Sustainable Development and Biodiversity

Volume 30

Series Editor

Kishan Gopal Ramawat, Botany Department, Mohanlal Sukhadia University, Udaipur, India

Sustainable Development Goals are best achieved by mechanisms such as research, innovation, and knowledge sharing. This book series aims to help researchers by reporting recent progress and providing complete, comprehensive, and broad subject-based reviews about all aspects of sustainable development and ecological biodiversity.

The series explores linkages of biodiversity with delivery of various ecosystem services and it offers a discourse in understanding the biotic and abiotic interactions, ecosystem dynamics, biological invasion, ecological restoration and remediation, diversity of habitats and conservation strategies. It is a broad scoped collection of volumes, addressing relationship between ecosystem processes and biodiversity. It aims to support the global efforts towards achieving sustainable development goals by enriching the scientific literature. The books in the series brings out the latest reading material for botanists, environmentalists, marine biologists, conservationists, policy makers and NGOs working for environment protection.

We welcome volumes on the themes --

Agroecosystems, Agroforestry, Biodiversity, Biodiversity conservation, Conservation of ecosystem, Ecosystem, Endangered species, Forest conservation, Genetic diversity, Global climate change, Hotspots, Impact assessment, Invasive species, Livelihood of people, Plant biotechnology, Plant resource utilization, Sustainability of the environment, Sustainable management of forests, Sustainable use of terrestrial ecosystems and plants, Traditional methods, Urban horticulture.

Swapna Thacheril Sukumaran • Keerthi T R
Editors

Conservation and Sustainable Utilization of Bioresources

Springer

Editors
Swapna Thacheril Sukumaran
Department of Botany
University of Kerala
Thiruvananthapuram, Kerala, India

Keerthi T R
School of Biosciences
Mahatma Gandhi University
Kottayam, Kerala, India

ISSN 2352-474X ISSN 2352-4758 (electronic)
Sustainable Development and Biodiversity
ISBN 978-981-19-5840-3 ISBN 978-981-19-5841-0 (eBook)
https://doi.org/10.1007/978-981-19-5841-0

This Springer imprint is published by the registered company Springer Nature Singapore Pte Ltd.
The registered company address is: 152 Beach Road, #21-01/04 Gateway East, Singapore 189721, Singapore

Foreword

Bioresources are essential for all life on, in, and above the Earth. They are the center of the economic development of a nation, and it is vital to maintain the health of people, grow food materials for future generations, generate energy, and manage the environment. Bioresources refer to the existence of wide diversity of living species on earth, including plants, animals, marine organisms, microbes, and their value-added products. In the current scenario, the industries have concentrated on a new revolution of bio-based economy to become more economic through the sustainable production of renewable sources from the underutilized biomass, agricultural wastes/effluents, and the available processed biomaterials by using advanced technologies. These resources are also exploited for generating functional consumer-friendly foods to increase the health benefits and protection against many chronic disorders. Recent discoveries in the utilization of lower plant forms have attracted industries to develop safe and effective nutraceuticals, pharmacologically active ingredients as well as economically valuable products. More emphasis has been made on consuming leafy greens, wild tubers/rhizomes, and underutilized fruits due to their potent therapeutic effects. The coastal and marine resources are also highlighted in the present world for the commercialization of food supplements and pharmaceuticals. Mangroves are also an irreplaceable component of the coastal ecosystem in providing rich biological and genetic diversity. They are considered natural carbon stores and good water purifying plants. Medicinal plants are always considered a huge source of drugs with special respect to the production of bioactive secondary metabolites. Various governments in different regions of the world have brought numerous initiatives to promote, maintain, and appraise the valuable traditional knowledge related to their uses of these medicinal plants. But many plants are threatened with extinction and belong to the RET group of IUCN. In vitro technologies are the most promising tool for the storage of germplasm of various plant species and production of diverse secondary metabolites. The botanical information among folk people has an important role in the development of new drug leads. The natural products derived from different plants/other organisms are crucial drug leads for the discovery of effective drugs for harmful diseases. Recently, computational methods are available to discover drugs with less time and low cost in contrast to conventional methods. Considering all these aspects of biological resources, it is very critical to conserve and consume the existing natural resources without overexploitation for maintaining the balance of our ecosystem. These diverse aspects are described in detail in this book.

I congratulate the editors for the presentation of this excellent book which, I am confident, will be of great value to the readers.

Centre for Innovation and Translational Research, CSIR-Indian Institute of Toxicology Research, Lucknow, India

Ashok Pandey

Preface

Bioresources include all kinds of biological materials that serve to support life on the earth, such as plants, animals, microbes, or any other biogenic products. Since they are rich in diversity, people utilize these resources to meet their requirements in the form of food, medicine, fuel, and other value-added products. Hence, bioresources have a significant role in various fields like agriculture, food, pharmaceutical, and fuel industries. The potential bioactive molecules of plant, animal, and microbial origin lead to high demand and increased utilization of these bioresources. At the same time, it is very necessary to conserve these resources for future generations. Hence, research is focused on developing various strategies for the conservation and sustainable utilization of these bioresources. Also, it is essential to discover novel and underutilized biological resources to meet the requirements of the increasing population.

For this, the younger generation should be aware of the various possibilities of proper and sustainable utilization of bioresources and their conservation. The book entitled *Conservation and Sustainable Utilization of Bioresources* is a collection of 25 chapters under four sections that deals with the diverse bioresources, various strategies for their conservation, the non-conventional or biotechnological methods of conservation, and the sustainable utilization of these bioresources. The chapters of this book are contributed by the scientists in Kerala, India, in collaboration with the scientist from many countries, and deliver information on various topics regarding bioresources and their diversity, various methods (conventional and non-conventional) for the conservation of bioresources, *in vitro* secondary metabolite production for the sustainable utilization of endangered medicinal plants, computational approaches for identifying the therapeutic potential of phytocompounds, etc.

This book is an effort from our side to make the younger generation aware of the need for sustainable utilization and conservation of bioresources. We thank all those who have contributed valuable information and made an effort to get this book published under Springer- Nature and hope the younger researchers and students in the science field will make use of the information in the book fruitfully.

Thiruvananthapuram, Kerala, India Swapna Thacheril Sukumaran
Kottayam, Kerala, India Keerthi T R

The original version of this book has been revised: Series Volume number 30 has been assigned. An correction to this book can be found at https://doi.org/10.1007/978-981-19-5841-0_26

Contents

Editors and Contributors

About the Editors

Swapna Thacheril Sukumaran is currently officiating as Professor and Head of the Department of Botany, University of Kerala. She completed her doctoral research in the Department of Biotechnology, Cochin University of Science and Technology, India. She has 26 years of teaching and research experience and published 13 books and over 120 research papers in journals and conference proceedings. Her research areas include *in vitro* secondary metabolite production, medicinal plant conservation, and phytochemistry. Dr. Swapna is a recipient of research grants from the Department of Environment and Climate Change, Government of Kerala, Western Ghats Cell Government of Kerala, State Medicinal Plant Board, Government of Kerala, University Grants Commission, SERB, Department of Science and Technology, Government of India. She is also serving as a Member of the Kerala State Biodiversity Board.

Keerthi T R has been awarded Ph.D. from Cochin University of Science and Technology, Cochin, Kerala, India. Currently, she is a Professor, School of Biosciences, Mahatma Gandhi University, Kerala, India. Her areas of interest are probiotics, prebiotics, microbiome, and bioprospecting of marine microbes. She has more than 20 years of experience in teaching and 25 years in research. She has published more than 35 papers in reputed peer-reviewed journals and 6 book chapters. She attended and presented her research findings at more than ten international conferences such as ACB 2011 held in Shanghai, China, International Biotechnology Symposium, "IBS 2012," September 16–21, 2012, Daegu, Korea (Chaired session), TSB International Forum-2013 BITEC Bang Na, Bangkok, Thailand, August 28–30, 2013; 4th Annual International Conference on Advances in Biotechnology (Biotech 2014) Dubai, BITS Pilani Campus 10–11th March 2014, 16th World Congress on Nutrition and Food Chemistry, Zurich, Switzerland September 18–20, 2017, Probiota Asia 2017, Singapore 11–13 October 2017. She is active in a number of academic and administrative bodies in her university and other universities. She serves as a reviewer of more than ten international journals as well as an editorial member of many journals. She has Membership in Scientific Societies such as International Probiotics Association (IPA), Zurich, Switzerland; Asian Federation of Biotechnology (AFOB), Korea; Indian Dairy Association (IDA), New Delhi.

Contributors

Athira V. Anand Department of Botany, University of Kerala, Thiruvananthapuram, Kerala, India

Anet Antony Department of Zoology, School of Biological Sciences, Central University of Kerala, Kasaragod, Kerala, India

Kizhakke Purayil Anupama Department of Zoology, School of Biological Sciences, Central University of Kerala, Kasaragod, Kerala, India

K. Arunkumar Department of Plant Science, School of Biological Sciences, Central University of Kerala, Kasaragod, Kerala, India

Anu Augustine Department of Biotechnology and Microbiology, Kannur University, Kannur, Kerala, India

Remya Babu Jawaharlal Nehru Tropical Botanic Garden and Research Institute, Thiruvananthapuram, Kerala, India

Vinitha S. Babu Department of Botany, University of Kerala, Thiruvananthapuram, Kerala, India

V. Balakrishnan Kerala State Biodiversity Board, Thiruvananthapuram, Kerala, India

Sarita G. Bhat Department of Biotechnology, Cochin University of Science and Technology, Cochin, India

Sajitha Biju School of Food and Agriculture, The University of Melbourne, Melbourne, VIC, Australia

E. S. Bindiya Department of Biotechnology, Cochin University of Science and Technology, Cochin, India

L. G. Campos-Muzquiz School of Chemistry, Universidad Autónoma de Coahuila, Saltillo, Coahuila, México

R. Castillo-Godina School of Chemistry, Universidad Autónoma de Coahuila, Saltillo, Coahuila, México

Amitha S. Chandra Department of Botany, University of Kerala, Thiruvananthapuram, Kerala, India

Sangeeth Chandran KSCSTE-Kerala Forest Research Institute, Thrissur, Kerala, India

P. Fahima Farook College (Autonomous), Kozhikode, Kerala, India

G. M. Greeshma CISSA Phytotech, Thiruvananthapuram, Kerala, India

Hunasanahally Puttaswamygowda Gurushankara Department of Zoology, School of Biological Sciences, Central University of Kerala, Kasaragod, Kerala, India

A. S. Hemanthakumar Biotechnology and Bioinformatics Division, Jawaharlal Nehru Tropical Botanic Garden and Research Institute, Kerala, India

Junaida M. Ibrahim Department of Computational Biology and Bioinformatics, University of Kerala, Thiruvananthapuram, Kerala, India

T. C. James Research and Information System for Developing Countries, New Delhi, India

Remya Krishnan Post Graduate Department and Research Centre of Botany, Mahatma Gandhi College, Thiruvananthapuram, Kerala, India

M. V. Lakshmi Department of Botany, University of Kerala, Thiruvananthapuram, Kerala, India

Bosco Lawrence Department of Botany and Biotechnology, Government of Arts College, Trivandrum, Kerala, India

G. P. Lekshmi Department of Botany, University of Kerala, Thiruvananthapuram, Kerala, India

S. Lekshmi Department of Botany, University of Kerala, Thiruvananthapuram, Kerala, India

Shanthala Mallikarjunaiah Center for Applied Genetics, Department of Studies in Zoology, Bangalore University, Bengaluru, Karnataka, India

K. P. Manoj Kumar Department of Plant Science, School of Biological Sciences, Central University of Kerala, Kasaragod, Kerala, India

Emi Mathew Bishop Chulaparambil Memorial College, Kottayam, Kerala, India
Faculty of Science, Istanbul University, Istanbul, Turkey

Jose Mathew Department of Botany, Sanatana Dharma College, Alappuzha, Kerala, India

Linu Mathew School of Biosciences, Mahatma Gandhi University, Kottayam, Kerala, India

V. Mithun KSCSTE-Malabar Botanical Garden and Institute for Plant Sciences, Kozhikode, Kerala, India

K. V. Mohanan Gregor Mendel Foundation, Calicut University, Kozhikode, Kerala, India

Jumana Muhammed Department of Biotechnology and Microbiology, Kannur University, Kannur, Kerala, India

K. Murugan CISSA Phytotech, Thiruvananthapuram, Kerala, India

Achuthsankar S. Nair Department of Computational Biology and Bioinformatics, University of Kerala, Thiruvananthapuram, Kerala, India

Bindu R. Nair Department of Botany, University of Kerala, Thiruvananthapuram, Kerala, India

Vinodkumar T. G. Nair Jawaharlal Nehru Tropical Botanic Garden and Research Institute, Thiruvananthapuram, Kerala, India

Rakhie Narayanan School of Biosciences, Mahatma Gandhi University, Kottayam, India

Thahiya Naushad Department of Botany, University of Kerala, Thiruvananthapuram, Kerala, India

Fundamental and Industrial Microbiology Division, Department of Biology, Faculty of Science, Istanbul University, Istanbul, Turkey

M. Navas Jawaharlal Nehru Tropical Botanic Garden and Research Institute, Thiruvananthapuram, Kerala, India

Shilpa Olakkaran Department of Zoology, School of Biological Sciences, Central University of Kerala, Kasaragod, Kerala, India

Kumudu Perera Faculty of Medicine, Department of Biochemistry, University of Peradeniya, Sri Lanka, Sri Lanka

Salim M. Pichan MS Swaminathan Research Foundation, Wayanad, Kerala, India

S. Pradeep Kumar KSCSTE-Malabar Botanical Garden and Institute for Plant Sciences, Kozhikode, Kerala, India

N. S. Pradeep KSCSTE-Malabar Botanical Garden and Institute for Plant Sciences, Kozhikode, Kerala, India

Sarath Praseetha Department of Biotechnology, University of Kerala, Thiruvananthapuram, Kerala, India
Microbiology Division, KSCSTE-Jawaharlal Nehru Tropical Botanic Garden and Research Institute, Thiruvananthapuram, Kerala, India

T. S. Preetha Department of Botany, University College, Thiruvananthapuram, Kerala, India

P. M. Radhamany Department of Botany, University of Kerala, Thiruvananthapuram, Kerala, India

A. V. Raghu KSCSTE-Kerala Forest Research Institute, Thrissur, Kerala, India

S. Rajasekharan Kerala State Biodiversity Board, Thiruvananthapuram, Kerala, India

Reshma Ramarajan Department of Zoology, School of Biological Sciences, Central University of Kerala, Kasaragod, Kerala, India

Prakashkumar Raveendran Jawaharlal Nehru Tropical Botanic Garden and Research Institute, Thiruvananthapuram, Kerala, India

Resmi Ravindran Department of Pathology and Laboratory Medicine, University of California, Sacramento, CA, USA

Jeena Elsa Reji School of Biosciences, Mahatma Gandhi University, Kottayam, Kerala, India

C. T. Riyas Department of Botany, University of Kerala, Thiruvananthapuram, Kerala, India

R. Rodriguez-Herrera School of Chemistry, Universidad Autónoma de Coahuila, Saltillo, Coahuila, México

E. A. Siril Department of Botany, University of Kerala, Thiruvananthapuram, Kerala, India

T. R. Sobha Farook College (Autonomous), Kozhikode, Kerala, India

M. J. Sreedevi Department of Botany, University of Kerala, Thiruvananthapuram, Kerala, India

P. M. Sreekanth Department of Biotechnology, Cochin University of Science and Technology, Cochin, India

R. Sreelekshmi Department of Botany, University of Kerala, Thiruvananthapuram, Kerala, India

S. Sreelekshmi Department of Botany, University of Kerala, Thiruvananthapuram, Kerala, India

L. S. Sreeshma Department of Botany, University of Kerala, Thiruvananthapuram, Kerala, India

Shiburaj Sugathan Department of Botany, University of Kerala, Thiruvananthapuram, Kerala, India

Suveena Sukumaran Department of Computational Biology and Bioinformatics, University of Kerala, Thiruvananthapuram, Kerala, India

T. S. Swapna Department of Botany, University of Kerala, Thiruvananthapuram, Kerala, India

Keerthi T R School of Biosciences, Mahatma Gandhi University, Kottayam, Kerala, India

Irfan Türetgen Fundamental and Industrial Microbiology Division, Department of Biology, Istanbul University, Istanbul, Turkey

Babu Valliyodan Department of Agriculture and Environment Science, Lincoln University, Jefferson City, MO, USA

C. P. Vibija Malabar Christian College, Kozhikode, Kerala, India

A. S. Vivek Department of Botany, University of Kerala, Thiruvananthapuram, Kerala, India

Chapter 1
Bioresources and Diversity

M. V. Lakshmi, S. Lekshmi, R. Castillo-Godina, L. G. Campos-Muzquiz, R. Rodriguez-Herrera, Sajitha Biju, Remya Krishnan, and T. S. Swapna

Abstract Bioresources are the biologically generated materials that support life on the earth. They include plants, animals, microorganisms, and the total biogenic products. The purposeful use of these resources involves the preparation of food, the discovery of value-added bio-products, and the generation of energy. So these resources have a significant role in agriculture, pharmacological industries, and the production of bioactive molecules with pharmaceutical and industrial potential, thus, contributing to the nation's overall economic development. Bioresources are generally diverse and abundant in nature. Due to this rich diversity, people depend on these resources to get enough goods and services to meet their needs. Microbial diversity is very vast and can have a wide range of applications in the agriculture, environment, and pharmaceutical industries. Legumes provide food security to millions of populations, and climate-smart agriculture can enhance sustainable utilization. Medicinal plants are storehouses of bioactive molecules with antimicrobial, anticancer, antidiabetic, and hepatoprotective potential. Validation of the ethnomedicinal properties of such plants can lead to the identification of many drug leads, which will be beneficial to mankind. The purposeful over-use of natural resources will lead to socio-economic and environmental issues. The conservation and sustainable consumption of biological resources are very critical for maintaining the balance of our ecosystem. There are many strategies to be followed for the prevention of overexploitation of these resources. A strong focus should be given to improving the resource efficiency of the available biomass to develop novel products with reduced costs. In this aspect, the multidisciplinary research studies on the

---

M. V. Lakshmi · S. Lekshmi · T. S. Swapna (✉)
University of Kerala, Thiruvananthapuram, Kerala, India

R. Castillo-Godina · L. G. Campos-Muzquiz · R. Rodriguez-Herrera
School of Chemistry, Universidad Autónoma de Coahuila, Saltillo, Coahuila, Mexico

S. Biju
School of Food and Agriculture, The University of Melbourne, Melbourne, VIC, Australia

R. Krishnan
Mahatma Gandhi College, Thiruvananthapuram, Kerala, India

S. T. Sukumaran, K. T R (eds.), *Conservation and Sustainable Utilization of Bioresources*, Sustainable Development and Biodiversity 30,
https://doi.org/10.1007/978-981-19-5841-0_1

utilization of different resources would be beneficial for the advancements in health systems and industrial areas through scientific and technological innovation.

Keywords Bioresources · Microbial diversity · Legumes · Medicinal plants · Bio-products · Multidisciplinary research

1.1 Introduction

The fundamental necessity of living organisms such as food, feed, medicine, minerals, shelter, and fuel is provided by bioresources which are renewable organic materials present on our earth. Bioresources are economically and industrially significant for the maintenance of the economy of a nation (Ingle et al. 2020). Bioactive molecules derived from plants, microbes, or marine organisms are indispensable prime elements for the maintenance of human health, which provides a wide scenario for research and development. The living organisms may vary in size, shape, habitat, nutritional requirements, and reproduction methods and are found in diverse ecosystems. The rich diversity of plants, animals, and microbes flourished on Earth through evolution, competition, and symbiosis (Wells and Varel 2011). They are highly exploited in industries, agriculture, forestry, and marine fields. Bioresources are more eco-friendly and economically available renewable energy sources (Ani 2016). Microorganisms have unlimited genetic diversity, and most of them present on the earth are still unknown (Vitorino and Bessa 2018). They include different kinds of unicellular organisms such as bacteria, archaea, protists, fungi, and viruses that live in diverse environments like air, water, soil, food, space, glaciers, and animal intestinal systems. The microorganisms are highly beneficial for the detoxification of the ecosystem, mineralization of organic substrates, and recovery of the destroyed ecosystem, thereby improving human livelihoods (Singh et al. 2019). The microbes like rhizobacteria and mycorrhizal fungi enhance the plant's defense systems and help resist abiotic stress (Pineda et al. 2013). As biological control agents, *Bacillus thuringiensis* and its products are widely used in the agricultural field (Usta 2013). Various microbes are used in biotechnology due to their applications as bio-fertilizers, bio-pesticides, bio-herbicides, and bio-insecticides (Ahirwar et al. 2020). The microbial community has a specific role in the organic and inorganic pollutant removal in floating treatment wetlands and plant growth promotion (Shahid et al. 2020). Therefore, these microbes are an irreplaceable resource for basic research studies and applications in the bioindustry, agriculture, health care, and the environment (Becker et al. 2019).

Biodiversity is the biological diversity of plants, animals, fungi, and microorganisms on earth and their genetic variation (Dirzo and Mendoza 2008). There are over 500,000 species of land plants (angiosperms, gymnosperms, ferns, lycophytes, and bryophytes) that show enormous diversity in the humid tropics (Corlett 2016). Algae are some of the most common and highly variable organisms inhabiting the Earth, which grow even in extreme conditions (Ścieszka and Klewicka 2019). Many algae have gained attraction due to their commercial importance in various industries.

They are good sources of biofuels, nutraceuticals, bio-fertilizers, bioactive metabolites, food supplements, and cosmetics (Sharma and Sharma 2017; Vidhyanandan et al. 2020). Angiosperms are most dominant in the earth's terrestrial ecosystems and represent nearly 90% of all extant plant species (Condamine et al. 2020). They are the most enormous and diverse autotrophic organisms on land (Armbruster 2014). There is an estimate of 369,434 flowering plants, 370,942 seed plants, 383,671 vascular plants and 403,911 land plants (Lughadha et al. 2016). The database on the World Checklist of Vascular Plants (WCVP, http://wcvp.science.kew.org/) included 1,383,297 plant names, 996,093 at the species level, representing 342,953 accepted vascular plant species up to 16th April 2021 (Govaerts et al. 2021). Gymnosperms are a group of woody plants, and about 850–1000 species presently inhabit our planet (Davis and Schaefer 2011). They have diversified more slowly than angiosperms as their lineages diverged in the Carboniferous (Crisp and Cook 2011). There are only 16% of all named species on earth are considered marine (Costello and Chaudhary 2017). Marine flora primarily includes brown, blue, green, blue-green, and red algae (Joseph 2016). Many pharmaceutical products developed from marine flora and fauna are available in the current market. Marine microbes provide plenty of non-toxic compounds with promising anti-biofouling/anti-biofilm properties. They can produce substances that inhibit the chemical components required for biofilm production and microbial growth and/or cell–to-cell communication (Adnan et al. 2018). Animal genetic resources (AnGR) such as animal species, breeds, strains, and their wild relatives are also widely exploited for food and agricultural production in the present or future (Rege and Gibson 2003).

The recent advancements in biotechnology and molecular biology helped scientists to produce modified user friendly natural resources on a low budget. Bioresources have a major role in the tribal culture and economy (Nimachow et al. 2008). The traditional/local information regarding any living organisms provides deep insight for the sustainable use of the existing natural resources, thus, preserving biodiversity with reduced exploitation. The overexploitation of natural resources will lead to genetic erosion and the transformation of habitable land into a desert (Uyoh et al. 2003). So it is very urgent to use our resources efficiently without wastage. In this perspective of increased use of biomass for food, energy, and materials, the concept of bioeconomy is developed for the conservation and production of renewable biological resources and also for the conversion of these resources into value-added products (Mougenot and Doussoulin 2022). Technology transfer and innovation support professionals are necessary for the bio-based economy to interconnect all parties involved in a particular value network and to understand the individual needs and motivation to shift from fossil to bio-based resources (Zörb et al. 2018). Bioeconomy covers the promotion, development, and establishment of biogenic resources in diverse kinds of industrial technologies, production processes, and products (Vogelpohl 2021). National Policy Strategy on Bioeconomy, initiated by the Federal Cabinet (2013), promotes the transition in industry and society toward a sustainable and bio-based economy in developed countries (Imbert et al. 2017). The German Federal Government published the new National Bioeconomy Strategy in January 2020 with the current relevance of the global bio-based economy (Birner

2018). These strategies will contribute to maintaining and preserving the bioresources with a balanced economy in the future.

1.2 Microbial Diversity: Generalities and Importance

Belowground diversity refers to the biological diversity, which includes the study of plants and animals, but the greatest diversity corresponds to microorganisms (bacteria, fungi, algae, nematodes, and viruses) in general. For its part, the biotechnological industry requires this diversity to transform it into innovative consumer goods and thus generate environmental wealth and incorporate it into the processes of sustainable development and the conservation of biodiversity itself.

The microbial diversity of the soil is the cause of its fertility. This phrase goes beyond the role previously attributed to the belowground diversity which was related to the degradation and recycling of organic matter, maintenance of the primary cycles of fixation, uptake, and release of some chemical elements and their main compounds (Cappello et al. 2015). The importance of the community of microorganisms and the flow of nutrients and energy in natural ecosystems have been extensively studied. Soil microorganisms are essential regulators of productivity, plant community dynamics, and plant diversity (Van der Heijden et al. 2008). Illarze et al. (2017) mentioned the benefits of microorganisms in the soil as nutrient providers. Knowledge of the dynamics of microbial communities in the soil, allows us to understand the role of microorganisms in climate change processes, propose and evaluate sustainable agronomical management practices, face the worldwide growing demand for food and biofuels, and discover new bioactive products (Soria 2016). Soil fertility, greenhouse gas flow, or the neutralization of pollutants in agroecosystems is increasingly studied to identify the role of microbial communities (Srivastava et al. 2017).

Different studies have been carried out on the diversity of soil microorganisms, from genetics to metabolism (microbiome), to understand their relationship with plant health and how it benefits or harms soil productivity (Erice et al. 2019). Microorganisms are generally associated with plant disease states; however, research has shown that beneficial soil microorganisms have the potential to improve soil conditions.

1.2.1 Culturable and Nonculturable Microorganisms

The predominant microbial communities in soils have been characterized by culture-dependent and culture-independent methods. For the characterization of the culture-dependent microbial community, classical microbiology techniques are used until pure strains are obtained, which are identified using taxonomic keys or at the molecular level by DNA sequencing (Sernaque Aguilar et al. 2019). On the other hand, some microorganisms cannot be studied easily due to their difficulty in growing and developing colonies under conventional culture media. If nonculturable microorganisms could be culturable, the normal vegetative cell changes morphologically and metabolically. The expression of their genes and their virulence potential are also modified (Li et al. 2014). On the other hand, due to their lack of development in culture media, the real density of Nonculturable-microorganism populations is often underestimated (Ramamurthy et al. 2014). According to Lara-Victoriano et al. (2011), most nonculturable microorganisms are beneficial, which makes them potentially useful for the biotechnological industry. The characterization of the microbial diversity present in a community represents the first stage in the process of industrialization and scaling of a microbial-derived product. This initial characterization must be followed by identifying the essential microorganisms in the process (Escalante et al. 2008).

1.2.2 Metagenomics

Metagenomics is a recent tool that uses mass DNA isolation and sequencing to study and characterize communities of microorganisms (Brenes-Guillén 2018). Metagenomics provides a broad overview of the study of the origin of microorganisms and their function in a given environment. These techniques are based on ecological diversity studies of cultivable and non-cultivable microorganisms using molecular biology tools (Lara-Victoriano et al. 2011). Metagenomics allows obtaining genome sequences of the different microorganisms that make up a community, extracting and analyzing their DNA globally, without the need to cultivate these microorganisms. Through metagenomics studies, valuable information has been obtained about the genes that code for enzymatic biocatalysts and biosynthetic pathways in non-cultured microorganisms faster than traditional molecular biology techniques (Wilson and Piel 2013). Metagenomics, thus, becomes a useful tool to assess the high microbial diversity of environmental samples and can take three routes: (a) PCR amplification of the 16S and 18S genes to determine microbial diversity, (b) digestion and cloning in expression vectors, and (c) direct sequencing of the sample (Hernández-León et al. 2010).

1.2.3 Metagenomic Tools for Microorganism Identification

Since classical microbiological tools do not allow the identification of strains of microorganisms that are not cultivable or that depend on very specific culture conditions, different genetic tools have been developed. Some of these procedures in the field of metagenomics are denaturing gradient gel electrophoresis (DGGE), amplified ribosomal DNA restriction analysis (ARDRA), and next-generation sequencing, including Illumina technology (van Hijum et al. 2013; Wilson and

Piel 2013). These technologies promise to expand the molecular approaches of ecological and evolutionary studies toward issues related to the conservation and management of biological diversity in the face of challenges such as climate change (Escalante et al. 2014). In addition, culture-independent molecular methods can lead to the identification of previously undetected microbial groups (Lappe-Oliveras et al. 2008).

PCR-DGGE is a culture-independent detection method that is denaturing gradient gel electrophoresis (DGGE), which consists of separating genes of the same size that differ in their fusion profile (denaturation), due to differences in the sequence of their nucleotides. This technique is fast compared to other classic microbial techniques (van Hijum et al. 2013).

ARDRA is a technique adopting restriction fragment length polymorphisms to create fingerprints of 16S rRNA or 18S rDNA genes that can be analyzed on agarose gels (van Hijum et al. 2013). Next-generation sequencing was used to study the entire genome, but now it has been used for defined regions of the genome. (Koboldt et al. 2013). New generation sequencing (NGS) tools allow the identification and characterization of cultivable and nonculturable microorganisms as well as those found in low abundances. (Escalante et al. 2014). Among the newest technologies, there is Illumina technology, which generates millions of sequences at a fraction of the cost of Sanger sequencing or Roche 454 pyrosequencing, from a sample of a microbial community (Bartram et al. 2011; Tanase et al. 2015). The DNA extraction method represents one of the factors that most influences the estimation of the biodiversity of different taxa present in the soil (Dopheide et al. 2019).

According to Lara-Victoriano et al. (2011), in metagenomics, the construction of genomic libraries with DNA isolated from soil microorganisms is required. The DNA quality is highly influenced by the different methods used during sampling, from soil sampling depth and soil conservation (Duque-Ortiz et al. 2017).

1.2.4 Sustainable Use

Sustainable use of belowground biodiversity is critical for improving agricultural management systems. Enhancing soil biodiversity and making targeted changes to soil microbial community composition can complement each other to increase overall ecosystem sustainability and stability in terms of long-term environmental impact (Bender et al. 2016). Soil ecosystems can be managed to improve agricultural productivity. For example, the addition of *Bacillus megaterium* var. phosphaticum and *Bacillus mucilaginosus* can solubilize the phosphorus present in animal bones to make it available for plant uptake through the release of organic acids (Kumar and Gopal 2015). Moreover, soil microbial communities can be manipulated to reduce nitrogen loss by controlling abundance, structure, and activities involved in N-cycling microorganisms. The sustainable technique used to achieve this is microbial inoculants; nonetheless, in situ microbiome engineering could provide an effective outcome (Hu and He 2018; Qiu et al. 2019). The most common

N-fixating microorganisms inoculated for nitrogen fixation are *Actinorhizobium* spp., *Azospirillum* spp., *Azotobacter* spp., and *Rhizobium* spp.(Arif et al. 2020).

Another use of belowground microorganisms is the removal of contaminants. For instance, *Azospirillum* sp. and *Pseudomonas* sp. degrade glyphosate residues (Travaglia et al. 2015), and arbuscular mycorrhizal fungi reduce plumb and cadmium (Wu et al. 2016). Also, various groups of pesticides such as carbamates, organochlorines, organophosphates, and pyrethroids can be degraded by microorganisms, mainly those belonging to phyla Actinobacteria, Ascomycota, Bacteroidetes, Basidiomycota, Chlorophyta, Cyanobacteria, Firmicutes, and Proteobacteria (Kumar et al. 2021).

Likewise, soil biodiversity can be managed to control soilborne diseases that affect plants. This can be achieved by anaerobic soil disinfestation (ASD). In this technique, a carbon source, such as wheat bran or ethanol, is incorporated into the soil to modify the microbiome, increasing the abundance of Firmicutes. This reduces the severity of diseases like root rot caused by *Pyrenochaeta lycopersici* (Testen and Miller 2018). Microbial competition is another way of controlling soilborne diseases. When *Trichoderma harzianum* is added to soil, the germination of *Fusarium* chlamydospore is reduced (Sivan and Chet 1989). In addition, it has been hypothesized that soil microorganisms could control soil-transmitted human diseases, like salmonellosis or campylobacteriosis, through competition for nutrients or due to the production of antimicrobial substances (Jeffery and Van Der Putten 2011).

Studies on the manipulation of soil microorganisms began a long time ago. However, metagenomic studies revolutionized the way of making approximations about the composition of the communities and the metabolic routes that community species carry out. Metagenomics allows knowing the mechanisms and interactions carried out by culturable and nonculturable soil microorganisms which help make decisions about their manipulation and usage. For example, Epihov and Beerling (2018) showed that N2-fixing trees enhance weathering due to an increase in microbial respiration, lithotripsy, and gluconic acid production. Using trees rich in N2 fixers during successional scenarios can be beneficial for ecosystem recovery. Also, metagenomics help identify soil microbial products that could be used in industry. For example, Lara-Victoriano et al. (2017) identified tannase allele variation in fungal strains isolated from soil and plants in five extreme areas of Coahuila, México. Tannase is an enzyme used to hydrolyze tannins in the food industry; thereby, these microorganisms found in soil could be a source of this enzyme.

1.3 Legumes—A Road Map for Sustainable Future Food Security

Food safety, climate change, and population growth are the three major challenges to the global food system. Providing food and nutritional security to nearly 9.7 billion by 2050 (UN 2017) will imply a huge pressure on the food production system in terms of efficient and sustainable strategies to provide nutritious food, mitigate climate change, and improve soil fertility. According to Food and Agriculture Organization (FAO), we will need to produce 60% more food to feed this population by 2050. This will definitely impart massive pressure on our natural resources. As food production systems are major sources of global greenhouse gas (GHG) emissions, deforestation, and water use, plant-based protein diets are encouraged to reduce GHG emissions, water use, and deforestation (Semba et al. 2021). Moreover, with the rapid rise in the world's population and food demand, existing trends in animal protein intake are not sustainable. Consequently, there is an urgent need to adopt legumes or pulses-based agricultural production systems due to their proven beneficial roles in nutrition. Legumes are the second most significant plant contributing to 27% of the global food production systems and providing one-third of the human dietary protein (Biju et al. 2021; Sita et al. 2017). Legumes are rich sources of protein, slow-release carbohydrates with low starch (55 or less) glycemic index (GI), dietary fiber, vitamins, and minerals (McCrory et al. 2010). Germinated legumes are rich in vitamin C, riboflavin, and niacin as the activity of the enzymes, amylase, protease, phytase, and lipase increase during germination (Swaminathan 1988). Moreover, legume seeds are considered poor man's meat and are the second most important food source after cereals like rice, wheat, maize, and barley. The nutritional demand for legumes is rising globally due to increased consumer perception of their nutritional and health gains.

Legumes can fix atmospheric nitrogen, which increases soil fertility, minimizes inorganic fertilizers, and improves crop production. Intercropping and crop rotation practices with legumes help mitigate diseases and pests in succeeding crops, reducing agricultural costs and protecting the environment. The significance of legumes in farming systems surpasses the production of cereal grains as they are also used as forage and green manure (Graham and Vance 2003). Pulse trade has been increased four-fold since 1980s which provides the growers an opportunity to get profit from their production (Joshi and Rao 2017). Undoubtedly, legumes can play a significant role in addressing global food safety, nutrition security, and unprecedented climate change. Acknowledging the varied roles of legumes in food security, human nutrition, and sustainable food production, the UN General Assembly nominated the year 2016 as the 'International Year of Pulses.' Despite these social, economic, and environmental benefits, along with increasing demand and global production volume (89.8 million metric tons in 2020), sustainable cultivation of legumes has not met the expectations, and the area of legume cultivation has significantly decreased worldwide in the last 50 years (Alexandratos and Bruinsma 2012; FAO 2020). Varied uses of legumes in human nutrition and agriculture to emerge as a sustainable food resource for the future are discussed here.

1.3.1 Legumes: Origin, Domestication, and Production

Legumes belong to the family of Leguminosae, the second largest family of seeded plants, including 700 genera and nearly 18,000 species (ILDIS 2017). Grain legumes are generally divided into two overlapping groups according to their consumption and seed composition. Pulses, such as the common bean (*Phaseolus vulgaris*) and cowpea (*Vigna unguiculata*), are grain legumes harvested mainly for dry seeds and are used for human and animal consumption (FAO 1994). Their seed composition includes higher levels of protein (>19%), carbohydrate (>60%), dietary fiber (>10%), and low lipid (<6%). Oilseed legumes, including peanut (*Arachis hypogaea*) and soybean (Glycine max), contain a high protein and lipid concentration but lower carbohydrate content. They are mainly utilized as a source of vegetable oil for culinary uses due to their high lipid content and are widely used in food preparations. Interestingly, when common bean (*Phaseolus vulgaris*) is harvested for dry seed, it is treated as a pulse, but when the same species are harvested as green beans, and not considered as a pulse. A list of legumes that are treated as pulses (FAO 1994) is given in Table 1.1.

Table 1.1 Plants that are treated as pulses according to FAO classification (FAO 2014)

Vernacular name	Scientific name
Common bean	*Phaseolus vulgaris* L
Lima bean	*Phaseolus lunatus* L
Scarlet brunner bean	*Phaseolus coccineus* L
Tepary bean	*Phaseolus acutifolius* A. Gray
Adzukibean	*Vigna angularis*(Willd) Ohwi & H. Ohashi
Mungbean	*Vigna radiata* (L) R. Wilczek
Mungo bean	*Vigna mungo* (L) Hepper
Rice bean	*Vigna umbellata* (Thunb) Ohwi & H. Ohashi
Moth bean	*Vigna aconitifolia* (Jacq) Maréchal
Bambarabean	*Vigna subterranea* (L) Verdc
Broadbean	*Viciafaba* L
Commonvetch	*Viciasativa* L
Pea	*Pisum sativum* L
Chickpea	*Cicer arietinum* L
Cowpea	*Vigna unguiculata* (L) Walp
Pigeonpea	*Cajanus cajan* (L) Huth
Lentil	*Lens culinaris* Medik
Lupine	Several *Lupinus* species
Hyacinthbean	*Lablab purpureus* (L) Sweet
Jackbean	*Canavalia ensiformis* (L) DC
Wingedbean	*Psophocarpus tetragonolobus* (L) DC
Guarbean	*Cyamopsis tetragonoloba* (L) Taub
Velvetbean	*Mucuna pruriens* (L) DC
African yam bean	*Sphenostylis stenocarpa* (Hochstex A. Rich) Harms

Table 1.2 Centre of domestication of major grain legumes (Adapted from Sinha 1977)

Common name	Scientific name	Centre of domestication
Peanut	*Arachis hypogaea*	Southern Bolivia, north-west Argentina
Chickpea	*Cicer arietinum*	Southeast Turkey
Soybean	*Glycine max*	China
Common bean	*Phaseolus vulgaris*	Mexico and Peru
Lima bean	*Phaseolus lunatus*	Peru, Central America, the Caribbean
Cowpea	*Vigna unguiculata*	Sub-Saharan Africa
Pigeon pea	*Cajanus cajan*	India
Lentil	*Lens culinaris*	Eastern Mediterranean, the Fertile Crescent (Iraq and Iran)
Field pea	*Pisum sativum*	The Fertile Crescent (Iraq and Iran), Turkey, Greece
Faba bean,	*Vicia faba*	Eastern Mediterranean, West Asia
Mung bean	*Vigna mungo*	India
Horse gram	*Macrotyloma uniflorum*	Southern India
Grass pea	*Lathyrus sativus*	Southern Europe, Southwest Asia
Lupinus albus; L.	*Lupin*	Greece, western Turkey, Eastern Mediterranean, Peru
Green gram	*Vigna radiata*	India, Southeast Asia

Legumes are believed to be one of the pioneer crops in the world, with their cultivation dating to pre-historic times and persisted as a staple food for millions across the globe (Table 1.2) (Yadav et al. 2010). Carbonized seeds of legumes (lentil, pea, and vetch) have been identified from the Neolithic age (7000–8000 years B.C.) in Turkey. These legume seeds were used as food in this region and spread to the north-west and south-west (Africa) and towards India. Over the years, domestication of the crops, selection of elite cultivars, and trade scenarios resulted in thousands of different species and a wide variety of legumes grown in different regions of the world, making them vital for both their economic and nutritional potential. In developing countries, legumes were the primary source of protein for various economic and social reasons. However, in most developed countries, the dry legume seeds were not used in diets as animal meat became the popular daily source of protein (Table 1.2).

Instead, fresh seeds and pods of legumes were used in the human diet. Major food legumes used for human consumption include lentil (*Lens culinaris* Medicum), chickpea (*Cicer arietinum* L.), soyabean (*Glycine max* L.), black gram (*Vigna mungo* L.), cowpea (Vigna unguiculata L.), lupin (*Lupinus* spp.), mung bean (*Vigna radiata* (L.) Wilczek), faba bean (*Vicia faba* L.), French bean (*Phaseolus vulgaris*), horse gram (*Macrotyloma uniflorum*), field pea (*Pisum sativum* L.), moth bean (*Vigna aconitifolia*), lathyrus (*Lathyrus sativus* L.), and pigeon pea (*Cajanus cajan* L.) (Fig. 1.1).

Legumes are ranked third in global crop production after oilseeds and cereals (Popelka et al. 2004). Asia is the largest producer with 45% global production, followed by Africa (30.4%) and Europe (23.5%) (Fig. 1.2a). Among Asian

Fig. 1.1 Main legume species for human consumption (**a**) Lentil; (**b**) Faba bean; (**c**) Pea nut; (**d**) Soyabean; (**e**) Pigeon pea; (**f**) Chickpea

countries, India is the largest producer (26% of global production), the consumer (27% of world consumption), and the importer (14%) of legumes in the world. On a global scale, soybeans rank first in production, followed by peanut, common beans, common peas, chickpea, cowpea, faba bean, lentil, pigeon pea, and lupin (FAO 2020) (Fig. 1.2b).

1.3.2 Legumes—Nutritional Properties and Health Benefits

Legumes play an important role in human nutrition due to their varied nutritional potential, especially in the diets of low-income people in developing countries (Table 1.3). Legumes are an excellent source of protein (20–45%) and rich in the essential amino acid lysine, which is lacking in cereal grains (Duranti 2006). When legumes are eaten combined with cereals, the protein value in the diet is considerably increased (Singh and Singh 1992). Legumes and cereals should be consumed in a ratio 35:65 to balance the nutritional level (Anonymous 2013). Table 1.4 shows the composition of protein, lipid, carbohydrates, and moisture content of grain legumes compared to other reference foods. The consumption of legumes may lower the risk of heart diseases as they may have a role in inhibiting diabetes and lowering breast cancer risk, and increased intake of legumes may also protect against obesity (Campos-Vega et al. 2010).

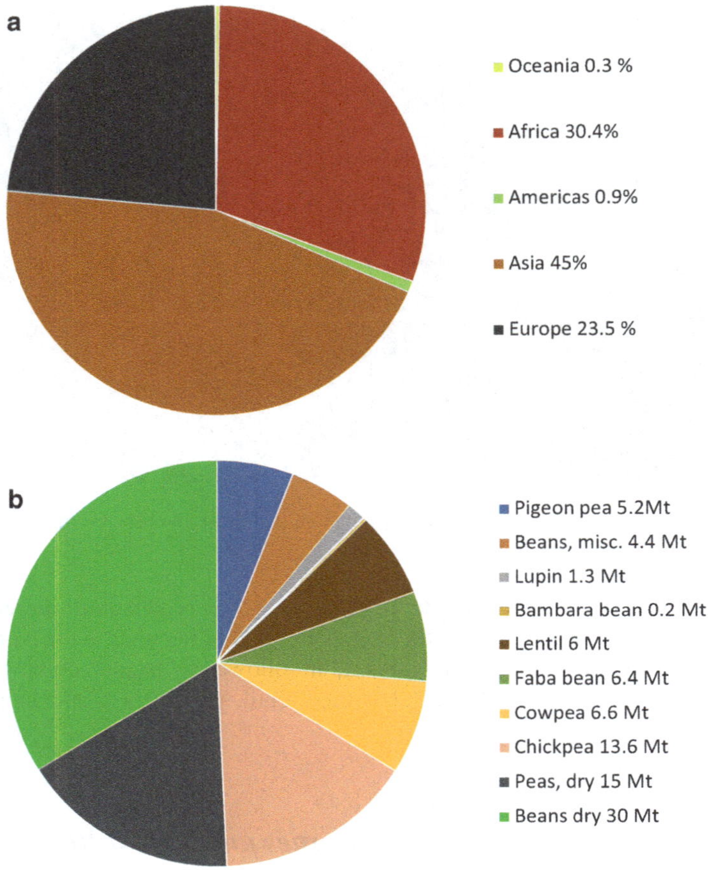

Fig. 1.2 (**a**) Global production of legumes (2014–2020) (*Mt* million tons) (FAOSTAT, 2020) (Adapted from Semba et al. 2021); (**b**) Global production of main legumes in metric tonnes

1.3.3 Proteins and Amino Acids

The major proteins found in legumes are globulins (legumin, 11S and vicilin, 7S) and albumins (enzymatic proteins, protease/ amylase inhibitors, inhibitors, lectins, and hormones). The minor proteins include prolamins and glutelins. These proteins also vary in their structure, molecular weight (MW) distribution, and solubility providing a basis for their extraction and isolation. The high protein content of legumes is supported by the interaction of the plant with rhizobium, nitrogen-fixing bacteria in their roots, which converts the unusable nitrogen gas into ammonium which the plant uses for protein synthesis. Compared to cereal grain production, the production of legume protein is more energy efficient and less energy intensive than the production of animal protein (Rockland and Radke 1981). The proteins and

Table 1.3 Nutritional analysis of 100 kilocalorie portions of the predominant pulses, undried legumes, and oilseed legumes

Chick pea	Cow pea	Dry bean	Snap bean	Dry pea	Green pea	Snap pea	Lentil	Peanut	Soybean	Edamame	
Approximate amount	~1/2	~1/2–1/3	~3	~1/2	~1	~2.5	~1/2	~2	~1/3	~1/2	
	Cup	Cup	Cup	Cups	Cup	Cup	Cups	Cup	Tbs	Cup	Cup
Protein (g)	5.4	6.7	6.7	5.4	7.2	6.4	5.7	7.8	4.6	10.6	7.8
Total lipid (g)	1.6	0.5	0.4	0.8	0.3	0.3	0.0	0.3	8.7	5.2	2.8
Carbohydrate (g)	16.7	17.9	18.0	22.5	17.7	18.6	17.2	17.4	2.8	4.9	11.1
Dietary fiber (g)	4.6	5.6	6.6	9.1	7.2	6.5	5.9	6.8	1.5	3.5	3.4
Folate (μg)	104.9	179.3	112.9	94.3	56.0	75.0	MV	156.0	42.3	31.4	MV
Iron (mg)	1.8	2.2	1.6	1.9	1.1	1.8	5.7	2.9	0.8	3.0	1.9
Potassium (mg)	177.4	239.7	268.9	417.1	312.1	322.6	MV	318.1	124.3	299.4	365.6
Calcium (mg)	29.9	20.7	20.5	125.7	12.1	32.1	107.3	16.4	16.2	59.3	66.4
Choline (mg)	26.1	27.8	24.7	48.3	28.3	35.4	MV	28.2	9.3	MV	MV
Magnesium (mg)	29.3	45.7	53.0	51.4	31.0	46.4	MV	31.0	29.6	50.0	MV
Vitamin A, RAE (μg)	0.6	0.9	0.0	100.0	0.0	47.6	MV	0.0	0.0	0.0	MV
Vitamin C (mg)	0.8	0.3	0.0	27.7	0.3	16.9	MV	1.3	0.0	1. 0	MV
Vitamin E (mg)	0.2	0.2	0.7	1.3	0.0	0.2	MV	0.1	1.5	0.2	MV

Table 1.4 Composition of protein, lipid, carbohydrates, and moisture content of grain legumes and other reference foods (g/100 g). (Adapted from Michaels 2015)

Food	Protein	Lipid	Carbohydrate	Moisture
Pulses				
Common bean	22.2	1.1	61.5	11.5
Lima bean	21.5	0.7	63.4	10.7
Chickpea	19.3	6.0	60.7	11.5
Green gram	23.9	1.1	62.6	9.0
Cowpea	23.8	2.1	59.6	11.1
Lupine	36.7	11.5	45.4	–
Oil seeds				
Soybean	36.5	19.9	30.2	8.5
Peanut	25.8	49.2	16.1	6.5
Cereals				
Wheat, hard white	11.3	1.7	75.9	9.6
Brown rice, raw	7.5	2.7	76.2	12.4
Animal products				
Beef, lean ground	20.0	10.0	0.0	69.5
Pork, ground	16.9	21.2	0.0	61.0
Fish, cod	17.8	0.7	0.0	81.2
Milk, cow's 3.25% fat	3.3	3.3	4.7	88.0
Egg. Hen's	12.5	10.0	1.2	75.3

peptides in legumes also offer many health benefits in addition to their nutritional properties. Peptides with angiotensin-1-converting enzyme (ACE) inhibitory properties were identified from Lentil, pea, bean, and chickpea (Barbana and Boye 2011; Rui et al. 2012). ACE plays a vital role in vasoconstriction that upshots in the elevation of blood pressure. Legume proteins are also used as components in flours and ingredients in soups, beverages, snacks, bakery, or meat products due to their varied functional properties such as solubility, water, and fat absorption capacity, foaming, gelation, and emulsification (Boye and Ma 2012; Ribéreau et al. 2018). Legumes are rich in amino acids like lysine, leucine, and arginine but are mostly deficient in S-containing amino acids (methionine, cystine, and cysteine) (Leonard 2012; Yao et al. 2015). The amino acid composition of selected legumes is given in Table 1.5.

1.3.4 Carbohydrates, Micronutrients, and Fats

Legumes are good sources of carbohydrates, with up to 60% carbohydrates (dry weight), due to their high dietary fiber content (5–37%) (Table 1.3). Legume starch with low GI can support the maintenance of blood glucose levels and reduce blood cholesterol levels, preventing obesity, diabetes, heart disorders, piles, and cancers

Table 1.5 Amino acid composition of selected legumes (g/100 g protein)

Amino acid	BGN	CP	SB	AB	LP	LB	LT	CK	BB	KB
Arginine	4.0	1.6	7.2	1.3	3.9	2.2	2.2	1.8	0.7	1.5
Aspartic acid	5.0	2.8	11.7	2.4	3.9	2.9	3.1	2.3	0.8	2.9
Histidine	2.2	0.7	2.5	0.5	1.0	0.6	0.8	0.5	0.2	0.7
Serine	3.2	1.2	5.1	1.0	1.9	1.1	1.3	1.0	0.3	1.3
Glutamic acid	16.5	4.5	18.7	3.1	8.7	4.2	4.4	3.4	1.3	3.6
Proline	3.2	1.1	5.5	0.9	1.5	1.0	1.2	0.8	0.3	1.0
Glycine	3.3	1.0	4.2	0.8	1.5	1.1	1.1	0.8	0.3	0.9
Alanine	3.5	1.1	4.3	1.2	1.3	1.1	1.2	0.8	0.3	1.0
Lysine	3.0	1.6	6.4	1.5	1.9	1.8	2.0	1.3	0.5	1.6
Threonine	2.5	0.9	3.9	0.7	1.3	0.9	1.0	0.7	0.3	1.0
Valine	3.8	1.1	4.8	1.0	1.5	1.2	1.4	0.8	0.3	1.2
Isoleucine	3.8	1.0	4.5	0.8	1.6	1.0	1.2	0.8	0.3	1.0
Leucine	6.8	1.8	7.8	1.7	2.7	1.8	2.0	1.4	0.6	1.9
Tyrosine	3.2	0.8	3.1	0.6	1.4	0.7	0.8	0.5	0.2	0.7
Phenylalanine	4.3	1.4	4.9	1.1	1.4	1.1	1.4	1.0	0.3	1.3
Tryptophan	0.7	0.3	1.3	0.9	0.3	0.3	0.3	0.2	0.1	0.3
Cystine	0.5	0.3	1.3	0.2	0.4	0.4	0.4	0.3	0.1	0.3
Methionine	2.0	0.3	1.3	0.2	0.3	0.3	0.2	0.3	0.1	0.4

BGN Bambara groundnut, *CP* Cowpea, *SB* soybean, *AB* Adzuki bean, *LP* Lupins, *LB* Lima beans, *LT* Lentils, *CK* Chickpea, *BB* Broad beans, *KB* Kidney beans
Adapted from (Leonard 2012; Yao et al. 2015)

(Leonard 2012; Esteban et al. 1998; Philips 1993). As legumes are gluten free, they are suitable for celiac disease patients or people allergic to proteins, gliadin, and glutenin (Mlyneková et al. 2014). Legumes also contain considerable amounts of resistant starch and oligosaccharides along with the monomer such as glucose, galactose, fucose, arabinose, rhamnose, xylose, and mannose, including raffinose which has prebiotic properties (Maphosa and Jideani 2017). Legumes are a good source of vitamins such as folic acid, thiamin, and riboflavin, which help in reducing the risk of neural tube disorders in newborn babies (Messina 2016). Legumes contain significant amounts of the essential minerals, calcium, selenium, chromium, phosphorus, zinc, iron, and magnesium, which play important vital roles in maintaining carbohydrate and lipid metabolism, iron metabolism, haemoglobin synthesis, bone health, and protein synthesis (Maphosa and Jideani 2017). Although iron is absent in legumes, if they are eaten together with vitamin C-rich foods, the absorption of iron is enhanced. Legumes are generally low in fat and contain considerable amounts of mono- and polyunsaturated fatty acids (PUFA), which are essential for human health (Kouris-Blazos and Belski 2016).

1.3.5 Legumes and Climate Smart Agriculture

Legumes offer a wide range of benefits for sustainable food production systems. An important aspect of legumes is their ability to biologically fix atmospheric nitrogen. They can convert atmospheric nitrogen into nitrogen compounds that plants in symbiosis can use with bacteria such as Rhizobium and Bradyrhizobium (Zahran 2001). The bacteria invade the root hairs and reside in nodules (swellings on roots) where atmospheric N is fixed for subsequent transport within the plant and later assimilation into protein compounds. They are inter-cropped with cereals or grown in rotation with other crops (crop rotation) and add nitrogen to the soil via N fixation, which promotes farm biodiversity and increases soil health (Blanchart et al. 2005; Nulik et al. 2013). Intercropping can offer subsistence farmers multiple foods from one farmland, yield higher protein per hectare, pool corresponding canopy crops, equalize consumption of available soil moisture, and lower competition by weeds and pests. Cultivation of legumes as a rotation crop offers numerous ecological and economic benefits in cropping systems as it has the potential to increase soil N, yield, and protein content for subsequent cereal crops. They also provide increased phosphorus (P) availability, and microbial biodiversity in the soil and break the chain of soilborne disease cycles.

Legumes are also used as green manure crops and can free phosphorus from soil to enhance soil fertility and maintains soil structure. Based on broad genetic diversity, legume species can tolerate extreme environmental conditions, survive in water limited environments, and possess a low carbon footprint. Moreover, more climate-smart varieties can be developed from this wide diversity. Numerous legume species have demonstrated a positive effect on climate change and adaptation to biotic and abiotic stresses (Biju et al. 2018; Liu et al. 2019; Rubiales et al. 2015). Legumes require less non-renewable energy inputs for their production compared to cereal plants and, hence, lowers GHG emissions. With the predicted rise in temperature to reach or exceed 1.50 °C in the next 20 years (IPCC 2021), food legumes will become an integral part of climate-smart food production systems. Legume crop residue is one of the major sources of carbon (C) in soils (Dhakal et al. 2016). Unlike other crops, the cultivation of legumes significantly increases the soil C sequestration due to their higher belowground biomass production (Abberton 2010). Legumes can accumulate 30% higher soil organic carbon when compared to other crop species due to their N-fixing property. The C sequestration potential of legumes depends mainly on specific legume species, root architecture, environmental conditions and the existing cropping system, and agronomic interventions during the crop growth period.

Feeding the world in a sustainable way is one of the major challenges faced by the global food system today. Legumes are an integral part of human nutrition, ecological sustainability, and nutritional security. Legume-based cropping systems are sustainable strategies to restore sol health and microbial biodiversity. The cultivation and consumption of legumes can be promoted as a sustainable food resource for the future, and awareness should be developed on the role of pulses in food security and nutrition. Increased production and improved productivity of pulses should be encouraged, and more focus should be given to research and development and extension services in legume based research.

Some of the other biological sources of natural neutraceutical, antimicrobial, antidiabetic, anticancer and hepatoprotective agents are discussed below.

1.4 Bioresources as Nutraceuticals

The term nutraceutical was derived from 'nutrition' and 'pharmaceutical' and was coined by Stephen Defelice in 1989 (Brower 1998). Nutraceuticals are food or food components that provide nutrients as well as health benefits, including the prevention and treatment of diseases (Trottier et al. 2010). They are bioactive molecules from natural sources that contribute positive effects on health promotion and disease prevention. They possess an advantage over medicines in lacking side effects. The development of nutraceuticals requires a connection between the food and pharmaceutical industries (Asif and Mohd 2019).

Nutraceuticals slightly differ from functional foods. Functional foods are those that contain adequate amounts of carbohydrates, proteins, fat, vitamins, etc. required for a healthy life. When a functional food accounts for disease prevention or treatment, it becomes a nutraceutical. Fortified dairy products and citrus fruits are some common examples of nutraceuticals (Kalra 2003; Cockbill 1994). As per AAFCO (1996), 'Nutrient' is a feed constituent that will support the life of humans or animals, whereas 'Nutraceutical' is any non-toxic food material that possesses scientifically proven health benefits (Stephen 2013). Nutraceuticals cover many therapeutic sectors such as antidiabetic, anti-arthritic, cold and cough, sleeping disorders, prevention of certain cancers, osteoporosis, blood pressure, cholesterol control, depression, etc. (Sami Labs 2002; Pandey et al. 2010).

The concept of nutraceuticals first arose in a survey conducted in the U.K., Germany, and France, where the consumers preferred diet over-exercise or hereditary factors for attaining healthy life (Pandey et al. 2010). Broadly, nutraceuticals can be classified into two major groups: potential nutraceuticals and established nutraceuticals regarding the promise of nutraceuticals. A potential nutraceutical is one that holds a promise of particular health benefits, and a potential nutraceutical will become an established one only when there is sufficient clinical evidence to support its effectiveness. The disappointing fact is that majority of the present nutraceuticals are included in the 'potential' category waiting to become established (Bickford et al. 2006).

1.4.1 Categories of Nutraceuticals

The natural food sources used as nutraceuticals can be classified as Dietary fiber, Probiotics, Prebiotics, Polyunsaturated fatty acids, Antioxidant vitamins, Polyphenols, and Spices (Kalia 2005; Kokate et al. 2002).

1.4.1.1 Dietary Fibre

Dietary fibres include plant food material that is not hydrolyzed by digestive enzymes of the body but by the gut microflora. Foods like oats, fruits, barley, beans etc., are rich sources of soluble fibers. Dietary fiber intake lowers the risk of obesity (Lairon et al. 2005), diabetes (Montonen et al. 2003), hypertension (Whelton et al. 2005), stroke (Steffen et al. 2003), and certain gastrointestinal disorders (Petruzziello et al. 2006), and also promotes immunity in humans (Watzl et al. 2005). The recommended dietary fiber intake for adults and children is reported to be 14 g/1000 kcal (Anderson et al. 2009).

1.4.1.2 Polyunsaturated Fatty Acids (PUFA)

Polyunsaturated fatty acids are also known as essential fatty acids, which indicate their significance in the proper functioning of the body. These are introduced into the body through diet and comprise omega-3-(n-3) fatty acids and omega-6-(n-6) fatty acids. Fatty fishes such as mackerel, salmon, herring, trout, blue fin tuna, and fish oils are rich in omega-3-fatty acids. Soybeans, canola, flax seeds, some nuts, and red/black currant seeds also constitute principal sources of omega-3-fatty acids. Omega-6-fatty acids mainly occur in vegetable oils (e.g., sunflower, soybean, corn, etc.) and animal products like eggs, meat, and poultry (Das et al. 2012). Omega-3-fatty acids are effective in cardiovascular diseases and have anti-arrhythmic (Leray et al. 2001; Stoll et al. 1999), hypolipidemic, and anti-thrombotic effects (Bucher et al. 2002).

1.4.1.3 Probiotics and Prebiotics

Probiotics are live microbial feed supplements that beneficially improve the intestinal microbial balance of the host organism (Hotel and Cordoba 2001; Fuller 1992). Probiotics include *Lactobacilli* such as *L. acidophilus*, gram-positive cocci such as *Lactococcus lactis*, and bifidobacteria such as *B. thermophilum*. Various probiotics are available in the market, such as capsules, powder, liquid, gel, or granule forms (Suvarna and Boby 2005). Reports suggest that consumption of probiotics reduces the risk of systemic conditions like allergy, asthma, cancer, and infections of the ear and urinary tract (Lenoir-Wijnkoop et al. 2007).

Prebiotics are dietary components that beneficially alter the composition or metabolism of the gut microbiota (Macfarlane et al. 2006). These are short-chain polysaccharides, especially fructose-based oligosaccharides that usually promote lactobacillus and bifidobacterial growth in the gut, thereby improving metabolism (Hord 2008). Tomatoes, Allium, Banana, and chicory roots are rich sources of fructo-oligosaccharides. Administration of adequate amounts of prebiotics improves lactose tolerance, intestinal immune system, neutralization of toxins, antitumor properties, reduction of constipation, and blood cholesterol levels (Fuller 1992; Sanders 1994).

1.4.1.4 Antioxidant Vitamins

Vitamin C, vitamin E, and carotenoids are usually known as antioxidant vitamins. They help reduce oxidative stress and prevent stress-related diseases such as cancer, cataracts, cardiovascular diseases, etc., through their radical scavenging mechanisms (Elliott 1999). Fruits and vegetables are the main sources of these vitamins. Studies suggest that high consumption of foods rich in vitamin E, vitamin C, and β- carotene is inversely associated with coronary artery disease (Badimon et al. 2010).

1.4.1.5 Polyphenols

Plants produce polyphenols as secondary metabolites, and polyphenols commonly occurring in food include flavanoids and phenolic acids. Polyphenols can regulate cellular processes like apoptosis, gene expression, intercellular signalling, and anticancer and anti-atherogenic effects (Duthie et al. 2003). They also possess anti-inflammatory, antioxidant, antimicrobial, antidiabetic, and cardioprotective activities (Scalbert et al. 2005). A triphenolic stilbene, namely reservatrol present in grapes, is responsible for the antioxidant and anti-inflammatory activities of red wine (Das et al. 2012).

1.4.1.6 Spices

Dietary spices in minute quantities can promote human health by their antioxidant, anti-inflammatory, immunomodulatory, chemopreventive, and antimutagenic effects on cells (Kochhar 2008). Curcumin (from turmeric) and gingerol (from ginger) are reported to have antimutagenic or anticancer potential. Curcumin also possesses antimicrobial, anti-inflammatory, anti-arthritic, and digestive stimulant properties (Das et al. 2012).

Nutraceuticals can be obtained from various natural sources such as plants, animals or microbes, or marine organisms. Some examples of nutraceuticals obtained from different bioresources are mentioned below based on their source of origin.

Plants produce various secondary metabolites as part of their defense mechanism. The majority of these phytochemicals are valuable bioactive molecules that exert many beneficial effects on humans and other animals. Numerous nutraceutical products derived from plants have been reported in various research articles. Curcumin is a well-known polyphenolic compound derived from *Curcuma longa* (Turmeric) that possesses antioxidant, anticancer, and anti-inflammatory properties (Ardalan and Rafieian-Kopaei 2014; Tavafi 2013). Beta sitosterol is a potential anticancer agent reported in the plant *Anaphyllum wightii* and many other Araceae plants (Lekshmi and Swapna 2020).

Bromelain present in the extracts of Pine apple, turmeric, and tea has anti-inflammatory activity. Gentianine, a constituent of Gentian root, is also an active anti-inflammatory agent (Nasri et al. 2014). Capsaicin conjugated linoleic acid, *Momordica charantia*, and *Psyllium* fiber is nutraceuticals having anti obese potential (Rubin and Levin 1994). Caffeine, ephedrine, chitosan, and green tea are effective for reducing body weight (Boozer et al. 2001). Some of the popular nutraceuticals available include botanicals such as ginseng, *Ginkgo biloba*, St. John's wort, and *Echinacea* (Chauhan et al. 2013).

Although a majority of natural nutraceuticals are of plant origin, there are a number of animal products that possess potential health-promoting effects. Some common examples for animal- derived nutraceuticals include dairy products, fish, meat, eggs, etc. Dairy products are rich in calcium, which can prevent osteoporosis and possibly colon cancer (Prates and Mateus 2002). Probiotics are fermented dairy products that have anticarcinogenic, hypo cholesterolemic effects and are effective against enteric pathogens and other intestinal organisms. Scientific reports support the role of probiotics in reducing the risk of cancers like colon cancer (Mitall and Garg 1995).

Lean red meat is a good source of iron, zinc, vitamin B12, n-3 fatty acids, and diets with high amounts of lean red meat can lower plasma cholesterol if it is fat trimmed (Mann 2000). In 1987, an anticarcinogenic fatty acid known as conjugated linoleic acid was isolated from grilled beef (Ha et al. 1987). Eggs are rich sources of many essential dietary components such as proteins, n-3 PUFA, sphingolipids, and choline that promote health (Hasler 2000). Fish oils are primary sources of n-3 fatty acids that are effective against cancer, cardiovascular diseases, etc. (Prates and Mateus 2002). L- carnitine is a betane derivative of β- hydroxy butyrate synthesized in animal liver, and it is an essential nutrient in infancy, pregnancy, breast feeding, etc. (Giovannini et al. 1991).

Many fungi and other microorganisms provide health-promoting nutraceuticals. The fungus *Colletotrichum capsici* is a good source of taxol, an anticancer drug (Kumaran et al. 2011). *Trichoderma viride* produces 3- beta hydroxyurs- 12-en-28-oic acid that possesses anticancer potential against HeLa cell line (Sheeba et al. 2020). The fugal biomass of *Aspergillus oryzae*, *Neurospora intermedia*, etc. is rich source of vegan protein (Souza Filho et al. 2018).

The commercial cultivation of the microalga *Chlorella* was started in 1960 in Japan as a source of single-cell protein (Varfolomeev and Wasserman 2011). *Chlorella* can reduce blood pressure and cholesterol level and boost the immune system (Bishop and Zubeck 2012). *Haematococcus pluvialis* is a freshwater green microalga used for the commercial production of astaxanthin. Astaxanthin is a nutritional supplement having anti-inflammatory, anticancer, antidiabetic, and immunity-enhancing activities (Guerin et al. 2003).

Fishes and other marine species are good sources of novel bioactive compounds that may contribute health-promoting effects for humans as well as other animals. Because of its high nutritional value, fishes constitute one of the essential food stuffs in the human diet. Fishes provide many nutritional components such as proteins, vitamins, minerals, higher quantities of PUFAs, and other micronutrients (Suleria et al. 2015). Macroalgae or seaweeds are commercially significant marine renewable bioresources that may provide novel drugs for inflammations, diabetes, and microbial infections (Bhagavathy et al. 2011). Marine microalgae like *Porphyridium cruentum*, *Isochrysis galbana*, and *Arthrospira* are rich sources of vitamin C, K, A, E, and alpha-carotene that possess strong antioxidant potential (Suleria et al. 2015).

A family of macrocyclic lactones known as bryostatins possess anticancer properties and are obtained from the marine bryozoan *Bugula neritina* (Davidson et al. 2001). Marine probiotics may be useful in fighting antibiotic resistance. *Lactobacillus* (L. casei, L. acidophilus), *Bifidobacterium* (B. bifidum, B. longum, B. infantis), *Leuconostoc* spp., and *Streptococcus* spp. constitute common marine probiotic bacteria (Ande et al. 2016).

1.5 Natural Antimicrobial Agents

Microbial infections lead to the death of millions of people every year worldwide. Even though several antibiotics are available nowadays, the evolution of resistance to such antibiotics in micro organisms leads to the inefficiency of these drugs (Geneva 2014; Baym et al. 2016). The use of a combination of other molecules along with these antibiotics may restore their efficacy of them (Brown 2015; Rana et al. 2019). Non-antibiotic molecules with potential antimicrobial activities can be used for such therapeutic approaches (Vandevelde et al. 2016). Hence, researchers are focused on isolating and identifying novel bioactive compounds with antimicrobial potential from natural sources such as plants, animals, microbes, etc. Such natural antimicrobial compounds can act either alone or in combination with antibiotics to enhance their activity against various microorganisms (Bazzaz et al. 2016; Bazzaz et al. 2018). Chemically complex compounds often possess lesser side effects than synthetic drugs, and the chance of developing resistance is also less (Lewis and Ausubel 2006).

Medicinal plants are rich sources of novel bioactive molecules with various pharmaceutical potential including antimicrobial properties. Studies suggest that the compounds spermidine, quercetin, rutin, tocopherol, and carotenoids isolated from Capparis sp. have antimicrobial and antiviral properties. The seed extracts of

C. decidua also showed antifungal, antibacterial and antileishmanial activities that may be due to glucosinolates and quaternary ammonium (Tlili et al. 2011).

When administered in combination with ciprofloxacin, the piperine alkaloid isolated from *Piper nigrum* and *P. longum* could reduce the minimal inhibitory concentration (MIC) value for a mutant *Staphylococcus aureus* significantly and inhibit its growth (Khan et al. 2006). A steroidal alkaloid tomatidine found mainly in solanaceous plants, including potato, tomato, and egg plants, exhibited antibacterial activity against S. aureus alone or on coadministration with aminoglycosides (Jiang et al. 2016). Reports suggest that tomatidine could be considered an antibiotic potentiator for various antibiotics like ciprofloxacin, gentamicin, ampicillin, and cefepime against both gram-positive and gram-negative bacteria (Soltani et al. 2017).

Baicalein and wogonin are the two active constituents reported from the root extracts of *Scutellaria baicalensis*, which are responsible for its antifungal activity against some dermatophytes, *Candida albicans* and *Aspergillus fumigates* (Da et al. 2019). The teasaponin isolated from tea is an effective antifungal agent against *Candida albicans,* and it can suppress the filament formation via intracellular cAMP and inhibit biofilm formation (Li et al. 2020). Magnoflorin, an alkaloid present in plants like *Magnolia officinalis*, is an antifungal agent effective against *Candida albicans*. It inhibits α- glucosidase action required for normal cell wall composition and virulence of *C. albicans* (Kim et al. 2018).

Nowadays, research is focused on antiviral compounds against SARS-CoV-2 that are derived from plant extracts. Plants like *Dipteryx odorata* and *Cinnamomum aromaticum* contain the phenolic compound coumarin, possessing antiviral activity by inhibiting the protease function of SARS-CoV-2. The molecular-docking studies in *Memecylon randerianum* clearly proved that the compound namely cosmosiin possesses significant inhibitory effects against herpes, chickenpox, and COVID-19 with significant binding energy (Lakshmi and Swapna 2021). Fruits such as grapes, citrus fruits, and tomatoes contain the flavanone naringenin, which can inhibit endolysosomal two-pore channels involved in COVID-19 infections (Khan et al. 2021). Rutaretin 1′-(6″- sinapoylglucoside), reported from the leaves of *Pittosporum dasycaulon*, has been suggested as a promising inhibitor of COVID-19 mpro catalytic dyad through in silico studies (Thodi et al. 2021).

Animal bioresources also provide potential antimicrobial agents. Arenicins, a group of peptides isolated from *Arenicola marina* (sand worm), possess antimicrobial potential against gram- negative bacteria. They are mainly of three types, arenicin 1, arenicin 2, and arenicin 3. Among these, arenicin 1 exhibited good antimicrobial activity against *Escherichia coli* and *Pseudomonas aeruginosa* (Orlov et al. 2019). Reports suggest that slightly modified arenicin-3 peptide showed potent antimicrobial activity against extensive drug resistant and multi-drug-resistant strains like *Escherichia coli, Klebsiella pneumoniae, Pseudomonas aeruginosa,* etc. (Elliott et al. 2020).

Poly (glycolide-co-lactide) is another antimicrobial peptide isolated from frog skin hemocytes that can prevent bacterial adhesion by causing a conformational change and elimination of bacterial pili (da Silva and Teschke 2003). Squalamine, a natural steroid polyamine isolated for the first time from a dog fish shark, exhibited antimicrobial potential against gram-negative and gram-positive bacteria. Squalamine interacts with the negatively charged phosphate groups in the outer membrane of gram-negative bacteria and sequentially disrupts the membrane. In Gram-positive bacteria, it leads to the depolarization of the cytoplasmic membrane, leakage of cytoplasmic contents, and finally, rapid death of cells (Alhanout et al. 2010).

About two thirds of currently used natural antibiotics are synthesized by *Actinobacteria*, most of them coming from the genus *Streptomyces* (Ait Barka et al. 2016; Chater 2016). Meijiemycin is an antifungal compound isolated from Streptomycetes strain that is effective against *Candida albicans*. This compound can inhibit hyphal growth by inducing ergosterol aggregation and interfering with the fungal plasma membrane structure (JieáLow et al. 2020). Garvicin KS is a bacteriocin, produced by *Lactococcus garvieae*, which can inhibit Acinetobacter spp. and has a synergistic effect with polymixin B against E. coli and Acinetobacter spp. (Chi and Holo 2018). Sonorensin is an antimicrobial peptide that is able to kill non-multiplying *Escherichia coli* cells and is produced by *Bacillus sonorensis* (Chopra et al. 2015).

Marine algae are one of the most inexhaustible sources of potential antimicrobial compounds. The sulfated polysaccharides such as fucoidans, laminarin, and alginates isolated from the marine alga *Chaetomorpha aerea* were reported to have potential antimicrobial activity against *Escherichia coli and Staphylococcus aureus* (de Jesus Raposo et al. 2015). Laminarin extracted from *Ascophyllum nodosum* and *Laminaria hyperborean* showed potential antimicrobial properties against gram-positive bacteria and gram-negative bacteria such as *S. aureus, L. monocytogenes, E. coli*, and *Salmonella typhimurium* (Kadam et al. 2015). Many of the sulfated polysaccharides obtained from marine algae can block the cell attachment and colonization of certain intestinal pathogens (Chua et al. 2015).

Seven sesterterpene sulfates isolated from the sponge Dysidea sp. exhibited antibacterial property against *Bacillus subtilis and Proteus vulgaris* by strongly inhibiting the enzyme isocitratelyase (Lee et al. 2008). The marine sponge *Hyrtiose erectus* contains the sesterterpenoidhyrtiosal that binds to HIV N-terminal domain at Ser17, Trp19, and Lys34, and can be applied in anti- HIV research (Qiu et al. 2004). Certain ring B aromatic steroids of the marine endophytic fungus Colletotrichum sp. have exhibited antimicrobial effects against E. coli, *Bacillus megaterium*, and M*icrobotryum violaceum* (Zhang et al. 2009).

1.6 Natural Anticancer Agents

Cancer is a dreadful global health issue and a leading cause of death. The mortality rate due to cancer is expected to rise to 13.1 million by 2030 (Gmeiner and Ghosh 2014). Global Cancer Observatory (GCO) (https://gco.iarc.fr/) is a web-based platform that focuses on global cancer statistics to provide information for cancer control and cancer research. High-quality research studies are concentrated on the development of affordable and effective anticancer drugs (Theofylaktou et al. 2021). Many natural products have been considered a vital source for reducing the complexity of a number of diseases, including cancer. The toxicity effect of various anticancer drugs may range from mild symptoms to lethal conditions. Widely used cancer treatments such as surgery, chemotherapy, and radiotherapy always cause serious adverse effects. The administration of natural products is the only safe remedy as part of disease prevention strategies. But the use of some chemoprotective drugs is still under debate. So the, detailed clinical trials are essential for the approval of these natural compounds as anticancer agents and their effective distribution among the public. Nature itself is an exemplary reservoir of promising chemotherapeutic agents (Khazir et al. 2014). Microbial diversity is recognized as a potential source of novel anticancer lead compounds especially symbiotic microorganisms that lives with other higher marine organisms (Dan and Sanawar 2017). Anthracyclines (such as doxorubicin), bleomycin, dactinomycin (actinomycin), and mitomycin C are some of the antitumor antibiotics from microbes (Kinghorn et al. 2009). More than 30% of the secondary metabolites were isolated from the most common ascomycetes, namely Aspergillus, Penicillium, and Fusarium species (Berdy 2005). Campothesins, taxanes, and vinca alkaloids are extensively used natural anticancer drugs from plants with excellent clinical efficiency and less toxicity (Chavda et al. 2021). Curcumin (diferuloylmethane) is the most well-known anticancer agent studied in the rhizome of *Curcuma longa* (Fridlender et al. 2015). Vinblastine and vincristine are the dominant anticancer drugs isolated from the plant *Catharanthus roseus* among its 30 different alkaloids. Topotecan and irinotecan are the two camptothecin analogs that received Food and Drug Administration (FDA)approval for treating various types of cancer, including ovarian, lung, breast, and colon cancers, and 10-hydroxycamptothecin, with reduced toxicity (Huang et al. 2021). In addition to numerous anticancer agents from plants, numerous compounds have been isolated from marine systems like ecteinascidin, halichondrin B, and dolastatins, from microbes such as bleomycin, doxorubicin, staurosporin, and epothilone B from slime molds (Cragg and Pezzuto 2016). Marine flora and fauna, such as algae, bacteria, sponges, fungi, seaweeds, corals, diatoms, ascidian, etc., are the different marine resources having anticancer potential (Wali et al. 2019). The natural compounds such as psammaplin, didemnin, dolastin, ecteinascidin, and halichondrin isolated from the marine plants and animals such as microalgae, cyanobacteria, heterotrophic bacteria, and invertebrates (e.g., sponges, tunicates, and soft corals) have been tested on various cell lines and studied in experimental animals and used in human chemotherapy (Lichota and Gwozdzinski 2018). Bacterial anticancer agents like antibiotics, bacteriocins, non-ribosomal peptides, polyketides, toxins, etc. have been considered a better solution for cancer treatment in recent years (Baindara and Mandal 2020). Saeed et al. (2021) have critically reviewed the features of marine-derived natural products for the development of anticancer drugs. They have tabulated important anticancer chemicals such as alkaloids, polyketides, polyphenols, terpenes, peptides, carbohydrates, and glycosides isolated from various marine sources and their biological activities. Paclitaxel, a chemotherapeutic agent marketed under the brand name Taxol, Abraxane is the first isolated anticancer natural drug. It is discovered from the bark of the *Taxus brevifolia*, the Pacific yew or western yew, which contains endophytic fungi that synthesize paclitaxel. Thus, the endophytes within the plants can produce bioactive compounds without causing any harm to the host plant. The natural products from various bioresources will contribute beneficial bioactive molecules for the discovery of many drugs. It is also necessary to understand both the in vitro and in vivo mechanisms of these drugs in various cancer cell lines.

1.7 Natural Sources of Hepatoprotective Agents

Liver diseases are a global health problem that occurs throughout the world, with an estimate of two million deaths per year worldwide, one million due to complications of cirrhosis, and one million due to viral hepatitis and hepatocellular carcinoma (Asrani et al. 2019). The hepatotoxicity caused by drugs is hard to detect and prevent in a wide range of patients (Paniagua and Amariles 2017). The food habit and changing lifestyles also cause these liver problems. Azab and Albasha (2018) gave a detailed description of the different hepatotoxic agents and prevention of hepatic disorders by using *Curcuma longa*, *Trigonella foenum graecum*, *Allium sativum*, *Coffea arabica*, *Petroselinum crispum*, *Olea europaea* leaves, and *Mentha piperita*. The microbial sources for the treatment of various liver disorders were scientifically validated through many research investigations. The bioactive compound Cordycepin isolated from the fungus, *Cordyceps militaris* has the ability to safeguard the liver against the accumulation of lipids (Quy et al. 2019). The phytoconstituents from algae such as *Chlorella vulgaris*, *Dunaliella salina*, *Ecklonia stolonifera*, *Hizikia fusiformis*, *Hypnea muciformis*, *Padina boergesenii,* and *Sargassum polycystum* show considerable hepatoprotective action as demonstrated by in vivo experiments conducted on animal models (Jędrejko et al. 2021). The utilization of *Spirulina,* a microscopic blue-green alga that belongs to cyanobacteria, has evidenced the hepatoprotective effect through CCl_4-induced assay (Agrawal et al. 2013). Brown seaweed, namely, *Sargassum ilicifolium*, has shown significant hepatoprotective activity comparable with standard drug silymarin (Hira et al. 2021). In recent literature, many plants were reported to have bioactive compounds, which prove the hepatoprotective activity. The hepatoprotective potential of many plant-derived compounds has been clinically studied.

1.8 Industrial Applications of Bioresources

Many of the natural bioresources and their byproducts have applications in various industries. Natural fibers especially plant-based fibers, and their composites, are of potential applicability in the biomedical field. Plant fibers such as coir, flax, kenaf, jute, ramie, pineapple, cotton, banana, bamboo, etc., and also those from animal sources like wool, silk, and chicken feather fibers are major sources of natural fibers (Mukhopadhyay and Fangueiro 2009). Natural fibers-based polymer composites have significant biomedical applications since they have mechanical properties and biocompatibility comparable to that of human tissues and often lack adverse effects, which is a major requirement for any material to be used in the biomedical field (Cheung et al. 2009).

The pharmaceutical industry is also dependent on natural product sources, predominantly plants, microbes, and marine macroorganisms (Wildman 1997). For example, the pharmaceutically significant compound taxol is obtained from the plant Taxus sp. (Wani et al. 1971) and also from certain fungal endophytes such as *Taxomyces andreane*, *Pestalotia bicilia*, *Fusarium lateritium*, etc. (Strobel et al. 1996). The use of microbes in traditional foods such as bread, wine, vinegar, cheese, soy sauce, and pickles has been reported for a long time. The recent industrial trend involves the applications of the microbiome in the production of biofuels (biogas, biohydrogen, butanol, ethanol, etc.), bio-based chemicals (1,3- propanediol), and biomaterials (polyhydroxyalkanoates) (Jiang et al. 2017).

The process of disposal of food wastes and their byproducts is a major problem faced by the food processing industry since it creates problems regarding environmental protection and sustainability (Russ and Pittroff 2004). Hence, researchers are focused on developing new ideas and techniques to utilize the byproducts of the food industry efficiently. For example, animal blood is a byproduct of meat, and it can be fractionated into several useful components. Certain blood products are used in tissue culture media as an ingredient of blood agar and peptones (Kurbanoglu and Kurbanoglu 2004). Similarly, gelatin can be extracted from animal skin and hides that can be used in the food industry (Cho et al. 2005). Khan et al. (2003) reported the use of waste fish scraps to produce fish protein hydrolysate by enzymatic treatment. This fish protein hydrolysate can be used as a cryoprotectant to avoid protein denaturation of lizardfish surimi during frozen storage (Ohba et al. 2003). Thus, diverse bioresources and their byproducts contribute significantly to various industries and, thereby, economic development of the nation.

The Convention on Biological Diversity (CBD) is the first global agreement that offers opportunities to address and solve issues related to the conservation and sustainable use of biodiversity (Chandra and Idrisova 2011). As per the Biological Diversity Act (2002), "bioresources means plants, animals and microorganisms or parts thereof, their genetic material and by-products (excluding value added products) with actual or potential use or value, but does not include human genetic material." Various factors such as increasing population, agricultural land degradation, and urbanization lead to loss of biodiversity, and the overexploitation of bioresources by the growing population causes various environmental problems (Dkhar et al. 2012). For the better living of the present and future generations, it is the duty of every citizen to be conscious of the available natural bioresources and their sustainable utilization and proper conservation. Each and every organism in the world, whether it is a single-celled microorganism or a complex multicellular one, has its own significance in the ecosystem and the conservation of bioresources is absolutely necessary for the existence of the universe (Pradeep et al. 2017).

1.9 Conclusion

Bioresources, including plants, animals, microorganisms, and other biogenic products, play a significant role in agriculture, food, and pharmaceutical industries. The sustainable use of these bioresources contributes significantly to the economic development of a nation. Factors such as exploding population, urbanization, and agricultural land degradation lead to the overexploitation of bioresources and consequent environmental issues that we are facing at present. The conservation and sustainable utilization of bioresources are essential for the existence of our ecosystem and also for future generations. Hence, along with the research studies going on to explore new bioactive compounds from the natural resources, the focus should also be given to develop new strategies for the conservation as well as sustainable use of these bioresources.

References

Abberton M (2010) Enhancing the role of legumes: potential and obstacles. In: Grassland carbon sequestration: management, policy and economics. Proceedings of the workshop on the role of grassland carbon sequestration in the mitigation of climate change. Rome, 2009

Adnan M, Alshammari E, Patel M, Ashraf SA, Khan S, Hadi S (2018) Significance and potential of marine microbial natural bioactive compounds against biofilms/biofouling: necessity for green chemistry. PeerJ 6:5049

Agrawal R, Soni K, Tomar JS, Saxena S (2013) Hepatoprotective activity of Spirulina species. Int J Sci Eng Res 4(10):1093–1101

Ahirwar NK, Singh R, Chaurasia S, Chandra R, Ramana S (2020) Effective role of beneficial microbes in achieving the sustainable agriculture and eco-friendly environment development goals: a review. Front Microbiol 5:111–123

Ait Barka E, Vatsa P, Sanchez L, Gaveau-Vaillant N, Jacquard C, Klenk HP (2016) The actinobacteria: taxonomy, physiology and their natural products. Microbiol Mol Biol Rev 80: 1–43

Alexandratos N, Bruinsma J (2012) World agriculture towards 2030/2050: the 2012 revision. Food and Agriculture Organization of the United Nations, Rome

Alhanout K, Malesinki S, Vidal N, Peyrot V, Rolain JM, Brunel JM (2010) New insights into the antibacterial mechanism of action of squalamine. J Antimicrob Chemother 65(8):1688–1693

Ande MP, Syamala K, SrinivasaRao P, MuraliMohan K, Lingam SS (2016) Marine nutraceuticals. In: Marine OMICS: principles and applications, pp 329–345

Anderson JW, Baird P, Davis RH, Ferreri S, Knudtson M, Koraym A, Waters V, Williams CL (2009) Health benefits of dietary fiber. Nutr Rev 67(4):188–205

Ani FN (2016) Utilization of bioresources as fuels and energy generation. In: Electric renewable energy systems. Elsevier, pp 140–155

Anonymous (2013) Grain composition. Lupin food Australia. Perth. http://www.lupinfoods.com.au/grain-composition/

Ardalan MR, Rafieian-Kopaei M (2014) Antioxidant supplementation in hypertension. J Renal Inj Prev 3(2):39–40
Arif I, Batool M, Schenk PM (2020) Plant microbiome engineering: expected benefits for improved crop growth and resilience. Trends Biotechnol 38(12):1385–1396
Armbruster WS (2014) Floral specialization and angiosperm diversity: phenotypic divergence, fitness trade-offs and realized pollination accuracy. AoB Plants 6:plu003
Asif M, Mohd I (2019) Prospects of medicinal plants derived nutraceuticals: a re-emerging new era of medicine and health aid. Progr Chem Biochem Res 2(4):150–169
Asrani SK, Devarbhavi H, Eaton J, Kamath PS (2019) Burden of liver diseases in the world. J Hepatol 70(1):151–171
Azab AE, Albasha MO (2018) Hepatoprotective effect of some medicinal plants and herbs against hepatic disorders induced by hepatotoxic agents. J Biotechnol Bioeng 2(1):8–23
Badimon L, Vilahur G, Padro T (2010) Nutraceuticals and atherosclerosis: human trials. Cardiovasc Ther 28(4):202–215
Baindara P, Mandal SM (2020) Bacteria and bacterial anticancer agents as a promising alternative for cancer therapeutics. Biochimie 177:164–189
Barbana C, Boye JI (2011) Angiotensin I-converting enzyme inhibitory properties of enzymatic lentil protein hydrolysates: determination of the kinetic of inhibition. Food Chem 127(1): 94–101
Bartram AK, Lynch MDJ, Stearns JC, Moreno-Hagelsieb G, Neufeld J (2011) Generation of multimillion-sequence 16S rRNA gene libraries from complex microbial communities by assembling paired-end illumina reads. Appl Environ Microbiol 77(11):3846–3852
Baym M, Stone LK, Kishony R (2016) Multidrug evolutionary strategies to reverse antibiotic resistance. Science 351(6268):3292
Bazzaz BSF, Sarabandi S, Khameneh B, Hosseinzadeh H (2016) Effect of catechins, green tea extract and methylxanthines in combination with gentamicin against Staphylococcus aureus and Pseudomonas aeruginosa: combination therapy against resistant bacteria. J Pharmacopuncture 19(4):312
Bazzaz BSF, Khameneh B, Ostad MRZ, Hosseinzadeh H (2018) *In vitro* evaluation of antibacterial activity of verbascoside, lemon verbena extract and caffeine in combination with gentamicin against drug-resistant *Staphylococcus aureus* and *Escherichia coli* clinical isolates. Avicenna J Phytomed 8(3):246
Becker P, Bosschaerts M, Chaerle P, Daniel HM, Hellemans A, Olbrechts A, Rigouts L, Wilmotte A, Hendrickx M (2019) Public microbial resource centers: key hubs for findable, accessible, interoperable, and reusable (FAIR) microorganisms and genetic materials. Appl Environ Microbiol 85(21):01444–01419
Bender SF, Wagg C, van der Heijden MGA (2016) An underground revolution: biodiversity and soil ecological engineering for agricultural sustainability. Trends Ecol Evol 31(6):440–452
Berdy J (2005) Bioactive microbial metabolites. J Antibiot 58(1):1–26
Bhagavathy S, Sumathi P, Bell IJS (2011) Green algae Chlorococcum humicola—a new source of bioactive compounds with antimicrobial activity. Asian Pac J Trop Biomed 1(1):S1–S7
Bickford PC, Tan J, Shytle RD, Sanberg CD, El-Badri N, Sanberg PR (2006) Nutraceuticals synergistically promote proliferation of human stem cells. Stem Cells Dev 15(1):118–123
Biju S, Fuentes S, Gupta D (2018) The use of infrared thermal imaging as a non-destructive screening tool for identifying drought-tolerant lentil genotypes. Plant Physiol Biochem 127: 11–24
Biju S, Fuentes S, Gonzalez Viejo C, Torrico DD, Inayat S, Gupta D (2021) Silicon supplementation improves the nutritional and sensory characteristics of lentil seeds obtained from drought-stressed plants. J Sci Food Agric 101(4):1454–1466
Biological Diversity Act (2002). http://www.nbaindia.org/act/act_ch3.htm
Birner R (2018) Bioeconomy concepts. In: Bioeconomy. Springer, Cham, pp 17–38
Bishop WM, Zubeck HM (2012) Evaluation of microalgae for use as nutraceuticals and nutritional supplements. Nutr Food Sci 2(5):1–6

Blanchart E, Villenave C, Viallatoux A, Barthès B, Girardin C, Azontonde A, Feller C (2005) Long-term effect of a legume cover crop (*Mucuna pruriens var utilis*) on the communities of soil macrofauna and nemato fauna under maize cultivation, in southern Benin. Eur J Soil Biol 42 (S1):136–144

Boozer CN, Nasser JA, Heymsfield SB, Wang V, Chen G, Solomon JL (2001) An herbal supplement containing Ma Huang-Guarana for weight loss: a randomized, double-blind trial. Int J Obes 25(3):316–324

Boye JI, Ma Z (2012) Finger on the pulse. Food Sci Technol 26:20–24

Brenes-Guillén L (2018) 4 bases de datos para realizar análisis bioinformático de comunidades microbianas. BLOGRBT—SERIE 1. Rev Biol Trop, 3p

Brower V (1998) Nutraceuticals: poised for a healthy slice of the healthcare market? Nat Biotechnol 16(8):728–731

Brown D (2015) Antibiotic resistance breakers: can repurposed drugs fill the antibiotic discovery void? Nat Rev Drug Discov 14(12):821–832

Bucher HC, Hengstler P, Schindler C, Meier G (2002) N-3 polyunsaturated fatty acids in coronary heart disease: a meta-analysis of randomized controlled trials. Am J Med 112(4):298–304

Campos-Vega R, Loarca-Piña G, Oomah BD (2010) Minor components of pulses and their potential impact on human health. Food Res Int 43(2):461–482

Cappello S, Volta A, Santisi S, Genovese L, Maricchiolo G (2015) Study of bacterial communities in mussel *Mytilus galloprovincialis* (Bivalvia: Mytilidae) by a combination of 16s crDNA and 16s rDNA sequencing. JSM Microbiol 3(1):1016

Chandra A, Idrisova A (2011) Convention on biological diversity: a review of national challenges and opportunities for implementation. Biodivers Conserv 20(14):3295–3316

Chater KF (2016) Recent advances in understanding Streptomyces. F1000Res 5:2795

Chauhan B, Kumar G, Kalam N, Ansari SH (2013) Current concepts and prospects of herbal nutraceutical: a review. J Adv Pharm Technol Res 4(1):4

Chavda VP, Ertas YN, Walhekar V, Modh D, Doshi A, Shah N, Anand K, Chhabria M (2021) Advanced computational methodologies used in the discovery of new natural anticancer compounds. Front Pharmacol 12(702611):1–15

Cheung HY, Ho MP, Lau KT, Cardona F, Hui D (2009) Natural fibre-reinforced composites for bioengineering and environmental engineering applications. Compos Part B 40(7):655–663

Chi H, Holo H (2018) Synergistic antimicrobial activity between the broad Spectrum bacteriocin Garvicin KS and nisin, farnesol and polymyxin B against Gram-positive and Gram-negative bacteria. Curr Microbiol 75(3):272–277

Cho SM, Gu YS, Kim SB (2005) Extracting optimization and physical properties of yellowfin tuna (*Thunnus albacares*) skin gelatin compared to mammalian gelatins. Food Hydrocoll 19(2): 221–229

Chopra L, Singh G, Kumar Jena K, Sahoo DK (2015) Sonorensin: a new bacteriocin with potential of an anti-biofilm agent and a food biopreservative. Sci Rep 5(1):1–13

Chua EG, Verbrugghe P, Perkins TT, Tay CY (2015) Fucoidans disrupt adherence of *Helicobacter pylori* to AGS cells *in vitro*. Evid Based Complement Alternat Med 2015:1–6

Cockbill CA (1994) Food law and functional foods. Br Food J 96(3):3–4

Condamine FL, Silvestro D, Koppelhus EB, Antonelli A (2020) The rise of angiosperms pushed conifers to decline during global cooling. Proc Natl Acad Sci U S A 117(46):28867–28875

Corlett RT (2016) Plant diversity in a changing world: status, trends, and conservation needs. Plant Divers 38(1):10–16

Costello MJ, Chaudhary C (2017) Marine biodiversity, biogeography, deep-sea gradients, and conservation. Curr Biol 27(11):R511–R527

Cragg GM, Pezzuto JM (2016) Natural products as a vital source for the discovery of cancer chemotherapeutic and chemopreventive agents. Med Princ Pract 25(2):41–59

Crisp MD, Cook LG (2011) Cenozoic extinctions account for the low diversity of extant gymnosperms compared with angiosperms. New Phytol 192(4):997–1009 da Silva JA, Teschke O (2003) Effects of the antimicrobial peptide PGLa on live Escherichia coli. Biochim Biophys Acta 1643(1–3):95–103

Da X, Nishiyama Y, Tie D, Hein KZ, Yamamoto O, Morita E (2019) Antifungal activity and mechanism of action of Ou-gon (*Scutellaria* root extract) components against pathogenic fungi. Sci Rep 9(1):1–12

Dan VM, Sanawar R (2017) Anti cancer agents from microbes. In: Bioresources and bioprocess in biotechnology. Springer, Singapore, pp 171–184

Das L, Bhaumik E, Raychaudhuri U, Chakraborty R (2012) Role of nutraceuticals in human health. J Food Sci Technol 49(2):173–183

Davidson SK, Allen SW, Lim GE, Anderson CM, Haygood M (2001) Evidence for the biosynthesis of bryostatins by the bacterial symbiont "*Candidatus Endobugula sertula*" of the bryozoan *Bugula neritina*. Appl Environ Microbiol 67(10):4531–4537

Davis CC, Schaefer H (2011) Plant evolution: pulses of extinction and speciation in gymnosperm diversity. Curr Biol 21(24):R995–R998

de Jesus Raposo MF, De Morais AMB, De Morais RMSC (2015) Marine polysaccharides from algae with potential biomedical applications. Mar Drugs 13(5):2967–3028

Dhakal Y, Meena RS, Kumar S (2016) Effect of INM on nodulation, yield, quality and available nutrient status in soil after harvest of green gram. Legum Res 39(4):590–594

Dirzo R, Mendoza E (2008) Biodiversity. In: Encyclopedia of ecology

Dkhar M, Tiwari BK, Tynsong H (2012) Traditional knowledge based management and utilization of bio-resources by war khasi tribe of Meghalaya, North-east India

Dopheide A, Xie D, Buckley TR, Drummond AJ, Newcomb RD (2019) Impacts of DNA extraction and PCR on DNA metabarcoding estimates of soil biodiversity. Methods Ecol Evol 10(1): 120–133

Duque-Ortiz A, Relova-Vento D, Pérez-Castillo A, Burger-Pulgarón Y, LoboRivero E (2017) Estabilidad del ADN deshidratado de Mycoplasma gallisepticum. Salud Anim 39(3):1–7

Duranti M (2006) Grain legume proteins and nutraceutical properties. Fitoterapia 77:67–82

Duthie GG, Gardner PT, Kyle JA (2003) Plant polyphenols: are they the new magic bullet? Proc Nutr Soc 62(3):599–603

Elliott JG (1999) Application of antioxidant vitamins in foods and beverages. Food Technol 53:46–48

Elliott AG, Huang JX, Neve S, Zuegg J, Edwards IA, Cain AK, Boinett CJ, Barquist L, Lundberg CV, Steen J, Butler MS (2020) An amphipathic peptide with antibiotic activity against multidrug-resistant Gram-negative bacteria. Nat Commun 11(1):1–13

Epihov DZ, Beerling DJ (2018) Discussion and conclusions the biogeochemistry and belowground metagenomics of N2-fixers: in search of fixer effects beyond nitrogen enrichment. In: Tropical legume trees and their soil-mineral microbiome: biogeochemistry and routes to enhanced mineral access, p 136

Erice P, Bonini V, Cirino G, Colla Y, Ruzzo M (2019) Influencia del microbioma del suelo en el rendimiento del cultivo de pimiento. Phytoma 309

Escalante A, Gattuso M, Pérez P, Zacchino S (2008) Evidence for the mechanism of action of the antifungal phytolaccoside B isolated from Phytolacca tetramera Hauman. J Nat Prod 71 (10):1720–1725

Escalante AE, Barbolla LJ, Ramírez-Barahona S, Eguiarte LE (2014) The study of biodiversity in the era of massive sequencing. Rev Mex Biodivers 85(4):1249–1264

Esteban RM, Mollá E, Valiente C, Jaime L, López-Andréu FJ, Martín-Cabrejas MA (1998) Dietary fibre: chemical and physiological aspects. Recent Res Dev Agric Food Chem 2:293–308

FAO (Food and Agriculture Organization of the United Nations) (1994) Definition and classification commodities, 4. Pulses and derived products. http://www.fao.org/es/faodef/fdef04e.htm. Accessed 29 Feb 2016

FAO (2014) Food and agricultural commodities production: country rank in the world, by commodity. http://faostat3.fao.org/browse/Q/QC/E

FAO (2020) Food and Agriculture Organisation of the United Nation

Fridlender M, Kapulnik Y, Koltai H (2015) Plant derived substances with anti-cancer activity: from folklore to practice. Front Plant Sci 6:799
Fuller R (ed) (1992) Probiotics: the scientific basis. Chapman and Hall, London
Geneva S (2014) Antimicrobial resistance: global report on surveillance
Giovannini C, Agostoni C, Salari PCPC, Acids F, Acidosis O (1991) Is carnitine essential in children? J Int Med Res 19(2):88–102
Gmeiner WH, Ghosh S (2014) Nanotechnology for cancer treatment. Nanotechnol Rev 3(2): 111–122
Govaerts R, Nic Lughadha E, Black N, Turner R, Paton A (2021) The world checklist of vascular plants, a continuously updated resource for exploring global plant diversity. Sci Data 8(1):1–10
Graham PH, Vance CP (2003) Legumes: importance and constraints to greater use. Plant Physiol 131(3):872–877
Guerin M, Huntley ME, Olaizola M (2003) *Haematococcus* astaxanthin: applications for human health and nutrition. Trends Biotechnol 21(5):210–216
Ha YL, Grimm NK, Pariza MW (1987) Anticarcinogens from fried ground beef: heat-altered derivatives of linoleic acid. Carcinogenesis 8(12):1881–1887
Hasler CM (2000) The changing face of functional foods. J Am Coll Nutr 19(5):499S–506S
Hernández-León R, Velázquez-Sepúlveda I, Orozco-Mosqueda MC, Santoyo G (2010) Metagenómica de suelos: grandes desafíos y nuevas oportunidades biotecnológicas. Phyton (Buenos Aires) 79(2):133–139
Hira K, Farhat H, Sohail N, Ansari M, Ara J, Ehteshamul-Haque S (2021) Hepatoprotective activity against acetaminophen-induced liver dysfunction and GC-MS profiling of a brown algae *Sargassum ilicifolium*. Clin Phytosci 7(1):1–11
Hord NG (2008) Eukaryotic-microbiota crosstalk: potential mechanisms for health benefits of prebiotics and probiotics. Annu Rev Nutr 28:215–231
Hotel ACP, Cordoba A (2001) Health and nutritional properties of probiotics in food including powder milk with live lactic acid bacteria. Prevention 5(1):1–10
Hu HW, He JZ (2018) Manipulating the soil microbiome for improved nitrogen management. Microbiology Australia 39(1):24
Huang M, Lu JJ, Ding J (2021) Natural products in cancer therapy: past, present and future. Nat Prod Bioprospect 11(1):5–13
ILDIS (2017) International legume database and information service. World Database of Legumes
Illarze G, Del Pino A, Riccetto S, Irisarri P (2017) Emisión de óxido nitroso, nitrificación, desnitrificación y mineralización de nitrógeno durante el cultivo del arroz en 2 suelos de Uruguay. Rev Argent Microbiol 50(1):97–104
Imbert E, Ladu L, Morone P, Quitzow R (2017) Comparing policy strategies for a transition to a bioeconomy in Europe: the case of Italy and Germany. Energy Res Soc Sci 33:70–81
Ingle AP, Philippini RR, Martiniano S, Marcelino PRF, Gupta I, Prasad S, da Silva SS (2020) Bioresources and their significance: prospects and obstacles. In: Current developments in biotechnology and bioengineering. Elsevier, pp 3–40
IPCC (2021) Climate change 2021: the physical science basis. Contribution of working group I to the sixth assessment report of the intergovernmental panel on climate change. Cambridge University Press, Cambridge
Jędrejko KJ, Lazur J, Muszyńska B (2021) Cordycepsmilitaris: an overview of its chemical constituents in relation to biological activity. Foods 10(11):2634
Jeffery S, Van Der Putten WH (2011) Soil borne human diseases. Publications Office of the European Union, Luxembourg. 49.10.2788: 37199
Jiang QW, Chen MW, Cheng KJ, Yu PZ, Wei X, Shi Z (2016) Therapeutic potential of steroidal alkaloids in cancer and other diseases. Med Res Rev 36(1):119–143
Jiang LL, Zhou JJ, Quan CS, Xiu ZL (2017) Advances in industrial microbiome based on microbial consortium for biorefinery. Bioresour Bioprocess 4(1):1–10
JieáLow Z, Tran H, LináWong S, MeiáPang L (2020) Discovery, biosynthesis and antifungal mechanism of the polyene-polyol meijiemycin. Chem Commun 56(5):822–825

Joseph A (2016) Oceans: abode of nutraceuticals, pharmaceuticals, and biotoxins. In: Joseph A (ed) Investigating seafloors and oceans. Candice Janco, Goa, pp 493–554

Joshi PK, Rao PP (2017) Global pulses scenario: status and outlook. Ann NY Acad Sci 1392:6–17

Kadam SU, O'Donnell CP, Rai DK, Hossain MB, Burgess CM, Walsh D, Tiwari BK (2015) Laminarin from Irish brown seaweeds *Ascophyllum nodosum* and *Laminaria hyperborea*: ultrasound assisted extraction, characterization and bioactivity. Mar Drugs 13(7):4270–4280

Kalia AN (2005) Textbook of industrial pharmacognocy. CBS Publisher and Distributor, New Delhi, pp 204–208

Kalra EK (2003) Nutraceutical-definition and introduction. AAPS Pharmsci 5(3):27–28

Khan MAA, Hossain MA, Hara K, Osatomi K, Ishihara T, Nozaki Y (2003) Effect of enzymatic fish-scrap protein hydrolysate on gel-forming ability and denaturation of lizard fish *Saurida wanieso surimi* during frozen storage. Fish Sci 69(6):1271–1280

Khan IA, Mirza ZM, Kumar A, Verma V, Qazi GN (2006) Piperine, a phytochemical potentiator of ciprofloxacin against *Staphylococcus aureus*. Antimicrob Agents Chemother 50(2):810–812

Khan N, Chen X, Geiger JD (2021) Possible therapeutic use of natural compounds against COVID-19. J Cell Signal 2(1):63–79

Khazir J, Riley DL, Pilcher LA, De-Maayer P, Mir BA (2014) Anticancer agents from diverse natural sources. Nat Prod Commun 9(11):1934578X1400901130

Kim J, Ha Quang Bao T, Shin YK, Kim KY (2018) Antifungal activity of magnoflorine against Candida strains. World J Microbiol Biotechnol 34(11):17

Kinghorn AD, Chin YW, Swanson SM (2009) Discovery of natural product anticancer agents from biodiverse organisms. Curr Opin Drug Discov Devel 12(2):189

Koboldt DC, Meltz KS, Larson DE, Wilson RK, Mardis ER (2013) The next-generation sequencing revolution and its impact on genomics. Cell 155:27–38

Kochhar KP (2008) Dietary spices in health and diseases: I. Indian J Physiol Pharmacol 52(2): 106–122

Kokate CK, Purohit AP, Gokhale SB (2002) Nutraceutical and cosmaceutical pharmacognosy, pp 542–549

Kouris-Blazos A, Belski R (2016) Health benefits of legumes and pulses with a focus on Australian sweet lupins. Asia Pacific J Clin Nutr 21(1):1–17

Kumar BL, Gopal DVRS (2015) Effective role of indigenous microorganisms for sustainable environment. 3 Biotech 5(6):867–876. https://doi.org/10.1007/s13205-015-0293-6

Kumar M, Yadav AN, Saxena R, Paul D, Tomar RS (2021) Biodiversity of pesticides degrading microbial communities and their environmental impact. Biocatal Agric Biotechnol 31:101883. https://doi.org/10.1016/j.bcab.2020.101883

Kumaran RS, Jung H, Kim HJ (2011) In vitro screening of taxol, an anticancer drug produced by the fungus, Colletotrichum capsici. Eng Life Sci 11(3):264–271

Kurbanoglu EB, Kurbanoglu NI (2004) Utilization as peptone for glycerol production of ram horn waste with a new process. Energy Convers Manag 45(2):225–234

Lairon D, Arnault N, Bertrais S, Planells R, Clero E, Hercberg S, Boutron-Ruault MC (2005) Dietary fiber intake and risk factors for cardiovascular disease in French adults. Am J Clin Nutr 82(6):1185–1194

Lakshmi MV, Swapna TS (2021) A computational study on Cosmosiin, an antiviral compound from Memecylon randerianum SM Almeida & MR Almeida. Med Plants Int J Phytomed Relat Ind 31:515–523

Lappe-Oliveras P, Moreno-Terrazas R, Arrizón-Gaviño J, Herrera-Suárez T, García-Mendoza A, Gschaedler-Mathis A (2008) Yeasts associated with the production of Mexican alcoholic non distilled and distilled Agave beverages. FEMS Yeast Res 8:1037–1052

Lara-Victoriano F, Castillo-Reyes F, Flores-Gallegos C, Aguilar CN, Rodríguez-Herrera R (2011) Metagenomics in plant pathology. In: Phytopathology in the omics era. Signpost Research, Kerala, 366p. isbn:978-81-308-0438-5

Lara-Victoriano F, Veana F, Hernández-Castillo FD, Aguilar CN, Reyes-Valdés MH, Rodríguez-Herrera R (2017) Variability among strains of Aspergillus section Nigri with capacity to degrade tannic acid isolated from extreme environments. Arch Microbiol 199(1):77–84. https://doi.org/10.1007/s00203-016-1277-6

Lee D, Shin J, Yoon KM, Kim TI, Lee SH, Lee HS, Oh KB (2008) Inhibition of Candida albicans isocitrate lyase activity by sesterterpenesulfates from the tropical sponge Dysidea sp. Bioorg Med Chem Lett 18(20):5377–5380

Lekshmi S, Swapna TS (2020) In vitro anticancer potential of Anaphyllum wightii Schott. against Dalton's lymphoma ascites cell lines and molecular docking studies of β-sitosterol. Ind J Exp Biol 58(8):522–526

Lenoir-Wijnkoop I, Sanders ME, Cabana MD, Caglar E, Corthier G, Rayes N, Sherman PM, Timmerman HM, Vaneechoutte M, Van Loo J, Wolvers DA (2007) Probiotic and prebiotic influence beyond the intestinal tract. Nutr Rev 65(11):469–489

Leonard E (2012) Cultivating good health. In: Grains and legumes nutrition council. Adelaide, Cadillac Printing, pp 3–18

Leray C, Wiesel ML, Freund M, Cazenave JP, Gachet C (2001) Long-chain n-3 fatty acids specifically affect rat coagulation factors dependent on vitamin K: relation to peroxidative stress. Arterioscler Thromb Vasc Biol 21(3):459–465

Lewis K, Ausubel FM (2006) Prospects for plant-derived antibacterials. Nat Biotechnol 24(12): 1504–1507

Li L, Mendis N, Trigui H, Oliver JD, Faucher SP (2014) The importance of the viable but non-culturable state in human bacterial pathogens. Front Microbiol 4:199. https://doi.org/10.3389/fmicb.2014.00258

Li Y, Shan M, Li S, Wang Y, Yang H, Chen Y, Gu B, Zhu Z (2020) Teasaponin suppresses Candida albicans filamentation by reducing the level of intracellular cAMP. Ann Transl Med 8(5):175

Lichota A, Gwozdzinski K (2018) Anticancer activity of natural compounds from plant and marine environment. Int J Mol Sci 19(11):3533

Liu Y, Li J, Zhu Y, Jones A, Rose RJ, Song Y (2019) Heat stress in legume seed setting: effects, causes, and future prospects. Front Plant Sci 10:938

Lughadha EN, Govaerts R, Belyaeva I, Black N, Lindon H, Allkin R, Magill RE, Nicolson N (2016) Counting counts: revised estimates of numbers of accepted species of flowering plants, seed plants, vascular plants and land plants with a review of other recent estimates. Phytotaxa 272(1):82–88

Macfarlane S, Macfarlane GT, Cummings JH (2006) Review article: prebiotics in the gastrointestinal tract. Aliment Pharmacol Ther 24:701–714

Mann N (2000) Dietary lean red meat and human evolution. Eur J Nutr 39(2):71–79

Maphosa Y, Jideani VA (2017) The role of legumes in human nutrition. In: Functional food-improve health through adequate food, vol. 1, pp 13

McCrory MA, Hamaker BR, Lovejoy JC, Eichelsdoerfer PE (2010) Pulse consumption, satiety, and weight management. Adv Nutr Int Rev J 1:17–30. https://doi.org/10.3945/an.110.1006

Messina MJ (2016) Legumes and soybeans: overview of their nutritional profiles and health effects. Asia Pacific J Clin Nutr 25(1):1–17. https://doi.org/10.1.1.847.8636

Michaels TE (2015) Grain legumes and their dietary impact: overview. The World of Food Grains, pp 265–273

Mitall BK, Garg SK (1995) Anticarcinogenic, hypocholesterolemic, and antagonistic activities of Lactobacillus acidophilus. Crit Rev Microbiol 21(3):175–214

Mlyneková Z, Chrenková M, Formelová Z (2014) Cereals and legumes in nutrition of people with celiac disease. Int J Celiac Dis 2(3):105–109. https://doi.org/10.12691/ijcd-2-3-3

Montonen J, Knekt P, Järvinen R, Aromaa A, Reunanen A (2003) Whole-grain and fiber intake and the incidence of type 2 diabetes. Am J Clin Nutr 77(3):622–629

Mougenot B, Doussoulin JP (2022) Conceptual evolution of the bioeconomy: a bibliometric analysis. Environ Dev Sustain 24(1):1031–1047

Mukhopadhyay S, Fangueiro R (2009) Physical modification of natural fibers and thermoplastic films for composites—a review. J Thermoplast Compos Mater 22(2):135–162

Nasri H, Ardalan MR, Rafieian-Kopaei R (2014) On the occasion of world hypertension day 2014. J Parathyr Dis 2(1):5–6

Nimachow G, Taga T, Tag H, Dai O (2008) Linkages between bio-resources and human livelihood: a case study of Adi tribes of Mirem Village, Arunachal Pradesh (India). Initiation 2(1):183–198

Nulik J, Dalgliesh N, Cox K, Gabb S (2013) Integrating herbaceous legumes into crop and livestock systems in eastern Indonesia. Australian Centre for International Agricultural Research, Canberra

Ohba R, Deguchi T, Kishikawa M, ARsyad F, Morimura S, Kida K (2003) Physiological functions of enzymatic hydrolysates of collagen or keratin contained in livestock and fish waste. Food Sci Technol Res 9(1):91–93

Orlov DS, Shamova OV, Eliseev IE, Zharkova MS, Chakchir OB, Antcheva N, Zachariev S, Panteleev PV, Kokryakov VN, Ovchinnikova TV, Tossi A (2019) Redesigning arenicin-1, an antimicrobial peptide from the marine polychaeta Arenicola marina, by strand rearrangement or branching, substitution of specific residues, and backbone linearization or cyclization. Mar Drugs 17(6):376

Pandey M, Verma RK, Saraf SA (2010) Nutraceuticals: new era of medicine and health. Asian J Pharm Clin Res 3(1):11–15

Paniagua AC, Amariles P (2017) Hepatotoxicity by drugs. In: Pharmacokinetics and adverse effects of drugs-mechanisms and risks factors. NtambweMalangu, IntechOpen

Petruzziello L, Iacopini F, Bulajic M, Shah S, Costamagna G (2006) Uncomplicated diverticular disease of the colon. Aliment Pharmacol Ther 23(10):1379–1391

Philips RD (1993) Starchy legumes in human nutrition and culture. Plant Foods Human Nutr 44(3): 195–211. https://doi.org/10.1007/BF01088314

Pineda A, Dicke M, Pieterse CM, Pozo MJ (2013) Beneficial microbes in a changing environment: are they always helping plants to deal with insects? Funct Ecol 27(3):574–586

Popelka JC, Terryn N, Higgins TJV (2004) Gene technology for grain legumes: can it contribute to the food challenge in developing countries? Plant Sci 167:195–206. https://doi.org/10.1016/j.plantsci.2004.03.027

Pradeep NS, Sugathan S, Abdulhameed S (2017) Bioresources: current status. In: Bioresources and bioprocess in biotechnology. Springer, Singapore, pp 3–23

Prates JM, Mateus CMRP (2002) Functional foods from animal sources and their physiologically active components. Rev Méd Vétér 153(3):155–160

Qiu Y, Deng Z, Pei Y, Fu H, Li J, Proksch P, Lin W (2004) Sesterterpenoids from the marine sponge Hyrtios erectus. J Nat Prod 67(5):921–924

Qiu Z, Egidi E, Liu H, Kaur S, Singh BK (2019) New frontiers in agriculture productivity: optimised microbial inoculants and in situ microbiome engineering. Biotechnol Adv 37(6): 107371. https://doi.org/10.1016/j.biotechadv.2019.03.010

Quy TN, Xuan TD, Andriana Y, Tran HD, Khanh TD, Teschke R (2019) Cordycepin isolated from Cordyceps militaris: its newly discovered herbicidal property and potential plant-based novel alternative to glyphosate. Molecules 24(16):2901

Ramamurthy T, Ghosh A, Pazhani GP, Shinoda S (2014) Current perspectives on viable but non-culturable (VBNC) pathogenic bacteria. Front Microbiol. https://doi.org/10.3389/fpubh.2014.00103

Rana R, Sharma R, Kumar A (2019) Repurposing of existing statin drugs for treatment of microbial infections: how much promising? Infect Disord Drug Targets 19(3):224–237

Rege JEO, Gibson JP (2003) Animal genetic resources and economic development: issues in relation to economic valuation. Ecol Econ 45(3):319–330

Ribéreau S, Aryee AN, Tanvier S, Han J, Boye JI (2018) Composition, digestibility, and functional properties of yellow pea as affected by processing. J Food Process Preserv 42(1):13375

Rockland LB, Radke TM (1981) Legume protein quality. Food Technology (USA)

Rubiales D, Fondevilla S, Chen W, Gentzbittel L, Higgins TJ, Castillejo MA, Singh KB, Rispail N (2015) Achievements and challenges in legume breeding for pest and disease resistance. Crit Rev Plant Sci 34(1–3):195–236

Rubin SA, Levin ER (1994) Clinical review 53: the endocrinology of vasoactive peptides: synthesis to function. J Clin Endocrinol Metabol 78(1):6–10

Rui X, Boye JI, Barbana C, Simpson BK, Prasher SO (2012) Electrophoretic profiles and angiotensin I-converting enzyme inhibitory activities of nine varieties of Phaseolus vulgaris protein hydrolysates. J Nutr Food Sci 2:1000156

Russ W, Pittroff RM (2004) Utilizing waste products from the food production and processing industries. Crit Rev Food Sci Nutr 44(2):57–62

Saeed AF, Su J, Ouyang S (2021) Marine-derived drugs: recent advances in cancer therapy and immune signaling. Biomed Pharmacother 134:111091

Sami Labs—pioneer in nutraceuticals (2002) The Hindu Newspaper. http://www.hinduonnet.com/thehindu/biz/2002/08/05/stories/2002080500040200.htm

Sanders ME (1994) Lactic acid bacteria as promoters of human health. In: Functional foods. Springer, Boston, pp 294–322

Scalbert A, Johnson IT, Saltmarsh M (2005) Polyphenols: antioxidants and beyond. Am J Clin Nutr 81(1):215S–217S

Ścieszka S, Klewicka E (2019) Algae in food: a general review. Crit Rev Food Sci Nutr 59(21):3538–3547

Semba RD, Ramsing R, Rahman N, Kraemer K, Bloem MW (2021) Legumes as a sustainable source of protein in human diets. Glob Food Sec 28:100520

Sernaque Aguilar YA, Cornejo La Torre M, Pierre Regard J, Mialhe Matonnier EL (2019) Caracterización molecular de bacteriascultivables y no cultivablesprocedentes de pozas de lixiviación con cianuro. Rev Peru Biol 26(2):275–282

Shahid MJ, Al-Surhanee AA, Kouadri F, Ali S, Nawaz N, Afzal M, Rizwan M, Ali B, Soliman MH (2020) Role of microorganisms in the remediation of wastewater in floating treatment wetlands: a review. Sustainability 12(14):5559

Sharma P, Sharma N (2017) Industrial and biotechnological applications of algae: a review. J Adv Plant Biol 1(1):1

Sheeba H, Ali MS, Anuradha V (2020) In-vitro anti-cancer activity of endophytic fungi isolated from Ziziphus mauritiana in cervical cancer cell line. Eur J Med Plants 31:38–48

Singh U, Singh B (1992) Tropical grain legumes as important human foods. Econ Bot 46(3):310–321

Singh AK, Sisodia A, Sisodia V, Padhi M (2019) Role of microbes in restoration ecology and ecosystem services. In: New and future developments in microbial biotechnology and bioengineering. Elsevier, pp 57–68

Sinha SK (1977) Food legumes: distribution, adaptability and biology of yield. FAO Plant Production and Protection Paper No. 3. FAO, Rome

Sita K, Sehgal A, Hanumantha Rao B, Nair RM, Vara Prasad PV, Kumar S, Gaur PM, Farooq M, Siddique KH, Varshney RK, Nayyar H (2017) Food legumes and rising temperatures: effects, adaptive functional mechanisms specific to reproductive growth stage and strategies to improve heat tolerance. Front Plant Sci 8:1658

Sivan A, Chet I (1989) The possible role of competition between Trichoderma harzianum and Fusarium oxysporum on rhizosphere colonization. Phytopathology 79(2):198–203

Soltani R, Fazeli H, Najafi RB, Jelokhanian A (2017) Evaluation of the synergistic effect of tomatidine with several antibiotics against standard and clinical isolates of Staphylococcus aureus, Enterococcus faecalis, Pseudomonas aeruginosa and Escherichia coli. Iran J Pharma Res 16(1):290

Soria MA (2016) Por qué son importanteslosmicroorganismos del suelo para la agricultura? Química Viva 15(2):3–10

Souza Filho PF, Nair RB, Andersson D, Lennartsson PR, Taherzadeh MJ (2018) Vegan-mycoprotein concentrate from pea-processing industry byproduct using edible filamentous fungi. Fungal Biol Biotechnol 5(1):1–10

Srivastava V, Sarkar A, Singh S, Singh P, de Araujo AS, Singh RP (2017) Agroecological responses of heavy metal pollution with special emphasis on soil health and plant performances. Front Environ Sci 5:64

Steffen LM, Jacobs DR Jr, Stevens J, Shahar E, Carithers T, Folsom AR (2003) Associations of whole-grain, refined-grain, and fruit and vegetable consumption with risks of all-cause mortality and incident coronary artery disease and ischemic stroke: the Atherosclerosis Risk in Communities (ARIC) Study. Am J Clin Nutr 78(3):383–390

Stephen D (2013) Nutraceuticals: what are they and do they work? 7(4). Kentucky Equine Research, Versailles, pp 1–50

Stoll AL, Severus WE, Freeman MP, Rueter S, Zboyan HA, Diamond E, Cress KK, Marangell LB (1999) Omega 3 fatty acids in bipolar disorder: a preliminary double-blind, placebo-controlled trial. Arch Gen Psychiatry 56(5):407–412

Strobel GA, Hess WM, Ford E, Sidhu RS, Yang X (1996) Taxol from fungal endophytes and the issue of biodiversity. J Ind Microbiol 17(5):417–423

Suleria HAR, Osborne S, Masci P, Gobe G (2015) Marine-based nutraceuticals: an innovative trend in the food and supplement industries. Mar Drugs 13(10):6336–6351

Suvarna VC, Boby VU (2005) Probiotics in human health: a current assessment. Curr Sci 88(11): 1744–1748

Swaminathan M (1988) Handbook of food science and experimental foods. Bangalore Printing and Publishing, Bangalore, pp 125–127

Tanase AM, Mereuta I, Chiciudean I, Ionescu R, Milea L, Petruta-Cornea C, Vassu T, Stoica I (2015) Comparison of total DNA extraction methods for microbial community form polluted soil. Agric Agric Sci Proc 6:616–622

Tavafi M (2013) Diabetic nephropathy and antioxidants. J Nephropathol 2(1):20

Testen AL, Miller SA (2018) Carbon source and soil origin shape soil microbiomes and tomato soilborne pathogen populations during anaerobic soil disinfestation. Phytobiomes J 2(3): 138–150. https://doi.org/10.1094/PBIOMES-02-18-0007-R

Theofylaktou D, Takan I, Karakülah G, Biz GM, Zanni V, Pavlopoulou A, Georgakilas AG (2021) Mining natural products with anticancer biological activity through a systems biology approach. In: Oxidative medicine and cellular longevity

Thodi RC, Ibrahim JM, Surendran VA, Nair AS, Sukumaran ST (2021) Rutaretin1′-(6 ″--sinapoylglucoside): promising inhibitor of COVID 19 mpro catalytic dyad from the leaves of Pittosporum dasycaulonmiq (Pittosporaceae). J Biomol Struct Dyn, 1–17

Tlili N, Elfalleh W, Saadaoui E, Khaldi A, Triki S, Nasri N (2011) The caper (Capparis L.): ethnopharmacology, phytochemical and pharmacological properties. Fitoterapia 82(2):93–101

Travaglia C, Masciarelli O, Fortuna J, Marchetti G, Cardozo P, Lucero M, Zorza E, Luna V, Reinoso H (2015) Towards sustainable maize production: glyphosate detoxification by Azospirillum sp. and Pseudomonas sp. Crop Prot 77:102–109. https://doi.org/10.1016/j.cropro.2015.07.003

Trottier G, Boström PJ, Lawrentschuk N, Fleshner NE (2010) Nutraceuticals and prostate cancer prevention: a current review. Nat Rev Urol 7(1):21–30

UN (2017) Probabilistic Population Projections Based on the World Population Prospects: the 2017 Revision. United Nations. https://esa.un.org/unpd/wpp/Download/Probabilistic/Population/

Usta C (2013) Microorganisms in biological pest control—a review (bacterial toxin application and effect of environmental factors). Curr Progr Biol Res 13:287–317

Uyoh EA, Nkang AE, Eneobong EE (2003) Biotechnology, genetic conservation and sustainable use of bioresources. Afr J Biotechnol 2(12):704–709

Van Der Heijden MG, Bardgett RD, Van Straalen NM (2008) The unseen majority: soil microbes as drivers of plant diversity and productivity in terrestrial ecosystems. Ecol Lett 11(3):296–310

van Hijum SA, Vaughan EE, Vogel RF (2013) Application of state-of-art sequencing technologies to indigenous food fermentations. Curr Opin Biotechnol 24:178–186

Vandevelde NM, Tulkens PM, Van Bambeke F (2016) Modulating antibiotic activity towards respiratory bacterial pathogens by co-medications: a multi-target approach. Drug Discov Today 21(7):1114–1129

Varfolomeev SD, Wasserman LA (2011) Microalgae as source of biofuel, food, fodder, and medicines. Appl Biochem Microbiol 47(9):789–807

Vidhyanandan LM, Kumar SM, Sukumaran ST (2020) Algal metabolites and phyco-medicine. In: Plant metabolites: methods, applications and prospects. Springer, Singapore, pp 291–316

Vitorino LC, Bessa LA (2018) Microbial diversity: the gap between the estimated and the known. Diversity 10(2):46

Vogelpohl T (2021) Transnational sustainability certification for the bioeconomy? Patterns and discourse coalitions of resistance and alternatives in biomass exporting regions. Energy Sustain Soc 11(1):1–13

Wali AF, Majid S, Rasool S, Shehada SB, Abdulkareem SK, Firdous A, Beigh S, Shakeel S, Mushtaq S, Akbar I, Madhkali H (2019) Natural products against cancer: review on phytochemicals from marine sources in preventing cancer. Saudi Pharm J 27(6):767–777

Wani MC, Taylor HL, Wall ME, Coggon P, McPhail AT (1971) Plant antitumor agents. VI. Isolation and structure of taxol, a novel antileukemic and antitumor agent from Taxus brevifolia. J Am Chem Soc 93(9):2325–2327

Watzl B, Girrbach S, Roller M (2005) Inulin, oligofructose and immunomodulation. Br J Nutr 93 (S1):S49–S55

Wells J, Varel V (2011) Symbiosis of plants, animals and microbes. Animal welfare in animal agriculture: husbandry, stewardship and sustainability in animal production. CRC Press, New York, pp 185–203

Whelton SP, Hyre AD, Pedersen B, Yi Y, Whelton PK, He J (2005) Effect of dietary fiber intake on blood pressure: a meta-analysis of randomized, controlled clinical trials. J Hypertens 23(3): 475–481

Wildman HG (1997) Pharmaceutical bioprospecting and its relationship to the conservation and utilization of bioresources. In: International conference on bioversity and bioresources: conservation and utilization, pp 23–37

Wilson MC, Piel J (2013) Metagenomic approaches for exploiting uncultivated bacteria as a resource for novel biosynthetic enzymology. Chem Biol 20:636–647

Wu Z, Wu W, Zhou S, Wu S (2016) Mycorrhizal inoculation affects pb and cd accumulation and translocation in pakchoi (Brassica chinensis L.). Pedosphere 26(1):13–26. https://doi.org/10.1016/S1002-0160(15)60018-2

Yadav SS, McNeil DL, Redden R, Patil SA (2010) Climate change and management of cool season grain legume crops. Springer Science & Business Media

Yao DN, Kouassi KN, Erba D, Scazzina F, Pellegrini N, Casiraghi MC (2015) Nutritive evaluation of the Bambara groundnut Ci12 landrace [Vigna subterranea (L.) Verdc. (Fabaceae)] produced in Côte d'Ivoire. Int J Mol Sci 16:21428–21441. https://doi.org/10.3390/ijms160921428

Zahran HH (2001) Rhizobia from wild legumes: diversity, taxonomy, ecology, nitrogen fixation and biotechnology. J Biotechnol 91(2-3):143–153

Zhang W, Draeger S, Schulz B, Krohn K (2009) Ring B aromatic steroids from an endophytic fungus, *Colletotrichum* sp. Nat Prod Commun 4(11):1934578X0900401101

Zörb C, Lewandowski I, Kindervater R, Göttert U, Patzelt D (2018) Biobased resources and value chains. In: Bioeconomy. Springer, Cham, pp 75–95

Chapter 2
Plant Resources and Functional Foods

Jeena Elsa Reji and Linu Mathew

Abstract A healthy diet can provide all the essential nutrients for the well-being of an individual. Functional foods can be defined as any modified food or food ingredient that may provide a health benefit beyond the traditional nutrients it contains. Consumer interest in functional foods has been increasing due to their potential to improve health or reduce the risk of diseases. This chapter discusses the bioactive components and the functional health benefits of plant-based foods. It focuses on some of the significant plant resources and their functional properties, which help reduce or minimize the risk of certain diseases.

Keywords Functional foods · Pigments · Anti-oxidants · Fruits and vegetables · Health benefits

2.1 Introduction

The primary role of food is to provide essential nutrients to maintain the life and growth of an individual. Affluence and urbanization are linked to a lifestyle where our daily routine with less physical activity and greater access to high-energy foods. Life-threatening diseases due to changes in lifestyle demand an increase in functional foods globally. The food industry is geared towards producing functional foods due to consumer demands for healthier foods. They fight against various physiological threats like oxidative stress, cardiovascular disease, cancer, and immune dysfunction (Shubha et al. 2020).

Functional foods provide benefits beyond basic nutrition and may play a role in reducing or minimizing the risk of certain diseases and other health conditions. The International Life Sciences Institute of North America (ILSI) has defined functional foods as 'physiologically active food components providing health benefits beyond

J. E. Reji · L. Mathew (✉)
School of Biosciences, Mahatma Gandhi University, Kottayam, Kerala, India
e-mail: linumathew@mgu.ac.in

S. T. Sukumaran, K. T R (eds.), *Conservation and Sustainable Utilization of Bioresources*, Sustainable Development and Biodiversity 30,
https://doi.org/10.1007/978-981-19-5841-0_2

basic nutrition.' The European Commission Concerted Action on Functional Food Science in Europe regards food as functional, 'if it is satisfactorily demonstrated to affect beneficially one or more target functions in the body, beyond adequate nutritional effects, in a way that is relevant to either an improved state of health and well-being and reduction of risk of disease' (Henry 2010). Functional foods are products that resemble traditional foods but have physiological benefits. Examples include conventional, fortified, enriched, or enhanced foods and dietary supplements (Bishnoi 2016). All foods are functional to some extent because all foods provide taste, aroma, and nutritional value. However, foods are now being examined intensively to identify physiologically active components from both plants and animals (known as phytochemicals and zoochemical) that could reduce the risk for various chronic diseases (Hasler 2002).

The functional product industry is in its infancy, characterized by dynamic growth, and new products are launched continually. These kinds of food offer many positive health-related actions, such as the potential to boost the immune system, reduce the risk of cardiovascular problems, osteoporosis, obesity, and cancer, and improve memory and fitness (Topolska et al. 2021). Functional foods occur in many forms. Some are conventional foods with bioactive components that can be identified and linked to positive health effects. Some are fortified or enhanced foods, specifically formulated to reduce disease risk. Consumers can choose from a wide variety of foods that contain functional components either inherently (e.g., soy protein, fish, olive oil) or through fortification (e.g., folate-fortified milk). As and when additional bioactive components are identified, the opportunities for developing functional foods will become broader. Foods that naturally provide a bioactive substance may be modified to enhance the food's level (e.g., eggs with increased levels of omega-3 fatty acids). Contrarily, foods that do not naturally contain a bioactive substance can be fortified with added nutrients (e.g., calcium-fortified orange juice). Food can also be altered by removing or reducing a harmful component and replacing it with another substance with beneficial effects (Lau et al. 2013). The food industry now focuses on achieving 'optimized' nutrition, maximizing life expectancy and quality by identifying food ingredients that improve the capacity to resist disease and enhance health when added to a' balanced' diet. As an outcome, functional foods are formulated as they can provide necessary health benefits beyond basic nutritional needs (Gibson and Williams 2000).

2.2 Roles of Functional Foods

All foods are essential as it provides nutrients and energy to sustain life. But, some of them can be termed as functional due to some active ingredients in them. The functional components in the food are the biologically active compounds like phytochemicals present in the plants that can prevent the onset of diseases. Foods with physiologically active compounds like alkaloids, flavonoids, carotenoids, polyphenolic compounds, and essential oils make the food functional, providing specific health benefits like antioxidative activity, detoxification, and anti-coronary disease activity, improving digestion, regulating metabolism, and enhancing immunity. Functional foods are not medicines, so they do not have therapeutic effects. The role of these foods in the disease is to reduce the risk of occurrence rather than prevent them. Thus, we can reduce the risk of illness and maintain our health and well-being through a healthy diet and lifestyle (Butnariu and Sarac 2019). Human diets of plant origin contain many hundreds of compounds that cannot be considered nutrients, but appear to play a role in maintaining health (Orzechowski et al. 2002). The major functional components in the food are as follows:

(a) Dietary fiber

These are non-starchy polysaccharides and structural components of plant-derived foods. It is composed of long, straight, and branched chains of carbohydrate molecules held together by bonds that human digestive enzymes cannot hydrolyze. Dietary fibers can be divided into water-soluble fibers and water-insoluble fibers. The water-soluble fibers are mainly β-glucans, gums, pectin, mucilage, and arabinoxylans, while the water-insoluble fibers are composed of lignin, cellulose, and hemicellulose (Abuajah et al. 2014). The consumption of dietary fibers has many potential health benefits, such as lowering the incidence of constipation and irritable bowel syndrome as it softens feces and promotes their elimination (Shandilya and Sharma 2017). It can also lower blood cholesterol, keep blood vessels unblocked, prevent arteriosclerosis, and the incidence of cardiovascular and cerebrovascular diseases (Xiao and Li 2020).

(b) Carotenoids

Carotenoids are fat-soluble pigments found in plants, algae, and photosynthetic bacteria. These pigments produce bright yellow, orange, and red colors for vegetables, fruits, and plants. Most carotenoids are hydrocarbons containing 40 carbon atoms and two terminal rings. Two classes of carotenoids are xanthophylls and carotenes. Xanthophylls are the oxygenated derivatives of carotenes such as lutein, violaxanthin, neoxanthin, and zeaxanthin. The carotenes, such as β-carotene, consist of linear hydrocarbons that can be cyclized at one end or both ends of the molecule (Botella-Pavía and Rodríguez-Concepción 2016). Carotenoids have beneficial properties such as strengthening the immune system, reducing the risk of degenerative diseases, anti-oxidant properties, and anti-inflammatory and anti-cancerous properties (Mezzomo and Ferreria 2016).

(c) Flavonoids

Flavonoids are a large family of polyphenolic plant compounds. Six major subclasses of flavonoids are anthocyanidins, flavan-3-ols, flavonols, flavanones, and isoflavones. Flavonoids are very effective anti-oxidants, and it has been proposed that they protect against cardiovascular disease by reducing the oxidation of low-density lipoproteins. Flavonoids help regulate cellular activity and fight off free radicals that cause oxidative stress on the body. They absorb dust and impurities remaining in the human digestive system and discharge them out of the body, thus, expelling toxins and clearing the stomach. It has anti-cancer effects by resisting oxidation and scavenging oxygen-free radicals, can treat hypertension, diabetes, and its complications by reducing blood sugar, blood fat, and urine sugar can also enhance the body's immune system function to play a role in liver protection. It can also treat abscess ulcers with anti-inflammatory, anti-allergy, and anti-virus effects (Xiao and Li 2020).

(d) Organosulphur compounds

These compounds are commonly found in cruciferous vegetables such as broccoli, cauliflower, Brussels sprouts, or allium vegetables. Organosulfur compounds contain sulfur atoms bound to a cyanate group or a carbon atom in a cyclic or non-cyclic configuration (Abuajah et al. 2014). Active sulfur-containing compounds induce apoptosis, promote chemoprevention by the induction of xenobiotic-metabolizing enzymes, ameliorate the detoxification of carcinogens, and are involved in cell cycle arrest, as well as in the blocking of the metabolism of hydrocarbons and nitrosamines, scavenging-free radicals, and modulating the enzymes responsible for DNA repair (Miekus et al. 2020).

(e) Phytosterol

Phytosterols are plant sterols with a chemical structure similar to cholesterol except for adding an extra methyl or ethyl group. Several hundreds of different phytosterol molecules have been identified in plant cells; the most common ones are beta-sitosterol, campesterol, stigmasterol, brassicasterol, and avenasterol (Poli et al. 2021). The food content in phytosterols is highest in oily fruits, oilseeds, and the oils obtained from them. Phytosterols can inhibit the uptake of both dietary and endogenously produced (biliary) cholesterol from intestinal cells. Such inhibition results in a decrease in serum total and LDL-cholesterol levels. Levels of HDL-cholesterol and triglycerides do not appear to be affected by dietary phytosterol consumption. Several studies suggest a protective role of phytosterols, especially beta-sitosterol, in colon, prostate, and breast cancer (Jones and AbuMweis 2009).

(f) Tocopherol

The molecular structure of tocopherols consists of a chromanol ring connected to a long carbon side chain. Variations in the number and position of the methyl groups on the ring result in different forms named α-, β-, γ-, and δ-tocopherol. Among these, α-Tocopherol is the preferred form of vitamin E, absorbed and accumulated in humans and other mammals. On the other hand, β-, γ-, and δ-tocopherols are referred to as having little vitamin E activity. Still, they have similar anti-oxidant activity and may convey additional health benefits (Delgado et al. 2020). It has anti-oxidant properties, reduces inflammation, and potentially promotes anti-cancer and anti-aging benefits. Tocopherols can also reduce that high cholesterol reduces the risk of cardiovascular diseases, osteoporosis, and several cancers.

2.3 Plants Resources as Functional Foods

Plants are the primary source of food that provides essential nutrients and energy to sustain life. Secondary metabolites produced in the plants are not directly involved in the normal growth and development, but they act as defense compounds against predators, diseases, and parasites. These bioactive molecules also have some health-promoting properties. A diet rich in plant foods may reduce the risk of degenerative diseases (Donno and Turrini 2020). Foods are considered functional if they contain bioactive components such as nutrients (vitamins, minerals, protein, etc.) or non-nutrient (phytochemicals including polyphenols, prebiotic dietary fibers, etc.) that affect one or more physiological functions in the body to improve health, reduce disease risk.

Carotenoid compounds, specifically beta-carotene, have been extensively studied for their influence on cancer development and cell growth. Based on epidemiological studies, beta-carotene was hypothesized to be among the most active compounds in fruit and vegetables responsible for the decreased incidence of lung cancer. More than 8000 phenolic compounds have been isolated from fruits and vegetables (Crowe and Coni 2013). Polyphenols are promising molecules for the prevention and treatment of neurodegenerative diseases, and anti-oxidant-based mechanisms of polyphenols can modulate several cell-signaling pathways and mediators. Plant-derived polyphenols such as curcumin, resveratrol, propolis, ginsenoside, rosmarinic acid, and polyunsaturated fatty acids have been applied to numerous neurodegenerative diseases for their antioxidative and anti-inflammatory effects. Several plant-derived foods contain these functional components and can be effectively included in the diet to reduce the risk of infections. Some potential plant-based functional foods are presented here.

2.3.1 Tomatoes

Tomatoes are rich in many bioactive compounds such as vitamin C and E and act as the source of strong anti-oxidants and carotenoids. Lycopene is the primary carotenoid present in these fruits playing a vital role in preventing prostrate cancer. The higher levels of lycopene in serum or tissue also reduce the risk of cancers like breast, digestive tract, cervix, bladder, and skin. The anti-oxidant activity of lycopene is the primary reason for its cancer preventive action (Nayak et al. 2019). It also can reduce cardiovascular disease development by reducing inflammation, inhibiting cholesterol synthesis, or improving immune function. Lycopene is known to have protective effects on oxidative stress, hypertension, atherosclerosis, and diabetes. The proposed mechanisms of action of lycopene in prostate cancer prevention include inhibition of proliferation, antiandrogen, and antigrowth factor effects. Foods rich in lycopene have decreased endogenous levels of DNA strand breaks in human lymphocytes (Canene-Adams et al. 2005).

2.3.2 Flaxseed

It is also known as linseed (Fa. Lineaceae); it is a blue-flowering annual herb that produces small flat seeds varying from golden yellow to reddish-brown color. Among the functional foods, flaxseed has emerged as a potential functional food as a good source of alpha-linolenic acid, lignans, high-quality protein, soluble fiber, and phenolic compounds (Kajla et al. 2015). Flaxseeds contain 51–55% alpha-linolenic acid (ALA), which inhibits cardiovascular disease, inflammation, blood pressure, cancer, skin diseases, and immune disorders such as renal failure, rheumatoid arthritis, and multiple sclerosis.

Consumption of flaxseed reduces total and LDL-cholesterol and platelet aggregation (Kaur and Das 2011). Flaxseed lignans are converted by beneficial gut flora into two hormone-like substances called enterolactone and enterodiol that have protective effects against breast cancer. Flaxseed oil eases symptoms of rheumatoid arthritis, lubricates joints, and lessens stiffness and joint pain as it is high in alpha-linolenic acid (ALA), a type of omega-3 fatty acid known as an anti-inflammatory (Alshafe et al. 2015).

2.3.3 Soybean

Soy protein products contain bioactive molecules called phytoestrogens or isoflavones. The three major isoflavones in soybeans are genistein, daidzein, and glycitein. Isoflavones may reduce the risk of several cancers, including breast, lung, colon, rectum, stomach, and prostate. Among the isoflavones, genistein is thought to act against cancers by interfering with cancer-promoting enzymes, blocking the activity of hormones in the body, and even interfering with the process by which tumors receive nutrients and oxygen. Isoflavones also play a major role in improving vascular functions and in reducing cholesterol. Consumption of soy protein reduces arterial stiffness, decreases LDL-cholesterol's susceptibility to oxidation, lowers total and LDL-cholesterol, and increases HDL-cholesterol, possibly reducing the risk of coronary artery disease. Soybeans have a very low glycaemic index and, thus, can be included in a diabetic diet. Regular consumption of soybean may help reduce symptoms associated with type 2 diabetes (Jooyandeh 2011).

Isoflavones have a chemical structure similar to the hormone estrogen, which allows them to bind to both estrogen receptors (ER)—ERα and ERβ, thus, exerting estrogen-like effects and so are referred to as phytoestrogens. They can reduce the rapid rate of bone loss associated with the onset of menopause (Messina 2016). Soy phytochemicals and their metabolites play a major role in preventing chronic diseases that arise due to oxidative stress, and therefore, it exhibits antioxidative and anti-inflammatory properties. The bioactive phytochemicals in soybean manifested complementary effects on renal function by improving renal flow, reducing proteinuria, renal histological damage, and retarding the development of kidney disease (Garg et al. 2016).

2.3.4 *Citrus Fruits*

Citrus fruits and juices are rich sources of bioactive compounds, like flavonoids, carotenoids, limonoids, coumarin-related compounds, folates, essential oils, pectins, and vitamin C. Citrus flavonoids exhibit anti-inflammatory, antioxidative, antiallergenic, antimicrobial, antiviral, antiulcer, and analgesic properties. Flavonoids protect DNA from oxidative damage and neutralize free radicals that promote mutation. Carotenoids have several biological properties, such as it reduces the risk of cancer and bone and cardiovascular diseases. Lutein and zeaxanthin also protect the eyes from age-related macular degeneration. Vitamin C is a water-soluble vitamin that plays a vital role in forming collagen, the main protein of connective tissue, essential for the proper resistance and strength of tendons, ligaments, skin, and other connective tissue tissues (Marti et al. 2009). Citrus fruits contain active secondary metabolites that help reduce the risk of cancer, including gastric cancer, breast cancer, lung tumorigenesis, colonic tumorigenesis, hepato-carcinogenesis, and hematopoietic malignancies. Citrus fiber possesses bioactive functions due to the presence of polyphenol-like components. Frequent consumption of dietary fiber is associated with a low risk of life-threatening chronic diseases such as bowel, and gastrointestinal disorders, obesity, diabetes, cardiovascular disease, and cancer. It promotes physiological functions, including reducing blood cholesterol levels and glucose attenuation (Rafiq et al. 2018).

2.3.5 *Broccoli*

Broccoli belongs to the family Brassicaceae. It contains vitamin C, soluble fiber, di indolyl methane, and selenium compounds. Di indolyl methane found in broccoli is a potent modulator of the innate immune system with antiviral, antibacterial, and anti-cancerous activity (Das et al. 2010). Broccoli is an excellent dietary source of phytochemicals, including glucosinolates, phenolics, other anti-oxidants like vitamins (A, B, C, K), and essential dietary minerals (Ca, P, S, K, Fe, I, etc.). Broccoli contains glucosinolates (GLS), the metabolic breakdown products of which are potent modulators of xenobiotic-metabolizing enzymes that protect DNA from damage. Isothiocyanates is, another GLS from broccoli, have the anti-oxidant capacity, cholesterol-lowering effects, and protective properties on cardiovascular disease.

Selenium is an essential mineral for the proper functioning of the immune system. Also, it has anti-cancerous properties that broccoli is known for its ability to accumulate high levels of Selenium in the form of semethylselenocysteine. It also contain polyphenols like kaempferol, quercetin, and traces of isorhamnetin. Broccoli is an excellent source of chromium, a trace mineral known to lower blood sugar (Shubha et al. 2020). The consumption of cruciferous vegetables has been associated with preventing lung, pancreas, bladder, prostate, thyroid, skin, stomach, and colon cancer. It can also lower blood glucose level and lipid profile and, thus, reduces cardiovascular diseases (Madhu and Kochhar 2013).

2.3.6 Green Tea

Green tea is solely made with the leaves of *Camellia sinensis*, which have undergone minimal oxidation during processing. Tea has catechins and epicatechin as primary functional foods, and they are known to possess chemo-preventive activities against prostate and ovarian cancers, anti-obesity, and anti-diabetic effects (Das et al. 2010). Consumption of green tea protects against stroke, liver disease, bacterial infection, cancer, and viral infection and lowers the risk of osteoporosis. It also reduces the risk of high blood cholesterol concentration and high blood pressure, thereby providing prevention against atherosclerosis and coronary heart disease (Adak and Abdel-Gabar 2011).

2.3.7 Almonds

Almonds are naturally high in vitamin E, riboflavin (vitamin B2), and the minerals calcium, magnesium, phosphorus, potassium, zinc, copper, and manganese. Almonds lack dietary cholesterol but are rich in chemically related phytosterols, a class of compounds that interfere with cholesterol absorption and, thus, helps maintain healthy blood cholesterol levels. Almonds are high in fiber and protein and have a low glycaemic index, which lowers the risk of type 2 diabetes. Almonds are considered an essential component of a healthy diet, and increased consumption has the potential to improve health. It can replace foods high in saturated fatty acids, sugars, and salt or those lacking in vitamins and minerals (Richardson et al. 2009). It contains a variety of anti-oxidant phytochemicals, including phenolic compounds and α-tocopherol, which reduce the risk factors for chronic diseases, such as cardiovascular diseases and diabetics. Consumption of almonds also helps control and maintain body weight because of their low energy, hunger suppression, and appealing taste (Martins et al. 2017).

2.3.8 Cereals

Cereals such as wheat, barley, psyllium, and oats, rich in many phytochemicals and other nutrients, are considered an excellent functional food. Beta-glucans in cereals are the sources of dietary fiber and have a therapeutic effect on coronary heart disease (CHD) by reducing cholesterol and glycemic control. Beta-glucans from oats and barley have been shown to lower serum cholesterol levels, thus, lowering the risk of CHD. Wheat, barley, and oats are also known to contain many other bioactive compounds such as lignans, phytosterols, isoflavones, resorcyclic acid, lactones, coumestans, unsaturated fatty acids, lutein, cryptoxanthin, zeaxanthin, tocopherols, tocotrienols, glutathione, and melatonin. These compounds protect against chronic diseases like coronary heart disease and some cancers. Flavonoids are present in almost all plants and are known to possess anticarcinogenic, anti-inflammatory, and anti-allergic properties. The most important flavonoids are quercetin, kaempferol, myricetin, and chrysin. Cereals have only small quantities of flavonoids, except that barley contains measurable amounts of catechin and some di- and tri-procyanidins (Sidhu et al. 2007). Cereals are rich in fiber and can lower systolic and diastolic hypertension and protect from strokes. Phenolics in cereals may inhibit platelet aggregation and LDL-cholesterol oxidation, two known risk factors in CVD. Vitamin E (tocotrienols and tocopherols) is another potent anti-oxidant group protecting cell membranes from oxidative damage. Phytic acid, vitamin E, and phenolics may all inhibit the formation of carcinogens from precursor compounds. Phytic acid can chelate various metals and inhibit iron- redox reactions. Phenolics can also induce detoxification systems (Borneo and León 2011).

2.3.9 Ginger

Ginger belongs to the family Zingiberaceae. Ginger root is used to attenuate and treat several common diseases, such as headaches, colds, nausea, and emesis. It can be categorized into volatiles and non-volatiles. Volatiles include sesquiterpene and monoterpenoid hydrocarbons, providing ginger's distinct aroma and taste. Contrarily, non-volatile aromatic compounds include gingerols, shogaols, paradols, and zingerone. Ginger has the potential for treating many ailments, including degenerative disorders (arthritis and rheumatism), digestive health (indigestion, constipation, and ulcer), cardiovascular disorders (atherosclerosis and hypertension), vomiting, diabetes mellitus, and cancer. It also has anti-inflammatory and antioxidative properties for controlling the process of aging. It also has antimicrobial potential, which can help treat infectious diseases (Mashhadi et al. 2013). The generation of free radicals or reactive oxygen species (ROS) during metabolism beyond the anti-oxidant capacity of a biological system results in oxidative stress. The bioactive molecules in ginger show anti-oxidant properties and are effective against oxidative stress. Experimental studies have demonstrated that ginger can prevent and treat several types of cancer, such as colorectal, prostate, breast, cervical, liver, and pancreatic cancer. The anti-cancer mechanisms mainly involve the induction of apoptosis and the inhibition of the proliferation of cancer cells (Mao et al. 2019).

Functional and nutraceutical ingredients in ginger rhizome have hypocholesterolemic and hypoglycaemic effects, especially anti-oxidants, fat-soluble vitamins, phytosterols, and some pyrazanol containing moieties. Gingerol shows anti-oxidant and anti-inflammatory behavior (Shoaib et al. 2016). Ginger also has several other health benefits, such as reducing blood glucose in Type 2 diabetes mellitus patients as an anti-pain cream, analgesic, uric acid, and muscle pain. Consuming ginger decoction, containing oleoresin and essential oils, can lower blood uric acid levels. Oleoresin and ginger essential oil reduce blood uric acid levels by inhibiting arachidonic acid metabolism and platelet aggregation and relieving pain by inhibiting the cyclooxygenase pathway to inhibit prostaglandin biosynthesis (Indiarto et al. 2020).

2.3.10 Cocoa

Cocoa is rich in flavanols, a class of polyphenols present in plants as nonconjugated molecules, including epicatechin and catechin, and as oligomers of these molecules, also named procyanidins. The concentration of flavanols in chocolates depends on both the flavanol content of the cacao plant and the procedures used for transforming the cocoa into chocolate (Fraga 2005). Polyphenol-rich foods have an important role in health preservation due to their anti-oxidant properties. It can act as proton donor-scavenging radicals, inhibitors of enzymes that increase oxidative stress, chelate metals, and bind carbohydrates and proteins. These properties serve as anticarcinogenic, anti-inflammatory, antihepatotoxic, antibacterial, antiviral, and antiallergenic compounds (Ackar et al. 2013).

Platelet dysfunction is another characteristic feature of atherosclerotic lesions. Cocoa has aspirin-like effects on platelet function, and the combined effects of cocoa and aspirin are additive, suggesting improved clot prevention. Chocolate has a dual effect on platelets. It decreases platelet aggregation and reduces platelet adhesion. Numerous studies have reported that flavanols in cocoa cause significant vasodilatation by increasing serum NO levels and endothelial NO bioavailability. Chocolate reduces stress levels by prompting serotonin production, a calming neurotransmitter. Cocoa intake results in increased cerebral blood flow, suggesting that cocoa might play a role in treating cerebral conditions such as dementia and stroke. Cocoa also has anti-diabetic, antihypertensive effects, eventually reducing the risk of cardiovascular diseases (Latif 2013).

2.3.11 Berries

Berries have been hailed as excellent reserves of health-restoring phytochemicals. The abundance of proanthocyanidins and anthocyanidins was responsible for enhancing the redox status of the body. Regular consumption of blueberries may reduce total cholesterol and triglyceride levels and increase HDL-cholesterol levels, leading to lower body weight. Blueberry has been shown effective against various food pathogens, owing to its abundance in phenolic components (Patel 2014). Raspberry is considered to be of highest value because it is a source of vitamin C, polyphenol compounds (anthocyanins), and mineral substances—potassium and iron. The prominent representatives of phenolic compounds in raspberry fruit are ellagic acid, gallic acid, caffeic acid, epicatechin, quercetin, lambertianin, kempferol (Akimov et al. 2021). It also have potent anti-oxidant, anti-inflammatory, and antiangiogenesis properties. Black raspberry contains many phytochemicals, including anthocyanins, ellagitannins, ferulic acid, β-sitosterol, bioflavonoids, fiber, vitamins, and minerals. Among them, anthocyanins and ellagitannins are considered the most potent anti-cancer components and are found in higher concentrations in black raspberry compared to other berries (Gu et al. 2013). Strawberries are a significant source of B-vitamins, vitamin C, vitamin E, potassium, folic acid, carotenoids, and specific flavonoids, such as pelargonidin, quercetin, and catechin. Strawberries also contain significant amounts of ellagic acid, tannins, and phytosterols. Ellagic acid and ellagitannins have been reported as significant contributors to strawberries' anti-oxidant and anticarcinogenic effects. Berry anthocyanidins have also been shown to inhibit monoamine oxidases A and B, and this has been implicated in protective effects against neurodegenerative disorders (Basu et al. 2014).

2.3.12 Olive

Natural olive oil contains highly monounsaturated fatty acids and natural anti-oxidant compounds such as vitamins, carotenoids, aliphatic, diterpenic and triterpene alcohols, and hydrocarbons, phytosterols, flavonoids, phenolic compounds, and secoiridoids. Polyphenols in olive oil have anti-inflammatory, antiproliferative, anti-oxidant, and antimicrobial properties and positively affect cancer, diabetes, skin diseases, and neurological and cardiovascular diseases. Olive oil intake is related to lowered mitochondrial oxidative stress and improvement of anti-oxidant capacity, which leads to healthier metabolic pathways (Akdas et al. 2020). The nutritional and anti-oxidant properties of EVOO are related to the presence and concentration of tocopherols, carotenoids, and phenolic compounds, which are of great importance for human health (Sinesi and Damato 2021). Olive oil is often considered the optimal dietary fat for preventing coronary heart disease. It has beneficial effects on hypertension, and high MUFA concentrations in olive oil reduce total and low-density lipoprotein cholesterol levels (Stark and Madar 2002).

2.3.13 Turmeric

Curcumin is the natural and safe bioactive compound of the turmeric (*Curcuma longa* L.) plant and possesses potent antimicrobial and immunomodulatory properties (Tripathy et al. 2021). Curcuma species have beneficial pharmacological properties, including antiproliferative, anti-inflammatory, anti-cancer, anti-diabetic, hypocholesterolemic, anti-thrombotic, antihepatotoxic, anti-diarrheal, carminative, diuretic, antirheumatic, hypotensive, antimicrobial, antiviral, anti-oxidant, larvicidal, insecticidal, antivenomous, and antityrosinase effects (Sharifi-Rad et al. 2020). The mechanisms of action of curcuminoids involved in the protection against cardiovascular diseases comprise the regulation of oxidative stress, suppression of apoptosis (programmed cell death), and anti-inflammatory activity. Curcuminoids display anti-inflammatory activity by increasing adipokine adiponectin and stress response proteins (heat shock proteins 70 and 90 and Sirtuin 1) and reducing inflammatory intermediate molecules such as interleukins and monocyte chemoattractant protein (Munekata et al. 2021). Curcumin can protect the kidneys from sclerosis, glomerular hypertrophy, tubulointerstitial changes, mononuclear cell interstitial infiltration, tubular lipid deposition, and in vitro fibroplasia (Hewlings and Kalman 2017).

2.4 Benefits

Functional foods exhibit health properties beyond the traditional nutrients it contains. They display physiological benefits and higher capacities to reduce the risk of chronic diseases beyond their basic nutritional functions, including maintaining gut health (Kaur et al. 2020). Functional foods improve well-being or health (Katan 2004). Several studies reported the beneficial effects of diets rich in vegetables and fruits on CVD risk. Soluble fibers, including pectins from apples and citrus fruits, β-glucan from oats and barley, and fibers from flaxseed and psyllium, are known to lower LDL-Cholesterol. The mechanisms of their cholesterol-lowering effects are the binding of bile acids and inhibition of cholesterol synthesis. Soy products contain many isoflavonoids (genistein, daidzein, glycitin) that are natural phytoestrogens that inhibit LDL oxidation, thus, decreasing the risk of atherosclerosis (Alissa and Ferns 2012). Cruciferous vegetables such as broccoli, cauliflower, and cabbage contain certain non-nutritional phytochemicals beneficial for cancer prevention. In contrast, onion and garlic contain certain biologically active compounds with anticarcinogenic and anticholinergic effects (Tiwary and Hussain 2021). Consumption of carotenoid-rich foods containing lutein is significantly related to a lowered risk of developing premenopausal breast cancer. Glucosinolates are found as the high amounts in cruciferous vegetables, assist in reducing the risk of developing different types of cancer such in lungs, breast, liver, esophagus, stomach, small intestine, and colon (Aghajanpour et al. 2017).

The traditional Indian diet, spices, and medicinal plants are rich sources of natural anti-oxidants. A high intake of this functional food can reduce health problems such as cardiovascular and inflammatory disease, cataracts, and cancer. Anti-oxidants prevent free radical-induced tissue damage by preventing the formation of radicals, scavenging them, or promoting their decomposition (Lobo et al. 2010). Phenolic compounds, including flavonoids, are extensively found in tea, nuts, fruits, vegetables, legumes, and cereals, possess anti-oxidant properties, and establish favorable effects on tumorigenesis and thrombosis. Similarly, phytoestrogens are present in flaxseed oil, vegetables, soy, fruits, and whole grains. They have shown promising results in controlling the further proliferation of cancer cell lines and various risk factors associated with cardiovascular diseases (Kaur et al. 2020).

Being overweight is one of the most critical risk factors for type 2 diabetes. The use of whole-grain foods, vegetables, fruits, low glycemic index starchy products, and low saturated fat foods can prevent diabetes (Riccardi et al. 2005). Functional foods also possess anti-aging properties. Anti-aging functional foods exert influence mostly through their anti-oxidant and anti-inflammatory effects, abrogating collagen degradation or increasing procollagen synthesis. Functional foods with some evidence of cutaneous anti-aging properties include carotenoids, polyphenols, vitamins C and E, red ginseng, squalene, omega-3 fatty acids, collagen peptides, and proteoglycans (Cho 2014). Osteoporosis is a health problem characterized by low bone mineral density, deterioration of bone microarchitecture, decreased bone mass, and increased bone fragility. It is considered an age-related disorder. It can be effectively reduced by consuming functional foods rich in anti-oxidants and calcium (Arnold et al. 2021).

Plant-based functional food diets can lower overall mortality and ischemic heart disease mortality. It can also reduce medication needs and support sustainable weight management, thus, reducing the incidence and severity of high-risk conditions, such as obesity and obesity-related inflammatory markers, hyperglycemia, hypertension, and hyperlipidemia. Plant foods exclusively contain two critical nutrients: fiber and phytonutrients. Fiber provides powerful protection to the gastrointestinal, cardiovascular, and immune systems, while phytonutrients, including glucosinolates, carotenoids, and flavonoids, work synergistically to reduce inflammation and oxidation, protecting from disease initiation and progression (Hever and Cronise 2017) (Table 2.1).

2.5 Conclusion

Recently, functional foods have been receiving global attention because of their nutritional value and the presence of constituents that are critical in regulating physiological processes. These foods are rich in specific minerals, vitamins, fatty acids, dietary fiber, anti-oxidants, prebiotics, and probiotics that balance the diet so that the body is provided with all the nutrients it needs and no longer found in the products offered by intensive farming (Butnariu and Sarac 2019). The bioavailability

Table 2.1 Functional components, sources, and their benefits (John and Singla 2021; Abuajah et al. 2014)

Bioactive components	Source	Potential benefits
Carotenoids		
Alpha-carotene/ beta-carotene	Carrots, fruits, vegetables	Neutralize free radicals which may cause damage to cells
Lutein	Green vegetables	Reduce the risk of muscular degeneration
Lycopene	Tomato	Reduce the risk of prostrate cancer
Non-starchy polysaccharide		
Fucoidan	Mushrooms, brown seaweeds	Apoptosis of cancer cells, stimulates brain development, anti-clotting effect, lower blood cholesterol levels, decrease high blood pressure, stabilize blood sugar
Insoluble dietary fiber	Wheat bran	Reduces risk of colon and breast cancer
Beta glucan	Oats, barley, rye	Reduces risk of cardiovascular disease, protects against heart disease and some cancers, lower LDL and total cholesterol
Soluble fiber	Psyllium	Reduces risk of cardiovascular disease, protects against heart disease and some cancers, lower LDL and total cholesterol
Phenolics		
Anthocyanidins	Fruits	Reduces the risk of cancers
Catechins	Tea, chocolates	Anti-oxidant activity and supports heart health
Flavonones	Citrus	Antioxidant activity and reduces the risk of cancers
Flavones	Fruits, vegetables	Neutralizes free radicals, reduces the risk of cancers
Flavonols—Quercetin	Onions, apples, broccoli	Neutralizes free radicals
Procyanidins and Proanthocyanidins	Cranberries, cocoa, tea, Cocoa, red wine, grapes	Supports urinary tract health and heart health
Lignans	Flax, rye, vegetables	Prevention of cancer, renal failure, supports heart and immune system.
Tannins	Cranberries	Improves urinary tract problem
Fatty acids		
Monounsaturated fatty acids (MUFAs)	Tree nuts, olive oil, canola oil	May reduce the risk of CHD
Polyunsaturated fatty acids (PUFAs)	Walnuts, Flaxseeds, flaxseed oil	Supports maintenance of heart, eye health and mental function
Omega-3 fatty acids—ALA		
Isothiocyanates		
Sulforaphane	Cauliflower, broccoli, cabbage, kale	

(continued)

Table 2.1 (continued)

Bioactive components	Source	Potential benefits
		May enhance detoxification of undesirable compounds, bolsters cellular antioxidant defenses
Tocopherols, ubiquitinol	Almonds, nuts	Decreases blood cholesterol, inhibits lipid peroxidation
Organosulphur compounds	Garlic, onion, broccoli	Reduces total and LDL cholesterol Decreases blood pressure
Phytosterols	Corn, soy, wheat, beverages	Antibacterial, antiviral, anticancer, and cholesterol lowering activity
Prebiotics	Banana, garlic, onion	Stimulate growth and activity of helpful bacteria in the colon, and thus, improves the host health
Soy protein	Soybeans and soy-based foods	May reduce the risk of CHD
Vitamins		
A	Carrots, sweet potato, Spinach	Supports maintenance of eye, immune and bone health, provides cell integrity
B1 (Thiamine)	Lentils, peas, brown or white rice, pistachios	Supports mental health, helps regulate metabolism
B2 (Riboflavin)	Green leafy vegetables	Supports cell growth, helps regulate metabolism
B3 (Niacin)	Fortified cereals	Regulates metabolism, supports cell growth
B5 (Pantothenic acid)	Sweet potato, soybean, lentils	Regulates metabolism and hormone synthesis
B6 (Pyridoxine)	Beans, legumes, citrus fruits, whole grains, nuts	Maintains immune health, regulates body metabolism
B9 (Folic acid)	Citrus fruits, cereals, legumes, beans	Reduces chance of birth defects, supports immune health
C	Guava, citrus fruits, kiwi, strawberry	Neutralizes free radicals, prevents scurvy, CHD, stroke and cancer
E	Almonds, hazelnuts, turnips, sunflower seeds	Neutralizes free radicals
Minerals		
Calcium	Spinach, beverages	Reduces the risk of osteoporosis
Magnesium	Pumpkin seeds, spinach, cereals, almonds	Supports maintenance of normal muscle and nerve function, immune health and bone health
Potassium	Banana, potatoes, cereals, Leafy greens	Reduces the risk of high blood pressure and stroke
Selenium	Whole grains, garlic	Neutralizes free radicals

of these components in the food and the level of its requirement in the human diet are the key factors required to optimize health. The market for functional foods is expanding worldwide due to their importance in the human diet.

Functional foods can be plant-based or animal-based. Plant-based functional foods are derived from natural or unprocessed plant foods and those modified via biotechnological techniques. They are products that have a relevant effect on well-being and health or reduce disease risk. Thus, there is growing interest in researching and developing plant-based functional foods. Health-conscious consumers are increasingly seeking functional foods to control their health and well-being. The functional food industry is in its growing stage, and information is scarce. There is a need to provide consumers with more information to effectively guide them in making wiser choices of diets that contain optimal levels of health-promoting functional food components. Rapid advance in science and technology, changes in lifestyle, and increased health concerns fuelled the interest in adding functional foods into the diet. A key challenge to ensuring the bright future of functional foods is to provide solid guarantees to consumers that they can trust the safety of functional foods and their promises about better health, performance, development, or growth (Weststrate et al. 2002).

References

Abuajah CI, Ogbonna AC, Osuji CM (2014) Functional components and medicinal properties of food: a review. J Food Sci Technol 52(5):2522–2529

Ackar D, Lendic KV, Valek M, Subaric D, Milicevic B, Babic J, Nedic I (2013) Cocoa polyphenols: can we consider cocoa and chocolate as potential functional food? J Chem 2013:1–7

Adak M, Abdel-Gabar M (2011) Green tea as a functional food for better health: a brief review. Res J Pharm Biol Chem Sci 2(2):645–664

Aghajanpour M, Nazer MR, Obeidavi Z, Akbari M, Ezati P, Kor NM (2017) Functional foods and their role in cancer prevention and health promotion: a comprehensive review. Am J Cancer Res 7(4):740–769

Akdas S, Kacar M, Erozkan C, Yazihan N (2020) Is olive oil a functional food with systemic benefits on pathological conditions and ageing by improving mitochondrial functions and antioxidant status? Acta Sci Nutr Health 4(2):1–5

Akimov MY, Koltsov VA, Zhbanova EV, Akimova OM (2021) Nutritional value of promising raspberry varieties. IOP Conf Series: Earth Environ Sci 640:022078

Alissa EM, Ferns GA (2012) Functional foods and nutraceuticals in the primary prevention of cardiovascular diseases. J Nutr Metab 2012:1–12

Alshafe MM, Kassem SS, Abdelkader MM, Hanafi EM (2015) Flaxseed as a functional food. Res J Pharm, Biol Chem Sci 64:1944–1951

Arnold M, Rajagukguk YV, Gramza-Michalowska A (2021) Functional food for elderly high in antioxidant and chicken eggshell calcium to reduce the risk of osteoporosis—A narrative review. Foods 10(3):656–682

Basu A, Nguyen A, Betts NM, Lyons TJ (2014) Strawberry as a functional food: an evidence-based review. Crit Rev Food Sci Nutr 54:790–806

Bishnoi S (2016) Herb as functional foods. In: Functional foods: sources & health benefits, pp 141–172

Borneo R, León AE (2011) Whole grain cereals: functional components and health benefits. Food and function

Botella-Pavía P, Rodríguez-Concepción M (2016) Carotenoid biotechnology in plants for nutritionally improved foods. Physiol Plant 126(3):369–381

Butnariu M, Sarac I (2019) Functional food. Int J Nutr 3(3):7–16

Canene-Adams K, Campbell JK, Zaripheh S, Jeffery EH, Erdman JW (2005) The tomato as a functional food. American Society for Nutritional Science

Cho S (2014) The role of functional foods in cutaneous anti-aging. J Lifestyle Med 4(1):8–16

Crowe KM, Coni F (2013) Functional foods. J Acad Nutr Diet 113(8):1096–1103

Das D, Vimala R, Das N (2010) Functional foods of natural origin—an overview. Indian J Nat Prod Resour 1(2):136–142

Delgado A, Al-Hamimi S, Ramadan MF, de Wit M, Durazzo A, Nyam KL, Issaoui M (2020) Contribution of Tocols to food sensorial properties, stability, and overall quality. J Food Qual 2020:1–8

Donno D, Turrini F (2020) Plant foods and underutilized fruits as source of functional food ingredients: chemical composition, quality traits, and biological properties. Foods 9(10):1474

Fraga CG (2005) Cocoa, diabetes, and hypertension: should we eat more chocolate? Am Soc Clin Nutr 81:341–342

Garg S, Lule VK, Malik RK, Tomar SK (2016) Soy bioactive components in functional perspective: a review. Int J Food Prop 19(11):2550–2574

Gibson GR, Williams CM (2000) Functional foods, concept to products. Woodhead Publishing Limited, Sawston

Gu J, Ahn-Jarvis JH, Riedl KM, Schwartz SJ, Clinton SK, Vodovotz Y (2013) Characterization of black raspberry functional food products for cancer prevention human clinical trials. J Agric Food Chem 62:3997–4006

Hasler CM (2002) Functional foods: benefits, concerns and challenges—A position paper from the American Council on Science and Health. J Nutr 132:3772–3781

Henry CJ (2010) Functional foods. Eur J Clin Nutr 64:657–659

Hever J, Cronise RJ (2017) Plant-based nutrition for healthcare professionals: implementing diet as a primary modality in the prevention and treatment of chronic disease. J Geriatr Cardiol 14(5): 355–368

Hewlings SJ, Kalman DS (2017) Curcumin: a review of its' effects on human health. Foods 6(10): 92

Indiarto R, Subroto E, Angeline, Selly (2020) Ginger rhizomes (Zingiber officinale) functionality in food and health perspective: a review. Food Res: 1–9

John R, Singla A (2021) Functional foods: components, health benefits, challenges, and major projects. DRC Sustain Future 2(1):61–72

Jones PJH, AbuMweis SS (2009) Phytosterols as functional food ingredients: linkages to cardiovascular disease and cancers. Curr Opin Clin Nutr Metab Care 12(2):147–151

Jooyandeh H (2011) Soy products as healthy and functional foods. Middle-East J Sci Res 7(1): 71–80

Kajla P, Sharma A, Sood DR (2015) Flaxseed—a potential functional food source. J Food Sci Technol 52(4):1857–1871

Katan MB (2004) Health claims for functional foods. BMJ 328(7433):180–181

Kaur S, Das M (2011) Functional foods: an overview. Food Sci Biotechnol 20(4):861–875

Kaur H, Agarwal S, Agarwal M, Agarwal V, Singh M (2020) Therapeutic and preventive role of functional foods in process of neurodegeneration. Int J Pharm Sci Res 11(6):2882–2891

Latif R (2013) Chocolate/cocoa and human health: a review. Neth J Med 71(2):63–68

Lau TC, Chan MW, Tan HP, Kwek CL (2013) Functional food: a growing trend among the health conscious. Asian Soc Sci 9(1):198–208

Lobo V, Patil A, Phatak A, Chandra N (2010) Free radicals, anti-oxidants and functional foods: impact on human health. Pharmacogn Rev 8(4):118–126

Madhu, Kochhar A (2013) Role of nutraceutical enriched broccoli in the management of lifestyle diseases. Int J Med Sci 6(2):69–76

Mao QQ, Xu XY, Cao SY, Gan RY, Corke H, Beta T, Li HB (2019) Bioactive compounds and bioactivities of ginger (*Zingiber ocinale* Roscoe). Foods 2(185):1–22

Marti N, Mena P, Canovas JA, Micol V, Saura D (2009) Vitamin C and the role of citrus juices as functional food. Nat Prod Commun 4(5):677–700

Martins IM, Q Chen, Oliver Chen CY (2017) Emerging functional foods derived from almonds. In: Wild plants, mushrooms, and nuts: functional food properties and applications, 1st edn, pp 445–469

Mashhadi NS, Askari G, Hariri M, Darvishi L, Mofid MR (2013) Anti-oxidative and anti-inflammatory effects of ginger in health and physical activity: review of current evidence. Int J Prev Med 4(1):S36–S42

Messina M (2016) Soy and health update: evaluation of the clinical and epidemiologic literature. Nutrients 8(12):754

Mezzomo N, Ferreria SRS (2016) Carotenoids functionality, sources and processing by supercritical technology: a review. J Chem 2016:1–16

Miekus N, Marszalek K, Podlacha M, Iqbal A, Puchalski C, Swiergiel AH (2020) Health benefits of plant-derived sulfur compounds, glucosinolates, and organosulfur compounds. Molecules 25(17):3804–3826

Munekata PES, Pateiro M, Zhang W, Dominguez R, Xing L, Fierro EM, Lorenzo JM (2021) Health benefits, extraction and development of functional foods with curcuminoids. J Funct Foods 79: 1–12

Nayak PK, Mohan CC, Radhakrishnnan K (2019) Food bioactives: functionality and applications in human health, functional foods from different sources, pp 37–58

Orzechowski A, Ostaszewski P, Jank M, Berwid SJ (2002) Bioactive substances of plant origin in food—impact on genomics. Reprod Nutr Dev 42(5):461–477

Patel S (2014) Blueberry as functional food and dietary supplement: the natural way to ensure holistic health. Mediterr J Nutr Metab 7:133–143

Poli A, Marangoni F, Corsini A, Manzato E, Marrocco W, Martini D, Medea G, Visioli F (2021) Phytosterols, cholesterol control, and cardiovascular disease. Nutrients 13:2810

Rafiq S, Kaul R, Sofi SA, Bashir N, Nazir F, Nayik GA (2018) Citrus peel as a source of functional ingredient: a review. J Saudi Soc Agric Sci 17:351–358

Riccardi G, Capaldo B, Vaccaro O (2005) Functional foods in the management of obesity and type 2 diabetes. Curr Opin Clin Nutr Metab Care 8:630–635

Richardson DP, Astrup A, Cocaul A, Ellis P (2009) The nutritional and health benefits of almonds: a healthy food choice. Food Sci Technol Bull Funct Foods 6(4):41–50

Shandilya UK, Sharma A (2017) Functional foods and their benefits: an overview. J Nutr Health Food Eng 7(4):350–356

Sharifi-Rad J, Rayess YE, Rizk AA, Sadaka C, Zgheib R, Zam W, Sestito S, Rapposelli S, Neffe-Skocińska K, Zielińska D, Salehi B, Setzer WN, Dosoky NS, Taheri Y, Beyrouthy ME, Martorell M, Ostrander EA, Hafiz Ansar Rasul Suleria HAR, Cho WC, Maroyi A, Martins N (2020) Turmeric and its major compound curcumin on health: bioactive effects and safety profiles for food, pharmaceutical, biotechnological and medicinal applications. Front Pharmacol 11(01021):1–23

Shoaib M, Shezhad A, Butt MS, Saeed M, Raza H, Niazi S, Khan IM, Shakeel A (2016) An overview: ginger a tremendous herb. J Glob Innov Agric Soc Sci 4(4):172–187

Shubha K, Reetu SA, Mukherjee A (2020) Broccoli: a potential functional food. Food Sci Report 1(5):26–28

Sidhu JS, Kabir Y, Huffman FG (2007) Functional foods from cereal grains. Int J Food Prop 10: 231–244

Sinesi A, Damato R (2021) Extra virgin olive oil (Evoo) as a valid functional food in the oral cavity. Novel Techn Nutr Food Sci 5(4):481–483

Stark AH, Madar Z (2002) Olive oil as a functional food: epidemiology and nutritional approaches. Nutr Rev 60(6):170–176

Tiwary S, Hussain S (2021) Functional foods for prevention and treatment of cancer. Asian J Pharm Clin Res 14(3):4–10

Topolska K, Florkiewicz A, Filipiak-Florkiewicz A (2021) Functional food—consumer motivations and expectations. Int J Environ Res Public Health 18(10):5327–5338

Tripathy S, Verma DK, Thakur M, Patel AR, Srivastav PP, Singh S, Gupta AK, Chavez-Gonzalez ML, Aguilar CN, Chakravorty N, Verma HK, Utama GL (2021) Curcumin extraction, isolation, quantification and its application in functional foods: a review with a focus on immune enhancement activities and COVID-19. Front Nutr 8(747956):1–29

Weststrate JA, van Poppel G, Verschurren PM (2002) Functional foods, trends and future. Br J Nutr 88(2):S233–S235

Xiao S, Li J (2020) Study on functional components of functional food based on food vitamins. J Phys Conf Ser 1549:032002

Chapter 3
Insight of Bioresources from Lower Plant Groups: Reconciling the Possibilities and Responsibilities

Remya Krishnan, G. M. Greeshma, Bosco Lawarence, and K. Murugan

Abstract Sustainable utilization of bioresources is the need of the hour that supports ecosystem and human wellbeing. Overexploitation of bioresource leads to catastrophic ecological processes like loss of habitat, which lead to irrecoverable loss of biological diversity. The biodiversity loss and unregulated ecosystem services result in sabotaging the declared UN Sustainable Development Goals. Many grassroots, and national and international studies provide mounting estimates of the overall gain delivered from the biological resources and ecological services. The global ecosystem services reveal a benefit of approximately 125–140 trillion USD/annum. There has been growing interest in lower group plant species in the last few decades as sources of novel nutraceuticals, pharmacologically active ingredients, or other economically valuable products. This chapter accounts for an insight into the utilization of bacteria, algae, fungi, lichens, bryophytes, pteridophytes, and gymnosperms and their bioactive molecules in ethnic medicine, trade of such bioresources with biopharmaceutical industries, and food or other industries. Unhealthy utilization accelerates the extinction rate of many living organisms. Peoples participatory policy formulations connected with biodiversity conservation and economic growth will helps to soothe environmental derailment and loss of biowealth from the earth.

Keywords Lower group plants · Bioresources · Active ingredients · Economic benefits

---

R. Krishnan
Post Graduate Department and Research Centre of Botany, Mahatma Gandhi College, Thiruvananthapuram, India

G. M. Greeshma · K. Murugan (✉)
CISSA Phytotech, Thiruvananthapuram, Kerala, India

B. Lawarence
Department of Botany, Government Womens College, Trivandrum, India

S. T. Sukumaran, K. T R (eds.), *Conservation and Sustainable Utilization of Bioresources*, Sustainable Development and Biodiversity 30,
https://doi.org/10.1007/978-981-19-5841-0_3

3.1 Introduction

Global biodiversity assessment and ecosystem services refer to the evidence of the importance of biodiversity to achieve the goal of Zero Hunger and meet the UN Sustainable Development Goals. Meanwhile, the population explosion and climate change resulted in to an increasing need for energy, food, water, and land significantly on the living planet. The scale of production and consumption, combined with unscientific usage of resources, has resulted in drastic and widespread loss of biodiversity. Pressures on biodiversity are blooming out. Diverse natural ecosystems have been partially/completely destroyed, limiting their services. A large number of plant and animal species are threatened by the altered or loss of habitat. In many countries, the endangered animal and plant species are increasing, particularly with high population density and high levels of human activities. As per the Global Risks Report, decision makers validated loss of biodiversity and habitat collapse as one of the ten most increased risks faced by the entire living organisms currently. In the present review, an insight into the utilization of tradable bioresources from lower plant groups such as bacteria, fungi, lichen, bryophytes, pteridophytes, and gymnosperms was recorded.

3.2 Tradable Bioresources and Products from Lower Groups

3.2.1 Bacteria

Microbial species are abundant, widespread, and longest-evolving forms of life on the earth. Recent studies suggest that the earth might abode 1 trillion (10^{12}) species (Louca et al. 2019). The present consumers around the world now increasingly prefer natural compounds over synthetic compounds. There is now an increased demand for the development of sustainable biotechnological processes in order to obtain new natural compounds that can eventually replace traditional synthetics. Recent development in the field of recombinant DNA technology has sparked the development of an array of biopharmaceutical products from natural products using microbes. During the early 1900s, about 80% of all medicines were obtained from plant sources (Siddiqui et al. 2014). The discovery of penicillin from *Penicillium notatum* in 1928 marked a significant transit from plants to microorganisms as a source of natural products (Pham et al. 2019). Many phytochemicals extracted from microorganisms have immense applications in the field of drugs, food products, and agriculture (Sanchez et al. 2012). Nowadays, large pharmaceutical companies have been entrusted to invest in this traditional domain (Dias et al. 2012). Approximately 60% of approved small molecule medicines are from natural products, and 70% of antibacterial agents originate mainly from natural products (Matsumura et al. 2018).

Common antibiotics produced from bacteria are streptomycin from *Streptomyces griseus*, chloramphenicol from *Streptomyces venezuelae*, chlortetracycline from *Streptomyces aureofaciens*, cephalosporin C from *Cephalosporium acremonium*, erythromycin from *Saccharopolyspora erythraea*, Amphotericin B from *Streptomyces nodosus*, Ieodoglucomide C from *Bacillus licheniformis*, Bleomycin from *Streptoalloteichus hindustanus* and vancomycin from *Amycolatopsis orientalis* (Pham et al. 2019). Similarly, immunosuppressant/Anti-inflammatory agents like Rapamycin (*Streptomyces rapamycinicus*), FK506 (*Streptomyces tsukubaensis*) and Anticancer and Antitumor agents like Bleomycin (*Streptoalloteichus hindustanus*), Daunorubicin (*Streptomyces peucetius*), Avermectins (*Streptomyces avermitilis*) (Shen 2015), and Mollemycin A 20 (*Streptomyces* sp.) (Blunt et al. 2016) were also commercially documented microbial products.

Antibiotic Pikromycin was the first reported polyketide molecule extracted from *S. venezuelae* and is effective against multi-drug resistant respiratory pathogens (Woo et al. 2014). Erythromycin A, produced by *S. erythraea* is prescribed to treat a wide variety of bacterial infections (Cobb et al. 2013). Streptomycin from *S. griseus* is the first aminoglycoside effective against pulmonary tuberculosis. Nisin A is a bacteriocin from *Lactococcus lactis* and Reuterin from *Lactobacillus reuteri* were also commercially important bacterial products (Gyawali and Ibrahim 2014). Amphotericin B is a traditional polyene antifungal product from *Streptomyces nodosus* (Tevyashova et al. 2013). Geldanamycin is a benzoquinone ansamycin antitumor compound from *Streptomyces hygroscopicus* var. *geldanus*, preventing ATPase activity (Singh et al. 2010). Examples of newly approved protein therapeutics from yeast were Recombumin®, Albucult®, human insulin (Actrapid®), and Pediarix®, all of which are obtained exclusively from *S. cerevisiae* (Nandy and Srivastava 2018). Ecallantide (Kalbitor®) is an FDA-approved recombinant peptide produced from *Pichia pastoris* for treating hereditary angioedema (Sheffer et al. 2011). Anakinra (Kineret®) was approved in 2001 in United States for treating rheumatoid arthritis (Baeshen et al. 2015). *Lactococcus lactis* has been used for centuries connected with the fermentation of food (Singh et al. 2018). Lactate is employed ideally as an emulsifying and moisturizing molecule in cosmetics and as a baseline material in drug fields (Papagianni 2012).

3.2.2 Algae

Algae refers to diverse taxonomic groups of eukaryotic organisms. It was estimated approximately to 30,000 to one million species. In India, the seaweeds were found abundantly in the southeast and west coastal areas comprising 271 genera and 1153 species. The thallus was ranged from unicellular, colonial, and filamentous to complex multicellular branched, reaching meters in length (kelps). The group includes microalgae and macroalgae. The bioresources from algae mainly include food, food additives, nutraceuticals, and feeds (Griffiths et al. 2016).

Microalgae have been used as a nutritional supplement in Asiatic countries like Japan, China, and Korea for years. They constitutes the rich source of carbohydrates, protein, enzymes, fibers, many vitamins (like vitamin A, C, B1, B2, B6), and minerals like niacin, iodine, potassium, iron, magnesium, and calcium. Microalgae are supplemented as food colorants or nutraceuticals in various preparations of pasta, snacks, or in soft drinks (Becker 2004). *Chlorella vulgaris*, *Haematococcus pluvialis*, *Dunaliella salina,* and the blue green *Spirulina maxima* are used as nutritional supplements for humans and as animal feed additives. *Spirulina platensis* is one of the most nutritious food and source of protein, polyunsaturated fatty acids, pigments, vitamins and phenolics, and phycocyanin. The products from *Spirulina* include powders (contain Provitamin A, β-carotene, Vitamin E, Thiamin B1, Riboflavin B2, Niacin B3, Vitamin B6, Inositol, Vitamin B12, Biotin, Folic acid, pantothenic acid, and Vitamin K) tablets, beverage ingredients, antioxidant capsules, chips, pasta, and liquid extracts [Hainan Simai Pharmacy Co.(China), Earthrise Nutritionals (California, USA), Cyanotech Corp. (Hawaii, USA), Myanmar Spirulina factory (Myanmar)] (Mobin and Alam 2017).

Chlorella was marketed in natural health groceries and also as food for fish (Priyadarshani and Rath 2012). *Chlorella* was also used to produce tablets, noodles, powders, etc. by Taiwan Chlorella Manufacturing Co. (Taiwan) and Klötze (Germany). *Dunaliella salina* is grown as a source of photosynthetic pigments and beta-carotene. Beta-carotene is used as an orange dye and a vitamin C supplement (Cognis Nutrition and Health, Australia). It was documented by the National Cancer Institute as an antimetastatic molecule and its efficacy in cholesterol regulation and also in reducing cardiac-related disorders (Priyadarshani and Rath 2012).

Some microalgal species are employed in the skincare market, such as *Arthrospira* and *Chlorella* (Stolz and Obermayer 2005). Major species used cosmetics were *Chondrus crispus*, *Mastocarpus stellatus*, *Ascophyllum nodosum*, *Alaria esculenta*, *Spirulina platensis*, *Nannochloropsis oculata*, *Chlorella vulgaris,* and *Dunaliella salina* (anti-aging cream, refreshing or regeneration care products, sun protection, and hair care products) (Priyadarshani and Rath 2012).

Polyunsaturated fatty acids (PUFA) like α-linolenic, eicosapentaenoic, docosapentaenoic, and docosahexaenoic acid were proven effective in preventing and treating several lifestyle disorders, including cardiovascular, cancer, type 2 diabetes, inflammatory bowel disorders, arthritis, asthma, kidney, skin allergies, and depression. Molecules to treat Schizophrenia was extracted from *Dunaliella*, *Chlorella,* and *Spirulina* species (Priyadarshani and Rath 2012). Fatty acids, DHA, EPA, PUFA from *Odontella aurita* were added to baby food, pharmaceuticals, and cosmetics ($650/kg) (Griffiths et al. 2016).

Haematococcus pluvialis (unicellular, biflagellate, freshwater microalga of Chlorophyta) is the principal compound commercially utilized as antioxidant astaxanthin (average price of $2500/kg) (Mobin and Alam 2017).

Approximate price value of food products like Wakame (*Undaria pinnatifida)* 169 US$/kg dried, Kombu (*Laminaria japonica)* 50–200 US$/kg dried, and Nori (*Porphyra)* 66–166 US$/kg sheets (Griffiths et al. 2016). In Indonesia, flours obtained from *Chlorella vulgaris, Arthrospira platensis,* and *Eucheuma cottonii* were added to wheat flour to produce noodles in which the contents of protein, fat, ash, and roughage were boosted; however, the sugar level was reduced (Kratzer and Murkovic 2021).

Small amounts of *Spirulina*, *Chlorella,* and *Scenedesmus* are incorporated into the animal feed, particularly to boost the immune response. *Spirulina* fed (5–10%) chickens displayed more carotene content in terms of bright yellow skin coloration and yolk content in the egg (Milledge 2011). *Spirulina* was proved for its immunity response, fertility, and skincare, and therefore, it was mixed with the diets of pet animals like cats, dogs, fish, ornamental birds, poultry, horses, cows, and pigs (Griffiths et al. 2016).

Hydrocolloids (alginate, carrageenans, agars, and agarose) extracted from macroalgae were used for thickening and gelling properties in a variety of food and industrial applications. The approximate price value of alginates (*Laminaria, Macrocystis,* and *Ascophyllum)* was $3–190/kg, Carrageenans (*Eucheuma cottonii, E.spinosum,* and *Chondrus crispus)* $5–140/kg powder, agars (*Gracilaria, Gelidium,* and *Pterocladia)* $ 5–100/kg powder and agarose up to $25,000/kg (Griffiths et al. 2016).

Chlamydomonas species with immense carbohydrates are ideal agents of soil moisturizing and water-holding properties to prevent soil leaching. Nitrogen (N)-fixing microalgae can absorb and transform nitrogen from the atmosphere into accessible forms to higher plants. *Aulosira, Anaebena, Nostoc, Tolypothrix,* and *Scytonema* species are traditionally applied as N-fixing agents in paddy cultivation by China, India and other Asian countries (Griffiths et al. 2016).

Dolastatin-10, Curacin-A, and Dolastatin-15 from Cyanobacteria are effective antimicrobial and antitumor agents (Griffiths et al. 2016). Antifungal products like Toyocamycin, Ciguatoxin, and Okadaic acid are produced from Cyanobacteria, *Gambierdiscus toxicus,* and *Prorocenrum lima,* respectively. Cosmetics produced from algal species like Protulines® of *Arthospira, Spirulina* are ideal anti-aging agents, Dermochlorella® can stimulate collagen synthesis (*Chlorella vulgaris),* Pepha®-Tight is used for skin tightening (*Nannochloropsis oculata),* Pepha®-Ctive, and Blue Retinol™ for stimulating cell proliferation (*Dunaliella salina),* and Remergent™ for repairing UV damaged skin (*Anacystis nidulans)* (Griffiths et al. 2016).

3.2.3 Fungi

Fungi are immensely applied in biotechnology-based industries for producing many lead molecules in developed countries. Screening fungi for antibiotics production by antagonistic culture testing has been reported, but it is unlikely to lead to industrial products because it will take over a decade to bring a given project based on a novel fungal metabolite into the preclinical studies. The Big Pharma industry has recently downsized its capacities for in-house research, meaning that academic sector organizations like the Bill and Melinda Gates Foundation and the Wellcome Trust have become more involved in the preclinical evaluation of new compounds (Hyde et al. 2019).

A wide variety of potentially active natural products have been isolated from fungi. Many macrofungi are frequently included in Traditional Chinese Medicine (TCM) formulas and are very commonly sold in the medicinal or local markets. *Ganoderma lingzhi, Wolfiporia cocos,* and *Ophiocordyceps sinensis* are very expensive as they are rare in Chinese markets (Hapuarachchi et al. 2019). Mevastatin and ML-236B are prepared from *Penicillium brevicompactum* as an antibiotic, and mevinolin is isolated from *Aspergillus terreus*. The first statin drug, lovastatin, is produced by several species of *Penicillium* (Jahromi et al. 2012). Another statin-containing product Xuezhikang or red yeast rice extract is produced by fermentation of *Monascus spp.*, and has been widely used in China for centuries for treating circulatory disorders. The active immunosuppressant principle of several marketed drugs, such as Myfortic® and CellCept®, is also now available commercially (Hyde et al. 2019).

Mycoinsecticides derived from *Lecanicillium, Beauveria, Isaria,* and *Metarhizium* species are found in the market. Most developed mycoinsecticdes were designed by European and American markets, whereas only a very few were aimed at the African and Asian markets (Mascarin et al. 2018). Some Representative mycoinsecticides in the market are Arizium, BioMetha GR, Plus, Metamax, Lı'quido, Metapremium, Metarfito, Metarhizium Probio, Opala, and Real from *Metarhizium anisopliae.* Met52 EC, Met52 Granular, and Bio1020 from *Metarhizium brunneum*. Adral, Agrivalle AUIN, Beauve Control, Bovemax EC, Broadband, and Ecobass from *Beauveria bassiana,* Challenger from *Isaria fumosorosea*, PreFeRal WG from *Isaria javanica*, Pae-Sin and Successor SC from *Isaria* spp. Mycotal from *Lecanicillium muscarium*, Verzam, Vercani WP, Lecafol from *Lecanicillium* spp. is common mycoinsecticides (Hyde et al. 2019).

Currently, biofertilizers are recommended over chemical fertilizers. Strains of *Alternaria, Aspergillus, Chaetomium, Fusarium, Penicillium, Serendipita (Piriformospora), Phoma*, and *Trichoderma* have been reported as plant growth-promoting fungi (Saldajeno et al. 2012). Numerous commercial fungal biofertilizer products have been manufactured and were marketed globally. Ketomium® has been created and modified from the strains of *Chaetomium* spp. as pellet and powder form. Several commercial ectomycorrhizal products like MycoRhiz®, Ectomycorrhiza Spawn®, Somycel PV, Mycobead® BioGrow Blend®, MycoApply®-Ecto, Ectovit®, and Mycor Tree® Ecto-Injectable-containing ectomycorrhizae spores are commercially available. The value-added products like MycoApply®-Endo/Ecto, BioOrganicTM Mycorrhizal Landscape Inoculant and Mycoke® Pro ARBOR·WP were produced by adding endo and ectomycorrhizal spores (Hyde et al. 2019). Different mycorrhizal fungi containing products include Oregonism XL (Six strains of endotrophic and six strains of ectotrophic mycorrhizae species), MycoStim (Eight strains of mycorrhizae and two species of *Trichoderma*), Patkar, and Naqvi (Patkar and Naqvi 2017).

Food and beverages from fungi are now a multi-million dollar business in Asia. The global production of mushrooms has increased from 30.2 million tons in 2010 to

48 million in 2017 (FAO). In China, production increased from 22.6 million tons in 2010 to 38.4 million in 2017, accounting for 75% of global production (CIRI 2017) (Hawksworth 2012). In China, there are six groups of mushrooms with a production capacity of over one million tons/year. *Auricularia*, *Lentinula edodes*, *Pleurotus species*, *Agaricus bisporu*, *Flammulina spp*, and *Pleurotus eryngii* species have high volume sales globally (approximately 8.2 million tons) (Zhang et al. 2015). Due to culinary, nutritional, and health benefits, the global market for mushrooms continues to grow, from $34.1 billion in 2015 to $69.3 billion by the end of 2024 (Bal 2018). Oyster mushrooms, morels (*Morchella spp.*), and truffles (*Tuber spp.*) have been commercially cultivated recently (Selosse et al. 2017).

Agaricus subrufescens, new Hitech hybrid of Thai and French strains, was engineered successfully by INRA, France, and Mae Fah Luang University and were enormously, marketed as edible fruiting bodies in tropical parts of the world. The *Auricularia cornea*, *Auricularia thailandica*, *Lepista sordida*, *Agaricus flocculosipes*, *A. subtilipes,* and *Macrolepiota dolichaula* white strains also have record trade in global markets (Thongklang et al. 2014). Food coloring products from filamentous fungi are widely used in food production to enhance the appearance of food colors. Red mold rice (red yeast rice) is an Asian traditional fermentation product of steamed rice fermented with filamentous fungi from the genus *Monascus* (Chen et al. 2015). Hong Qu glutinous rice wine provides a bright-red color and fine sweet flavor (Liu et al. 2018), which is a commercial breakthrough for fungi.

3.2.4 Lichens

Lichens are complex symbionts of algae and fungi and consist of about 19,500 species on the earth. Nearly 8% of the land surface includes lichens distributed from the Arctic to the Antarctic, deserts to the tropics, and littoral zones to mountain peaks. Indian biota comprises 2900 species which is 14.8% of the world's known species. Lichens produce unique secondary metabolites and have considerable biological activities. Many lichens are edible and often toxic. Lichen compounds are aliphatic lichen substances (including acids, zeorin compounds, and polyhydric alcohols) and aromatic components such as fulvic acid, depsides, depsidones, quinones, xanthone, diphenyleneoxide, nitrogen-containing derivatives, triterpenes, and tetronic acids. The third group includes polysaccharides. The most commonly reported molecules are lactones (protolichessterinic acid), phenolic compounds (atranol and resorcinol), depsides (diffractic acid), fulvinic acid derivative (vulpinic acid), dibenzofurans, usnic acids (usnic acid), atranorin, stictic acid, lecanoric acids, and pannarin.

Many commercial pharmacological products are developed. Usnic acid, the antiseptic products of Germany (Camillen 60 Fudes spray and nail oil) and Italy (Gessato™ shaving) have high-value trade, in the drug industries. Lichens of icelandic were traded for cold remedies formulation with the brand name of Isla-

Moos® (Engelhard Arzneimittel GmbH & Co. KG, Germany) and Broncholind® (MCM Klosterfrau Vertriebsgesellschaft mbH, Germany). Lichen extracts were an ingredient in cosmetics, pharmaceuticals, and nutraceutical products of Japan. The riminophenazine antibiotics, exemplified by clofazimine (Lamprene®), were created as antimycobacterial drugs. The antituberculous potential of these drugs was due to the active compounds diploicin and depsidone extracted from the Irish species *Buellia canescens* (Zambare and Christopher 2011).

Lichen harvest and trade are closely associated with the livelihood of most of the rural people in high-range zones. The lichens collected from West Nepal were mainly used in international trade, while those in East Nepal were used locally for food. Approximately 20 commercial lichen species were recorded from five centers and one local market. During 2000–2011, legal estimation recorded export of 2020 tons and collected USD 240,000. The mean annual quantity of turnover was 168 tons, though it is referred that much was illegally exported. Meanwhile, the legislation lacks an effective strategy and sustainable harvest of lichen resources. But, based on scientific data, in the future, better service can be given to local livelihood and lichen conservation globally (Devkota et al. 2017).

3.2.5 Bryophytes

Bryophytes represent the second largest group in the plant world, yet they have not properly evaluated their chemical fingerprints. Ethnic knowledge of bryophytes revealed its importance in curing many human ailments, including lifestyle diseases, wound healing, antiseptic, etc., due to the presence of many potential natural products such as polysaccharides, lipids, amino acids, terpenoids, phenylpropanoids, quinones, and other conjugated metabolites. Globally bryophytes comprise approximately about 20,000 species of liverworts, hornworts, and mosses. India recorded 2800 species (642 endemic). Compiled data about therapeutic features of documented bryophytes are lacking till now (Sabovljevic et al. 2016).

3.2.5.1 Tradable Species from Bryophytes

Anticancer diterpene and paclitaxel (Taxol™) were isolated profoundly from *Physcomitrella patens* by transferring the taxadiene synthase gene from Taxus brevifolia. The yield of taxa-4(5), 11(12)-diene from the in vivo plant was marginal, i.e., approximately 0.05% only (Horn et al. 2021). The antimalarial drug artemisinin (sesquiterpene lactone) was extracted from *P. patens* by engineering five artemisinin biosynthetic pathway genes. The yield was 0.21 mg/g dry weight which is almost at par with the plant *Artemisia annua* (KhairulIkram et al. 2017). The moss *Physcomitrella patens* were also engineered for the commercial production of valuable ingredients for the perfume industry, like the sesquiterpenoids patchoulol and β-santalene. The yield of patchoulol from the moss was 1.34 mg/g dry weight

(Decker and Reski 2020). The diterpenoid sclareol, another valuable component used in fragrances, was obtained from *P. patens* (2.84 mg/g dry weight after 18 days of cell culture) (Decker and Reski 2020). Similarly, the species was utilized by Mosspiration Biotech as a cell factory for the production of air fresheners (Horn et al. 2021). MossCellTec™ No. 1 of the Mibelle Biochemistry unit was the first moss species (*P. patens*) based cosmetic agent marketed globally (Wandrey et al. 2018).

Further, Greenovation, a biopharmaceutical company in Heilbronn, Germany, started the first clinical trial for moss-aGal recombinant form of human alpha-galactosidase as an enzyme replacement therapy (ERT) for patients suffering from deficient or defective alpha-galactosidase A, resulting in lysosomal storage disease Morbus Fabry. Moss-aGal one dosage was powerful in terms of reducing globotriaosylceramide (Gb3) levels even after 28 days, and the patients could recover from the disorders without side effects. This is the first moss-based clinical product to be tested in humans. Fabry disease is a rare genetic disease causing a deficiency of the enzyme alpha-galactosidase A (a-Gal A), hence, the name Moss-aGal for the moss compound (Decker and Reski 2020). Recent in vivo experiments also tested the antifeedant and antifungal activities of extracts from several bryophytes species, indicating their potential value of promised biopesticide against synthesized pesticides. This was already traded as a commercial product in Germany (Lu et al. 2019).

A sesquiterpenoid ricciocarpin, isolated from the liverwort *Ricciocarpos natans,* proved molluscicidal activity against the freshwater snail *Biomphalaria glabrata* (Asakawa and Ludwiczuk 2018). Chinese traditional medicine uses *Marchantia polymorpha* for external ailments (burns and cuts), *Sphagnum teres* for eye diseases and skin irritation, and *Rhodobryum giganteum* for minor heart problems (Horn et al. 2021). Transgenic *Physcomitrella* is now being used to produce blood-clotting factor IX for the treatment of hemophilia B (Saxena and Harinder 2004).

Bryophytes possess a high amount of polyunsaturated fatty acids of long chain (LC-PUFAs). LC-PUFAs are a valuable component of the human diet and are mainly obtained from algal oils as compared to fish (limited availability), which stresses the scope of sustainable harvest of these compounds for human needs (Lu et al. 2019). Metabolically engineered *P. patens* produce LC-PUFAs by encoding D5-elongase from the marine algae *Pavlova* species (Chodok et al. 2012). The synthesis of docosatetraenoic or adrenic acid (ADA) and n-3 docosapentaenoic acid (DPA) was reported from *P. patens*. The transgenic moss produces docosahexaenoic acid (DPA), a new source for the human diet (Chodok et al. 2012). Similarly, transgenic liverwort *Marchantia polymorpha* synthesizes prostaglandin F2a, prostaglandin E2, and prostaglandin D2 through heterologous expression of cyclooxygenase gene from the red alga *Gracilaria vermiculophylla* (Takemura et al. 2013). In addition, the bioproduction of prostaglandins was increased using an in vitro reaction system, the first bioproduction of PGs from plant species (Takemura et al. 2013).

In Sri Lanka, eco-friendly products such as coir pots, coir fiber pith (coco-peat), moss sticks, hanging wire baskets, and basket liners were made from bryophytes. In USA, *Sphagnum* species was used for the production of insulator sheets for houses.

In France, moss industries manufacture moss fancy carpets in diverse sizes, which are easy to implement along the road, lawns, and playgrounds (Saxena and Harinder 2004).

Liverworts and mosses have been used as fuel in developed countries like Finland, Sweden, Ireland, West Germany, Poland, and the Soviet Union. Peat is suitable for the production of low and intermediate BTU gas as well as hydrogen, ethylene, natural gas, methanol, and Fisher Tropsch gasoline. In Germany, *Sphagnum* mixed with wool has been used to prepare cheap clothes. In India, *Sphagnum, Hypnum cupressiforme, Macrothamnium submacrocarpum, Neckera crenulata, Trachypodopsis crispatula,* and *Thuidium tamariscellum* were used for the packing of apples and plum fruits from western Himalayas (Saxena and Harinder 2004).

3.2.6 Pteridophytes

Pteridophytes are primitive vascular plants that comprise 13,000 species distributed along the tropical and temperate zones of the earth. India holds 300 genera with 12,000 species, and 47 are endemic (Fraser-Jenkins 2008). Traditional and ethnobotanical medicinal knowledge of plants plays a significant role in scientific research. But, pteridophytes were unexplored, and their usage aspects are mostly ignored. Twenty-three species of 18 genera in 15 families were used in treating 16 different pregnancy (antenatal) issues during and after delivery (post-partum) health-related diseases by the tribal women of Gond, Korku, Bharia, Bhil, and Mabasi communities, which contribute to about 18.66% of pteridophytic diversity (134 species) of Pachmarhi Biosphere Reserve. Khine and Schneider (2020) recorded 603 species of pteridophytes with multiple uses in Myanmar and also carried their conservation. Traditional knowledge of Tharu tribes spread across Indo-Nepal border of Uttar Pradesh, and its sustainable utilization of 14 pteridophyte species for their livelihood was documented by Singh and Johari (2016). Ojha and Devkota (2021) documented that 26 species of pteridophytes were used as food, 43 species as traditional medicines, and 14 species were used for both in Nepal. *Diplazium esculentum, Diplazium maximum, Dryopteris cochleata,* and *Ophioglossum reticulatum* were edible ferns, and *Aleuritopteris albomarginata, Equisetum ramosissimum, Nephrolepis cordifolia,* and *Tectaria coadunata* were medicinal. Most of them were growing in natural habitat and the potential utilization in the market, which has supported the livelihood of local inhabitants. Some of them were also employed in the preparation of ethnic foods like fermented edible items and as pickles. *Pteridium aquilinum* is the most common edible fern in China and Nigeria (Liu et al. 2012). *Asplenium unilateral, Blechnum orientale, Coniogramme intermedia, Coniogramme denticulatoserrata, Matteuccia intermedia, Osmunda japonica, Pteris wallichiana, Thelypteris prolifera,* and *Woodwardia unigemmmata* were used as edible species in China. Similarly, *Cyathea gigantean, Hypodematium crenatum, Leucostegia truncata, Microsorum punctatum, Pitryrogramma calomelanos, Psilotum nudum, Pteris cretica, Pteris vittata,*

Pyrrosia lanceolata, Lygodium microphyllum, Huperzia phlegmaria, Huperzia squarrosa, and Thelypteris interrupta were used as medicine in India and other parts of the world (Minarchenko et al. 2017). Species like *Athyrium atkinsoni* and *A. strigillosum* were consumed as vegetables in Nepal. Baskaran et al. (2018) reviewed the usage of ferns for treating many human ailments. MannarMannan et al. (2008) also reviewed the ethnobotanical potentialities of ferns.

Dryopteridaceae, Pteridaceae, and Polypodiaceae were popular fern families recorded in nursery trade industries. Twenty-nine alien fern species (especially *Dryopteris erythrosora, D. cycadina, Polystichum polyblepharum,* and *Cyrtomium falcatum*) have high market values in Australia, Canada, New Zealand, South Africa, the United Kingdom, and the USA. Three hundred eighty-two traded ferns, including alien and native species, were documented. Generalized linear models were designed to analyze the market and species traits that influenced the possibility of establishment success in the country in terms of trade. The final model highlighted that different market traits positively affected the likelihood of establishment success in alien ferns, i.e., the number of varieties and cultivars available for a species, high market presence, and trade via e-commerce (McCulloch-Jones et al. 2021).

Scientific validation of its nutritional beneficiaries or medicinal potentialities of the ferns/fern allies was less documented. Similarly, no record of the international market of the species was recorded. The overexploitation practices at local levels in different zones of India, China, Nepal, Sri Lanka, and Japan result in the decline of many species in the wild habitats. Therefore, initiatives for conservation and cultivation (medicinal and decorative species) for income generation and economic empowerment of local communities are to be initiated. The uses of some Pteridophytes are listed in Tables 3.1 and 3.2.

3.2.7 Gymnosperms

Gymnosperms are non-flowering medium-size/tall trees consisting of naked seeds. One thousand one hundred seven gymnosperm species were reported globally (Wang and Ran 2014). Among these, *Ginkgo* was represented by one, gnetophytes 112, cycads 366, and conifers 628 species. Gymnosperms serve as a staple food for low-income or indigenous peoples (Ginkgo, Pinus, Cycas etc). In America and Middle East areas, the leaves of the species were eaten as raw green leafy vegetables and also employed in wine synthesis and other food products. Starch was isolated from the stem of *Cycas revoluta, C. circinalis* (sago), *Zamia,* and *Microzamia* (India and other Asian countries). Seeds of *Pinus girardiana* (Chilgoza) were roasted and eaten in India, Tibet, and Afghanistan, and similarly, *P. ninea* seeds were used in soups and desserts in Italy and Spain. In North America, *Pinus* seeds were used in the production of nut coffee, caramels, candies, and other sweets. Seeds of *Araucaria* and *Ginkgo* were roasted and used as value-added products in Chile and Japan.

Edible oil from the seeds of *C. revoluta, Macrozamia riedlei, Pinus cembra*, and *Cephalotaxus drupacea* possess high market values. Red cedar oil from the core skin

Table 3.1 Checklist of Pteridophyte species recorded as ethnically important wild edible/medicinal plant

Sl. no.	Plant name	Sustainable utilization
1.	*Adiantum capillus-veneris* L.	Stipe and rachis are used for piercing the ears & as ear studs by girls and women. The entire plant is used in witchery, Jadu-Tona and cough syrup. Leaves + honey for treating catarrh, throat, and bronchial disorders. Decoction of the leaves mixed with tea for curing irregularity in the menstrual period. Extract of fronds + honey for eye ailments, respiratory, and menstrual disorders
2.	*Adiantum philippense* L.	Stipe and rachis are used as ear ornaments (studs). Paste of the plant with mustard oil as an ointment for the cure of the boils. Fresh leaf is used for the cure of fits. Rhizome is used for anti-fertility, fronds as a decoction for the pulmonary infections, dysentery and glandular swelling, leaf paste in the treatment of leprosy and hair fall to remove obsession, rhizome to women for sterility, whole plant crushed and applied around navel region in flatulence, fresh leaf (2 g) paste taken orally on an empty stomach twice a day for 10 days for relief from indigestion
3.	*Ampelopteris prolifera* (Retz.) Copel.	Entire plant but usually new frond as a delicious vegetable. Juice of rhizome diluted with to cure sexual disorders in male. Diluted juice of rhizome as aphrodite to the tensile male sex organ
4.	*Ceratopteris thalictroides* (L.) Brongn.	Paste of the entire plant is boiled in mustard oil is used as an ointment for burn, fresh wounds and to stop bleeding. Cooked leaves are eaten as food. Fronds are used as a poultice in skin diseases; leaf powder along with turmeric is applied to unhealed wounds
5.	*Christella dentata* (Forssk.) Brownsey & Jermy in Brit.	New and juvenile fronds are used as a vegetable. The rhizome and leaf paste are used for boil treatment. Rhizome and sporophyll are used as an antibacterial agent
6.	*Christella parasitica* (L.) Lev.	Juvenile fronds are used as a vegetable and the entire plant as fodder. Paste of rhizome is to get rid of evil spirits. Fresh rhizome (5 g) along with fresh root (1 g) of *Asparagus racemosus* and sugar (5 g) are boiled in water (250 ml). The decoction is orally administered for 10 days to cure spermatorrhoea, gout, and rheumatism
7.	*Diplazium esculentum* (Retz.) Sw. in Schrad.	New fronds are used as a vegetable and entire plant as fodder for cows and goats, Crozier or tender leaf as salad and pickles. Young and fresh frond is boiled with salt and taken to maintain all round health. Rhizome as an insect repellent. Decoction of rhizome +2 mL of honey in empty stomach to cure spermatorrhoea

(continued)

Table 3.1 (continued)

Sl. no.	Plant name	Sustainable utilization
8.	*Equisetum ramosissimum* Desf. subsp. debile Hauke	Macerated plant mixed with red mud was topically for treatment and joining of the fractured bone. Plant paste is prepared in water and is applied twice a day in bone fracture. It is used for polishing wood and brass. Shoot and rhizomes are used for gonorrhea. Plant paste is topically applied to cure scabies, itches, and skin infections. Powdered stem is dissolved in water for enema in children
9.	*Helminthostachys zeylanica* (L.) Hook. Gen.	Strobilus of the plant is used as a vegetable. Extract of rhizome and the entire plant are used as Aphrodite, and rhizome mixed with other plants as a tonic for the cure of waist pain. Plant part-based herbal formulation is developed for sexual disorders. It has anodyne properties and is prescribed as a tonic. The rhizome is used as a memory enhancer to promote strength and vitality and cure impotency or erectile dysfunction
10.	*Lygodium flexsuosum* (L.) Sw., in Schrad.	Juvenile part of plants is used as a vegetable. Rhizome extract is boiled with mustard oil as a thick paste, which is topically applied for the treatment of arthritis and sore discharging water. Decoction of leaf cures jaundice. Fresh rhizome boiled with mustard oil is good for rheumatism, sprains, scabies, ulcers, eczema, and cut wounds; however, an aqueous extract is used to cure spermatorrhoea. Leaf paste in skin diseases; however, rhizome powder mixed with cow urine is a potent formulation of skin diseases. Rhizome and black pepper paste is used twice a day for dysmenorrhoea. Rachis tied over forehead reduces headache. Leaf powder mixed in milk to enhance memory teaspoonful of plants juice is given twice a day to relieve fever. Rhizomes extract (100 mL) twice daily for 2 weeks given orally for premature ejaculation
11.	*Marsilea minuta* L.	Leaves mixed with mint are macerated to prepare extract for the pacification of heatstroke and the plant juice for the treatment of eye disease. Stalk and leaves are used as a vegetable, consumed as a tonic after fever, and in insomnia and mental problems. Cakes of sporocarp called 'nardoo' are eaten. Decoction of leaves + ginger for cough and bronchitis. Juvenile leaves juice dropped in the nostrils twice a day can cure migraine
12.	*Microlepia speluncae* (L.) Moore	The plant is used as fodder for cow feed. Dried plants are also sprayed in the cattle shed as supplementary fodders and also to protect the animals from extreme cold, acting as an absorbent of urinal excreta
13.	*Ophioglossum reticulatum* L.	Leaves as a delicious vegetable, a remedy against headache, and fresh plant as a tonic for the treatment

(continued)

Table 3.1 (continued)

Sl. no.	Plant name	Sustainable utilization
		of worms and inflammation. Fresh leaf along with rice into a cake, and the boiled cake is taken orally on an empty stomach for 15–20 days against menstrual disorders. Paste of fresh leaves and tubers can applied topically for treatment of boils, burns, and as a cooling agent
14.	*Pteris biaurita* L	Entire plant as fodder. Dried plants are sprayed in the cattle shed as supplementary fodders and to protect the animals from extreme cold, acting as an absorbent of urinal excreta. Paste of the plants can be used on cuts and bruises

MannarMannan et al. (2008)

of *Juniperus virginiana* was used to clean small objects and oil-soaked lenses. Perfumed oils of *Cyptomeria, Cedrus deodara, Cupressus, japonica*, and serum peruvians are high-value products. Gymnosperm timbers were widely used in furniture and other construction industries. The woods are softwood, and hence, their durability is long lasting. The largest timber-producing tree on the earth is *Agathis australis*. The heavy and durable coniferous wood of *Pinus, Taxus, Sequoia, Cupressus,* etc., is used as a building material in furniture, railway sleepers, packing cases, poles, etc. The wood of Cedar is durable, oily, sweet scented, and generally without resin ducts. *C. deodara* (deodar) is a unique timber of North India. The resinous wood is resistant to pests and pathogens and, therefore, has high market values in carpentry industries. The heavy wood of *Araucaria* is used in making doors, bus chassis, carpentry, etc. The Abies wood is used in the manufacture of the soundboard of musical instruments, boxes, planks, cabinets, paper pulp, etc. *Abies concolor* wood is suitable for keeping dairy products. Agathis wood is stronger and used in building construction, boats, wooden machinery, etc. The *Podocarpus totara* is soft and is resistant to mine borer, hence, used for dock and sea work as well as bridges and shipbuilding.

Gymnosperms are used successfully to treat infectious diseases and other allergies, including colds, coughs, asthma, respiratory congestion, etc. *Taxus* yields taxol, an anticancer drug. Different kinds of cycas plants are used for producing hair care items, including shampoo, oil, lotion, etc. Alkaloid ephedrine from *Ephedra* is an ingredient in the cough syrup due to its dilating action on the bronchial tube also contracts mucous membranes and is used in nasal drops and inhalants. Extract of *Ginkgo biloba* is used for the treatment of cerebral insufficiency and vertigo. Leaves of Taxus are used in asthma, bronchitis, hiccough, epilepsy, and indigestion.

Table 3.2 Pteridophyta species used for gynecological/reproductive health-related issues by native people

Gynecological/reproductive health-related issues	Botanical name	Part used
Irregular menstruation	1. *Adiantum capillus-veneris* 2. *Adiantum philippense* 3. *Asplenium trichomanes* 4. *Cheilanthes farinosa* 5. *Nephrolepis exaltata* 6. *Ophioglosum reticulatum*	Fronds Fronds Whole plant Leaves Rhizome Leaves
Dysmennorrhea (painful menstruation)	1. *Parahemionitis cordata* 2. *Lygodium flexuosum*	Whole plant Rhizome
Amenorrhoea/Emmenagogue (absence or suppression of normal menstrual flow)	1. *Adiantum capillus-veneris* 2. *Selaginella ciliaris* 3. *Selaginella involvens*	Fronds Leaves Leaves
Menorrhagia (heavy menstrual bleeding)	1. *Lygodium flexuosum*	Whole plant
Female contraception/Abortifacient	1. *Adiantum capillus-veneris* 2. *Adiantum philippense* 3. *Osmunda regalis* 4. *Pleopeltis macrocarpa*	Fronds Rhizome Leaves Whole plants
Total sterility	1. *Blechnum orientale* 2. *Nephrolepis cordifolia*	Fronds Rhizome
For conception/remove infertility	1. *Actiniopteris radicata* 2. *Dicranopteris linearis.* 3. *Equisetum ramosissimum* 4. *Hypodematium crenatum* 5. *Lygodium flexuosum* 6. *Nephrolepis exaltata*	Leaves Fronds Rhizome Leaves Whole plant Rhizome
Uterine hemorrhage	1. *Ophioglosum reticulatum*	Whole plants
Gonorrhea	1. *Asplenium yoshinagae* var. *planicaule* 2. *Dryopteris cochleata* 3. *Equisetum ramosissimum* ssp. *debile* 4. *Lygodium flexuosum* 5. *Selaginella bryopteris*	Rhizome Whole plant Rhizome Rhizome Leaves
Leucorrhea	1. *Actiniopteris radicata* 2. *Ophioglosum reticulatum* 3. *Selaginella bryopteris*	Leaves Whole plants Leaves
Cold-imposed problems of the uterus	*Adiantum capillus-veneris*	Fronds

(continued)

Table 3.2 (continued)

Gynecological/reproductive health-related issues	Botanical name	Part used
Birth-aid in parturition/easy child birth	1. *Diplazium esculentum* 2. *Nephrolepis exaltata*	Leaves Rhizome
White discharge	1. *Alsophila gigantea*	Rhizome
Post-partum care/strengthening	1. *Actiniopteris radicata* 2. *Selaginella bryopteris* 3. *Ophioglosum reticulatum*	Leaves
Abscess of uterus	1. *Asplenium trichomanes*	Whole plants
Aphrodisiac	1. *Actiniopteris radicata* 2. *Adiantum capillus-veneris* 3. *Pteridium aquilinum*	Leaves Leaves Whole plants

Baskaran et al. (2018)

3.2.7.1 Industrial Uses of Gymnosperms

Cycas gum is used as an adjuvant, a remedy for snake bites and for dangerous sores. Tannins from the bark of Araucaria, Pinus, Sequoia, etc. are used in the leather industry. The turpentine of *Abies balsamea* is used as a balm in organic preparations. Fossil resin of *Pinus succinifera* was used for doors, posts, beams, wagon flooring, etc. Plywood from podocarpus, papers such as newsletters, stationery items, and print are prepared from the wood pulp of *Pinus, Picea, Abeis, Gnetum* etc. Cycad leaves are used to fix baskets, mats, hats, brooms, etc. Macrozamia and Cycas leaf fibers are employed to make pillows and mattresses for smooth sleep. Species of Abies, Cedrus, Juniperus, and Thuja are used for making soaps, perfumes, room fresheners, deodorants, disinfectants, medicines for cough, cold, rheumatism, cuts, wounds, eczema, and other skin smoothening. Craft paper of Pinus is valuable in the market. In India, *Pinus roxburghii* yields quality pulp, while craft paper was obtained from *Cryptomeria japonica*. Rayon, transparent and photographic film, lacquers etc., are produced from the wood pulp of Picea and Tsuga (Jan et al. 2009).

Cycas species are cultivated for decorative purposes. Ginkgo, the 'virgin' tree, is planted as an ornamental plant in Chinese and Japanese temples and is worshiped. *Thuja plicata, Biota orientalis* and *Juniperus* species are cultivated as ornamental throughout India. Pinus and Aurocaria are also cultivated in mass scale as ornamental in northern India. Gnetum and Ephedra varieties are also cultivated as ornaments in many garden nureseries.

Acts and regulations recognize the local community involvement in managing and protecting bioresources available in their locality. Their mode of life is interconnected with biological resources, which provide a source of food and earnings. Access and Benefit Sharing (ABS) in the Biodiversity Act empowers such native people to generate resources for local biodiversity management.

Recording and enlisting the bioresources are the need of the hour for its conservation and future sustainable utilization.

References

Asakawa Y, Ludwiczuk A (2018) Chemical constituents of bryophytes: structures and biological activity. J Nat Prod 81:641–660

Baeshen MN, Al-Hejin AM, Bora RS, Ahmed MM, Ramadan HA, Saini KS (2015) Production of biopharmaceuticals in *E. coli*: current scenario and future perspectives. J Microbiol Biotechnol 25:953–962. https://doi.org/10.4014/jmb.1412.12079

Bal C (2018) Benefits and uses of mushroom. J Bacteriol Mycol 6:155–156

Baskaran X, Geo Vigila A, Zhang S, Feng S, Liao WB (2018) A review of the use of pteridophytes for treating human ailments. J Zhejiang Univ Sci B 19:85–119

Becker W (2004) Microalgae in human and animal nutrition. In: Richmond A (ed) Handbook of microalgal culture. Blackwell, Oxford, pp 312–351

Blunt JW, Copp BR, Keyzers RA, Munro MH, Prinsep MR (2016) Marine natural products. Nat Prod Rep 33:382–431

Chen W, He Y, Zhou Y, Shao Y (2015) Edible filamentous fungi from the species Monascus early traditional fermentations, modern molecular biology, and future genomics. Compr Rev Food Sci Food Saf 14:555–567

Chodok P, Cove DJ, Quatrano RS, Kanjana-Opas A, Kaewsuwan S (2012) Metabolic engineering and oil supplementation of *Physcomitrella patens* for activation of C22 polyunsaturated fatty acid production. J Am Oil Chem Soc 89:465–476

Cobb RE, Luo Y, Freestone T, Zhao H (2013) Drug discovery and development via synthetic biology. In: Zhao H (ed) Synthetic biology. Academic, Boston, pp 183–206. https://doi.org/10.1016/b978-0-12-394430-6.00010-8

Decker EL, Reski R (2020) Mosses in biotechnology. Curr Opin Biotechnol 61:21–27

Devkota S, Chaudhary RP, Werth S, Scheidegger C (2017) Trade and legislation: consequences for the conservation of lichens in the Nepal Himalaya. Biodivers Conserv 26(4):2491–2250. https://doi.org/10.1007/s10531-017-1371-3

Dias DA, Urban S, Roessner U (2012) A historical overview of natural products in drug discovery. Metabolites 2:303–336

Fraser-Jenkins CR (2008) Taxonomic revision of three hundred Indian subcontinental pteridophytes with a revised census-list (a new picture of fern-taxonomy and nomenclature in the Indian subcontinent). Bishen Singh Mahendra Pal Singh, Dehra Dun

Griffiths M, Harrison STL, Smit M (2016) Major commercial products from micro- and macroalgae. In: Bux F, Chisti Y (eds) Algae biotechnology. Green Energy Technology, pp 269–300

Gyawali R, Ibrahim SA (2014) Natural products as antimicrobial agents. Food Control 46:412–449. https://doi.org/10.1016/j.foodcont.2014.05.047

Hapuarachchi KK, Karunarathna SC, McKenzie EHC, Wu XL, Kakumyan P, Hyde KD, Wen TC (2019) High phenotypic plasticity of *Ganoderma sinense* (Ganodermataceae, Polyporales) in China. Asian J Mycol 2:1–47

Hawksworth DL (2012) Global species numbers of fungi: are tropical studies and molecular approaches contributing to a more robust estimate. Biodivers Conserv 21:2425–2433

Horn A, Pascal A, Loncarevic I, Marques RV, Lu Y, Miguel S, Bourgaud F, Thorsteinsdottir M, Cronberg N, Becker JD, Reski R, Simonsen HT (2021) Natural products from bryophytes: from basic biology to biotechnological applications. Crit Rev Plant Sci 40(3):191–217

Hyde KD, Xu J, Rapior S (2019) The amazing potential of fungi: 50 ways we can exploit fungi industrially. Fungal Divers 97:1–136

Jahromi MF, Liang JB, Ho YW, Mohamad R, Goh YM, Shokryazdan P (2012) Lovastatin production by *Aspergillu sterreus* usingagro-biomass as substrate in solid state fermentation. J Biomed Biotechnol 2012:1–11

Jan G, Khan MA, Jan F (2009) Traditional medicinal and economic uses of gymnosperms of Dir Kohistan valleys, NWFP, Pakistan. Ethnobot Leafl 13:1509–1521

KhairulIkram NKB, Beyraghdar Kashkooli A, Peramuna AV, van der Krol AR, Bouwmeester H, Simonsen HT (2017) Stable production of the antimalarial drug artemisinin in the moss *Physcomitrella patens*. Front Bioeng Biotechnol 5:47–48

Khine PK, Schneider H (2020) First assessment of pteridophytes' composition and conservation status in Myanmar. Glob Ecol Conserv 22:1–9

Kratzer R, Murkovic M (2021) Food ingredients and nutraceuticals from microalgae: main product classes and biotechnological production. Foods 10:1–15

Liu Y, Wujisguleng W, Long C (2012) Food uses of ferns in China: a review. Acta Soc Bot Pol 81(4):263–270

Liu Z, Wang Z, Lv X, Zhu X (2018) Comparison study of the volatile profiles and microbial commnities of Wuyi Qu and Gutian Qu, two major types of traditional fermentation starters of Hong Qu glutinous rice wine. Food Microbiol 69:105–115

Louca S, Mazel F, Doebeli M, Parfrey LW (2019) A census-based estimate of earth's bacterial and archaeal diversity. PLos Biol 17(2):1–30

Lu Y, Eiriksson FF, Thorsteinsdottir M, Simonsen HT (2019) Valuable fatty acids in bryophytes-production, biosynthesis, analysis and applications. Plants 8:524–533

MannarMannan M, Maridass M, Victor B (2008) A review on the potential uses of ferns. Ethnobot Leafl 12:281–285

Mascarin MG, Lopes R, Delalibera I, Fernandes E (2018) Current status and perspectives of fungal entomopathogens used for microbial control of arthropod pests in Brazil. J Invertebr Pathol 165: 46–53. https://doi.org/10.1016/j.jip.2018.01.001

Matsumura E, Nakagawa A, Tomabechi Y, Ikushiro S, Sakaki T, Katayama T (2018) Microbial production of novel sulphated alkaloids for drug discovery. Sci Rep 8:79–80

McCulloch-Jones E, Kraaij T, Crouch NH, Fritz D (2021) The effect of horticultural trade on establishment success in alien terrestrial true ferns (Polypodiophyta). Biol Invasions 23:3583–3596. https://doi.org/10.1007/s10530-021-02599-0

Milledge JJ (2011) Commercial application of microalgae other than as biofuels: a brief review. Rev Environ Sci Biotechnol 10:31–41

Minarchenko V, Tymchenko I, Dvirna T, Makhynia L (2017) A review of the medicinal ferns of Ukraine. Scr Sci Pharm 4(1):7–15

Mobin S, Alam F (2017) Some promising microalgal species for commercial applications: a review. Energy Procedia 110:510–551

Nandy SK, Srivastava RK (2018) A review on sustainable yeast biotechnological processes and applications. Microbiol Res 207:83–90. https://doi.org/10.1016/j.micres.2017.11.013

Ojha R, Devkota HP (2021) Edible and medicinal pteridophytes of Nepal: a review. Ethnobot Res Appl 22(16):1–16

Papagianni M (2012) Metabolic engineering of lactic acid bacteria for the production of industrially important compounds. Comput Struct Biotechnol J 3(4):1–8. https://doi.org/10.5936/csbj.201210003

Patkar RN, Naqvi NI (2017) Fungal manipulation of hormone regulated plant defense. PLoS Pathol 13(6):e1006334. https://doi.org/10.1371/journal.ppat.1006334

Pham JV, Yilma MA, Feliz A, Majid MT, Maffetone N, Walker JR, Kim E, Cho HJ, Reynolds JM, Song MC, Park SR, Yoon YJ (2019) A review of the microbial production of bioactive natural products and biologics. Front Microbiol 10:1–27

Priyadarshani I, Rath B (2012) Commercial and industrial applications of micro algae—a review. J Algal Biomass Utln 3(4):89–100

Sabovljevic M, Sabovljevic A, Ikram N, Peramuna A, Bae H, Simonsen H (2016) Bryophytes—an emerging source for herbal remedies and chemical production. Plant Genet Res 14(4):314–327

Saldajeno MGB, Ito M, Hyakumachi M (2012) Interaction between the plant growth-promoting fungus Phoma sp. GS8-2 and the arbuscular mycorrhizal fungus *Glomus mosseae*: impact on biocontrol of soil-borne diseases, microbial population, and plant growth. Aus Plant Pathol 41: 271–281

Sanchez S, Guzman-Trampe S, Avalos M, Ruiz B, Rodriguez-Sanoja R, Jimenez-Estrada M (2012) Microbial natural products. In: Civjan N (ed) Natural products in chemical biology. Wiley, Hoboken, pp 65–108

Saxena DK, Harinder (2004) Uses of bryophytes. Resonance 9:56–65

Selosse MA, Schneider-Maunoury L, Taschen E, Rousset F, Richard F (2017) Black truffle, a hermaphrodite with forced unisexual behaviour. Trends Microbiol 25:784–787

Sheffer AL, Campion M, Levy RJ, Li HH, Horn PT, Pullman WE (2011) Ecallantide (DX-88) for acute hereditary angioedema attacks: integrated analysis of 2 double-blind, phase 3 studies. J Allergy Clin Immunol 128:153–159. https://doi.org/10.1016/j.jaci.2011.03.006

Shen B (2015) A new golden age of natural products drug discovery. Cell 163:1297–1300

Siddiqui AA, Iram F, Siddiqui S, Sahu K (2014) Role of natural products in drug discovery process. Int J Drug Dev Res 6:172–204

Singh AP, Johari D (2016) Sustainable utilization of biodiversity: role of pteridophytes in livelihoods of Tharu tribes of Uttar Pradesh. Uttar Pradesh Biodiversity Board. https://doi.org/10.13140/RG.2.2.32813.79841

Singh SB, Genilloud O, Pelaez F (2010) Terrestrial microorganisms—filamentous bacteria. In: Liu H-W, Mander L (eds) Comprehensive natural products II. Elsevier, Oxford, pp 109–140. https://doi.org/10.1016/b978-008045382-8.00036-8

Singh SK, Tiendrebeogo RW, Chourasia BK, Kana IH, Singh S, Theisen M (2018) *Lactococcuslactis* provides an efficient platform for production of disulfide-rich recombinant proteins from *Plasmodium falciparum*. Microb Cell Fact 17:55–60

Stolz P, Obermayer B (2005) Manufacturing microalgae for skin care. Cosmetics Toiletries 120:99–106

Takemura M, Kanamoto H, Nagaya S, Ohyama K (2013) Bioproduction of prostaglandins in a transgenic liverwort, *Marchantia polymorpha*. Transgenic Res 22:905–911

Tevyashova AN, Olsufyeva EN, Solovieva SE, Printsevskaya SS, Reznikova MI, Trenin AS (2013) Structure-antifungal activity relationships of polyene antibiotics of the amphotericin B group. Antimicrob Agents Chemother 57:3815–3822. https://doi.org/10.1128/aac.00270-13

Thongklang N, Sysouphanthong P, Callac P, Hyde KD (2014) First cultivation of *Agaricus flocculosipes* and a novel Thai strain of *A. subrufescens*. Mycosphere 5:814–820

Wandrey F, Henes B, Zulli F, Reski J (2018) Biotechnologically produced moss active improves skin resilience. SOFW J 144:34–37

Wang XQ, Ran JH (2014) Evolution and biogeography of gymnosperms. Mol Phylogen Evol 75: 24–40

Woo MW, Nah HJ, Choi SS, Kim ES (2014) Pikromycin production stimulation through antibiotic down-regulatory gene disruption in *Streptomyces venezuelae*. Biotechnol Bioprocess Eng 19: 973–977

Zambare VP, Christopher P (2011) Biopharmaceutical potential of lichens. Pharma Biol 50(6): 778–798

Zhang JX, Chen Q, Huang CY, Gao W, Qu JB (2015) History, current situation and trend of edible mushroom industry development. Mycosystema 34:524–540

Chapter 4
Nutraceutical Potential of Underutilized Wild Edible Fruits Endemic to Western Ghats in Southern India

Amitha S. Chandra, Vinitha S. Babu, and P. M. Radhamany

Abstract Fruits are a quintessential part of a healthy eating pattern and are a source of many vital nutrients. Heed in understanding the nutraceutical potential of underutilized fruits is a sore subject. Globally, a myriad of research has been carried out on many underutilized fruits, and these studies have highlighted their nutritional, ethnobotanical, and ethnomedicinal value. The present chapter focuses on the nutraceutical potential of underutilized wild edible fruits endemic to the south Western Ghats of India. The region is one of the hottest biodiversity hot spots, and the prevailing climatic conditions there, including the topography, wide rainfall variation, and varied temperature, have made it home to many flora and fauna. In recent years, rain-fed orchards have grabbed attention at a national level, and these wild fruits have gained further importance because of their availability all around the year. Apart from this, these serve as a source of food and medicine to native dwellers. The continuous depletion of forest areas due to human intervention has led to the loss of their habitat. The present chapter discusses the nutraceutical potential of selected underutilized fruits endemic to the southern Western Ghats of India.

Keywords Underutilized · South-western Ghats · Nutraceuticals · Biodiversity · Endemic

4.1 Introduction

'Let food be thy medicine, and medicine be thy food,' the principle advocated by Hippocrates (460–377 BC), father of medicine, highlighted the association between nutrition and human health. He also conceptualized the relationship between the use of appropriate foods for health and their therapeutic benefits (Palthur et al. 2010).

A. S. Chandra · V. S. Babu · P. M. Radhamany (✉)
Department of Botany, University of Kerala, Thiruvananthapuram, Kerala, India
e-mail: drradhamany@keralauniversity.ac.in

S. T. Sukumaran, K. T R (eds.), *Conservation and Sustainable Utilization of Bioresources*, Sustainable Development and Biodiversity 30,
https://doi.org/10.1007/978-981-19-5841-0_4

However, the present scenario of rapid development has led to an imbalance not only in their nutritional habits but also in the environment where they live. World Health Organization (WHO) defines malnutrition as deficiencies, excesses, or imbalances in a person's intake of energy and/or nutrients. As per WHO, worldwide, around 1.9 billion and 462 million adults, respectively, are overweight and underweight. Recent estimates have suggested that approximately 41 million children under five years of age are obese, whereas around 159 million show stunted growth. Adding to this burden, around 29% of women of reproductive age group are prone to anemia around the world, for which approximately half would be open for iron supplementation (Muller and Krawinkel 2005). Even though several initiatives are being carried out to enhance food production, nutritional security is a major concern worldwide. Hence, there is a need for the availability of inexpensive and healthy foods. Vegetables and fruits are 'treasure houses' for a repertoire of nutritional as well as nutraceutical molecules. Since the dawn of civilization, wild edible fruits have been an important source of food for mankind. Fruits are often called defensive food because of their innumerable accumulation of phytoconstituents, nutrients, and minerals. They are also a rich wellspring in dissolving dietary fibers that can lower cholesterol as well as fat deposition from the body, thereby helping in smooth defecation and aids in boosting the immune system. Research findings suggest that the educated and well-informed population of the world is aware of nutraceuticals' importance. Currently, a surge in non-communicable or chronic diseases has led to a burgeoning contribution to the global burden of disease. This scenario has been exacerbated by the overwhelming westernization apart from rapid industrialization and urbanization. Unfortunately, all the countries do not have the potential to increase their resource allocation to health. Even though the high-income countries have access to equitable and efficient health care systems, the situation is strange for the low-income and third-world countries. Thus, in this scenario, the role of underutilized fruits comes into action. Underutilized fruits are those which have market value but are not widely grown in the field and are rarely found in the market. They have been linked with the cultural heritage of the locality, traditional crops in localized areas, and neglected by agricultural research organizations. Underutilized fruits are also stated as those fruits which are less available, less utilized or rarely used, or are region specific. They are often categorized as underutilized because of their non-availability of their complete botanical information and inadequate research on their commercial exploitation. Apart from this, there is a lack of knowledge on their food nutrition value and potential. There is also a stigma that these are the food for the poor, irrespective of the nutritional benefits in them.

Fruits contain rich sources of nutraceuticals. They are an important source of trace elements, vitamins, pigments, and metabolites that are necessary for growth, normal physiological functioning, and maintenance of health. Currently, there is an increased global interest in nutraceuticals as they play a major role in health enhancement. Also, nutraceutical as a field of science and formulations has emerged in recent times and is progressing as an interdisciplinary science. The term nutraceutical was first defined by Dr. Stephen L. Defelice as a product isolated or purified from foods and sold in medicinal forms. Nutraceuticals which inhibit the oxidation of organic molecules are very important for the defense of living systems against oxidative stress. Hence, an insight into the importance of the nutritional significance of underutilized fruits is essential.

4.2 Underutilized Fruits in India

India, located at 8°N–38°N latitude and 68°E–93.5°E longitude, exhibits extreme diversities of edaphic, climatic conditions in its agro-climatic regions and ecosystems. The country also shows vivid altitudinal variations ranging from below sea level to more than 3500 m above mean sea level. The quintessential climatic conditions seen here are conducive to tropical plants and are the centers of origin of many fruit trees. India is also home to the world's most useful plants thriving in her diverse agro-ecological zones and altitudes—in the monsoon tropics of the south to temperate and alpine north-western Himalayas, from the extremely arid and semi-arid north-western plains to the humid tropics of the east. The several less-known fruit species which have the potential for commercial exploitation are yet to be utilized to their potential. Many of these plants have wide adaptability as well as a high degree of tolerance and hence can thrive even under most adverse situations. India is bestowed with a rich diversity of flora and fauna and is one of the major agro-diversity centers (Vavilov 1951). The nation covers over5% of the world's diversity even though it covers only 2% of the earth's surface, and it is considered one of the richest and most highly endangered ecosystems of the world (Myers et al. 2000). Various literature has documented the dietary use of wild edible fruits (Billore and Hemadri 1969). The importance of wild edible resources for the food and nutrient security of rural poor and tribal communities in India is also well recognized. It is estimated that more than 600 species of plants are known to have food value. Several researchers have documented wild edibles as well as their importance as domestic food among tribal peoples in India (Nair and Jayakumar 1999). According to the reports of Arora and Pandey, there are approximately 647 species of wild edible fruits under 112 families in India. The plethora of literature available on the ethnobotany of wild edible products thrusts the importance of forest and forest products (Vartak 1980). Apart from being a vital role in the livelihood of tribal and rural people, forests are endowed with value added products such as wax, honey, gum, resins, etc.

4.3 Western Ghats: Hottest Hot-Spot

The Western Ghats are, also known as '*Sahyadris*' composed of the province of Malabar, running parallelly through the western coast of India (8°N–21°N latitudes, 73°E–77°E longitudes), transversing through the states of Maharashtra, Goa,

Karnataka, Tamil Nadu, and Kerala over a distance of 1600 km at an average width of 200 km. The region has several mountain peaks that deter monsoon winds, thereby resulting in heavy rainfall. This further allows several plant species to inhabit the region. These mountain chains are very rich in diverse edible fruit yielding plants. They are also considered to be one of the hottest megadiversity hotspots that directly and indirectly support the livelihood of millions of human beings. The area is one of the biodiversity hotspots of the world and is home to 7402 species of angiosperms and 1814 species of non-flowering plants; it is likely that many undiscovered species live in the Western Ghats. This hotspot is rich in several minor and underutilized fruits that are rarely eaten and unfamiliar to kinfolks. As these fruits comprise a wide spectrum of essential nutrients, vitamins, and minerals as well as metabolites, they are contemplated for cultivation, consumption, and utilization. Unlike the cultivated tropical fruits, these wild fruits may not taste good but holds good proportions of antioxidants, essential nutrients, and bioactive molecules. Moreover, these are also a source of income for local communities in rural areas. Over the past, information regarding these fruits has been limited due to the lack of suitable and efficient processing techniques and increased deforestation. Furthermore, deprived knowledge about wild fruit identification is yet another hurdle to overcome. Also, there is a lack of documentation on these plants. Hence, cultivation, promotion, and conservation of these can be crucial for nutritional, medicinal, and economic purposes. Consequently, there requires a need for the popularization of these fruits by creating awareness among the locals.

The South Western Ghats belongs to an ecoregion in South India, occupying the portions of Karnataka, Kerala, and Tamil Nadu at an elevation ranging from 1000 to 2695 m. This region receives an annual rainfall exceeding 2800 mm (Fig. 4.1). Out of the 1500 species of flowering plants reported from India, more than 5094 are exclusive to the Southern Western Ghats (Sasidharan 2011). The Southern Western Ghats is endemic to numerous species. It has an area of 22,600 square kilometers and is estimated to have covered two thirds of the original forests. They contain the highest peaks, notably 'Anamudi' in Kerala. They are also the wettest portion of peninsular India and are surrounded by dry ecoregions (Wikramanayake et al. 2002). The cool and moist edaphic factors, accompanied by heavy rainfall, have supported lush and diverse forests. About 35% of the plant species are endemic to this region. A moist evergreen montane forest is the predominant habitat type. The region thereby supports a great diversity of flora and fauna. The major canopy tree forms include *Mesua ferrea, Cullenia exarillata, Gluta travancorica, Palaquium ellipticum, Nageia wallichiana,* etc. The area has also evergreen tree species like *Calophyllum austroindicum, Garcinia travancorica, G. rubro-echinata, Diospyros barberi, Memecylon subramanii, M. gracile, Goniothalamus rhynchantherus,* and *Vernonia travancorica*. The shola-grassland complex is yet another major feature characterized by small trees like *Linociera ramiflora, Mahonia napaulensis, Pygeumgardneri, Actinodaphne bourdillonii, Schefflera racemose,* etc. (Wikramanayake et al. 2002). Even though much information regarding the plant distribution is abundant, there is little documentation of endemic fruit trees available (Table 4.1). Hitherto, there has been a decline in the availability of these fruits

Fig. 4.1 Map of political boundaries of southern Western Ghats region

because of the change in land-use patterns and overexploitation. Hence, this chapter gives the documentation of selected edible endemic fruits of Southern Western Ghats (Figs. 4.1 and 4.2).

4.3.1 Spondias pinnata *(L. f.) Kurz*

Spondias pinnata, often known as hog plum or wild mango, is a member of the Anacardiaceae family. It is a deciduous, glabrous tree with edible fruit that can reach a height of 25 m. After ripening, the fruits turn a greenish yellow color. Ripe fruits are typically spherical to ovoid or ellipsoidal in shape, measuring around 5 cm long and 3.5 cm wide. The fruits have a sour taste and are high in vegetable protein, chitin, fiber, nutrients, and minerals and low in calories. Fruits could be eaten raw or

Table 4.1 Wild edible fruits of Western Ghats Region

Scientific name	Common name	Family	Habit	Flowering and fruiting	Ethnobotany	References
Annona muricata L.	Soursop	Annonaceae	Tree	April–October	Fruit juice is given orally to treat hematuria, liver problems, urethritis, and to increase mother's milk. The fruit juice and crushed seeds are used as a vermifuge and anthelmintic	Badrie and Schauss (2010)
Artocarpus hirsutus Lam.	Wild jack	Moraceae	Tree	December–March	The unripe fruits are sour, astringent, and sweet in taste, having an aphrodisiac, constipating properties. Fruit juice stimulates appetite and relieves hemorrhage pains	Thakur and Vidyasagaran (2014), Hari et al. (2014)
Aegle marmelos (L.) Correa	Bael	Rutaceae	Tree	June–July	The unripe, dried fruit is astringent, digestive, and stomachic. The roots and the bark of the tree are used in the treatment of fever by making a decoction of them	Kirtikar and Basu (1932)
Averrhoa carambola L.	Star fruit	Oxalidaceae	Tree	All year round	The fruit is consumed raw and is used in the preparation of pickles	
Berberis tinctoria Lesch.	Nilgiri barberry	Berberidaceae	Shrub	February–October	The fruit has antioxidative, anticarcinogenic, antihypertensive, antiinflammatory, antibacterial, immune-stimulating, and cholesterol-lowering qualities	Tabeshpour et al. (2017)
Carissa spinarum L.	Wild Karanda	Apocynaceae	Shrub	April–July	The fruits are eaten along with the seeds, and they have a sweet flavor and are of good quality. The ripe fruits are consumed as snacks to alleviate diarrhea, diabetes, heart disease, and obesity	Chauhan (2015)

Elaeagnus conferta Roxb.	Wild olive	Elaeagnaceae	Shrub	January–March	Ripe fruits are consumed. Dry fruit powder helps in blood alcohol removal. The fruits are used as medicine for the treatment of indigestion	Wu et al. (2011), Patil et al. (2012)
Elaeocarpus tectorius (Lour.) Poir	Bikkannu	Elaeocarpaceae	Tree	April–November	Fruits are good for treating microbial infections and are a potential source of antioxidants and diseases including diabetes, rheumatism, and piles	Sharvani and Devaki (2014), Manoharan et al. (2019), Keerthana and Chitra (2020)
Flacourtia indica (Burm. f.) Merr.	Governor's plum Madagascar plum	Salicaceae	Shrub	November–March	The ripe fruits are consumed or used to make jelly or jam. Fruits are useful appetizing and digesting food as well as a diuretic and in the treatment of jaundice and enlarged spleen	Kirtikar and Basu (1932)
Flacourtia jangomas (Lour.) Raeusch	Indian plum	Salicaceae	Tree	March–October	Fruits are used for making dessert, juice, syrup, marmalade, pickles, and sauces. Alleviate digestive disorders, biliousness, nausea, diarrhea, and allay thirst	Srivastava et al. (2009)
Flacourtia montana J. Grahm.	Moutain Sweet Thorn	Salicaceae	Tree	April–June	Proteins, antioxidants, macro, and micronutrients are abundant in fruits. Eaten raw and made into jelly or wine. Catechin, chlorogenic acid, and epicatechin are the major polyphenols	Mundaragi et al. (2019), Mundaragi and Thangadurai (2015)

(continued)

Table 4.1 (continued)

Scientific name	Common name	Family	Habit	Flowering and fruiting	Ethnobotany	References
Garcinia gummi-gutta (L.) Roxb.	Malabar tamarind	Clusiaceae	Tree	January–September	Fruits are commonly used as a souring agent for curries and as a fish preservative. Fruits have the potential to cure parasite infections, hemorrhoids, rheumatism, and menstrual irregularities. Hydroxycitric acid (HCA) content in the fruit act as a weight loss agent	Tharachand and Avadhani (2013), Onakpoya et al. (2011)
Garcinia indica (Thouars) Choisy	Kokum	Clusiaceae	Tree	November–August	Fruits are eaten raw, and the juice is used for curing stomach and liver disorders. Hydroxy citric acid in the fruit boost fat burning. Potent anti-oxidant, antiinflammatory, antimi-crobial and anticancer agent	Krishnamurthy et al. (1982), Jena et al. (2002), Padhye et al. (2009), Khatib et al. (2010)
Gardenia gummifera L.f	Cumbi gum tree	Rubiaceae	Tree	December–May	Consumed fresh. antimicrobial, antiradical, and insecticidal activity	Kekuda et al. (2017)
Glycosmis pentaphylla (Retz.) DC.	Orangeberry	Rutaceae	Shrub	September–April	Rich in proteins, carbohydrates, fiber, energy content, and a suffi-cient amount of minerals. The fruits are used by rural people for their food and medicine	Kuladip and Varsha (2018)

Limonia acidissima L.	Wood apple	Rutaceae	Tree	December–June	Fruits can be consumed raw or stored like preserves and beverages. Fruits pulp possesses antiinflammatory, antipyretic, and analgesic properties. Psoralene, xanthotoxin, 2, 6-dimethoxybenzoquinone, and osthenol are the antifungal compounds found in the fruit shells and used to exfoliate the skin	Ahamed et al. (2008), Ilango and Chitra (2010)
Passiflora edulis Sims	Passion fruit	Passifloraceae	Climber	July–November	Rich in minerals and contain provitamin A, ascorbic acid, riboflavin, and niacin. The pulp obtained after scooping is used to make fruit salads, ice cream, juices, jelly, jam, and squash. Fruits have antioxidant, antibacterial, antiinflammatory, antihypertensive, hepatoprotective, anti-diabetic, sedative, and antidepressant properties	Menzel et al. (1993), Silva et al. (2015), Dzotam et al. (2016), Zhang et al. (2016)
Rhodomyrtus tomentosa (Aiton) Hassk.	Rose myrtle Ceylon hill gooseberry	Myrtaceae	Shrub	February–May	Fruits are with antioxidant, anticancer, antiinflammatory, and anti-obesity properties and are consumed fresh	Ovesna et al. (2006), Kita et al. (2012), Son et al. (2010), Kwon et al. (2012)
Rubus ellipticus Smith	Indian yellow raspberry Himalayan raspberry	Rosaceae	Shrub	November–March	Fruits are consumed fresh and can enhance their antioxidant properties, balancing blood glucose levels	Deighton et al. (2000), Halvorsen et al. (2002), Reyes-Carmona et al. (2005), Zhang et al. (2010), Sharma and Kumar (2011)
Rubus racemosus Roxb.	Black raspberry	Rosaceae	Shrub	December-May	Fruits help in digestion and have anti inflammatory potential by inhibiting the expression of COX-1 and COX-2 enzymes	Bowen-Forbes et al. (2010), Sathyavathi and Janardhanan (2014)

(continued)

Table 4.1 (continued)

Scientific name	Common name	Family	Habit	Flowering and fruiting	Ethnobotany	References
Rubus rugosus Smith.	Indian Raspberry	Rosaceae	Shrub	June–July	Consumed fresh and its juice is effective for curing fistula	Singh et al. (2012), Sharma and Kumar (2013)
Schleichera oleosa (Lour.) Oken.	Ceylon oak	Sapindaceae	Tree	March–June	The ripe fruit is eaten raw and is a good appetizer. *Tribal people have used Kusum oil'* for curing a cold, itch, and ulcers for both humans and cattle	Meshram et al. (2015)
Spondias pinnata (L. f.) Kurz	Hog plum	Anacardiaceae	Tree	March–December	Fruits are eaten raw or processed into jams, jellies, juices, or pickles. Pulp is used to treat rheumatism, bilious dyspepsia	Muhammad et al. (2011)
Syzygium calophyllifolium (Wight) Walp	Pretty-Leaved Plum	Myrtaceae	Tree	February–May	Fruits are edible and are efficient antioxidant and antibacterial agent	Sathyanarayanan et al. (2018)
Syzygium cumini (L.) Skeels	Jamun	Myrtaceae	Tree	April–July	They were consumed at the ripe stage. Fruit juice for treating diabetics, chronic diarrhea, enteric illnesses, and antibacterial infections. The alkaloid Jambosin and the glycoside Jambolin or Antimellin found in the seeds restrict the diastatic conversion of starch to sugars	Migliato (2005), Benherlal and Arumughan (2007), Giri et al. (1985)
Syzygium zeylanicum (L.) DC	Spicate eugenia	Myrtaceae	Tree	January–April	Fruits are edible and fermented to make wine. They are rich in vitamins, proteins, minerals, and sugars	Shilpa and Krishnakumar (2015)

Terminalia bellerica (Gaertn.) Roxb.	Bastard myrobalan, Bedda nut tree	Combretaceae	Tree	January–April	The matured kernel is consumed. Fruits have antiinflammatory, antihelmintic, antidiabetic, antibacterial, expectorant, and anti-pyretic properties. They are used to treat asthma, bronchitis, dropsy, dyspepsia, heart disorders, skin ill-nesses, leprosy, and ulcers. One of the ingredients in Triphala medicine of Ayurveda are also effective in cataract, glaucoma, progressive myopia, and conjunctivitis	Nadkarni (1954), Kumar et al. (2010), Latha and Daisy (2010)
Terminalia chebula Retz.	Black myrobalan	Combretaceae	Tree	February–August	The fruit extract is rich in phenolic compounds and shows antioxidant, antihyperglycemic, and wound healing potential.	Murali et al. (2004), Li et al. (2011)
Toddalia asiatica (L.) Lam.	Wild Orange tree	Rutaceae	Climber	September–July	The fruits are used to treat coughs and sore throat. Aside from that, the fruits are used for making pickles, and the unripe fruit is rubbed with oil to make stimulant liniment for arthritis	Rajkumar et al. (2008), Reddy et al. (1998)
Ziziphus jujuba Mill.	Red date	Rhamnaceae	Tree	February–April	Active components in the fruit include phenolics, flavonoids, triterpenic acids, vitamin C, and polysaccharides. These compounds have anti-aging, antiinflammatory, and antibacterial qualities as well as reducing glucose levels, which pre-vents extra weight gain, particularly belly fat. They also have neuro-protective benefits and increase memory and learning	Gao et al. (2013), Chen and Tsim (2020), Tahergorabi et al. (2015)

(continued)

Table 4.1 (continued)

Scientific name	Common name	Family	Habit	Flowering and fruiting	Ethnobotany	References
Ziziphus oenopolia (L.) Mill.	Jackal jujube	Rhamnaceae	Climber	November–March	The fruits are rich in polyphenols like flavonoids and tannin . Commonly used to treat diabetes mellitus	Thirugnanasampandan et al. (2017), Goyal et al. (2021)
Ziziphus rugosa Lam.	Zunnaberry	Rhamnaceae	Shrub	November–May	The deseeded pulp is used to make dosa and juice. Fruit is used topically to cure boils and for the treatment of throat and bronchial irritation. The methanolic extract of pericarp and seed revealed to be an effective anticancer agent	Hooker (1875), Kaur et al. (2011)

Fig. 4.2 Selected underutilized and edible fruits of Western Ghats in Southern India

processed into jams, jellies, juices, or pickles and used as pig fodder, thus, the popular name 'hog plum' (Florido and Cortiguerra 2004). The fruits are used in worship, as a vegetable, and for flavoring in curry. In Ayurveda, the unripe fruit is thought to eradicate '*vata,*' enrich the blood and heal rheumatism. The fruit juice exhibits antiscorbutic properties. The pulp is used to treat rheumatism and bilious dyspepsia, and a home-made cure for biliousness is made with 10 g of fruit juice, 50 g of candy sugar, and 0.6–0.8 g of black pepper powder (Arif and Fareed 2010).

4.3.2 Artocarpus hirsutus *Lam.*

Artocarpous hirsutus commonly known as wild jack or 'Anjili chakka' belongs to the Moraceae family. This endemic tree grows only in the Southern Western Ghats of peninsular India and Maharashtra (Matthew et al. 2006). The tall evergreen tree grows to a height of 20–25 m and a diameter of up to 5 m. It bears edible ovoid fruits that are bright yellow and covered with spines, having ovoid white seeds. The smaller size of spherical fruits distinguishes *A. hirsutus* from the common jack fruit (*Artocarpus heterophyllus*). It is an essential foodstuff that has received a lot of attention as the natural nutrition of the people of Kerala state; its fruits and roasted seeds are consumed as a highly nutritious cuisine. The fruit pulp has a high nutritional value due to its high level of antioxidants, thiamine, ascorbic acid, riboflavin, niacin, amino acids, and lipids. The unripe fruits are sour, astringent, and sweet in taste having anaphrodisiac, constipating properties (Thakur and Vidyasagaran 2014). The fruit juice stimulates appetite and relieves hemorrhage pains (Hari et al. 2014). The juice from cooked fruits is thought to stimulate hunger and, when applied to the anus, can reduce hemorrhage pains. Methanolic extract of the fruit has significant DPPH and reduces power-scavenging activity (Narayanaswamy and Balakrishnan 2011).

4.3.3 Flacourtia indica *(Burm. f.) Merr.*

Flacourtia indica (Burm. f.) Merr. is a branched dioecious shrub or small tree of the Salicaceae (previously Flacourtiaceae) family that grows up to 5 m tall. Due to its resemblance to small plums, it is also called as 'Indian plum,' 'Governor's plum,' and 'Madagascar plum.' The fruit is globular in shape, reddish to reddish black in color, juicy, and contains 4–10 seeds. The soft, juicy pulp of the fruit is acidic to sweet taste. The ripe fruits are consumed or used to make jelly or jam. Wine can be made by fermenting it. From ancient times, numerous tribal societies have employed the herb as an effective cure for various ailments. Fruits are useful appetizing and digesting food as well as a diuretic. It is also useful in treating jaundice and enlarged spleen (Kirtikar and Basu 1932). Nutritive value: Crude fat (0.21%), Crude Fiber (2.96%), Crude protein (2.42%), Carbohydrate (18.20%), P (11.1 mg/100 g), K (142.8 mg/100 g), Mg (18.91 mg/100 g), Ca (51.12 mg/100 g), and Na (44.12 mg/100 g) (Mundaragi et al. 2017).

4.3.4 Carissa spinarum *L.*

Carissa spinarum is commonly known as 'wild Karanda,' or 'bush plum' and is closely related to *Carissa carandas* L. The erect thorny shrub belongs to the family

Apocynaceae with forked branches and grows to a height of 2–3 m. These drought-resistant plants are commonly growing in forests and wastelands up to 1500 m in height. Oval green berries develop from white star-shaped blossoms. When the berries become mature, they turn black or dark in color and are consumed by everyone. They are also sold in many places. *Carissa spinarum* L. is also known as 'Magic Shrub' in some African countries because it is used for treating a variety of diseases and illnesses (Ansari and Patil 2018). The fruits are eaten along with the seeds, and they have a sweet flavor and are of good quality. Fruits contain acids, sugars, reducing sugars, non-reducing sugars, tannins, pectin and vitamin C, carissol (an epimer of α-amyrin), lupeol, oxalic, tartaric, citric, malic, malonic, and glycolic acids, glycine, alanine, phenyl alkaline, cerine, glucose, and galactose (Siyum and Meresa 2021). The ripe fruits are consumed as snacks to alleviate diarrhea, diabetes, heart disease, and obesity (Chauhan et al. 2015). Sahreen et al. (2010) found that chloroform and aqueous fractions of *C. spinarum* fruits have significant antioxidant activity. The phenolic and flavonoid composition of these fractions is thought to be responsible for their activity. The plant has been employed as a treatment for a variety of ailments in ethnobotany, and its fruits are a rich source of vitamin C and iron (Mundaragi and Thangadurai 2017).

4.3.5 Ziziphus rugosa *Lamk.*

Ziziphus rugosa Lamk. is a huge evergreen thorny shrub belonging to the family Rhamnaceae. It is grown widely in the deciduous and semi-evergreen forest of Western Ghats up to 6000 feet in height. 3–6 m tall tree bears orange to black-colored fruits having 9–12 mm length and 8–10 mm breadth. The fruits are commonly known as Cheruthudali, Mullanu, Kottimullu and are the known famine, medicinal, and edible fruits sold by locals. The Kodava community in the Western Ghats' Kodagu region eats raw and ripened fruit for nutritional sources (Greeshma and Sridhar 2016). The deseeded pulp is used to make dosa and juice, and it contains macronutrients like nitrogen, potassium, calcium, magnesium, and micronutrients like zinc, copper, manganese, and iron (Krishnamurthy and Sarala 2012). The fruit of *Z. rugosa* is used by Sylhet and Moulvibazar folk medicinal healers to treat cancers (Rahmatullah et al. 2015). Villages in the upper Ghat (Salkani and Killara) and two villages in the coastal zone (Murur and Kallabbe) in the central Western Ghats, Karnataka, India, used the fruit as a coolant and to keep the body hydrated. Rural communities in Tiruchirappalli District, Tamil Nadu, South India, use the fruit to treat wounds and diarrhea (Hegde et al. 2015). The fruit is used topically to cure boils and for the treatment of throat and bronchial irritation (Hooker 1875). The pericarp extract has the potential for curing bacterial infections, free radical damage, and arboviral diseases like chikungunya, dengue, etc. The phytoconstituents present in the extract like alkaloids, saponins, flavonoids, and glycosides may be responsible for the tested biological efficacies of the extract (Prashith et al. 2011). When tested against human melanoma cells, the methanolic extract of pericarp and seed was revealed to be an effective anticancer agent (Kaur et al. 2011).

4.3.6 Berberis tinctoria *Lesch.*

Berberis tinctoria Lesch. is an evergreen shrub endemic to the Southern Western Ghats, most abundant in the upper elevations of the Nilgris, Tamil Nadu. It comes under the Family Berberidaceae and is commonly known as Nilgiri Barberry (locally called as 'Oosikala' or 'JakkalaHannu'). In the forest, the plant grows to 2–3 feet tall and can reach a height of 15 feet. It bears blue blackberries with few seeds and short, sturdy styles in a glaucous spindle form. The Berberis genus has a number of well-known medicinal plants that are used to treat a variety of illnesses, including gallstones, eye disease, jaundice, rheumatism, diabetes, fever, kidney stone, vomiting during pregnancy, and others (Srivastava et al. 2015). Its leaves and fruits are primarily taken uncooked by tribal and rural people in the Nilgiris for a variety of ailments that have not been formally documented. Secondary metabolites in barberry fruit extract, such as phenol (410 0.082 mg (GAE)/100 g) and flavonoid (320 0.120 mg (QE)/100 g), have the potential to scavenge DPPH, O_2, NO, OH, and $ABTS^+$ radicals as well as a chelated ferrous ion (Sasikumar et al. 2012). Compared to the leaf and stem, the fruit contains more vital minerals such as Na (20.296 0.544 mg/100 g), Mg (75.477 12.845 mg/100 g), Al (52.393 15.986 mg/100 g), and K (177.898 38.722 mg/100 g) (Vignesh et al. 2021). The high nutritional value of carbohydrates, amino acids, vitamins, and protein minerals in the fruit shows that it has antioxidative, anticarcinogenic, antihypertensive, antiinflammatory, antibacterial, immune-stimulating, and cholesterol-lowering qualities (Tabeshpour et al. 2017).

4.3.7 Elaeagnus conferta *Roxb.*

Elaeagnus conferta Roxb. is a thorny climber shrub or small bushy deciduous plant that can reach a height of 12 m. It is commonly called 'Nerli' and in Malayalam '*Kattumunnthiringa.*' It is a member of the Elaeaganaceae family and can be found in evergreen and semi-evergreen forests. The edible fruit is long and elliptical in shape, about 4 cm long, reddish with whitish dots, and has a sour taste. Fruit contains macroelements like nitrogen, phosphorus, potassium, calcium, magnesium, sodium, and microelements like ferrous, zinc, copper, and manganese (Valvi and Rathod 2011). Pulp extract contains a higher quantity of ascorbic acid (8.2 mg/100 g) and total carotene (16.00 mg/100 g), so the consumption of fresh fruit provides an adequate source of vitamin C, provitamin A, and other antioxidants (Patil et al. 2012). They also found that 100 g of *E. conferta* berries contains 8.2 mg of ascorbic acid, 6.80 mg of total phenolics, and 11.68 mg of flavonoids. These small molecules serve a crucial role in the prevention of oxidative damage, scavenging free radicals, and alleviating various human diseases. Dry fruit powder helps in blood alcohol removal by boosting the action of ADH and ALDH (Wu et al. 2011). The fruits are used as medicine for the treatment of indigestion (Patil et al. 2012). GC-MS analysis revealed the presence of 9-Octadecanoicacid, n-hexadecenoic acid, Cyclopentanone, 2-methyl, and 9-Octadecanoic acid as the major compounds (Valvi et al. 2014).

4.3.8 Rhodomyrtus tomentosa *(Alt.)*

Rhodomyrtus tomentosa (Alt.) Hassk. also known as rose myrtle, which grows at higher altitudes in the Western Ghats region. The flowering plant comes under Myrtaceae family and grows up to 5 m in height. It bears an ellipsoid berry with a persistent calyx. The immature fruits have a green skin that matures to a purple black color. The pulp is purplish in color, soft, and sweet when fully mature. It is a popular ornamental plant in tropical and subtropical gardens, planted for its abundant flowers and sweet, edible fruit. The fruit can be used in salads or made into pies and jellies. The unripe fruits of the plant are used to treat diarrhea in Vietnamese traditional medicine, while the ripe ones are used to stimulate the immune system (Agro Forestry Tree Database 1992; Institute of Chinese Medicine 2010). The fruit contains nutrient substances such as proteins, carbohydrates, lipids, vitamins, minerals, dietary fiber, essential oil, and trace elements (Huang et al. 2010; Lai et al. 2015; Wu et al. 2004). The major amino acid was reported to be tryptophan. Fatty acids like linoleic and palmitic acids constituted 75.36% and 10.45% of total fatty acids. Lai et al. (2013) identified 19 phenolic compounds found in the fruit, including ellagitannins, stilbenes, anthocyanins, flavonols, and phenolic acids. Piceatannol, a stilbene, has antioxidant (Ovesná et al. 2006), anticancer (Kita et al. 2012), antiinflammatory (Son et al. 2010), and anti-obesity properties (Kwon et al. 2012).

4.3.9 Rubus ellipticus *Smith*

Rubus ellipticus Smith is a thorny evergreen shrub that grows up to 100–300 cm in height in higher altitude regions of Western Ghats. It is a member of the Rosaceae family, commonly known as Indian yellow raspberry or Himalayan raspberry. It bears yellow spherical berries that look similar to a raspberry having 10 mm in diameter. Fruits are sweet and slightly acidic in taste and can be consumed raw or cooked. The fruits are highly nutritious and rich in vitamins, sugars (Parmar and Kaushal 1982), minerals (Saklani et al. 2012; Ahmad et al. 2015), and amino acids like L-Hydroxyproline, DL Iso-leucine, DL Valine, DL-2-Aminobutyric acid, L-Cystein hydroxyl, DL-Nor-leucine, DL Alanine, L-Glutamic acid, L-Arginine, DL-Aspartic acid, L-Cysteinhydroxychloride, L-Leucine, L-Lysine monochloride, DL-Methionine, etc. 200 g of fruit is sufficient to meet an individual's daily nutritional requirements due to its high level of fiber, lipids, minerals, and proteins (Saklani et al. 2012). Secondary metabolites like flavonoids, glycosides, steroids, phenols, tannins, anthocyanin, and resin are all found in fruit (Saklani et al. 2012; Karuppusamy et al. 2011). The richness of phenolic components enhances the antioxidant properties (Deighton et al. 2000; Halvorsen et al. 2002; Reyes-Carmona et al. 2005; Zhang et al. 2010) as well as balancing blood glucose levels (Sharma and Kumar 2011).

4.3.10 Syzygium cumini *(L.) Skeels*

Syzygium is the largest genus in the Myrtaceae family, includes more than 1200 species, and a widely distributed across the Indian subcontinent (Parnell et al. 2007; Govaerts et al. 2008). Syzygium with a high rate of endemism in the evergreen and shola forests of the Western Ghats and the north-eastern ghats. When compared to the middle and northern Western Ghats, the Southern Western Ghats, particularly in Kerala, have a higher diversity and endemism of Syzygium. Fruits of many species, such as *S. cumini* (L.) Skeels, *S. aqueum* (Burm. f.) Alston, *S. jambos* (L.) Alston, *S. malaccense* (L.) Merr. & L. M. Perry, *S. caryophyllatum* (L.) Alston, and *S. zeylanicum* (L.) DC. are nutritionally rich and are edible. *Syzygium cumini* (L.) Skeels are locally known as '*Njaval.*' Ripe fruits are mostly eaten raw and used to make jam, wine, or pickles. Fresh fruit cleanses the blood, prevents bad breath, and strengthens the gums and teeth. Gargling with diluted fruit juice relieves throat pain (Gordon et al. 2011). Fruit juice or vinegar is very effective for treating diarrhea, spleen enlargement, and urine retention. Jamun fruit extract shows a significant antioxidant activity as compared to the other non-traditional fruits, which can be attributed to constituents such as anthocyanins, tannins, and flavonols (Gordon et al. 2011) and for treating diabetics, chronic diarrhea, enteric illnesses, and antibacterial infections (Migliato 2005; Benherlal and Arumughan 2007). The alkaloid Jambosin and the glycoside Jambolin or Antimellin found in the seeds restrict the diastatic conversion of starch to sugars. The seed powder is often used to immediately reduce the amount of sugar in the urine (Giri et al. 1985).

4.4 Conclusion

The tropical and subtropical tracts of India are bestowed with a wide range of diversity in several fruits, which are growing wild/semi-wild, are unattended and underutilized. Most of these species have wide adaptability as well as a high degree of tolerance and, hence, can thrive well under most adverse situations. In spite of the rich germplasm existing in India, for most of the underutilized fruits, no standard variety has been developed so far. Many of these fruits are nutritionally very rich and are of great medicinal value. These fruits hold promise for sustainable agriculture, particularly for small farmers, by augmenting their income with the least risk. Our urgent task would be to develop/select suitable variety/genotype and to standardize production protocol and popularize these fruits.

References

Agro Forestry Tree Database (1992) A tree species reference and selection guide: Rhodomyrtus tomentosa. http://www.worldagroforestrycentre.org/sea/Products/AFDbases/af/asp/SpeciesInfo.asp?SpID=18093

Ahamed SM, Swamy SK, Jayaverra KN, Rao JV, Kumar VS (2008) Antiinflammatory, antipyretic and analgesic activity of methanolic extract of *Feronia limonia* fruit pulp. Pharmacologyonline 3:852–857

Ahmad M, Masood S, Sultana S, Hadda TB, Bader A, Zafar M (2015) Antioxidant and nutraceutical value of wild medicinal Rubus berries. Pak J Pharm Sci 28(1):241–247

Ansari I, Patil DT (2018) A brief review on phytochemical and pharmacological profile of *Carissa spinarum* L. Asian J Pharm Clin Res 11:12–18

Arif M, Fareed S (2010) Pharmacognostic investigation and authentication of potentially utilized fruit *Spondias mangifera* (willd). Int J Curr Pharm Res 2(1):31–35

Badrie N, Schauss AG (2010) Soursop (*Annona muricata* L.): composition, nutritional value, medicinal uses, and toxicology. In: Bioactive foods in promoting health. Academic, pp 621–643

Benherlal PS, Arumughan C (2007) Chemical composition and in vitro antioxidant studies on *Syzygium cumini* fruit. J Sci Food Agric 87(14):2560–2569

Billore KV, Hemadri K (1969) Observations on the Flora of Harishchandragarh, Sahyadri range, Maharashtra. Nelumbo 11(3–4):335–346

Bowen-Forbes CS, Zhang Y, Nair MG (2010) Anthocyanin content, antioxidant, antiinflammatory and anticancer properties of blackberry and raspberry fruits. J Food Compos Anal 23(6): 554–560

Chauhan AT, Beenu A, Intelli (2015) Influence of processing on physiochemical, nutritional and phytochemical composition of *Carissa spinarum* (karonda) fruit. Asian J Pharm Clin Res 8(6): 254–259

Chen J, Tsim KW (2020) A review of edible jujube, the *Ziziphus jujuba* fruit: a heath food supplement for anemia prevalence. Front Pharmacol 11:593655

Deighton N, Brennan R, Finn C, Davies HV (2000) Antioxidant properties of domesticated and wild Rubus species. J Sci Food Agric 80(9):1307–1313

Dzotam JK, Touani FK, Kuete V (2016) Antibacterial activities of the methanol extracts of *Canarium schweinfurthii* and four other *Cameroonian* dietary plants against multi-drug resistant Gram-negative bacteria. Saudi J Biol Sci 23(5):565–570

Florido HB, Cortiguerra FF (2004) Research information series on ecosystems. J Trop For Sci 3:1–10

Gao QH, Wu CS, Wang M (2013) The jujube (*Ziziphus jujuba* Mill.) fruit: a review of current knowledge of fruit composition and health benefits. J Agric Food Chem 61(14):3351–3363

Giri J, Sathidevi T, Dushyanth N (1985) Effect of jamun seed extract on alloxan induced diabetes in rats. J Diabetic Assoc India 25(4):115–119

Gordon A, Jungfer E, da Silva BA, Maia JGS, Marx F (2011) Phenolic constituents and antioxidant capacity of four underutilized fruits from the Amazon region. J Agric Food Chem 59(14): 7688–7699

Govaerts R, Sobral M, Ashton P, Barrie F, Holst BK, Landrum LL, Matsumoto K, Mazine FF, Lughadha EN, Proneça C, Soares-Silva LH (2008) World checklist of Myrtaceae. Royal Botanic Gardens

Goyal PK, Jeyabalan G, Singh Y (2021) In-vitro free radical scavenging and hypoglycemic evaluation of fruit extract and solvent fractions of *Ziziphus oenoplia* mill (Rhamnaceae). Int J Pharmacogn 8(5):216–223

Greeshma AA, Sridhar KR (2016) Ethnic plant-based neutraceutical values in Kodagu region of the Western Ghats. Biodivers India 8:299–317

Halvorsen BL, Holte K, Myhrstad MC, Barikmo I, Hvattum E, Remberg SF, Wold AB, Haffner K, Baugerødn H, Andersen LF, Moskaug O (2002) A systematic screening of total antioxidants in dietary plants. J Nutr 132(3):461–471

Hari A, Revikumar KG, Divya D (2014) *Artocarpus*: a review of its phytochemistry and pharmacology. J Pharma Search 9(1):7–12

Hegde VM, Vasudeva R, Kamatekar SL, Sthapit BR, Parthasarathy VA, Rao VR (2015) Traditional knowledge associated with tropical fruit tree genetic resources: comparison of upper-Ghat and coastal situation of Central Western Ghats, India. Indian J Plant Genet Resour 28(1):95–105

Hooker JD (1875) The flora of british India (vol. 1). L. Reeve

Huang WY, Cai YZ, Corke H, Sun M (2010) Survey of antioxidant capacity and nutritional quality of selected edible and medicinal fruit plants in Hong Kong. J Food Compos Anal 23(6):510–517

Ilango K, Chitra V (2010) Wound healing and antioxidant activities of the fruit pulp of *Limonia acidissima* Linn (Rutaceae) in rats. Trop J Pharm Res 9(3):223–230

Institute of Chinese Medicine (2010). Encyclopedia on contemporary medicinal plants: *Rhodomyrtus tomentosa* (Downy Rose Myrtle). http://www.hkjcicm.org/cm_database/plants/detail_e.aspx?herb_id=489. Last accessed 24 Sept 2011

Jena BS, Jayaprakasha GK, Singh RP, Sakariah KK (2002) Chemistry and biochemistry of (−)-hydroxycitric acid from *Garcinia*. J Agric Food Chem 50(1):10–22

Karuppusamy S, Muthuraja G, Rajasekaran KM (2011) Antioxidant activity of selected lesser known edible fruits from Western Ghats of India

Kaur R, Kapoor K, Kaur H (2011) Plants as a source of anticancer agents. J Nat Prod Plant Resour 1(1):119–124

Keerthana M, Chitra P (2020) Antidiabetic activity of chemical constituents in *Elaeocarpus tectorius* fruits-an in silico study. J Univ Shanghai Sci Technol 22(12):342–358

Kekuda PTR, Raghavendra HL, Shilpa M, Pushpavathi D, Petkar T, Siddiqha A (2017) Antimicrobial, antiradical and insecticidal activity of *Gardenia gummifera* Lf (Rubiaceae). Int J Pharm Pharm Sci 9(10):265–272

Khatib NA, Pawase K, Patil PA (2010) Evaluation of anti inflammatory activity of *Garcinia indica* fruit rind extracts in wistar rats. Int J Res Ayurveda Pharm 1(2):449–454

Kirtikar KR, Basu BD (1932) Indian medicinal plants. Lalit Mohan Basu, Allahabad

Kita Y, Miura Y, Yagasaki K (2012) Antiproliferative and anti-invasive effect of piceatannol, a polyphenol present in grapes and wine, against hepatoma AH109A cells. J Biomed Biotechnol 2012:1–7

Krishnamurthy SR, Sarala P (2012) Determination of nutritive value of *Ziziphus rugosa* Lamk.: a famine edible fruit and medicinal plant of Western Ghats. Ind J Nat Prod Resour 3(1):20–27

Krishnamurthy N, Lewis YS, Ravindranath B (1982) Chemical constituents of kokam fruit rind. J Food Sci Technol 19(3):97–100

Kuladip G, Varsha JR (2018) Proximate and mineral analysis of fruit of *Zanthoxylum rhetsa* DC. and *Glycosmis pentaphylla* (Retz.) DC.: most useful ethnomedicinal plants in Kolhapur district. Int J Life Sci 9:48–52

Kumar B, Divakar K, Tiwari P, Salhan M, Goli D (2010) Evaluation of anti-diarrhoeal effect of aqueous and ethanolic extracts of fruit pulp of *Terminalia belerica* in rats. Int J Drug Dev Res 2(4):769–779

Kwon JY, Seo SG, Heo YS, Yue S, Cheng JX, Lee KW, Kim KH (2012) Piceatannol, natural polyphenolic stilbene, inhibits adipogenesis via modulation of mitotic clonal expansion and insulin receptor-dependent insulin signaling in early phase of differentiation. J Biol Chem 287(14):11566–11578

Lai TNH, Herent MF, Quetin-Leclercq J, Nguyen TBT, Rogez H, Larondelle Y, Andre CM (2013) Piceatannol, a potent bioactive stilbene, as major phenolic component in *Rhodomyrtus tomentosa*. Food Chem 138(2–3):1421–1430

Lai TNH, Andre C, Rogez H, Mignolet E, Nguyen TBT, Larondelle Y (2015) Nutritional composition and antioxidant properties of the sim fruit (*Rhodomyrtus tomentosa*). Food Chem 168: 410–416

Latha RCR, Daisy P (2010) Parameters in streptozotocin diabetic rats. Int J Pharmacol 6(2):89–96

Li K, Diao Y, Zhang H, Wang S, Zhang Z, Yu B, Huang S, Yang H (2011) Tannin extracts from immature fruits of *Terminalia chebula* Fructus Retz. promote cutaneous wound healing in rats. BMC Complement Alternat Med 11(1):1–9

Manoharan AL, Thamburaj S, Muniyandi K, Jagadeesan G, Sathyanarayanan S, Nataraj G, Thangaraj P (2019) Antioxidant and antimicrobial investigations of *Elaeocarpus tectorius* (Lour.) Poir. fruits against urinary tract infection pathogens. Biocatal Agric Biotechnol 20: 101260

Matthew SP, Mohandas A, Shareef SM, Nair GM (2006) Biocultural diversity of the endemic 'wild jack tree' on the Malabar coast of South India. Ethnobot Res Appl 4:25–40

Menzel CM, Winks CW, Simpson DR (1993) Passion fruit production and research in Australia. Tropical Fruits Newsletter (IICA), 8:3–4

Meshram N, Ojha M, Singh A, Alexander A, Sharma M (2015) Significance and traditional medicinal properties of *Schleichera oleosa*. Asian J Pharma Res 5(1):61

Migliato KF (2005) Standardization of the extract of *Syzygium cumini* (L.) skeels fruits through the antimicrobial activity. Caderno de Farmacia 21(1):55–56

Muhammad A, Rahman M, Kabir ANM, Kabir S, Hossain M (2011) Antibacterial and cytotoxic activities of *Spondias pinnata* (Linn. f.) Kurz fruit extract. Ind J Nat Prod Resour 2(2):265–267

Muller O, Krawinkel M (2005) Malnutrition and health in developing countries. Can Med Assoc J 173(3):279–286

Mundaragi A, Thangadurai D (2015) Proximate composition, nutritive value and antioxidant activity of *Flacourtia montana* J. Graham. (Salicaceae). Vegetos 28(4):181–187

Mundaragi A, Thangadurai D (2017) Process optimization, physicochemical characterization and antioxidant potential of novel wine from an underutilized fruit *Carissa spinarum* L. (Apocynaceae). Food Sci Technol 38:428–433

Mundaragi A, Thangadurai D, Bhat S, Sangeetha J (2017) Proximate analysis and mineral composition of potential minor fruits of Western Ghats of India. Sci Papers Ser A Agron 60:340–346

Mundaragi A, Thangadurai D, Appaiah KAA, Dandin CJ, Sangeetha J (2019) Physicochemical characterization, antioxidant potential and sensory quality of wine from wild edible fruits of *Flacourtia montana* J. Graham. Braz Arch Biol Technol 62:1–12

Murali YK, Chandra R, Murthy PS (2004) Antihyperglycemic effect of water extract of dry fruits of *Terminalia chebula* in experimental diabetes mellitus. Indian J Clin Biochem 19(2):202–204

Myers N, Mittermeier RA, Mittermeier CG, Da Fonseca GA, Kent J (2000) Biodiversity hotspots for conservation priorities. Nature 403(6772):853–858

Nadkarni A (1954) Nadkarni's Indian Materia Medica

Nair KKN, Jayakumar R (1999) Ethnobotany of Hill-Pulaya tribe in the context of biodiversity rehabilitation at Chinnar Wildlife Sanctuary, Western Ghats of India. J Econ Taxon Bot 23(2): 431–449

Narayanaswamy N, Balakrishnan KP (2011) Evaluation of some medicinal plants for their antioxidant properties. Int J Pharmtech Res 3(1):381–385

Onakpoya I, Hung SK, Perry R, Wider B, Ernst E (2011) The use of garcinia extract (hydroxycitric acid) as a weight loss supplement: a systematic review and meta-analysis of randomised clinical trials. J Obes 2(1):11–18

Ovesná Z, Kozics K, Bader Y, Saiko P, Handler N, Erker T, Szekeres T (2006) Antioxidant activity of resveratrol, piceatannol and 3, 3′, 4, 4′, 5, 5′-hexahydroxy-trans-stilbene in three leukemia cell lines. Oncol Rep 16(3):617–624

Padhye S, Ahmad A, Oswal N, Sarkar FH (2009) Emerging role of garcinol, the antioxidant chalcone from Garcinia indica Choisy and its synthetic analogs. J Hematol Oncol 2(1):1–13

Palthur MP, Palthur SS, Chitta SK (2010) Nutraceuticals: a conceptual definition. Int J Pharm Pharm Sci 2(3):19–27

Parmar C, Kaushal MK (1982) Rubus ellipticus. Wild fruits. Kalyani Publishers, New Delhi, pp 84–87

Parnell JA, Craven LA, Biffin E (2007) Matters of scale: dealing with one of the largest genera of angiosperms. In: Reconstructing the tree of life: taxonomy and systematics of species rich taxa. CRC Press, Boca Raton

Patil RP, Pai SR, Pawar NV, Shimpale VB, Patil RM, Nimbalkar MS (2012) Chemical characterization, mineral analysis, and antioxidant potential of two underutilized berries (*Carissa carandus* and *Eleagnus conferta*) from the Western Ghats of India. Crit Rev Food Sci Nutr 52(4):312–320

Prashith KT, Vinayaka KS, Mallikarjun N, Bharath AC, Shailendra KB, Rakesh KM, Vinod KH (2011) Antibacterial, insecticidal and free radical scavenging activity of methanol extract of *Ziziphus rugosa* Lam.(Rhamnaceae) fruit pericarp. Pharm J 2(18):65–69

Rahmatullah M, Khairuzzaman M, Saleem SM, Sattar F, Rahman I, Yesmin MS, Malek I, Bashar (2015) Documentation of some folk medicinal practices in Sylhet &Moulavibazar districts, Bangladesh. World J Pharm Pharm Sci 4:176–186

Rajkumar M, Chandra R, Asres K, Veeresham C (2008) *Toddalia asiatica* (Linn.) Lam.—a comprehensive review. Pharmacogn Rev 2(4):386

Reddy MH, Reddy RV, Raju RV (1998) Rutaceous plants from tribal medicine of Andhra Pradesh, India. Anc Sci Life 17(4):251

Reyes-Carmona J, Yousef GG, Martínez-Peniche RA, Lila MA (2005) Antioxidant capacity of fruit extracts of blackberry (Rubus sp.) produced in different climatic regions. J Food Sci 70(7):s497–s503

Sahreen S, Khan MR, Khan RA (2010) Evaluation of antioxidant activities of various solvent extracts of *Carissa opaca* fruits. Food Chem 122(4):1205–1211

Saklani S, Chandra S, Badoni PP, Dogra S (2012) Antimicrobial activity, nutritional profile and phytochemical screening of wild edible fruit of *Rubus ellipticus*. Int J Med Aromat Plants 2(2): 269–274

Sasidharan N (2011) Flowering plants of Kerala ver. 2.0 (DVD). Kerala Forest Research Institute, Peechi

Sasikumar JM, Maheshu V, Smilin AG, Gincy MM, Joji C (2012) Antioxidant and antihemolytic activities of common Nilgiri barberry (*Berberis tinctoria* Lesch.) from south India. Int Food Res J 19(4):1601–1607

Sathyanarayanan S, Chandran R, Thankarajan S, Abrahamse H, Thangaraj P (2018) Phytochemical composition, antioxidant and antibacterial activity of *Syzygium calophyllifolium* Walp. fruit. J Food Sci Technol 55(1):341–350

Sathyavathi R, Janardhanan K (2014) Wild edible fruits used by Badagas of Nilgiri District, Western Ghats, Tamilnadu, India. J Med Plants Res 8(2):128–132

Sharma US, Kumar A (2011) Anti-diabetic effect of *Rubus ellipticus* fruit extracts in alloxan induced diabetic rats. J Diabetol 2(2):4

Sharma M, Kumar A (2013) Traditional medicinal plants of Rajasthan used in tribal medicine—a review. Int J Life Sci Pharma Res 3(2):45–49

Sharvani KA, Devaki NS (2014) Distinct characters of *Elaeocarpus*, a conservation dependent endemic genus of Western Ghats. Acta Biol Ind 3:663–667

Shilpa KJ, Krishnakumar G (2015) Nutritional, fermentation and pharmacological studies of *Syzygium caryophyllatum* (L.) Alston and *Syzygium zeylanicum* (L.) DC. fruits. Cogent Food Agric 1(1):1018694

Silva RO, Damasceno SR, Brito TV, Dias JM, Fontenele AM, Braúna IS, Júnior JS, Maciel JS, de Paula RC, Ribeiro RA, Souza MH (2015) Polysaccharide fraction isolated from *Passiflora edulis* inhibits the inflammatory response and the oxidative stress in mice. J Pharm Pharmacol 67(7):1017–1027

Singh B, Sinha BK, Phukan SJ, Borthakur SK, Singh VN (2012) Wild edible plants used by Garo tribes of Nokrek Biosphere Reserve in Meghalaya, India

Siyum ZH, Meresa TA (2021) Physicochemical properties and nutritional values of *Carissa spinarum* L. "Agam" fruit. Int J Fruit Sci 21(1):826–834

Son PS, Park SA, Na HK, Jue DM, Kim S, Surh YJ (2010) Piceatannol, a catechol-type polyphenol, inhibits phorbol ester-induced NF-κB activation and cyclooxygenase-2 expression in human breast epithelial cells: cysteine 179 of IKKβ as a potential target. Carcinogenesis 31(8): 1442–1449

Srivastava D, Prabhuji SK, Rao GP (2009) Taxonomic and ethno-biological status of *Flacourtia jungomas* (Lour.) Raeus.: an endemic nutraceutical plant of eastern UP. Med Plants 1(1):49–53

Srivastava S, Srivastava M, Misra A, Pandey G, Rawat A (2015) A review on biological and chemical diversity in Berberis (Berberidaceae). EXCLI J 14:247

Tabeshpour J, Imenshahidi M, Hosseinzadeh H (2017) A review of the effects of *Berberis vulgaris* and its major component, berberine, in metabolic syndrome. Iran J Basic Med Sci 20(5):557

Tahergorabi Z, Abedini MR, Mitra M, Fard MH, Beydokhti H (2015) "Ziziphus jujuba": a red fruit with promising anticancer activities. Pharmacogn Rev 9(18):99

Thakur S, Vidyasagaran K (2014) Influence of zonal variation on physical characteristics of Artocarpus hirsutus fruit and size wise variation in biochemical composition of the fruit. J Recent Adv Agric 2(6):263–270

Tharachand SI, Avadhani M (2013) Medicinal properties of Malabar tamarind *Garcinia cambogia* (Gaertn) DESR. Int J Pharm Sci Rev Res 19(2):101–107

Thirugnanasampandan R, Ramya G, Bhuvaneswari G, Aravindh S, Vaishnavi S, Gogulramnath M (2017) Preliminary phytochemical analysis and evaluation of antioxidant, cytotoxic and inhibition of lipopolysaccaride-induced NOS (iNOS) expression in BALB/c mice liver by *Ziziphus oenoplia* Mill. fruit. J Complement Integr Med 14(2):122–129

Valvi SR, Rathod VS (2011) Mineral composition of some wild edible fruits from Kolhapur district. Int J Appl Biol Pharm Technol 2(1):392–396

Valvi SR, Jadhav VD, Gadekar SS, Yesane DP (2014) Assessment of bioactive compounds from five wild edible fruits, *Ficus racemosa*, *Elaegnus conferta*, *Grewia tillifolia*, *Scleichera oleosa* and *Antidesma ghasembilla*. Acta Biol Ind 3(1):549–555

Vartak VD (1980) Observations on wild plants from hilly regions of Maharashtra and Goa resume and future prospects. In: Jain SK (ed) Glimpses of Indian ethnobotany

Vavilov NI (1951) The origin, variation, immunity and breeding of cultivated plants, vol. 72, no. 6. LWW, p 482

Vignesh A, Pradeepa Veerakumari K, Selvakumar S, Rakkiyappan R, Vasanth K (2021) Nutritional assessment, antioxidant, antiinflammatory and antidiabetic potential of traditionally used wild plant, Berberis tinctoria Lesch. Trends Phytochem Res 5(2):71–92

Wikramanayake ED, Dinerstein E, Loucks CJ (2002) Terrestrial ecoregions of the Indo-Pacific: a conservation assessment, vol 3. Island Press, Washington

Wu X, Beecher GR, Holden JM, Haytowitz DB, Gebhardt SE, Prior RL (2004) Lipophilic and hydrophilic antioxidant capacities of common foods in the United States. J Agric Food Chem 52(12):4026–4037

Wu MC, Hu HT, Yang L (2011) Proteomic analysis of up-accumulated proteins associated with fruit quality during autumn olive (*Elaeagnus umbellata*) fruit ripening. J Agric Food Chem 59(2):577–583

Zhang L, Li J, Hogan S, Chung H, Welbaum GE, Zhou K (2010) Inhibitory effect of raspberries on starch digestive enzyme and their antioxidant properties and phenolic composition. Food Chem 119(2):592–599

Zhang YJ, Zhou T, Wang F, Zhou Y, Li Y, Zhang JJ, Zheng J, Xu DP, Li HB (2016) The effects of *Syzygium samarangense*, *Passiflora edulis* and *Solanum muricatum* on alcohol-induced liver injury. Int J Mol Sci 17(10):1616

Chapter 5
Microgreens: A Future Super Food

Lekshmi G P and Bindu R. Nair

Abstract Leafy greens are nutrient-packed leaves of herbs, shrubs, or trees, consumed along with tender petioles and shoots as vegetables and complete a balanced diet. They are high in dietary fiber, eaten raw or cooked, and appreciated for their bland to tangy taste. The leafy greens of herbaceous plants, more so in the recent past, are being marketed in the form of microgreens. The microgreens are miniature seedlings of herbs, very tender, crunchy, fresh, and fragrant. Moreover, they are highly nutritional compared to mature leafy greens. Microgreens contain a considerably higher concentration of vitamins and carotenoids than mature plants, but the type and quantity differ among the microgreens originating from the different plant types. The current chapter focuses on the different aspects that should be considered while growing microgreens, as well as the information about seeds that are suitable to be raised as microgreens. Also, there is an account of the growth media, light, temperature, nutrient requirements, and other methods that may be adopted to cultivate microgreens either at home or on small-scale commercial farms. Preliminary studies have irrevocably proved the nutritional and health benefits offered by microgreens, and so the recent findings in this area are updated for the interested readers. Microgreens have a short growth period, and there is every chance that they will be contaminated during harvest, storage, and transport. Therefore, awareness about the necessity of axenic conditions for microgreen cultivation is also provided. Many pre and postharvest practices may be adopted to obtain fresh and healthy produce. This would enable us to bring variety to our food palette. Thus, it is clearly evident that if proper care is taken, microgreens can become a future food choice. Hopefully, we will be able to witness the impending microgreen revolution.

Keywords Microgreen · Leafy green · Vitamin-rich food · Super food

L. G. P · B. R. Nair (✉)
Department of Botany, University of Kerala, Kariavattom, Thiruvananthapuram, India
e-mail: bindunair@keralauniversity.ac.in

S. T. Sukumaran, K. T R (eds.), *Conservation and Sustainable Utilization of Bioresources*, Sustainable Development and Biodiversity 30,
https://doi.org/10.1007/978-981-19-5841-0_5

5.1 Introduction

Green leafy vegetables are considered to be rich sources of several vital nutrients (Kamble and Jadhav 2013). The diversity of leafy greens and their usage varies widely both with respect to the geographical area and cultural affiliations of the residents (Welbaum 2015). Among the known leafy crops worldwide (~50,000), only a very limited (~1000) number is commonly used as vegetables. It is estimated that in India, only about 125 plant species are consumed as edible greens (Bandopadhyay and Mukherjee 2009). However, it should be understood that there are many more potential leafy edibles. Many lesser-known, underutilized leafy greens can be utilized for nourishing the ever-increasing human population due to their promising nutritive value, diversity in taste, and fragrance (Sheela et al. 2004). Underutilized plant resources are the foundation of diversity in developing nation societies, ancient, and autochthonous food systems (Sahoo et al. 2021).

Botanically, leafy greens are defined as young, edible, active growth-phased plants with fresh leaves. The leaves are usually consumed along with tender petiole and shoot. They are mostly short-lived and rapidly growing plants with herbaceous nature. They are marketed within a short period of time (Sahu et al. 2020). The leafy greens are blessed with an array of bioactive compounds, including antioxidants, vitamins, and minerals, also called nature's anti-aging wonders. They are low in carbohydrates but rich in folic acid and dietary fiber content. The nutritional value of these underutilized greens is 20 times more than in other market vegetables (Sudha and Mathanghi 2012). Cultivating these greens will be a great way to increase food production and ensure balanced nutrition, food security, health security, and poverty alleviation in the deprived sections of society (Buragohain et al. 2013).

5.2 Microgreens

Leafy greens may be consumed as miniature forms, also referred to as the 'microgreens.' Microgreens have gained acceptance in society in very recent times for almost a decade or so (Bulgari et al. 2021; Turner et al. 2020). Consequently, several leafy greens are being marketed as microgreens, which are smaller, very tender, and more nutritional (in terms of vitamins and minerals) than mature leafy greens.

Microgreens are described as young and tender edible seedlings with fully developed cotyledonary leaves along with the emerging first true leaves. Microgreens are commonly termed as 'Vegetable Confetti,' portrayed as soft juvenile greens raised from the seeds of grains, vegetables, or herbs (Sharma et al. 2020). They have three basic parts: a central stem, cotyledon leaves, and typically the first pair of true leaves. The size of the harvested microgreens may vary depending upon the specific plant type. When the greens grow beyond the above-mentioned size, they are called petite greens. The seeds of a wide variety of herbs (e.g., basil and cilantro), vegetables (e.g., radish, broccoli), and even flowers (e.g., sunflower) are grown as microgreens.

Microgreens are primarily used in the restaurant industry to embellish the cuisine and are most commonly consumed fresh in salads, soups, and sandwiches. Traditionally only a very few plant varieties are offered as microgreens, such as Argula, Basil, Beets, Kale, and Cilantro in the South East Asian countries (Xiao et al. 2012). Today, it is getting more popular, and in the US, there is a boom in the microgreen industry as a variety of seed companies and growers are involved in the growing and marketing of microgreens. These are also provided as a mixture called Rainbow Mix. Mustard cress, Radish green, Spinach, Basil, Coriander, Fennel, Argula, Beet greens, and Kale microgreens are commonly available in the markets (Fig. 5.1). Scientific data on the nutritional contents of microgreens are limited. Still, research has shown that microgreens contain high concentrations of many nutrients when compared with mature, fully grown vegetables or herbs. Data on nutrients such as carbohydrates and proteins are scarce. However, several studies have demonstrated the high level of phytonutrients, antioxidants, vitamins, and minerals that the microgreens contain (Xiao et al. 2012).

Microgreens are considered potential functional foods with nutritional benefits and health improving or ailment prevention characteristics. Microgreens contain a considerably higher concentration of vitamins and carotenoids than mature plants; however, they differ in the type of plant. The maximum value of vitamins C, E, and K was seen in red cabbages, garnet amaranth, and green daikon radish compared to many others studied (Xiao et al. 2012). Cilantro microgreens showed the highest concentrations of lutein, zeaxanthin, violaxanthin, and beta-carotene, but Popcorn or Maize shoots and golden pea tendrils have relatively low vitamins carotenoids compared to their mature forms (Xiao et al. 2012). All the vitamins and carotenoids are usually concentrated in the cotyledon leaves. The macro and micro-minerals were found to be high in almost all microgreens. However, the mineral content varied according to the growth media used for culturing them. The mineral concentration of compost-grown broccoli microgreens had higher amounts of P, K, Mg, Mn, Zn, Fe, Ca, Na, and Cu than the others, while the hydroponic cultures showed larger quantities of Mg, Mn, Cu, and Zn compared to control (Weber 2017). Microgreens are valuable nutritional supplements to the human diet. Therefore, microgreens should be produced sufficient to ensure availability to all.

Light exposure during the sprouting, growing, and storing period had an effect on quality. For example, light exposure during storage had an effect on the phenolic and alpha-tocopherol content. It also increases the amount of ascorbic acid. Dark storage results in higher hydroxyl radical scavenging capacity and carotenoids retention, i.e., light exposure accelerates the deterioration of radish microgreens, while dark storage maintains quality (Xiao et al. 2014). Golden pea tendrils, which were grown in the absence of light, showed much lower vitamin and carotenoid content. Hence, it appears that light plays a vital role in nutritional value (Xiao et al. 2012).

They are attractive with their colors, textures, and flavors. Microgreens can be grown and harvested within 7–14 days after germination in a very simple way; hence, they are considered suitable for urban agriculture (Renna and Paradiso 2020).

Fig. 5.1 Microgreens of local vegetables in India

Microgreens are also ideal for indoor production and controlled environmental agriculture. Short harvest time heightens its market values and categorizes them as important Controlled Environment Agriculture crops (Wood 2019).

Although interest in microgreens has expanded, the main market clients continue to be restaurant chefs as well as health food store owners. The most likely successful marketing strategy for producers interested in growing microgreens is to work in association with a restaurateur or chef, growing and delivering microgreens at their requests and preferences. Microgreens are typically purchased and used by restaurants in small amounts, and the quick growing and harvest time may make this a more attractive crop for very small growers interested in developing nearby, high-end specialty markets for fresh produce. Even though microgreens are super greens, they face many challenges in markets due to their speedy degradation and short storage life.

However, as the microgreens perish fast, they should be washed, stored properly, and subjected to good handling practices. They can create substantial marketing challenges, particularly for inexperienced growers. Due to these difficulties, they are usually cultivated in small batches on specialized farms. There are innumerable websites dedicated to microgreen cultivation protocols (https://www.allthatgrows.in/collections/micro-greens-seeds). It appears that many people have taken up microgreen farming as a serious profession as the demand for these rises around the world. Supermarkets are now stocking them in small fresh batches. These baby greens are valued and popular as they are entirely organic and also as they are nutritional and crunchy, and their display looks wonderful on a dish.

5.3 Plant Seeds Used for Microgreen Cultivation

Microgreens can be produced from the seeds of many vegetables, herbs, and agronomic crops. Microgreens are cultivated in a variety of environments, specifically indoor, outdoor, and controlled environments, i.e., greenhouse and developing frameworks (growing systems) that is either containing soil or soilless substrate, depending on the scale of production. Containerized production, adaptable both to micro-scale urban and large-scale commercial operations, allows for the commercialization of the product while growing on the media, to be harvested directly by the end-user (Bhatt and Sharma 2018).

Microgreens are often marketed as specialty mixes and tagged with adjectives such as 'sweet,' 'mild,' 'colourful,' or 'spicy.' Crops like lettuces are not usually used as microgreens because they are too delicate and wilt easily. Growers should evaluate various crop varieties to determine their value as microgreens. The kinds of crops that are selected for production and sale as microgreens have value in terms of color, unique textures, or distinct flavors. Certain crops, including cabbage, beet, kale, kohlrabi, mizuna, mustard, radish, swiss chard, and amaranth of microgreens, germinate easily and grow quickly. Soaking of seeds prior to sowing helps facilitate germination.

Almost 80–100 crop varieties are used as microgreens. According to Parida (2020), seeds of Anise, Basil, Beet, Cabbage, Carrot, Cauliflower, Celery, Cucumber, Chickpea, Coriander, Corn, Dwarf copperleaf, Edible amaranth, Fennel, Fenugreek, Finger millet, Flax, Foxtail millet, Great Millet, Green gram, Green amaranth, Lemongrass, Little millet, Marigold, Mint, Mustard, Oats, Onion, Pea shoots, Pearl millet, Pendant amaranth, Pigweed, Pumpkin, Purple amaranth, Radish, Sesame, Spinach, Spiny amaranth, Sunflower, Turnip, and Wheatgrass can be used for innovative farming at home. Most commonly, seeds used for microgreen cultivation include carrot, cress, arugula, basil, onion, chive, broccoli, fennel, lemongrass, popcorn, buckwheat, spinach, sweet pea, and celery. Prices for microgreens generally range from $30 to $50 per pound. The product is packed in plastic clamshell containers that are typically 4–8 oz. by weight but can be sold in 1 lb. containers as well (Treadwell et al. 2020).

5.4 Mode of Cultivation of Microgreens

Growing small quantities of microgreens at home is very easy, but growing and marketing high-quality microgreens commercially are a bit difficult. Commercial growers should select crops that have a similar growth rate so that the produce from each growth tray can be harvested at once. Alternatively, growers can seed the various crops singularly and mix them after harvest (Treadwell et al. 2020).

Cultivation of microgreens is a simple process that does not require much time, energy, and experience (Franks and Richardson 2009). However, little care should be needed in each step, and further research is required to improve productivity, nutritional quality, and the cost of production. Even if microgreens' life cycle is very short, seed germination velocity should be improved to determine a more rapid stabilization and to promote vigorous seedlings (Delian et al. 2015). Microgreen production depends on factors such as water, substrate, and energy intake. Energy consumption mainly depends on the lighting since the optimization of light emission in terms of quality and intensity can improve the system's sustainability. The effect of substrates such as vermiculite, coconut fiber, and jute fabric was tested on microgreens' production and quality traits to identify the best cultivation conditions (Bulgari et al. 2021).

Indigenous landraces, underutilized crops, and wild-edible plants constitute a vast repository for the selection of genetic material for microgreens. Microgreens can be grown in single-planting flat containers, mats, or lining. The substrata should be placed in the bottom of a tray or longer trough using standard, sterile, loose, soilless germinating media, including peat, vermiculite, perlite, coconut fiber, or any others. The substrata should be filled in a tray, and irrigation programs should be devised accordingly. Overhead mist irrigation is used only through the germination stage in these media systems.

According to Gayathree et al. (2019), seed germination of most of the species was >75%, with the time taken to reach 75% germination within 14 days. Lettuce and carrot were found to be the most preferred microgreens, followed by peas and amaranth. Poor germination can be attributed to unfavorable temperature, quality and quantity of light, lack of moisture, and inherent factors. After germination, trays should be sub-irrigated to avoid excess moisture in the plant canopy. Seeding density is difficult to recommend. Most growers want to grow microgreens as densely as possible to maximize production. However, it should be noted that when dense, crowding leads to elongation of stems and increases the risk of diseases (Treadwell et al. 2020).

Pre-sowing treatments are good for standardizing and shorten the production cycle. Seed surface sterilization with non-chemical treatments having antimicrobial action is more effective for yielding healthy greens. Crop-specific information on sowing rate along with details on its yield and quality need further attention (Kyriacou et al. 2016). Modular light exposure may fortify the bioactive content of microgreens and augment their sensorial attributes. Most crops require little or no fertilizer because the seed provides adequate nutrition for the young crop. Some longer-growing microgreen crops, such as micro-carrot, dill, and celery, may benefit from light exposure applied to the tray bottom. Some of the faster-growing greens, such as mustard cress and chard, may also benefit from the light because they germinate quickly and exhaust their self-contained nutrient supply quickly. Light exposure is best achieved by floating each tray of microgreens for 30 s in a prepared nutrient solution of approximately 80 ppm nitrogen (Treadwell et al. 2020).

Microgreens are smaller than baby greens and larger than sprouts; therefore, their harvesting stages are also in between those two stages; hence, special attention is required (Xiao et al. 2012; Ebert 2013). Harvesting at the right stage is an important production strategy as it varies greatly from crop to crop. Some growers cultivate mixed crops of microgreens having similar growth rates so that the whole crops can be harvested at once (Bhatt and Sharma 2018).

It is mostly recommended that seedling height should be the harvested index due to its easiness in determination. However, the leaf area can also be used as a harvesting stage index (Gayathree et al. 2019). Microgreens are ready for harvest when they reach the first true leaf stage, usually at about two in. tall. The time from seeding to harvest can vary greatly by crop, from 7 to 21 days. Production in small trays will likely require harvesting with scissors. This is a very time-consuming part of the production cycle and is often mentioned by growers as a major drawback. The seeding mat type of production system has gained popularity with many growers because it facilitates faster harvesting. The mats can be picked up by hand and held vertically while an electric knife or trimmer is used for harvesting, allowing cut microgreens to fall from the mat into a clean harvest container.

Harvested microgreens are highly perishable and immediately washed and cooled as soon as possible, using good handling practices for food safety (Treadwell et al. 2020). Microgreens are generally packed in polyethylene packages and cooled to recommended temperatures before supplying to the market or consumers (Bhatt and Sharma 2018). To improve quality, some chefs ask growers to deliver microgreens in the trays or mats so that they can cut the microgreens as needed.

Pre and postharvest select-waveband, intensity, and photoperiod combinations can elicit compound-specific improvements in functional quality and in shelf life. Research is needed to identify effective sanitization and drying methods (do not compromise on quality) and shelf life for the commercialization of ready-to-eat packaged microgreens. Genotypic variability in postharvest chilling sensitivity and the interactions of temperature, light conditions, and packaging gas permeability should be further examined to establish environments that are suppressive on respiration but preventive of off-odor development (Kyriacou et al. 2016). Aloe vera gel spray coating was suggested as an eco-friendly ergonomic preharvest treatment along with PET–CS for the enhancement of postharvest quality and shelf life in radish and rosella microgreens, with a high potential to be extended to other microgreens.

Scientists have explored the preharvest and postharvest interventions, such as calcium treatments, modified atmosphere packaging, temperature control, and light, to maintain quality, augment nutritional value, and extend shelf life. However, more research is to be conducted to optimize production and storage conditions to improve safety and quality and increase the shelf life of microgreens, thereby expanding their potential markets (Turner et al. 2020).

5.5 Nutritional Composition and Health Benefits of Microgreens

Microgreens are miniature greens that can be recommended as nutraceuticals and functional foods due to their health-promoting and disease-preventing properties and their nutritional value. It is obvious that microgreens are good sources of phytonutrients. Nowadays, microgreens are becoming the emerging alternative to fortified and genetically modified food, as it provides a sufficient amount of nutrition. Microgreens are low in fat, calories, and antinutrients but rich in moisture content, vitamins, carotenoids, and other phytochemicals. The microgreens were found to be moderate to good sources of protein, dietary fiber, and essential elements. They are excellent sources of ascorbic acid, Vitamin E, and beta-carotene (provitamin A) (Ghoora et al. 2020). Microgreens may have 4–40 times the amount of some nutrients and vitamins as the vegetables a mature plant would produce (Weber 2017).

During seed germination, biochemical changes within the seeds lead to the activation of various enzymes, which in turn result in the degradation of macromolecules into micro-molecules that the body can absorb easily. These changes result in the rapid elevation of the amount of bioactive constituents such as vitamins and antioxidants that are easily absorbable and are beneficial to the human body (Kowitcharoen et al. 2021). Thiamine, riboflavin, and niacin were high in microgreens of *Amaranthusviridis, Vignaradiata,* and *Allium cepa*, respectively. However, vitamin content was higher in sprouts than in microgreens and mature edible parts of selected plants, even though some exceptions were found. Antinutrients analyzed in microgreens were below the toxic levels. Microgreens have low carbohydrates and antinutrient contents; but high vitamin contents, so these can be recommended as a dietary supplement, especially for those who prefer less carbohydrate-containing food supplements (Nair and Lekshmi 2019). According to Xiao (2013), among the 25 microgreens assayed, red cabbage, cilantro, garnet amaranth, and green daikon radish had the highest concentrations of ascorbic acids, carotenoids, phylloquinone, and tocopherols, respectively when compared to the nutrient concentrations in mature leaves recorded in USDA National Nutrient Database, microgreens possessed higher nutrient density.

Research has shown that microgreens possess antioxidants and a number of polyphenols in contrast to their fully grown vegetable counterparts (Xiao et al. 2019). Microgreens of mung beans and lentils had higher carbohydrate and protein contents than others. Lentil microgreens had the highest total chlorophyll, carotenoid, and ascorbic acid contents. Buckwheat microgreens showed the highest total phenol content (TPC) and maximum DPPH scavenging activity. Red cabbage has a higher anthocyanin content than purple radish (Kowitcharoen et al. 2021). The highest concentration of vitamin C, carotenoids, phylloquinone, and tocopherols was found in red cabbage, cilantro, garnet amaranth, and green daikon radish. The study concluded that microgreen cotyledon leaves possess higher nutritional value than mature leaves. Researchers also found about five times greater vitamins in microgreens than in their mature plant counterparts (Xiao et al. 2012). Broccoli and cauliflower microgreens had higher concentrations of carotenoid contents than mature florets (Xiao et al. 2019). Niroula et al. (2019) found that carotenoid content increased in wheat and barley microgreens over the course of the 16-day growth period studied.

The concentration of bioactive compounds from microgreens of wheat, lentils, radishes, and sunflowers showed that they had richer saturated fatty acids (palmitic acid) than unsaturated fatty acids, whereas alfalfa microgreens have significantly higher amounts of unsaturated fatty acids (oleic and linoleic acids) than other microgreens. For radish sprouts, it was noticed that the content of glucosinolates increased after germination. Glucosinolates are beneficial against human cancers. The studies evaluated that microgreens are nutritionally superior to sprouts in terms of polyamine content. Alfalfa microgreens showed the highest levels of agmatine, lentil microgreens had the highest content of spermidine, while fenugreek microgreens showed the highest content of spermine. Moreover, the nutritionally beneficial polyamines (agmatine, spermidine, and spermine) were accumulated in microgreens and a lower cadaverine content (Renna and Paradiso 2020). There was a substantial increase in vitamin C content from amaranth sprouts to microgreens (2.7-fold). Both provitamins A, α-carotene, and β-carotene were considerably increased from sprouts to microgreens. The content of α-carotene was the same at the microgreen and fully developed stage (Kyriacou et al. 2016). Brassicaceae microgreens are good sources of macroelements (e.g., potassium and calcium) and microelements (e.g., iron and zinc) (Xiao et al. 2016). Ebert et al. (2014) reported that fully grown *Amaranthus tricolor* showed higher protein, iron, vitamin c, b-carotene, lutein, and violaxanthin contents than sprouts and microgreens. Microgreens had a higher content of α-carotene, βcarotene, violaxanthin, lutein, and neoxanthin than sprouts.

According to Sun et al. (2013), who profiled the polyphenols in five microgreen cultivars of the genus Brassica, there were about 165 phenolic compounds comprising many highly glycosylated acylated quercetin, kaempferol, cyanidin glycones, and complex hydroxycinnamic and benzoic acids. They reported more complex polyphenol profiles and a greater variety of polyphenols in the microgreens than in their mature plant counterparts. The protective effect against the oxidative stress shown by Brassicaceae (broccoli, Brussels sprouts, cabbage, kale, and cauliflower) is given by glucosinolates which are sulfur-containing glucosides. For example, in broccoli, there are sinigrin, glucoraphanin, and progoitrin; in Chinese cabbage, there is indolyl glucosinolate gluco brassicin and glucoraphanin (one of the most abundant glucosinolates present in broccoli).

Lettuce contains various health-promoting phytochemicals, including vitamins and phenolic compounds with antioxidant properties. Oh et al. (2010) found that young lettuce (*Lactucasativa*) seedlings, after 7 days of germination, had the highest total phenolic concentration and antioxidant capacity in comparison to the mature leaves.

Regardless of how they were grown, microgreens had larger quantities of Mg, Mn, Cu, and Zn than vegetables. However, compost-grown (C) microgreens had higher P, K, Mg, Mn, Zn, Fe, Ca, Na, and Cu concentrations than the vegetables (Weber 2017). Analysis of microgreens of *Brassica* revealed that they are excellent sources of the macroelements, K and Ca, and the microelements, Fe and Zn (Xiao et al. 2016). Additionally, they were found to be moderate to excellent sources of ascorbic acid, phylloquinone, carotenoids, tocopherols, glucosinolates, and polyphenols (Xiao et al. 2019). Cauliflower, rapini, red radish, China rose radish, and ruby radish microgreens were found to have the greatest contents of total ascorbic acid, phylloquinone, total tocopherols, total glucosinolates, and TPC, respectively. Ruby radish microgreens also had the greatest DPPH radical scavenging capacity. Radish and mustard were found to have the highest bioaccessible fraction (BF) for ascorbic acid, total carotenoids, and total isothiocyanates, while broccoli, kale, and radish all had comparable high BF (Body Fat) or total polyphenols. Broccoli and mustard showed the lowest and highest BF values, respectively, for potassium and magnesium, while kale had the highest BF value for calcium (de la Fuente et al. 2019). Pinto et al. (2015) showed that microgreen lettuce (*Latucasativa* var. *capitata*; 2 week old) had a higher content of most minerals (Ca, Mg, Fe, Mn, Zn, Se, and Mo) than mature lettuce (10 week old). Kyriacou et al. (2019) reported that basil and swiss chard microgreens were good sources of K and Mg, and purple basil was high in ascorbic acid, while green basil and coriander were especially excellent sources of beta-carotene and total polyphenols. Lupin microgreen bread retained high levels of genistein which has anticarcinogenic properties, especially in women (Romognolo et al. 2017). Polash et al. (2018) demonstrated that bioactive components and antioxidant activity in mustard, radish, and cabbage microgreens degraded rapidly after harvest so that to obtain substantial health benefits from eating microgreens, they should be consumed soon after harvest. Microgreens possess a higher content of most minerals (Ca, Mg, Fe, Mn, Zn, Se, and Mo) and a lower NO_3^- content than mature lettuces. Therefore, microgreens can be considered a good source of minerals in the human diet, and their consumption could be an important strategy to meet children's dietary mineral requirements without exposing them to harmful NO_3^- (Pinto et al. 2015). The wild greens (*Sanguisorba minor* Scop, *Sinapis arvensis* L., and *Taraxacum officinale* Weber ex F. H. Wigg.) showed high amounts of Mg, P, Zn, Mn, Mo, and Fe content. However, the wild greens also showed high amounts of nitrate and traces of some metals, potentially detrimental to health. This cautions against the use of wild species for producing microgreens and baby leaves (Lenzi et al. 2019). Studies indicated that the nutrient data of Brassicaceae microgreens could be valuable to dieticians, nutrition policymakers, health-conscious consumers, and growers in the selection of nutrient-dense vegetables. For instance, reports state that broccoli microgreens have higher contents of bioactive compounds and potent antioxidants and exhibit higher anti-inflammatory and anticancer activities than their corresponding mature plants (Kowitcharoen et al. 2021).

In addition to their high nutritional value, microgreens are considered functional foods with particular health-promoting or disease-preventing properties. Sprouts and microgreens can be easily produced in urban or peri-urban settings where land is often a limiting factor, by either specialized vegetable farmers or the consumers themselves. Sprouts and microgreens can be grown without soil and without external inputs like fertilizers and pesticides, around or inside residential areas considering their short growth cycle. Moreover, sprouts and microgreens are usually consumed raw; hence, there is no loss or degradation of micronutrients through food processing.

Microgreens are gaining increasing popularity as an innovative horticultural product. Six microgreen species were evaluated by researchers: Dijon mustard (*Brassica juncea* L. Czern.), opal basil (*Ocimum basilicum* L.), bull's blood beet (*Beta vulgaris* L.), red amaranth (*Amaranthus tricolor* L.), peppercress (*Lepidium bonariense* L.), and China rose radish (*Raphanus sativus*L.). They found a strong correlation between the total phenolic content of the microgreens and their flavor attributes. The chemical analysis revealed that China rose radish, opal basil, and red amaranth have the highest concentrations of total ascorbic acid, phylloquinone, carotenoids, and tocopherols, while the highest concentrations of total phenolics were found in China rose radish and opal basil. The study also concluded that microgreen growers could use pH and total phenolic values as indicators and predictors of consumer acceptability. Unfortunately, the study results reflected the reality that people reject some vegetables because of their unpleasant flavor (bitterness, astringency), even if they have a beneficial effect on human health (Stoleru et al. 2016).

Microgreens play a vital role in health-promoting diets. Microgreens are a nutrient-rich food that contains more digestible vitamins, minerals, and other phytonutrients than the corresponding mature plants. They are excellent sources of nutritional and bioactive compounds, hence, having the potential for the prevention of malnutrition and chronic diseases (Teng et al. 2021). In a comparative study of fresh microgreen and radish vegetables, the concentration of five minerals, i.e., F, Mg, K, Ca, and total phosphorous, vitamin B9, vitamin C, and Omega 3 fatty acid concentration are 1–3 times higher than the fresh radish (Rani et al. 2018).

Microgreens of *Brassica oleracea* L. var. *italica* have a substantial amount of bioactive constituents, including glucosinolates, phenolic compounds, vitamins, and essential minerals. These secondary metabolites are positively associated with health benefits. In vitro and in vivo studies demonstrated that broccoli seedlings possess various biological properties, including antioxidant, anticancer, antimicrobial, anti-inflammatory, antiobesity, and antidiabetic activities. Hence, it can serve as a potential reference for food selections of consumers and applications in functional food and nutraceutical industries (Le et al. 2020).

Both in vitro and in vivo studies showed that microgreens can act as the best functional food with potential benefits to human health because of their anti-inflammatory, anticancer, antibacterial, and antihyperglycemia properties, prevention, and treatment of chronic diseases (Zhang et al. 2021). According to Huang et al. (2016), red cabbage microgreen lowered the weight gain induced by high-fat diet and reduced circulating LDL levels and expression of hepatic inflammatory cytokines in mice fed a high-fat diet. This is probably due to the effect of polyphenol contents in red cabbage microgreens with antioxidant and anti-inflammation properties.

Chlorophyll and Carotenoids contents in the cereal microgreens correlate with its health-promoting components like phenolics and antioxidants, play a role against clinical conditions like thalassemia and hemolytic anemia, and reduce the risk of some chronic diseases, such as cancer, cardiovascular diseases, skin diseases, and age-related eye diseases (Niroula et al. 2019).

According to Tan et al. (2020), the soil-grown farm broccoli microgreens possessed a significantly higher vitamin C content than hydroponically grown and commercial microgreens. The farm grown broccoli microgreens possessed much higher chlorophyll content and total phenolic content as compared to many species of mature vegetables. Regardless of the growing method or environment, microgreens have the potential to serve as functional foods. Microgreen supplementation has the ability to lower circulating LDL levels in animals fed with a high-fat diet and reduce hepatic cholesterol ester, triacylglycerol levels, and expression of inflammatory cytokines in the liver. Microgreens of red cabbages can modulate weight gain and cholesterol metabolism and may protect against CVD by preventing hypercholesterolemia (Huang et al. 2016). According to Di Gioia et al. (2019), Brassicaceae species grown in soilless systems are good targets for producing high-quality Zn and Fe-biofortified microgreens by simply manipulating nutrient solution composition.

So, in general, we can say that microgreens can be recommended as a functional food to enhance and reduce the dietary-related problems in humans. But elaborate research is required to underline and highlights its health benefits.

5.6 Food Quality of Microgreens and Safety in Consumption

The rapidly developing microgreen industries face many challenges. One major limitation is the rapid quality deterioration of microgreens due to their short shelf life. Once harvested, microgreens quickly dehydrate, wilt, decay, and rapidly lose their tender texture, color, and certain nutrients. Deterioration occurs soon after harvest, restricting to local sale units. Second, there is a chance of carrying foodborne pathogens, and hence, steps should be taken during cultivation and harvest to reduce the pathogen attacks (Turner et al. 2020). Warriner et al. (2003) reported that microgreens are more susceptible to bacterial internalization than mature vegetable plants.

The seed releases a mixture of carbohydrates and peptides that can attract surrounding bacteria in the rhizosphere during seed germination. Bacteria can enter via germinating radicals or secondary roots and can persist in localized sites. In immature plants, protective structures are not fully formed, which enables the entry of bacteria into the xylem, thereby spreading it. For microgreen production, seeds are soaked in water to enhance germination without light, in very high humidity and warm temperatures, often with recirculating water. Any microbial contamination on seeds or production equipment, introduced by insects, or lack of hygienic practices by workers will lead to contamination of the entire batch mainly due to the rapid growth of microbes. Therefore, seeds should be certified, and microgreens should be produced by good agricultural practices. Seed decontamination is another crucial step.

5.7 Best Pre and Postharvest Practices for Microgreens

Microgreens' industrial production and marketing are limited by their short shelf life and quick deterioration in product quality. It raises prices and confines trade to local sales. Microgreens that have been harvested are readily dehydrated, wilt, rot, and lose some nutrients, as well as their soft texture and color. Microgreens' shelf life is now being extended using a variety of pre and postharvest procedures.

Microgreens can have a longer shelf life if they are sold still rooted in the growing medium or if they are picked and kept cool. According to previous research, quality can be maintained for over 14 days (Turner et al. 2020). Berba and Uchanski (2012) suggested that the age of the seedlings influences the shelf life of microgreens at harvest. For example, radish is harvested at 7 days, arugula at 9 days, and red cabbage at 11 days.

Major preharvest factors of microgreen production, such as crop selection, fertilization, biofortification, lighting, and growth stage at harvest, are addressed with respect to crop physiology and quality, as well as postharvest handling and applications, temperature, atmospheric composition, lighting, and packaging technology which influence shelf life and microbial safety (Kyriacou et al. 2016).

The effects of temperature, relative humidity, packing material, and microbial load significantly differ in the shelf life. It is advised that products are stored in the dark at a low temperature throughout distribution and retail. According to Kou et al. (2013), temperatures between 5 and 10 °C with moderately high O_2 (14.0–16.5 kPa) and moderately low CO_2 (1.0–1.5 kPa) conditions can be used to maximize the shelf life of buckwheat microgreens. Likewise, researchers from the USDA-ARS Food Quality Laboratory identified that the usage of film bags with high porosity to oxygen content helps daikon radish microgreens to maintain their nutritional value. The US FDA (1999) recommended that all seeds used for sprout/microgreen production should be treated to reduce the pathogens. Alfalfa and clover seeds are mainly used for sprouting but are not typically grown for microgreens. In contrast, radish and broccoli seeds are widely used for microgreen production. Plants in microgreen production are generally anchored by rooting them in a solid medium. After germination, they are exposed to light and moving air, which causes water evaporation from the growth matrix and reduces humidity. Here, water is not usually recirculated but frequently grown in hydroponic systems. Microgreens are generally harvested by cutting stems above the root at the emergence of the first pair of true leaves and the fully developed cotyledons (Berba and Uchanski 2012). All these are much safer, but care should be needed for proper cutting.

Clean water should be used to reduce the chance of contamination while growing. Riggio et al. (2019) suggested that the interaction of the harvesting implement with the cut edge of the stem may be another source of contamination. Instead of worker handling, mechanization of harvesting can reduce the chance of further contamination of microgreens, or education can improve worker hygiene practices. Studies have demonstrated that microgreens growing in hydroponic systems are vulnerable to pathogen proliferation when seeds are contaminated. Hence, seed sanitation is necessary (Reed et al. 2018; Wright and Holden 2018; Xiao et al. 2015). Overall, microgreens are harvested by cutting the stems and, hence, are highly susceptible to dehydration and quality deterioration. So, refrigeration and proper packaging are essential to maintain their quality (Turner et al. 2020). The US FDA has not defined commodity-specific guidelines for microgreens (Wang 2016).

Microgreens may be washed after harvest to remove soil particles, which reduces the initial bacterial load to provide a clean product for packaging. Washing prior to packaging creates a humid environment which promotes microbial growth. Hence, removal of excess water is essential. Many growers choose not to wash, as the additional handling by washing and dewatering can damage the delicate microgreens, which leads to a greater susceptibility to microbial growth. Removing excess moisture after washing without causing damage is a challenge. Thus, a delicate balance is required to maintain temperature, moisture, and atmosphere that optimize microgreens' quality retention and shelf life while discouraging the growth of spoilage microbes and human pathogens (Turner et al. 2020).

The sensory attributes and overall acceptability and liking of microgreens are primarily influenced by their phytochemical content. Microgreens have become used commonly only in the recent past, and research on microgreens is in its infancy. Studies should optimize the pre and postharvest practices for nutrient enhancement and retention. Therefore, increasing public awareness and acceptance of microgreens may be an important consideration in the promotion of this new functional food to the general population (Zhang et al. 2021).

For the commercialization of ready-to-eat packaged microgreens, more research is needed to identify effective sanitizers and drying methods that are not harmful to quality or shelf life (Kyriacou et al. 2016). The microgreens are usually washed clean postharvest. Improved wash/drying methods should be available for ready-to-eat microgreens with greater quality and shelf life. The postharvest wash phase can be avoided if microgreens are produced in a controlled environment with little microbial contamination. Microgreens may be cultivated indoors, so the propagation materials may be properly decontaminated (Kou et al. 2015).

Chandra et al. (2012) reported the effects of different sanitizers such as chlorine, citric acid, ascorbic acid, and ethanol spray on the quality and microbial population of *Brassica campestris var. farinose* microgreens. The study found that a spray of citric acid and ethanol might successfully substitute the chlorine normally used to wash microgreens. Simply washing the microgreens with chlorinated water (without affecting taste and flavor) can prevent microbial growth in the postharvest stage (Mir et al. 2017). Kou et al. (2014) investigated the positive role of preharvest calcium application on the quality of broccoli microgreens. Studies on the effects of various pre/postharvest cleaning may include calcium treatments on the quality and shelf life of broccoli microgreens. It was noticed that $CaCl_2$ without postharvest wash was the most efficient treatment as compared with Ca lactate and Calcium amino acid chelate. Since broccoli microgreens are so fragile in nature, dipping/washing and drying them reduces their quality drastically. The postharvest quality and shelf life could be extended to 21 days. The addition of 10 mM calcium chloride to the microgreens not only tripled the calcium concentration but also, there was an increase in biomass synthesis, superoxide dismutase activity, and peroxidase activity in microgreens as well as a substantial reduction in microbial growth during storage.

Key areas which required to be addressed are (Bandyopadhyay et al. 2002) testing and validation of microgreens' health-promoting effects in both animal models and human studies; (Berba and Uchanski 2012) determining the bioavailability of bioactive components from microgreens; (Bhatt and Sharma 2018) identifying the mechanism of action of microgreen components on cellular pathways in inflammatory processes to also include the microbiome; and (Bulgari et al. 2021) determining the optimum growing conditions and postharvest processing and the effects of these factors on the nutrient content of microgreens. Therefore, further studies are necessary to fully realize the value of microgreens in human health (Choe et al. 2018).

Shelf life extension of microgreens can be achieved using the relatively modern approach of microencapsulation. The vitamin-rich *Vigna radiata* microgreens could be locked within microspheres made from inert materials using microencapsulation technologies (Fig. 5.2) such as ionotropic gelation and lyophilization (Nair and Lekshmi 2020).

Fig. 5.2 Vitamin-rich microcapsules developed from various microgreens via microencapsulation

Future research aimed at improving the quality and shelf life of microgreens, on the other hand, should be encouraged in order for the microgreen sector to thrive.

5.8 Microgreen Market Challenges

Microgreens are gaining popularity across the world since they are 40 times more nutritious than mature veggies. Microgreen cultivation has both environmental and economic benefits. Consumption patterns and income levels influence the progression of the microgreens market to its true development potential. Climate and societal emergencies, like the COVID-19 epidemic, compel us to reconsider how we create food fortified with bioactive components that enhance health and maintain the immune system. Under such circumstances, microgreen culture should be considered and implemented as a remedy (Paraschivu et al. 2021).

According to recently available data and market research, the microgreens industry is on the rise throughout the world, particularly in cities owing to the rapid use of indoor and vertical farming systems; however, lack of knowledge about shelf life, food safety, and packaging of microgreens act as a primary challenge in many developing and underdeveloped areas. In such a setting, the promotion of microgreens to the general population must take precedence. The popularity and sales of microgreens will rise as more people become aware of the market. Furthermore, shelf life, food safety, and packaging of microgreens should be addressed, in addition to marketing or farmers who cultivate microgreens at home or on the farm. Customers must, thus, have a thorough understanding of microgreens before the industry can be established in a particular location (Van Rooyen et al. 2021).

5.9 Microgreens: The Future Food

NASA scientists are investigating the difficulties and advantages of cultivating microgreens in space. Microgreens are suited for space travel because they can be grown on a static shallow substrate with little to no nutrients, alleviating poor crop performance in microgravity hydroponics due to low oxygen and nutrient solubility. Hence it is an exciting, novel area of research (Dey and Chakraborty 2021). Plants are prized in space for their ability to renew oxygen, fix nitrogen, offer essential nutrients and fresh ingredients, and boost astronaut morale over long periods away from Earth (Kyriacou et al. 2017).

Microgreens are perfect for salads due to their small size, minimal nutrition, growth medium requirements, and short growing period. Seed germination, watering plants, anchoring plant roots, and restricted resources are just a few of the obstacles that come with growing food in space. Padgett and Smith (2018) outlines how films are made and tested to keep seeds in place during growing. Microgravity, according to Vandenbrink and Kiss (2016), may impact epigenetic processes and, as a result, gene expression in plants.

Furthermore, food safety problems are crucial for astronauts, who have limited access to medical treatment and are confined together in a tiny area during space travel. To guarantee that seeds do not retain human infections, they must be cleaned (Padgett and Smith 2018).

Novel applications, such as meals or ingredients made from discarded microgreens or microgreens nearing the end of their shelf life, are the last topic of study that has yet to be fully explored. Despite the importance of extending the shelf life of microgreens, which has been described in this study to decrease waste, innovative processing, and reformulation of discarded microgreens into new products is an area of future research.

References

Bandopadhyay S, Mukherjee SK (2009) Wild edible plants of Koch Bihar district, West Bengal. Nat Prod Radiance 8(1):64–72

Bandyopadhyay U, Biswas K, Chattopadhyay I, Banerjee RK (2002) Biological activities and medicinal properties of neem (*Azadirachta indica*). Curr Sci 82(11):1336–1345

Berba KJ, Uchanski ME (2012) Postharvest physiology of microgreens. J Young Invest 24:1–5

Bhatt P, Sharma S (2018) Microgreens: a nutrient rich crop that can diversify food system. Int J Pure Appl Biosci 6:182–186

Bulgari R, Negri M, Santoro P, Ferrante A (2021) Quality evaluation of indoor-grown microgreens cultivated on three different substrates. Horticulturae 7(5):96

Buragohain J, Singh VB, Deka BC, Jha AK, Wanshnong K, Angami T (2013) Collection and evaluation of some underutilized leafy vegetables of Meghalaya. Indian J Hill Farming 26(2): 111–115

Chandra D, Kim JG, Kim YP (2012) Changes in microbial population and quality of microgreens treated with different sanitizers and packaging films. Hortic Environ Biotechnol 53(1):32–40

Choe U, Yu LL, Wang TT (2018) The science behind microgreens as an exciting new food for the 21st century. J Agric Food Chem 66(44):11519–11530
de la Fuente B, López-García G, Máñez V, Alegría A, Barberá R, Cilla A (2019) Evaluation of the bio accessibility of antioxidant bioactive compounds and minerals of four genotypes of Brassicaceae microgreens. Foods 8(7):250
Delian E, Chira A, Bădulescu L, Chira L (2015) Insights into microgreens physiology. Sci Pap Ser B Hortic 59:447–454
Dey S, Chakraborty AP (2021) Microgreens: food for the future. Science Reporter, pp 44–46
Di Gioia F, Petropoulos SA, Ozores-Hampton M, Morgan K, Rosskopf EN (2019) Zinc and iron agronomic biofortification of Brassicaceae microgreens. Agronomy 9(11):677
Ebert AW (2013) Sprouts, microgreens, and edible flowers: the potential for high value specialty produce in Asia. SEAVEG 2012: high value vegetables in Southeast Asia: production, supply and demand, pp 216–227
Ebert AW, Wu TH Yang RY (2014) Amaranth sprouts and microgreens-a homestead vegetable production option to enhance food and nutrition security in the rural-urban continuum. In: Proceedings of the Regional Symposium on Sustaining Small-Scale Vegetable Production and Marketing Systems for Food and Nutrition Security (SEVEG 2014), Bangkok, Thailand, pp 25–27
Franks E, Richardson J (2009) Microgreens: a guide to growing nutrient-packed greens. Gibbs Smith, Utah
Gayathree PK, Samarasinghe D (2019) Green stimuli characteristics and green self identity towards ethically minded consumption behavior with special reference to mediating effect of positive and negative emotions. Asian Soc Sci 15(7):77–102
Ghoora MD, Babu DR, Srividya N (2020) Nutrient composition, oxalate content and nutritional ranking of ten culinary microgreens. J Food Compos Anal 91:103495
Huang H, Jiang X, Xiao Z, Yu L, Pham Q, Sun J, Wang TT (2016) Red cabbage microgreens lower circulating low-density lipoprotein (LDL), liver cholesterol, and inflammatory cytokines in mice fed a high-fat diet. J Agric Food Chem 64(48):9161–9171
Kamble VS, Jadhav VD (2013) Traditional leafy vegetables: a future herbal medicine. Int J Agric Food Sci 3(2):56–58
Kou L, Luo Y, Yang T, Xiao Z, Turner ER, Lester GE, Camp MJ (2013) Postharvest biology, quality and shelf life of buckwheat microgreens. LWT Food Sci Technol 51(1):73–78
Kou L, Yang T, Luo Y, Liu X, Huang L, Codling E (2014) Pre-harvest calcium application increases biomass and delays senescence of broccoli microgreens. Postharvest Biol Technol 87:70–78
Kou L, Yang T, Liu X, Luo Y (2015) Effects of pre-and post-harvest calcium treatments on shelf life and post-harvest quality of broccoli microgreens. HortScience 50(12):1801–1808
Kowitcharoen L, Phornvillay S, Lekkham P, Pongprasert N, Srilaong V (2021) Bioactive composition and nutritional profile of microgreens cultivated in Thailand. Appl Sci 11(17):7981
Kyriacou MC, Rouphael Y, Di Gioia F, Kyratzis A, Serio F, Renna M, Santamaria P (2016) Micro-scale vegetable production and the rise of microgreens. Trends Food Sci Technol 57:103–115
Kyriacou MC, De Pascale S, Kyratzis A, Rouphael Y (2017) Microgreens as a component of space life support systems: a cornucopia of functional food. Front Plant Sci 8:1587. https://doi.org/10.3389/fpls.2017.01587
Kyriacou MC, El-Nakhel C, Graziani G, Pannico A, Soteriou GA, Giordano M, Rouphael Y (2019) Functional quality in novel food sources: genotypic variation in the nutritive and phytochemical composition of thirteen microgreens species. Food Chem 277:107–118
Le TN, Chiu CH, Hsieh PC (2020) Bioactive compounds and bioactivities of *Brassica oleracea* l. var. italica sprouts and microgreens: an updated overview from a nutraceutical perspective. Plants 9(8):946
Lenzi A, Orlandini A, Bulgari R, Ferrante A, Bruschi P (2019) Antioxidant and mineral composition of three wild leafy species: a comparison between microgreens and baby greens. Foods 8(10):487
Mir SA, Shah MA, Mir MM (2017) Microgreens: production, shelf life, and bioactive components. Crit Rev Food Sci Nutr 57(12):2730–2736

Nair BR, Lekshmi GP (2019) Nutritional and anti-nutritional analysis of some selected microgreens. Appl Biol Res 21(2):137–144. https://doi.org/10.5958/0974-4517.2019.00014.4

Nair BR, Lekshmi GP (2020) Preparation, characterization and evaluation of a vitamin-enriched dietary supplement from *Vignaradiata* L. microgreens. Nutrafoods 1:159–168. https://doi.org/10.17470/NF-020-0022

Niroula A, Khatri S, Timilsina R, Khadka D, Khadka AandOjha P (2019) Profile of chlorophylls and carotenoids of wheat (*Triticumaestivum* L.) and barley (*Hordeumvulgare* L.) microgreens. J Food Sci Technol 56(5):2758–2763

Oh MM, Carey EE, Rajashekar CB (2010) Regulated water deficits improve phytochemical concentration in lettuce. J Am Soc Hortic Sci 135(3):223–229

Padgett N, Smith T (2018) Researching seeds: films, sanitation methods, microbiological growth, viability, and selection for new crops (No. KSC-E-DAA-TN55349)

Paraschivu M, Cotuna O, Sărățeanu V, Durău CC, Păunescu RA (2021) Microgreens-current status, global market trends and forward statements. Sci Pap Ser Manag Econ Eng Agric Rural Dev 21(3):633–639

Parida S (2020) Innovative farming of edible microgreens at home and their nutritional composition. Test Eng Manag 83:17630–17640

Pinto E, Almeida AA, Aguiar AA, Ferreira IM (2015) Comparison between the mineral profile and nitrate content of microgreens and mature lettuces. J Food Compos Anal 37:38–43

Polash MAS, Sakil MA, Hossain MA (2018) Post-harvest biodegradation of bioactive substances and antioxidant activity in microgreens. J Bangladesh Agric Univ 166(2):250–253

Rani S, Singh N, Maurya SB (2018) The comparative nutrients assessment of spicer salad: radish microgreens, pp 107–111

Reed E, Ferreira CM, Bell R, Brown EW, Zheng J (2018) Plant-microbe and abiotic factors influencing *Salmonella* survival and growth on alfalfa sprouts and Swiss chard microgreens. Appl Environ Microbiol 84(9):e02814–e02817

Renna M, Paradiso VM (2020) Ongoing research on microgreens: nutritional properties, shelf-life, sustainable production, innovative growing and processing approaches. Foods 9(6):826

Riggio GM, Wang Q, Kniel KE, Gibson KE (2019) Microgreens—a review of food safety considerations along the farm to fork continuum. Int J Food Microbiol 290:76–85

Romognolo DF, Donovan MG, Papoutsis AJ, Doetschman TC, Selmin OI (2017) Genistein prevents BRCA1 CpG methylation and proliferation in human breast cancer cells with activated aromatic hydrocarbon receptor. Curr Dev Nutr 1(6):e000562

Sahoo G, Swamy S, Rout S, Wani A, Mishra A (2021) Exploitation of wild leafy vegetables and under-utilized fruits: consequences for food and nutritional security. Ann Roman Soc Cell Biol 25(6):5656–5668

Sahu S, Mhedhbi A, Salihoglu S, Lin J, Özsu MT (2020) The ubiquity of large graphs and surprising challenges of graph processing: extended survey. VLDB J 29(2):595–618

Sharma P, Sharma A, Rasane P, Dey A, Choudhury A, Singh J, Kaur D (2020) Optimization of a process for microgreen and fruit-based functional beverage. An Acad Bras Cienc 92:e20190596

Sheela K, Nath KG, Vijayalakshmi D, Yankanchi GM, Patil RB (2004) Proximate composition of underutilized green leafy vegetables in southern Karnataka. J Hum Ecol 15(3):227–229

Stoleru T, Ioniță A, Zamfirache M (2016) Microgreens—a new food product with great expectations. Roman J Biol 61:7–16

Sudha K, Mathanghi SK (2012) Traditional underutilized green leafy vegetables and its curative properties. Int J Pharm 2:786–793

Sun J, Xiao Z, Lin LZ, Lester GE, Wang Q, Harnly JM, Chen P (2013) Profiling polyphenols in five *Brassica* species microgreens by UHPLC-PDA-ESI/HRMS n. J Agric Food Chem 61(46): 10960–10970

Tan L, Nuffer H, Feng J, Kwan SH, Chen H, Tong X, Kong L (2020) Antioxidant properties and sensory evaluation of microgreens from commercial and local farms. Food Sci Human Wellness 9(1):45–51

Teng J, Liao P, Wang M (2021) The role of emerging micro-scale vegetables in human diet and health benefits-an updated review based on microgreens. Food Funct 12(5):1914–1932

Treadwell D, Hochmuth R, Landrum L, Laughlin W (2020) Microgreens: a new specialty crop: HS1164, rev. 9/2020. Edis 2020(5)

Turner ER, Luo Y, Buchanan RL (2020) Microgreen nutrition, food safety, and shelf life: a review. J Food Sci 85(4):870–882

Van Rooyen J, Shrestha P, Shrestha A, Tamang PL, Nepali PP, Lala OS (2021) Impacts of lack of awareness that is restricting the microgreen market of Nepal. Int J Innov Sci Res Rev 3(1): 640–644

Vandenbrink JP, Kiss JZ (2016) Space, the final frontier: a critical review of recent experiments performed in microgravity. Plant Sci 243:115–119

Wang Q (2016) Interaction of human norovirus and its surrogates with fresh produce. University of Delaware

Warriner K, Ibrahim F, Dickinson M, Wright C, Waites WM (2003) Internalization of human pathogens within growing salad vegetables. Biotechnol Genet Eng Rev 20:117–136

Weber CF (2017) Broccoli microgreens: a mineral-rich crop that can diversify food systems. Front Nutr 4:7

Welbaum GE (2015) Vegetable production and practices. CABI, Wallingford

Wood L (2019) World-wide indoor farming market outlook 2019–2024-the decrease in cultivable land is driving growth. Research and Markets

Wright KM, Holden NJ (2018) Quantification and colonisation dynamics of Escherichia coli O157: H7 inoculation of microgreens species and plant growth substrates. Int J Food Microbiol 273:1–10

Xiao Z (2013) Nutrition, sensory, quality and safety evaluation of a new specialty produce: microgreens (Doctoral dissertation, University of Maryland, College Park)

Xiao Z, Lester GE, Luo Y, Wang Q (2012) Assessment of vitamin and carotenoid concentrations of emerging food products: edible microgreens. J Agric Food Chem 60(31):7644–7651

Xiao Z, Lester GE, Luo Y, Xie ZK, Yu LL, Wang Q (2014) Effect of light exposure on sensorial quality, concentrations of bioactive compounds and antioxidant capacity of radish microgreens during low temperature storage. Food Chem 151:472–479

Xiao Z, Lester GE, Park E, Saftner RA, Luo Yand Wang Q (2015) Evaluation and correlation of sensory attributes and chemical compositions of emerging fresh produce: microgreens. Postharvest Biol Technol 110:140–148

Xiao Z, Codling EE, Luo Y, Nou X, Lester GE, Wang Q (2016) Microgreens of Brassicaceae: mineral composition and content of 30 varieties. J Food Compos Anal 49:87–93

Xiao Z, Rausch SR, Luo Y, Sun J, Yu L, Wang Q, Stommel JR (2019) Microgreens of Brassicaceae: genetic diversity of phytochemical concentrations and antioxidant capacity. LWT 101:731–737

Zhang Y, Xiao Z, Ager E, Kong L, Tan L (2021) Nutritional quality and health benefits of microgreens a crop of modern agriculture. J Future Foods 1(1):58–66

Chapter 6
Wild-Edible Tubers and Rhizomes of South Western Ghats, India

V. Balakrishnan, Jose Mathew, and Salim M. Pichan

Abstract Wild tubers can play a pivotal role in eradicating malnourishment and food insecurity. Besides, such crops also offer economic advantages due to their uniqueness, suitability to the environments in which they are grown and low input requirements. Hills and valleys of south Western Ghats are the habitats of several indigenous tubers and roots with edible potential. In this context, documentation of the diversity of potential wild-edible tubers of SWG viz., *Alocasia, Amorphophallus, Asparagus, Arisaema, Ceropegia, Chlorophytum, Curcuma, Dioscorea, Leea, Nelumbo* and *Nymphaea* are attempted. Scientific names along with their synonyms and common names, taxonomic description, distribution, etc., are dealt with in detail. Traditional botanical information on tuber crops integrated with scientific studies will validate their medicinal/nutraceutical benefits and support further research on the bioprospecting of these ethnomedicinal plants.

Keywords Wild edibles · Western Ghats · Endemism · Domestication · Conservation

6.1 Wild-Edible Tubers and Rhizomes

The Food and Agricultural Organization points out that nearly one billion people are undernourished, hungry and living without adequate daily calories, and they denote the scope of wild edibles to meet the everyday healthful prerequisites (FAO 2010). It

V. Balakrishnan
Kerala State Biodiversity Board, Thiruvananthapuram, Kerala, India

MS Swaminathan Research Foundation, Wayanad, Kerala, India

J. Mathew (✉)
Department of Botany, Sanatana Dharma College, Alappuzha, Kerala, India
e-mail: josemathew@sdcollege.in

S. M. Pichan
MS Swaminathan Research Foundation, Wayanad, Kerala, India

S. T. Sukumaran, K. T R (eds.), *Conservation and Sustainable Utilization of Bioresources*, Sustainable Development and Biodiversity 30,
https://doi.org/10.1007/978-981-19-5841-0_6

is estimated that about four billion people of the world's population depend on wild-edible herbs/herbal medicines for their primary healthcare requirements (Food and Agriculture Organization 1990). According to food and agricultural organisation (Food and Agriculture Organization 2003), an estimated total of 7000 plants have been identified as crop plants at some point in human history. However, global population depends on three plants—rice, maize and wheat, for 60% of their daily energy requirement. This trend in agriculture has led to the marginalization of locally important crops with adaptability to marginal farming conditions, relevance to local food culture and diverse nutritional values.

On the other hand, our crops all have their starting points in wild biodiversity and have wild predecessors. To some extent, these wild species are still used within the local, national, or even international communities and make a significant contribution to the diet of rural households, particularly during drought, famine and the dry season. They are suggested to have high nutritional value, but their role in nutrition security is not adequately understood, and they have not been mainstreamed into existing policies and programmes on nutrition.

Tubers and root crops are significant sources of several compounds, namely, saponins, phenolic compounds, glycol alkaloids, phytic acids, carotenoids and ascorbic acid. Many bioactivities, namely, antioxidant, immune modulator, antimicrobial, antidiabetic, antiobesity and hypocholesterolemic activities, are reported for tubers and root crops. An important agronomic advantage of root and tuber crops as staple foods is their favourable adaptation to diverse agro-climatic conditions and various farming systems with minimum agricultural inputs (Chandrasekara and Kumar 2016).

Presently underutilized food sources, including root and tuber crops, can possibly safeguard against inner and outside market disturbances and environmental vulnerabilities and lead to better biological system capacities and administrations, along these lines upgrading supportability. High energy worths of roots and tubers guarantee nourishment security through adjusting food things in families' food containers. Underutilized tubers have drawn in little exploration discoveries in spite of the way that they are adjusted to a wide scope of developing circumstances, add to food security, particularly under pressure conditions, and are significant for a healthful well-balanced diet. The information and comprehension of wild tuber crops can be incredibly valuable for the human culture since wild species are connected with crops and are a significant wellspring of agronomically significant qualities for plant rearing, expanding interest as a wellspring of versatile attributes that can assist our yields with adapting better despite the environmental change.

The forests and valleys along the southern Western Ghats are the most species-rich ecological region in peninsular India with respect to species diversity and endemism. About 80% of the flowering plant species of the entire Western Ghats are distributed in this ecological region (Mathew et al. 2019). This study aims to document the checklist and diversity of edible tuber crops and rhizomes of the southern Western Ghats based on extensive field studies. In this array, domesticated tuber crops and wild species are considered separately. It is expected that intensive floristic studies give due emphasis to correct identity, distribution, present status and extent of the threat, if any, endemism and the dynamism have tremendous significance to safeguard biodiversity and food security.

6.2 Methodology in Brief

Intensive floristic explorations and traditional/ethnobotanical studies were conducted during the period 2005–2018 by frequent field visits and specimen collections. Forests and farm fields of Wayanad, Idukki, Kollam and Thiruvananthapuram districts of Kerala were the study area. Tribal groups viz., *Kattunayikka, Cholanayikka, Paniya, Kurichyar, Malampandaram, Kani, Malayarayar* and traditional farmers are the main informers. Plant specimens were taxonomically identified using the pertinent literature and different herbaria. The herbarium specimens were prepared as per the standard specifications (Bridson and Forman 1991; Fosberg and Sachet 1965). The voucher specimens were deposited in the herbarium of MSSRF (MS Swaminathan Research Foundation, Kalpetta, Wayanad, Kerala). The occurrence and distribution of endemic species were verified and analysed with the help of standard publications (Ahmedullah and Nayar 1987; IUCN 2021; Sasidharan 2013). The systematic documentation of the identified species was done according to the classification system of Bentham and Hooker with necessary alterations as suggested by Hutchinson (Bentham and Hooker 1883; Hutchinson 1926, 1934, 1956, 1973). Artificial diagnostic (Dichotomous parallel) keys were prepared for the identification of varieties. Name of the authors and citation of the original publication of the generic name are given based on IPNI (IPNI 2021).

6.3 Results

The botanical investigations conducted among the farmers and tribes have brought to light new insights regarding the utility of tubers and rhizomes, including their wild crop relatives. The major domesticated tubers and rhizomes are enlisted in Table 6.1. Consumption of wild-edible tubers constitutes an integral part of the tribal life of the study area. Their living surroundings are rich with edible roots/tubers, greens, piths and fruits. Digging of tubers and the collection of edibles were done by both men and women.

6.3.1 Wild-Edible Tubers and Rhizomes of South Western Ghats

During the present study, 24 wild-edible tubers and rhizomes belonging to 11genera were recorded. *Dioscorea* was the largest genus in the family Dioscoreaceae with 13 taxa, followed by *Amorphophallus* (2 spp.). Eight species among the enlisted plants are coming under endemic status (Fig. 6.1).

Table 6.1 Domesticated tubers and rhizomes in Kerala part of south Western Ghats

Sl. no.	Scientific name	Common name	Family	Uses
1	*Daucus carota* L.	Carrot	Apiaceae	Food
2	*Amorphophallus paeoniifolius* (Dennst.) Nicolson	*Chena* (Mal.); Elephant foot yam	Araceae	Food, medicine
3	*Colocasia esculenta* (L.) Schott	*Chembu* (Mal.); Taro		Food, medicine
4	*Raphanus raphanistrum* subsp. *sativus* (L.) Domin	Radish	Brassicaceae	Food, value-added products
5	*Ipomoea batatas* (L.) Lam.	*Cheenikizhang*; *Madurakizhangu* (Mal.); Sweet Potato	Convolvulaceae	Food, sugar, value-added products
6	*Dioscorea alata* L.	Kaachil (Mal.); Asiatic yam	Dioscoreaceae	Food, medicine
7	*Dioscorea esculenta* (Lour.) Burkill	Nanakizhangu (Mal.); White yam		Food
8	*Manihot esculenta* Crantz	*Kappa; Marachini* (Mal.); Topioca plant; Cassava	Euphorbiaceae	Food, Sugar, value-added products
9	*Plectranthus rotundifolius* Spreng.	Koorka (Mal.); Chinese Potato	Lamiaceae	Food
10	*Maranta arundinacea* L.	*Koova* (Mal.); Arrow root	Marantaceae	Food, value-added products
11	*Solanum tuberosum* L.	*Urulakizhangu* (Mal.); Sweet Potato	Solanaceae	Food, Sugar, Value-added products
12	*Curcuma longa* L	*Manjal* (Mal.); Turmeric	Zingiberaceae	Food ingredient, medicine, preservative, value-added products
13	*Zingiber officinale* Rosc.	*Inji* (Mal.); Ginger		Food ingredient, medicine, preservative, value-added products

6.3.1.1 *Alocasia longiloba* Miq., Fl. Ned. Ind. II. 207. 1855. (Araceae) (Fig. 6.1a)

Synonym: *Alocasia amabilis* W.Bull; A. argyrea Sander; *A. cochinchinensis* Pieree ex Engl. & K.Krause; *A. curtisii* N.E.Br.; *A. cuspidata* Engl.; *A. denudata* Engl.; *A. denudata* var. *elongata* Engl.; *A. eminens* N.E.Br.; *A. korthalsii* Schott; *A. lowii* Hook.f.; *A. lucianii* Pucii ex Rodigas; *A. pucciana* Andr.; *A. putzeysii* N.E.Br.; *A. singaporensis* Linden; *A. spectabilis* Engl. & K. Krause; *A. thibantiana* Mast.; *A. veitchii* (Lindl.) Schott; *A. watsoniana* Sander; *Caladium lowii* Lem.; *C. veitchii* Lind.

Fig. 6.1 (**a**) *Alocasia longiloba* Miq. (**b**) *Amorphophallus nicolsonianus* Sivad. (**c**) *Asparagus racemosus* Willd. (**d**, **e**) *Arisaema subulatum* Manudev & Nampy. (**f**) *Ceropegia fimbriifera* Bedd. (**g**) *Chlorophytum indicum* (Willd.) Dress

Ethnic Name: Neelanchembu (Kattunaykkar Tribe of Kerala).

Description: Herbs, terrestrial to 150 cm tall. Stem rhizomatous, elongate, 8–60 × 2–8 cm. Leaves 1–3, subtended by lanceolate papery-membranous cataphylls; petiole terete; leaf blade is pendant, green to dark green, dark green often with major venation grey- green adaxially, narrowly hastate-sagittate, 27–85 × 14–40 cm. Posterior lobes pointed. Spathe 7–17 cm, abruptly constricted 1.5–3.5 cm from the base, stipitate, female zone 1.5–2.5 cm, pistil green, subglobose, stigma white, sterile zone 7–10 mm, narrower than fertile zone, synandrodes mostly rhombic hexagonal, flat topped, male zone is ivory and subcylindric. Fruit ripening orange red.

Flowering and Fruiting: June–October.

Distribution: In rain forests and regrowth understory and on rocks of Cambodia, Laos, Vietnam, China, Malaysia, Borneo, Java, Sulawsi and India (World). In India, it is only found in rain forests of Wayanad and Thiruvananthapuram Districts of Kerala, part of South Western Ghats.

Status in Southern Western Ghats: Rare.

6.3.1.2 *Amorphophallus bonaccordensis* Sivad. & N. Mohanan, Blumea 39: 295. 1994 (Araceae)

Ethnic Name: *Cheruchena* (*Kani* Tribe).

Description: Herbs, corms subglobose. Leaf is trichotomously decompound with petiole cylindric, 30–75 cm long, smooth, green with dark greenish brown mottles; rachis of the segments 15–20 cm long, shallowly channelled above and with decurrent leaf bases; leaflets sessile, ovate to oblong, acuminate at apex, acute at base, base unequal and decurrent on rachis, greenish above, 6–13 × 3–6.5 cm; primary veins 15–20 pairs, closely parallel, united below the margin forming an intramarginal vein. Stolons are cylindric, 4–5 × 0.4–0.7 cm, nodes with thin-scale leaves and few root primordia. Inflorescence with peduncle, 30–75 cm high, 1–1.6 cm diameter at the base, gradually narrowly to the tip; Spadix stipitate. Pistillate flowers subspirally arranged; ovary sessile, subglobose, 0.18–0.2 × 0.2–0.3 cm, greenish, 2 or 3 loculed, each with one sub-basal ovule; style very short, cylindric, 0.5–0.8 × 0.8–1 mm; stigma is 2 or 3-lobed, covered with short unicellular papillae; neuter flowers in 1–3 rows, creamy white, obovoid, 0.2–0.3 cm diameter, few appear to be transitional to the staminate flowers. Staminate flowers closely arranged, each compressed of 2–5 stamens borne on a white cushion-like tissue of 0.8–1 mm thickness; stamen 0.1–0.13 cm high, inconspicuously 2-lobed; thecae dehisce by narrow apical horizontal slits.

Flowering and Fruiting: March–April.

Distribution: Evergreen forests of South Western Ghats. Endemic to Agasthyavam Biological Park.

Status in Southern Western Ghats: Rare.

6.3.1.3 *Amorphophallus nicolsonianus* Sivad., Pl. Syst. Evol. 153:165. 1986. (Araceae) (Fig. 6.1b)

Ethnic Name: *Kuzhichena* (*Malapanadaram* Tribe).

Description: Corm oblong or subcylindrical. Leaves 25–30 cm long; leaflets 13–15 × 4–5 cm, ovate or oblong, acuminate, base acute or obtuse, glabrous, glossy above. Peduncle terete, 17–20 cm long, spathe ovate lanceolate, acuminate, 8–10 × 2.5–3 cm, greenish brown, basally convolute; spadix sessile, slightly longer than the spathe, flowers confined towards the base; sterile appendix narrowly cylindric, tapering to the apex.

Flowering and Fruiting: January–March.

Habitat and Distribution: In evergreen forests of South Western Ghats (Endemic).

Status in Southern Western Ghats: Rare.

6.3.1.4 *Arisaema subulatum* Manudev & Nampy, Rheedea 29 (2): 166.2019 (Araceae) (Fig. 6.1d, e)

Ethnic Name: *Thavittuchena* (*Malayarayan* Tribes).

Description: Corm 3–10 cm across, brown in colour, depressed globose. Leaf one; leaflets 7–12, to 25 × 7 cm, lanceolate, acuminate; nerves many, close, parallel, forming an intramarginal vein, prominent, glabrous; petiole to 70 cm long, 1–3 cm thick, brownish. Spadix peduncled, below the level of the leaf; spathe 10–22 cm long, tube 5–15 cm long, 1–3.5 cm broad, with thick greenish brown streaks; limb ovate, apex curved down, finely acuminate, green with brown streaks. Spadix 11–15 cm long, clavate at apex, female flowers many in lower 1–3 cm long; neuter flowers filiform, simple or bifurcate.

Flowering and Fruiting: May–December.

Habitat and Distribution: Margins of evergreen forests, sholas of South Western Ghats, found in Thrissur and Idukki districts of Kerala.

Status in Southern Western Ghats: Rare.

6.3.1.5 *Asparagus racemosus* Willd., Sp. Pl., ed. 4 [Willdenow] 2 (1): 152 (1799) (Asparagaceae) (Fig. 6.1c)

Synonym: *Asparagopsis abyssinica* Kunth; *A. acerosa* Kunth; *A. rownei* Kunth; *A. decaisnei* Kunth; *A. floribunda* Kunth; *A. hohenackeri* Kunth; *A. javanica* Kunth; *A. retrofracta* Schweinf. ex Baker; *A. sarmentosa* Dalzell & A.Gibson; *A. subquadrangularis* Kunth; *Asparagus acerosus* Roxb.; *A. dubius* Decne.; *A. fasciculatus* R.Br.; *A. jacquemontii* Baker; *A. penduliflorus* Zipp. ex Span.; *A. petitianus* A.Rich.; *A. stachyoides* Spreng. ex Baker; *A. tetragonus* Bresler; *A. zeylanicus* (Baker) Hook.f.; *Protasparagus acerosus* (Kunth) Kamble; *P. jacquemontii* (Baker) Kamble; *P. racemosus* (Willd.) Oberm.; *P. racemosus* var. *javanicus* (Kunth) Kamble; *P. racemosus* var. *subacerosus* (Baker) Kamble; *P. zeylanicus* (Hook.f.) Kamble; *Geitonoplesium scandens* Hassk.

Ethnic Name: *Chathavalli, Sathavali, Sathavari, Thalicheria, Thaliperiya, Thannivayan-kizhangu.*

Common Name: Asparagus, Wild asparagus.

Description: Woody perennial climbers. Stem often spinescent, terete, green; rootstock with fascicled tuberous roots. Cladodes from the axils of scale leaves in clusters of 2–6, 0.8–1.5 × 0.1–0.3 cm, linear falcate, slightly triquetrous, base narrow, apex acute. Racemes 2.5–5 cm long, slender, axillary, solitary, or clustered. Flowers bisexual, 5–6 mm across; bracts triangular; pedicel c. 1 mm long. Perianth lobes 6, white, c. 3 × 0.5 mm, oblong, acute. Stamens 6, adnate to the perianth lobes; filaments subulate. Ovary 2–3 mm long, globose to slightly 3-gonous, 3-celled; ovules 2 per cell; stigma-3, recurved. Berry 4–6 mm diam., globose, purple on ripening. Seeds 2–5, c. 2 mm across, globose.

Flowering and Fruiting: July–August.

Distribution: Paleotropics, In Kerala, this plant grows in the shades of dry deciduous forest to montane evergreen forests.

Status in Southern Western Ghats: Common.

6.3.1.6 *Ceropegia fimbriifera* Bedd., Madras J. Lit. Sci. ser. 3, 1: 53. 1864 (Apocynaceae) (Fig. 6.1f)

Ethnic Name: *Urulankizhangu* (*Malampandaram* Tribes of Kerala).

Description: Erect tuberous rooted herbs; stem terete, glabrous, internodes 2–4 cm long. Leaves simple, linear, to 12 × 0.5 cm, base and apex acute, margin entire, 1-nerved, glabrescent; petiole 0 or very minute. Flower(s) greenish yellow, solitary; pedicel c. 5 mm long, glabrous; bracts subulate, 1–2 mm long; calyx lobes 5, linear, c. 3 mm long; corolla tube more or less 2 cm long, inflated below, tubular portion c. 1 cm long, glabrous inside, lobes 5, lanceolate, anthers 1 cm long, connate at tip, villous hairy inside at base; corona lobes 10, 2-seriate, outer lobes bifid, subulate, divergent, inner lobes spathulate; anther without appendages, pollen masses waxy; ovaries 2, style-apex truncate.

Flowering and Fruiting: June–September.

Habitat and Distribution: Margins of sholas and grasslands of South Western Ghats, found in Kollam, Wayanad and Idukki districts of Kerala.

Status in Southern Western Ghats: Vulnerable (Sasidharan 2013).

6.3.1.7 *Chlorophytum indicum* (Willd.) Dress, Baileya 9: 43. 1961. (Asparagaceae) (Fig. 6.1g)

Synonym: *Ornithogalum indicum* Willd. ex Schult. & Schult.f.; *Phalangium indicum* (Willd. ex Schult. & Schult.f.) Kunth.; *Liliago indica* (Willd. ex Schult. & Schult.f.) C. Presel.; *Chlorophytum attenuatum* Baker.

Description: Perennial geophytes 200–450 mm tall, grows solitary, rhizome reduced to disc, covered with fibrous remnant of leaves; roots fibrous, tuberous; tubers very few, ellipsoied to oblong 1–2.0 × 0.5–1 cm, towards the end of roots. Leaves all rosettes, elliptic oblong, sessile, rarely pseudopetiolate, 20–40 × 1.5–3.0 cm, glaucous green below, 19–21 veined, apex acute; margin wavy, hyaline. Scape 20–45 cm tall, solitary, unbranched, smooth, naked, upper 1/2 part fairly dense raceme; sterile bracts absent, rarely present, linear, caducous, green, floral bracts ovate or triangular, green to translucent. Flowers white, 1.2–1.7 cm across, pedicelate, in alternate to subopposite 5–8 flowered clusters, Pedicels 0.5–1.2 cm long, cylindrical, white, smooth, jointed above middle, swollen at articulation, pedicel portion below the joint 0.3–0.8 cm long, cylindrical; pedicel portion above the joint 0.3–0.6 cm long, angled. Perianth segments 6, in two whorls of 3 each, rarely recurved; outer perianth segments 0.7–0.9 × 0.3–0.4 cm, white, elliptic, 5-nerved, apex acute, margins hyaline; inner perianth segments 0.6–0.8 × 0.3–0.4 cm, white, elliptic lanceolate, smaller than the outer, 3-nerved, apex acute, margins hyaline. Stamens 6, 0.8–1.0 cm long, erect, introse; filaments 0.4–0.5 cm long, white, papillose, inflated at middle; anther 0.4–0.5 cm long, basifixed, yellow to mustard coloured, dehiscing by longitudinal slits, longer than or equal to filaments. Ovary 0.2–0.4 cm in diam., sessile, green, globose triquetrous; style 0.8–1.0 cm long, white, longer than stamens, bends out from base; stigma minutely papillose. Capsule green, triquetrous, 3-sulcate, c. 1.5 × 1.0 cm. Seeds 3–5 in each cell.

Flowering and Fruiting: August–December.

Distribution: India, dry hills of Andhra Pradesh, Karnataka and Tamil Nadu states. In Kerala, *Chlorophytum indicum* grows in the rock crevices under shades in the dry deciduous forest.

Status in Southern Western Ghats: Common in montane grasslands.

6.3.1.8 *Curcuma amada* var. *glabra* Velay., Unnikr., Asha & Maya, J. Econ. Taxon. Bot. 33(1): 164 (2009). (Zingiberaceae) (Fig. 6.2a)

Ethnic Name: *Koova.*

Description: Annual herbs. Spike arising laterally, sterile bracts, 4–5 in number enclosing the peduncle base, whitish at the base, green above; lower sterile bracts ovate, 2.5 × 2.3 cm, upper ones oblong lanceolate, 10.0–10.5 × 3.0–3.2 cm, acute, cuspidate to obtuse at tip, minutely hairy, peduncle 10.0 cm long; spike 10.5 cm long, 7.0–7.5 cm wide; fertile bracts 12–15 in numbers, ovate, tip obtuse, light green with light purple tint towards tip, hairy on both sides, 3.8 × 2.1 cm in size; coma bracts 5–6, oblong lanceolate, tip acute, whitish at the base, light purple violet/lilac from middle to tip, minutely hairy on both sides, spreading, showy, 5.7 × 2.1 cm. Flowers slightly exserted or as long as the bract with lip visible outside, 4.5 cm long, 2–3 flowers per bract, each enclosed by ovate/elliptic inner emarginated bract. Calyx transparent white, glabrous, 0.8–1.0 cm long, minutely toothed. Corolla tube

Fig. 6.2 (**a**) *Curcuma amada* var. *glabra* Velay. (**b**, **c**) *Dioscorea belophylla* (Prain) Voigt ex Haines. (**d**) *Dioscorea bulbifera* L. (**e**) *Dioscorea hamiltonii* Hook.f. (**f**) *Dioscorea hispida* Dennst

whitish; upper lobe beaked, 1.7 × 1.6 cm; side lobes oblong, 2.3 × 0.9 cm. Staminodes three lobed; lip obscurely three lobed and midlobe slightly emarginate, pale yellow, median yellow, side lobes obovate to oblong. Stamens oblique, filaments constricted at the tip; anthers versatile, 2 mm long, basal spurs white, converging; ovary 3 mm long, hairy, white, stigma exserted.

Flowering and Fruiting: April–September.

Habitat and Distribution: Scrub jungles and wastelands of Kerala, found throughout Kerala.

Note: Tubers used to prepare powder after processing for making various foods.

Status in Southern Western Ghats: Common in lowlands.

6.3.1.9 *Dioscorea belophylla* (Prain) Voigt ex Haines, Forest Fl. Chota Nagpur 530 (1910). (Dioscoreaceae) (Fig. 6.2b, c)

Synonym: *Dioscorea nummularia* var. *belophylla* Prain; *D. sagittata* Royle.

Ethnic Name: *Hekku (Kattunaika* Tribe).

Description: Tall climbers. Stem unarmed, greenish, 1–2.5 cm dia., terete, glabrous. Bulbils axillary, solitary, rarely 2, rootlets absent. Tuber deep, 1–4, stalked; stalk 20–50 cm long, tuber cylindrical, 25–150 cm long, rarely branched, 5–10 cm dia., 0.5–15 kg weight, surface smooth, rootlets small, thin, abundant, skin pale brown or yellowish, flesh white to lemon yellow inside, tender, very mucilaginous. Leaves simple, lower leaves often alternate, the upper usually opposite; lamina ovate, sagitate acuminate, base deeply cordate or conical, apex long acuminate, upper surface dark green, lower surface pale green; petiole 4–12 cm, channelled above, glabrous, green or pinkish. Male inflorescence simple paniculate, 8–20 cm long; spike 1–3 in a node, zig-zag, alternate on the axis; flowers sessile, dense, opens from the lowest upward, greenish in colour, sweetly scented; bracts ovate, acuminate, tepals 6, biseriate, glabrous, stamens 6, half as long as the perianth, anther as long as filaments. Female inflorescence simple spicate, axillary, 1–2 from an axil, pendulous, 8–17 cm long, bracts long, acuminate, bracts and bracteoles ovate, flower 4–6 mm long, tepals 6, biseriate, glabrous, staminodes minute; ovary 3 ridged, ca.3 mm long, stigma-3 partite. Capsules apex truncate or retuse, base rounded, margin and axis darker, green and silky glaucous when ripe; wings long, evenly rounded, broader than long, 2.5–3.7 × 2–3 cm, torus and tepals persisting; Seeds winged, reddish brown all around.

Flowering and Fruiting: September–January.

Distribution: India, Bangladesh and Nepal. In India, on very well drained mountain slopes from Himalaya of Kashmir and the salt range of the Punjab, eastwards to the Khasia Hills and southwards to the Nilgiri Hills and in Kerala. It flourishes upon the rock crests of steep hilly slopes. In Kerala the plant grows best at altitudes between 1300 and 1500 m above MSL on hill slopes.

Status in South Western Ghats: Rare.

6.3.1.10 *Dioscorea bulbifera* L., Sp. Pl. 2: 1033 (1753) (Dioscoreaceae) (Fig. 6.2d)

Synonym: *Dioscorea anthropophagorum* A. Chev.; *D. bulbifera* var. *albotuberosa* Y. F. Zhou, Z. L. Xu & Y. Y. Hang; *D. bulbifera* var. *brachybotryum* Y. Y. Hang & Y. F. Zhou; *D. bulbifera* var. *elongata* (F.M.Bailey) Prain & Burkill; *D. crispata* Roxb.; *D. heterophylla* Roxb.; *D. hoffa* Cordem.; *D. hofika* Jum. & H. Perrier; *D. korrorensis* R. Knuth; *D. latifolia* Benth.; *D. longipetiolata* Baudon; *D. perrieri* R. Knuth; *D. pulchella* Roxb.; *D. rogersii* Prain & Burkill; *D. sativa* var. *elongata* F. M. Bailey; *D. sativa* var. *rotunda* F. M. Bailey; *D. tamifolia* Salisb.; *D. tenuiflora* Schltdl.; *D. violacea* Baudon; *Helmia bulbifera* (L.) Kunth; *Polynome bulbifera* (L.) Salisb.

Ethnic Name: *Alanthal, Kattukachil, Nukappa, Pannikizhangu.*

Description: Large climbers. Stem twining to left, internodes 10–30 cm long, unarmed, pinkish, 1–3 cm dia., ribs shortly winged, glabrous. Bulbils axillary, solitary, discoid, spherical or ovoid, 1.5–6 cm across, dark brown outside, reddish yellow inside, surface tuberculate, rootlets thick. Tuber solitary, very variable in shape, globose, obovoid to pyriform, rarely branched, 10–40 cm dia., 0.5–10 kg weight, surface smooth, rootlets long, dense, skin purplish balck or muddy brown, flesh white to lemon yellow, sometimes marked with purple flecks, very mucilaginous. Leaves alternate, simple, lamina broadly ovate cordate, petiole 5–20 cm, channelled above, broadly auricled at base, auricle 0.5–1 cm broad, glabrous, green or pinkish. Male inflorescence simple, paniculate, pendulous, flowers sessile, dense, opens from the lowest upward, white or pale rose in colour, sweetly scented, torus very small, tepals 6, biseriate, glabrous, 1.5–4 mm, outer linear lanceolate, apex acute, inner ones small, apex rounded; stamens 6, half as long as the perianth, anther as long as filaments. Female inflorescence simple spicate, axillary, flower 4–7 mm long, tepals 6 biseriate, glabrous, 1.5–4 mm, outer tepals broader; ovary 3 ridged, 4–6 mm long, reflexed after flowering, stigmas as three pairs of recurved hooks. Capsules oblong to elliptic or oblong obovate, margin and axis darker, base rounded to retuse, apex obtuse, Seeds winged on the basal side only; wing oblong.

Flowering and Fruiting: July–December.

Distribution: The species is a native of the tropics of the old world and occurs in rain forests extending from the west coast of Africa to the islands in the Pacific. It is common throughout India and growing up to 6000 ft. In Southern Western Ghats, it is common in all the Districts and usually seen in openings of deciduous, semi-evergreen forests. Its habitat varied from marshy areas of the forest to low land and coastal area.

Notes: Tubers are used mostly as a famine food. They are very acrid and bitter. These tubers are used as a famine food for *Paniya* tribal community in the Wayanad District of Kerala by coursing with ashes and steeping in cold water. Bulbils are used as edible after processing and found in marketing in the Goa vegetable market. The tubers are used for the preparation of starch in Japan. *D. bulbifera* tubers are used in Kashmir for washing wool and as fish bait. Dried and pounded tubers are used as an application for ulcers, piles, dysentery and syphilis. Bulbils of this species are used as an application for sores. Paste of bulbils used externally as a remedy for abdominal pain. It is also used in dysentery, bone fracture and jaundice.

Status in South Western Ghats: Common.

6.3.1.11 *Dioscorea hamiltonii* Hook.f., Fl. Brit. India [J. D. Hooker] 6(18): 295 (1892). (Dioscoreaceae) (Fig. 6.2e)

Synonym: *Dioscorea persimilis* Prain & Burkill; *D. persimilis* var. *pubescens* C. T. Ting & M. C. Chang; *D. raishaensis* Hayata.

Ethnic name: Venni kilangu (*Kattunaika, Then kuruma, Cholanaika*), Noolvenni kilangu, Arikilangu, Kalu Venni.

Description: Large climbers. Stem twining to right, unarmed, greenish or pinkish, angular, 4–8 ribbed, glabrous. Bulbils present. Tuber 1 or 2, cylindrical, very deeply buried, long stalked, tender, 3–5 cm dia., 2–6 kg weight, surface smooth, rootlets small, thin, sparsely arranged, skin whitish or pale brown, flesh white, less mucilaginous. Leaves opposite at base, alternate towards the apex, simple, lamina cordate or often sub-sagittate at the base, apex acuminate, petiole slightly channelled above, glabrous, green, pinkish at both ends. Male inflorescence paniculate, axillary or terminal, spike 1–3 in a node, whorled on the axis, rachis markedly zig-zag with a flower at each angle; flowers sessile, dense, yellow, 16–38 per spike; buds globose ca. 1.8 mm long, tepals 6, biseriate, glabrous, inner tepals cuneate, obovate, apex rounded, shorter than outer tepals, concave; stamens 6, perfect. Female inflorescence simple, spicate, axillary, 1–2 from an axil, 15–25 cm long; tepals six biseriate, glabrous, 2–2.5 mm, outer tepal ovate acute, thick; stigmas-3 partite, branch sickle like. Capsules winged, thin, reniform, glabrous, with distinctly margined, angles triangular, yellow coloured. Seeds laterally winged or all around.

Flowering and Fruiting: September–December.

Distribution: Bangladesh, China South-Central, China Southeast, East Himalaya, Hainan, India, Myanmar, Nepal, Taiwan, Thailand, Vietnam. In Southern Western Ghats of India it is rarely seen in moist deciduous forests and semi-evergreen forests.

Status in South Western Ghats: Rare.

6.3.1.12 *Dioscorea hispida* Dennst., Schlüssel Hortus Malab. 15, 20, 33 (1818). (Dioscoreaceae) (Fig. 6.2f)

Synonym: *Dioscorea daemona* Roxb.; *D. daemona* var. *reticulata* Hook.f.; *D. hirsuta* Blume; *D. hispida* var. *neoscaphoides* Prain & Burkill; *D. hispida* var. *reticulata* (Hook.f.) Prain & Burkill; *D. hispida* var. *scaphoides* Prain & Burkill; *D. lunata* B.Heyne ex Roth; *D. mollissima* Blume; *D. virosa* Wall. ex Klotzsch & Garcke; *Helmia daemona* (Roxb.) Kunth; *H. hirsuta* (Blume) Kunth.

Ethnic name: *Podava kelengu Rheede; Bolkande* (Naika Tribe, Kasaragode); *Podava kizhangu* (Paniya and Kuruma tribes); *Podukkilangu* (Paniya tribe), *Venni Nangu* (Cholanaika tribe).

Description: Large climbers. Stem twining to the left, armed with green recurved prickles, pinkish on base, terete, glabrous. Bulbils are not produced normally. Tuber solitary, globose, lobed, occasionally slightly elongated, 10–50 cm dia., 4–35 kg weight, surface smooth, rootlets long, dense, skin straw coloured to light grey, flesh white to lemon yellow, very mucilaginous, very poisonous. Leaves alternate, 3-foliolate, very rarely simple leaves found in the leafy inflorescence axis, middle leaflet elliptic or elliptic oblong, rarely obovate or tri-partite, petiole 15–32 cm, longer than the middle leaflet, channelled above, prickly, glabrous, green; petiolule 1–2 cm. Male inflorescence simple, paniculate, pendulous, flowers sessile, dense, 50–60 per spikes, greenish, fragrant tepals 6, biseriate, glabrous, stamens 6, anthers subsessile; pistillode very small. Female inflorescence simple spicate, axillary or solitary, terminal, axis pubescent, 1–2 from an axil, pendulous, flower 0.5–1.5 cm long, spaced, tepals 6, biseriate, outer tepals, ovate lanceolate, pubescent; inner tepals slightly smaller; ovary 3–3.5 mm long, stigmas broad and short. Capsules oblong to elliptic or oblong obovate, the wings rounded at both ends, 2–5 × 1.5–2.5 cm, torus and tepals persisting. Seeds 6 from a capsule, winged on the basal side only, wing oblong, 2–3 × 0.5–1.3 cm, brown.

Flowering and Fruiting*:* May–September.

Distribution: Tropics of Asia from India through Malaysia, Philippine Islands to New Guinea. In the southern Western Ghats, the species is seen from sea level up to an altitude of 1200 m in semi-evergreen forests.

Status in South Western Ghats: Rare.

Notes: Poisonous, Edible for a few tribal groups after a series of processing.

6.3.1.13 *Dioscorea intermedia* Thwaites, Enum. Pl. Zeyl. [Thwaites] 326 (1864). (Dioscoreaceae) (Fig. 6.3a)

Synonym: *Dioscorea spicata* Hook.f.

Ethnic Name: *Eyyar.*

Description: Small climbers. Stem twining to the right, internodes 8–10 cm long, unarmed, green, 5–7 mm dia., terete, glabrous. Bulbils absent. Tuber two or three, cylindrical, ca. 4 cm dia., 30–50 cm long, 0.5–1 kg weight, surface smooth, rootlets small, sparse, skin whitish, flesh white, very mucilaginous. Leaves opposite rarely sub opposite, simple, lamina narrowly ovate or elliptic or oblanceolate, 9–13 × 3–5 cm, base acute, apex acuminate, acumen ca. 1 cm long, primary veins 3, all are reaching up to the apex, margin entire, glabrous, coriaceous, secondary nerves not prominent; petiole 0.5–1 cm, glabrous, green or pinkish. Male inflorescence simple, axillary paniculate spikes, 3–4 cm long, spike 1–3 in an axil, bracts and bracteoles ovate, apex acuminate, ca. 1 mm long, flowers sessile, dense, ca. 20 per spike, white or pale green in colour; tepals 6, biseriate, glabrous, ca. 1 mm, outer widely ovate, apex acute; inner ones small, narrower, obovate,

Fig. 6.3 (**a**) *Dioscorea intermedia* Thwaites. (**b**) *Dioscorea kalkapershadii* Prain & Burkill. (**c**) *Dioscorea oppositifolia* L. (**d**) *Dioscorea pentaphylla* L. (**e**) *Dioscorea pubera* Blume

apex rounded; stamen 6, broader at the base, free; pistillode small. Female inflorescence not obtained.

Flowering and Fruiting: October–January.

Distribution: Sri Lanka and Southern India. Semi-evergreen forests of south Western Ghats.

Status in South Western Ghats: Rare.

6.3.1.14 *Dioscorea kalkapershadii* Prain & Burkill, J. Proc. Asiat. Soc. Bengal 10: 24, hybr. (1914). (Dioscoreaceae) (Fig. 6.3b)

Ethnic Name: *Nara* (Kattunaika Tribe).

Description: Large climbers. Stem twining to the left, internodes 15–28 cm long, armed, spines hard, greyish, 1–2 cm dia., terete, glabrous. Bulbils axillary, sometimes present in the node of spike axis, solitary, globose, ovoid, spherical, varied in shape, 0.5–1 cm dia., rootlets absent. Tuber solitary, lobed, very variable in shape, rarely branched, 30–150 cm long, 3–15 kg weight, surface rough, rootlets long, thick, hard, dense, skin reddish or dark muddy brown, flesh reddish to orange yellow, sometimes marked with purple flecks, inner flesh with hard fibres, less mucilaginous. Leaves alternate, compound, 3–7 leaflets; lamina oblanceolate, petiole 10–20 cm, spinous, pubescent, green or brownish. Male inflorescence simple, paniculate, flowers sessile, dense, rusty red in colour, tepals and stamens crowded together, tepals 6, biseriate, glabrous, 1.5–4 mm, outer linear lanceolate, apex acute, inner ones small, apex rounded; stamens 3, anthers small, filament short; staminodes 3, alternating with fertile stamens; pistillode rudimentary and dome shaped. Female inflorescence simple spicate, axillary, flower ca. 60 per spike, tepals as in staminate flowers but slightly larger, fleshy, the outer tepals 1–2.3 × 0.5–1.4 mm; stigmas 2-partiate, recurved at apex. Capsules oblong, basally rounded, apically truncate to retuse, becoming glossy grey and glabrate at maturity; wings more or less elliptic, 6–7 mm broad; seeds winged, basely cuneate, wing 5–16 mm long dark brown.

Flowering and Fruiting: September–December.

Distribution: Sheveroy hills, in Southern Western Ghats reported from Wayanad, Kasaragode and a collection from Goa. Restricted in distribution.

Status in South Western Ghats: Rare.

6.3.1.15 *Dioscorea oppositifolia* L., Sp. Pl. 2: 1033 (1753). (Dioscoreaceae) (Fig. 6.3c)

Synonym: *Dioscorea opposita* Thunb.

Ethnic Name: *Kachil, Kanji, Kanjirakizhangu, Kavala, Narankizhangu, Noorankizhangu, Vettilavally, Vellakkizhangu.*

Description: Medium sized climbers. Tubers solitary or branched, cylindrical penetrating deeply in the soil; stem terete, twine to the right, purplish or green, glabrous. Leaves all opposite or sub opposite, elliptic to ovate, margin cartilaginous and hyaline, base acute or rounded or slightly cordate, apex acuminate or cuspidate, glabrous or rarely pubescent,3–5 nerved. Bulbils absent. Female inflorescence axillary, solitary or clustered, glabrous to pubescent; capsule twice or more as broad as long. Male inflorescences are usually on axillary spikes or on leafless branches, glabrous or pubescent; stamens 6. Female Flowers distant; tepals 1 mm long, orbicular, glabrous. Capsule 20 × 30 mm, glabrous.

Flowering and Fruiting*:* August–November.

Distribution: Indo-Malaysia and China. In south Western Ghats, moist deciduous, semi-evergreen and evergreen forests are the habitat.

Status in South Western Ghats: Common.

Note: In South Western Ghats, 4 distinct varieties of *Dioscorea oppositifolia* have been observed.

Key to the varieties of *Dioscorea oppositifolia*

1: Axis of male inflorescence pubescent	**2**
1: Axis of male inflorescence glabrous	**3**
2. Male flowers on leafless branches	var. ***linnaei***
2: Male flowers on axillary spikes	var. ***thwetesii***
3. Leaves ovate	var. ***dukhunensis***
3: Leaves acuminate or cuspidate	var. ***oppositifolia***

6.3.1.16 *Dioscorea pentaphylla* L., Sp. Pl. 2: 1032 (1753). (Dioscoreaceae) (Fig. 6.3d)

Synonym: *Botryosicyos pentaphyllus* (L.) Hochst.; *Dioscorea hangjiangensis* F. W. Xing & Z. X. Li; *D. codonopsidifolia* Kamik.; *D. digitata* Mill.; *D. globifera* R. Knuth; *D. jacquemontii* Hook.f.; *D. kleiniana* Kunth; *D. pentaphylla* var. *papuana* Burkill; *D. spinosa* Burm.; *D. sumbawensis* R. Knuth; *D. triphylla* L.; *Hamatris triphylla* (L.) Salisb.; *Ubium quadrifarium* J.F.Gmel.; *Ubium scandens* J. St.-Hil.

Ethnic Name: *Chaval; Henthikkorana; Kattunurunnakizhangu; Kattukizhangu; Kornapidan; Nurunnakizhangu; Noorakizhangu; Nooran; Nuranchaavl; Narakkizhangu; Mulluvalykizhangu; Vellachikizhangu.*

Common Name: Fiji yam; Kawan yam.

Description: Large climber Leaves alternate; petiole up to 17 cm long, deep rusty red or dirty white pubescence, petiolules rusty, up to 2 cm long; leaflets 3–5, the middle one longer than the rest; lateral leaflets unequal sided with one additional primary nerve outside the midrib, pubescent on both surface but soon glabrescent above, more pubescent and glandular beneath. Stem up to 1 cm dia. and up to 30 m long, usually prickly, sparingly hairy when young, glabrous when old. Bulbils many, globose, linear, cylindrical, or shortly ellipsoid, sometimes lobed, or irregular in shape, 1–9 cm long and 0.5–4 cm diameter, skin brown, flesh yellow. Tubers are usually simple, sometimes branched, variable in size and long. In some varieties elongated and burying deeply, in others not so and then globose or pyriform, in some cultigens palmately lobed, never stalked, generally coated if short by bristly roots, or if long with such roots on the upper parts; flesh white or lemon yellow, sometimes with purple flecks in it. Staminate inflorescence paniculate or racemose or both in the same or separate axils; panicles terminal or 1–5 per axil, 10–88 flowered; the axis densely pubescent. Male flowers 1–2 × 2–3 mm; tepals more or less erect but slightly incumbent at apices often scarious on margins, the outer tepals ovate or triangular, stamens free or inserted at the base of the outer tepals, the filaments ca. 0.2 mm long, the anther 0.2–0.4 mm long; staminodes up to 0.4 (0.8) mm long, free or inserted at the base of the inner tepals; pistillode 0.4–0.8 mm long, stumpy or conical. Pistillate inflorescence in axillary spike, raceme or panicles, pendulous, 1–4 per axil of a spike, raceme or panicles, the axis 5–48 cm long, densely pubescent tepals as in staminate flowers but slightly larger, fleshy, the outer tepals 1–2.3 × 0.5–1.4 mm; staminodes ca. 1 mm long, inserted at base of the tepals; ovary 2–8 × 0.7–5 mm, densely or often woolly pubescent with a short or indistinct neck, styles connate, 0.3–0.4 mm long, stigmas 2-partiate, recurved at apex. Capsules loosely imbricate, oblong, basally rounded, apically truncate to retuse, 2.5–3.8 × 1.5–1.8 cm, becoming glossy blackish and glabrate at maturity; wings more or less elliptic, 6–7 mm broad; stipe 3–4 mm long. Seeds winged, up to 20 × 5 mm including the wing, basally cuneate; wing 5–15 mm long, smoky brown.

Distribution: From upper India through Malaysia to the remoter islands of the Pacific. In India, found in degraded deciduous forests and waste places, also in sacred groves.

Status in South Western Ghats: Common.

Note: *Dioscorea pentaphylla* is the most confusing species among *Dioscorea* since its aerial morphological parts varied considerably in different agro-climatic and physiographic conditions. All the tribal community uses this species as edible food. "Noora" the folk name of this species, is commonly used by the tribal community in Kerala part of Western Ghats. It shows its distribution in all the agro-climatic conditions. A number of morphotypes or varieties can be seen in the wild as well as in the forest openings. The habitat of this species varies from the marshy stream side to the hilly rock crevices. Some of the varieties of *D.pentaphylla* grow in the wet area and even up to the dry deciduous forest and even in the rocky area. Locating both male and female plants of the same variety is a much more difficult task or not even identified so far. Mainly five varieties vis. *Dioscorea pentaphylla L. var. pentaphylla, D.pentaphylla var. communis, D.pentaphylla var. jacquemontii, D. pentpahylla var.rheedei and D.pentpahylla var.linnaei* reported during this study but a number of variations could be seen among these varieties.

This is a very variable plant. Some recent publication like flora of Kerala (Sasidharan 2013), denotes the varieties of *Dioscorea pentaphylla* were merged into variety *pentaphylla* proper. In this study, it is revealed that the varieties such as *Dioscorea pentaphylla var. pentaphylla, D. pentaphylla var. communis, D. pentaphylla var. jacquemontii, D. pentpahylla var. rheedei, D. pentpahylla var. linnaei, D. pentpahylla var.chenakorana* and *D. pentpahylla var. manalkorana* shows variations in tuber characters, bulbils, leaves and stem, etc. Moreover, ethnobotanical uses and season of tuber maturation show many variations from variety to variety. Hence the varieties are treated separately in this study.

Key to the varieties of *D. pentpahylla* in South Western Ghats

1: Tubers not elongated; rootlets abundant; male flowers with brown pubescence	**2**
1: Tubers elongated; rootlets scattered or absent; male flowers with white pubescence	**5**
2. Leaflets hairy, apex acute	var. ***communis***
2. Leaflets glabrous, apex abruptly cuspidate	**3**
3. Bulbils elongated	var. ***rheedei***
3. Bulbils not elongated, mostly spherical	**4**
4. Stem spinous; leaf lethery	var. ***jacquemontii***
4: Stem without spines if present on the base; leaf not leathery	var. ***chenakorana***
5. Leaves with few rusty red pubescence on both surface but soon glabrescent	var. ***pentaphylla***
5. Leaves silvery pubescent below	**6**
6. Tuber flesh white, starchy; stem dark brown, spinous	var. ***linnaei***
6. Tuber flesh yellowish, fibrous; stem greenish, without spines, sometime basely spinous	var. ***manalkorana***

6.3.1.17 *Dioscorea pubera* Blume, Enum. Pl. Javae 1: 21 (1827). (Dioscoreaceae) (Fig. 6.3e)

Synonym: *Dioscorea anguina* Roxb.; *D. combilium* Buch.-Ham. ex Wall.; *D. cornifolia* Kunth.

Ethnic name: Boojikavala (Kattunaika tribe of Wayanad District).

Description: Medium climbers. Stem twining to the right; internodes 10–15 cm long, unarmed, greenish, 0.5–1 cm dia., terete, softly pubescent. Bulbils axillary, solitary, ovoid, 1–3 cm across, skin thin, silvery green outside, light yellow inside, surface tuberculate; rootlets absent. Tuber one or two, columnar, stalked, 4–8 cm dia., 1–5 kg weight, 1–2 m long, grown deep into the soil, surface smooth, rootlets small, dense, skin tawny orange, flesh lemon yellow, fibrous in the upper part of the tuber, less mucilaginous. Leaves often alternate, rarely opposite, simple; lamina broadly ovate cordate or suborbicular, base deeply cordate, petiole 5–9 cm, as long as blade, usually livid in colour at the pulvini, pubescent, green. Male inflorescence simple, axillary or on special leafless branches, 1–2, tepals 6, biseriate, glabrous, outer tepals shortly ovate, much incurved, obtuse, thin, ca. 2 mm long, pubescent; inner tepal elliptic ovate; obtuse, almost glabrous, shorter than the outer tepal; stamens 6, perfect, ca. 1 mm long, adnate with tepals, filaments short; pistillode of 3 small points. Female inflorescence simple spicate, axillary, 1–3 from an axil, 10–16 cm long; bracts ovate acuminate, tepals six biseriate, glabrous, 0.5–0.7 mm, outer tepals broader, inner tepals shorter; staminodes minute, pubescent. Capsules apex retuse, base almost truncate, the wings rounded at both ends, ca. 15 × 18 mm, torus and tepals persisting. Seeds winged all around, orbicular, dark brown.

Flowering and Fruiting: August–December.

Distribution: South East Asia and Malaysia, Sumatra coast, Java. In Kerala the inhabited in marshy sandy soil in wet deciduous forest of Wayanad and Idukki.

Note: Tubers and bulbils used as a famine food after processing (Kattunayikka) and is also used as an oral contraceptive.

Status in south Western Ghats: Rare.

6.3.1.18 *Dioscorea spicata* Roth, Nov. Pl. Sp. 371 (1821). (Dioscoreaceae)

Ethnic Name: *Athikizhangu, Arathikizhzngu, Kavalakizhzngu (Kattunayikkar).*

Description: Small climbers. Stem twining to the right; internodes 4–6 cm long, small prickle at the base, green with mottling of pinkish or purplish, terete, glabrous. Bulbils absent. Tuber solitary, cylindrical, 5–7 cm dia., ca 1 m. long, 0.5–5 kg weight, surface smooth, rootlets small, dense, skin whitish or brown, flesh white. Leaves opposite to alternate, simple; lamina broadly elliptic or elliptic lanceolate, 5–8 × 3–4 cm, base acute, apex acuminate or cuspidate, acumen ca. 1 cm long, primary veins 3, all are reaching up to the apex, margin entire, glabrous, cartilaginous, secondary nerves prominent; petiole 3–5 cm, glabrous, green. Male inflorescence simple, axillary spikes, 5–15 cm long; spike 1–3 in an axil, bracts and bracteoles ovate, apex acuminate, ca. 1.5 mm long; flowers sessile, dense, 10–25 per spike, white or pale green in colour, tepals 6, biseriate, glabrous, ca. 1.5 mm, outer widely ovate, apex acute; inner ones small, obovate, apex rounded; stamens broader at the base, anther oblong; pistillode conical Female inflorescence not located.

Flowering and Fruiting: September–December.

Distribution: Sri Lankan tropical rain forests and southern India. In Southern Western Ghats, it is very rare and found about 1800 m above MSL.

Status in south Western Ghats: Rare.

6.3.1.19 *Dioscorea tomentosa* Spreng., Pl. Min. Cogn. Pug. 2: 92 (1815). (Dioscoreaceae)

Synonym: *Helmia tomentosa* (J.Koenig ex Spreng.) Kunth.

Ethnic Name: *Chavali, Inckkachil, Inthikachil, Nuli, Noolamkizhangu, Chavankizhangu, Pindi.*

Description: Large climbers. Stem twining to left, internodes 10–25 cm long, prickly; prickles straight or upcurved, greenish, 0.5–1 cm dia., 20–40 m long, terete, pubescent. Bulbils not produced. Tuber several, cylindrical rarely branched, dense, long roots on the upper part of the tuber, skin brown, flesh white or pale yellow, soft and fibrous, very mucilaginous. Leaves alternate, very variable, 5-foliate below, simple leaves near the end of the branches; margin entire, membraneous, secondary nerves prominent, scalariform; petiole 6–12 cm; petiolules 2–6 mm, upper surface more or less pubescent, lower surface densely white tomentose, greenish. Male inflorescence terminal or axillary racemes, 20–45 cm long; spike 1–4 in a node, alternate on the axis; bracts ovate to broadly ovate, placed just below the flower, ca. 1 mm long, densely pubescent outside, margins scarious; bracteoles similar but smaller; pubescent; inner tepals smaller, oblong ovate, rounded apically, less pubescent; stamens 3, anthers as long as the filament; staminodes 3, pistillode short, columnar. Female inflorescence simple spicate, axillary, 1–4 from an axil, pendulous, 5–50 cm long, tomentose; bracts ovate, acuminate, ca. 2 mm long; flower 40–60 per spike, tepals six biseriate, outer tepals pubescent, inner glabrous, fleshy; ovary ca. 6 mm long; stigmas as three pairs of short rays, staminodes 3. Capsules oblong to elliptic, base rounded, apex truncate to rounded, brownish to blackish; wings rounded at both ends, broadening upwards, 20–25 × 10–15 mm, pubescent. Seeds winged towards the base of the locules, 15–20 mm long.

Flowering and Fruiting: July–November.

Distribution: Native to Bangladesh, India and Sri Lanka. In India, South of Gengetic plains, Sri Lanka and in one locality in Bangladesh. The distribution is throughout in Southern Western Ghats, up to 1000 m., forest broader and open areas, usually seen along open bamboo patches in dry deciduous forest.

Status in South Western Ghats: Common.

Notes: No report of using this tuber as food from the study area but reports says that it is edible in north and also used as medicine against the poison of other tubers.

6.3.1.20 *Dioscorea wallichii* Hook.f., Fl. Brit. India [J. D. Hooker] 6(18): 295, in syn. (1892). (Dioscoreaceae)

Ethnic Name: *Kattukizhangu, Purakilangu, Narukizhangu, Vali, Vara-kilangu (Kattunayikkar).*

Description: Very large woody climbers. Stem twining to the right; internodes 10–15 cm long, armed at the base, greenish, 0.5–1.5 cm dia., terete, glabrous. Bulbils are absent. Tuber numerous, fascicled, cylindrical, 0.5–4 m. long, 4–9 cm dia., 2–8 kg weight, surface smooth, rootlets small, dense, skin brown, flesh white to lemon yellow, highly fibrous, mucilage absent. Leaves alternate, simple; lamina broadly cordate, secondary nerves prominent, scalariform; petiole 6–9 cm, channelled above, glabrous, green, pulvinate, lower pulvinus tinged with purplish red. Male inflorescence simple on the axillary spike or paniculate on a leafless branch, pyramid shape, flowers sessile, 20–36 in an axis, whitish or greenish in colour, tepals 6, biseriate, glabrous, outer tepals ovate, ca. 1 mm long, incurved, glabrous; inner tepals obovate, ca. 1 mm long, glabrous; Stamens 6, half as long as the perianth; pistillode rudimentary. Female inflorescence paniculate, axillary, flower 4–6 mm long, tepals 6 biseriate, glabrous, outer tepals broader, stigmas-3 partite and short sickle like organs. Capsules broadly obovate, apex rounded or truncate, base obtuse or rounded, whitish coloured when ripe, 18–20 × 14–17 mm, stipe ca. 4 mm long. Seeds winged all round, brown.

Flowering and Fruiting: October–January.

Distribution: India (Bombay coast to the Bengal plans, the lower Himalayas and Hills of Assam–Burma frontier) to Indo-China with the discontinuous distribution. In Southern Western Ghats area, it is reported from all the forest types but common in low lands and wet deciduous forests and open forests and reported from the entire District in the study area.

Note: The name 'Nara" is because of the presence of hard fibre in the flesh of the tuber. It is consumed by some of the tribal communities in the study area. The tuber is used as medicine after delivery by the tribal communities.

Status in Southern Western Ghats: Common.

6.3.1.21 *Dioscorea wightii* Hook.f., Fl. Brit. India [J. D. Hooker] 6(18): 291 (1892). (Dioscoreaceae) (Fig. 6.4a)

Description: Small climbers. Stem twining to the right; internodes 6–8 cm long, unarmed in the upper part and armed in the base, pale green, 1–1.5 cm dia., terete, glabrous. Bulbils absent. Tuber solitary, cylindrical, 15–30 cm dia., 0.5–2 m. long, 0.5–18 kg weight, surface smooth, rootlets small, dense, skin pale yellow; flesh white. Leaves opposite to alternate, simple; lamina ovate, 10–12 × 5–6 cm, base minutely cordate, apex acuminate, sinus vide; acumen ca. 2 cm long, primary veins

Fig. 6.4 (**a**) *Dioscorea wightii* Hook.f. (**b**) *Leea macrophylla* Roxb. ex Hornem. (**c**) *Nelumbo nucifera* Gaertn. (**d**) *Nymphaea nouchali* Burm.f

7, 3 reaching up to the apex, margin entire, glabrous, cartilaginous, secondary nerves not prominent; petiole 3–5 cm, pulvinus on both side, glabrous, purple to pinkish. Male inflorescence simple, axillary spikes, slender, 4–7 cm long; spike 1–3 in an axil, bracts and bracteoles ovate, apex acuminate, ca. 1 mm long; flowers sessile, 14–18 in an axis, white or pale green in colour; tepals 6, biseriate, glabrous, ca. 2 mm, outer widely ovate, apex acute; inner ones small, obovate, apex acute, reddish brown; stamens 6, longer than broad; pistillode small. Capsules margin and axis darker, oblong to ovate, straw coloured when ripe, the wings rounded at both ends, broader than long, ca. 1–2.5 cm long, torus and tepals not persisting. Seeds winged all around, wing rounded, pale brown.

Flowering and Fruiting: October–January.

Distribution: This species is imperfectly known in India. Distribution is from east coast, Courtallum of Tamilnadu, near the borders of Kerala state. In Kerala occurs only in certain areas, up to 900 m altitude.

Status in the south Western Ghats: Data Deficient (IUCN 2021); Critically Endangered (Sasidharan 2013).

6.3.1.22 *Leea macrophylla* Roxb. ex Hornem., Hort. Hafn. 1: 231. 1813 (Leeaceae) (Fig. 6.4b)

Synonym: *Leea angustifolia* M. A. Lawson; *L. cinerea* M. A. Lawson; *L. coriacea* M. A. Lawson; *L. diffusa* M. A. Lawson; *L. integrifolia* Roxb.; *L. latifolia* Wall.; *L. macrophylla* Roxb.; *L. pallida* Craib; *L. parallela* Wall.; *L. robusta* Roxb.; *L. simplicifolia* Griff.; *L. talbotii* King ex Talbot; *L. venkobarowii* Gamble.

Ethnic Name: *Njallu* (*Malampandarm* Tribe).

Common Name: *Hathikana.*

Description: Shrubs to 2 m high; young branches, rachises, petioles, petiolules and inflorescences hairy to mealy pubescent. Leaves alternate, 1 or 3-foliolate or 1–3 pinnate; if 1 or 3-foliolate; petioles to 20 cm long; rachises 10–15 cm long; stipules obovate, 2–6 × 1–4 cm; leaflets oblique, broadly ovate, oblong ovate, elliptic or rhomboid, 15–60 × 10–50 cm, cordate at base, acute or short acuminate at apex; if 1–3 pinnate leaflets 7–21 or more, oblong, ovate lanceolate or elliptic, 9–30 × 4–9 cm, rounded at base, serrate at margin, acuminate to caudate at apex, chartaceous to subcoriaceous, glabrous to sparsely hairy above, sparsely to densely hairy sometimes mealy pubescent beneath; lateral nerves to 14 pairs, pubescent to hairy; petiolules to 25 mm long. Inflorescences 12–45 cm long, much branched; peduncles to 25 cm long; bracts deltoid to narrowly 3-angled, to 6 mm long; pedicels 1–2 mm long, pubescent. Flowers greenish white. Calyx 1.5–3 × 2.5–4 mm, 5-lobed, mealy pubescent; lobes 3-angled, 0.75–1 × 0.8–1 mm. Corolla tube with staminal lobes 3–4 mm long; corolla lobes 5, linear ovate, thick, 2–4 × 0.8–1 mm, greyish pubescent to papillose. Staminal lobes slightly retuse or shallowly cleft; stamens free, in between staminal lobes; staminal column 2–3 mm long; filaments ca 1 mm long; anthers oblong, ca 1 mm long, medifixed. Ovary globose, 1–1.5 mm,

6-loculed; style to 2 mm long. Fruits subglobose or globose depressed, 10–15 mm across, green; seeds usually 6, 3-gonous, ca 4 × 3 mm.

Flowering and Fruiting: November–December.

Habitat and Distribution: South and southeast Asia; in Kerala, Margins of sholas of south Western Ghats.

Status in Southern Western Ghats: Rare.

6.3.1.23 *Nelumbo nucifera* Gaertn., Fruct. Sem. Pl. i. 73 (1788). (Nelumbonaceae) (Fig. 6.4c)

Synonym: *Nelumbium album* Bercht. & J. Presl; *N. asiaticum* Rich.; *N. caspicum* Fisch. ex DC.; *N. caspium* Eichw.; *N. discolor* Steud.; *N. indicum* Poir.; *N. javanicum* Poir.; *N. marginatum* Steud.; *N. nelumbo* (L.) Druce; *N. rheedii* C. Presl; *N. speciosum* Willd.; *N. tamara* (DC.) Sweet; *N. transversum* C. Presl; *N. turbinatum* Blanco; *N. venosum* C. Presl; *Nelumbo caspica* (Fisch.) Schipcz.; *N. indica* Pers.; *N. komarovii* Grossh.; *N. nelumbo* (L.) H. Karst.; *N. speciosa* G. Lawson; *N. speciosa* var. *alba* F. M. Bailey; *Tamara alba* Roxb. ex Steud.; *T. hemisphaerica* Buch.-Ham. ex Pritz.; *T. rubra* Roxb. ex Steud.; *Nymphaea nelumbo* L.

Common Name: *Aravindam;Thamara; Chenthamara;Venthamara* (Mal.); Sacred lotus; Indian lotus; Chinese water lily.

Description: Perennial aquatic rhizomatous herbs. Leaves alternate, orbicular, entire or wavy at margin, 20–90 cm across, glaucous on both surfaces, dark green above, flat when floating, concave or bowl shaped when emersed; petioles 60–180 cm long, weak, often minutely prickled. Flowers projecting above water level, 10–25 cm across, rose pink, white or white at base and pink at tips, rarely creamy yellow; peduncles to 180 cm long, minutely prickly. Sepals ovate or elliptic, concave, 2–5 × 1–3.5 cm, green when petals white or pinkish green when petals rose pink. Petals numerous, obovate or elliptic oblong, obtuse or subacute, concave, (2-) 8–10 × (1-)3–5 cm. Receptacle obconic, 9–11 cm high, 2–4 cm across at top, yellow. Stamens ca 200, 2–4.5 cm long; anthers yellow or orange; connective appendages recurved, white or yellow. Carpels many, oblong cylindric, ca 8 × 3 mm, sunk in the cavities of the spongy receptacle; stigmas are protruding from the receptacle. Fruiting carpels (nuts) ellipsoid, 1.5–2 × ca 1 cm; seeds ovoid oblong, black.

Flowering and Fruiting: August–January.

Habitat and Distribution: South and East Asia to far eastern Russia and to Australia; In Kerala, it is found in freshwater ponds and lakes.

Note: Tubers are used as a famine food and to prepare powder after processing for making various foods.

Status in Southern Western Ghats: Common.

6.3.1.24 *Nymphaea nouchali* Burm.f., Fl. Ind. (N. L. Burman) 120 (1768). (Nymphaeaceae) (Fig. 6.4d)

Synonym: *Nymphaea nouchali* var. *caerulea* (Savigny) Verdc.; *N. nouchali* var. *mutandaensis* Verdc.; *N. nouchali* var. *ovalifolia* (Conard) Verdc.; *N. nouchali* var. *petersiana* (Klotzsch) Verdc.; *N. nouchali* var. *versicolor* (Sims) R. Ansari & Jeeja; *N. nouchali* var. *zanzibariensis* (Casp.) Verdc.

Common Name: *Ampala; Ambel; Neerambel; Periambel; Poothali; Vellampel* (Mal.); Indian waterlily; Indian bllue waterlily.

Description: Rhizomatous aquatic herbs. Rhizomes often produce long runners. Leaves alternate, elliptic or orbicular, elliptic sagitate when young, entire or irregularly sinuate at margin, rounded acute at apex, (6-) 10–35(−45) × (5-) 8–28(−40) cm, reddish purple beneath; main nerves 7–15, palmate; midnerve grooved above, prominently angled beneath; secondary nerves 4–7 pairs, slightly grooved above; petioles terete, 2–5 mm thick, glabrous. Flowers 6–14 cm across, slightly fragrant. Sepals are lanceolate to oblong lanceolate, obtuse, 2.5–7.5 × 0.7–2.5 cm, green, streaked purple outside. Petals 8–15, elliptic lanceolate, 2.5–6 × 0.5–1.2 cm, mauve. Stamens 20–60, yellow; outer ones 1.5–2.5 cm long; the inner shorter; connective appendages 1.5–3 mm long; anthers 0.5–1.5 cm long. Ovary urceolate, sunken, 10–20 loculed; ovules numerous on superficial placentae; stigmas 8–20, radiating, connate at base. Fruits globose, 1.5–3.5 cm, with remnants of sepals, petals and stamens; seeds ellipsoid globose, 8–10 mm long, vertically fine lined, ciliate on ribs, becoming glabrate with the growth of aril.

Flowering and Fruiting: Throughout the year.

Habitat and Distribution: Indo-Malesia and Tropical Africa; Ponds and pools in plains.

Note: Tubers are used as a famine food and to prepare powder after processing for making various foods.

Status in Southern Western Ghats*:* Common.

6.4 Summary

Consumption of wild-edible tubers and rhizomes constitutes an integral part of tribal/traditional life. Based on the field investigation, wild relatives of 24 species are the main source of food in this category. Among them, eight species are endemic to the south Western Ghats. It is expected that the data generated from this study would provide a guideline for future domestication, breeding, conservation and bioprospecting scopes of these wild plants and lead to food and nutritional security.

Acknowledgements The authors are thankful to the Kerala Forest Department for permission to conduct research. The corresponding author is grateful to the Principal and Management of the Sanatana Dharma College, Alappuzha, Kerala for support and to Mohamed bin Zayed Species Conservation Fund (project 190521457); SERB (DST–INDIA) Project No. NPDF-PDF/2017/000135; DOEACC, Kerala (head no. 3435–04–104–98(P) –1) and NMPB (Project No. HG/KE–01/2019–2020) for financial assistance.

References

Ahmedullah M, Nayar M (1987) Endemic plants of the Indian region. I. Peninsular India. Botanical Survey of India, Calcutta

Bentham G, Hooker JD (1883) Genera plantarum, vol 1–3. L. Reeve & Co., London, (1862–1883)

Bridson DM, Forman L (1991) The herbarium handbook. Richmond, Royal Botanic Gardens, Kew

Chandrasekara A, Kumar TJ (2016) Roots and tuber crops as functional foods: a review on phytochemical constituents and their potential health benefits. Int J Food Sci 3631647:1–15

Food and Agricultural Organization of the United Nations (2010) Forests for improved nutrition and food security. FAO

Food and Agriculture Organization (1990): Roots, tubers, plantains and bananas in human nutrition, vol. 24 of Food and nutrition series, Food and Agriculture Organization, Rome

Food and Agriculture Organization of the United Nations (2003) Wild food plants: agrobiodiversity strategies to combat food insecurity and HIV/AIDS impact in rural Africa, Food and Agriculture Organization

Fosberg FR, Sachet M (1965) Manual for tropical herbaria (Reg. Veg. 39). IAPT, Utrecht

Hutchinson J (1926) The families of flowering plants, vol I. Macmillan & Co. Ltd., London

Hutchinson J (1934) The families of flowering plants, vol II. Macmillan & Co. Ltd., London

Hutchinson J (1956) The families of flowering plants, vol I and II, 2nd edn. Clarendon Press, Oxford

Hutchinson J (1973) The families of flowering plants, Monocotyledons, vol II, 3rd edn. Clarendon Press, Oxford

IPNI International plant names index (2021) Published on the internet: http://www.ipni.org

IUCN (2021) IUCN red list of threatened species. http://www.iucnredlist.org

Mathew J, Salim PM, Radhamany PM, George KV (2019) Araceae of Agasthyamala biosphere reserve, South Western Ghats, India. In: Ethnopharmacology and biodiversity of medicinal plants. Apple Academic Press, Palm Bay

Sasidharan N (2013) Flowering plants of Kerala (CD) 2.0. Kerala Forest Research Institute, Thrissur

Chapter 7
Medicinal Plants as Control for Prevalent and Infectious Diseases

Sarath Praseetha, Swapna Thacheril Sukumaran, Resmi Ravindran, and Shiburaj Sugathan

Abstract The greatest loss to the world's economy has resulted from the multidrug resistance (MDR) strains and other prevalent infectious diseases like leishmania, diarrheal diseases, malaria, tuberculosis, parasitic infections, pneumonia, and trypanosomiasis. Antibiotics have played the core role of wonder drugs in treating a variety of diseases. However, the irregular, unsuitable, and irrational uses of antibiotics have led to the emergence of antibiotic resistance. This has paved the way to an increased usage of medicinal plants as currently, many nutraceuticals and pharmaceuticals are plant derived. This chapter describes the huge repositories of secondary metabolites present in medicinal plants that may provide novel antibiotics to tackle all the infectious disease-causing pathogens and other prevalent diseases and reclaim the currently used antibiotics. The wide spectrum of phytochemicals in medicinal plants is to be explored as potential therapeutics for the sustainable use of plant resources. This chapter focuses on the diverse efficacy and effects of plant-derived multiple compounds for the development of bioactive therapeutics to identify novel antibiotics in controlling many infectious pathogens and finding a cure for the most common prevalent diseases. More research into unexplored plants is the need of the hour for global health benefits to identify novel antibiotics.

Keywords Bio-molecules · Antimicrobials · Anti-dengue · Anti-diarrheal · Biofilms · Leishmaniasis

---

S. Praseetha
Department of Biotechnology, University of Kerala, Thiruvananthapuram, Kerala, India

Microbiology Division, KSCSTE-Jawaharlal Nehru Tropical Botanic Garden and Research Institute, Thiruvananthapuram, Kerala, India

S. T. Sukumaran · S. Sugathan (✉)
Department of Botany, University of Kerala, Thiruvananthapuram, Kerala, India
e-mail: drshiburaj@karalauniversity.ac.in

R. Ravindran
Department of Pathology and Laboratory Medicine, University of California, Davis Medical Center, Sacramento, CA, USA

S. T. Sukumaran, K. T R (eds.), *Conservation and Sustainable Utilization of Bioresources*, Sustainable Development and Biodiversity 30,
https://doi.org/10.1007/978-981-19-5841-0_7

7.1 Introduction

Human beings have been dependent on nature for all their requirements from food, shelter, medicines, clothing, fragrances, fertilizers, and much more since time immemorial. For years, medicinal plants have continued to play a pivotal role in the healthcare system in large proportions of the world, mainly in the developing countries where herbal formulations have an ancient history of usage. The recognition of medicinal plants and bioactive compounds from them as therapeutic agents is emerging in both Industrialized and developing countries (Dar et al. 2017). For thousands of years, a traditional system of medicine formed from medicinal plants existed. Plants have offered many novel drugs to mankind and still are a potent source of new therapeutic agents or drug leads. The earliest reports of usage of plant material to cure diseases of about 2600 BC include oils of *Cedrus* species (Cedar), *Papaver somniferum* (Poppy juice), *Commiphora* species (Myrrh), *Cupressus sempervirens* (Cypress), and *Glycyrrhiza glabra* (Licorice) are still being used today to cure coughs, colds, inflammations, and parasitic infections.

Plants have contributed remarkably in diversified ways to various industries such as pharmaceuticals, agriculture, cosmetics, and industrial raw materials. A dynamic part has been played by medicinal plants in drug discovery. Medicinal plants have proved their influential role in eliminating many diseases like diabetes, atherosclerosis, malaria, cancer, AIDS, and Hepatitis. In the US drug market, approximately a hundred new plant-based drugs were released from 1950 to 1970, including vincristine, deseridine, reseinnamine, reserpine, and vinblastin. From 1971 to 1995, many other plant-derived drugs like ginkgolides, teniposide, artmisinin, lectinam, irinotecan, E-guggulsterone, toptecan, plaunotol, gomishin, ectoposide, paclitaxel, nabilone, and Zguggulsterone, appeared all around the world (Dar et al. 2017).

The strongest and most promising agents of plants are their secondary metabolites, on which we depend (Robinson and Zhang 2011). The food and Drug Administration (FDA) has approved many natural products and their derivatives as drugs (Chavan et al. 2018). Over the last two decades, many attempts have been made to discover novel therapeutics to combat microbes, MDR strains, and other infectious pathogens, especially from plant products. There are several relationships between matter and life, which are beautifully covered by natural products. Such interactions pave the way for using secondary metabolites and their derivatives for improvement in the health and pharmaceutical sector (Anand et al. 2019).

Due to the emergence of more Multi-Drug-Resistant (MDR) microbes and other infectious diseases and due to these pandemics caused by COVID-19, there is a warning signal for the pharmaceutical industry to provide new weapons in the form of biomolecules for the development of novel drugs to combat these problems (Newman and Cragg 2016; Luepke et al. 2017). A special branch of traditional medicines is herbal medicines dealing with the body and mind (Blair et al. 2015). The majority of the global population depends upon traditional medicines. There are lots of concerns being raised on the development of MDR and the side effects caused by drug intake on the public health, warning the medical researchers to return to the pre-antibiotic era (Atanasov et al. 2015). To date, no one particular antimicrobial agent has been successfully developed that could cure all bacterial infections.

Novel biomolecules from medicinal plants uninfluenced by these resistance mechanisms can be developed. To overcome or prolong the ongoing increasing resistance shown by various infectious agents, targeting resistance by new biomolecules in a novel method is the best way.

7.2 Plant Metabolites: Key Target Players

The secondary metabolites or small organic molecules are not essential for plants' normal growth or development yet have a powerful property of being used as antimicrobial agents and explored further to be formulated as drugs (Boy et al. 2018; Mawalagedera et al. 2019). Secondary metabolites are naturally occurring heterogeneous groups of compounds used for the treatment of various diseases (Moloney 2016). Traditional natural drugs have made an enormous contribution to the development of affordable medicines worldwide with their known biochemistry. Secondary metabolites are divided into three main categories, namely terpenes, alkaloids, and phenolics. These phytochemicals belong to phenolics, flavonoids, alkaloids, quinines, polyphenols, lectins, coumarins, terpenes, saponins polypeptides, etc. that play a prominent role as antimicrobials, anti-virulent agents, and for treatment of other diseases (Anand et al. 2019). Some of the pharmacological activities of main phytochemicals from medicinal plants are as follows:

7.2.1 Flavonoids

Flavonoids are aromatic compounds found in seeds, nuts, flowers, fruits, vegetables, and honey. All flavonoid compounds have a basic 2-phenylbenzopyrane or flavane nucleus, comprising two benzene rings linked by a heterocyclic pyrane ring. Preparations with these compounds have been used to treat many human diseases. Fowler et al. (2011) have suggested that the antimicrobial properties of flavonoids are due to their capacity to form complexes with bacterial membranes and both extracellular and soluble proteins. The compounds produced by some flavonoids by hydroxylating the prenyl groups of stipulin, like the bartericin A and angusticornin B, have shown profound antimicrobial activity (Kuete 2010). 6-hydroxy-7-methoxyluteolin and the xanthone 8-carboxymethyl-1,5,6-trihydroxy 3-methoxyxanthone, the two flavonoids purified from the leaves of *Leiothrix spiralis*, exhibited potent activity against *Escherichia coli* and *Pseudomonas aeruginosa* (De Freitas Araújo et al. 2011). Many synergistic activities of flavonoids and chemotherapeutics have been established (Cushnie and Lamb 2011). Quercetin, a flavonoid, has antibacterial action because it inhibits DNA gyrase. The mechanism of many flavonoids' action has been established, and Epigallocatechin gallate and sophoraflavone G inhibit the cytoplasmic membrane, whereas licochalcones A and C inhibit energy metabolism (Savoia 2012). Flavonoids have also shown their action in various stages of the cancer cell cycle system, thus, proving their involvement in anticancer mechanisms as well. Quercetin has shown antibacterial activity against a wide range of bacteria like *Bacillus subtilis, E. coli, Bacillus cereus, Staphylococcus aureus,* and *Pseudomonas fluorescens.* Apigenin was found to be active against *Streptococcus pyogenes, Streptococcus jaccalis, Vibrio cholera, E. coli, Klebsiella pneumoniae, B. subtilis,* and *Streptococcus viridans.* Kaempferol showed activity against *V. cholera and Enterococcus faecalis* (Ahmad et al. 2015b). Apart from antibacterial activity, many flavonoids like 3-O-β-D-galactoside, rutin, quercetin, and kaempferol have been reported for anti-cholinesterase activity for the treatment of Alzheimer's disease. Flavonoids are known for their antiviral activity as well. It shows action at various stages of a viral attack, right from viral entry to replication to translation and packaging. It exhibited activity against many viral diseases such as Herpes, AIDS, Hepatitis, and SARS to SARS-CoV-2, which has become a prevalent disease now and causing much chaos (Badshah et al. 2021). It was also found to be useful in the treatment of H1N1 viruses also (Panche et al. 2016).

7.2.2 Alkaloids

Alkaloids play a prime role in the defense against many pathogens. It has revealed antimicrobial properties and anti-HIV and antiparasitic activities (Ekpenyong et al. 2015). Alkaloids have a heterocyclic nitrogen group with various antimicrobial properties that include membrane disruption of microbes and are found to be a DNA intercalator, which targets DNA gyrase, topoisomerase IV, and RNA polymerase. Morphine was the first medically useful alkaloid isolated from *Papaver somniferum* (Kim et al. 2002). Berberine is an isoquinoline alkaloid widely used in traditional medicines. It exhibited a wide spectrum of activity against viruses, bacteria, fungi, and protozoa (Savoia 2012). Berberine was also found to be active against FtsZ protein, an important protein involved in bacterial cell division (Boberek et al. 2010). Most of the alkaloids act by inhibiting efflux pumps, as proved by using ethidium bromide (EtBr) as substrate. Lysergol, an ergoline alkaloid, causes inhibition of ATPases efflux pump (Cushnie and Lamb 2011). Carpaine isolated from the leaves and seeds of *Carica papaya* L., exhibited major antimicrobial activity. Harmine, a beta-carboline alkaloid, exhibited many pharmacological properties like antiplasmodial, antioxidant, antitumor, antimutagenic, cytotoxic, and antimicrobial action(Patel et al. 2012). There is an urgent need to develop antileishmanial agents against Leishmaniasis, one of the most neglected tropical diseases in the world. Pleiocarpine, purified from the stem bark of *Kopsia griffithii,* berberine, Isoguattouregidine isolated from barks of *Guatteria foliosa,* Anonaine from the trunk of *Annona spinescens* and Corynantheine and corynantheidine purified from the bark of *Corynanthe pachyceras,* all showed antileishmanial activities (Mishra et al. 2009).

7.2.3 Terpenes

These are isoprenoids, and their derivatives containing extra elements, usually oxygen, are terpenoids. Very potent antimicrobial activity of terpenoids' various types like monoterpenes (C10), diterpenoids, sesquiterpenes (C15), and triterpenoids have been reported (Kurek et al. 2011). The mechanism of action of terpenoids is thought to be because of its membrane disruption ability owing to its lipophilic character (Savoia 2012). Interaction studies of terpenes with bacterial cell membranes were done by evaluating intracellular K^+ leakage. Leakage of K^+ ions from the cells showed the antibacterial potency of terpenes. Carvone is an important terpene used in anti-infective therapy. It was found to be effective against *Campylobacter jejuni, E. faecium, E. coli, and Listeria monocytogenes* (Khameneh et al. 2019). Thymol alone and in combination with fluconazole exhibited antifungal activity against *Candida krusei, C. albicans,* and *C. glabrata* (Cheesman et al. 2012). Eugenol and cinnamaldehyde showed activity against a major human pathogen, *Helicobacter pylori.* These compounds inhibited the growth of *H. pylori*, without allowing the bacteria to develop resistance against it. Many terpenoids like reynosin, elatol, santamarine, elisapterosin B, aureol, totarol, debromolaurinterol, and costunolide exhibited activity against *M. tuberculosis* also. Their antimycobacterial activity is due to the high lipophilic nature of these terpenes that eased their way into the mycobacterial cell wall (Khameneh et al. 2019). Nerolidol, a sesquiterpene, exhibited antileishmanial activity by blocking the early stages of mevalonate pathways (Arruda et al. 2005). Different terpenes like nerolidol, linalool, and limonene were found to inhibit *P. falciparum.* They inhibited the dolichol biosynthetic pathway of both the trophozoite and schizont stages (Goulart et al. 2004).

7.2.4 Phenolics and Polyphenols

They are a huge group of aromatic compounds that includes flavonoids, flavones, and flavonols with one carbonyl group; quinones contain two carbonyl groups; coumarins are phenolic substances with fused pyrone and benzene groups; and tannins are polymeric phenolic compounds (Fowler et al. 2011). They protect plants from microbial infections and find a variety of applications as antibacterial agents, antioxidative agents, and anti-infectives. They have the ability to destroy bacterial cell walls (Cazarolli et al. 2008). Flavin and epicatechin have shown synergistic activity against many nosocomial bacterial pathogens (Betts et al. 2011). Anthraquinones, in turn, cause the inactivation of bacterial proteins like cell wall proteins, adhesins, and membrane proteins, thereby having a greater spectrum of antibacterial properties (Kurek et al. 2011). Gallotannin has a strong affection for iron and destroys membrane-bound proteins of bacteria (Savoia 2012). Proanthocyanidins, a condensed tannin, obtained from flavanols, inhibit uropathogenic *E. coli*, by an anti-peroxidation mechanism (Cimolai and Cimolai 2007). Blackberry juice is a rich source of anthocyanins and has been used to treat infections of the eyes and mouth for decades (Savoia 2012). Essential oils are rich in monoterpenes and sesquiterpenes and have shown various activities like antifungal, antibacterial, antioxidant, antibiofilm, insecticidal, and antiviral. Owing to its medicinal properties, essential oil from Cinnamomum (Lauraceae), such as *C. zeylanicum and C. cassia*, is used in traditional medicines as antidiabetic, diuretic, astringent, antiseptic, digestive, stimulant, tonic, vasodilator, bronchitis, anti-diarrheal, gastric ulcers, and skin infections. They also demonstrated antifungal activity against *C. albicans* (Firmino et al. 2018). P-caumaric acid and gentisic acid are proved to cure cutaneous leishmaniasis (Monzote et al. 2016).

7.3 Mechanism of Phytochemicals' Action

The heart of natural science is Ethnobotany, which connects plants with human beings. Pharmaceutical companies are exploring the existing knowledge of medicinal plants for the development of new drugs for the betterment of the health sector. The knowledge of ethnobotany is crucial to identifying new biologically and chemically active natural compounds from plant species (Garnatje et al. 2017). Table 7.1 gives an insight into the medicines developed from different plants.

Synthetic chemical drugs and natural bioactive compounds differ in their configuration of different radicals (Szychowski et al. 2014). The naturally occurring compounds exhibited enhanced scaffold, molecular complexity, diversity in a ring structure, and have less nitrogen, halogens, and phosphorous and carbohydrate contents (Anand et al. 2019). Plant acts as powerful modulators of immune response, signal transduction, and apoptosis by inhibiting or modifying protein–protein interactions (Vadhana et al. 2015). The biological potential of traditional medicines clearly shows how it kills and stops the advance of pathogens and diseases. Due to the multiple bioactive phytochemicals present in plant extracts, developing resistance will be difficult for the pathogens. Thus, with the traditional techniques and methodologies, new researches focused on to isolate and purify bioactive compounds from plant extracts. Secondary metabolites affect the microbial cells and infectious parasites in many ways:

1. Disruption of cell structure.
2. Interference with cell membrane functions and metabolism (Chitemerere and Mukanganyama 2014).
3. Interference with DNA/RNA synthesis (Zhao and Chuncheng 2020).
4. Interruption in cell communication (Radulovic et al. 2013).
5. Disruption of cell architecture by the destruction of cytoskeletons in the case of parasites (Wink 2012).

Table 7.1 Plant-derived natural compounds are used as drugs for the treatment of prevalent diseases

Sl. no.	Plant-derived drugs/ molecules	Plant species	Disease/mechanism of action	References
1.	Colchicine	*Colchicum autumnale* L.	Antileukemic activity, Gout treatment (inhibits tubulin polymerization)	Anand et al. (2019)
2.	Artemisinin	*Artemisia annua* L.	Treatment of Malaria (generate free radicals that attach proteins)	Itokawa et al. (2008)
3.	Arglabin	*Artemisia glabella* Kar. & Kir.	Cancer chemotherapy (inhibition of farnesyl transferase)	Atanasov et al. (2015)
4.	Digoxin and digitoxin	*Digitalis purpurea L.*	Anticancer agent (inhibits cell proliferation)	Anand et al. (2019)
5.	Cannabidiol	*Cannabis sativa L.*	Chronic neuropathic pain (CB1 and CB2 receptor activation)	Anand et al. (2019); Atanasov et al. (2015)
6.	Berberine	*Coptis chinensis* Franch.	Treatment of various diseases like diarrhea, gastrointestinal infections, anticancer (activates AMP-activated protein kinase (AMPK)	Qiu et al. (2014)
7	Salbutamol and Salmetrol	*Ephedra sinica* Stapf.	Anti-asthmatic (stimulate beta-2 receptors in the bronchial musculature)	Cragg and Newman (2013); Ullman et al. (1990)
8	Capsaicin	*Capsicum annum* L., or *C. minimum* Mill.	Postherpetic neuralgia (TRPV1 activator)	Atanasov et al. (2015)
9	Papaverine	*Papaver somniferum L.*	Relaxation of smooth muscles	Anand et al. (2019)
10	Aspirin	*Filipendula ulmaria* (L.) Maxim	Inhibits cyclooxygenase (COX)	Anand et al. (2019); Vane and Botting (2003)
11	Vinblastine and vincristine	*Catharanthus roseus* (L.) G. Don	Treat Hodgkin's lymphoma and acute childhood lymphoblastic leukemia (targets tubulin and inhibits mitosis)	Anand et al. (2019); Itokawa et al. (2008)
12	Zosteric acid	*Zostera marina* L.	Anti-dengue	Abd Kadir et al. (2013)
13	Quinine	*Cinchona officinalis L.*	Antimalarial (inhibits nucleic acid and protein synthesis)	Cragg and Newman (2013)
14	Galanthamine	*Galanthus caucasicus* (Baker) Grossh	Alzheimer's disease (ligand of human nicotinic acetylcholine receptors)	Atanasov et al. (2015)

(continued)

Table 7.1 (continued)

Sl. no.	Plant-derived drugs/ molecules	Plant species	Disease/mechanism of action	References
15	Verapami	*Papaver somniferum L.*	Hypertension (calcium channel blocker)	Cragg and Newman (2013)
16	Allicin (diallylthiosulfonate)	*Allium sativum* L.	Antioxidants	Capasso (2013)
17	Fucoidan	*Cladosiphon okamuranus*	Anti-dengue	Abd Kadir et al. (2013)
18	Reserpine	*Rauwolfia serpentine* Benth.	Hypertension (inhibits the uptake of norepinephrine)	Cragg and Newman (2013)
19	Camptothecin.	*Camptotheca acuminate* Decne	Treatment of Tumors (prevent DNA re-legation)	Anand et al. (2019)
20	Paclitaxel	*Taxus brevifolia* Nutt.	Cancer treatment, (antimitotic agent)	Itokawa et al. (2008)
21	Chromolyn	Ammi visnaga (L) Lamk.	Bronchodilator (prevents mast cells from releasing histamines)	Cragg and Newman (2013)
22	Metformin	Galega officinalis L.	Antidiabetic (decreases hepatic glucose production, decreases intestinal absorption of glucose)	Cragg and Newman (2013)

Bacterial cell membranes are a potential target for the development of new antibacterial drugs. Plants produce secondary metabolites containing steroids, triterpenoids aglycon attached to sugar chains that exhibit cell membrane permeabilization effects. (Chitemerere and Mukanganyama 2014). There is a direct relationship between permeability and enhancement activity; permeability enhancers decrease cellular resistance and allow drug passages.

Many plant-derived compounds that damage DNA have proved to be antiparasitic. These are usually DNA-alkylating compounds that form covalent bonds with DNA, producing mutations in essential protein-coding genes and, thus, causing the death of parasites. Aristolochic acid from Aristolochia, furanocoumarins from Fabaceae, and cycasin from Cycadaceae are some of the DNA-alkylating compounds (Wink 2012). The major proteins of the cytoskeleton are the actin filaments and the microtubules, which are responsible for the cell architecture in eukaryotes. These microtubules are essential for the assembly of the mitotic spindle for cell division. A number of natural products that inhibits the polymerization of tubulin have been reported. They are vinblastine from *Catharanthus roseus,* colchicine from *Colchicum* spp., sanguinarine from *Sanguinaria canadensis*, and chalcones from *Combretum caffrum* and podophyllotoxin from *Podophyllum* Spp. (Stanton et al. 2011).

7.4 New Therapeutic Strategies for the Development of Drugs

7.4.1 *Antimicrobial Activity*

The new antibacterial therapeutics are designed in novel ways, which do not cause bacteria to develop resistance to the drug. We here are discussing new therapeutic strategies for the development of drugs:

7.4.1.1 Efflux Pump Inhibitory Activity

Apart from possessing acquired resistance, many of the microorganisms now have developed increased resistance resulting from mutations that alter the intrinsic expression of genes and, thus, contribute to resistance to antibiotics (Nikaido and Pagès 2012). One of the major causes of intrinsic resistance in gram-negative bacteria is efflux pumps. Thus, an alternative approach to many diseases could be based on the molecules that interfere with the process of efflux. It was proved that biomolecules from plants are active against gram-positive bacteria by acting as effective efflux pump inhibitors (Holler et al. 2012). Gram-negative bacteria owing to its efflux pumps; particularly, the AcrAB-TolC efflux system shows multidrug resistance to many antimicrobials (Piddock et al. 2006). Many plant alkaloids, flavones, isoflavones, and berberine have been shown to interfere with the efflux pumps (Stavri et al. 2007). The accumulation of berberine was increased in the cells in the presence of a multidrug pump inhibitor, 5′-methoxyhydnocarpin, present in chaulmoogra oil extracted from *Hydnocarpus wightianus*. The presence of some efflux pumps like MexAB-OprM and AcrAB-TolC intricates the resistance development in gram-negative bacteria to natural products. Gram-negative efflux pump inhibitors are present in the extracts of several herbal plants (Savoia 2012). Falcarindol from *Levisticum officinale* showed synergistic activity in combination with ciprofloxacin. A synergistic effect of *Dichrostachys glomerata* extracts with some antibiotics was noted (Lacmata et al. 2012). Plant-derived alkaloids, which have already proved themselves as efflux pump inhibitors, are now proving themselves as a potent synergistic agent against anti-TB drugs. Such alkaloids are: reserpine, piperine, and berberine, all derived from plants (Pule et al. 2016).

7.4.1.2 Plant Extracts with Bacterial Quorum Sensing (QS) Inhibitory Activity

Both gram-positive and gram-negative bacterial species are communicated by producing diffusible molecules to perform diverse functions like swarming, toxin production, fluorescence, and biofilm formation. As many pathogenic bacteria are using the QS system to regulate their virulence, this makes them an antimicrobial therapy target. Thus, QS inhibitors are novel therapeutics that interrupt the bacterial communication system and are new therapeutic prospects (Savoia 2012). The basic strategy adopted for QS inhibitors is the receptor antagonistic approach. QS inhibitors should ideally be low-molecular mass molecules that can reduce the expression of QS-controlling genes without exerting toxic side effects on bacteria and host (Yarmolinsky et al. 2015). Many extracts from herbs, fruits, and spices and their bioactive compounds have shown QS inhibitory activities. Natural compounds, a group of halogenated furanones, isolated from the marine red algae, *Delisea pulchra*, showed bacterial QS inhibition activity. The mechanism of action was found to be due to the inactivation of an enzyme, namely, LuxS (S-ribosylhomocysteine lyase), which produces autoinducers-2, which are responsible for interspecies QS communications (Vikram et al. 2010). Rasmussen et al. have identified many plant-derived compounds from chamomile, garlic, yellow pepper, bean sprouts, water lily, and habanero with anti-QS activity against *P. aeruginosa* (Rasmussen et al. 2005). It was further confirmed through GeneChip analysis that garlic extract reduced the biofilm tolerance of *P. aeruginosa* to tobramycin and reduced its pathogenicity in *Caenorhabditis elegans*. An antifungal agent, resveratrol (3,5,4′-trihydroxystilbene), found in many plants and grapes, inhibited QS of *P. aeruginosa* in vitro. Many plant extracts have shown inhibition of LasA protease and pyoverdine, QS-controlling virulence factors in *P. aeruginosa*. Hamamelitannin (2′,5-di-O-galloyl-d-hamamel-ose), is a compound isolated from the bark of *Hamamelis virginiana*, was found to inhibit QS virulence factors like δ hemolysisn and RNA III in *Staphylococcus epidermidis* and *S. aureus* (Savoia 2012). Iberin isolated from *Armoracia rusticana* was found to downregulate rhlR gene of *P. aeruginosa* (Jakobsen et al. 2012). Kaempferol isolated from *Centella asiatica* was found to inhibit violacein production in Chromobacterium violaceum (Vasavi et al. 2014). L-canvanine the bioactive compound present in *Medigo sativa* inhibited violacein production in *Sinochrizobium melioti* (Keshavan et al. 2005). Thus, it can be summarized that the plant extracts and the phytochemicals isolated from them inhibit QS using the following mechanisms: (Yarmolinsky et al. 2015)

(a) Enzymatic inactivation of QS signals.
(b) Inducer analogs.
(c) Inhibition of the production of QS signals.
(d) Inhibition of inducer receptors.

7.4.2 Plant Extracts with Biofilm Inhibitory Activity

Biofilms formation is a default mode of life for many microorganisms, and its eradication with classic antibiotic therapies is difficult. Two natural compounds, proAc (proanthocyanidin A2-phosphatidylcholine) isolated from *A. hippocastanum* and CH (chelerythrine) isolated from *Macleya cordata*, showed de novo inhibition of biofilm formation by Staphylococci, without any bactericidal activity (Artini et al. 2012). proAc inhibits the iron-binding protein, blocking the transition from planktonic to the sessile state of bacteria and blocking autolysin (penicillin-binding protein), thereby inhibiting biofilm formation. Both sanguinarine and CH are found to act on proteins of heat shock response, methoxy-mycolic acid synthase, and surface-exposed lipids, thus, inhibiting protein synthesis. Carvacrol, a natural monoterpenic phenol, was found to inhibit biofilm formation by *S. aureus* and *Salmonella enterica* serovar *typhimurium*. These molecules prevented the proteins involved in the developmental stages of biofilm formation and, thus, arresting the microcolony stage in the biofilm development (Savoia 2012). Carvacrol, together with thymol, makes structural and functional changes in the cytoplasmic membranes of bacteria, thus, making them lose their integrity. These compounds are also found to destabilize the polysaccharide matrix of biofilm by diffusing through it.

A polyphenolic extract isolated from the bark of *Hamamelis virginiana* is Hamamelitannin. It was found to reduce biofilm formation in various microbes (Cobrado et al. 2012). 1-deoxynoijirimycin isolated from *Morus alba* reduced bacterial extracellular polysaccharide secretions in *S. mutans*, the major causative agent of dental caries (Islam et al. 2008). 5- Dodecanolide (DD) is a natural aromatic lactone found in a few fruits. It is found to inhibit biofilm formation and alter the virulent gene expression of *S. aureus* and *S. pyogenes* (Valliammai et al. 2020). 3Furancarboxaldehyde (3FCA), a honey-derived compound from a flower, has shown antibiofilm activity against *S. pyogenes*. It showed downregulated covR gene, which is a main target for antibiofilm activity (Subramenium et al. 2015). A naturally available compound found in many plant species and mostly in the bark of birch trees is a triterpenoid. Betulin (Lup-20(29)-ene-3-3β,28-diol) showed antivirulence properties by suppressing ropB core regulon, sagA, and dltA gene expression and, thus, inhibited the biofilm formation capacity of *S. pyogenes* (Viszwapriya et al. 2016). A biflavonoid, Fukugicide, isolated from leaves of *Garcinia travancoria* exhibited a concentration-dependent biofilm inhibition against many serotypes of *S. pyogenes*. Differential regulation of genes like dltA, ropB, srv, speB showed that it lowered cell surface hydrophobicity, thereby destabilizing the biofilm matrix and, thus, inhibiting biofilm formation in *S. pyogenes*. Many virulences encoded genes like slo, hasA, and col370 were found to be downregulated, suggesting diminished virulence (Nandu et al. 2018).

7.4.3 Antiparasitic Activity

New aspects of designing antiparasitic therapeutics are by understanding the target of the drugs and by understanding the mechanism of their action to avoid possible resistant development.

1. Resisting oxidative stress: A key factor for the survival of parasites within the host is their ability to detoxify the reactive oxygen intermediates. Thus, targeting their ability to resist oxidative stress is a suitable way. One of the potential targets to be exploited for drug development is trypanothione reductase (Renslo and McKerrow 2006).

2. Developing Protease Inhibitors: Proteases are validated druggable targets and hold good as potential targets for developing a drug. Cysteine and aspartyl proteases are the two closely related enzymes used by the parasitic organism to carry out their functions in vertebrates. Cysteine proteases play a significant role in the proteolysis of extracellular parasites. *Schistoma mansoni* and *P. falciparum* use a network of Cysteine and aspartyl proteases for destroying host proteins. This means that drugs that act as inhibitors of these classes of enzymes will inhibit replication in the parasite (Renslo and McKerrow 2006).
3. Targeting parasitic carbohydrate metabolism: African trypanosomatids, especially the bloodstream forms, completely rely on glycolysis for ATP synthesis. A parasite epimerase has been identified by the University of Dundee group that converts glucose to galactose. This is essential because parasites have no means of hunting for galactose from the surrounding. Thus, targeting this epimerase would provide an interesting and potential target for the new therapy (Verlinde et al. 2001).

7.4.4 Medicinal Plants with Anti-infectious Activity

A very valuable source of secondary metabolites is plants, finding their application in many sectors, especially as pharmaceuticals, biopesticides, flavors, colors, agrochemicals, fragrances, and food additives. 75% of people around the world are using plants for therapeutic purposes. In the USA, 25% of pharmaceuticals are based on plant-derived compounds (Al-Snafi 2015). Medicinal plants have an enormous amount of properties to be used for the treatment of many infectious diseases:

7.4.5 Medicinal Plants with Antimalarial Activity

Malaria is one of the leading causes of death in developing countries. As the malarial parasites are showing resistance to the existing antimalarial drugs, the focus is now more on natural products from medicinal plants (Basco et al. 1994). Some plants with antimalarial activity are being reviewed below:

7.4.5.1 *Cryptolepis sanguinolenta* (Lindl.) Schlechter

Cryptolepis sanguinolenta has much ethnomedicinal importance and is traditionally used for the treatment of malaria, diarrhea, upper respiratory tract infections, and wounds. A tea bag formulation of the root of this plant has proved to clear the *Plasmodium falciparum* parasitemia from 50% of patients with uncomplicated malaria within 72 h (Bugyei et al. 2011).

7.4.5.2 Terminalia ivorensis A. Chev.

Various parts of this plant are used traditionally to treat malaria, stomach ulcers, wounds, and yellow fever. An in vitro antiplasmodial activity against P. falciparum chloroquine-sensitive (3D7) and resistant (W2) strains from the aqueous extract of this plant was proved by Komlaga et al. (Komlaga et al. 2016).

7.4.5.3 *Syzygium aromaticum* (L)

Commonly known as clove, apart from being a flavoring agent, it has many medicinal properties like antiasthma, antiparasitic, anthelmintic, antiviral, and anti-inflammatory (Mittal et al. 2014). The methanolic extract from the flower buds of this plant showed activity against *P. falciparum* CQ-sensitive (3D7) and CQ-resistant (Dd2 and INDO) strains (Bagavan et al. 2011).

7.4.6 Medicinal Plants with Activity Against Dengue

Dengue is caused by an arthropod-borne flavivirus named dengue virus (DENV), which is transmitted by Aedes aegypti mosquitoes. World Health Organization (WHO) states that 80% of people in Asian and African countries use traditional medicines for their primary health care (Abd Kadir et al. 2013). Following are a few plant species that are used for the treatment of Dengue:

7.4.6.1 Boesenbergia rotunda

It belongs to the family Zingiberaceae. Some compounds isolated from this plant showed the inhibition of dengue virus proteases. Panduratin A derived from *B. rotunda* proved to inhibit DENV-2 NS3 protease (Abd Kadir et al. 2013).

7.4.6.2 Euphorbia hirta

Water decoction of leaves from this plant is used to treat dengue fever traditionally. It was found to stop internal hemorrhaging and reduce fever. The tea made from the boiled leaves of *E. hirta* is used to cure dengue fever as well (Abd Kadir et al. 2013).

7.4.6.3 Psidium guajava

The leaf extract of *P. guajava* showed potent inhibitory activity against the dengue virus. Water boiled with leaves of guava was found to increase the platelet count and reduce bleeding in Dengue hemorrhagic fever. Also, the ripe fruit and its juice relieved dengue fever by reducing the platelet levels (Pink Roses 2011).

Apart from the above-mentioned plants, many have shown potent activity against the dengue virus-like *Alternanthera philoxeroides*, *Cladosiphon okamuranus*, *Rhizophora apiculate*, *Lippia citriodora*, *Leucaena leucocephala*, *Quercus lusitanica*, and *Cladogynos orientalis* (Abd Kadir et al. 2013).

7.4.6.4 *Allium sativum* (Garlic)

Garlic extract has been reported to show a broad spectrum of antibacterial activity against both gram-negative and gram-positive microorganisms. Its action is not only limited to bacteria, its spectrum of action is also reported to be against viruses, fungus, and protozoans, but it also has useful impacts on immune systems. It is also used to treat various types of cancers like stomach cancer, lung cancer, colon cancer, rectal cancer, and breast cancer. It proved to be useful in the treatment of various cardiovascular diseases showing potent antihypertensive and antilipemic activity. Apart from the above beneficiaries, *Allium* is proved to be useful in the treatment of traveler's diarrhea as well (Rivlin et al. 2006) .

7.4.7 Antimicrobial Activity

The aqueous and alcoholic extract, as well as its essential oil, inhibited the growth of *Clostridium*, *Staphylococcus aureus*, *Mycobacterium tuberculosis*, *Shigella sonnei*, *Streptococcus faecalis*, *Escherichia coli*, *Bacillus* sp., *Pseudomonas aeruginosa*, *Proteus* sp., *Candida* sp., *Aspergillus niger*, *Cryptococcus* sp., *and Trichosporon pullulans* (Al-Snafi 2015; Harris et al. 2001). It also showed a miraculous cure rate for tuberculosis as well. Almost 30 strains of Mycobacterium were inhibited by lower concentrations of garlic extract. Garlic extracts can also arrest the production of different types of enterotoxin produced by *Staphylococcus* (Al-Snafi 2015). It was also found to be effective against Helicobacter pylori (Cellini et al. 1996). The bioactive compound allicin purified from garlic extract has shown many antibacterial activities by inhibiting bacterial enzymes such as phosphotransacetyl-CoA synthetase acetate kinase. Allicin also inhibited protein, DNA, and RNA synthesis mechanisms (Al-Snafi 2015).

7.4.8 Antiviral Activity

The flavonoid (quercetin) and organosulfur (allicin) components present in garlic are responsible for the immunomodulatory effects of garlic. The various organosulfur compounds showed activity against a wide range of viruses like para-influenza, herpes simplex virus, vaccinia, and influenza (Khorshed Alam et al. 2016), and recently, it showed activity against COVID-19 virus also. The bio-actives of garlic formed hydrogen bonds with the serine proteases of COVID-19 and thus inhibited COVID-19 virus (Khubber et al. 2020).

7.4.9 Anthelminthic and Antiparasitic Activity

Worldwide, garlic extract is used for the treatment of helminthes infections and intestinal disorders. It possessed antiparasitic activity against human intestinal parasites like *Giardia lamblia*, *Entamoeba histolytica*, and *Ascaris lumbricoides* (Khorshed Alam et al. 2016).

7.4.9.1 *Achillea santolina*

They have a wide range of applications as antispasmodic, anti-inflammatory, antimicrobial, insecticidal, insect repellent, antidiabetic, antiulcer, antitumor, and used in the treatment of pneumonia, hemorrhage, and for wound healings (Saeidnia et al. 2011; Al-Snafi 2013). It showed potent activity with low MIC against *Staphylococcus aureus*, *Candida albicans,* and *Pseudomonas aeruginosa*. The stem and leaf extract showed more activity than the flower extract (Saeidnia et al.). The essential oil showed the presence of 54 components, the major being fragranyl acetate, 1,8-cineole, terpin-4-ol, and fragranol (Al-Snafi 2013).

7.4.9.2 *Cymbopogon citratus* (DC.) Stapf. (Lemongrass)

The infusions and decoctions of this plant are used for the treatment of many diseases like antihypertensive, stomachic, and antispasmodic. It also has antibacterial activity against *Streptococcus agalactiae, Bacillus cereus, and Staphylococcus aureus and Escherichia coli,* and *Salmonella enterica* (Ortega-Ramirez et al. 2014). The major essential compound was citral and geraniol, which was found to inhibit many bacteria. It also showed antibiofilm activity against S. aureus (Aiemsaard et al. 2011). The essential oil from this plant effectively inhibited *P. berghei*. It also exhibited antiviral activity against the herpes simplex virus and HIV (Ekpenyong et al. 2015). The essential oil was also shown to be a promising agent in the treatment of *Trypanosoma cruzi* (Santoro et al. 2007).

7.4.9.3 *Ailanthus altissima* (Mill.)

Both water and methanolic extracts from the leaves showed efficient activity against gram-positive bacteria. Stigmasterols isolated from the fruits of this plant showed potent activity against many bacteria (Al-Snafi 2015). Quassinoids isolated from *Ailanthus altissima* showed the anti-tuberculosis property. (De Martino and De Feo 2008). The ethanolic extract of the root bark of this plant inhibited the replication of the tobacco mosaic virus (Tamura et al. 2006). A Quassinoid, namely ailanthone, isolated from the extract of this plant, was found to be active against the various strains of chloroquine-resistant and chloroquine-sensitive *P. falciparum* (Dhanapal and Ratna 2012).

7.4.9.4 *Asparagus officinalis* L.

The herb Asparagus is known as "Shatavari" or "Queen of herbs" (De Martino and De Feo 2008). Desoukey et al. (2020) showed the antibacterial activity of the ethanolic extract of this plant against *B. subtilis* and *Ralstonia solanacearum.* The antibacterial potential was also shown against *E. coli* and *P. aeruginosa, S. aureus,* and *B. cereus.* Asparagus is rich in polyphenols. The aqueous extracts of the plant leaves showed antioxidant and tyrosinase inhibition activity. With increasing concentration of asparagus aqueous extract above 1.21 mg/mL showed inhibition of melanin formation. The crude extracts of asparagus showed antitumor activity. (De Martino and De Feo 2008). The antifungal activity was proven against *Alternaria tenuissima, Fusarium oxysporum,* and *Rhizoctonia solani.* This plant is a remedy for cancer, toothache, and face acne lesions (Hildah Mfengwana and Sitheni Mashele 2020).

7.4.9.5 Alhagi maurorum

The leaves and flower extracts showed potent antibacterial activity against *S. aureus, S. typhimurium, B. subtilis, P. aeruginosa,* and *C. albicans.* The methanolic extract of the fresh aerial part of this plant showed inhibition against *B cereus and L. ivanovii.* The methanolic extract also showed anti-helicobacter activity as well. The ethanolic extract of this plant showed antifungal activities against *A. alternate, Rhizoctonia solani,* and *F. oxysporum.* This plant also revealed anti-diarrheal, anti-inflammatory, anti-diuretic, and anti-ulcerogenic properties as well (Ahmad et al. 2015a, b).

7.5 Conclusion

Drug development and its perspective are changing internationally towards the development of natural products. Botanical drugs, the basis of which is a traditional medication with proven safety and clinical effects, were used to treat many infectious diseases. Plants are reservoirs from which many therapeutic potential medicines are developed. Many potential drugs can be developed from natural products and their analogs by using activity-guided fractionation and isolation. Regular developments in bioassay methods coupled with the discovery of new biological targets have benefitted the development of world-class medicines. All plants discussed here exhibited potent activity in their ability to treat many diseases.

A systematic network between countries needs to be established for the utilization of natural resources. With government's support and leading investment by the companies can lead to the development of botanical based research and pharmaceutical industries to full fledge.

References

Abd Kadir SL, Yaakob H, Mohamed Zulkifli R (2013) Potential anti-dengue medicinal plants: a review. J Nat Med 67(4):677–689. https://doi.org/10.1007/s11418-013-0767-y

Ahmad N, Bibi Y, Saboon RI, Zahara K, Idrees S, Khalid N, Bashir T, Tabassum S, Mudrikah (2015a) Traditional uses and pharmacological properties of Alhagi maurorum: a review. Asian Pac J Trop Dis 5(11):856–861. https://doi.org/10.1016/S2222-1808(15)60945-8

Ahmad A, Kaleem M, Ahmed Z, Shafiq H (2015b) Therapeutic potential of flavonoids and their mechanism of action against microbial and viral infections—a review. Food Res Int 77:221–235. https://doi.org/10.1016/j.foodres.2015.06.021

Aiemsaard J, Aiumlamai S, Aromdee C, Taweechaisupapong S, Khunkitti W (2011) The effect of lemongrass oil and its major components on clinical isolate mastitis pathogens and their mechanisms of action on *Staphylococcus aureus* DMST 4745. Res Vet Sci 91(3):e31–e37

Al-Snafi AE (2013) Chemical constituents and pharmacological activities of milfoil (Achillea santolina) a review. Int J Pharmtech Res 5(3):1373–1377

Al-Snafi AE (2015) Therapeutic properties of medicinal plants: a review of their effect on reproductive systems. Indian J Pharma Sci Res 5(4):240–248

Anand U, Jacobo-Herrera N, Altemimi A, Lakhssassi N (2019) A comprehensive review on medicinal plants as antimicrobial therapeutics: potential avenues of biocompatible drug discovery. Metabolites 9(11):1–13

Arruda DC, Alexandri FL, Katzin AM, Uliana SRB (2005) Antileishmanial activity of the terpene nerolidol. Antimicrob Agents Chemother 49(5):1679–1687. https://doi.org/10.1128/AAC.49.5.1679

Artini M, Papa R, Barbato G, Scoarughi GL, Cellini A, Morazzoni P, Bombardelli E, Selan L (2012) Bacterial biofilm formation inhibitory activity revealed for plant derived natural compounds. Bioorganic Med Chem 20(2):920–926. https://doi.org/10.1016/j.bmc.2011.11.052

Atanasov AG, Waltenberger B, Linder T, Wawrosch C, Uhrin P, Temml V, Schwaiger S, Heiss EH, Rollinger JM, Schus D, Breuss JM, Bochkov V, Mihovilovic MD, Bauer R, Dirsch VM, Stuppner H, Thomas L, Christoph W, Pavel U, Rudolf B (2015) Discovery and resupply of pharmacologically active plant-derived natural products: a review. Biotechnol Adv 33(8):1582–1614. https://doi.org/10.1016/j.biotechadv.2015.08.001

Badshah SL, Faisal S, Muhammad A, Poulson BG, Emwas AH, Jaremko M (2021) Antiviral activities of flavonoids. Biomed Pharmacother 140:111596. https://doi.org/10.1016/j.biopha.2021.111596

Bagavan A, Rahuman AA, Kaushik NK, Sahal D (2011) In vitro antimalarial activity of medicinal plant extracts against Plasmodium falciparum. Parasitol Res 108(1):15–22. https://doi.org/10.1007/s00436-010-2034-4

Basco LK, Mitaku S, Skaltsounis AL, Ravelomanantsoa N, Tillequin F, Koch M, Le Bras J (1994) In vitro activities of furoquinoline and acridone alkaloids against Plasmodium falciparum. Antimicrob Agents Chemother 38(5):1169–1171. https://doi.org/10.1128/AAC.38.5.1169
Betts JW, Kelly SM, Haswell SJ (2011) Antibacterial effects of theaflavin and synergy with epicatechin against clinical isolates of Acinetobacter baumannii and Stenotrophomonas maltophilia. Int J Antimicrob Agents 38(5):421–425. https://doi.org/10.1016/j.ijantimicag.2011.07.006
Blair JMA, Webber MA, Baylay AJ, Ogbolu DO, Piddock LJV (2015) Molecular mechanisms of antibiotic resistance. Nat Rev Microbiol 13(1):42–51. https://doi.org/10.1038/nrmicro3380
Boberek JM, Stach J, Good L (2010) Genetic evidence for inhibition of bacterial division protein FtsZ by berberine. PLoS One 5(10):1–9. https://doi.org/10.1371/journal.pone.0013745
Boy HIA, Rutilla AJH, Santos KA, Ty AMT, Yu AI, Mahboob T, Tangpoong J, Nissapatorn V (2018) Recommended medicinal plants as source of natural products: a review. Digital Chin Med 1(2):131–142. https://doi.org/10.1016/s2589-3777(19)30018-7
Bugyei K, Boye G, Addy M (2011) Clinical efficacy of a tea-bag formulation of *Cryptolepis sanguinolenta* root in the treatment of acute uncomplicated falciparum malaria. Ghana Med J 44(1):3–9. https://doi.org/10.4314/gmj.v44i1.68849
Capasso A (2013) Antioxidant action and therapeutic efficacy of Allium sativum L. Molecules 18(1):690–700. https://doi.org/10.3390/molecules18010690
Cazarolli L, Zanatta L, Alberton E, Bonorino Figueiredo MS, Folador P, Damazio R, Pizzolatti M, Barreto Silva FR (2008) Flavonoids: prospective drug candidates. Mini Rev Med Chem 8(13):1429–1440. https://doi.org/10.2174/138955708786369564
Cellini L, Di Campli E, Di Masulli M, Bartolomeo S, Allocati N (1996) Inhibition of Helicobacter pylori by garlic extract (*Allium sativum*). FEMS Immunol Med Microbiol 13(4):273–277
Chavan SS, Damale MG, Devanand B, Sangshetti JN (2018) Antibacterial and antifungal drugs from natural source: a review of clinical development, vol 1. Natural Products in Clinical Trials, Sharjah, p 114
Cheesman L, Nair JJ, Van Staden J (2012) Antibacterial activity of crinane alkaloids from Boophone disticha (Amaryllidaceae). J Ethnopharmacol 140(2):405–408. https://doi.org/10.1016/j.jep.2012.01.037
Chitemerere TA, Mukanganyama S (2014) Evaluation of cell membrane integrity as a potential antimicrobial target for plant products. BMC Complement Altern Med 14:1–8. https://doi.org/10.1186/1472-6882-14-278
Cimolai N, Cimolai T (2007) The cranberry and the urinary tract. Eur J Clin Microbiol Infect Dis 26(11):767–776. https://doi.org/10.1007/s10096-007-0379-0
Cobrado L, Azevedo MM, Silva-Dias A, Ramos JP, Pina-Vaz C, Rodrigues AG (2012) Cerium, chitosan and hamamelitannin as novel biofilm inhibitors? J Antimicrob Chemother 67(5):1159–1162. https://doi.org/10.1093/jac/dks007
Cragg GM, Newman DJ (2013) Natural products: a continuing source of novel drug leads. Biochim Biophys Acta Gen Subj 1830(6):3670–3695. https://doi.org/10.1016/j.bbagen.2013.02.008
Cushnie TPT, Lamb AJ (2011) Recent advances in understanding the antibacterial properties of flavonoids. Int J Antimicrob Agents 38(2):99–107. https://doi.org/10.1016/j.ijantimicag.2011.02.014
Dar RA, Shahnawaz M, Qazi PH (2017) General overview of medicinal plants: a review. J Phytopharmacol 6(6):349–351
De Freitas Araújo MG, Hilário F, Nogueira LG, Vilegas W, Dos Santos LC, Bauab TM (2011) Chemical constituents of the methanolic extract of leaves of leiothrix spiralis ruhland and their antimicrobial activity. Molecules 16(12):10479–10490. https://doi.org/10.3390/molecules161210479
De Martino L, De Feo V (2008) Chemistry and biological activities of Ailanthus altissima swingle: a review. Pharmacogn Rev 2(4):339–350. citeulike-article-id:9347292
Desoukey SF, Nahas SE, Sabh AZ, Taha Z (2020) Antimicrobial effect of Asparagus officinalis L. extracts. Plant Arch 20(2):9253–9264

Dhanapal R, Ratna JV (2012) Nanosuspensions technology in drug delivery—a review. Int J Pharm Rev Res 2(1):46–52

Ekpenyong CE, Akpan E, Nyoh A (2015) Ethnopharmacology, phytochemistry, and biological activities of Cymbopogon citratus (DC.) Stapf extracts. Chin J Nat Med 13(5):321–337. https://doi.org/10.1016/S1875-5364(15)30023-6

Firmino DF, Cavalcante TTA, Gomes GA, Firmino NCS, Rosa LD, De Carvalho MG, Catunda FEA (2018) Antibacterial and antibiofilm activities of Cinnamomum Sp. essential oil and cinnamaldehyde: antimicrobial activities. ScientificWorldJournal 2018:7405736. https://doi.org/10.1155/2018/7405736

Fowler ZL, Baron CM, Panepinto JC, Koffas MA (2011) Melanization of flavonoids by fungal and bacterial laccases. Yeast 28(3):181–188

Garnatje T, Peñuelas J, Vallès J (2017) Ethnobotany, phylogeny, and 'omics' for human health and food security. Trends Plant Sci 22(3):187–191. https://doi.org/10.1016/j.tplants.2017.01.001

Goulart HR, Kimura EA, Peres VJ, Couto AS, Duarte FAA, Katzin AM (2004) Terpenes arrest parasite development and inhibit biosynthesis of isoprenoids in Plasmodium falciparum. Antimicrob Agents Chemother 48(7):2502–2509. https://doi.org/10.1128/AAC.48.7.2502-2509.2004

Harris JC, Cottrell SL, Plummer S, Lloyd D (2001) Antimicrobial properties of *Allium sativum* (garlic). Appl Microbiol Biotechnol 57(3):282–286

Hildah Mfengwana PMA, Sitheni Mashele S (2020) Medicinal properties of selected Asparagus species: a review. Phytochemicals in human health. https://doi.org/10.5772/intechopen.87048

Holler JG, Slotved HC, Mølgaard P, Olsen CE, Christensen SB (2012) Chalcone inhibitors of the NorA efflux pump in *Staphylococcus aureus* whole cells and enriched everted membrane vesicles. Bioorg Med Chem 20(14):4514–4521

Islam B, Khan SN, Haque I, Alam M, Mushfiq M, Khan AU (2008) Novel anti-adherence activity of mulberry leaves: inhibition of Streptococcus mutans biofilm by 1-deoxynojirimycin isolated from Morus alba. J Antimicrob Chemother 62(4):751–757. https://doi.org/10.1093/jac/dkn253

Itokawa H, Morris-Natschke SL, Akiyama T, Lee KH (2008) Plant-derived natural product research aimed at new drug discovery. J Nat Med 62(3):263–280. https://doi.org/10.1007/s11418-008-0246-z

Jakobsen TH, Bragason SK, Phipps RK, Christensen LD, Van Gennip M, Alhede M, Skindersoe M, Larsen TO, Høiby N, Bjarnsholt T, Givskov M (2012) Food as a source for quorum sensing inhibitors: iberin from horseradish revealed as a quorum sensing inhibitor of Pseudomonas aeruginosa. Appl Environ Microbiol 78(7):2410–2421. https://doi.org/10.1128/AEM.05992-11

Keshavan ND, Chowdhary PK, Haines DC, González JE (2005) L-canavanine made by Medicago sativa interferes with quorum sensing in Sinorhizobium meliloti. J Bacteriol 187(24):8427–8436. https://doi.org/10.1128/JB.187.24.8427-8436.2005

Khameneh B, Iranshahy M, Soheili V, Sedigheh B, Bazzaz F (2019) Review on plant antimicrobials a mechanistic viewpoint. Antimicrob Resist Infect Control 8(1):1–28

Khorshed Alam M, Obydul Hoq M, Shahab Uddin M (2016) Medicinal plant Allium sativum: a review. J Med Plants Stud 4(6):72–79

Khubber S, Hashemifesharaki R, Mohammadi M, Gharibzahedi SMT (2020) Garlic (Allium sativum L.): a potential unique therapeutic food rich in organosulfur and flavonoid compounds to fight with COVID-19. Nutr J 19(1):20–22. https://doi.org/10.1186/s12937-020-00643-8

Kim SH, Lee SJ, Lee JH, Sun WS (2002) Antimicrobial activity of 9- O -Acyl- and 9- O -alkylberberrubine derivatives, pp 1–5

Komlaga G, Cojean S, Dickson RA, Beniddir MA, Suyyagh-Albouz S, Mensah MLK, Agyare C, Champy P, Loiseau PM (2016) Antiplasmodial activity of selected medicinal plants used to treat malaria in Ghana. Parasitol Res 115(8):3185–3195. https://doi.org/10.1007/s00436-016-5080-8

Kuete V (2010) Potential of Cameroonian plants and derived products against microbial infections: a review. Planta Med 76(14):1479–1491. https://doi.org/10.1055/s-0030-1250027

Kurek A, Grudniak AM, Kraczkiewicz-Dowjat A, Wolska KI (2011) New antibacterial therapeutics and strategies. Pol J Microbiol 60(1):3–12. https://doi.org/10.33073/pjm-2011-001

Lacmata ST, Kuete V, Dzoyem JP, Tankeo SB, Teke GN, Kuiate JR, Pages JM (2012) Antibacterial activities of selected cameroonian plants and their synergistic effects with antibiotics against bacteria expressing MDR phenotypes. Evid Based Complement Alternat Med 2012:623723. https://doi.org/10.1155/2012/623723

Luepke KH, Suda KJ, Boucher H, Russo RL, Bonney MW, Hunt TD, Mohr JF (2017) Past, present, and future of antibacterial economics: increasing bacterial resistance, limited antibiotic pipeline, and societal implications. Pharmacotherapy 37(1):71–84. https://doi.org/10.1002/phar.1868

Mawalagedera SMUP, Callahan DL, Gaskett AC, Rønsted N, Symonds MRE (2019) Combining evolutionary inference and metabolomics to identify plants with medicinal potential. Front Ecol Evol 7(July):1–11. https://doi.org/10.3389/fevo.2019.00267

Mishra BB, Kale RR, Singh RK, Tiwari VK (2009) Alkaloids: future prospective to combat leishmaniasis. Fitoterapia 80(2):81–90. https://doi.org/10.1016/j.fitote.2008.10.009

Mittal M, Gupta N, Parashar P, Mehra V, Khatri M (2014) Phytochemical evaluation and pharmacological activity of syzygium aromaticum: a comprehensive review. Int J Pharm Pharm Sci 6(8):67–72

Moloney MG (2016) Natural products as a source for novel antibiotics. Trends Pharmacol Sci 37(8):689–701. https://doi.org/10.1016/j.tips.2016.05.001

Monzote L, Perera Córdova WH, García M, Piñón A, Setzer WN (2016) In-vitro and in-vivo activities of phenolic compounds against cutaneous leishmaniasis. Records Nat Prod 10(3): 269–276

Nandu TG, Subramenium GA, Shiburaj S, Viszwapriya D, Iyer PM, Balamurugan K, Rameshkumar KB, Pandian SK (2018) Fukugiside, a biflavonoid from garcinia travancorica inhibits biofilm formation of Streptococcus pyogenes and its associated virulence factors. J Med Microbiol 67(9):1391–1401. https://doi.org/10.1099/jmm.0.000799

Newman DJ, Cragg GM (2016) Natural products as sources of new drugs from 1981 to 2014. J Nat Prod 79(3):629–661. https://doi.org/10.1021/acs.jnatprod.5b01055

Nikaido H, Pagès JM (2012) Broad-specificity efflux pumps and their role in multidrug resistance of Gram-negative bacteria. FEMS Microbiol Rev 36(2):340–363. https://doi.org/10.1111/j.1574-6976.2011.00290.x

Ortega-Ramirez LA, Rodriguez-Garcia I, Leyva JM, Cruz-Valenzuela MR, Silva-Espinoza BA, Gonzalez-Aguilar GA, Siddiqui MW, Ayala-Zavala JF (2014) Potential of medicinal plants as antimicrobial and antioxidant agents in food industry: a hypothesis. J Food Sci 79(2):129–137. https://doi.org/10.1111/1750-3841.12341

Panche AN, Diwan AD, Chandra SR (2016) Flavonoids: an overview. J Nutr Sci 5:e47. https://doi.org/10.1017/jns.2016.41

Patel K, Gadewar M, Tripathi R, Prasad SK, Patel DK (2012) A review on medicinal importance, pharmacological activity and bioanalytical aspects of beta-carboline alkaloid "Harmine". Asian Pac J Trop Biomed 2(8):660–664. https://doi.org/10.1016/S2221-1691(12)60116-6

Piddock LJ (2006) Clinically relevant chromosomally encoded multidrug resistance efflux pumps in bacteria. Clin Microbiol Rev 19(2):382–402

Pink Roses (2011) Guava leaf extract potential cure dengue fever. http://pinkroses.info/guava-leaf-extract-potential-cure-dengue-fever

Pule CM, Sampson SL, Warren RM, Black PA, Van Helden PD, Victor TC, Louw GE (2016) Efflux pump inhibitors: targeting mycobacterial efflux systems to enhance TB therapy. J Antimicrob Chemother 71(1):17–26. https://doi.org/10.1093/jac/dkv316

Qiu S, Sun H, Zhang AH, Xu HY, Yan GL, Han Y, Wang XJ (2014) Natural alkaloids: basic aspects, biological roles, and future perspectives. Chin J Nat Med 12(6):401–406. https://doi.org/10.1016/S1875-5364(14)60063-7

Radulovic NS, Blagojevic PD, Stojanovic-Radic ZZ, Stojanovic NM (2013) Antimicrobial plant metabolites: structural diversity and mechanism of action. Curr Med Chem 20(7):932–952

Rasmussen TB, Skindersoe ME, Bjarnsholt T, Phipps RK, Christensen KB, Jensen PO, Andersen JB, Koch B, Larsen TO, Hentzer M, Eberl L, Hoiby N, Givskov M (2005) Identity and effects of quorum-sensing inhibitors produced by Penicillium species. Microbiology 151(5):1325–1340. https://doi.org/10.1099/mic.0.27715-0

Renslo AR, McKerrow JH (2006) Drug discovery and development for neglected parasitic diseases. Nat Chem Biol 2(12):701–710. https://doi.org/10.1038/nchembio837

Rivlin RS, Budoff M, Amagase H (2006) Significance of garlic and its constituents in cancer and cardiovascular disease. J Nutr 136(3):716–725. https://doi.org/10.1093/jn/136.3.v

Robinson MM, Zhang X (2011) The World medicines situation 2011 traditional medicines: global situation, issues and challenges. World Health Organization, Geneva 3rd edn: 1–14

Saeidnia S, Gohari AR, Mokhber-Dezfuli N, Kiuchi F (2011) A review on phytochemistry and medicinal properties of the genus Achillea. DARU 19(3):173

Santoro GF, Cardoso MG, Guimarães LGL, Freire JM, Soares MJ (2007) Anti-proliferative effect of the essential oil of Cymbopogon citratus (DC) Stapf (lemongrass) on intracellular amastigotes, bloodstream trypomastigotes and culture epimastigotes of Trypanosoma cruzi (protozoa: Kinetoplastida). Parasitology 134(11):1649–1656. https://doi.org/10.1017/S0031182007002958

Savoia D (2012) Plant-derived antimicrobial compounds: alternatives to antibiotics. Future Microbiol 7(8):979–990

Stanton RA, Gernert KM, Nettles JH, Aneja R (2011) Drugs that target dynamic microtubules: a new molecular perspective. Med Res Rev 31(3):443–481

Stavri M, Piddock LJ, Gibbons S (2007) Bacterial efflux pump inhibitors from natural sources. J Antimicrob Chemother 59(6):1247–1260

Subramenium GA, Viszwapriya D, Iyer PM, Balamurugan K, Pandian SK (2015) CovR mediated antibiofilm activity of 3-furancarboxaldehyde increases the virulence of group a Streptococcus. PLoS One 10(5):1–19. https://doi.org/10.1371/journal.pone.0127210

Szychowski J, Truchon JF, Bennani YL (2014) Natural products in medicine: transformational outcome of synthetic chemistry. J Med Chem 57(22):9292–9308. https://doi.org/10.1021/jm500941m

Tamura S, Fukamiya N, Okano M, Koike K (2006) A new quassinoid, ailantinol H from Ailanthus altissima. Nat Prod Res 20(12):1105–1109. https://doi.org/10.1080/14786410600825907

Ullman A, Hedner J, Svedmyr N (1990) Inhaled salmeterol and salbutamol in asthmatic patients. An evaluation of asthma symptoms and the possible development of tachyphylaxis. Am Rev Respir Dis 142(3):571–575. https://doi.org/10.1164/ajrccm/142.3.571

Vadhana P, Singh BR, Bharadwaj M, Singh SV (2015) Emergence of herbal antimicrobial drug resistance in clinical bacterial isolates. Pharm Anal Acta 6(10):434

Valliammai A, Selvaraj A, Sangeetha M, Sethupathy S, Pandian SK (2020) 5-Dodecanolide inhibits biofilm formation and virulence of Streptococcus pyogenes by suppressing core regulons of virulence. Life Sci 262(September):118554. https://doi.org/10.1016/j.lfs.2020.118554

Vane JR, Botting RM (2003) The mechanism of action of aspirin. Thromb Res 110(5–6):255–258. https://doi.org/10.1016/S0049-3848(03)00379-7

Vasavi HS, Arun AB, Rekha PD (2014) Anti-quorum sensing activity of Psidium guajava L. flavonoids against Chromobacterium violaceum and Pseudomonas aeruginosa PAO1. Microbiol Immunol 58(5):286–293. https://doi.org/10.1111/1348-0421.12150

Verlinde CLMJ, Hannaert V, Blonski C, Willson M, Périé JJ, Fothergill-Gilmore LA, Opperdoes FR, Gelb MH, Hol WGJ, Michels PAM (2001) Glycolysis as a target for the design of new anti-trypanosome drugs. Drug Resist Updat 4(1):50–65. https://doi.org/10.1054/drup.2000.0177

Vikram A, Jesudhasan PR, Jayaprakasha GK, Pillai BS, Patil BS (2010) Grapefruit bioactive limonoids modulate E. coli O157:H7 TTSS and biofilm. Int J Food Microbiol 140(2–3): 109–116. https://doi.org/10.1016/j.ijfoodmicro.2010.04.012

Viszwapriya D, Subramenium GA, Prithika U, Balamurugan K, Pandian SK (2016) Betulin inhibits virulence and biofilm of Streptococcus pyogenes by suppressing ropB core regulon, sagA and dltA. FEMS Pathog Dis 74(7):88. https://doi.org/10.1093/femspd/ftw088
Wink M (2012) Medicinal plants: a source of antiparasitic secondary metabolites. Molecules 17(11):12771–12791. https://doi.org/10.3390/molecules171112771
Yarmolinsky L, Bronstein M, Gorelick J (2015) Review: inhibition of bacterial quorum sensing by plant extracts. Israel J Plant Sci 62(4):294–297. https://doi.org/10.1080/07929978.2015.1067076
Zhao X, Chuncheng W (2020) Recent advances in peptide nucleic acids for rapid detection of foodborne pathogens. Food Anal Methods 13(10):1956–1972

Chapter 8
Coral Reef: A Hot Spot of Marine Biodiversity

T. R. Sobha, C. P. Vibija, and P. Fahima

Abstract Coral reefs are the largest reservoirs of biodiversity on earth: they host 32 of 34 recognized phyla and approximately one third of all marine biodiversity. The center of global coral reef biodiversity is the "Coral Triangle" (CT), including eastern Indonesia and Malaysia, the Philippines, Timor-Lest e, Papua New Guinea, and the Solomon Islands. Coral reefs support human life and livelihoods and are important economically. Nearly 500 million people depend directly and indirectly on coral reefs for their livelihoods, food, and other resources. The current serious and further deteriorating status of coral reefs around the world is directly due to damaging stresses that arose during the Anthropocene effectively since the mid-eighteenth century, and particularly since 1950 when human pressures ramped up to destructive levels. The major threats include extractive activities, pollution, sedimentation, physical destruction, and the effects of anthropogenic action as well as climate change. It has been estimated that 33% of all reef-building corals could become extinct due to damage from local threats combined with climate change impacts. The Coral reefs are considered as fascinating hot spots of life in the ocean, which stabilize marine ecosystems as key underwater biodiversity hot spots. With the earth's coral reefs being destroyed at an alarming rate, it is important to take action mainly by proper management and to restore these underwater biodiversity hot spots.

Keywords Coral reefs · Marine biodiversity · Hot spots · Coral triangle · Extractive activities · Pollution · Sedimentation · Physical destruction · Anthropogenic action · Climate change

---

T. R. Sobha (✉) · P. Fahima
Farook College (Autonomous), Kozhikode, Kerala, India

C. P. Vibija
Malabar Christian College, Kozhikode, Kerala, India

S. T. Sukumaran, K. T R (eds.), *Conservation and Sustainable Utilization of Bioresources*, Sustainable Development and Biodiversity 30,
https://doi.org/10.1007/978-981-19-5841-0_8

8.1 Introduction

The oceans are marvelous creations of beauty with a blend of stunning underwater gardens, coral reefs, and marine life. Coral reefs are some of the most diverse and magnificently beautiful ecosystems in the world. Coral reefs can be referred to as the wonderland underwater. They are large underwater structures that resemble ridges or mounds composed of tiny, colonial marine invertebrates called corals. The term 'Coral' has been originated from the ancient Greek word 'Korallion' which refers to the valuable red coral of the Mediterranean known as *Corallum rubrum* (Etnoyer and Morgan 2003; Williams 1986).

The reefs are formed by hard corals or Scleractinians that belong to the phylum Cnidaria (Goreau et al. 1979). Being the members of phylum Cnidaria, they are radially symmetrical organisms that lack a head, usually having a crown of tentacles around their mouth, and they live in the skeleton secreted by their tissues. The coral species that build reefs are known as hermatypic or hard corals because they extract calcium carbonate from seawater to create a hard-durable exoskeleton forming the stony framework of the reef. And these exoskeletons protect their soft sac-like bodies https://www.livescience.com/40276-coral-reefs.html (n.d.).

Each individual coral is called a polyp. Coral polyp lives on the calcium carbonate exoskeletons of their ancestors by adding their own exoskeleton to the existing coral structure. As the years pass, the coral reef gradually grows from a tiny form to a massive feature of the marine environment. According to CORAL, there exist hundreds of different species of corals. They can be found in various shapes and colors, ranging from round, folded brain corals to tall and graceful sea fans that resemble tangled trees. A coral polyp can be as small as the head of a pin (*Montipora*) to as large as a foot in diameter as in the case of *Fungia* (Parasharya et al. 2012). Despite their sizes, they have a stunning impact on the marine ecosystem.

Coral reefs are typically categorized into three groups; Fringing reefs, barrier reefs, and atolls. Fringing reefs are the most common, and they grow in shallow water near coastlines. Barrier reefs are parallel to the coast but are separated from the coastlines by a deeper channel of water. Atolls are rings of the coral island that encloses a central lagoon.

Corals are some of the oldest animals on the planet; Fossil records date back 450 to 500 million years to the Ordovician period of the Paleozoic Era. And the modern-day corals evolved during the Mesozoic Era 245 million years ago, and they diversified into today's reef-building corals in the marine environment (Williams 1986). According to the coral reef alliance, most coral reefs found today are between 5000 and 10,000 years old.

A fascinating feature of corals is their symbiotic relationship with the unicellular algae, Zooxanthellae (*Symbiodinium microadriaticum*) (Velloth and Nayak 2013). These algae live inside the coral polyp's body, where they conduct photosynthesis to produce nutrients, many of which they pass to coral cells, and the polyps, in turn, provide a home and carbon dioxide for the algae. The corals also expel waste products in the form of ammonium, which the algae takes up as food. This relationship maintains the nutrient cycling within the corals. Symbiosis also plays a role in building reefs. The corals that host the algae can deposit calcium carbonate up to 10 times faster than the non-symbiotic corals. In addition to this, the Zooxanthellae is also responsible for the lively, vibrant colors of the coral polyp bodies. This interconnection of algae and coral, which is crucial for the health of the coral reefs, began more than 210 million years ago (Zandonella 2016).

They are most often found in warm shallow waters (ocean waters of depth less than 100 m) where there is plenty of sunlight to nurture the algae, and they are disfavored by extreme temperatures, high turbidity, and low salinity and pH changes. Owing to the diversity of life found in the habitats created by the corals, reefs are often entitled "rainforests of the sea" (Reaka-Kudla 1997; Knowlton et al. 2010).

Coral reefs are home to a substantial variety of organisms. About 25 percent of all known marine species rely on coral reefs for food, shelter, and for their breeding (Wilson and Kaufmann 1987; Rogers 1994). They serve as the breeding grounds for 10–20% of the world's fisheries. As reported by CORAL, coral reefs are the primary habitat for more than 4000 species of fishes, 700 species of corals, and thousands of other plant and animal species.

8.2 Major Reef Areas

Worldwide, coral reefs cover an estimated 110,000 square miles (284,300 square kilometers), which is less than 0.2% of the global ocean area and about 15% of the shallow sea areas within 0–30 m depth (Venkataraman et al. 2003; Lalli and Parsons 1997). Coral reefs can be found in tropical destinations around the world, mostly in areas close to the equator where the water is warmer. Most reefs are located between the Tropics of Cancer and Capricorn, in the Pacific Ocean, the Indian Ocean, the Caribbean Sea, the Red Sea, and the Persian Gulf. Around 54% of the coral reefs lie in the Asiatic Mediterranean and the Indian Ocean. Pacific reefs account for 25% of the remaining, Caribbean reefs for 9%, Atlantic for 6%, Red Sea reefs 4%, and Persian Gulf reefs for 2% (Lalli and Parsons 1997). More than 100 countries have a coral reef within their borders, and over half of the world's coral reefs are found in six countries: Australia, Indonesia, Philippines, Papua New Guinea, Fiji, and the Maldives. Even though coral reefs only cover 0.2 percent of the ocean floor, they support the life of about 25 percent of marine life, which makes the decline of these ecosystems more impactful.

Even though various types of corals can be found from the water's surface to depths of 19,700 ft. (6000 m), reef-building corals are generally restricted to depths of less than 150 ft. (46 m), where sunlight penetrates in and can be taken up by the symbiotic algae. Reef-building corals require warm ocean temperatures (68–82 °F, or 20–28 °C). Warm water flows along the eastern shores of major landmasses. Corals are also found in places farther from the equator where warm currents flow out of the tropics, such as in Florida and southern Japan https://seaworld.org/animals/all-about/coral-and-coral-reefs/habitat/ (n.d.). Coral reefs are generally absent on the west coast of tropical continents due to the presence of cold coastal currents on the west coast of the America (Peru cold current), west coast of Africa (Benguela cold current and Canary current), and West coast of Australia (west Australian cold current). Coral reefs do not flourish in the cold currents due to the upsurge of the cold water from depths that cools the warm surface water. The optimum temperature for the survival of the corals is 20 °C. Tremendous coral growth can be observed in clear water. Clear water allows light to reach the symbiotic algae living within the coral polyp's tissue. Light-absorbing adaptations enable some reef-building corals to live in dim blue light. Reef development is generally more plentiful in areas that are subject to strong wave action. Waves carry food, nutrients, and oxygen to the reef; distribute coral larvae, and prevent sediment from settling on the coral reef. Precipitation of calcium from the water is necessary to form a coral polyp's skeleton. This precipitation takes place when water temperature and salinity are high, and carbon dioxide concentrations are low. These conditions are typical of shallow, warm tropical waters.

The highest distribution of the coral reefs can be found in the northern Carrebian reefs. The coral reefs of Gulf of Kutch are also considered to be the northern most reef of the Asian continent. At the same time, the southernmost distribution of coral reefs can be observed in the south of Australia and also in the Southern Africa. The western Atlantic and the Indo-Pacific are the two main coral reef regions in the world. From the biodiversity point of view, the Indo-Pacific reef is roughly ten times more diverse than the western Atlantic reef. For example, there are approximately 60 species of hermatypic corals reported from the coral reefs of the western Atlantic, whereas 500–600 species are reported from the Indo-Pacific reefs (Smith 1978).

More than 100 countries are endowed with coral reefs worldwide; most of these are developing countries (Venkataraman et al. 2003). The Great Barrier Reef (GBR) is the largest reef settlement in the world, covering an area of more than 344,000 square kilometers. The reef begins in the north at Australia's Cape York Peninsula and ends midway down the eastern coast at Lady Elliot Island, located just 90 km northeast of Bundaberg. The reef's catchment area is made up of 35 river basins and, together with the GBR, totals 424,000 km in area (Stoeckl et al. 2011). The GBR holds the position of the most intensively managed marine park in the world. In 1975, the region gained protection through the creation of the Great Barrier Reef Marine Park (GBRMP) and later designated as a World Heritage Area (GBRWHA) in 1981 (Hoegh-Guldberg and Hoegh-Guldberg 2004). Around 600 species of corals have been reported from here. Sea grass, mangroves, estuaries, and other marine habitats help to host 100 species of jellyfish, 3000 varieties of molluscs, 500 species of worms, 1625 types of fish, 133 varieties of sharks and rays, more than 30 species of whales and dolphins, and various turtles and crocodiles Great Barrier Marine Park Authority (2015).

The species diversity of corals varies in different regions. The maximum coral species are found in South East Asia at a region known as the Coral Triangle (CT) where 605 species of coral species have been reported. The Coral Triangle is a geographical term referring to the tropical marine waters of Indonesia, Malaysia, Papua New Guinea, Philippines, Solomon Islands, and Timor-Leste. The highest coral diversity resides in the Bird's Head Peninsula of Indonesian Papua, which hosts 574 species, with individual reefs supporting up to 280 species per hectare. Within the Bird's Head Peninsula, the reef of Raja Ampat is the world's best coral diversity spot, with 553 species (Venkataraman et al. 2003).

8.2.1 Coral Reefs in South Asia

South Asian coastlines possess diverse marine and coastal habitats, including coral reefs, mangroves, and seagrass. With 19,210 km^2 of reef area, South Asia contributes 6% to the world's coral reef area (Wilkinson 2008). Bangladesh, India, Maldives, Pakistan, and Sri Lanka are the maritime countries of south Asia with diverse coral reefs (Velloth and Nayak 2013).

1. **Bangladesh:**The country has only one coral reef island, the St. Martin's island in the Bay of Bengal. The coral reefs of Bangladesh cover about 50 square kilometers of area. The live coral cover of this area is generally 5%, but it is 10–15% at Boro Shiler Bandh. The coral genera evident here are Porites, followed by Favites, Goniopora, Cyphastrea, and Goniastrea. The presence of 37 species of fish and 46 species of algae has been reported in this area. A total of 66 species of scleractinian corals belonging to 22 genera and 15 families are found here (Wilkinson 2008). The area was declared an Ecologically Critical Area (ECA) in 1996 under the Bangladesh Environment Conservation Act.
2. **Maldives:** The Maldives is in the center of the Lacadive–Chagos ridge. The reef area of Maldives covers about 3.14% of the global reef area. It has 1190 coral islands within an area of 8920 km, including 23 atolls. Coral reefs serve as a resource base for major sectors of tourism and fisheries. A total of 187 species of corals belonging to 33 genera have been recorded from the Maldives. The western atolls of Maldives exhibit a higher coral cover than the eastern atolls. The "Atoll Ecosystem Conservation Project" is one of the international initiatives that have been carried out in the Maldives to enhance the conservation efforts for this ecosystem.
3. **Pakistan:** The total reef area is estimated to be less than 50 square kilometers. Here the high turbidity and sedimentation limit the reef development. The coral communities along the Baluchistan coast have been poorly studied, although surveys around Jiwani and the area near Astola Island have identified 25 coral species and 77 species of reef fishes (Wilkinson 2008). Recently, isolated patches of corals on the hard substratum have been reported along Baluchistan coast in Pakistan.
4. **Sri Lanka:** The reef area of Sri Lanka covers about 680 square kilometers along the coast. The most extensive coral reefs are offshore in the Gulf of Mannar region. The southwestern coast of Sri Lanka has fringing coral reefs. Along the

eastern coast of the country, better development of fringing reefs has been noted, both on the leeward side of the mainlands and on offshore rocks and islands. The reef types include fringing, barrier, and patch reefs. It also includes sandstone or limestone and rocky reef habitats. The coral community represents a total of 183 species belonging to 68 genera. The Government of Sri Lanka has declared 5 Sanctuaries since 1973 and 3 National parks since 1969 for the conservation of corals and turtles.

5. **India:** India which is bounded by a coastline on three sides, has four major coral reef areas that cover about square kilometers, equal to 2.04% of the global reef area. The four nationally recognized coral reefs include the Andaman and Nicobar group of Islands, Lakshadweep Islands, the Gulf of Mannar, and the Gulf of Kutch. Scattered patches of corals are found in the intertidal areas and occasionally at subtidal depths down to a few meters along the west coast of India, notably at Ratnagiri, Malwan, Rede port, and Vizhingam. The Indian coral reefs show fringing and atolls, and submerged reef platforms. India has more than 208 species of hard corals belonging to 60 genera and 15 families. Among 208 species in the reefs of India, the family Acroporidae and Faviidae are the most dominant. In 2002, India added new legislation to protect corals and included corals in Schedule-I of the Wildlife Protection Act 1972 so that corals are ensured maximum protection.

8.2.2 Coral Reefs of India

The total coral reef area in India is 5790 square kilometers, distributed between 4 major regions: Lakshadweep, Gulf of Mannar, Gulf of Kutch, and Andaman and the Nicobar Islands. Considering the area of reefs, Andaman and Nicobar contribute to 41% of the Indian reefs, followed by Lakshadweep (35%), the Gulf of Kutch (20%), and the Gulf of Mannar (4%) (Wilkinson 2008). All of these reefs are Fringing reefs, except Lakshadweep, which are Atolls. Patchy corals are present along with the intertidal areas of the central west coast like the intertidal regions of Ratnagiri, Gaveshani Bank, etc. There is also scattered coral growth on submerged banks along the East and West coasts of the mainland. The family Acroporidae has the maximum representation (74 species—34% of scleractenian fauna of India). This is followed by the family Faviidae (22 species—11%). The family Trachyphyllidae is represented by a single species (Venkataraman et al. 2003).

Reef structure and species diversity depend on the region and its environmental conditions. The environmental conditions in India have definitely affected the development of coral reefs in India. The seasonal monsoons, equatorial calm, tropical cyclones, and trade winds are the major factors that regulate the reef distribution in the Indian Ocean regions Wells (1988). Exposure to the atmosphere and desiccation limits the growth of corals, algae, and other associated organisms in the reef zones in India. Lakshadweep is an archipelago of 12 atolls surrounded by deep water on the northern end of the Laccadive–Chagos ridge. In the Gulf of Kutch, shallow patchy reefs are growing on sandstone platforms surrounding 34 islands. The reefs experience high salinity, frequent emersion, high temperature fluctuations, and heavy sedimentation. In the Gulf of Mannar, coral reefs are found mainly around 21 islands between Rameshwaram and Tuticorin. The Andaman and Nicobar Islands consist of 530 islands with extensive fringing reefs in good condition.

Andaman and Nicobar Islands

Geographic Location of Andaman Islands: 10°30′–14°N 92–93°E

Geographic Location of Nicobar Islands: 6°30′–9°30′ N 93–94° E

There are about 321 islands in the Andaman and Nicobar group, with a total land area of 8293 sq. km.

Reef type: Fringing type.

Corals found: *Acropora, Porites, Pocillopora, Montipora, Heliopora, Tubipora,* and *Favia.*

Lakshadweep Islands

The geographic location of Lakshadweep islands: 19–12°N 72–74°E

The Lakshadweep group of 12 atolls, three reefs, five submerged banks, and 36 islands form the Lakshadweep archipelago.

Reef type: Atolls.

Corals found: *Acropora* species, *Porites* species, D*iploastrea, Heliopora, Goniastrea retiformis,* and *Lobophyllia corymbosa.*

Gulf of Kutch

Geographic Location of Gulf of Kutch: 22°15′–23°40′N 68°20′–70°40′E

The Gulf of Kutch is the northernmost reef of India. There are about 42 islands on the southern flank of the Gulf.

Reef type: Fringing reefs.

Corals found: All corals of massive form.

Branching corals like Acropora, Pocillopora, Stylophora, and Seritopora are totally absent.

Gulf of Mannar

Geographic location of Gulf of Mannar:8 ° 48’ N, 78 ° 9′ E (Tuticorin).

9°14′N, 79°14′ E (Rameshwaram)

Reef type: Fringing reefs.

Coral reserves in India

- Gulf of Mannar biosphere reserve
- Gulf of Kutch Marine National Park
- Mahatma Gandhi Marine National Park, Andaman and Nicobar
- Lakshadweep
- Rani Jhansi Marine National Park in Richie’s Archipelago, South Andaman Islands

8.3 Coral Reefs: The Marine Hot Spots

Hot spots are regions that harbor a great diversity of endemic species and, at the same time, have been significantly impacted and altered by human activities. Coral reefs are often referred to as the rain forests of the sea that account for nearly one-quarter of the total marine biodiversity, despite only covering 0.2% of the total seafloor by area (Reaka-Kudla 1997). They are widely regarded as one of the top conservation ecosystems and prioritize globally due to their extraordinary biodiversity, countless ecosystem services, and global anthropogenic stress (Crosby et al. 2002; Klein et al. 2014). Coral reefs, the richest of tropical marine habitats, are at risk of disappearing at an incredibly fast rate. Worldwide, already 25% of coral reefs have been destroyed or badly degraded. Some scientist reckons that by 2020 up to 70% might be permanently lost (Allen et al. 2003).

8.3.1 World Distribution of Coral Reef Hot Spots

Coral reefs reach the highest diversity in the Indo-Australian archipelago, with secondary peaks in the southwestern Indian Ocean and West Atlantic. Diversity declines into the Central Pacific, is lower in the East Pacific than in the Caribbean for most (not all) taxa, and is lowest in the East Atlantic. Endemism varies with scale—local endemism shows no significant geographic pattern, but regional endemism is high in the Indo-Australian archipelago, moderate in the western Indian Ocean and Central Pacific, and highest in the Americas, especially the East Pacific (Reaka and Lombardi 2011).

8.3.2 World List of Coral Reef Hot Spots

In 2002, researchers identified global priority areas for coral reef conservation and prepared a list of the world's top 10 coral reef hot spots. These are areas rich in marine species, which are found only in a small areas. Therefore, they are highly vulnerable to extinction. These ten hot spots contain just 24% of the world's coral reefs, or 0.017% of the oceans, but claim 34% of restricted-range species. An interesting fact is that eight of the ten coral reef hot spots are adjacent to a terrestrial hot spot. Those are regions of the world that harbor the highest concentrations of species on land and are also at the greatest risk. The tropical reef ecosystems include "wilderness" areas, which remain far less impacted by people, are rich in species, and relative to degraded areas, still contain abundant populations of reef species that have already disappeared from overexploited reefs (Allen et al. 2003).

The global environmental group Conservation International presented a strategy to concentrate conservation efforts on these hot spots. This strategy prioritizes and targets conservation investments where they have the greatest impact. They named 25 Biodiversity Hot spots (on land and in the sea) and divided the so-called Coral Triangle (Indonesia, Philippines, Papua New Guinea) into three different regions (Philippines, Sundaland, Wallacea). Accordingly, there are 11 coral reef hot spots ranked by the degree of threat (Allen et al. 2003).

8.3.2.1 The Philippines

The coral reefs of the Philippines are the second largest in Southeast Asia, covering 10,000 square miles (26,000 square kilometers) (Burke et al. 2002). The reefs support extraordinary biodiversity, including more than 400 species of hard coral—12 of which are unique to the area—and more than 900 species of reef fish (Licuanan and Gomez 2000; Wilkinson 2000). They are already endangered by direct human activity, and these reefs are under additional pressure from global warming. These are ecologically unique in as far that there are a lot of small regions and areas (such as an island) that are highly diverse (Allen et al. 2003). Destructive fishing methods using explosives and poison (cyanide fishing for the aquarium trade), excessive fishing, coastal development, farming, and fish farming, pollution runoff from logging, agriculture, and urban development pose a growing threat to Philippine coral reefs. This is one of the most-threatened hot spots due to its population density. Filipino residents depend on these tremendous natural resources for fishing, tourism, and storm protection, valued at more than U.S.$2 billion annually (Wilkinson 2008; Ming 1998).

8.3.2.2 The Sundaland Hot Spot

It encompasses some 1,600,000 km^2 and covers the western half of the Indonesian archipelago (Bali, Java, Sumatra, Borneo), Malaysia, and a small part of Thailand. It is part of the so-called Coral Triangle, which is probably the most diverse coastal area on the planet, having a richness of marine species and a large occurrence of endemism. Major threats are pollution from land-based sources, intensive destructive fishing (dynamite), and a growing live reef fish trade (for the aquarium trade). Indonesia is one of the favorite places to go for scuba diving. The reefs on this 5000-km-long archipelago are the best. There are said to be over 15,000 islands with more than 80,000-km coastline—that is a lot of potential area for coral reefs! However, these are also one of the most endangered reef areas in the world. Major threats are destructive fishing methods using explosives and poison (cyanide fishing for the aquarium trade), excessive fishing, pollution runoff from logging, agriculture, and urban development. Wallacea and Sundaland are some of the most-threatened hot spots (Allen et al. 2003).

8.3.2.3 The Wallacea Hot Spot

It encompasses some 346,782 km^2 and covers Nusa Tenggara (Lombok, Sumbawa, Komodo, Flores, Sumba, Savu, Roti and Timor), the Mollucas, and Sulawesi. Wallacea is divided from Sundaland by the Wallace's Line. Nusa Tenggara is the Indonesian name for the over 500 islands east of Bali, running from Lombok in the west to Timor in the east. Nusa Tenggara stretches 1300 km and lies just a few degrees south of the equator. The northern islands (Lombok, Sumbawa, Flores to Alor) are volcanic, and the southern islands (Sumba, Savu, Roti and Timor) are uplifted coral limestone and sediment. There are over 40 volcanoes, with half of them still active. Major threats are pollution from land-based sources, sediment pollution from logging, intensive destructive fishing (dynamite), and live reef fish trade (for the aquarium trade) (Allen et al. 2003).

8.3.2.4 Gulf of Guinea

This hot spot encompasses the four islands (Annobón, Bioco, São Tomé, and Príncipe) of the Gulf of Guinea, off the West African coast. Príncipe is part of the marine biodiversity hot spot of the Gulf of Guinea due to its high levels of endemism despite low overall species richness. Coral cover is limited in the tropical eastern Atlantic, as this area is at the extreme margins of the ecological tolerances of hard corals, and this reef hot spot is among the ten most threatened in the world. Nevertheless, interesting diversity of species can be found underwater (Felipe Spina 2016). The exact area of the reef is unknown but is likely to be less than 200 km^2. There are rivers nearby, so the water is not very salty. Major threats from coastal development, sediment pollution from logging, over-fishing, and a proposed coral harvesting business (Allen et al. 2003).

8.3.2.5 Southern Mascarene Islands

This hot spot encompasses approximately 1000 km^2 of the reef surrounding the islands of Mauritius, La Reunion, and Rodriguez in the southern Indian Ocean. The Mascarene Islands are surrounded by approximately 750 square km of coral reef. Rodrigues has nearly continuous fringing reefs bounding an extensive lagoon, 7–20 km wide, with deep channels. Mauritius is also surrounded by a fringing reef (Strahm 1996).

In contrast, Réunion has very short stretches of narrow fringing reefs along the western and southwestern coasts only. The islets of the Cargados Carajos Shoals, which have a very depauperate terrestrial biota owing to being so low-lying and swamped during cyclones, are bound to the east by an extensive arc of fringing reef, which accounts for ~30% of the reefs of the Mascarene Islands (Allen et al. 2003). Lagoon reefs and reef flats are dominated by scleractinian corals such as branching and tubular *Acropora*, *Porites* massives, foliaceous *Montipora* and *Pavona*, and sand consolidated with beds of seagrass such as *Halophila* spp. (Hydrocharitaceae). Among coral reef fishes, wrasses (Labridae), damselfish (Pomacentridae), carnivorous groupers (Serranidae), and surgeonfishes (Acanthuridae) have many species (Thébaud et al. 2009). Major threats are the rapidly growing human population, pollution from intensive sugar cane production, coastal and agricultural development, and over-fishing.

8.3.2.6 Eastern South Africa

This hot spot lies adjacent to Cape Floristic and encompasses less than 200 km^2. The coral reefs that stretch from Kosi Bay to St Lucia along South Africa's east coast have been identified as among the most-threatened habitats of their kind in the world. According to the United Nations Environment Programme (UNEP), the region ranks fifth on a list of the planet's 10 most-threatened coral reef 'hotspot' areas. The coral reefs along South Africa's east coast are about 4000 years old, and formed on top of the remnants of ancient sand dunes. The summer flooding of rivers limits the growth of coral. Major threats are land-based sources of pollution, fishing, and the development of tourism. Ahead of South Africa on the list, ranked according to the degree of threat, there are the Philippines, the Gulf of Guinea Islands, the Sunday Islands (Indonesia), and the Southern Mascarene Islands (near Madagascar) (Allen et al. 2003).

8.3.2.7 North Indian Ocean

This hotspot encompasses a total of 10,000 km^2. Ten percent of Western and Northern Indian Ocean coral species are endemic, and the global warming in 1998 increased the sea surface temperatures and resulted in severe coral bleaching. Global climate change continues to pose a threat, as do coral mining, over-fishing, and ornamental fish collection (Allen et al. 2003).

8.3.2.8 Southern Japan, Taiwan, and Southern China

Over 3000 km^2 of reefs extend from Kyushu in Japan, through Taiwan to the coast of southern China. Major threats are shoreline development (proposed airport) and conversion for agriculture and aquaculture, rapidly growing human population and global climate change, sea warming, and plagues of coral-eating Crown-of-Thorns starfish (Allen et al. 2003).

8.3.2.9 Cape Verde Islands

It is approximately 200 km^2 in the mid-Atlantic off the West African coast. Major threats are coastal development, pollution from land clearing and agriculture, and over-fishing (Allen et al. 2003). The Cabo Verde archipelago comprises ten islands and five islets located about 300 nautical miles WNW off the coast of Dakar, Senegal (Monteiro et al. 2008). Volcanic in origin, the islands rise from the deep abyssal plain beyond the African continental shelf. These islands are considered to be an important hotspot of tropical reef biodiversity and one of the top ten priority locations worldwide for the conservation of reef habitats (Roberts et al. 2002). Despite the absence of major coral reef structures and of the relatively low richness of coral species and reef-associated fauna, these islands' coral communities constitute unique habitats as many species display a small distributional range (Roberts et al. 2002). These coral communities are of major importance for coastal ecosystems (providing habitat and supporting other fauna and flora) and play a key role on the sustainability of local fishing and economy (Wilkinson 2004a, b).

8.3.2.10 Western Caribbean

The Western Caribbean is recognized as the most species-rich sea in the Atlantic regarding its coral reefs (Miloslavich et al. 2010). This hotspot encompasses the Caribbean islands and coastal reefs from the Mexican Yucatan Peninsula to Colombia, more than 4000 km^2 of reefs. During the last three decades, reef corals of this region have experienced unprecedented thermal stress (Aronson et al. 2000). Such extreme thermal stress has led to extensive coral bleaching, coral mortality, and shifts in coral community composition. In addition, there has been a purported increase in the number of coral disease outbreaks in the Caribbean, which also has contributed to extensive declines in coral populations (Altizer et al. 2013; Loh and Pawlik 2014). Although it has been proposed that thermal stress is a driver of several coral diseases. Major threats are epidemic diseases and coral bleaching resulting from global warming and coastal development for tourism (Allen et al. 2003).

8.3.2.11 The Red Sea and Gulf of Aden

The Red Sea and Gulf of Aden hotspot extend for 2500 km from north to south, including the Gulfs of Aqaba and Suez. Major threats are coastal and industrial developments, tourism, and oil spills from tankers. The western and southern coasts are less threatened (Allen et al. 2003). The Red Sea is a unique body of water, hosting some of the most productive and richest coral reef ecosystems with a coral reef framework along its entire coastline (DiBattista et al. 2016). Clear blue water of the Red Sea and a profusion of life, as well as its proximity to Europe, have always been a source of attraction to generations of explorers, and naturalists have been reviewed (Berumen et al. 2013) and reported the high diversity and endemism (Stehli and Wells 1971). DiBattista et al. (2016) reported 365 scleractinian coral species from the Red Sea, including 19 (5.5%) endemic species (DiBattista et al. 2016).

8.4 Significance of Coral Reefs

Coral reefs are extremely productive ecosystems and provide humans with many services.

8.4.1 Provisioning Services

Coral reefs support human life and livelihoods and are important economically. Nearly 500 million people depend—directly and indirectly—on coral reefs for their livelihoods, food, and other resources (Wilkinson 2004a, b). Further, it is estimated that nearly 30 million of the poorest human populations in the world depend entirely on coral reefs for their food (Wilkinson 2004a, b). A km^2 of well-managed coral reef can yield an average of 15 tons of fish and other seafood every year http://www.panda.org/about_wwf/what_we_do/marine/blue_planet/ (n.d.). In 1985, the world export value of the marine aquarium trade was estimated at 25–40 million USD per year. In 1996, the world export value was about 200 million USD. The annual export of marine aquarium fish from Southeast Asia alone is estimated to be between 10–30 million fish, with a retail value of up to 750 million USD (Bruckner and Bruckner 2006).

8.4.2 Medicinal Values

Many coral species and species associated with coral reefs have medicinal values. Several species are used in Traditional Chinese Medicine (TCM), and many are now providing novel resources for allopathic medicine. Most of these species are used in Asia (Hunt and Vincent 2006). Some hard-coral species are used in bone grafts. Others contain chemicals that might be used as natural sunscreen products (Demers et al. 2002). There are some 500 species of cone snails that live in and around coral reefs. These species have a range of venoms that are currently being investigated for use as non-addictive pain killers (Chivian 2006).

8.4.3 Regulating Services

Coral reefs protect the shoreline and reduce flooding. Very importantly, coral reefs protect the shoreline, providing a physical barrier—a wall—against tidal surges, extreme weather events, ocean currents, tides, and winds. In doing so, they prevent coastal erosion, flooding, and loss of infrastructure. Because of this, they serve to reduce huge costs involved with destruction and displacement due to extreme weather events. The value of this protective service of coral reefs is estimated at 314 million USD in Indonesia (Burke et al. 2002).

8.4.4 Supporting Services

Coral reefs are an essential part of land accretion. The natural action of waves breaks pieces of calcified coral, and these are washed up onto beaches. Through the process of natural physical breakdown, these larger pieces are broken into smaller and smaller pieces and eventually become part of the rubble, building these beaches. Corals, therefore, contribute, in part, to the process of accretion—which is the opposite of erosion (Miththapala 2008).

Coral reefs are very diverse. Corals do not even cover 1% of the Earth's surface, but they are extremely diverse. In fact, they are dubbed the rain forests of the sea because of this immense diversity. The nooks and crannies formed within reefs by the constant beating of waves provide shelter to many species. They are the home (they provide shelter and nursery grounds) of 25% of marine fish (Burke et al. 2002). Thirty two out of the 34 described groups of organisms are found in coral reefs. (As a comparison, only nine groups are found in tropical rain forests.) (Wilkinson 2002). Coral reefs support a complex and interdependent community of photosynthesizing organisms and animals. There is an incredible diversity of life on coral reefs, such as algae, corals (there may be as many as 750 species on one coral reef), sponges, marine worms, echinoderms (sea stars and their relatives), molluscs (snails, mussels and their relatives), crustaceans (crabs, shrimps and their relatives), and fish http://assets.panda.org/ (n.d.).

Coral reefs have high primary productivity. Zooxanthellae photosynthesize and produce their own food (like green plants do on land), and corals benefit from this association. Because of the immense diversity of coral reefs, a great deal of exchange of nutrients and primary productivity (food production) is very high. The primary productivity of coral reefs is estimated at 5–10 g $C/m^2/day$ (Sorokin 1995). This productivity is derived mainly from algae.

8.4.5 Cultural Services

Coral reefs have intrinsic, esthetic, and recreational values. The beauty of coral reefs and their diversity are essential parts of many cultures in different parts of the world. Because of their easy access, visiting coral reefs is an important recreation for snorkelers, scuba divers, recreational fishermen, and beach lovers. In Seychelles, tourism was estimated to have generated one fifth of GDP and over 60% of foreign exchange earnings in 1995 (Mathieu et al. 2000). In the Maldives, 'tourism contributes more than 60% of foreign exchange receipts, over 90% of government tax revenue comes from import duties and tourism-related taxes, and almost 40% of the workforce is employed in the industry' (Emerton 2006).

8.5 Threats to Coral Reefs

Despite their significance to humans, coral reefs are subjected to various types of large-scale natural and anthropogenic disturbances, which results in widespread mortality of these reef-building corals (Burke et al. 2011). These threats can be either local or global. As cited by Reefs at Risk Revisited (RRR), six primary stressors cause a decline in coral population: over-fishing, watershed-based pollution, marine-based pollution and damage, coastal development, coastal development, thermal stress, and ocean acidification (Burke et al. 2011).

8.5.1 Seawater Temperature Increase and Bleaching

Corals rely on the symbiotic relationship between corals and intracellular zooxanthellae. When corals are subjected to heat stress, the symbiotic relationship between corals and dinoflagellate breaks, the loss of zooxanthellae or loss/degradation of pigments from zooxanthellae from corals leaves the coral paler in appearance, hence, known as coral bleaching (Douglas 2003). The increasing temperature of the ocean due to human-induced global warming plays a greater role in heat stress.

Mass coral bleaching happened in the past years is posing a serious threat to the future of corals worldwide (Hughes et al. 2017). Various reports suggest a vast decline in these corals, up to 90% of corals had been found to be dead in many sites around the Indian Ocean, and 16% of reef-building corals of the world died (Sheppard 2003). In 2017, The Great Barrier Reef (Australia) experienced the largest bleaching event in the history of corals. Coral bleaching events are occurring more frequently and becoming longer in duration. Although corals can recover in weeks to months by reacquiring the symbionts, they grow slower and have decreased fecundity than unbleached corals (Heron et al. 2016).

8.5.2 Ocean Acidification

The increasing concentration of CO_2 in the sea waters and decreasing pH of the sea due to the formation of carbonic acid are another side effect of global warming. From 1751 to 1996, the average surface ocean pH has decreased from 8.25 to 8.14, which resulted in the shifting of chemical equilibria toward certain species over the others. A 35% increase in H^+ ion concentration in world's oceans was reported in 2005 (Jacobson 2005). Calcium carbonate is an important constituent of exoskeletons and shells of many corals, and as pH continues to decline, the concentration of carbonic acid in seawater increases, which in turn reacts with solid calcium carbonate and causes the dissolving of calcium, affecting the coral's fitness and growth (Mora et al. 2013).

Coldwater corals found at a depth of 3000 m are most vulnerable to acidification. When corals are once exposed to these corrosive waters, they will lose their skeletons and shell, and the reefs will collapse. Several studies report that, by the end of this century, 70% of the present deep-sea corals will be exposed to these corrosive waters (Guinotte et al. 2006).

8.5.3 Sedimentation

Sedimentation can result from coastal development works, terrestrial runoff, forestry, and agriculture. Sediments can smother corals and other reef organisms, and it greatly reduces the recruitment and survival of early life stages in corals. Sediments that are enriched with organic matter are more damaging to corals as they facilitate bacterial growth, resulting in the formation of muddy marine snow and anoxia and also reduced seawater pH in the boundary layer (Stafford-Smith and Ormond 1992). Severe exposure to terrestrial runoff leads to increased bioerosion, reduced reef calcification, shallower photosynthetic compensation points, changed coral community structure, and reduced species richness (Stafford-Smith and Ormond 1992).

8.5.4 Crown-of-Thorns (COT's) Starfish Outbreaks

The crown-of-thorns (COT's) sea star, *Acanthaster planci*, is a "corallivore" that feeds on coral tissues and is widely distributed throughout the tropical reefs across the planet. The outbreaks of COTs are considered a major threat to coral reef ecosystems. COTs commonly show cyclical oscillations in populations, with long periods of low density and scarcely distributed individuals and short periods of unsustainably high densities commonly called 'outbreaks.' What causes outbreaks of COT's is not still fully understood, but an increase in the phytoplankton availability based on nutrient delivery and warmer seawater temperature could be an important driver in promoting COT's outbreaks (Hutchings et al. 2019).

The COT's outbreaks, though found to be more or less localized to isolated reefs, their impact on the reef is more devastating. For instance, in Moorea, the outbreak which occurred in 2006, impacted the foreshore reef all around the island and wiped out all living corals in the outbreak area. COTs commonly described as 'cleaners of the reef,' are considered a part of a natural evolution of the reef, but the frequency of occurrence of COT's outbreaks poses a significant threat to the corals (Brodie et al. 2005).

8.5.5 Nutrients

The surface runoff mostly carries pesticides (herbicides, insecticides, and fungicides) which can disrupt photosynthesis in marine plants like mangroves, seagrasses, coral zooxanthellae, and crustose coralline algae. These have been found to directly affect the physiology of corals, especially their reproductive, larval, and juvenile stages (Hutchings et al. 2019). Shallow nearshore and lagoon coral reefs are considered the first to be impacted by land-based pollutants. Research across the coral reefs confirms the presence of herbicides like atrazine, simazine and alachlor, and organochlorine insecticides (Bambridge et al. 2019).

Studies demonstrate that herbicides and inorganic nutrients can induce coral bleaching and promote phytoplankton bloom and algal growth, which cause an increase in the crown-of-thorns starfish population, hence, constituting a major threat to coral reefs. In coral reef lagoons, nutrients like nitrogen and phosphorus generally act as a major limiting factor in primary production, but in lagoons with ocean water inputs, nitrogen is the essential limiting factor (Salvat et al. 2016).

8.5.6 Extreme Climatic Events

Natural events like the occurrence of Tsunami waves can generate a remarkable influence on marine ecosystems. These extremely powerful waves can affect corals, including whole or partial damage of corals by destruction, rotation, sedimentation of sands on corals, and local slides of reef slopes. These waves have the capability to destroy the whole coral communities, and they can even change the coral habitability radically (Atwater et al. 2017).

Cyclones, too, can destroy the whole coral communities; although they are brief disturbances, they have long-term impacts as they damage the three-dimensional framework of coral constructions. The recovery potential of corals following a cyclone is much more compared to a tsunami, especially for coral species which have the capacity for asexual propagation (Kayal et al. 2015).

8.5.7 *Tourism*

Activities like SCUBA diving and snorkeling as a part of promoting tourism, when not properly organized, could lead to the destruction of corals. Destruction of corals could result from infrastructure constructions like resorts, roads, boardwalks, etc., and water contamination through wastewater runoff carrying sedimentation into these reef waters (Juhasz et al. 2010). Physical damage to corals could also occur due to standing, walking, kicking, touching, trampling, and when their equipment damages these corals.

Boating and anchoring can also break or fragment these corals by their movement. Anchoring could damage the habitats near reefs that serve as nurseries and habitats for juveniles of different coral reef organisms. Chemicals added to paints used on boats and fishnets could also pollute coral reef waters. Several studies also suggest that the developing whale shark tourism has negatively impacted the coral reefs (Wong et al. 2019).

8.5.8 *Fishery Activities*

The coral reef supports a wide variety of fishes that depend on corals for their survival, and millions of people living along the world's coasts depend on reef fisheries for livelihood. Reef fishery activities are very difficult to assess as a wide variety of gears are used for fishing, lack of centralized markets, and there is an overlapping cultural and economic motivations to catch fish (Rassweiler et al. 2020).

The capturing of large fishes from coral reefs alters the composition of benthic fauna and increases the population of smaller fishes at lower trophic levels. The exclusion of larger fishes also led to an increase in the algal population as overfishing removes grazing fish species, which keep the coral clean by removing the algal growth on them (Daniela et al. 2006).

8.5.9 *Microplastics*

Microplastics (less than 5 mm in length sized plastics) also pose a potential threat to all marine organisms, including the corals. The most abundant plastic polymer found is polystyrene, followed by polyethylene and polypropylene. These microplastics come from the usage of Styrofoam, food and beverage packages, and fishing devices. Most of these plastics enter the oceans through land-based activities and a small portion by fishing and aquaculture activities. Microplastics in coral reef ecosystems threaten living organisms as they vary in color, making marine organisms like fishes mistaken for planktons (Muhammad et al. 2018).

8.6 Strategies for Conservation

Coral reef ecosystems suffered a drastic loss in the past few decades (Jackson et al. 2014). Reefs are subjected to several anthropogenic disturbances along with climate change. Although coral reefs have an innate capacity for natural recovery, the frequency of coral destruction is rapidly increasing, diminishing the time and capacity for recovery between catastrophic events (Montefalcone et al. 2018). Local-scale restoration action, along with the large-scale action on climate change and indirect management actions, could help restore coral reefs.

Coral restoration projects are mainly focused on five species of corals mostly (*Acropora cervicornis, Pocillo pora danicornis, Stylophora pistillata, A. palmate,* and *Porites cylindrical).* Of these, *A. cervicornis* and *A. palmate* are listed as endangered on International Union for Conservation of Nature Red List of Endangered Species (IUCN 2018). Coral restoration studies are dominated by short-term projects with less than 18 months of monitoring (Boström-Einarsson et al. 2020).

There are two types of restoration: Active restoration and Passive restoration. Passive restoration is the process of natural regeneration or an indirect restoration. It is implemented in the sites with less damage and basically relies on increasing the individual coral species without direct planting or seeding. Active restoration is the reintroduction or augmentation of coral species and is implemented in areas where unassisted natural recovery is unlikely (Boström-Einarsson et al. 2020).

There are different ways of restoration, of which some are described here.

8.6.1 Direct Transplantation

It is one of the most common methods of coral restoration in which coral fragments from a donor reef are directly transplanted. For this purpose, mostly fast-growing corals are selected, but their efficacy is not much (Boström-Einarsson et al. 2020).

8.6.2 Coral Gardening

Direct transplantation can have detrimental effects on donor corals populations. In coral gardening, small coral fragments are raised in nurseries prior to planting them on restoration sites. This will protect corals from harsh environmental conditions during their post-transplantation period. Corals are grown either in in situ or ex situ conditions. Some previous studies do report a high survival rate (Putchim et al. 2008) than direct transplantation method.

(a) *Micro-fragmentation*: This method is mostly based on massive corals where a diamond blade saw is used to cut massive corals into small fragments. These fragments can be further broken down into micro fragments, which can readily fuse with reef substrates or dead coral skeletons to form large colonies (Forsman et al. 2015). It has a very high survival rate.

(b) *Larval enhancement:* This method targets to increase the rate of coral fertilization, larval survival, and recruitment. A large number of coral larvae was found to be swept away from reefs, hence, failing to settle on the substratum (Jones et al. 2009). There are two types of larval enhancement procedures: (1) settlement of larvae on artificial structures where harvested ex situ reared gametes are used to improve post settlement survival rates and (2) larva collected during spawning was settled directly onto the reef or reared in holding tanks during the settlement period. Recently these methods were implemented successfully on highly degraded reef areas in many places (Dela Cruz and Harrison 2017).

(c) *Artificial Reefs:* Artificial reefs are human-made underwater structure which is built to enhance marine life. For creating such a substratum, substances are placed on the seabed, which closely resembles the natural reef, thereby promoting the habitat for reef fauna, fisheries production, and diving opportunities. It is found mostly in association with the project of coral transplantation. These have shown a 66% survival rate compared to other methods (Ferse 2009).

(d) *Stabilization of substratum:* Damaged substratum must be physically restored to increase the coral population. This involves stabilizing areas that have been affected by storms or ships by placing rubble in that area (Boström-Einarsson et al. 2020). A most common method to stabilize is to prevent the movement of rubble by installing mesh or net over the rubble. Such stabilization of reefs was reported from the works based on US territorial waters (Boström-Einarsson et al. 2020).

(e) *Substratum enhancement with electricity:* This process is done with the aim of mimicking the physical and chemical properties of reef limestone by enhancing calcium and magnesium precipitation on artificial substrata. When a direct electric current is established between electrodes, precipitation of calcium and magnesium occurs, which increases the calcification of coral polyps and increases their growth. Various scientific studies reported both positive and negative outcomes for these types of coral restoration (Romatzki 2014).

8.7 Conclusion

This chapter has summarized the coral reefs—their types, important hotspots, significance, causes of their depletion, and some of the coral restoration projects undertaken to preserve these valuable reefs. Corals play a prominent role in the formation of other ecosystems. They protect the shores from strong waves and currents and also protect and shelter many species of fish. Several restoration projects have been carried out to preserve them, but due to short-term nature of the projects, poor design, lack of adequate monitoring system, lack of outcomes, and increasing climatic variations play a prominent role in the failures of these projects. Therefore, the future of coral reefs will depend on our actions to save them. All of us can do our part to protect coral reefs: for example, eating less coral reef fish, recycling plastic waste so it does not end up in the ocean, or reducing energy consumption to slow climate change. This way, coral reefs of the future may remain colorful oases full of life in a marine desert.

References

Allen GR, Kinch JP, McKenna SA, Seeto P, eds (2003) A marine rapid assessment of Milne Bay Province, Papua New Guinea, RAP bulletin of biological assessment, second survey 29. Center for Applied Biodiversity Science, Conservation International, Washington, DC

Altizer S, Ostfeld RS, Johnson PTJ, Kutz S, Harvell CD (2013) Climate change and infectious diseases: from evidence to a predictive framework. Science 341(6145):514–519. https://doi.org/10.1126/science.1239401

Aronson R, Precht W, Macintyre I (2000) Coral bleach-out in Belize. Nature 405:36. https://doi.org/10.1038/35011132

Atwater BF, ten Brink USUS, Cescon AL, Feuillet N, Fuentes Z, Halley RB, Nuñez C, Reinhardt EG, Roger JH, Sawai Y, Spiske M, Tuttle MP, Wei Y, Weil-Accardo J (2017) Extreme waves in the British Virgin Islands during the last centuries before 1500 CE. Geosphere 13(2):301–368

Bambridge T, Chlous F, D'Arcy P, Claudet J, Pascal N, Reynaud S, Rodolfo-Metalpa R, Tambutte S, Thomassin A, Recuero Virto L (2019) Society- based solutions to coral reef threats in French pacific territories. Reg Stud Mar Sci 29:2352–4855

Berumen ML, Hoey AS, Bass WH, Bouwmeester J, Catania D, Cochran JE, Saenz-Agudelo P (2013) The status of coral reef ecology research in the Red Sea. Coral Reefs 32(3):737–748

Boström-Einarsson L, Babcock RC, Bayraktarov E, Ceccarelli D, Cook N, Ferse SCA, Hancock B, Harrison P, Hein M, Shaver E, Smith A, Suggett D, Stewart-Sinclair PJ, Vardi T, McLeod IM (2020) Coral restoration—A systematic review of current methods, successes, failures and future directions. PLoS One 15(1):e0226631. https://doi.org/10.1371/journal.pone.0226631

Brodie J, Fabricus K, Death G, Okaji K (2005) Are increased nutrient inputs responsible for more outbreaks of crown- of – thorns starfish? An appraisal of the evidence. Mar Pollut Bull 51(1–4): 266–278

Bruckner AW, Bruckner RJ (2006) The recent decline of Montastraea annularis (complex) coral populations in western Curaçao: a cause for concern. Rev Biol Trop 54:45–58

Burke L, Selig E, Spalding M (2002) Reefs at risk in Southeast Asia. World Resources Institute. http://pdf.wri.org/rrseasia_full.pdf

Burke L, Reytar K, Spalding M, Perry A (2011) Reefs at risk revisited. World Resources Institute, Washington, DC

Chivian E (2006) Medicines from natural sources: how human health depends on nature. In: Miththapala S (ed) Conserving medicinal species: securing a healthy future. Ecosystems and Livelihoods Group Asia, IUCN, Colombo, pp 1–7, 184 pp

Crosby MP, Brighouse G, Pichon M (2002) Priorities and strategies for addressing natural and anthropogenic threats to coral reefs in Pacific Island nations. Ocean Coastal Manag 45(2–3): 121–137. https://doi.org/10.1016/S0964-5691(02)00051-0

Roberts CM, McClean CJ, Veron JEN, Hawkins JP, Allen GR, McAllister DE, Mittermeier CG, Schueler FW, Spalding M, Wells F, Vynne C, Werne TB (2002) Marine biodiversity hotspots and conservation priorities for tropical reefs. Science 295(5558):1280–1284. https://doi.org/10.1126/science.1067728

Klein CJ, Ban NC, Halpern BS, Beger M, Game ET, Grantham HS, Green A, Klein TJ, Kininmonth S, Treml E, Wilson K, Possingha HP (2014) Prioritizing land and sea conservation investments to protect coral reefs. Mar Policy 49:146–154. https://doi.org/10.1371/journal.pone.0012431

Zandonella C (2016) When corals met algae: symbiotic relationship crucial to reef survival dates to the Triassic. Princeton University, Princeton

Ming CL (1998) Status of southeast Asian coral reefs. In: Wilkinson C (ed) Status of coral reefs of the world, 1998. Australian Institute of Marine Science, Townsville

Wilkinson C (2004a) Status of coral reefs of the world, 1:301 Global Coral Reef Monitoring Network

Dela Cruz DW, Harrison PL (2017) Enhanced larval supply and recruitment can replenish reef corals on degraded reefs. Sci Rep 7:1–13. https://doi.org/10.1038/s41598-016-0028-x

Daniela MC, Hughes TP, McCook LJ (2006) Impacts of simulated overfishing on the territoriality of coral reef damselfish. Mar Ecol Prog Ser 309:255–262. https://doi.org/10.3354/meps309255

Demers C, Hamdy CR, Corsi K, Chellat F, Tabrizian M, Yahia LH (2002) Natural coral exoskeleton as a bone graft substitute: a review. Biomed Mater Eng 12(1):15–35

DiBattista JD, Roberts MB, Bouwmeester J, Bowen BW, Coker DJ, Lozano-Cortés DF, Choat JH, Gaither MR, Hobbs JPA, Khalil MT, Kochzius M, Myers RF, Paulay G, Robitzch VSN, Saenz-Agudelo P, Salas E, Sinclair-Taylor TH, Toonen RJ, Westneat MW, Tand WS, Berumen ML (2016) A review of contemporary patterns of endemism for shallow water reef fauna in the Red Sea. Biogeography 43(3):423–439. https://doi.org/10.1111/jbi.12649

Douglas AE (2003) Coral bleaching- how and why? Mar Pollut Bull 46(4):385–392

Emerton LA (2006) Counting coastal ecosystems as an economic part of development infrastructure. Ecosystems and Livelihoods Group, Asia, IUCN, Colombo, 12 pp

Etnoyer P, Morgan L (2003) Occurrences of habitat-forming deep-sea corals in the Northeast Pacific Ocean. A report to NOAA's office of habitat protection. Marine Conservation Biology Institute, Silver Spring, pp 33

Spina F (2016) Príncipe Island: a tiny biodiversity wonder in the Gulf of Guinea. flora and fauna international. https://www.fauna-flora.org/news/principe-island-a-tiny-biodiversity-wonder-in-the-gulf-of-guinea/

Ferse SCA (2009) Multivariate responses of the coral reef fish community to artificial structures and coral transplants. In: Riegl BM, Dodge RE (eds) National Coral Reef Institute, Nova Southeastern University, Fort Lauderdale

Forsman ZH, Page CA, Toonen RJ, Vaughan D (2015) Growing coral larger and faster: micro-colony-fusion as a strategy for accelerating coral cover. PeerJ 3:e1313. https://doi.org/10.7717/peerj.1313. PMID: 26500822

Goreau TF, Goreau NI, Goreau TJ (1979) Corals and coral reefs. Sci Am

Great Barrier Marine Park Authority (2015) Facts about the great barrier reef. https://www.nature.com/climate/2009/0906/full/climate.2009.52.html

Guinotte JM, Orr J, Cairns S, Freiwald A, Morgan L, George R (2006) Will human induced changes in seawater chemistry alter the distribution of deep- sea scleractinian corals? Front Ecol Environ 4:141–146

Heron S, Maynard J, van Hooidonk R (2016) Warming trends and bleaching stress of the world's coral reefs 1985–2012. Sci Rep 6:38402. https://doi.org/10.1038/srep38402

Hoegh-Guldberg O., Hoegh-Guldberg H.(2004) Great barrier reef 2050, implications of climate change for the Australia's great barrier reef

http://assets.panda.org/ (n.d.)

http://www.panda.org/about_wwf/what_we_do/marine/blue_planet/ (n.d.)

https://seaworld.org/animals/all-about/coral-and-coral-reefs/habitat/ (n.d.)

https://www.livescience.com/40276-coral-reefs.html (n.d.)

Hughes TP, Kerry JT, Álvarez-Noriega M, Álvarez-Romero JG, Anderson KD, Baird AH (2017) Global warming and recurrent mass bleaching of corals. Nature 543:373–377. https://doi.org/10.1038/nature21707

Hunt B, Vincent AC (2006) Scale and sustainability of marine bioprospecting for pharmaceuticals. AMBIO 35(2):57–64

Hutchings P, Kingsford M, Hoegh-Guldberg O (2019) The great barrier reef: biology, environment and management. Csiro Publishing, 488pp

Monteiro J, Almeida C, Freitas R, Delgado A, Porteiro F, Santos RS (2008) Coral assemblages of Cabo Verde: preliminary assessment and description. In: Proceedings of the 11th international coral reef symposium, Ft. Lauderdale, Florida, pp 1416–1419

Jackson J, Donovan M, Cramer K, Lam V (eds) (2014) Status and trends of Caribbean coral reefs: 1970–2012. IUCN, Gland

Jacobson MZ (2005) Studying ocean acidification with conservative, stable numerical schemes for nonequilibrium air- ocean exchange and ocean equilibrium chemistry. J Geophys Res Atmos 110(D7):1–17. https://doi.org/10.1029/2004JD005220

Jones GP, Almany GR, Russ GR, Sale PF, Steneck RS, Van Oppen MJH, Willis BL (2009) Larval retention and connectivity among populations of corals and reef fishes: history, advances and challenges. Coral Reefs 28:307–325

Juhasz A, Ho E, Bender E, Fong P (2010) Does use of tropical beaches by tourists and island residents result in damage to fringing coral reefs? A case study in moorea French Polynesia. Mar Pollut Bull 60:2251–2256

Kayal M, Vercelloni J, Wand MP, Adjeroud M (2015) Searching for the best bet in life- strategy: a quantitative approach to individual performance and population dynamics in reef- building corals. Ecol Complex 23:73–84

Knowlton N, Brainard RE, Fisher R, Moews M, Plaisance L, Caley MJ (2010) Coral reef biodiversity. In: McIntyre AD (ed) Life in the world's oceans: diversity, distribution, and abundance. Wiley-Blackwell, Hoboken, pp 65–77

Lalli CM, Parsons TR (1997) Biological oceanography, an introduction, 2nd edn. Buttersworth-Heinmann, Oxford

Licuanan WY, Gomez ED (2000) Philippine coral reefs, reef fishes, and associated fisheries: status and recommendations to improve their management. In: Wilkinson C (ed) Status of coral reefs of the world: 2000. Australian Institute of Marine Science, Townsville

Loh TL, Pawlik JR (2014) Chemical defenses and resource trade-offs structure sponge communities on Caribbean coral reefs. Proc Natl Acad Sci U S A 111(11):4151–4156

Mathieu L, Langford IH, Kenyon W (2000) Valuing marine parks in a developing country: a case study of the Seychelles. CSERGE working paper GEC 2000–27

Miththapala S (2008) Coral reefs, coastal ecosystems series 1:1–36 + iii. Ecosystems and Livelihoods Group Asia, IUCN, Colombo

Montefalcone M, Morri C, Bianchi CN (2018) Long-term change in bioconstruction potential of Maldivian coral reefs following extreme climate anomalies. Global Change Biol 24:5629–5641. https://doi.org/10.1111/gcb.14439. PMID: 30194747

Mora C, Wei CL, Rollo A, Amaro T, Baco AR, Billett D, Bopp L, Chen Q, Collier M, Danovaro R (2013) Biotic and human vulnerability to projected changes in ocean biogeochemistry over the 21st century. PLoS Biol 11:e1001682. https://doi.org/10.1371/journal.pbio.1001682

Muhammad RC, Hadi TA, Prayudha B (2018) Occurrence and abundance of microplastics in coral reef sediment: a case study in Sekotong, Lombok-Indonesia. AES Bioflux 10:23–29. https://doi.org/10.5281/zenodo.1297719

Miloslavich P, Díaz JM, Klein E, Alvarado JJ, Díaz C (2010) PLoS One. journals.plos.org

Parasharya, Piyushkumar D, Padate, Geeta (2012) Study of corals and some associates in the marine national park and sanctuary Jamnagar. PhD thesis, Maharaja Sayajirao University of Baroda

Putchim L, Thongtham N, Hewett A, Chansang H (2008) Survival and growth of Acropora spp. in mid-water nursery and after transplantation at Phi Phi Islands, Andaman Sea, Thailand, Proceeding of the 11th international coral reef symposium, Ft. Lauderdale, Florida, 7–11 July 2008 session number 24, pp 1258–1261

Rassweiler A, Lauer M, Lester SE, Holbrook SJ, Schmitt RJ, Madi Moussa R, Munsterman KS, Lenihan HS, Brooks AJ, Wencelius J, Claudet J (2020) Perceptions and responses of Pacific Island fishers to changing coral reefs. Ambio 49(1):130–143

Reaka-Kudla ML (1997) The global biodiversity of coral reefs: a comparison with rain forests. In: Reaka-Kudla ML, Wilson DE, Wilson EO (eds) Biodiversity II: understanding and protecting our biological resources. Joseph Henry Press, Washington, DC, pp 83–108

Reaka M, Lombardi SA (2011) In: Zachos FE, Habel JC (eds) Biodiversity hotspots. Springer, Berlin, pp 471–501. https://doi.org/10.1007/978-3-642-20992-5_24

Rogers AD (1994) The biology of seamounts. Adv Mar Biol 30:305–350. https://doi.org/10.1016/s0065-2881(08)60065-6

Romatzki SBC (2014) Influence of electrical fields on the performance of Acropora coral transplants on two different designs of structures. Mar Biol Res 10:449–459

Salvat B, Roche H, Ramade F (2016) On the occurrence of a widespread contamination by herbicides of coral reef biota in French Polynesia. Environ Sci Pollut Res 23:49–60

Sheppard CRC (2003) Predicted recurrences of mass mortality in the Indian Ocean. Nature 425: 294–297

Smith SV (1978) Coral reef area and the contribution of reefs to processes and resources of the world oceans. Nature 273:225–226

Sorokin YI (1995) Coral reef ecology. Springer, Berlin, p 465

Stafford-Smith MG, Ormond RFG (1992) Sediment rejection mechanisms of 42 scleractinian corals. Aust J Marine Freshwater Sci 43:638–705

Stehli FG, Wells JW (1971) Diversity and age patterns in Indian Ocean corals. Syst Zool 20:115–126

Stoeckl N, Hicks C, Mills M, Fabricius K, Esparon M, Kroon F, Kaur K, Costanza R (2011) The economic value of ecosystem services in the great barrier reef: our state of knowledge. Ann N Y Acad Sci 1219(1):113–133

Strahm W (1996) Mascarene Islands an introduction. Curtis's Botanical Magazine 13(4):182–185. https://doi.org/10.1111/j.1467-8748.1996.tb00567.x.JSTOR45066025

Thébaud C, Ben W, Cheke A, Strasberg D (2009) Mascarene Islands biology. In: Encyclopedias of the natural world, pp 612–619

Velloth S, Nayak S (2013) Habitat classification of coral reefs of India using multispectral and hyperspectral remote sensing data. Department of marine geology. Mangalore University, 18, p 160

Venkataraman K, Satyanarayana C, Alfred JRB, Wolstenholme J (2003) Handbook on hard corals of India. Zoological Survey of India, Kolkata, 1–266pp

Wells SM (1988) Coral reefs of the world, vol II. Indian Ocean, UNEP, IUCN Centre, Townsville

Wilkinson C (2000) Status of coral reefs of the world. Australian Institute of Marine Science, Townsville

Wilkinson C (2002) Status of coral reefs of the world 2002. Australian Institute of Marine Science (AIMS), Global Coral Reef Monitoring Network (GCRMN), Townsville

Wilkinson C (2004b) Status of the coral reefs of the world: 2004. Australian Institute of Marine Science, Townsville

Wilkinson C (2008) Status of coral reefs of the world. Global Coral Reef Monitoring Network, and Reef and Rainforest Research Centre, Townsville

Williams GC (1986) What are corals? Sagittarius—magazine of the south African museum. Cape Town 1(2):11–15

Wilson RR, Kaufmann RS (1987) Seamount biota and biogeography. In: Keating BH, Fryer P, Batiza R, Boehlert GW (eds) Seamounts, islands, and atolls geophysical monograph, vol. 43. American Geophysical Union, Washington, pp 355–378. https://doi.org/10.1029/gm043p0355

Wong CWM, Conti- Jerpe I, Raymundo LJ, Dingle C, Araujo G, Ponzo A, Baker DM (2019) Whale shark tourism: impacts on coral reefs in the Philippines. Environ Manag 63:282–291. https://doi.org/10.1007/s00267-018-1125-3

Chapter 9
The Seaweed a Gold Mine for Drugs

K. P. Manoj Kumar and K. Arunkumar

Abstract The earth's crust is mainly composed of water (75%), and it is estimated that roughly 50–80% of the earth's oxygen production comes from the oceans. Most products are made from seaweed. Half of the world's biodiversity consists of marine life, which provides numerous new compounds that can be evaluated for medicinal activity. The macroalgae include numerous marine species-rich in bioactive compounds that are available to humans and animals. In the last three decades, seaweeds have been known for the high diversity of their metabolites which are used for antioxidant, anti-inflammatory, anti-aging, antiviral antihyperlipidemic, anticoagulant, antifungal, antibacterial, and tissue healing properties. At first, the seaweed was consumed as food; later, some discoveries in new species led the way to yield industrial, medicinal, pharmaceutical, and commercial products. In the proposed chapter, we discussed different types of seaweed species used for drug development and their medicinal properties.

Keywords Seaweeds · Marine algae · Macroalgae · Nutraceuticals

9.1 Introduction

Aquatic algae thrive in freshwater, brackish water, saltwater (shallow or coastal), lakes, ponds, and estuaries. In some cases, algae are also found in air and soil habitats. Algae are a primitive crypt group that evolved as microscopic organisms that do not have true roots, stems, and leaves but have pseudoroots known as fixatives/rhizomes and fronds for photosynthesis (Surender Reddy et al. 2018). In the early days, the importance of marine aquatic plants was not recognized, so it was eventually called seaweed (Kumar 2018). The importance of these seaweeds in modern times has been well studied and popular among scientists due to the

K. P. Manoj Kumar · K. Arunkumar (✉)
Department of Plant Science, School of Biological Sciences, Central University of Kerala, Kasaragod, Kerala, India

S. T. Sukumaran, K. T R (eds.), *Conservation and Sustainable Utilization of Bioresources*, Sustainable Development and Biodiversity 30,
https://doi.org/10.1007/978-981-19-5841-0_9

bioactive compounds. Seaweed plays a vital role in the aquatic food web and is a huge source of vitamins, minerals, carbohydrates, proteins, fats, and essential micronutrients that humans consume directly (Blunt et al. 2009). Therefore, seaweed is called the "healing food" of the twenty-first century (Surender Reddy et al. 2018). The composition of algae depends on geographic location, season, and temperature. Seaweed is an economical and renewable source used to convert solar energy into biomass because of its simple cellular structure and living in the water. Commercial use of this algae did not begin until after World War II. Because there was not enough protein for the vast world population, this historic incident paved the way for the development of industrial seaweed production. Countries, like India, China, Japan, and Republic of Korea, are the largest seaweed producers in the world.

Seaweeds are classified based on the pigments they produce: Rhodophyta (red algae), Chlorophyta (green algae), and Phaeophyceae (brown algae) (Menaa et al. 2020). They have numerous pharmaceutical and biomedical properties that are used in pharmaceutical, biomedical, biotechnological, agricultural, aquacultural, and energy industries (Menaa et al. 2020). In the last decades, many pharmaceutical and biomedical companies started investing in marine organisms, including seaweeds, because of their vast benefits in making novel drugs. Seaweeds, such as Phaeophyceae, which are rich in fucoidan, have antibacterial, antiviral, anti-inflammatory, anticoagulant, and antithrombotic properties. Rhodophyta (red algae) contains large amounts of carrageenan and agar. Such carrageenan and agar are widely used in medicine and pharmaceuticals, and in particular, agar used in biopharmaceuticals is also used as an anticoagulant. Ulvan is found in Ulva (green algae), a sulfated polysaccharide used primarily in the cosmetic and pharmaceutical industries and is used as an emulsifier, stabilizer, and thickener in foods. In one way or another, algae are used directly or indirectly by humans, animals, and agriculture (Lomartire et al. 2021). Here, in this proposed chapter, we discuss the novel and current uses of seaweeds as drugs that play an important role in pharmaceutical industries.

9.2 Distribution of Seaweeds

Seventy-one percent of the globe is dominated by photosynthetic macrophytes known as seaweeds, namely, Chlorophyta (green algae), Phaeophyta (brown algae), and Rhodophyta (red algae), occurring in tidal, intertidal, and subtidal regions, respectively. Some of the common green seaweeds, like *Ulva* (sea lettuce), *Enteromorpha* (green string lettuce), *Chaetomorpha, Codium,* and *Caulerpa* are found in the intertidal zone. The tidal zone is inhabited by common brown seaweeds, such as *Sargassum, Laminaria, Turbinaria, and Dictyota*. Red seaweeds, like *Gracilaria, Gelidiella, Eucheuma, Ceramium, and Acanthophora,* grow in the subtidal waters. Commercial seaweed activities are processed in 42 countries throughout the world (distribution in bhava doc). Among that 221 species of seaweeds are utilized commercially. The coastline of India is about 7500 km, including those of islands, including Andaman & Nicobar and Lakshadweep. This coastline displays much marine biological diversity, including seaweeds, which play a prominent role in exhibiting various properties and economic purposes. These seaweeds are mostly distributed along India's coastline (Bhagyaraj and Kunchithapatham 2016). More than 200 seaweed species support an international economy in primarily phycocolloid (algins, agars, and carrageenans) and food products valued at over the U.S. $6.2 billion (Williams and Smith 2007).

India is one of the mega biodiversity nations in the world. The coastline of India is 8100 km long and has an Exclusive Economic Zone (EEZ) of 2.17 million km^2 (equal to 66% of the total mainland area). Nearly thirty percent of the human population is dependent on rich exploitable coastal and marine resources (Athulya and Anitha 2019). The Indian coastline supports the luxuriant growth of diverse seaweed populations, which increases economic importance. The first report on seaweed diversity in India was published by Iyengar (1927) on the flora of Krusadai Island, on the Southeast coast of India. Subsequently, Thivy (1948), Krishnamurthy (1957), and Umamaheswara Rao (1969) made substantial contributions to increasing our knowledge of India's seaweed diversity. Central Salt and Marine Chemicals Research Institute (CSMCRI) conducted a comprehensive survey from 1980 to 1990 and estimated the total seaweed standing stocks as 97,400 tons wet wt. from the coast of Tamil Nadu, 7500 tons wet wt. from the coast of Andhra Pradesh, and 19,345 tons wet wt. from the Lakshadweep islands (Kaliaperumal and Kalimuthu 1997). Likewise, the CSMCRI team researched the seaweed diversity of the Gujarat and Tamil Nadu coasts (Jha et al. 2009), known for their relative abundance of seaweeds. The results show that these two maritime states are the home to 366 seaweed species that represent nearly half of India's total seaweed diversity. Gujarat has a coastline of 1600 km long and is located in northwest India. A survey was conducted in Gujarat about intertidal seaweed diversity. It states that 198 species represent all three significant seaweed groups: Rhodophyta 109 species from 62 genera; Chlorophyta 54 species from 23 genera; and Phaeophyceae 35 species from 16 genera. Likewise, Tamil Nadu's coastline is 1076 km long, with 13 coastal districts and 15 major ports and harbors. There is a variation in the nature of substratum in littoral and sublittoral regions in Tamil Nadu. The northern coastal area has long and wide sandy beaches intermittently populated with flat rocks, boulders, and stones in the lower and upper intertidal regions, experiencing severe surf energy. Species belonging to the genera *Ulva, Chaetomorpha, Bryopsis, and Grateloupia* inhabit the rocks and boulders. The region around pal bay falls within the Cauvery deltaic region, filled with muddy intertidal and subtidal coast is due to the confluence of many rivers flowing to the sea. Several species, like *Gracilaria*, including *G. edulis, G. foliifera, G. verrucosa, and G. Salicornia,* are dominant and commercially harvested in this region. Subsequently, in the areas of Mannar, all the three groups of seaweeds that grow in intertidal and subtidal are found abundantly. Species, like *Sargassum, Acanthophora,* and *Hypnea,* are predominant and cultivated in this region (Ganesan et al. 2019).

Coastline of Kerala has 580 km, which is extended in nine districts of the State from Poovar, Thiruvananthapuram district in the South to Thalappady, Kasaragod district in the north. It is the third-largest coast of India after Gujarat with 1600 km (Jha et al. 2009) and Maharashtra with 720 km (Sakhalkar and Mishra 2014). A study of seaweeds was conducted by SK Yadavan Mookkan Palanisamy, in 2015 along the coastlines of Kerala. The results of the study were the identification of a total of 42 economically important seaweed species from the Kerala coast. Among these, 29 species are edible for humans, 24 species are suitable for industries to extract the phycocolloids (agar-agar, agaroids, algin, carrageenan, etc.), 14 species as fodder for domestic animals, 11species for the production of manures, and seven species are medicinal (Athulya and Anitha 2019).

9.3 Classification of Seaweeds

Approximately 500,000 species populate the oceans, which means that seawater is native to nearly three-quarters of all known species. Since algae were the precursors to photosynthesis, there are theories explaining that life might have originated from these unique creatures on our planet. Photosynthesis may have begun in some blue-green prokaryotic microorganisms that were previously thought to be algae and belong to the phylum Cyanobacteria, now part of the Kingdom Monera. More than 30,000 algal species have been described, and their scientific study is known as phycology. According to the current definition of algae, blue-green varieties are not considered algae as they are prokaryotes, and only eukaryotes (unicellular, such as phytoplankton microalgae, or multicellular, such as macroalgae) fall into this category. The taxonomic classification of algae is complicated by the variety of species present and the numerous classification criteria used. They are a polyphyletic group, which means they are related to multiple kin groups. As a result, the classification is not well defined and may vary depending on the authors, but they are currently classified as Protista, with the exception of macroalgae, which are classified as Plantae. According to modern Linnaean taxonomy, the marine plants described belonging to two empires and four kingdoms. The Empire Eukaryota is made up of three kingdoms, Plantae, Chromista, and Protozoa, while the Empire Prokaryota is made up of the Kingdom Eubacteria. Seagrasses (Phylum Tracheophyta or marine flowering plants), green algae (Phylum Chlorophyta), and red algae (Phylum Rhodophyta) are all members of the Kingdom Plantae. Brown algae are members of the Kingdom Chromista (Phylum Ochrophyta, Class Phaeophyceae, and Bacillariophyceae). Protozoa is a kingdom of unicellular algae-dinoflagellates (Phylum Myzozoa). Blue-green algae make up the Kingdom Eubacteria (Phylum Cyanobacteria). The marine plant classification is given in Latin. Representatives of different Phyla (Divisions) differ primarily in their somatic and reproductive structures, life histories, photosynthetic pigments, biochemistry of their storage products, and cell wall ultrastructure (Titlyanov et al. 2017). Macroalgae are also a very diverse group, ranging in size from a few centimeters to specimens up to 100 m in length. About 15,000 species in this group have been reported. Because they are also autotrophs and photosynthetic organisms, their habitat is limited to a certain depth, usually up to 60 m mostly within the intertidal zone, and their growth is usually vertical, seeking sunlight. This macroalgae group was classified based on the pigment that makes up the composition and allows it to perform photosynthesis to feed autotrophically (Gomez-Zavaglia et al. 2019).

The Rhodophyta is characterized by a predominance of phycoerythrin and phycocyanin pigments, which give this group a red color and mask the pigments, like chlorophylls a and c. Other pigments found in their cells are carotene and xanthophylls (such as lutein and zeaxanthin) (Titlyanov et al. 2017). Rhodophytes are divided into four subtypes and eight classes. The two most important classes in terms of hydrocolloid production are Bangiophyceae and Florideophyceae, representing 161 and 6224 species, respectively. Species of the genera *Porphyra, Gelidium, Gloiopeltis, Eucheuma, and Gracilaria* are the most widely distributed worldwide (Rioux and Turgeon 2015). They are predominantly marine in distribution; only approximately 3% of over 5000 species occur in truly freshwater habitats. In North America, 25 genera are recognized in inland habitats. Freshwater red algae are largely restricted to streams and rivers and can occur in other inland habitats, such as lakes, hot springs, soils, caves, and even sloth hair (Sheath and Wehr 2015).

Phaeophyta (brown algae) has unique characteristics. Photosynthetic pigments include chlorophyll, β-carotene, fucoxanthin, violaxanthin, diatoxanthin, and other xanthophylls, and typically contain excess carotenoids than chlorophyll pigments. Fucoxanthin is present in sufficient amounts to mask the green color of the chlorophyll and give the algae its characteristic brown color (Sheath and Wehr 2015). Brown algae (Ochrophyta) mostly found in marine waters are classified into 20 classes. More than 1800 species of brown algae are found in the class Phaeophyceae. *Dictyotales, Ectocarpales, Fucales, and Laminariales Ecklonia, Eisenia, Fucus, Laminaria, Lessonia, Macrocystis, Sargassum, and Undaria* are the orders that include most species (Rioux and Turgeon 2015). Exceptionally there are six genera of freshwater brown algae, of which four were identified from North America. Brown algae are mainly distributed in the sea. Less than 1% of the species live in freshwater. Inland species are benthic, inhabiting lakes or streams, with a scattered distribution.

Phylum Chlorophyta (Green algae) are named after their green chloroplasts. They are characterized by a predominance of green pigments (chlorophylls a and b), masking carotene, xanthophylls (such as lutein, zeaxanthin, and cyphonoxanthin), and other pigments (Titlyanov et al. 2017). This includes 44 flagellate genera, at least 129 cocci-like and immobile colonies, 81 filamentous and plant-like genera, and 48 conjugating genera and desmids. Some members of the green algae (most filamentous and plant species, such as Chlorophyta) attach to the solid surface of stagnant or flowing water, but some may float or exist in soil or other aquatic habitats. It is often found in the stagnant waters of ditches and ponds. Desmids are more common in ponds and streams with low conductivity and moderate nutrient levels and are often mixed with macro flora (Sheath and Wehr 2015).

9.4 Morphology and Anatomy of Seaweeds

The diversity of the frond structures of algae contrasts sharply with the uniformity of vascular plants. In the latter case, parenchymal meristems (e.g., the tips of shoots and roots) form tissues with different shapes. In marine algae, the parenchyma predominates only in brown algae. Larger algae, particularly *Laminariales* and *Fucales*, have several distinct tissues and cell types, including photosynthetic epidermis, cortex, medulla, phloem, and mucous ducts. The parenchymal ontology of Dictyotales was followed extensively by Katsaros and Galatis (1988), citing many sources. Cell division may occur throughout the algae or be localized to meristem areas. If localized, it is most often at the apex, but may be at the base or somewhere in between (intercalary). A simple filament consists of an unbranched chain of cells attached by their end walls and results from cell division only in the plane perpendicular to the axis of the filament. Unbranched filaments are uncommon among seaweeds; examples are Ulothrix and Chaetomorpha. Usually, some cell division takes place parallel to the filament axis to produce branches (Cladophora, Ectocarpus, Antithamnion). The adhesion of filaments can also produce a pseudoparenchymatous crust (Peyssonnelia, Neoralfsia) or blade (Anadyomene). Many siphon green algae, including Halimeda and Codium, are formed by intertwining numerous filaments. Although there is an interaction between the morphology of whole algae and the environment, the physiological responses to the environment and the mechanisms by which the overall morphology is generated occur within individual cells (Niklas and Ulrich, 2009). Cells are protected by walls and membranes and separated by membrane organelles, and through these membranes and walls, they must come into contact with the environment. Thus, the structure and organization of cellular components provide the necessary background for physiological and ecological studies. Different cellular components are different in algae. These include the composition and structure of the cell wall, the flagellar apparatus, the cytoskeleton, and the structure of the thylakoid photosystem. Algal cells also have unique structures, many of which contain bioactive secondary metabolites. (Phaeophyta 1979) In red algae plants, the foliar greens of red algae have thylakoids that occur alone (not stacked like green algae and brown algae). The cell walls are low in cellulose gelatinous or amorphous sulfuric acid, such as agar and carrageenan. It contains chemicalized galactan polymer, and the stored nutrient is Floridian starch. Red algae are reproduced in various ways. Their reproductive method and life history are the most complex, including both sexual and asexual modes. Most of the advanced genera have a gamete production stage (sexual stage) and two sporulation stages (asexual stage) (Titlyanov et al. 2017). The structure of brown algae fronds varies from atypical filamentous to pseudoparenchymal uniaxial morphology to true parenchymal morphology. No single-celled or simple filamentous brown algae are known, and the simplest frond tissue in this phylum is a branched heterotrichous filament. Their cell walls are mainly composed of cellulose and alginic acid. Laminarin and mannitol are the main forms of nutrient storage for brown algae. Laminarin is a polymer of glucose and mannitol, and mannitol is a

6-carbon sugar alcohol. All motile reproductive cells in brown algae are laterally bifurcated, except for those in Dictyotales, which produce uniflagellates and non-flagellates. Each brown alga has both sexual and asexual reproduction (Titlyanov et al. 2017). Green algae vary widely in morphology, from microscopic unicellular algae to macroscopic multicellular algae (Lewis and McCourt 2004). Also, giant algae are rarely longer than a meter. Green algae are photosynthetic eukaryotes that transport plastids containing chlorophyll an and b as well as starch (Rioux and Turgeon 2015). The cell wall of green algae is composed of a pectin layer and an inner cellulose layer. Starch is a reserve nutrient. As a general rule, green algae reproduce both sexually and asexually. In most cases, flagellated spores are produced, and non-flagellated spores are produced to a lesser extent (Titlyanov et al. 2017).

9.5 Ethnobotanical Study of Seaweeds

Ancient people have been using seaweed as medicine and food for thousands of years. In recent years, seaweed has been widely used in the pharmaceutical, food, and cosmetic industries (Liu et al. 2020). Seaweed has been used for medicinal purposes since ancient times. The Romans benefited from seaweed and used it to heal wounds and burns. Basic seaweed compounds, such as alginate and agar, are already used in medicine as binders and carrier materials for medical tablets or dressings (Smit 2004). According to recent reports, seaweed is still used as a folk remedy. Compounds extracted from seaweed have high nutritional and medicinal value, so interest in the development of seaweed-based pharmaceuticals is growing. Seaweed has also been used to treat goiter and other gland problems and has been practiced in Japan and China since 300 BC. Red algae, also known as *Digenea simplex*, has proven to be very effective in treating acariasis. Species, such as *Chondrus, Gracilaria, Pterocladia, and Gelidium,* are used to treat the intestines, such as constipation, stomach pain, and ulcers (Abdussalam 1990). The chemical composition of seaweeds is characterized by a wide range of molecules with antibiotic activity, in particular, halogenated molecules (haloforms, halogenated alkanes, alkenes, alcohols, aldehydes, hydroquinones, ketones, polysaccharides, fatty acids, phlorotannins, pigments, lectins, alkaloids, terpenoids), sterols, and some heterocyclic and phenolic compounds (Pérez et al. 2016; Smit 2004). All seaweed algae extracts are tested worldwide for antibacterial and antiviral activity. For example, algae extracts of seaweeds, such *as Gelidiella acerosa, Gracilaria verrucosa, and Hypnea musciformis,* were dissolved in various solvent extraction methods, and excellent results were obtained for *paratyphi, Enterococcus aerogenes, Staphylococcus epidermidis, Salmonella typhi, and Shigella flexneri.* (Pal et al. 2014).

9.6 Nutraceutical and Biomedical Aspects of Seaweeds

Seaweed is a source of minerals, vitamins, protein, carbohydrates, and fiber. Seaweed contains many macro- and microelements. Brown algae are known for their high iodine content. Likewise, red and green algae are rich in protein (Kolanjinathan et al. 2014).

A study of the high protein content of algae off the coast of Mandapam found that brown algae contained more protein than red and green algae (Thirunavukkarasu et al. 2019). The content of biochemical elements, such as proteins, lipids, carbohydrates, vitamins, and minerals, varies depending on the season and terrain conditions. Seeds, such as brown algae, contain biologically active compounds that are used as drugs for their antibacterial, antiviral, anti-inflammatory, and antithrombotic properties. However, the algae belonging to the class Rhodophyta are used in the medical and pharmaceutical industries, especially in the production of agar and carrageenan. In addition, green algae contain a large amount of ulvan, a major physiologically active ingredient used as an emulsifier, stabilizer, and thickener in biopharmaceuticals, cosmetics, pharmaceuticals, and foods. Consuming algae as food or natural medicine will contribute to a healthier lifestyle for mankind. The nutraceutical, biomedical, and pharmaceutical uses, including bioactive seaweed compounds, were analyzed in detail. The beneficial properties and biological compounds of seaweed have aroused interest among researchers. These compounds can be used in pharmaceutical and biomedical applications to obtain new drugs of natural origin and create new and functional food (Lomartire et al. 2021).

9.7 Applications of Seaweeds in Pharmaceuticals and Medicine

9.7.1 Antiviral Activity

Viral infections are the number one cause of death worldwide, and the current novel coronavirus breakout shatters the world. Aqueous, methanol, chloroform–methanol, and dichloromethanol extracts of 16 Moroccan red algae were tested for herpes simplex virus type 1 (HSV1) using the cell viability method, with some positive results showing increased antiviral activity. Plouguerné et al. (2013) isolated the antiherpes compounds from the Brazilian kelp *Sargassum vulgare*. This bioactive compound was purified from the lipid fraction of the crude extract and was identified as sulfoquinovosyl diacylglycerol (SQDG). SQDG should have high antiviral activity against the herpes simplex virus (HSV1 and HSV2). *Chondrus crispus* (Rhodophyta) has long been used as a mucus remedy for diarrhea, dysentery, and stomach ulcers and has been used as an ingredient in numerous medicinal teas, such as colds. Sulfated polysaccharides, such as galactan sulfate, have been laboratory tested for effectiveness against HIV and herpes simplex virus (HSV). For example, sulfated polysaccharides from *Schizymenia pacifica* have been shown for the first time to possess anti-HIV reverse transcriptase in vitro. The red alga *Solieria chordalis* lives in coastal areas of France. *S. chordalis* extract showed potential antiviral activity. J. B. Hudson Kim examined 16 British Columbia seaweed extracts for antiviral activity. Among them, in *Analipus japonicus*, the antiviral activity was mainly virucidal (direct virus effect). For the antiviral properties of seaweed, polysaccharides were a common ingredient (Hudson et al. 1998). Marine polysaccharides, in particular, are seaweed with a unique structure, and it has a virucidal effect that interferes with different stages. Therefore, the viral infection process has attracted great interest in pharmaceutical discovery and the development of antivirals drugs (Shi et al. 2017).

Hui WANG extracted six brown seaweed species with hot water for their antiviral properties. He found that the extracts of *Hydroclathrus clathratus and Lobophora variegata* showed more potential anti-HSV activities and had average anti-respiratory syncytial virus (RSV) activities but could not inhibit influenza A virus (Wang et al. 2008). Likewise, green algae *Ulva fasciata* and *Codium decorticatum* both showed the highest activity (99.9%) against HSV-1, with triacylglycerols and fatty acids as the major components. The red alga *Laurencia dendroidea* showed good activity against HSV-1 (97.5%). (Soares et al. 2012) The extract of marine alga *Codium fragile* was active against all three test viruses (herpes simplex, HSV; Sindbis, SINV; polio virus), whereas the waste of *Enteromorpha linza, Colpomenia bullosa, Scytosiphon lomentaria, and Uncaria pinnatifid* was active against HSV and SINV, but not poliovirus (Hudson et al. 1999). The red seaweed *Sebdenia polydactyla extract* has sulfated polysaccharides, particularly xylomannan sulfate, showing strong antiviral properties (Ghosh et al. 2009). Furthermore, a polysaccharide called fucopyranosyl is considered very important for antiherpetic activity (Mandal et al. 2007).

9.7.2 Antibacterial Activity

Seaweeds are a potentially abundant source of bioactive secondary metabolites and provide clues to help develop new pharmaceutical compounds. These seaweeds are partially unexplored in terms of potential bioactivity. Organic extracts of *Amphiroa anceps, Corallina officinalis, and Sargassum filipendula* were tested for antibacterial activity against the three pathogens *Salmonella typhimurium, Staphylococcus aureus, and Escherichia coli. C. officinalis* extract showed the highest inhibition of *Staphylococcus aureus* among the above seaweeds. Seven bacterial isolates, such as *Escherichia coli, Klebsiella pneumoniae, Pseudomonas aeruginosa, Proteus mirabilis* (as Gram-negative bacteria), *Bacillus subtilis*, *Staphylococcus aureus,* and *Streptococcus aureus,* were selected and antibacterial isolate was tested for *Sargassum linear*. Among them, *Staphylococcus aureus* was sensitive to *Sargassum and cystoseiraclinite*, and the growth of *Bacillus subtilis* was strongly influenced by *cystoseiraclinite* (Abu-Khudir et al. 2021). Bhuyarm et al. conducted an antibacterial experiment against *B. cereus* using the disk diffusion method on *Kappaphycus alvarezii.* The results indicate that *Kappaphycus alvarezii* was more effective against the bacteria *B. cereus* (Bhuyar et al. 2020). Numerous seaweeds possess bioactive components that are capable of inhibiting the growth of some Gram-positive and Gram-negative bacterial pathogens. The algal extracts were used as a curative and inhibitory agent for various diseases. Species of Rhodophyta (*Laurencia okamurai, Dasya scoparia, Grateloupia filicina, and Plocamium telfairiae)* showed a wide spectrum of antibacterial activity. Antibacterial activity was exhibited in 45 extracts of nine algae, namely, *Sargassum wightii, fragilisima, Gracilaria edulis,* and *Enteromorpha,* against the bacterial species that were Gram positive and Gram negative, such as *Haemophilus, Streptococcus pneumoniae, Staphylococcus aureus, and Moraxella* (De Corato et al. 2017).

Choudhury et al. screened for the biological activity of solvent extracts of three marine algae *Gracilaria tikvahiae, Ulva lactuca, and Ulva fasciata,* which showed activity against virulent strains of bacteria pathogenic to fish, viz., *Edwardsiella tarda, Vibrio alginolyticus, Pseudomonas fluorescens,* and *Pseudomonas aeruginosa* (Carneiro et al. 2020; Kolanjinathan et al. 2014). Antagonist compounds, such as fatty acids, bromophenols, tannins, phloroglucinols, and terpenoids, are used as antibiotics to kill bacteria, fungi, and viruses (Reichelt and Borowitzka 1984). Certain seaweed extracts from various regions have been shown to have antibacterial properties (Anand et al. 2016).

9.7.3 Antifungal Activity

Fungal infections are a growing threat to public health. Opportunistic infections, such as *aspergillosis, candidiasis, and cryptococcosis,* have become a major problem worldwide for cancer patients, transplant recipients, and other immunocompromised individuals, including those with acquired immunodeficiency syndrome (AIDS), despite an increase in the number of drugs available to treat infections over the past decade. Since the discovery of penicillin, mushrooms have been considered a good source of natural antibacterial products. Researchers have created many antifungal drugs from natural sources. Seaweed contains many bioactive compounds that scientists have not studied. P. Kumar studied the biosynthesis of silver nanoparticles using *Gracilaria corticata* for antifungal activity against Candida spp. The results showed that Ag nanoparticles biosynthesized from *G. corticata* showed antifungal activities against *Candida albicans* and *C. glabrata*. Two lectins recently isolated from the galactose-binding green alga *C. isthmocladum* were found to inhibit up to 60% biofilm formation of *Staphylococcus aureus* and *Staphylococcus epidermidis* compared to controls (Carneiro et al. 2020). The effect of fluorotannins on dermatomygal pathogens which were extracted from *Cystoseira nodicaulis, Crassiphycususneoides, and Fucus Spiralis* using n-hexane followed by extraction with acetone:water (7:3) were evaluated. These phlorotannins exhibit antifungal activity against *C. albicans, E. floccosum, and T. Mentagrophytes*

(Negara et al. 2021). Floretol and fucofloretol extracted from *kelp digitalis* are effective against fungal plant pathogens, such as *B. cinerea and M. laxa,* by 100% inhibition of mycelial growth (De Corato et al. 2017). The extracts of brown algae belonging to *Sargassum muticum, Dictyota bartyrensiana, Padina gymospora, Chnoospora implexa, and S. wightii* have been reported by Jayaraj et al. They are active against leaf rot in Indian rice caused by *Rhizoctonia solani* (Jayaraj et al. 2008). Methanol and aqueous extracts of *Cladophora glomerata, Ulva lactuca, Ulva reticulata, Gracilaria corticata, Kappaphycus alvarezii, and Sargassum wightii* have been used in antifungal studies against fungal pathogens, such as *Aspergillus niger, Aspergillus cervisum, Aspergillus flavus, and Mucor indicus.* The green algae, *U. lactuca,* exhibited high activity, followed by the brown algae, *S. wightii, green algae C. glomerata, and red algae K. alvarezii.* It exhibits antifungal activity and inhibits the growth of mold (Aruna et al. 2010).

9.7.4 Antioxidant Activity

Seaweeds are known to contain compounds with antioxidant or free radical scavenging properties, such as glutathione peroxidase, catalase, superoxide dismutase, and polyphenols. Methods, such as beta-carotene depigmentation (ABTS) radical depigmentation assay, DPPH radical scavenging assay, and iron-reducing antioxidant (FRAP) assay, are used to determine antioxidant activity. Antioxidant activity was evaluated by Lee Je-Hyeok et al. using the extract of *Sargassum muticum, Polysiphonia morrowii, Dictyopteris undulata Holmes, Sargassum micracanthum, and Sargassum macrocarpum.* All of the above seaweeds showed antioxidant activity, in DPPH assay (Lee and Kim 2015). Antioxidant activity has been reported in many seaweeds, including Colpomenia, *Gracilaria, Halymenia, Hydroclathrus, Laurencia, Padina, Polysiphonia, and Turbinaria* (Cornish and Garbary 2010). Reactive oxygen species (ROS) include a group of oxygen-containing compounds that are highly reactive with important biomolecules and pose a threat to cellular integrity (Shanura Fernando et al. 2016). ROS, as well as reactive nitrogen species (collectively RS), have been identified as causative agents of various pathogenic diseases and adverse clinical conditions relevant to human health. These include cancer, cardiovascular disease, atherosclerosis, hypertension, ischemia/reperfusion, diabetes mellitus, hyperoxaluria, neurodegenerative diseases, such as Alzheimer's and Parkinson's diseases, arthritis rheumatism, and aging. Polyphenols are derived from certain types of brown algae provide high protection against skin carcinogenesis caused by UVB rays (Hwang et al. 2006). Yasantha athukorala evaluated the antioxidant potential of crude extracts of *G. filicina* for their ability to scavenge ROS (DPPH, DH, H2O2, and O2-) and inhibit lipid peroxidation. Using the tested methanolic extract of *G. filicina,* 82% of DPPH radicals were scavenged, nearly three times more than using BHT. Similarly, the same methanol extract recovered 65% of superoxide anion, which is near twice the amount of superoxide anion recovered from BHT and cxtocopherol (Athukorala et al. 2003). Additionally, De

Souza et al. observed higher antioxidant capacity in sulfated polysaccharides, where fucoidan and fucans polysaccharides from *Fucus vesiculosus and Padina gymnospora,* respectively, inhibited the formation of hydroxyl radicals and superoxide radicals. A positive correlation exists between the number of polysaccharides in macroalgae and their antioxidant activity, as demonstrated by Rocha de Souza et al. (El-Shafei et al. 2021). Several studies have shown that marine sulfated polysaccharides are antioxidants. Sulfated polysaccharides are macromolecules that have an array of biological functions. Researchers have studied seaweeds most widely for their antioxidants. Furthermore, antioxidants have been demonstrated to play a significant role in human health, thus raising consumer demand for such products. A variety of nutritional products from seaweeds may contain antioxidants that are commonly used in health food or nutraceutical supplements (Balakrishnan et al. 2014).

9.7.5 Antibiotics

Active antibiotic molecules from the algae and their characteristics associated with various pathogenic microorganisms were studied by Broadbent (1966). Manuel L. Lemos studied bacteria that produce antibiotics from the microbial flora attached to algae, and 224 bacterial strains were isolated from five green and brown seaweeds were tested for antibiotic production. A total of 38 strains showed antibiotic activity, and *Enteromorpha intestinalis* was the source of most production strains. All epiphytic bacteria with antibiotic activity were assigned to the *Pseudomonas alteromonas* group (Lemos et al. 1985). Fifty-five seaweed extracts from 11 species of seaweed were studied against antibiotic-resistant postoperative infectious drug-resistant pathogens, namely, *E. coli, Klebsiella pneumoniae, Pseudomonas aeruginosa, Streptococcus pyogenes, and Staphylococcus aureus.* Among the seaweed extracts, the acetone extract of *Caulerpa cupressoides* showed the greatest inhibitory activity against E. coli, and the propanol extract of *Gracilaria edulis* showed the greatest inhibitory activity against *K. pneumoniae*. Acetone extracts of *Padina tetrastromatica and Laurencia cruciata* show maximum inhibitory activity against *P. aeruginosa*, and butanol extracts of *Hypnea musciformis, Caulerpa cupressoides, and Chaetomorpha linoides* show maximum inhibitory effect against *S. aureus* (Ravikumar et al. 2002).

9.8 Novel Drugs Developed from Seaweed Extract

Nosocomial diseases or nosocomial infections occurs in patients during their stay in the hospital. Recent studies have shown that 5–19% of patients in acute care hospitals have in-hospital infectious conditions. Bacteria are the leading cause of disease in hospitals, with viruses as the second cause. Sometimes fungi cause the disease but rarely are protozoa involved. Mostly nosocomial diseases are caused by Gram-negative bacilli, like E. coli (Ravikumar et al. 2002). Likewise, Neglected Tropical Disease (NTD) is a group of parasites, bacteria, viruses, and fungal infections and is one of the most common causes of illness in the poorest people in developing countries. These diseases are widespread in tropical and subtropical situations in 149 countries around the world, and affected individuals are often affected by multiple parasites or infectious diseases. NTDs include leishmaniosis, Chagas disease, and dengue fever, which affect about 1.4 billion people each year, and in developing countries, billions of dollars were utilized to prevent this. It is estimated that these diseases cause 35,000 deaths daily worldwide (Cohen et al. 2017). Unfortunately, NTD is overshadowed by three major diseases: the human immunodeficiency virus (HIV), tuberculosis, and malaria, which receive more funding for research. However, coinfection with NTDs can also be fatal to these diseases. As a result, NTD was also considered and included in the Global Fund for Fighting HIV, tuberculosis, and malaria. Trypanosomiasis is a contagious NTD caused by Kinetoplastid parasites and transmission of the genus Trypanosome from tsetse flies (genus Glossina) or triatomin to human beetle bites. Despite the prevalence of Chagas disease and its risk of spreading, the number of new cases of African human t*rypanosomiasis r*eported to the WHO has declined in recent years due to national control programs, bilateral agreements, and NGOs (Freile-Pelegrín and Tasdemir 2019). In all cases, natural drugs are used to treat the diseases. Natural products of marine life are widely researched as a source of new bioactive molecules, but despite seaweed having great potential, no seaweed-based molecules are in drug discovery. Currently, seaweed extracts are used in herbal medicine to correct nutritional imbalances, and numerous pharmaceuticals incorporate algae colloids in their formulations (as excipients in syrups and capsules); seaweed is one of the richest sources of new bioactive compounds known. In fact, thousands of molecules have been identified. Polysaccharides and lipids and small secondary metabolites, such as polyphenols, halogenated phenols, and monoterpenes. In seaweeds, terpenes, a class of chemical compounds, represent one of the major classes of metabolites. Some of these agents are effective in the treatment of cancer, malaria, and heart disease. They have also been reported to possess insecticidal properties (Cabral et al. 2021). In western countries, the coronavirus disease 2019 (COVID-19) pandemic has been very devastating. In Japan, it appears to be less severe and seaweed consumption is possibly one reason. SARS-CoV-2 shares 79.6% of its genome with the SARS-CoV genome. Both viruses bind to angiotensin-converting enzyme 2 (ACE2) via the coronavirus spike (S) protein and use ACE2 as an entry receptor. Various bioactive components found in these edible seaweeds, like *Saccharina* (kombu), *Undaria pinnatifida (*wakame*), Cladosiphon okamuranus (*mozuku*), and Sargassum fusiforme (*hijiki*); Porphyra spp (*nori*); and Ulva spp* (sea lettuce) have a wide variety of health benefits, including antihypertensive, antioxidant, and anti-inflammatory effects. The consumption of seaweed may provide protection against COVID-19 through a variety of mechanisms. ACE inhibitory components may prevent ACE dominance from SARS-CoV-2 infection (Tamama 2021). Despite the evidence that marine macroalgae have a lot of potential, few clinical trials for marine-derived compounds have been conducted. Polysaccharides, polyphenols, and peptides in seaweed are biologically active components that can modulate the diversity and abundance of bacterial populations in the gut. It has been suggested that prebiotics increase bacterial populations and often their ability to produce short-chain fatty acids, which are used by gastrointestinal epithelial cells, protect them against pathogens, and help induce apoptosis in colon cancer cells. The bioactive compounds found in seaweeds might provide some potential for the prevention and, in some instances, treatment of diseases in humans. However, large-scale clinical trials are required to confirm these potential therapeutic effects (Shannon et al. 2021). Human papillomavirus (HPV) infection is one of the major risk factors for the development of premalignant lesions that will undergo cancerous transformation. Cervical cancer represents one of the leading causes of cancer-related death in women all over the world. Studies have shown that sulfated polysaccharides, such as dextran sulfate, heparan, and cellulose sulfate, inhibit HPV colonization and carcinogenesis. The compound fucoidans inhibited the growth of HeLa cervical cells in vitro. Terpenoids obtained from brown algae are also promising agents for preventing cervical cancer. The human papillomavirus (HPV) is a type of deoxyribonucleic acid (RNA) virus that penetrates genital and urinary tracts, throat, and mouth in humans and other vertebrates. HPV infection is found in 43% of vulvar tumors, 70% of vaginal tumors, and 100% of cervical tumors. HPV is almost always transmitted by sexual contact, but other routes, such as perinatal vertical transmission or physical contact, have also been documented (Moga et al. 2021). A marine heparinoid polysaccharide derived from brown algae has been shown to inhibit HPV infection in vitro, with an IC50 value of 90%. Carrageenans are found in red seaweed and are mainly made of D-linked galactose residues, which possess antiviral, anticoagulant, and anticancer properties; their anti-HPV properties have also been documented in numerous studies (Moga et al. 2021). Even though fucoidan is not as effective as carrageenan, it has the potential to be a novel chemotherapy agent in the future for cervical cancer and HPV infection (Buck et al. 2006). The laminarins reported in the gastrointestinal tract have bioactive properties and affect intestinal mucus structure, pH, and short-chain fatty acid production. In addition, laminarin prevents infection by bacterial pathogens and is very protective against radiation, increases the numbers of B cells and T helpers, and reduces systolic blood pressure, so it interferes with typical metabolic abnormalities. (Keleszade et al. 2021). It is believed that *Undaria pinnatifida* (Phaeophyceae) acts as an anti-inflammatory and a diuretic. The algae can also be used to treat fever and edema in women after childbirth (Lomartire et al. 2021). In a study by Celikler et al., the extracts from the chlorophyte ulva lidar were found to exert antigen toxic activity in in vitro human lymphocytes. The antigenic activity itself was not observed, but these extracts displayed chemotherapeutic activity against mitomycin C in human lymphocytes (Celikler et al. 2008). *Gracilaria edulis* extract exhibited diverse properties that are applicable to pharmaceutical formulations, including antidiabetic, antioxidant, antimicrobial, anticoagulant, anti-inflammatory, and anti-proliferative activities. Red algae have been found to possess hypoglycemic properties by inhibiting carbohydrate-digesting enzymes, reducing glucose absorption, and forming antiglycating end products (Lomartire et al. 2021). Neurodegenerative diseases generally strike older adults in midlife, causing steady declines in motor or cognitive functions and shortening life expectancy. They are the result of several environmental factors and genetic causes. In the series of diseases leading to extensive loss of neurons, Parkinson's and Alzheimer's are well-defined examples of neurodegenerative diseases. There might be some neuroprotective macro algal compounds that can be used to prevent and treat neurodegenerative diseases, including Parkinson's disease and Alzheimer's disease (Bauer et al. 2021).

In addition to carbohydrates, polysaccharides and monosaccharides constitute the majority of seaweed components. Moreover, red algae do not typically contain monosaccharides or oligosaccharides, though many of them do. There has been great interest in extracting polysaccharides from seaweed because of their neuroprotective and neuro-repair properties. These polysaccharides may prove to be the next significant breakthrough in neurodegenerative disease treatment. Recent research by Bauer et al. suggests that fucoidan and its derivatives may be potential agents in treating Alzheimer's disease (Bauer et al. 2021). Multiple Sclerosis (MS) is one of the most common chronic, inflammatory, nerve demyelination diseases of the brain, and it is believed to be the result of an autoimmune response. In recent years, many studies worldwide have focused on multiple sclerosis, making it possible to continuously improve the quality of life for patients. The quality of diet and lifestyle factors can influence the course of multiple sclerosis; therefore, dietary recommendations for people with multiple sclerosis may help reduce the severity of certain symptoms. Red algae Dulse (*Palmaria palmata*) and brown algae Kombu *(Saccharina latissima)* have polyphenols, vitamins A, B12, and C, high levels of fiber, protein, minerals, and arginine and are low in saturated fatty acids and various polyunsaturated fatty acids, such as omega-6 or EPA linoleic and arachidonic acids. Therefore, it is a good seaweed for general strengthening in anemia and asthenia conditions. It also strengthens eyesight (high-level vitamin A) and is recommended for the treatment of gastrointestinal and intestinal disorders, and regeneration of mucous membranes (respiratory, stomach, and vaginal) because of the high content of Vitamin B12. This red alga prevents cardiovascular disease since this vitamin reduces the level of homocysteine in the blood when large amounts are deposited in the blood vessels (Pereira and Valado 2021).

9.9 Conclusion

Our discussion in this chapter has focused on the pharma uses of seaweed and its sustainability in the development of drugs. Regardless of their abundance, marine algae are a valuable resource, containing a variety of functional metabolites, such as polysaccharides, proteins, peptides, lipids, amino acids, polyphenols, and mineral salts. Usually, there is a relationship between their traditional use and in vitro or in vivo experimental data. For example, brown and red seaweeds contain unique secondary metabolites with potential application to human infectious diseases. These seaweeds contain numerous bioactive compounds, and their use for the preparation of new functional ingredients in food is promising, as well as treatment and prevention of chronic disease. In addition to being rich in minerals, vitamins, and fiber, they are also low in lipids, so they are a suitable choice for weight loss regimens, easing intestinal transit, lowering blood cholesterol levels, and reducing cancer.

References

Abdussalam S (1990) Drugs from seaweeds. Med Hypotheses 32(1):33–35. https://doi.org/10.1016/0306-9877(90)90064-L

Abu-Khudir R, Ismail GA, Diab T (2021) Antimicrobial, antioxidant, and anti-tumor activities of Sargassum linearifolium and Cystoseira crinita from Egyptian Mediterranean Coast. Nutr Cancer 73(5):829–844. https://doi.org/10.1080/01635581.2020.1764069

Anand N, Rachel D, Thangaraju N, Anantharaman P (2016) Potential of marine algae (seaweeds) as a source of medicinally important compounds. Plant Genet Resour: Characterisation Util 14(4): 303–313. https://doi.org/10.1017/S1479262116000381

Aruna P, Mansuya P, Sridhar S, Suresh Kumar J, Babu S (2010) Pharmacognostical and antifungal activity of selected seaweeds from Gulf of Mannar region. Recent Res Sci Technol 2(1): 2076–5061. www.recent-science.com

Athukorala Y, Lee KW, Song C, Ahn CB, Shin TS, Cha YJ, Shahidi F, JeonY J (2003) Potential antioxidant activity of marine red alga grateloupia filicina extracts. J Food Lipids 10(3): 251–265. https://doi.org/10.1111/j.1745-4522.2003.tb00019.x

Athulya K, Anitha T (2019) Algal biodiversity along southern coasts of india : a review. 6(2): 93–101.

Balakrishnan D, Kandasamy D, Nithyanand P (2014) A review on antioxidant activity of marine organisms. Int J ChemTech Res 6(7):3431–3436

Bauer S, Jin W, Zhang F, Linhardt RJ (2021) The application of seaweed polysaccharides and their derived products with potential for the treatment of Alzheimer's disease. Mar Drugs 19(2):89. https://doi.org/10.3390/md19020089

Bhagyaraj I, Kunchithapatham VR (2016) Diversity and distribution of seaweeds in the shores and water lagoons of Chennai and Rameshwaram coastal areas, South-Eastern coast of India. Biodivers J 7(4):923–934

Bhuyar P, Rahim MH, Sundararaju S, Maniam GP, Govindan N (2020) Antioxidant and antibacterial activity of red seaweed; Kappaphycus alvarezii against pathogenic bacteria. Glob J Environ Sci Manag 6(1):47–58. https://doi.org/10.22034/gjesm.2020.01.04

Blunt JW, Copp BR, Hu WP, Munro MHG, Northcote PT, Prinsep MR (2009) Marine natural products. Nat Prod Rep 26(2):170–244. https://doi.org/10.1039/b805113p

Broadbent D (1966) Imperial Chemical Industries Limited, Pharmaceuticals Division, Research Department, Alderley Park, Macclesfield, Cheshire, England. 32:219–242.

Buck CB, Thompson CD, Roberts JN, Müller M, Lowy DR, Schiller JT (2006) Carrageenan is a potent inhibitor of papillomavirus infection. PLoS Pathog 2(7):671–680. https://doi.org/10.1371/journal.ppat.0020069

Cabral EM, Oliveira M, Mondala JRM, Curtin J, Tiwari BK, Garcia-Vaquero M (2021) Antimicrobials from seaweeds for food applications. Mar Drugs 19(4):211. https://doi.org/10.3390/md19040211

Carneiro RF, Duarte PL, Chaves RP, da Silva SR, Feitosa RR, de Sousa BL, da Silva Alves AW, de Vasconcelos MA, da Rocha BAM, Teixeira EH, Sampaio AH, Nagano CS (2020) New lectins from Codium isthmocladum Vickers show unique amino acid sequence and antibiofilm effect

on pathogenic bacteria. J Appl Phycol 32(6):4263–4276. https://doi.org/10.1007/s10811-020-02198-x
Celikler S, Yildiz G, Vatan O, Bilaloglu R (2008) In vitro antigenotoxicity of Ulva rigida C. Agardh (Chlorophyceae) extract against induction of chromosome aberration, sister chromatid exchange and micronuclei by mutagenic agent MMC. Biomed Environ Sci 21(6):492–498. https://doi.org/10.1016/S0895-3988(09)60008-8
Cohen AJ, Brauer M, Burnett R, Anderson HR, Frostad J, Estep K, Balakrishnan K, Brunekreef B, Dandona L, Dandona R, Feigin V, Freedman G, Hubbell B, Jobling A, Kan H, Knibbs L, Liu Y, Martin R, Morawska L, Pope CA, Shin H, Straif K, Shaddick G, Thomas M, Dingenen R, Donkelaar A, Vos T, Murray CJL, Forouzanfar MH (2017) Estimates and 25-year trends of the global burden of disease attributable to ambient air pollution: an analysis of data from the Global Burden of Diseases Study 2015. Lancet 389(10082):1907–1918
Cornish M, Garbary DJ (2010) Antioxidants from macroalgae: potential applications in human health and nutrition. Algae 25(4):155–171. https://doi.org/10.4490/algae.2010.25.4.155
De Corato U, Salimbeni R, De Pretis A, Avella N, Patruno G (2017) Antifungal activity of crude extracts from brown and red seaweeds by a supercritical carbon dioxide technique against fruit postharvest fungal diseases. Postharvest Biol Technol 131(January):16–30. https://doi.org/10.1016/j.postharvbio.2017.04.011
El-Shafei R, Hegazy H, Acharya B (2021) A review of antiviral and antioxidant activity of bioactive metabolite of macroalgae within an optimized extraction method. Energies 14(11):3092. https://doi.org/10.3390/en14113092
Freile-Pelegrín Y, Tasdemir D (2019) Seaweeds to the rescue of forgotten diseases: a review. Bot Mar 62(3):211–226. https://doi.org/10.1515/bot-2018-0071
Ganesan M, Trivedi N, Gupta V, Madhav SV, Radhakrishna Reddy C, Levine IA (2019) Seaweed resources in India—Current status of diversity and cultivation: Prospects and challenges. Bot Mar 2016(Fao 2018):1–21. https://doi.org/10.1515/bot-2018-0056
Ghosh T, Pujol CA, Damonte EB, Sinha S, Ray B (2009) Sulfated xylomannans from the red seaweed Sebdenia polydactyla: structural features, chemical modification and antiviral activity. Antiviral Chem Chemother 19(6):235–242. https://doi.org/10.1177/095632020901900603
Gomez-Zavaglia A, Prieto Lage MA, Jimenez-Lopez C, Mejuto JC, Simal-Gandara J (2019) The potential of seaweeds as a source of functional ingredients of prebiotic and antioxidant value. Antioxidants 8(9):406. https://doi.org/10.3390/antiox8090406
Hudson JB, Kim JH, Lee MK, DeWreede RE, Hong YK (1998) Antiviral compounds in extracts of Korean seaweeds: evidence for multiple activities. J Appl Phycol 10(5):427–434. https://doi.org/10.1023/A:1008004117305
Hudson JB, Kim JH, Lee MK, Hong YK, DeWreede RE (1999) Multiple antiviral activities in extracts of seaweeds from British Columbia. Pharma Biol 37(4):300–306
Hwang H, Chen T, Nines RG, Shin HC, Stoner GD (2006) Photochemoprevention of UVB-induced skin carcinogenesis in SKH-1 mice by brown algae polyphenols. Int J Cancer 119(12):2742–2749. https://doi.org/10.1002/ijc.22147
Jayaraj J, Wan A, Rahman M, Punja ZK (2008) Seaweed extract reduces foliar fungal diseases on carrot. Crop Protect 27(10):1360–1366. https://doi.org/10.1016/j.cropro.2008.05.005
Jha B, Reddy CRK, Thakur MK, Rao MU (2009) Seaweeds of India. The diversity and distribution of Seaweeds in Gujarat Coast. CSMCRI, Bhavnagar, p 215
Kaliaperumal N, Kalimuthu S (1997) Seaweed potential and its exploitation in India. Seaweed Res Util 19(1&2):33–40
Katsaros C, Galatis B (1988) Thallus development in *Dictyopteris membranacea* (Phaeophyta, Dictyotales). Br Phycol J 23:71–88
Keleszade E, Patterson M, Trangmar S, Guinan KJ, Costabile A (2021) Clinical efficacy of brown seaweeds ascophyllum nodosum and fucus vesiculosus in the prevention or delay progression of the metabolic syndrome: a review of clinical trials. Molecules 26(3):714. https://doi.org/10.3390/molecules26030714

Kolanjinathan K, Ganesh P, Saranraj P (2014) Pharmacological importance of seaweeds: a review. World J Fish Marine Sci 6(1):1–15. https://doi.org/10.5829/idosi.wjfms.2014.06.01.76195
Kumar KS (2018) Seaweeds: distribution, production and uses
Lee JH, Kim GH (2015) Evaluation of antioxidant activity of marine algae-extracts from Korea. J Aquat Food Prod Technol 24(3):227–240. https://doi.org/10.1080/10498850.2013.770809
Lemos ML, Toranzo AE, Barja JL (1985) Antibiotic activity of epiphytic bacteria isolated from intertidal seaweeds. Microb Ecol 11(2):149–163. https://doi.org/10.1007/BF02010487
Lewis LA, Mccourt RM (2004) Green algae and the origin of land plants. Am J Bot 91(10):1535–1556
Liu J, Luthuli S, Wu Q, Wu M, Choi J II, Tong H (2020) Pharmaceutical and nutraceutical potential applications of Sargassum fulvellum. Biomed Res Int 2020:13–15. https://doi.org/10.1155/2020/2417410
Lomartire S, Marques JC, Gonçalves AMM (2021) An overview to the health benefits of seaweeds consumption. Mar Drugs 19(6):341. https://doi.org/10.3390/md19060341
Mandal P, Mateu CG, Chattopadhyay K, Pujol CA, Damonte EB, Ray B (2007) Structural features and antiviral activity of sulphated fucans from the brown seaweed Cystoseira indica. Antiviral Chem Chemother 18(3):153–162. https://doi.org/10.1177/095632020701800305
Menaa F, Wijesinghe PAUI, Thiripuranathar G, Uzair B, Iqbal H, Khan BA, Menaa B (2020) Ecological and industrial implications of dynamic seaweed-associated microbiota interactions. Mar Drugs 18(12):1–25. https://doi.org/10.3390/md18120641
Moga MA, Dima L, Balan A, Blidaru A, Dimienescu OG, Podasca C, Toma S (2021) Are bioactive molecules from seaweeds a novel and challenging option for the prevention of HPV infection and cervical cancer therapy?—a review. Int J Mol Sci 22(2):1–18. https://doi.org/10.3390/ijms22020629
Negara BFSP, Sohn JH, Kim JS, Choi JS (2021) Antifungal and larvicidal activities of phlorotannins from brown seaweeds. Mar Drugs 19(4):1–11. https://doi.org/10.3390/MD19040223
Niklas KJ, Ulrich K (2009) The evolutionary development of plant body plans. Funct Plant Biol 36:682–695
Pal A, Kamthania MC, Kumar A (2014) Bioactive compounds and properties of seaweeds—a review. OALib 1(4):1–17. https://doi.org/10.4236/oalib.1100752
Pereira L, Valado A (2021) The seaweed diet in prevention and treatment of the neurodegenerative diseases. Mar Drugs 19:1–25
Pérez MJ, Falqué E, Domínguez H (2016) Antimicrobial action of compounds from marine seaweed. Mar Drugs 14(3):1–38. https://doi.org/10.3390/md14030052
Phaeophyta T (1979) Range of thallus organization. I.
Plouguerné E, de Souza LM, Sassaki GL, Cavalcanti JF, Villela Romanos MT, da Gama BA, Pereira RC, Barreto-Bergter E (2013) Antiviral Sulfoquinovosyldiacylglycerols (SQDGs) from the Brazilian brown seaweed Sargassum vulgare. Mar Drugs 11(11):4628–4640. https://doi.org/10.3390/md11114628. PMID: 24284427; PMCID: PMC3853750
Ravikumar S, Anburajan L, Ramanathan G, Kaliaperumal N (2002) Screening of seaweed extracts against antibiotic resistant post operative infectious pathogens. Seaweed Res Util 24:95–99
Reichelt JL, Borowitzka MA (1984) Antimicrobial activity from marine algae: results of a large-scale screening programme. In: Bird CJ, Ragan MA (eds) Eleventh International Seaweed Symposium. Developments in hydrobiology, vol 22. Springer, Dordrecht. https://doi.org/10.1007/978-94-009-6560-7_26
Rioux LE, Turgeon SL (2015) Seaweed carbohydrates. In: Seaweed sustainability: food and non-food applications. Elsevier, pp 141–192. https://doi.org/10.1016/B978-0-12-418697-2.00007-6
Sakhalkar SS, Mishra RL (2014) Biodiversity of Marine Benthic Algae from Intertidal Zone of Konkan Coast (Maharashtra). Indian J Appl Res 4(2):1–3
Shannon E, Conlon M, Hayes M (2021) Seaweed components as potential modulators of the gut microbiota. Mar Drugs 19(7):1–50. https://doi.org/10.3390/md19070358

Shanura Fernando IP, Kim M, Son KT, Jeong Y, Jeon YJ (2016) Antioxidant activity of marine algal polyphenolic compounds: a mechanistic approach. J Med Food 19(7):615–628. https://doi.org/10.1089/jmf.2016.3706

Sheath RG, Wehr JD (2015) Introduction to the freshwater algae. In: Freshwater algae of North America: ecology and classification, pp 1–11. https://doi.org/10.1016/B978-0-12-385876-4.00001-3

Shi Q, Wang A, Lu Z, Qin C, Hu J, Yin J (2017) Overview on the antiviral activities and mechanisms of marine polysaccharides from seaweeds. Carbohydr Res 453–454:1–9. https://doi.org/10.1016/j.carres.2017.10.020

Smit AJ (2004) Medicinal and pharmaceutical uses of seaweed natural products: a review. J Appl Phycol 16(4):245–262. https://doi.org/10.1023/B:JAPH.0000047783.36600.ef

Soares AR, Robaina MCS, Mendes GS, Silva TSL, Gestinari LMS, Pamplona OS, Yoneshigue-Valentin Y, Kaiser CR, Romanos MTV (2012) Antiviral activity of extracts from Brazilian seaweeds against herpes simplex virus. Rev Bras Farma 22(4):714–723. https://doi.org/10.1590/S0102-695X2012005000061

Surender Reddy K, Abraham A, Afewerki B, Tsegay B, Ghebremedhin H, Teklehaimanot B (2018) Extraction of agar and alginate from marine seaweeds in Red Sea region. Int J Mar Biol Res 3(2):1–8. https://doi.org/10.15226/24754706/3/2/00126

Tamama K (2021) Potential benefits of dietary seaweeds as protection against COVID-19. Nutr Rev 79(7):814–823. https://doi.org/10.1093/nutrit/nuaa126

Thirunavukkarasu R, Mary Shamya A, Joseph J (2019) Screening of phytochemical, antioxidant activity and antibacterial activity of marine seaweeds. Int J Pharm Pharm Sci 11(1):61. https://doi.org/10.22159/ijpps.2019v11i1.29119

Titlyanov EA, Titlyanova TV, Li X, Huang H (2017) Marine plants of coral reefs. In: Coral reef marine plants of Hainan island. https://doi.org/10.1016/b978-0-12-811963-1.00002-0

Wang H, Ooi EV, Ang PO (2008) Antiviral activities of extracts from Hong Kong seaweeds. J Zhejiang Univ Sci B 9(12):969–976. https://doi.org/10.1631/jzus.B0820154

Williams SL, Smith JE (2007) A global review of the distribution, taxonomy, and impacts of introduced seaweeds. Annu Rev Ecol Evol Syst 38:327–359. https://doi.org/10.1146/annurev.ecolsys.38.091206.095543

Chapter 10
Mangroves: An Underutilized Gene Pool to Combat Salinity

Anu Augustine, Jumana Muhammed, and Babu Valliyodan

Abstract Salinity is a global problem, being aggravated by climate change, scanty rainfall, poor irrigation systems, salt ingression, water contamination, and other environmental factors. The salinity stress tolerance mechanism is a very complex phenomenon, and stress pathways are co-ordinately linked to impart salt tolerance. Although a number of salt-responsive genes have been reported from the halophytes, there is always a quest for promising stress-responsive genes that can modulate plant physiology according to salt stress. Several known genes, like antiporters, antioxidant encoding genes, and some novel genes, were isolated from halophytes and explored for developing stress tolerance in the crop plants (glycophytes). We provide here a comprehensive update on salinity-induced adverse effects on soils and plants. In this chapter, the physiological and biochemical adaptation strategies that help mangroves and crop plants grow and survive in salinity-affected areas are reviewed. In this review, mangroves are discussed as an underutilized gene pool of salt-responsive genes that can be utilized for developing salinity tolerance in crop plants using strategies, like genetic engineering and molecular breeding by marker-assisted breeding phenotyping technologies, GWAS, etc.

Keywords Salt stress · Mangroves · Mechanisms of tolerance · Salinity tolerance genes · Physiology · Molecular breeding methods · QTLs · Phenotyping · GWAS

A. Augustine (✉) · J. Muhammed
Department Biotechnology and Microbiology, Kannur University, Kannur, Kerala, India
e-mail: anua@kannuruniv.ac.in

B. Valliyodan
Department of Agriculture and Environment Science, Lincoln University, Jefferson City, MO, USA

S. T. Sukumaran, K. T R (eds.), *Conservation and Sustainable Utilization of Bioresources*, Sustainable Development and Biodiversity 30,
https://doi.org/10.1007/978-981-19-5841-0_10

10.1 Introduction

Salt stress is a grave concern limiting crop production and causing damage amounting to 27.3 billion dollars annually (Qadir et al. 2014), forcing to look for expansion of irrigated agriculture to saline water. A 50% increase in the demand for food and fresh water is predicted by 2050 to sustain the expected 10–13 billion people (Gleeson et al. 2012; UN Water and Energy 2014; FAO 2017). The scarcity of food is bound to increase more than ever before, especially in sub-Saharan Africa, the Middle East, and South Asia, where population growth far exceeds agricultural productivity (FAO 2017). If we achieve the attainable productivity in these regions in rice, maize, and wheat, the productivity can be increased by 230%, 135%, and 224%, respectively, in these crops in the sub-Saharan region and by 129%, 83%, and 61% in South Asia (Mueller et al. 2012). The situation requires us to expand the irrigated agriculture to saline water which forms 98% of global water resources (Morton et al. 2019). Morton et al. (2019) define salt tolerance as the ability to maintain desirable plant performance traits under salt stress relative to control conditions.

Salt tolerance is the ability of the plant to reverse the deleterious effects of high salt concentration and grow to complete its lifecycle. Halophytes or salt-loving plants are further classified as obligate and facultative based on their dependence on salt for optimal growth. Studies on selected halophytes to investigate the salt requirement for growth and development showed that salts are not compulsorily required for the development of halophytic species, but the availability of water and nutrients is the key limiting factor for growth in natural saline habitats (Grigore et al. 2012). It is also hypothesized that halophytes got distributed to saline areas to avoid competition with glycophytic species (Mishra and Tanna 2017).

Mangroves are facultative halophytes tolerant to fluctuating salinity levels. Downton (1982) and Clough (1984) have included mangroves under obligate halophytes. Mangroves grow optimally at a salt concentration ranging from 5 to 25% (Downton 1982; Clough 1984; Ball 1988; Burchett et al. 1989; Ball and Pidsley 1995). Species-specific variations in the range of tolerance to salinity are reported (Ball 1988). Total absence and excessive NaCl are found to inhibit the growth of several mangrove species (Downton 1982; Clough 1984; Burchett et al. 1989; Pezeshki et al. 1990; Ball and Pidsley 1995). Mangroves being adapted to high salinity serve as models for studies on plant responses to salinity

Studies on salt tolerance mechanisms of model systems, like *Arabidopsis thaliana* and the facultative halophyte *Mesembryanthemum crystallinum* (Apse et al. 1999; Nanjo et al. 1999; Liu et al. 2000; Quesada et al. 2000; Shi et al. 2000; Zhu 2000; Elphick et al. 2001; Ratajczak et al. 1994; Low et al. 1996; Golldack and Dietz 2001; Su et al. 2001; Agarie et al. 2007), are many. Downton (1982) and Clough (1984) stressed the need to study mangroves as salt stress adaptors. The use of mangroves for breeding salt-tolerant crops has been reinstated by several authors (Meera et al. 2013).

Soil salinization is the accumulation of soluble salts in the soils (Bockheim and Gennadiyev 2000). Arid and semi-arid climates particularly favor this, with evapotranspiration volumes being greater than precipitation volumes throughout the year. Saline soils are defined as those with an electrical conductivity (EC) of the saturation soil paste extract of more than 4 dS/m at 25 °C, which corresponds to approximately 40 mM NaCl, producing an osmotic pressure of approximately 0.2 MPa (Munns and Tester 2008; USDA-ARS 2008). Salts may include chlorides, sulfates, carbonates, and bicarbonates of sodium, potassium, magnesium, and calcium. The diverse ionic composition of salt-affected soils results in a wide range of physiochemical properties. A significant reduction in yield is observed in crops grown on soils with an EC value above 4.

Saline-sodic soils are those with a combination of high alkalinity, high Na+, and high salt concentration (Eynard et al. 2005), whereas sodicity refers to the amount of Na+ retained in the soil. High sodicity (more than 5% of Na+ of the overall cation content) causes the clay to swell excessively when wet, causing limited air and water movements and poor drainage. Salts may arise naturally in subsoil (primary salinization) or maybe be introduced (secondary salinization) by soil amendments, inorganic fertilizers, and irrigation with brackish water (Carillo et al. 2011). These have led to the salinization of more than 6% of the world's total land area (800 million hectares). Of the current 230 million ha of irrigated land, 45 million ha (19.5%) has already been damaged by salt (FAO 2017). Among the salinity-affected soils, saline, sodic, and saline-sodic account for 60, 26, and 14%, respectively. Approximately 736 (65%), 227 (20%), 53 (5%), and 114 (10%) Mha of all salinity-affected soils are further categorized as slightly, moderately, highly, and extremely salinity-affected soils, respectively; the severity, varying across the world. Middle East, Australia, North Africa, and the former Soviet Union account for 176, 169, 161, and 126 Mha of land, respectively (Rahman et al. 2021). NaCl is the most soluble and widespread salt; plants have evolved mechanisms to tolerate/exclude it while allowing the acquisition of other nutrients available at low concentrations, such as phosphate, potassium, and nitrate (Hanin et al. 2016)

In glycophytes (most crop plants), growth is inhibited or even completely prevented by NaCl concentrations of 100–200 mM, resulting in plant death (Munns and Termaat 1986), even by short-term exposure (Hernández and Almansa 2002). On the other hand, halophytes survive in high NaCl concentrations (300–500 mM) as they are better equipped with characteristic salt resistance mechanisms (Hernández and Almansa 2002; Parida and Das 2005). Euhalophythes (plants growing in saline habitats) regulate their salt content by salt exclusion by preventing the entry of salts into the vascular system, salt elimination via. salt-secreting glands and hairs that eliminate salts, salt succulence i.e. increasing the storage volume cells progressively with the uptake of salt by uptake of water and thereby maintaining salt concentration constant and salt redistribution, by Na+ and Cl− being readily translocated through phloem so as to redistribute the increase in concentration caused by actively transpiring leaves (Acosta-Motos et al. 2017)

Based on their capacity to tolerate salt concentrations, halophytes are named obligate halophytes exhibiting low morphological and taxonomical diversity with optimal growth rates in highly saline habitats. Facultative halophytes are found inhabiting less saline habitats along the border between saline and non-saline upland and are characterized by broader physiological diversity enabling them to thrive under saline and non-saline conditions. Mangroves are generally considered facultative halophytes, though some authors include them under the obligate category, such as Downton 1982 and Clough 1984 (Meera et al. 2013).

Santos et al. (2016) developed a database of halophytes, namely, eHALOPH (http://www.sussex.ac.uk/affiliates/halophytes/), to accumulate the information on plant species that survive under 80 mM or higher salt concentrations. They classified the halophytes into seven groups, typically growing in aquatic conditions or on wet soil (hydrohalophytes); clifftops, rocky and sandy seashores, and saltmarshes (chasmophytes); deep-rooted plants obtaining water from a deep underground source that may or may not be saline (phreatophytes); sandy soils (psammophiles); adapted to extreme drought-prone areas (xerophytes); species predominantly invading and colonizing highly disturbed sites or regions (weedy halophyte); and adapted to inland salt desert and saline habitats (Xerohalophytes). In addition, based on the physiological basis of salt tolerance as well as the accumulation and transportation of ions, Breckle (1995) classified halophytes into three major categories, including recretohalophytes (with salt-secreting structures), euhalophytes (dilute absorbed salts, succulents), and salt-exclusion halophytes (also referred to as salt excluders—shed leaves with toxic levels of salt) (Rahman et al. 2021).

Mangroves represent a group of plants (woody halophytes) adapted to withstand harsh coastal conditions, like high salinity, extreme temperatures, and anaerobic soils. The molecular mechanism behind the origin, diversification, and adaptation of these stress-tolerant trees are ever so relevant with respect to climate change. The mangroves comprise around 80 species derived from 20 plant families with genetic, physiological, and anatomical adaptations to withstand extreme environmental conditions (Duke et al. 1998; Duke 2013). Mangroves serve as models for stress tolerance and as reservoirs for novel genes and proteins involved in salt tolerance. With the onset and development of salt/ drought stress within a plant, major processes, such as photosynthesis, protein synthesis, and lipid metabolisms, are affected. Biochemical and physiological mechanisms involving water relations, photosynthesis, and accumulation of various inorganic ions and organic metabolites contribute to salt tolerance. The ability of the plant to react to high salinity depends on the genes that are expressed during stress (Sreeshan et al. 2014)

Rajalakshmi and Parida (2012) reported mangroves as a resource for abiotic stress genes, like high lactate dehydrogenase activity in the roots of mangrove *Suaeda maritima* (Weston et al. 2012). Mangroves being unique grow equally well under extreme saline conditions and under freshwater conditions and can be used as source material for isolation of stress-tolerant genes (Tomlinson 1986). *Avicennia marina*, *Bruguiera cylindrica*, *Bruguiera gymnorhiza*, *Aegiceras corniculatum*, *Rhizophora stylosa*, and *Rhizophora mucronata* have been used to source genes concerned with stress tolerance. EST sequencing of a salt-stressed *Avicennia marina* leaf library revealed that 30% novel cDNAs with no homology to previously characterized genes in public databases (Mehta et al. 2005), of which 26 clones were highly upregulated and five clones were downregulated during salt stress. The suppression subtractive hybridization (SSH) approach led to the isolation of a number of salt-tolerant genes from the root of mangrove *Bruguiera cylindrica*; 81 salt-induced cDNAs were identified, of which 42 did not have homology to previously reported cDNAs (Wong et al. 2007). *Bruguiera gymnorhiza*, cDNA libraries constructed from the leaves and roots revealed 44 putative salt tolerance genes, including Bg70 and cyc02 homologs (Miyama et al. 2006). Transgenic Arabidopsis plants overexpressing these two genes and lipid transfer and ankyrin genes exhibited increased tolerance to NaCl (Ezawa and Tada 2009). SSH analysis of salt-stressed leaves *Aegiceras corniculatum* led to the isolation of 577 ESTs, of which 30 ESTs showed no homology to any previously reported sequence (Fu et al. 2005). Suppressive subtraction was used to isolate salt-responsive genes from the roots of *Rhizophora stylosa* (Mohammad et al. 2011).

It is suggested that the majority of mangrove plants are angiosperms which first appeared in the fossil record as sparse montane shrubs during the Early Cretaceous (Takhtajan 1980). Fossils of mangroves have been found since the end of the Palaeozoic, as coastal vegetation, before the origin of angiosperms (Tomlinson 1986). True mangroves consisting of three genera of mangroves are found to have originated during the Palaeocene Eocene thermal maximum (PETM), ~55-5 Myr ago, when the sea level rose due to the melting of ice. Some of the land plants developed specialized structures and characters to combat the salty environment, prior to whole-genome duplication (WGD). Mangroves are found to strongly prefer 4 AAS and avoid five others across the genomes as in Convergent evolution (He et al. 2020).

Geological and fossil records based on pollen, fruits, and wood of Nypa, Avicennia, Sonneratia, Rhizophoraceae, and mangrove associates have led Srivastava and Prasad (2019) to trace the origin of mangroves to Late Cretaceous till Miocene. The oldest geological record of Nypa palm (late Cretaceous) exemplifies the broad ecological tolerance and pantropical distribution of mangroves. Sea-level rise and humid climate led to the development of 12 genera of mangroves in nine families and their expansion and multiplication in newer regions during early to middle Eocene (~50–40 Ma). Neogene latitudinal contraction, extinction, and migration, the Himalayan uplift, and the establishment of the Asian summer monsoon caused the mangrove distribution of the Indian subcontinent. Three genera, Avicennia, Nypa, and Rhizophora, found along the Tethys Sea coasts may be the ancestors of Old World mangroves, and four genera adding Pelliciera to the above three flourishing along America's coasts may have descended to form the New World mangroves.

10.2 Salt Stress Tolerance

Salt-resistant plants show several adaptations in their morphological, physiological, and biochemical characteristics, which help them adapt to the deleterious effects caused by salinity. Plants inhabiting saline environments confront challenges, like water loss, decrease in photosynthetic activity, leaf ion toxicity, etc. These plants show adaptations, like changes in leaf anatomy, increase in chlorophyll content, and root/canopy ratio, and evolve certain mechanisms to protect photosynthetic machineries, like photorespiration pathway, water–water cycle, and xanthophyll cycle. Studies have revealed the interaction of salt-induced oxidative stress at the subcellular level and its impact on antioxidant systems in salt-tolerant and salt-sensitive plants, thereby revealing the effect of salinity on physiological processes (Acosta-Motos et al. 2017).

An elevation in soil salinity adversely affects plant growth and development in several ways. A highly saline soil creates water paucity and this, in turn, lowers the osmotic potential of soil water, which interrupts imbibition in plant seeds and severely affects seed germination (Ashraf and Harris 2013; Acosta-Motos et al. 2015a, b). As a consequence of disruption in seed imbibition, plant shows alteration in enzymatic activities, protein metabolism, hormonal imbalance, a decline in the seed reserve utilization, and eventually damage to the ultrastructure of cells, tissues, and organs (Acosta-Motos et al. 2015a, b; Mazher et al. 2007). The elevated soil salinity primarily has a negative impact on plant growth and biomass after seed germination, which varies with salinity levels, duration of salt exposure, and the type of plant species (Ashraf and Harris 2013). When a plant cell is exposed to salinity, within seconds, the cell gets shrunk by dehydration (Passioura 1988). A shrunken cell can regain its size over hours, but the cell elongation rates remain reduced (Vamerali et al. 2003). Cell division rates are also hindered, leading to an overall decrease in plant growth (Franco et al.) Salinity can also produce obvious damage to the older leaves, as well as the mortality of older and younger leaves, leading to plant death before seed maturity (Passioura 1988; Vamerali et al. 2003). The roots are the most exposed to saline soils and serve as a primary conduit for mineral translocation to aerial portions of the plant. Root responses are primarily responsible for the lower growth rate of plants under salinity (Franco et al. 2011a, b). Salt stress reduces the water potential of the soil solution, interfering with root water conductivity, thereby reducing water and mineral influx to the plant body (Ma et al. 2010). High concentrations of Na+ and Cl in soil solution limit micronutrient uptake and accumulation, causing signal transduction, membrane stability, chlorophyll production, and stomatal opening to be disrupted (Franco et al. 2011a, b; Steudle 2000). Furthermore, the detrimental effects of salinity on photosynthesis also lead to the breakdown of photosynthetic pigments. Plants exposed to salt stress had lower stomatal density, stomatal conductance, photosynthetic activity, and carbon assimilation, but higher mesophyll resistance, lowering photosystem (PS)-I, and PS-II light absorption efficiency (Ranathunge et al. 2010). Salt stress also reduces the activity of ribulose-1,5-bisphosphate carboxylase/oxygenase (RuBisCO), a crucial enzyme involved in the assimilation of carbon dioxide (CO2) during photosynthesis' dark phase (Miteva et al. 1992). Salinity also stimulates the generation of reactive oxygen species (ROS), such as the hydroxyl radical (OH·), superoxide (O_2·), hydrogen peroxide (H_2O_2), and singlet oxygen ($^1O_2$), in cellular organelles, such as chloroplasts, mitochondria, peroxisomes, and the endoplasmic reticulum (Munns 2005; Koyro 2006). Increased ROS levels can harm plant growth and development by reducing protein synthesis, destroying cell membranes, generating genomic instability, and damaging the photosynthetic system (Mostofa et al. 2021). Despite the fact that Cl^- is the most prevalent anion in salt soils, it has received less attention than Na+ in terms of its effects on plant performance, as well as its uptake and transport processes inside plants (Whiley et al. 2002). Cl^- is an important micronutrient, thought to have functions in enzyme activation in the cytoplasm, oxygen evolution during photosynthesis, membrane potential stability, and plant cell turgor pressure, and pH maintenance (Whiley et al. 2002). Cl^-, on the other hand, has far more harmful effects than Na^+ in disturbing the normal activities of cells once it has accumulated excessively (Whiley et al. 2002). All of the abovementioned salinity-induced negative impacts on plants result in a reduction in crop production; however, the severity and extent of the damage vary between plant species (Whiley et al. 2002).

To summarize, salt stress produces a considerable decrease in plant growth in salt-sensitive plants, especially in the aerial part. The accumulation of phytotoxic ions in leaves results in a nutrient imbalance, causing decreased K+ and Ca2+ levels. Salinity also topples over the plant's water relations, leading to lower leaf relative water content (RWC) and leaf water potential (ψl). When photosynthesis rate (PN) decreases, chlorophyll content and chlorophyll fluorescence parameters decrease in parallel. Oxidative stress at the subcellular level takes place, the synthesis of ethylene and abscisic acid (ABA) increases, and a decrease occurs in indol-3-acetic acid (IAA) and cytokinin (CK) synthesis, causing senescence in salinized leaves. Due to the stress, chloroplast gets damaged, and starch accumulates. The accumulation of salts in the root zone causes a decrease in osmotic potential, root water potential (ψr), and the root hydraulic conductivity (Lp), diminishing the water available to the root zone and causing a nutrient imbalance in roots (Acosta-Motos et al. 2017).

Euhalophytes (salt-loving plants) may also store salt in their cell sap to the point that their osmotic potentials are lower than the soil solution. Aside from salts, soluble carbohydrates play a crucial function in maintaining a low osmotic cell sap potential. The protoplasm's capacity to tolerate high salt concentrations is dependent on the compartmentalization of ions entering the cell. The vacuoles gather the majority of the salt ions (includer mechanisms). Salt stress affects plant development in the short term by producing osmotic stress owing to decreased water availability and in the long term by salt-induced ion toxicity due to nutritional imbalance in the cytosol (Munns 2005; Koyro 2006). Salinity leads to the excessive uptake of salts, like sodium and chloride as well as deficiency of calcium and potassium, also the production of ROS. All of these salinity responses contribute to the negative impacts of salt stress on plants (Hernandez et al. 1993, 1995; Hernández et al. 2003; Mittova et al. 2003). To deal with the stress caused by saline environments, plants should activate several physiological and biochemical systems. Morphology, anatomy, water relations, photosynthesis, the hormonal profile, toxic ion distribution, and biochemical adaptability (such as the antioxidative metabolism response) are examples of such systems (Rahman et al. 2021).

Among glycophytes and halophytes and between different species of the same family and genus, differences in response to saline water irrigation are found. Plant responses also depend upon the salt concentration in irrigation water and the time of exposure. Munns 1992 found that while salts do not directly affect development, they do alter turgor, photosynthesis, and/or the activity of certain enzymes. Vacuoles serve as the storage unit of incoming salt. With the increase in the concentration of and the subsequent rise of salt in the cell walls or cytoplasm, vacuoles may no longer sequester incoming salt (ionic phase), resulting in a drop-in soil water potential (osmotic phase), followed by salt damage in leaves, eventually leading to decline in growth. Munns 1992 discovered that salt accumulation in old leaves causes them to die, thereby reducing the availability of carbohydrates and/or growth hormones in the meristematic areas causing restricted growth. The development of particular metabolites that directly impede growth is influenced by the fact that plant growth is limited by the drop-in photosynthetic rate and excessive uptake of salts (Mazher et al. 2007; Rahman et al. 2021)

10.3 Morphology of Roots and Aerial Parts

Root performance is determined by the anatomy of the root system (length, root diameter, and so on), which allows plants to collect water and nutrients and hence enhance the replacement of water lost from the plant (Passioura 1988). Roots act as an interface between the plant and soil, and optimal root systems can enhance shoot growth and improve plant yield (Vamerali et al. 2003). Plants with a more extensive root system appear to be healthier since it helps them reach deeper layers of soil for water and nutrients (Franco et al. 2011a, b). Recent studies have found that alternative root characteristics, such as tiny roots, can be more beneficial for shoot development (Ma et al. 2010). Even a few roots in moist soil can supply significant amounts of water, regardless of the number of roots. The permeability of roots to water is also determined by other root features, such as the number and diameter of xylem vessels, the breadth of the root cortex, the number of root hairs, and the suberin deposition in both the root exodermis and endodermis (Steudle 2000; Ranathunge et al. 2010). Furthermore, changes in soil environmental conditions (such as temperature, lack of oxygen, mechanical impedance, and salt) can have a significant impact on root structure. The cell walls of salinized plants' root cells are frequently thickened and twisted unevenly (Shannon et al. 1994). Salts encourage the suberization of the hypodermis and endodermis in woody tree roots, resulting in the development of a well-developed Casparian strip closer to the root apex (Walker et al. 1984). Furthermore, certain plants' morphology reveals their salinity sensitivity. Avocado trees, for example, have a shallow root structure with little ramification, which reduces their water and nutrient absorption capacity (Schaffer et al. 2013) and makes them more sensitive to soil salt (Bernstein et al. 2004). These physical characteristics limit the crop's distribution to places with good irrigation water. The root system morphology of *Callistemon citrinus* plants has also been observed to be altered by saline water irrigation (Álvarez and Sanchez-Blanco 2014). Furthermore, Álvarez et al. (2011) and Álvarez and Sánchez-Blanco (2013) reported similar reactions in *C. citrinus* under deficit irrigation, implying that the effect of water stress on root system morphology in *C. citrinus* is very similar to that caused by irrigation with salty water. The same behavior was found in *Euonymus japonica* plants irrigated with a NaCl solution and reclaimed water having various salt concentrations, which resulted in a reduction in total root length, particularly in thin (0.5 mm) and medium thickness (0.5–2.0 mm) roots (Gómez-Bellot et al. 2013a, b). Franco et al. and Croser et al. showed an increase in root diameter (hypertrophy) in response to salt in various plant species (*Picea sp., Pinus banksiana, Portulaca oleracea*). The higher root density observed in these plants suggests greater robustness and, presumably, a higher accumulation of reserves (Franco et al. 2011a, b; Gómez-Bellot et al. 2013a, b; Croser et al. 2001), which would improve plant resistance to saline environments and speed up the establishment of plants, particularly ornamental plants for gardening and landscaping. High sodium uptake ratio values cause sodicity, which increases soil resistance, reduces root growth, and reduces water movement through the root with a decrease in hydraulic conductivity (Rengasamy and Olsson 1993). In general, when plants are irrigated with low-quality water, their root hydraulic conductivity decreases. In reality, one of the most important criteria influencing the long-term viability of a recycled water irrigation system is root hydraulic conductivity. Root hydraulic conductance is usually stated in terms of the dry weight (DW) of the entire root, ignoring the effect of root design on water uptake capacity. However, the number of fine roots, which determine the root length and surface area, can vary substantially for any given root dry weight value, altering the water absorption level (Fitz Gerald et al. 2006; Zobel et al. 2007). Furthermore, the rootstock qualities impact the plant response to salinity in assays using grafts. Abiotic stress is mitigated more effectively by salt-tolerant rootstocks than by salt-sensitive rootstocks. It was found that *Clemenules mandarin* trees grafted on Carrizo (salt-sensitive rootstock) had lower fruit output and quality than Cleopatra (salt-tolerant rootstock) when both were irrigated with NaCl solution (30 mM) (Navarro et al. 2010). Salt-tolerant rootstocks were found to boost the output of a commercial pepper cultivar under saline irrigation in trials conducted by Penella et al. (2015, 2016). Different nutritional and physiological responses, such as ionic regulation, optimum photosynthetic performance, and sink strength maintenance, were observed. Inoculating the roots with arbuscular mycorrhizal fungi (AMF), which are important bioameliorators for saline soils, is another way to relieve or protect plants from saline stress. Navarro et al. (2014) investigated the impact of AMF in alleviating the detrimental effects of salt stress in citrus seedlings inoculated with a combination of AMF (*Rhizofagus irregularis* and *Funneliformis mosseae*). This protective effect, or synergic interaction, is enhanced when salt-tolerant rootstocks are used. When the roots develop in a salty environment, good AMF–soil–plant interactions make it viable to utilize reclaimed water. Salt stress causes a general drop-in fresh weight (FW) or DW in all plant tissues, although it is most visible in the aerial section. A drop-in FW or DW has been linked by many writers to a reduction in the number of leaves and leaf abscissions. Some experiments have indicated that a specific Cl build-up in the leaves of salt-stressed plants triggers high-efficiency 1-aminocyclopropane-1-carboxylic acid (ACC) synthesis and conversion to ethylene, releasing enough hormone to trigger leaf abscission as seen in citrus leaves and other plants (Tudela and Primo-Millo 1992). In the halophyte *Allenrolfea occidentalis*, salt and osmotic stress aided the conversion of ACC to ethylene (Chrominski et al. 1988). Early events during the osmotic phase of salt stress induce leaf senescence prior to the huge build-up of harmful ions, according to experiments conducted on tomato. The advancement of senescence in salinized leaves may be aided by an indirect impact associated to early ABA build-up and a reduction in IAA and CK levels. The ethylene precursor ACC, on the other hand, is the key hormonal signal that is temporally linked to the initiation of oxidative damage, chlorophyll fluorescence reductions, and significant Na+ build-up (Albacete et al. 2008; Ghanem et al. 2008). A decrease in total leaf area is another common reaction to salt stress. The earliest response of glycophytes exposed to salt stress is reduced leaf growth (Munns and Termaat 1986). When the stomata are closed, the observed reduction in canopy area could be interpreted as an avoidance mechanism that reduces water loss through transpiration (Savé et al. 1994; Ruiz-Sánchez et al. 2000). This action may favor the preservation of harmful ions in the roots, reducing their build-up in the plant's aerial section (Munns and Tester 2008; Colmer et al. 2005). Cell wall characteristics alter under salinity, and leaf turgor and photosynthetic rates (PN) decrease, resulting in a decrease in total leaf area (Franco et al. 1997; Rodrıguez et al. 2005). Furthermore, excessive salt concentrations inhibit stem development (which is a component of the aerial portion). Reduced leaf and stem diameters result in a reduction in the size of all aerial parts as well as the plant's height. The most frequent response to salt stress is an increase in root to shoot ratio or a decrease in the shoot to root ratio, which is due to characteristics associated with water stress (osmotic impact) rather than a salt-specific effect (Hsiao and Xu 2000). Under salt stress, a higher root percentage may favor the preservation of harmful ions in this organ, limiting their transfer to the aerial portions. This reaction might be a common plant resistance/survival strategy under saline conditions (Cassaniti et al. 2009, 2012). Reduced plant growth (Munns and Tester 2008), restricting leaf expansion (Cramer 2002a, b), and disrupting the interaction between the aerial and root portions (Tattini et al. 1995) are all effects of high salt concentrations in irrigation water. Some plant species have shown a higher dry root mass than shoot dry mass under high salinity, resulting in an increased root to shoot ratio, which is thought to improve the water and nutrient source/sink ratio under such conditions (Zekri and Parsons 1989).

Plants react to salt and drought stress in similar ways, and the processes involved are similar. Salt and drought stress have an impact on every part of plant physiology and metabolism. These are good candidates for genetic modification to increase salt and drought resistance. Ionic and osmotic stress signaling for re-establishing cellular homeostasis under stress conditions, detoxification signaling to control and repair stress damages, and signaling to coordinate cell division and expansion to levels appropriate for the particular stress condition are the three functional categories of salt and drought stress signaling (Zhu 2002). Single-function genes and regulatory genes are two types of genes that are triggered by salt stress. Osmolytes, transporters/ channel proteins, antioxidative enzymes, lipid biosynthesis genes, polyamines, and other protective metabolites are all produced by the first group of genes. Regulatory proteins, such as basic leucine zipper (bZIP), drought-responsive element-binding protein (DREB), myelocytomatosis/myeloblastoma (MYC/MYB), and no apical meristem, ATAF 1, 2, and cup-shaped cotyledon (NAC), affect the expression of several downstream salt stress tolerance genes (Shinozaki and Yamaguchi-Shinozaki 2007; Agarwal and Jha 2010). These two types of genes interact in the signaling pathways indicated above to provide abiotic stress tolerance.

Under salt stress, secondary metabolites, such as the products of phenylpropanoid biosynthesis, are among the molecules accumulating in plants to combat it (Korkina et al. 2007). In salt-stressed *Kandelia candel*, the total concentration of phenols showed no significant difference between 0 and 200 mM NaCl treatments, but it increased 17% under 500 mM NaCl stress. Anthocyanins and proanthocyanidins levels also increased significantly when NaCl concentration increased from 200 to 500 mM. The lignin concentration also increased, although it nearly peaked at 200 mM NaCl. The most abundant amino acid under all three conditions was glutamine (Glu) which accounted for >50% of the total free amino acid concentration in the leaves of *K. candel*. Threonine and histidine increased fivefold and 34-fold under 500 mM NaCl stress. Proline did not show an increase in concentration. GABA, a four-carbon non-protein amino acid, also showed an increase. The concentration of free Glu, serine, valine, and lysine also increased significantly in response to salt stress. The GSH (an antioxidant) levels in the leaves of *K. candel* also showed 27% increase at 200 mM and 15% increase at 500 mM NaCl (Wang et al. 2016a, b).

10.4 Chromatin Modifications and Epigenetics in Salt Tolerance

A plant's ability to adapt to varied environmental pressures is thought to be aided by chromatin alteration, also known as epigenetic modifications (Gómez-Bellot et al. 2013a, b; Fernández-García et al. 2014). The role of chromatin remodeling in plant stress physiology has been understood through multiple studies. Plants, when hyper osmotically primed and later grown under control conditions (showing apparently no changes compared to untreated plants), were noticed to have accumulated less Na+ ions when salt was added and thus found much tolerant than untreated ones. Further, it was observed that HKT1expression in pretreated plants was more intense than in control, which may be the reason for variation in Na+ deposition. No conclusive evidence of epigenetic inheritance of salt tolerance from one generation to the next has been found in research on salinity to date. The ploidy status of a plant also has a significant impact. In Arabidopsis, auto polyploidy has been demonstrated to account for enhanced salt resistance, which results in more effective potassium accumulation (Zhen et al. 2010; Deinlein et al. 2014).

Seed priming using phytohormones has emerged as a viable technique of abiotic stress management in plants from stresses by increasing antioxidant enzyme activity, reducing oxidative damage, and improving plant growth. This method enhances abiotic stress tolerance in crop plants and can be used to maintain sustainable crop production in drought, salinity, and flood-prone areas around the world (Rahman et al. 2021).

10.5 Small RNAs

Small RNAs in plants are complex as they are derived through several pathways and play a vital role in abiotic stress regulation. One of the candidates in regulatory RNA, viz., small non-coding RNAs (ncRNAs), regulate distinct biological activities in plants as well as other organisms. MicroRNA (miRNA) and siRNA, which are ncRNAs, participate in the regulation of the expression of vital genes in distinct biological processes by RNA interference (RNAi). Hence, crop improvement utilizes the strategy of siRNA and miRNA interference. Various studies have confirmed the role of miRNA in salinity stress response. Enhanced salt stress tolerance has been shown by the transgenic Arabidopsis and creeping bent grass, which overexpressed miR393, a resistant form of TIR1, and rice miR528, respectively. In Arabidopsis, it was observed that water loss was reduced and germination rate decreased, whereas in the other, tiller number has increased (Chen et al. 2015; Yuan et al. 2015). Pan et al. 2016 has found that miR172a participates in the salt stress tolerance of soybean by downregulating the expression of an AP2/EREBP type transcription factor SSAC1 by acting as a long-distance signal. Microarray studies by Ding et al. (2009) found that miR156, miR164, miR167, and miR396 family members of maize were downregulated and miR162, miR168, miR395, and miR474 were upregulated during salt stress. sRNAs that regulate salt stress and impart salinity tolerance in the plant have been identified. A 24 nt long siRNA was found to regulate the expression of MYB74; an Arabidopsis salinity-induced transcription factor, by RNA-directed DNA methylation (Xu et al. (2015). Silencing of the rice floral meristem-specific cytokinin oxidase (OsCKX2) using siRNA-based approach, provided salinity tolerance without any yield loss (Joshi et al. 2018a, b). RNAi is a promising technology for obtaining crop varieties tolerant to different types of abiotic stresses and with improved yield since they are relatively safer than transgenics. RNAi has great scope since it can engineer the plant by controlling the regulation of vital genes and transcription factors involved in stress responses. Identification of sRNAs and their corresponding targets enhance the scope of engineering plants for stress tolerance. sRNA-based engineering of plants is mainly done through overexpression and short-term tandem mimicry (STTM) for the regulation of sRNAs. STTM approach has been performed for the knockdown of miR166 maize, resulting in abiotic stress tolerance (Li et al. 2020). In RNAi technique, transformed plants that express dsRNA show genetic modification and confer resistance against abiotic stresses. Moreover, deregulation of target genes using artificial miRNA (amiRNA) and ta-siRNA (ata-siRNAs) can also be done (Cisneros and Carbonell 2020). Overexpression of miRNA can be obtained by expressing pre-miRNA in expression vectors. The latest clustered regularly interspaced short palindromic repeats (CRISPR)/CRISPR-associated protein (Cas) technology has high potential in developing abiotic stress-tolerant varieties by editing MIR genes. The utilization of micro peptides, a translational product of pri-miRNA, is yet to be deciphered (Iborra et al. 2001; Prasad et al. 2020; Tiwari and Rajam 2022)

10.6 Nanoparticles

Nanoparticles (NPs) may be described as materials with diameters between 1 and 100 nm in at least one dimension. It has been observed that NPs minimize the uptake of heavy metals by modifying the expression of genes responsible for metal uptake and thereby reduce HM bioaccumulation. Furthermore, NP treatment improves the physiological and biochemical parameters of the plants, such as enhancing the synthesis of defense enzymes (SOD, POX, CAT, APX, etc.); augmenting nutrient uptake; decreasing the loss of electrolytes; improving pigments and soluble proteins; reducing peroxidation; and causing a rise in the levels of proline, glutathione, and phytochelatins. These attributes are primarily responsible for the overall increase in the abiotic stress tolerance of the crops by NP treatment and the effect may vary slightly according to the genotype (Rajput et al. 2021). NPs can be successfully used in plant genetic engineering as a genome editor also. All of these factors make NPs a perfect candidate to boost plant immunity against abiotic stress (Khalil et al. 2021).

Both drought and cadmium stresses were reduced by the application of iron oxide NPs (IONPs) and hydrogel nanoparticles (HGNPs) simultaneously in rice (Ahmed et al. 2021). Levels of both hydrogen peroxide and lipid peroxidation after drought were reduced when γ-Fe_2O_3 NPs were delivered by irrigation in a nutrient solution to *Brassica napus* plants grown in soil (Palmqvist et al. 2017).

10.7 Species and Mechanisms

Many species which are conventionally regarded as halophytes are sensitive to salinity and several species among glycophytes show resistance to salinity and other abiotic stresses. Halophytes are the best models for studying salinity tolerance mechanisms in plants (Shabala 2013; Flowers and Colmer 2015; Himabindu et al. 2016). Several reports suggest that both glycophytes and halophytes possess similar mechanisms to resist salinity stress, but the difference lies in the quantitative aspect, not the qualitative (Anjum et al. 2012; Rai et al. 2012; Bartels and Dinakar 2013; Sreeshan et al. 2014; Joshi et al. 2015; Volkov 2015; Muchate et al. 2016). Functional validation of salinity tolerance in halophytes is done in transgenic systems. Salt tolerance in halophytes is possibly due to the difference in the expression level of important genes that participate in salt tolerance or the production of comparatively active proteins, which are involved in stress tolerance (Anjum et al. 2012; Das and Strasser 2013; Himabindu et al. 2016; Muchate et al. 2016). *Thellungiella salsuginea*, a halophyte closely related to the model plant *Arabidopsis thaliana* is regarded as a model system for studying halophyte genes responsive to salinity (Amtmann 2009; Bartels and Dinakar 2013). Genomic studies in *T. salsuginea* provide insights into the genetic basis of halophyte salinity tolerance mechanisms (Wu et al. 2012). Comparative microarray studies on *T. salsuginea* showed that when compared to Arabidopsis, only a few genes were induced in *Thellungiella* on exposure to salt (Taji et al. 2004). However, Wong et al. (2006) reported that compared to *Arabidopsis*, *Thellungiella* differentially expressed 154 genes under salt stress. A random gene transfer from *Lepidium crassifolium*, a halophyte related to *Arabidopsis*, resulted in increased salinity tolerance in *Arabidopsis* (Rigó et al. 2016). Most of the halophytic genes transferred to crop species belong in the category of antiporters, ROS scavangers, K+ transporters, ion channels, protein candidates in signal transduction, antioxidants, and novel genes from extreme halophytes, like Salicornia *brachiata* (Udawat et al. 2014, 2017; Singh et al. 2016a, b; Mishra and Tanna 2017).

Studies indicate that transformation of salt sensitive plants with halophytic genes can impart salinity tolerance to them. A variety of transporters encoding genes from different halophytic species are transferred to salt sensitive species, showing enhancement in salt tolerance. For example, *N. sibirica* NsNHX1 gene transformed to poplar plant, SbHKT1 gene from *S. bigelovii* to cotton, ZxNRT1.5 of *Zygophyllum xanthoxylum* in Arabidopsis, resulted in increased salt tolerance in transgenic lines (Mwando et al. 2020; Nanjo et al. 1999; Navarro et al. 2010; Rahman et al. 2021). Mangroves accumulate low molecular mass compatible solutes, such as pinitol, mannitol, and proline in the cytoplasm to maintain ionic balance (Hasegawa et al. 2000; Ashihara et al. 2003). Major osmoregulatory compounds are different in different mangrove species. Pinitol is the osmoregulator for *Bruguiera gymnorrhiza*, *Kandelia candel*, and *Rhizophora stylosa* (Hibino et al. 2001), mannitol for *Sonneratia alba* and *Lumnitzera racemosa* (Yasumoto et al. 1999; Ashihara et al. 2003), and carbohydrates for *Acanthus ilicifolius*, *H. littoralis*, and *H. tiliaceus* (Popp et al. 1985). Among these osmoregulators, the amino acid proline is regarded as a good osmoregulator, and its concentration is positively correlated with resistance to abiotic stresses (Matysik et al. 2002). Moreover, it can also act as a reactive oxygen scavenger and provide membrane and protein stability (Alia et al. 1991; Hanson and Burnet 1994). Along with proline, glycine betaine also acts as an osmoregulator, and they are accumulated through betaine/proline transporters AmT1, 2, and 3 (Waditee et al. 2002).

Salicornia is also a functional food since it contains unique oligosaccharides (Mishra et al. 2013), metabolites (Mishra et al. 2015), and sulfur-rich seed storage proteins (Jha et al. 2012). *Porteresia coarctata*, a wild rice relative with great salinity and submergence tolerance, had 152,367 distinct transcript sequences discovered by transcriptomics (Garg et al. 2014). To decipher important metabolic pathways, researchers identified 15,158 genes implicated in salinity and submergence tolerance. Salinity and submergence tolerance in rice are engineered using genes from halophytes (Garg et al. 2014).

10.8 Promoters of Halophytic Genes and Transgene Expression

Crop plants must be engineered with a powerful and well-regulated promoter. According to a comparative transcriptome study, many stress-related genes were constitutively expressed at greater levels in T. halophila than in their A. thaliana homologs (Taji et al. 2004, 2010). This study reveals that halophytes have a well-functioning transcriptional regulatory network for stress-responsive genes. The presence of several stress-inducible motifs was discovered in the cis-regulatory regions of different stress-responsive genes from certain halophytes recently (Tiwari et al. 2014, 2016). Yin et al. (2002) discovered that salt stress substantially induces the promoter of the AcBADH gene in *Atriplex centralasiatica*, which has two salt-responsive enhancer areas (located from 1115 to 890 and 462 to 230) and one silencer region (placed between 890 and 641). Under salt stress (400 mmol/L NaCl), the SlBADH gene promoter segment (300 bp alone) from *Suaeda liaotungensis* displayed roughly 6.3-fold more expression when compared to the control (Zhang et al. 2008). Under salt stress, the TsVP1 gene promoter from the halophyte *T. halophila* featured a 130-bp unique cis-acting region, which resulted in increased GUS expression in transgenic Arabidopsis (Sun et al. 2010). Under NaCl stress (200 mmol/L), an 897-bp promoter region of the SlPEAMT gene (*S. liaotungensis*) showed an 18.6-fold increase in GUS activity (Li et al. 2016). These findings demonstrate that even a short section of promoter might include critical cis-acting elements that influence gene expression in stressful situations. Basic elements were also found in the promoters of CMO genes from *S. liaotungensis* and *Salicornia europaea*, indicating that they were salt inducible (Li et al. 2007; Wu et al. 2011). In the halophyte *M. crystallinum*, Schaeffer et al. (1995) found enhancer and silencer areas involved in the transcriptional activation of salt-responsive expression of CAM (Crassulacean Acid Metabolism) genes. AlSAP is a promoter found in *Aeluropus littoralis* that is age dependent, abiotic stress inducible, organ specific, and tissue specific (Saad et al. 2011). Furthermore, in transgenic rice, gusA expressed at the same level as AlSAP transcript under the control of the AlSAP gene promoter (Ben-Saad et al. 2015). They also discovered that the regulatory domains of two orthologs, AlSAP and OsSAP9 (from rice), regulate and induce stress in distinct ways in rice. Sun et al. (2010) discovered a 130-bp unique cis-acting element in the promoter region of the halophyte *T. halophila* (TsVP1) that promotes GUS production in transgenic Arabidopsis under salt stress. Under abiotic and biotic stress conditions, the CBL1 gene promoter identified from *Ammopiptanthus mongolicus* regulated the production of the reporter gene (Guo et al. 2010). The inclusion of enhancer and repressor binding sites in the cis-regulatory elements, as well as stress-inducible motifs, was shown in a model for transcriptional regulation of the SbpAPX gene (from *S. brachiata*) (Tiwari et al. 2014). Similarly, a variety of abiotic stress sensitive cis-regulatory motifs was found in the SbGSTU promoter, which governs the expression of the GSTU gene in *S. brachiata* (Tiwari et al. 2016). As evident from these studies, the halophytic promoters appear to be a good choice for developing abiotic stress tolerance in crops by high-level transgene expression.

10.9 Salt-Responsive Genes in Halophytes and Salt Tolerance in Transgenics

The most prevalent method of using antiporters, regulated by a multigene family, includes Na+ efflux, compartmentalization of Na+ in vacuoles, and inhibition of Na + input (Rajendran et al. 2009; Kronzucker and Britto 2011). Several antiporters from glycophytes and halophytes have been functionally described (Kronzucker and Britto 2011; Sreeshan et al. 2014). Under the control of the non-specific CaMV35S promoter, overexpression of glycophytic transporters producing genes (NHX, SOS, HKT, ATPase, etc.) exhibited tolerance in the range of 150–250 mM NaCl, although their halophytic homologs offered a tolerance of up to 400 mM NaCl (Kronzucker and Britto 2011; Sreeshan et al. 2014; Volkov 2015). Under salt stress treatments, the effects of overexpression of halophytic genes were commonly observed, but under control (unstressed) conditions, there was negligible difference between wild-type plants and transgenic lines (Jha et al. 2011a, b; Joshi et al. 2012; Volkov 2015; Tiwari et al. 2015; Singh et al. 2016a, b; Udawat et al. 2016). In many crops, including tomato, brassica, maize, and wheat, the glycophytic NHX gene from *A. thaliana* was used to acquire salt tolerance (Zhang et al. 2001; Xue and Loveridge 2004; Yin et al. 2004). Other glycophytic NHX1 genes, such as BnNHX1 (*Brassica napus*), GhNHX1 (*Gossypium hirsutum*), and HbNHX1 (*Hordeum brevisubulatum*), have also been shown to induce salt tolerance in tobacco (Wang et al. 2004; Wu et al. 2004; Lü et al. 2005). The NHX1 genes from halophytes and glycophytes demonstrated salt tolerance activity, although the salt tolerance intensity differed. In *Oryza sativa*, the antiporter AgNHX1 (from the halophyte *Atriplex gmelini)* has a 75% amino acid sequence similarity to AtNHX1 (from *Arabidopsis thaliana*) and a greater salt tolerance (Hamada et al. 2001; Ohta et al. 2002). Compared to glycophytic counterparts, transgenic plants overexpressing AgNHX1 (*A. gmelini*), SaNHX1 (*Spartina anglica*), or SsNHX1 (*Suaeda salsa*) demonstrated resistance to 300–400 mM NaCl (Ohta et al. 2002; Zhao et al. 2006; Lan et al. 2011). In transgenic tobacco, overexpression of the SbNHX1 gene resulted in 200 mM salt tolerance; however, only 100 mM NaCl tolerance was reported in transgenic jatropha and castor plants (Joshi et al. 2013; Patel et al. 2015). Other halophytic genes, including as SbpAPX, SbUSP, and SbGSTU, were overexpressed in transgenic plants and demonstrated superior salinity tolerance (200–300 mM NaCl) than their glycophytic homologs (Jha et al. 2011a, b; Udawat et al. 2016; Singh et al. 2014a, b). When compared to the identical gene from the glycophyte, *Panax ginseng*, transgenic Arabidopsis plants overexpressing the TIP1 gene from the halophyte *T. salsuginea* showed higher salt tolerance (Peng et al. 2007). In the transgenic plants, APX and GST from rice showed lesser tolerance up to 150–200 mM than the same genes (200–300 mM NaCl) from the *halophyte S. brachiata* (Lu et al. 2007a, b; Jha et al. 2011a, b; Sharma et al. 2014; Singh et al. 2014a). Overexpression of a stress-associated protein gene (AlSAP) from *A. littoralis* has recently been shown to increase abiotic stress tolerance in tobacco, wheat, and rice (Ben-Saad et al. 2015). They also discovered that several abiotic conditions activate AlSAP transcripts, the rice OsSAP9 ortholog gene being predominantly stimulated by cold and heat treatments. Antiporter SOS1 gene expression was found to be greater in *Thellungiella species* than in Arabidopsis, according to a transcript expression study (Oh et al. 2010). Similarly, numerous genes involved in Na+ excretion, compartmentation, and diffusion, including SOS2, NHX1, and HKT1, were expressed at greater levels in Thellungiella than in Arabidopsis (Taji et al. 2010). To examine the Na+ hypersensitive response, Arabidopsis lines overexpressing either AtHKT1 (*A. thaliana*) or TsHKT1 (*T. salsuginea*) were studied, and AtHKT1 lines showed slower root development than TsHKT1 lines (Ali et al. 2012). In transgenic lines expressing AtHKT1, shoot sensitivity was detected. They also discovered that salt stress causes a substantial upregulation of TsHKT1 but a strong inhibition of AtHKT1 expression (Ali et al. 2012). These findings show that halophytes are to be investigated further in order to source novel genes and to produce transgenic crop plants with a better level of salt tolerance than their glycophytic counterpart genes.

10.10 Mangrove Genes for Salt Tolerance

Salt tolerance was induced in various plants in a variety of ways, including transformation and overexpression of genes associated with abiotic stress tolerance (Singla-Pareek et al. 2001; Vinocur and Altman 2005; Bhatnagar-Mathur et al. 2008; Ashraf and Akram 2009). Many reviews on various areas of salt stress tolerance have recently been published (Zhu 2003; Agarwal et al. 2006; Umezawa et al. 2006; Apse and Blumwald 2007; Gaxiola et al. 2007; Shinozaki and

Yamaguchi-Shinozaki 2007; Century et al. 2008; Shao et al. 2008a, b; Rodríguez-Rosales et al. 2009; Agarwal and Jha 2010).

The application of mangrove genes has been found to be effective in elevating the salt tolerance capacity of plants, and individual genes are exploited for the same. Sultana et al. (2012) overexpressed AeMDHAR genes in rice to improve salt tolerance. AeMDHAR is an enzyme that confers salt tolerance by scavenging ROS. A mangrove candidate gene(s) for salt tolerance is MYB. MYB1 gene was isolated from salt-secreting mangrove *Avicennia marina* and constitutively expressed in tobacco for enhanced tolerance to salt (Ganesan et al. 2012).

Salt-tolerant plants overcome environmental stress through antioxidant defense systems. Many enzymes are involved in this strategy of defense, among which superoxide dismutase functions as the first line of defense. Prashanth et al. (2008) studied the superoxide dismutase genes in *Avicennia marina*, and the cDNA was transferred into the rice. The transformed rice showed elevated defense against methyl viologen-mediated oxidative stress as well as salinity stress. *Avicennia marina* CSRG1 gene was seen to elevate the salt defense mechanisms against a plethora of ions in the transformed tobacco plant. Reported to encode 197 amino acids with no homologous sequence found in the gene bank (Hantao et al. 2004). MDAR (Monodehydroascorbate reductase) is another salt-induced gene that induces resistance to salinity in transformed tobacco (Kavita and Alka 2010). Such studies prove that mangroves are a hub of genes that give tolerance against a combination of environmental stresses by activating different types of pathways involved in antioxidant protection, homeostasis, and regulatory pathways (Shinozaki and Yamaguchi-Shinozaki 2007; Agarwal and Jha 2010).

One hundred and seven salinity-tolerant candidate genes were identified and isolated from a mangrove plant, *Acanthus ebracteatus Vahl,* and verified in *E. coli* host (Nguyen et al. 2007). Huang et al. 2003 isolated ten cDNAs from genes of *Kandelia candel* and studied their expression. Of these, two were unknown; three belonged to small heat shock proteins (sHSPs) and ADP ribosylation factor categories, five genes were repressed under NaCI stress, two encoded cyclophilins, and three tonoplast intrinsic proteins, early light-induced protein, and 60S ribosomal protein, respectively.

cDNA library constructed from *A. marina* leaves and screened for betaine aldehyde dehydrogenase genes (BADH) which efficiently catalyze the oxidation of betaine aldehyde, one of which showed higher expression under salinity in *E. coli* expression systems (Hibino et al. 2001). *Rhizophora mucronata* leaves were used to analyze free amino acids and the expression of selected genes. Chromatograms showed the accumulation of free amino acids, like proline, glycine, aspartic acid, valine, leucine, and glutamic acid, in the presence of salt. Among the genes studied (P5CS, BADH, NHX1), BADH gene was found to be expressed more during salt stress (Sreeshan et al. 2018). Several stress-related gene homologs, such as chaperonin-60, clpP protease of the clp/Hsp100 family of chaperones, ubiquitin, eukaryotic elongation factor 1 A (eEF1A), drought-induced AtDi19 gene of *Arabidopsis thaliana*, and secretory peroxidase, were successfully isolated from *A. marina* (Tanaka et al. 2002).

Table 10.1 Over expression of mangrove gene for salt tolerance

Gene	Mangrove source	Transgenic plant	References
AcMDHAR	*Acanthus ebracteatus*	Rice	Sultana et al. (2012)
AmMYB	*Avicennia marina*	Tobacco	Ganesan et al. (2012)
Cu/ZnSOD	*Avicennia marina*	Rice	Prashanth et al. (2008)
CSRG1	*Avicennia marina*	Tobacco	Hantao et al. (2004)
Am MDAR		Tobacco	Kavitha et al. (2010)

Provided by (Meera et al. 2013)

The role of mangrove-specific allene oxide cyclase (mangrin) in salt tolerance was revealed by functional screening for cDNAs in the mangrove *Brugiera sexangula* (Yamada et al. 2002). Meera et al. 2013 reviewed a list of the reported salt-responsive mangrove genes (Table 10.1) and a list for future use (Table 10.2). Some of the genes were reported to induce salt/ drought tolerance to include Na^+ H^+ antiporter (Apse et al. 1999; Apse et al. 2003; Jha et al. 2010), dehydrin (DHN) (Brini et al. 2007), late embryogenesis abundant protein (LEA)-bZip (Qu et al. 2012), phytoene synthase (Han et al. 2008), DREB (Bhatnagar-Mathur et al. 2007), aquaporin (Peng et al. 2007), glutathione S-transferase (Chen et al. 2012), fatty acid desaturase (Zhang et al. 2012), trihelix transcription factor genes GmGT-2A and GmGT-2B (Xie et al. 2009), and mannose-1-phosphate guanyl transferase (Kumar et al. 2012) as reviewed by Meera et al. (2013).

Mangroves form an ideal system to unveil the genetic background of salt tolerance. Genome and transcriptome sequence information form the basis of gene characterization in mangrove species.

Salt stress-induced transcriptome of *Bruguiera gymnorhiza* root tissues was profiled using the oligo microarray (NCBI GEO accession no. GSE10942). The study came out with a vivid observation that emphasizes the importance of tissue-specific transcriptome. The expression profiles were more or less similar in the lateral and main roots. The nucleotide sequence data were also generated for the screening of ESTs (DDBJ accessions AB429341-AB429362) (Yamanaka et al. 2009). The first mangrove de novo transcriptome analyses were carried out in *Rhizophora mangle* (Rhizophoraceae) and *Heritiera littoralis* (Malvaceae) Dassanayake et al. 2009). *Rhizophora mangle* transcriptome is the foundation for MTDB. The transcriptome analysis of another mangrove species, *Sonneratia alba,* was performed in 2011. Also, genes responsible for salt tolerance of *Sonneratia alba* were identified through in silico annotations (Chen et al. 2011). Transcriptome analysis was performed and compared between *Acanthus ilicifoli*us and its terrestrial relative *Acanthus leucostachyus* (Yang et al. 2015a). Little is known about the origin and evolution of species of Ceriops. In a recent study, Illumina-based transcriptome sequence analysis of two Ceriops species (*Ceriops zippeliana* and *Ceriops tagal*) and one of their terrestrial congenator (Pellacalyx yunnanensis) was successfully conducted (Yang et al. 2015b). *Nypa fruticans* is the one and only monocot species among true mangroves. De novo transcriptome profiling of *Nypa fruticans* was conducted so as to compare the simple sequence repeats with other palm trees (He et al. 2015). Leaf tissue-specific transcriptome sequence and de novo assembly

Table 10.2 List of salt responsive mangrove genes suggested to be used in future transgenic works

Gene	Mangrove source	References
Glutathione S- transferase Z1	*Acanthus*	Nguyen et al. (2007)
Glutathione transferase	*ebracteatus*	Nguyen et al. (2007)
Manganese superoxide dismutase	*Acanthus*	Nguyen et al. (2007)
Myo-insitol 1-phosphate synthase	*ebracteatus*	Nguyen et al. (2007)
Homogentisate	*Acanthus*	Nguyen et al. (2007)
phytylprenyltransferase	*ebracteatus*	Nguyen et al. (2007)
Carotenoid cleavage dioxygenase1–1	*Acanthus*	Nguyen et al. (2007)
Glutamate synthase (ferredoxin)	*ebracteatus*	Nguyen et al. (2007)
Glutamine synthetase	*Acanthus*	Nguyen et al. (2007)
Glutamine synthetase GS58	*ebracteatus*	Nguyen et al. (2007)
Glycine dehydrogenase	*Acanthus*	Nguyen et al. (2007)
Putative alanine aminotransferase	*ebracteatus*	Nguyen et al. (2007)
Putative quinone reductase	*Acanthus*	Nguyen et al. (2007)
NAD malate dehydrogenase	*ebracteatus*	Nguyen et al. (2007)
NADP-malic enzyme	*Acanthus*	Nguyen et al. (2007)
Plastidic aldolase	*ebracteatus*	Nguyen et al. (2007)
Plastidic aldolase NPALDP1	*Acanthus*	Nguyen et al. (2007)
Latex plastidic aldolase	*ebracteatus*	Nguyen et al. (2007)
Phospholipid/ glycerol	*Acanthus*	Nguyen et al. (2007)
acyltransferase	*ebracteatus*	Nguyen et al. (2007)
Chloroplast protease	*Acanthus*	Nguyen et al. (2007)
Amino acid permease-like protein	*ebracteatus*	Nguyen et al. (2007)
Plasma membrane intrinsic protein	*Acanthus*	Nguyen et al. (2007)
Ubiquitin-conjugating enzyme fam-	*ebracteatus*	Nguyen et al. (2007)
ily protein	*Acanthus*	Nguyen et al. (2007)
E3 ubiquitin ligase SCF	*ebracteatus*	Nguyen et al. (2007)
33 kDa polypeptide of water oxidiz-	*Acanthus*	Nguyen et al. (2007)
ing	*ebracteatus*	Nguyen et al. (2007)
Germin-like protein	*Acanthus*	Nguyen et al. (2007)
Proline rich protein	*ebracteatus*	Nguyen et al. (2007)
Putative salt tolerance protein	*Acanthus*	Nguyen et al. (2007)
Universal stress protein (USP)	*ebracteatus*	Nguyen et al. (2007)
14-3-3 protein	*Acanthus*	Nguyen et al. (2007)
GTP binding protein	*ebracteatus*	Nguyen et al. (2007)
Small GTP binding protein Sar1BNt	*Acanthus*	Nguyen et al. (2007)
Zinc finger (DNL type) family pro-	*ebracteatus*	Huang et al. (2003)
tein	*Acanthus*	Huang et al. (2003)
Calcium binding EF hand family	*ebracteatus*	Huang et al. (2003)
Cyclic nucleotide and calmodulin	*Acanthus*	Hibino et al. (2001)
regulated	*ebracteatus*	Tanaka et al. (2002)
DNAJ Heat shock protein	*Acanthus*	Tanaka et al. (2002)
cytosolic class I loss molecular mass	*ebracteatus*	Tanaka et al. (2002)
HSP	*Acanthus*	Tanaka et al. (2002)
cytosolic class II low molecular mass	*ebracteatus*	Tanaka et al. (2002)Tanaka et al.
HSP	*Acanthus*	(2002)
ADP-ribosylation factor (ARF)	*ebracteatus*	Fan et al. (2009)
Betaine aldehyde dehydrogenase	*Acanthus*	Yamada et al. (2002)
(BADH)	*ebracteatus*	
Chloroplast chaperonin-60	*Acanthus*	
Clp protease (clpP)	*ebracteatus*	

(continued)

Table 10.2 (continued)

Gene	Mangrove source	References
Ubiquitin	*Acanthus*	
Translation elongation factor-1 a	*ebracteatus*	
Drought-induced 19 (AtDi119)	*Acanthus*	
Secretory peroxidase	*ebracteatus*	
fructose-1,6-bisphosphate aldolase	*Acanthus*	
Mangrin(Allene oxide cyclase)	*ebracteatus*	
	Acanthus	
	ebracteatus	
	Acanthus	
	ebracteatus	
	Acanthus	
	ebracteatus	
	Acanthus	
	ebracteatus	
	Acanthus	
	ebracteatus	
	Acanthus	
	ebracteatus	
	Acanthus	
	ebracteatus	
	Kandelia candel	
	Kandelia candel	
	Kandelia candel	
	Avicennia marina	
	Avicennia marina	
	Avicennia marina	
	Avicennia marina	
	Avicennia marina	
	Avicennia marina	
	Avicennia marina	
	Sesuvium	
	portulacastrum	
	Brugiera sexangula	

Provided by (Meera et al. 2013)

of datasets of *Asiatic mangrove Rhizophora mucronata* Lam. were performed to develop a reference sequence database to decode gene candidates involved in stress tolerance (Meera et al. 2020).

Plants have a system developed to detoxify this MG consisting of two major enzymes: glyoxalase I (Gly I) and glyoxalase II (Gly II), and hence known as the glyoxalase system. Recently, a novel glyoxalase enzyme, named glyoxalase III (Gly III), has been detected in plants, providing a shorter pathway for MG detoxification. Glyoxalase systems have strong interactions in conferring abiotic stress tolerance in plants through the detoxification of ROS and MG. Upregulation of both Gly I and Gly II, as well as their overexpression in plant species, showed enhanced tolerance to various abiotic stresses, including salinity and drought (Espartero et al. 1995; Veena et al. 1999; Skipsey et al. 2000; Jain et al. 2002; Singla-Pareek et al. 2003, 2008;

Saxena et al. 2005; Yadav et al. 2005; Hossain and Fujita 2009; Lin et al. 2010; Mustafiz et al. 2011; Tuomainen et al. 2011; Wu et al. 2013; Hasanuzzaman et al. 2017a, b).

Meera and Augustine (2020) reported and characterized the GLY I, II, and III in *R. mucronata,* establishing their role in the salinity response of the plant. Glutathione is the key regulator of routine glyoxalase pathway, and relative expression of some glutathione catabolic enzymes and 14 glutathione-dependent transporter proteins under saline conditions was analyzed. Also suggested the feasibility of using mangrove glyoxalases to induce higher salt tolerance in crop plants. Salt-induced changes in gene expression levels of glyoxalase enzymes in the mangrove species were evaluated. The expression status of glutathione synthetic and catalytic enzymes, as well as glutathione regulated transporter protein, was estimated. Rm GLY I was found to be more responsive to salinity stress than the other two glyoxalases. A lead for probable structural stability of mangrove glyoxalases under salinity is also suggested.

10.11 Salinity Tolerance in Crop Species

Barley makes an excellent model for salinity tolerance in cereals as it is the most tolerant among the cultivated cereals (Munns and Tester 2008). A significant QTL on chromosome number 2 was found to be associated with salt tolerance by studies on NILs differing only in the target QTL (Xu et al. 2012) by biochemical and proteomic analyses. Tolerant NILs maintained lower Na+ content in leaves and higher K+ content in the roots when compared to sensitive lines under salt conditions. H2O2, MDA, and proline accumulation was noticed in salinity-sensitive NILs. Proteomic studies using two-dimensional gel electrophoresis and tandem mass spectrometry analyses identified 53 and 51 differentially expressed proteins (associated with photosynthesis, ROS scavenging, and ATP synthase) in the leaves and roots, respectively; Zhu et al. (2020).

Three hundred seventy-seven two-row spring barley cultivars during both the vegetative, in a controlled environment, and the reproductive stages, in the field were used to set up a diversity panel for association mapping for the different components of salinity tolerance. It was observed that the two developmental stages did not share genetic regions associated with the components of salinity tolerance, suggesting that different mechanisms play distinct roles throughout the barley life cycle. Genetically defined regions containing known flowering genes (Vrn-H3, Vrn-H1, and HvNAM-1) were found responsive to salt stress. A salt-responsive locus, such as 7H, 128.35 cM, was associated with grain number per ear, and a gene encoding a vacuolar H+-translocating pyrophosphatase, HVP1, was identified as a candidate. A QTL on chromosome 3H (139.22 cM), significant for ear number per plant, and a locus on chromosome 2H (141.87 cM), which was associated with a yield component, interacted with salinity stress.

HVP1 could be the candidate vacuolar H+-translocating pyrophosphatase underlying the 7H locus. Arabidopsis vacuolar pyrophosphatase AVP1 showed increased shoot biomass production and yield under saline field conditions (Han et al. 2015). Therefore, HVP1 merits further investigation as the causal gene underlying the 7H locus. The means for ear number per plant and grain number per ear for each condition (control or saline) and genotype at these three loci (2H, 141.87 cM; 3H, 139.22 cM; and 7H 128.35 cM). The condition, the genotype, and the condition-by-genotype interaction showed all three as possible candidates for salinity tolerance in barley (Saade et al. 2020).

The genetics of salinity tolerance in barley during germination is reviewed by summarizing reported quantitative trait loci (QTLs) and functional genes by Mwando et al. (2020). Gene candidates for salinity tolerance in Arabidopsis, soybean, maize, wheat, and rice have been blasted and mapped on the barley reference genome. The genetic diversity of three reported functional gene families for salt tolerance during barley germination, namely, dehydration-responsive element-binding (DREB) protein, somatic embryogenesis receptor-like kinase and aquaporin genes, is discussed. The DREB gene family is found to be more diverse in barley than in wheat and rice.

Salinity tolerance in barley is grouped into four classes based on their function (Wu et al. 2011; Walia et al. 2006; Wu et al. 2011; Yin et al. 2018): (1) osmotic protectors, like HvPIP2;5 (Alavilli et al. 2016), HVA1 (Lal et al. 2008), HvDREB1, HvCBF4, HvWRKY38 (Gürel et al. 2016), and ROS scavengers, like trehalose synthesis, mannitol-1-phosphate dehydrogenase (M1PD), and pyrroline-5-carboxylase synthetase (P5CS); (2) Na+ and K+ transporters, like the high affinity potassium transporter (HKT) family (e.g., HvHKT1;5 (Hazzouri et al. 2018; Huang et al. 2019), HvHKT1;1 (Han et al. 2018), HvHKT2;1 (Mian et al. 2011; Assaha et al. 2017), HvHAK1 (Mangano et al. 2008), HvHKT1, HvHKT2 (Qiu et al. 2011), the Na+ /H+ exchanger (NHX) family (HvNax4 (Rivandi et al. 2011), and salt overly sensitive (SOS) engaged Na+/H+ antiporters (HvSOS1 (HvNHX7), HvSOS2 (HvCIPK24), HvSOS3 (HvCBL4), HvNHX1, HVA) (Yousefirad et al. 2018; Wu et al. 2019); (3) Regulatory proteins, like the CBF/ DREB (C-repeat-binding protein/ dehydration-responsive element-binding protein) family (e.g., HvRAF (Jung et al. 2007), HvAP2/ERF (ethylene response factor) (Guo et al. 2016), HvDREB1 (Xu et al. 2009), HvCBF4, HvWRKY38 (Gürel et al. 2016), HvDRF1 (Xue and Loveridge 2004)) in the signaling pathways of long-distance and downstream gene expression. Wu et al. (2011) identified twenty CBF genes have in barley, which enhances tolerance to drought, salinity, and low temperature; and (4) Jasmonate inducers, like (JA) biosynthesis, late embryogenesis abundant (LEA) protein genes (e.g., HVA1 expressed in response to water and salinity stresses, HVA22 expressed in response to dehydration, extreme temperatures, ABA secretion, and salinity stress (al-Yassin and Khademian 2015).

Wheat genes showed the highest sequence similarity to barley (90.93%), followed by maize (83.00%), rice (67.58%), soybean (59.43%), and Arabidopsis (57.94%). The aquaporin-like superfamily protein (HORVU2Hr1G096360.13) had the highest expression levels in the first three development stages of embryo, embryo shoot, and embryo root. PEG treatments showed a high positive correlation has been to salinity stress, indicating that salt stress in germinating seeds is mostly osmotic (Mano et al. 1996). Dehydrogenases (Jörnvall et al. 1984) in embryos catalyze the oxidation of D-glucose to synthesize sugars for osmoprotective functions (Witzel et al. 2010).

Dehydrins, a subfamily of LEA proteins from *Hordeum vulgare* (aba2), enhanced seed germination and water uptake in transgenic lines of Arabidopsis under salinity stress (Calestani et al. 2015; Hara 2010); a role in cellular detoxification also suggested as they contain lipids and has metal-binding capacity preventing lipid peroxidation (Cheng et al. 2002; Krüger et al. 2002; Alsheikh et al. 2003; Koag et al. 2003, 2009)

Somatic embryogenesis receptor-like kinase (SERK) genes (HvSERK1/2/3) isolated from barley were induced in microspore derived embryogenic callus under salt stress (Li et al. 2016), indicating their protective role for developing embryos during salinity stress.

A gene encoding DREB proteins in *H. vulgare* (HvDREB1) induced by exogenous ABA to enhance germination and early root growth in Arabidopsis plants under salinity stress was reported (Xue and Loveridge 2004). TF HvDREB1 is a member of the AP2 group of the DREB subfamily that is vital for regulating responses to various stresses (Agarwal et al. 2006; Xu et al. 2008) and reportedly improves salt, drought, and cold responses in transgenic plants (Oh et al. 2007).

Barley ERF-type TF HvRAF improved seed germination and root growth under salinity stress but was not induced by ABA treatment in transgenic Arabidopsis plants (Jung et al. 2007).

Overexpression of a barley aquaporin gene HvPIP2;5 in yeast enhanced salt and osmotic stress tolerance. In transgenic *Arabidopsis* the gene showed better seed germination and root growth than the wild type under salinity stress (Alavilli et al. 2016)

Three genes' families—DREB protein, SERKs, and aquaporin gene—were considered for further analysis based on their high numbers of hits. The proteins from the three genes were blasted on the Phytozome (https://phytozome.jgi.doe.gov/pz/portal.html) to download matched homologs (>30%) in 40 different plant species. Gene alignments showed three distinct regions across the sequences of the gene families. The start and end sections of the sequences were more divergent than the middle segments. The phylogenetic tree shows that DREB proteins are no more diverse in barley than in wheat or rice (Mwando et al. 2020).

10.12 Phenotyping Technologies for Comprehensive Salt Stress Phenotyping

Several high throughput (HTP) systems have emerged, the majority of which rely on regular, non-invasive, automated imaging to the phenotype of tens, hundreds, or even thousands of plants in a short period of time (Furbank and Tester 2011; Ghanem et al. 2015; Junker et al. 2015). In both controlled and outdoor situations, these methods make phenotyping large numbers of genotypes easier. Red green blue (RGB), thermal infrared (TIR), chlorophyll fluorescence (ChlF), and, more recently, multispectral and hyperspectral imaging are all used by HTP systems (Humplík et al. 2015; Fahlgren et al. 2015).

10.13 Harnessing the Genetic Diversity of Exotic Germplasm

To supplement the genetic scarcity of modern elite cultivars, several sources of genetic variation are accessible. Because variations can be directly produced in commercial germplasm, genetic diversity generated artificially through mutagenesis has proven to be a beneficial resource for a variety of agricultural species (Caldwell et al. 2004; Mba 2013; Nikam et al. 2015; Gulfishan et al. 2016; Çelik and Atak 2017; Pando and Deza 2017; Wang et al. 2017; Tu Anh et al. 2018). Given that salt tolerance is most likely the result of the combined actions of multiple pathways, the ability to develop salt-tolerant variations intentionally is definitely restricted. On the other hand, as seen in landraces or wild cousins of crops, these are quite likely to evolve through natural selection in plants that are subjected to adverse environmental conditions (Mayes et al. 2012). As a result, it is critical to research the naturally diverse germplasm that is available.

The rapid advancement of DNA sequencing and genome assembly capabilities over the last decade has made genomic characterisation of novel and uncharacterized germplasm—potential salt tolerance allele and gene repositories easier. Over 300 plant species now have reference genome sequences (as of March 2018; http://www.plabipd.de/), with chromosome-level assemblies for several major crops, like rice, maize, and wheat, as well as less well-known species like quinoa; Schnable 2012; Mayer et al. 2014; Jackson 2016; Jarvis et al. 2017.

The capacity of these genomic resources to facilitate the capture and description of genetic diversity is their greatest value. Indeed, from well-established genotyping chips to more modern genotyping-by-sequencing methods, such as various reduced representation sequencing and whole-genome re-sequencing approaches, high-quality genomes breed high-quality genotyping (Bevan et al. 2017; Dwivedi et al. 2017; Scheben et al. 2017). With these methods, large numbers of accessions may now be genotyped fast and cheaply, with the potential for millions of single-nucleotide polymorphisms (SNPs) to be discovered. Genomic resources, such as diversity panels and mapping populations (e.g., bi-parental, nested association), are now available for all sequenced main crop species, a number of wild relatives, and increasingly for the aforementioned orphan crops (Rasheed et al. 2018).

10.14 Single-Step GWAS as New GWAS Technique (ssGWAS)

The most comprehensive information for genetic evaluation comes from the ssGWAS approach. The ssGWAS approach allows researchers to assess all available Table 10.2. Data in one step, including genetic markers, phenotype records, and pedigree information. Much recent research has verified this method and successfully implemented ssGWAS in pigs (Howard et al. 2015; Wu et al. 2018) and other species (Silva et al. 2017), yielding higher power and more exact estimates than other models. Thus, in a single model, genetic markers, repeat measurements of each variable, spatial and design elements, and marker–treatment interaction may all be considered. The statistical power of the conclusions from this procedure is projected to increase because each individual plant is used as input data rather than the genotype means. Because of the inclusion of the interaction model in the ssGWAS, allows finding important loci particularly related to salt stress, which could be advantageous for **complicated quantitative salt tolerance traits** with multiple components that can be successfully deconstructed by a ssGWAS model.

10.15 Conclusions and Future Perspectives

Climate change and global warming necessitate the cultivation of plant species that can survive under ambient salinity and extremes of temperature. Mangroves are extremophiles that have evolved their own mechanisms to cope with extremes of temperature and salinity. They still remain as an untapped and underutilized gene pool for genes providing tolerance to abiotic stress. The emergence of a mangrove species as a model plant for the molecular elucidation of salinity/abiotic stress tolerance can enlighten our understanding of the salt tolerance mechanisms.

Many a time, expressed genes in transgenic systems have proven to be salt resistant. As abiotic stress tolerance is a multigenic trait, researchers aim to pyramid more than one type of gene involved in stress tolerance for enhanced transgenic system efficacies. While every researcher's ambition is to find the elusive "master control" that activates the complete cascade of genes in the abiotic stress pathway, current methodologies have made significant progress in understanding the fundamental mechanism of abiotic stress tolerance in plants. In the future, mangroves, including facultative and obligate halophytes, have the potential to be a rich repository for genes for QTL-based techniques used in abiotic stress tolerance breeding, which includes salinity tolerance. Enormous progress has been made in understanding halophyte abiotic stress response systems with the aid of techniques, like transcriptomic profiling and QTL mapping (Flowers 2004). At the molecular level, however, there are several gaps in the pathways implicated in salt stress responses. For a complete knowledge of plant cell adaptation to salt stress, a far deeper understanding of protein–protein interactions and intracellular alterations is required.

Multiple studies have revealed that both biotic and abiotic stresses influence the expression of several common genes that work together to impart stress tolerance. Stress tolerance cannot be altered by manipulating a single trait as there are many component traits. Also, it is clear that overexpression or suppression of genes has far-reaching consequences. Overexpression of some of the abiotic stress-responsive genes from halophytes in crop systems has been encouraging. Apart from being a Na +/H+ vacuolar antiporter, AtNHX1 mutant plants displayed altered expression of genes related to intravesicular trafficking (Sottosanto et al. 2004). To uncover the genes linked with a QTL area, current methodologies employ DNA arrays in conjunction with QTL mapping. In order to solve the problem of abiotic stress tolerance, QTL mapping and marker-assisted breeding may be the best option. The ability to pyramid or combine QTLs for diverse stress tolerances is a key benefit of this strategy (Takeda and Matsuoka 2008). While overexpression of genes can result in greater stress tolerance, it is critical to test these transgenic systems in open field circumstances, where the crops are exposed to a range of stressors at different phases of plant growth (Rajalakshmi and Parida 2012). As 21 of the 28 crops cultivated are largely glycophytes, it is crucial to emphasize the molecular distinctions between these two plant groups, which is indispensable for glycophyte transformation utilising halophytic candidate genes, opening up a new line of research into how to prepare crops for increased soil salt. Despite their significant implications, knowledge of the discovery, metabolic activities, and isolation of key genes linked with salt tolerance in halophytes is relatively restricted. As a result, identifying and characterization novel salt tolerance-associated genes from halophytes, as well as their subsequent use in genetic engineering, are critical areas of plant research that require more attention in order to sustainably increase crop production and achieve global food security in the face of climate change. Future research should concentrate on identifying halophytic plants that have numerous salt tolerance mechanisms and can generate more biomass in order to remove the greatest amount of salts from polluted soils. Furthermore, after more than a century of research on halophytes, numerous basic physiological, biochemical, and molecular pathways remain unknown. As a result, further research is needed to understand the underlying salt tolerance processes of halophytes in order to improve soil conservation. Furthermore, new genetic and omic research has revealed that salt tolerance is a complex feature that is influenced by whole-plant responses rather than a single gene or protein. Exploration of the salt tolerance capability of diverse halophytes, as well as the production of transgenic plants with improved remediation properties, is therefore critical for sustainable agriculture in salinity-affected areas, as well as for mitigating the effects of global climate change (Rahman et al. 2021). While examining salt tolerance, yield is usually the most significant factor, but biomass, fruit or grain quality, and other factors may also be relevant. Other qualities (e.g., transpiration, photosynthetic activity, metabolite profiles) should be evaluated as possible variables contributing to salt tolerance in terms of these traits of real-world importance rather than as measures of salt tolerance in and of themselves (Mitchell et al. 2019). Correlation studies may be used to acquire insight into the eventual contribution of component qualities to salt tolerance in terms of yield using a suitably detailed phenotyping approach. Despite the fact that each of these components is complicated in and of itself, this disambiguation aids in the dissection of salt tolerance for yield into basic portions that are more genetically traceable than the total. The processes and subcomponents that contribute to salt tolerance for yield can change across and within species, as well as during development and across different environmental circumstances. Genotype, salt stress concentration and distribution (e.g., temporal dynamics, topography), developmental stage at which stress is imposed, weather conditions, cultivation practices, soil type, and water availability are all important factors to consider (Greenway and Munns 1980; Tal and Shannon 1983; Mano and Takeda 1997; Foolad 1999; Yamaguchi and Blumwald 2005). To supplement the genetic scarcity of current elite cultivars, several sources of genetic diversity are accessible. Because variations may be directly created in commercial germplasm, genetic diversity generated artificially by mutagenesis has proved to be a beneficial resource for a variety of agricultural species (Caldwell et al. 2004; Mba 2013; Nikam et al. 2015; Gulfishan et al. 2016; Çelik and Atak 2017; Pando and Deza 2017; Wang et al. 2017; Tu Anh et al. 2018). Given that salt tolerance is most likely the result of the combined actions of many pathways, the ability to develop salt-tolerant variations intentionally is clearly restricted (Mitchell et al. 2019)

Rapidly growing genome-wide association models in plants are promising and extremely valuable tools for finding alleles and markers. In crops like rice, combining genomic selection with information from population structure, GWAS, and the known genetic architecture of certain traits has resulted in greater breeding efficiency (Spindel et al. 2015). Plant species are yet to be bred for salt tolerance through genomic selection. Lower genotyping costs and better genomic models for predicting breeding values will help speed up the delivery of superior breeding lines with salt tolerance. Selection of high-yield stress resistant crops that boost agricultural productivity can be made by identifying the link between genotype and phenotype. Furthermore, the use of genome editing technologies will allow for the confirmation of causative gene activity. CRISPR-Cas9, for example, is predicted to reveal causal links between candidate genes discovered by GWAS and phenotypic features in order to conduct tailored molecular-assisted breeding programs (Yin et al. 2017). Genome editing, on the other hand, will not only aid in the functional validation of loci of interest but will also allow for the simultaneous change of many genetic loci in elite crops, advancing breeding programs, and the translation of discoveries to the field. Novel genetic resources, such as understudied crops, and new imaging techniques, such as HTP, combined with new statistical association models, will speed up the discovery of new stress tolerance pathways leading to the development of enhanced salt tolerance in crop plants.

References

Acosta-Motos JR, Diaz-Vivancos P, Álvarez S, Fernández-García N, Sánchez-Blanco MJ, Hernández JA (2015a) NaCl-induced physiological and biochemical adaptative mechanisms in the ornamental Myrtus communis L. plants. J Plant Physiol 183:41–51

Acosta-Motos JR, Diaz-Vivancos P, Alvarez S, Fernández-García N, Sanchez-Blanco MJ, Hernández JA (2015b) Physiological and biochemical mechanisms of the ornamental Eugenia myrtifolia L. plants for coping with NaCl stress and recovery. Planta 242(4):829–846

Acosta-Motos JR, Ortuño MF, Bernal-Vicente A, Diaz-Vivancos P, Sanchez-Blanco MJ, Hernandez JA (2017) Plant responses to salt stress: adaptive mechanisms. Agronomy 7(1):18

Agarie S, Shimoda T, Shimizu Y, Baumann K, Sunagawa H, Kondo A, Ueno O, Nakahara T, Nose A, Cushman JC (2007) Salt tolerance, salt accumulation, and ionic homeostasis in an epidermal bladder-cell-less mutant of the common ice plant Mesembryanthemum crystallinum. J Exp Bot 58(8):1957–1967

Agarwal PK, Jha B (2010) Transcription factors in plants and ABA dependent and independent abiotic stress signalling. Biol Plant 54(2):201–212

Agarwal PK, Agarwal P, Reddy MK, Sopory SK (2006) Role of DREB transcription factors in abiotic and biotic stress tolerance in plants. Plant Cell Rep 25(12):1263–1274

Ahmed T, Noman M, Manzoor N, Shahid M, Abdullah M, Ali L, Wang G, Hashem A, Al-Arjani ABF, Alqarawi AA, Abd-Allah EF (2021) Nanoparticle-based amelioration of drought stress and cadmium toxicity in rice via triggering the stress responsive genetic mechanisms and nutrient acquisition. Ecotoxicol Environ Saf 209:111829

Alavilli H, Awasthi JP, Rout GR, Sahoo L, Lee BH, Panda SK (2016) Overexpression of a barley aquaporin gene, HvPIP2; 5 confers salt and osmotic stress tolerance in yeast and plants. Front Plant Sci *7*:1566

Albacete A, Ghanem ME, Martínez-Andújar C, Acosta M, Sánchez-Bravo J, Martínez V, Lutts S, Dodd IC, Pérez-Alfocea F (2008) Hormonal changes in relation to biomass partitioning and shoot growth impairment in salinized tomato (*Solanum lycopersicum* L.) plants. J Exp Bot 59(15):4119–4131

Ali Z, Park HC, Ali A, Oh DH, Aman R, Kropornicka A, Hong H, Choi W, Chung WS, Kim WY, Bressan RA (2012) TsHKT1; 2, a HKT1 homolog from the extremophile Arabidopsis relative Thellungiella salsuginea, shows K+ specificity in the presence of NaCl. Plant Physiol 158(3): 1463–1474

Alia PSP, Saradhi PP, Mohanty P (1991) Proline enhances primary photochemical activities in isolated thylakoid membranes of Brassica juncea by arresting photo inhibitory damage. Biochem Biophys Res Commun 181(3):1238–1244. https://doi.org/10.1016/0006-291x(91) 92071-q. PMID: 1764073

Alsheikh MK, Heyen BJ, Randall SK (2003) Ion binding properties of the dehydrin ERD14 are dependent upon phosphorylation. J Biol Chem 278(42):40882–40889

Álvarez S, Sánchez-Blanco MJ (2013) Changes in growth rate, root morphology and water use efficiency of potted Callistemon citrinus plants in response to different levels of water deficit. Sci Hortic 156:54–62

Álvarez S, Sanchez-Blanco MJ (2014) Long-term effect of salinity on plant quality, water relations, photosynthetic parameters and ion distribution in Callistemon citrinus. Plant Biol *16*(4): 757–764

Álvarez S, Navarro A, Nicolás E, Sánchez-Blanco MJ (2011) Transpiration, photosynthetic responses, tissue water relations and dry mass partitioning in Callistemon plants during drought conditions. Sci Hortic 129(2):306–312

Al-Yassin A, Khademian R (2015) Allelic variation of salinity tolerance genes in barley ecotypes (natural populations) using EcoTILLING: a review article. Am Eur J Agric Environ Sci 15(4): 563–572

Amtmann A (2009) Learning from evolution: Thellungiella generates new knowledge on essential and critical components of abiotic stress tolerance in plants. Mol Plant 2(1):3–12

Anjum NA, Gill SS, Ahmad I, Tuteja N, Soni P, Pareek A, Umar S, Iqbal M, Pacheco M, Duarte AC, Pereira E (2012) Understanding stress-responsive mechanisms in plants: an overview of transcriptomics and proteomics approaches. In: Improving crop resistance to abiotic stress, pp 337–355

Apse MP, Blumwald E (2007) Na+ transport in plants. FEBS Lett 581(12):2247–2254

Apse MP, Aharon GS, Snedden WA, Blumwald E (1999) Salt tolerance conferred by overexpression of a vacuolar Na+/H+ antiport in Arabidopsis. Science 285(5431):1256–1258

Apse MP, Sottosanto JB, Blumwald E (2003) Vacuolar cation/H+ exchange, ion homeostasis, and leaf development are altered in a T-DNA insertional mutant of AtNHX1, the Arabidopsis vacuolar Na+/H+ antiporter. Plant J 36(2):229–239

Ashihara H, Wakahara S, Suzuki M, Kato A, Sasamoto H, Baba S (2003) Comparison of adenosine metabolism in leaves of several mangrove plants and a poplar species. Plant Physiol Biochem 41(2):133–139

Ashraf M, Akram NA (2009) Improving salinity tolerance of plants through conventional breeding and genetic engineering: an analytical comparison. Biotechnol Adv 27(6):744–752

Ashraf M, Harris PJ (2013) Photosynthesis under stressful environments: an overview. Photosynthetica 51(2):163–190

Assaha DVM, Ueda A, Saneoka H, Al-Yahyai R, Yaish MW (2017) The role of Na+ and K+ transporters in salt stress adaptation in glycophytes. Front Physiol 8:1–19

Ball MC (1988) Ecophysiology of mangroves. Trees 2(3):129–142

Ball MC, Pidsley SM (1995) Growth responses to salinity in relation to distribution of two mangrove species, Sonneratia alba and S. lanceolata, in northern Australia. Funct Ecol 9:77–85

Bartels D, Dinakar C (2013) Balancing salinity stress responses in halophytes and non-halophytes: a comparison between Thellungiella and Arabidopsis thaliana. Funct Plant Biol 40(9):819–831

Ben-Saad R, Meynard D, Ben-Romdhane W, Mieulet D, Verdeil JL, Al-Doss A, Guiderdoni E, Hassairi A (2015) The promoter of the AlSAP gene from the halophyte grass Aeluropus littoralis directs a stress-inducible expression pattern in transgenic rice plants. Plant Cell Rep 34(10):1791–1806

Bernstein N, Meiri A, Zilberstaine M (2004) Root growth of avocado is more sensitive to salinity than shoot growth. J Am Soc Hortic Sci 129(2):188–192

Bevan MW, Uauy C, Wulff BB, Zhou J, Krasileva K, Clark MD (2017) Genomic innovation for crop improvement. Nature 543(7645):346–354

Bhatnagar-Mathur P, Devi MJ, Reddy DS, Lavanya M, Vadez V, Serraj R, Yamaguchi-Shinozaki-K, Sharma KK (2007) Stress-inducible expression of At DREB1A in transgenic peanut (Arachis hypogaea L.) increases transpiration efficiency under water-limiting conditions. Plant Cell Rep 26(12):2071–2082

Bhatnagar-Mathur P, Vadez V, Sharma KK (2008) Transgenic approaches for abiotic stress tolerance in plants: retrospect and prospects. Plant Cell Rep 27(3):411–424

Bockheim JG, Gennadiyev AN (2000) The role of soil-forming processes in the definition of taxa in soil taxonomy and the world soil reference base. Geoderma 95(1–2):53–72

Breckle SW (1995) How do halophytes overcome salinity? In: Biology of salt tolerant plants, vol. 23, pp 199–203

Brini F, Hanin M, Lumbreras V, Amara I, Khoudi H, Hassairi A, Masmoudi K (2007) Overexpression of wheat dehydrin DHN-5 enhances tolerance to salt and osmotic stress in Arabidopsis thaliana. Plant Cell Rep 26(11):2017–2026

Burchett MD, Clarke CJ, Field CD, Pulkownik A (1989) Growth and respiration in two mangrove species at a range of salinities. Physiol Plant 75(2):299–303

Caldwell DG, McCallum N, Shaw P, Muehlbauer GJ, Marshall DF, Waugh R (2004) A structured mutant population for forward and reverse genetics in Barley (*Hordeum vulgare* L.). Plant J 40(1):143–150

Calestani C, Moses MS, Maestri E, Marmiroli N, Bray EA (2015) Constitutive expression of the barley dehydrin gene aba2 enhances Arabidopsis germination in response to salt stress. Int J Plant Biol 6(1):5826

Carillo P, Annunziata MG, Pontecorvo G, Fuggi A, Woodrow P (2011) Salinity stress and salt tolerance. In: Abiotic stress in plants-mechanisms and adaptations, vol. 1, pp 21–38

Cassaniti C, Leonardi C, Flowers TJ (2009) The effects of sodium chloride on ornamental shrubs. Sci Hortic 122(4):586–593

Cassaniti C, Romano D, Flowers TJ (2012) The response of ornamental plants to saline irrigation water. In: Irrigation: water management, pollution and alternative strategies, pp 131–158

Çelik Ö, Atak Ç (2017) Applications of ionizing radiation in mutation breeding. In: New insights on gamma rays, pp 111–132

Century K, Reuber TL, Ratcliffe OJ (2008) Regulating the regulators: the future prospects for transcription-factor-based agricultural biotechnology products. Plant Physiol 147(1):20–29

Chen S, Zhou R, Huang Y, Zhang M, Yang G, Zhong C, Shi S (2011) Transcriptome sequencing of a highly salt tolerant mangrove species Sonneratia alba using Illumina platform. Mar Genomics 4(2):129–136

Chen JH, Jiang HW, Hsieh EJ, Chen HY, Chien CT, Hsieh HL, Lin TP (2012) Drought and salt stress tolerance of an Arabidopsis glutathione S-transferase U17 knockout mutant are attributed to the combined effect of glutathione and abscisic acid. Plant Physiol 158(1):340–351

Chen Z, Hu L, Han N, Hu J, Yang Y, Xiang T, Zhang X, Wang L (2015) Overexpression of a miR393-resistant form of transport inhibitor response protein 1 (mTIR1) enhances salt tolerance by increased osmoregulation and Na+ exclusion in Arabidopsis thaliana. Plant Cell Physiol 56(1):73–83

Cheng Z, Targolli J, Huang X, Wu R (2002) Wheat LEA genes, PMA80 and PMA1959, enhance dehydration tolerance of transgenic rice (Oryza sativa L.). Mol Breed 10(1):71–82

Chrominski A, Bhat RB, Weber DJ, Smith BN (1988) Osmotic stress-dependent conversion of 1-aminocyclopropane-1-carboxylic acid (ACC) to ethylene in the halophyte, Allenrolfea occidentalis. Environ Exp Bot 28(3):171–174

Cisneros AE, Carbonell A (2020) Artificial small RNA-based silencing tools for antiviral resistance in plants. Plants 9(6):669

Clough BF (1984) Growth and salt balance of the mangroves Avicennia marina (Forsk.) Vierh. and Rhizophora stylosa Griff. in relation to salinity. Funct Plant Biol 11(5):419–430

Colmer TD, Munns R, Flowers TJ (2005) Improving salt tolerance of wheat and barley: future prospects. Aust J Exp Agric 45(11):1425–1443

Cramer GR (2002a) Response of abscisic acid mutants of Arabidopsis to salinity. Funct Plant Biol 29(5):561–567

Cramer GR (2002b) Sodium-calcium interactions under salinity stress. In: Salinity: environment-plants-molecules. Springer, Dordrecht, pp 205–227

Croser C, Renault S, Franklin J, Zwiazek J (2001) The effect of salinity on the emergence and seedling growth of Picea mariana, Picea glauca, and Pinus banksiana. Environ Pollut 115(1): 9–16

Das AB, Strasser RJ (2013) Salinity-induced genes and molecular basis of salt-tolerant strategies in Mangroves. In: Molecular stress physiology of plants. Springer, India, pp 53–86

Dassanayake M, Haas JS, Bohnert HJ, Cheeseman JM (2009) Shedding light on an extremophile lifestyle through transcriptomics. New Phytol 183(3):764–775

Deinlein U, Stephan AB, Horie T, Luo W, Xu G, Schroeder JI (2014) Plant salt-tolerance mechanisms. Trends Plant Sci 19(6):371–379

Ding D, Zhang L, Wang H, Liu Z, Zhang Z, Zheng Y (2009) Differential expression of miRNAs in response to salt stress in maize roots. Ann Bot 103(1):29–38

Downton WJS (1982) Growth and osmotic relations of the mangrove Avicennia marina, as influenced by salinity. Funct Plant Biol 9(5):519–528

Duke NC (2013) Mangrove floristics and biogeography. In: Robertson AI, Alongi DM (eds) Tropical mangrove ecosystems. American Geophysical Union, Washington, pp 63–100

Duke N, Ball M, Ellison J (1998) Factors influencing biodiversity and distributional gradients in mangroves. Glob Ecol Biogeogr Lett 7(1):27–47

Dwivedi SL, Scheben A, Edwards D, Spillane C, Ortiz R (2017) Assessing and exploiting functional diversity in germplasm pools to enhance abiotic stress adaptation and yield in cereals and food legumes. Front Plant Sci 8:1461
Elphick CH, Sanders D, Maathuis FJ (2001) Critical role of divalent cations and Na+ efflux in Arabidopsis thaliana salt tolerance. Plant Cell Environ 24(7):733–740
Espartero J, Sánchez-Aguayo I, Pardo JM (1995) Molecular characterization of glyoxalase-I from a higher plant; upregulation by stress. Plant Mol Biol 29(6):1223–1233
Eynard A, Lal R, Wiebe K (2005) Crop response in salt-affected soils. J Sustain Agric 27(1):5–50
Ezawa S, Tada Y (2009) Identification of salt tolerance genes from the mangrove plant Bruguiera gymnorhiza using Agrobacterium functional screening. Plant Sci 176(2):272–278
Fahlgren N, Gehan MA, Baxter I (2015) Lights, camera, action: high-throughput plant phenotyping is ready for a close-up. Curr Opin Plant Biol 24:93–99
FAO (2017) The future of food and agriculture—trends and challenges. Food and Agriculture Organization of the United Nations, Rome
Fan W, Zhang Z, Zhang Y (2009) Cloning and molecular characterization of fructose-1, 6-bisphosphate aldolase gene regulated by high salinity and drought in Sesuvium portulacastrum. Plant Cell Rep 28(6):975–984
Fernández-García N, Olmos E, Bardisi E, García-De la Garma J, López-Berenguer C, Rubio-Asensio JS (2014) Intrinsic water use efficiency controls the adaptation to high salinity in a semi-arid adapted plant, henna (Lawsonia inermis L.). J Plant Physiol 171(5):64–75
Fitz Gerald JN, Lehti-Shiu MD, Ingram PA, Deak KI, Biesiada T, Malamy JE (2006) Identification of quantitative trait loci that regulate Arabidopsis root system size and plasticity. Genetics 172(1):485–498
Flowers TJ (2004) Improving crop salt tolerance. J Exp Bot 55(396):307–319
Flowers TJ, Colmer TD (2015) Plant salt tolerance: adaptations in halophytes. Ann Bot 115(3): 327–331
Foolad MR (1999) Genetics of salt and cold tolerance in tomato quantitative analysis and QTL mapping. Plant Biotechnol 16(1):55–64
Franco JA, Fernandez JA, Bañón S, González A (1997) Relationship between the effects of salinity on seedling leaf area and fruit yield of six muskmelon cultivars. Hortscience 32(4):642–644
Franco JA, Bañón S, Vicente MJ, Miralles J, Martínez-Sánchez JJ (2011a) Root development in horticultural plants grown under abiotic stress conditions—a review. J Hortic Sci Biotechnol 86(6):543–556
Franco JA, Cros V, Vicente MJ, Martínez-Sánchez JJ (2011b) Effects of salinity on the germination, growth, and nitrate contents of purslane (Portulaca oleracea L.) cultivated under different climatic conditions. J Hortic Sci Biotechnol 86(1):1–6
Fu X, Huang Y, Deng S, Zho R, Yang G, Ni X, Li W, Shi S (2005) Construction of a SSH library of Aegiceras corniculatum under salt stress and expression analysis of four transcripts. Plant Sci 169(1):147–154
Furbank RT, Tester M (2011) Phenomics–technologies to relieve the phenotyping bottleneck. Trends Plant Sci 16(12):635–644
Ganesan G, Sankararamasubramanian HM, Harikrishnan M, Ashwin G, Parida A (2012) A MYB transcription factor from the grey mangrove is induced by stress and confers NaCl tolerance in tobacco. J Exp Bot 63(12):4549–4561
Garg R, Verma M, Agrawal S, Shankar R, Majee M, Jain M (2014) Deep transcriptome sequencing of wild halophyte rice, Porteresia coarctata, provides novel insights into the salinity and submergence tolerance factors. DNA Res 21(1):69–84
Gaxiola RA, Palmgren MG, Schumacher K (2007) Plant proton pumps. FEBS Lett 581(12): 2204–2214
Ghanem ME, Albacete A, Martínez-Andújar C, Acosta M, Romero-Aranda R, Dodd IC, Lutts S, Pérez-Alfocea F (2008) Hormonal changes during salinity-induced leaf senescence in tomato (Solanum lycopersicum L.). J Exp Bot 59(11):3039–3050

Ghanem ME, Marrou H, Sinclair TR (2015) Physiological phenotyping of plants for crop improvement. Trends Plant Sci 20(3):139–144

Gleeson T, Wada Y, Bierkens MF, Van Beek LP (2012) Water balance of global aquifers revealed by groundwater footprint. Nature 488(7410):197–200

Golldack D, Dietz KJ (2001) Salt-induced expression of the vacuolar H+-ATPase in the common ice plant is developmentally controlled and tissue specific. Plant Physiol 125(4):1643–1654

Gómez-Bellot MJ, Álvarez S, Bañón S, Ortuño MF, Sánchez-Blanco MJ (2013a) Physiological mechanisms involved in the recovery of euonymus and laurustinus subjected to saline waters. Agric Water Manag 128:131–139

Gómez-Bellot MJ, Alvarez S, Castillo M, Bañón S, Ortuño MF, Sánchez-Blanco MJ (2013b) Water relations, nutrient content and developmental responses of Euonymus plants irrigated with water of different degrees of salinity and quality. J Plant Res 126(4):567–576

Greenway H, Munns R (1980) Mechanisms of salt tolerance in non-halophytes. Annu Rev Plant Physiol 31(1):149–190

Grigore MN, Villanueva Lozano M, Boscaiu Neagu MT, Vicente Meana Ó (2012) Do halophytes really require salts for their growth and development? An experimental approach. Not Sci Biol 4(2):23–29

Gulfishan M, Bhat TA, Oves M (2016) Mutants as a genetic resource for future crop improvement. In: Al-Khayri JM, Jain SM, Johnson DV (eds) Advances in plant breeding strategies: breeding, biotechnology and molecular tools. Springer International Publishing, Cham, pp 95–112

Guo L, Yu Y, Xia X, Yin W (2010) Identification and functional characterisation of the promoter of the calcium sensor gene CBL1 from the xerophyte Ammopiptanthus mongolicus. BMC Plant Biol 10(1):1–16

Guo W, Chen T, Hussain N, Zhang G, Jiang L (2016) Characterization of salinity tolerance of transgenic rice lines harboring HsCBL8 of wild barley (*Hordeum spontanum*) line from Qinghai-Tibet plateau. Front Plant Sci 7:1678

Gürel F, Öztürk ZN, Uçarlı C, Rosellini D (2016) Barley genes as tools to confer abiotic stress tolerance in crops. Front Plant Sci 7:1137

Hamada A, Shono M, Xia T, Ohta M, Hayashi Y, Tanaka A, Hayakawa T (2001) Isolation and characterization of a Na+/H+ antiporter gene from the halophyte Atriplex gmelini. Plant Mol Biol 46(1):35–42

Han H, Li Y, Zhou S (2008) Overexpression of phytoene synthase gene from Salicornia europaea alters response to reactive oxygen species under salt stress in transgenic Arabidopsis. Biotechnol Lett 30(8):1501–1507

Han JS, Park KI, Jeon SM, Park S, Naing AH, Kim CK (2015) Assessments of salt tolerance in a bottle gourd line expressing the Arabidopsis H+-pyrophosphatase AVP 1 gene and in a watermelon plant grafted onto a transgenic bottle gourd rootstock. Plant Breed 134(2):233–238

Han Y, Yin S, Huang L, Wu X, Zeng J, Liu X, Qiu L, Munns R, Chen ZH, Zhang G (2018) A sodium transporter HvHKT1; 1 confers salt tolerance in barley via regulating tissue and cell ion homeostasis. Plant Cell Physiol 59(10):1976–1989

Hanin M, Ebel C, Ngom M, Laplaze L, Masmoudi K (2016) New insights on plant salt tolerance mechanisms and their potential use for breeding. Front Plant Sci 7:1787

Hanson AD, Burnet M (1994) Evolution and metabolic engineering of osmoprotectant accumulation in higher plants. In: Biochemical and cellular mechanisms of stress tolerance in plants. Springer, Berlin, pp 291–302

Hantao Z, Qingtong L, Wen P, Yuanyuan G, Pan C, Xu C, Bo L (2004) Transformation of the salt tolerant gene of Avicennia marina into tobacco plants and cultivation of salt-tolerant lines. Chin Sci Bull 49:456–461

Hara M (2010) The multifunctionality of dehydrins: an overview. Plant Signal Behav 5(5):503–508

Hasanuzzaman M, Nahar K, Anee TI, Fujita M (2017a) Glutathione in plants: biosynthesis and physiological role in environmental stress tolerance. Physiol Mol Biol Plants 23(2):249–268

Hasanuzzaman M, Nahar K, Hossain M, Mahmud JA, Rahman A, Inafuku M, Oku H, Fujita M (2017b) Coordinated actions of glyoxalase and antioxidant defense systems in conferring abiotic stress tolerance in plants. Int J Mol Sci 18(1):200

Hasegawa PM, Bressan RA, Zhu JK, Bohnert HJ (2000) Plant cellular and molecular responses to high salinity. Annu Rev Plant Biol 51(1):463–499

Hazzouri KM, Khraiwesh B, Amiri K, Pauli D, Blake T, Shahid M, Mullath SK, Nelson D, Mansour AL, Salehi-Ashtiani K, Purugganan M (2018) Mapping of HKT1; 5 gene in barley using GWAS approach and its implication in salt tolerance mechanism. Front Plant Sci 9:156

He Z, Zhang Z, Guo W, Zhang Y, Zhou R, Shi S (2015) De novo assembly of coding sequences of the mangrove palm (Nypa fruticans) using RNA-Seq and discovery of whole-genome duplications in the ancestor of palms. PLoS One 10(12):e0145385

He Z, Xu S, Zhang Z, Guo W, Lyu H, Zhong C, Boufford DE, Duke NC, International Mangrove Consortium, Shi S (2020) Convergent adaptation of the genomes of woody plants at the land–sea interface. National science review 7(6):978–993

Hernández JA, Almansa MS (2002) Short-term effects of salt stress on antioxidant systems and leaf water relations of pea leaves. Physiol Plant 115(2):251–257

Hernandez JA, Corpas FJ, Gomez M, del Rio LA, Sevilla F (1993) Salt-induced oxidative stress mediated by activated oxygen species in pea leaf mitochondria. Physiol Plant 89(1):103–110

Hernandez JA, Olmos E, Corpas FJ, Sevilla F, Del Rio LA (1995) Salt-induced oxidative stress in chloroplasts of pea plants. Plant Sci 105(2):151–167

Hernández JA, Aguilar AB, Portillo B, López-Gómez E, Beneyto JM, García-Legaz MF (2003) The effect of calcium on the antioxidant enzymes from salt-treated loquat and anger plants. Funct Plant Biol 30(11):1127–1137

Hibino T, Meng YL, Kawamitsu Y, Uehara N, Matsuda N, Tanaka Y, Ishikawa H, Baba S, Takabe T, Wada K, Ishii T (2001) Molecular cloning and functional characterization of two kinds of betaine-aldehyde dehydrogenase in betaine-accumulating mangrove Avicennia marina (Forsk.) Vierh. Plant Mol Biol 45(3):353–363

Himabindu Y, Chakradhar T, Reddy MC, Kanygin A, Redding KE, Chandrasekhar T (2016) Salt-tolerant genes from halophytes are potential key players of salt tolerance in glycophytes. Environ Exp Bot 124:39–63

Hossain MA, Fujita M (2009) Purification of glyoxalase I from onion bulbs and molecular cloning of its cDNA. Biosci Biotechnol Biochem 73(9):2007–2013

Howard JT, Jiao S, Tiezzi F, Huang Y, Gray KA, Maltecca C (2015) Genome-wide association study on legendre random regression coefficients for the growth and feed intake trajectory on Duroc Boars. BMC Genet 16(1):1–11

Hsiao TC, Xu LK (2000) Sensitivity of growth of roots versus leaves to water stress: biophysical analysis and relation to water transport. J Exp Bot 51(350):1595–1616

Huang W, Fang XD, Li GY, Lin QF, Zhao WM (2003) Cloning and expression analysis of salt responsive gene from Kandelia candel. Biol Plant 47(4):501–507

Huang L, Kuang L, Wu L, Wu D, Zhang G (2019) Comparisons in functions of HKT1; 5 transporters between Hordeum marinum and Hordeum vulgare in responses to salt stress. Plant Growth Regul 89(3):309–319

Humplík JF, Lazár D, Husičková A, Spíchal L (2015) Automated phenotyping of plant shoots using imaging methods for analysis of plant stress responses—a review. Plant Methods 11(1):1–10

Iborra FJ, Jackson DA, Cook PR (2001) Coupled transcription and translation within nuclei of mammalian cells. Science 293(5532):1139–1142

Jackson SA (2016) Rice: the first crop genome. Rice 9:1–3

Jain M, Choudhary D, Kale RK, Bhalla-Sarin N (2002) Salt-and glyphosate-induced increase in glyoxalase I activity in cell lines of groundnut (Arachis hypogaea). Physiol Plant 114(4): 499–505

Jarvis DE, Ho YS, Lightfoot DJ, Schmöckel SM, Li B, Borm TJ, Ohyanagi H, Mineta K, Michell CT, Saber N, Kharbatia NM (2017) The genome of Chenopodium quinoa. Nature 542(7641): 307–312

Jha D, Shirley N, Tester M, Roy SJ (2010) Variation in salinity tolerance and shoot sodium accumulation in Arabidopsis ecotypes linked to differences in the natural expression levels of transporters involved in sodium transport. Plant Cell Environ 33(5):793–804

Jha A, Joshi M, Yadav NS, Agarwal PK, Jha B (2011a) Cloning and characterization of the Salicornia brachiata Na+/H+ antiporter gene SbNHX1 and its expression by abiotic stress. Mol Biol Rep 38(3):1965–1973

Jha B, Sharma A, Mishra A (2011b) Expression of SbGSTU (tau class glutathione S-transferase) gene isolated from Salicornia brachiata in tobacco for salt tolerance. Mol Biol Rep *38*(7): 4823–4832

Jha B, Singh NP, Mishra A (2012) Proteome profiling of seed storage proteins reveals the nutritional potential of Salicornia brachiata Roxb., an extreme halophyte. J Agric Food Chem 60:4320–4326

Jörnvall H, von Bahr-Lindström H, Jany KD, Ulmer W, Fröschle M (1984) Extended superfamily of short alcohol-polyol-sugar dehydrogenases: structural similarities between glucose and ribitol dehydrogenases. FEBS Lett 165(2):190–196

Joshi M, Mishra A, Jha B (2012) NaCl plays a key role for in vitro micropropagation of Salicornia brachiata, an extreme halophyte. Indus Crops Prod 35(1):313–316

Joshi M, Jha A, Mishra A, Jha B (2013) Developing transgenic Jatropha using the SbNHX1 gene from an extreme halophyte for cultivation in saline wasteland. PLoS One 8:e71136. https://doi.org/10.1371/journal.pone.0071136

Joshi R, Mangu VR, Bedre R, Sanchez L, Pilcher W, Zandkarimi H, Baisakh N (2015) Salt adaptation mechanisms of halophytes: improvement of salt tolerance in crop plants. In: Elucidation of abiotic stress signaling in plants. Springer, New York, pp 243–279

Joshi A, Kanthaliya B, Arora J (2018a) Halophytes of Thar desert: potential source of nutrition and feedstuff. Int J Bioassays 8:5674–5683

Joshi R, Sahoo KK, Tripathi AK, Kumar R, Gupta BK, Pareek A, Singla-Pareek SL (2018b) Knockdown of an inflorescence meristem-specific cytokinin oxidase–OsCKX2 in rice reduces yield penalty under salinity stress condition. Plant Cell Environ 41(5):936–946

Jung J, Won SY, Suh SC, Kim H, Wing R, Jeong Y, Hwang I, Kim M (2007) The barley ERF-type transcription factor HvRAF confers enhanced pathogen resistance and salt tolerance in Arabidopsis. Planta 225(3):575–588

Junker A, Muraya MM, Weigelt-Fischer K, Arana-Ceballos F, Klukas C, Melchinger AE, Meyer RC, Riewe D, Altmann T (2015) Optimizing experimental procedures for quantitative evaluation of crop plant performance in high throughput phenotyping systems. Front Plant Sci *5*:770

Kavita K, Alka S (2010) Assessment of salinity tolerance of Vigna mungo var. Pu-19 using ex vitro and in vitro methods. Asian J Biotechol 2(2):73–85

Kavitha K, George S, Venkataraman G, Parida A (2010) A salt-inducible chloroplastic monodehydroascorbate reductase from halophyte Avicennia marina confers salt stress tolerance on transgenic plants. Biochimie 92(10):1321–1329

Khalil R, ElSayed N, Hashem HA (2021) Nanoparticles as a new promising tool to increase plant immunity against abiotic stress. In: Sustainable agriculture reviews, vol 53. Springer, Cham, pp 61–91

Koag MC, Fenton RD, Wilkens S, Close TJ (2003) The binding of maize DHN1 to lipid vesicles. Gain of structure and lipid specificity. Plant Physiol 131(1):309–316

Koag MC, Wilkens S, Fenton RD, Resnik J, Vo E, Close TJ (2009) The K-segment of maize DHN1 mediates binding to anionic phospholipid vesicles and concomitant structural changes. Plant Physiol 150(3):1503–1514

Korkina LG, Mikhal'Chik E, Suprun MV, Pastore S, Dal Toso R (2007) Molecular mechanisms underlying wound healing and anti-inflammatory properties of naturally occurring biotechnologically produced phenylpropanoid glycosides. Cell Mol Biol 53(5):84–91

Koyro HW (2006) Effect of salinity on growth, photosynthesis, water relations and solute composition of the potential cash crop halophyte Plantago coronopus (L.). Environ Exp Bot 56(2): 136–146

Kronzucker HJ, Britto DT (2011) Sodium transport in plants: a critical review. New Phytol 189(1): 54–81

Krüger C, Berkowitz O, Stephan UW, Hell R (2002) A metal-binding member of the late embryogenesis abundant protein family transports iron in the phloem Ofricinus communis L. J Biol Chem *277*(28):25062–25069

Kumar R, Mustafiz A, Sahoo KK, Sharma V, Samanta S, Sopory SK, Pareek A, Singla-Pareek SL (2012) Functional screening of cDNA library from a salt tolerant rice genotype Pokkali identifies mannose-1-phosphate guanyl transferase gene (OsMPG1) as a key member of salinity stress response. Plant Mol Biol 79(6):555–568

Lal S, Gulyani V, Khurana P (2008) Overexpression of HVA1 gene from barley generates tolerance to salinity and water stress in transgenic mulberry (Morus indica). Transgen Res 17(4):651–663

Lan T, Duan Y, Wang B, Zhou Y, Wu W (2011) Molecular cloning and functional characterization of a Na+/H+ antiporter gene from halophyte Spartina anglica. Turk J Agric Forest 35(5): 535–543

Li Q, Yin H, Li D, Zhu H, Zhang Y, Zhu W (2007) Isolation and characterization of CMO gene promoter from halophyte Suaeda liaotungensis K. J Genet Genomics 34:355–361. https://doi.org/10.1016/S1673-8527(07)60038-1

Li QL, Xie JH, Ma XQ, Li D (2016) Molecular cloning of Phosphoethanolamine N-methyltransferase (PEAMT) gene and its promoter from the halophyte Suaeda liaotungensis and their response to salt stress. Acta Physiol Plant 38(2):1–8

Li N, Yang T, Guo Z, Wang Q, Chai M, Wu M, Li X, Li W, Li G, Tang J, Tang G (2020) Maize microrna166 inactivation confers plant development and abiotic stress resistance. Int J Mol Sci 21(24):9506

Lin F, Xu J, Shi J, Li H, Li B (2010) Molecular cloning and characterization of a novel glyoxalase I gene TaGly I in wheat (Triticum aestivum L.). Mol Biol Rep 37(2):729–735

Liu J, Ishitani M, Halfter U, Kim CS, Zhu JK (2000) The Arabidopsis thaliana SOS2 gene encodes a protein kinase that is required for salt tolerance. Proc Natl Acad Sci U S A 97:3730–3734

Low R, Rockel B, Kirsch M, Ratajczak R, Hortensteiner S, Martinoia E, Luttge U, Rausch T (1996) Early salt stress effects on the differential expression of vacuolar H+-ATPase genes in roots and leaves of Mesembryanthemum crystallinum. Plant Physiol 110(1):259–265

Lü SY, Jing YX, Shen SH, Zhao HY, Ma IQ, Zhou XJ, Ren Q, Li YF (2005) Antiporter gene from Hordum brevisubulatum (Trin.) link and its overexpression in transgenic tobaccos. J Integr Plant Biol 47(3):343–349

Lu PL, Chen NZ, An R, Su Z, Qi BS, Ren F, Chen J, Wang XC (2007a) A novel drought-inducible gene, ATAF1, encodes a NAC family protein that negatively regulates the expression of stress-responsive genes in Arabidopsis. Plant Mol Biol 63(2):289–305

Lu Z, Liu D, Liu S (2007b) Two rice cytosolic ascorbate peroxidases differentially improve salt tolerance in transgenic Arabidopsis. Plant Cell Rep 26(10):1909–1917

Ma SC, Li FM, Xu BC, Huang ZB (2010) Effect of lowering the root/shoot ratio by pruning roots on water use efficiency and grain yield of winter wheat. Field Crop Res 115(2):158–164

Mangano S, Silberstein S, Santa-María GE (2008) Point mutations in the barley HvHAK1 potassium transporter lead to improved K+-nutrition and enhanced resistance to salt stress. FEBS Lett 582(28):3922–3928

Mano Y, Takeda K (1997) Mapping quantitative trait loci for salt tolerance at germination and the seedling stage in barley (Hordeum vulgare L.). Euphytica 94(3):263–272

Mano Y, Nakazumi H, Takeda K (1996) Varietal variation in and effects of some major genes on salt tolerance at the germination stage in barley. Japan J Breed 46(3):227–233

Matysik J, Alia BB, Mohanty P (2002) Molecular mechanisms of quenching of reactive oxygen species by proline under stress in plants. Curr Sci 82:525–532

Mayer KF, Rogers J, Dole˘zel J, Pozniak C, Eversole K, Feuillet C, Gill B, Friebe B, Lukaszewski AJ, International Wheat Genome Sequencing Consortium (IWGSC) (2014) A chromosome-based draft sequence of the hexaploid bread wheat (Triticum aestivum) genome. Science 345: 1251788

Mayes S, Massawe FJ, Alderson PG, Roberts JA, Azam-Ali SN, Hermann M (2012) The potential for underutilized crops to improve security of food production. J Exp Bot 63(3):1075–1079
Mazher AA, El-Quesni EF, Farahat MM (2007) Responses of ornamental and woody trees to salinity. World J Agric Sci 3(3):386–395
Mba C (2013) Induced mutations unleash the potentials of plant genetic resources for food and agriculture. Agronomy 3(1):200–231
Meera SP, Augustine A (2020) De novo transcriptome analysis of Rhizophora mucronata Lam. furnishes evidence for the existence of glyoxalase system correlated to glutathione metabolic enzymes and glutathione regulated transporter in salt tolerant mangroves. Plant Physiol Biochem 155:683–696
Meera SP, Sreeshan A, Augustine A (2013) Functional screening and genetic engineering of mangrove salt responsive genes: a review. Ann Plant Sci 02(12):535–542
Meera SP, Sreeshan A, Augustine A (2020) Leaf tissue specific transcriptome sequence and de novo assembly datasets of Asiatic mangrove Rhizophora mucronata Lam. Data Brief 31:105747
Mehta PA, Sivaprakash K, Parani M, Venkataraman G, Parida AK (2005) Generation and analysis of expressed sequence tags from the salt-tolerant mangrove species Avicennia marina (Forsk) Vierh. Theor Appl Genet 110(3):416–424
Mian A, Oomen RJ, Isayenkov S, Sentenac H, Maathuis FJ, Véry AA (2011) Over-expression of an Na+-and K+-permeable HKT transporter in barley improves salt tolerance. Plant J 68(3): 468–479
Mishra A, Joshi M, Jha B (2013) Oligosaccharide mass profiling of nutritionally important Salicornia brachiata, an extreme halophyte. Carbohydr Polym 92(2):1942–1945
Mishra A, Patel MK, Jha B (2015) Non-targeted metabolomics and scavenging activity of reactive oxygen species reveal the potential of Salicornia brachiata as a functional food. J Funct Foods 1(13):21–31
Mishra A, Tanna B (2017) Halophytes: potential resources for salt stress tolerance genes and promoters. Front Plant Sci 8:829
Mitchell N, Campbell LG, Ahern JR, Paine KC, Giroldo AB, Whitney KD (2019) Correlates of hybridization in plants. Evolution Let 3(6):570–585
Miteva T, Zhelev NZ, Popova L (1992) Effect of salinity on the synthesis of ribulose-1,5-bisphosphate carboxylase/oxygenase in barley leaves. J Plant Physiol 140:46–51
Mittova V, Tal M, Volokita M, Guy M (2003) Up-regulation of the leaf mitochondrial and peroxisomal antioxidative systems in response to salt-induced oxidative stress in the wild salt-tolerant tomato species Lycopersicon pennellii. Plant Cell Environ 26(6):845–856
Miyama M, Shimizu H, Sugiyama M, Hanagata N (2006) Sequencing and analysis of 14,842 expressed sequence tags of Burma mangrove, Bruguiera gymnorrhiza. Plant Sci 171(2): 234–241
Mohammad B, Yuji K, Shigeyuki B, Naoya S, Hironori I, Edy BMS, Hirosuke O (2011) Isolation of salt stress tolerance genes from roots of mangrove plant, rhizophora stylosa Griff., using PCR based suppression subtractive hybridization. Plant Mol Biol Rep 29(3):533–543
Morton MJ, Awlia M, Al-Tamimi N, Saade S, Pailles Y, Negrão S, Tester M (2019) Salt stress under the scalpel–dissecting the genetics of salt tolerance. Plant J 97(1):148–163
Mostofa MG, Rahman M, Ansary MU, Keya SS, Abdelrahman M, Miah G, Tran LSP (2021) Silicon in mitigation of abiotic stress-induced oxidative damage in plants. Crit Rev Biotechnol 41:918–934
Muchate NS, Nikalje GC, Rajurkar NS, Suprasanna P, Nikam TD (2016) Plant salt stress: adaptive responses, tolerance mechanism and bioengineering for salt tolerance. Bot Rev 82(4):371–406
Mueller ND, Gerber JS, Johnston M, Ray DK, Ramankutty N, Foley JA (2012) Closing yield gaps through nutrient and water management. Nature 490(7419):254–257
Munns R (1992) A leaf elongation assay detects an unknown growth inhibitor in xylem sap from wheat and barley. Funct Plant Biol 19(2):127–135
Munns R (2005) Genes and salt tolerance: bringing them together. New Phytol 167(3):645–663
Munns R, Termaat A (1986) Whole-plant responses to salinity. Funct Plant Biol 13(1):143–160

Munns R, Tester M (2008) Mechanisms of salinity tolerance. Annu Rev Plant Biol 59:651–681

Mustafiz A, Singh AK, Pareek A, Sopory SK, Singla-Pareek SL (2011) Genome-wide analysis of rice and Arabidopsis identifies two glyoxalase genes that are highly expressed in abiotic stresses. Funct Integr Genomics 11(2):293–305

Mwando E, Angessa TT, Han Y, Li C (2020) Salinity tolerance in barley during germination—homologs and potential genes. J Zhejiang Univ Sci B 21(2):93–121

Nanjo T, Kobayashi M, Yoshiba Y, Kakubari Y, Yamaguchi-Shinozaki K, Shinozaki K (1999) Antisense suppression of proline degradation improves tolerance to freezing and salinity in Arabidopsis thaliana. FEBS Lett 461(3):205–210

Navarro JM, Gomez-Gomez A, Pérez-Pérez JG, Botia P (2010) Effect of saline conditions on the maturation process of Clementine Clemenules fruits on two different rootstocks. Spanish J Agric Res 8:21–29

Navarro JM, Pérez-Tornero O, Morte A (2014) Alleviation of salt stress in citrus seedlings inoculated with arbuscular mycorrhizal fungi depends on the rootstock salt tolerance. J Plant Physiol 171(1):76–85

Nguyen PD, Ho CL, Harikrishna JA, Wong ML, Abdul Rahim R (2007) Functional screening for salinity tolerant genes from Acanthus ebracteatus Vahl using Escherichia coli as a host. Trees 21(5):515–520

Nikam AA, Devarumath RM, Ahuja A, Babu H, Shitole MG, Suprasanna P (2015) Radiation-induced in vitro mutagenesis system for salt tolerance and other agronomic characters in sugarcane (Saccharum officinarum L.). Crop J 3(1):46–56

Oh SJ, Kwon CW, Choi DW, Song SI, Kim JK (2007) Expression of barley HvCBF4 enhances tolerance to abiotic stress in transgenic rice. Plant Biotechnol J 5(5):646–656

Oh DH, Dassanayake M, Haas JS, Kropornika A, Wright C, d'Urzo MP, Hong H, Ali S, Hernandez A, Lambert GM, Inan G (2010) Genome structures and halophyte-specific gene expression of the extremophile Thellungiella parvula in comparison with Thellungiella salsuginea (Thellungiella halophila) and Arabidopsis. Plant Physiol 154(3):1040–1052

Ohta M, Hayashi Y, Nakashima A, Hamada A, Tanaka A, Nakamura T, Hayakawa T (2002) Introduction of a Na+/H+ antiporter gene from Atriplex gmelini confers salt tolerance to rice. FEBS Lett 532(3):279–282

Palmqvist NM, Seisenbaeva GA, Svedlindh P, Kessler VG (2017) Maghemite nanoparticles acts as nanozymes, Improving growth and abiotic stress tolerance in Brassica napus. Nano Res Lett 12(1):1–9

Pan WJ, Tao JJ, Cheng T, Bian XH, Wei W, Zhang WK, Ma B, Chen SY, Zhang JS (2016) Soybean miR172a improves salt tolerance and can function as a long-distance signal. Mol Plant 9(9): 1337–1340

Pando GL, Deza P (2017) Development of advanced mutant lines of native grains through radiation-induced mutagenesis in Peru. Hortic Int J 1(3):93–96

Parida AK, Das AB (2005) Salt tolerance and salinity effects on plants: a review. Ecotoxicol Environ Saf 60(3):324–349

Passioura JB (1988) Water transport in and to roots. Annu Rev Plant Physiol Plant Mol Biol 39(1): 245–265

Patel MK, Joshi M, Mishra A, Jha B (2015) Ectopic expression of SbNHX1 gene in transgenic castor (Ricinus communis L.) enhances salt stress by modulating physiological process. Plant Cell Tissue Organ Cult 122(2):477–490

Penella C, Nebauer SG, Quinones A, Bautista S, López-Galarza S, Calatayud A (2015) Some rootstocks improve pepper tolerance to mild salinity through ionic regulation. Plant Sci 230:12–22

Penella C, Landi M, Guidi L, Nebauer SG, Pellegrini E, San Bautista A, Remorini D, Nali C, López-Galarza S, Calatayud A (2016) Salt-tolerant rootstock increases yield of pepper under salinity through maintenance of photosynthetic performance and sinks strength. J Plant Physiol 193:1–11

Peng Y, Lin W, Cai W, Arora R (2007) Overexpression of a Panax ginseng tonoplast aquaporin alters salt tolerance, drought tolerance and cold acclimation ability in transgenic Arabidopsis plants. Planta 226(3):729–740

Pezeshki SR, DeLaune RD, Patrick WH Jr (1990) Flooding and saltwater intrusion: potential effects on survival and productivity of wetland forests along the US Gulf Coast. Forest Ecol Manag *33*:287–301

Popp M, Larher F, Weigel P (1985) Osmotic adaption in Australian mangroves. In: Ecology of coastal vegetation. Springer, Dordrecht, pp 247–253

Prasad A, Sharma N, Prasad M (2020) Noncoding but coding: pri-miRNA into the action. Trends Plant Sci *26*(3):204–206

Prashanth SR, Sadhasivam V, Parida A (2008) Over expression of cytosolic copper/zinc superoxide dismutase from a mangrove plant Avicennia marina in indica rice var Pusa Basmati-1 confers abiotic stress tolerance. Transgen Res 17(2):281–291

Qadir M, Quillérou E, Nangia V, Murtaza G, Singh M, Thomas RJ, Drechsel P, Noble AD (2014) Economics of salt-induced land degradation and restoration. Nat Resour Forum 38(4):282–295

Qiu L, Wu D, Ali S, Cai S, Dai F, Jin X, Wu F, Zhang G (2011) Evaluation of salinity tolerance and analysis of allelic function of HvHKT1 and HvHKT2 in Tibetan wild barley. Theor Appl Genet 122(4):695–703

Qu GZ, Zang L, Xilin H, Gao C, Zheng T, Li KL (2012) Co-transfer of LEA and bZip genes from Tamarix confers additive salt and osmotic stress tolerance in transgenic tobacco. Plant Mol Biol Report 30(2):512–518

Quesada V, Ponce MR, Micol JL (2000) Genetic analysis of salt-tolerant mutants in Arabidopsis thaliana. Genetics 154(1):421–436

Rahman M, Mostofa MG, Keya SS, Siddiqui M, Ansary M, Uddin M, Das AK, Tran LSP (2021) Adaptive mechanisms of halophytes and their potential in improving salinity tolerance in plants. Int J Mol Sci *22*(19):107

Rai V, Tuteja N, Takabe T (2012) Transporters and abiotic stress tolerance in plants. In: Improving crop resistance to abiotic stress, pp 507-522

Rajalakshmi S, Parida A (2012) Halophytes as a source of genes for abiotic stress tolerance. J Plant Biochem Biotechnol *21*(1):63–66

Rajendran K, Tester M, Roy SJ (2009) Quantifying the three main components of salinity tolerance in cereals. Plant Cell Environ 32(3):237–249

Rajput VD, Minkina T, Kumari A, Singh VK, Verma KK, Mandzhieva S, Sushkova S, Srivastava S, Keswani C (2021) Coping with the challenges of abiotic stress in plants: new dimensions in the field application of nanoparticles. Plants 10(6):1221

Ranathunge K, Shao S, Qutob D, Gijzen M, Peterson CA, Bernards MA (2010) Properties of the soybean seed coat cuticle change during development. Planta *231*(5):1171–1188

Rasheed A, Mujeeb-Kazi A, Ogbonnaya FC, He Z, Rajaram S (2018) Wheat genetic resources in the post-genomics era: promise and challenges. Ann Bot 121(4):603–616

Ratajczak R, Richter J, Lüttge U (1994) Adaptation of the tonoplast V-type H+-ATPase of Mesembryanthemum crystallinum to salt stress, C3–CAM transition and plant age. Plant Cell Environ 17(10):1101–1112

Rengasamy P, Olsson KA (1993) Irrigation and sodicity. Soil Research 31(6):821–837

Rigó G, Valkai I, Faragó D, Kiss E, Van Houdt S, Van de Steene N, Hannah MA, Szabados L (2016) Gene mining in halophytes: functional identification of stress tolerance genes in Lepidium crassifolium. Plant Cell Environ 39:2074–2084

Rivandi J, Miyazaki J, Hrmova M, Pallotta M, Tester M, Collins N (2011) A SOS3 homologue maps to HvNax4, a barley locus controlling an environmentally sensitive Na+ exclusion trait. J Exp Bot 62(3):1201–1216

Rodrıguez P, Torrecillas A, Morales MA, Ortuño MF, Sánchez-Blanco MJ (2005) Effects of NaCl salinity and water stress on growth and leaf water relations of Asteriscus maritimus plants. Environ Exp Bot 53(2):113–123

Rodríguez-Rosales MP, Gálvez FJ, Huertas R, Aranda MN, Baghour M, Cagnac O, Venema K (2009) Plant NHX cation/proton antiporters. Plant Signal Behav 4(4):265–276

Ruiz-Sánchez MC, Domingo R, Torrecillas A, Pérez-Pastor A (2000) Water stress preconditioning to improve drought resistance in young apricot plants. Plant Sci 156(2):245–251

Saad RB, Romdhan WB, Zouari N, Azaza J, Mieulet D, Verdeil JL, Guiderdoni E, Hassairi A (2011) Promoter of the AlSAP gene from the halophyte grass Aeluropus littoralis directs developmental-regulated, stress-inducible, and organ-specific gene expression in transgenic tobacco. Transgen Res 20(5):1003–1018

Saade S, Brien C, Pailles Y, Berger B, Shahid M, Russell J, Waugh R, Negrão S, Tester M (2020) Dissecting new genetic components of salinity tolerance in two-row spring barley at the vegetative and reproductive stages. PLoS One 15(7):e0236037

Santos J, Al-Azzawi M, Aronson J, Flowers TJ (2016) eHALOPH a database of salt-tolerant plants: helping put halophytes to work. Plant Cell Physiol 57(1):e10

Savé R, Olivella C, Biel C, Adillón J, Rabella R (1994) Seasonal patterns of water relationships, photosynthetic pigments and morphology of Actinidia deliciosa plants of the Hayward and Tomuri cultivars. Agronomie 14(2):121–126

Saxena M, Bisht R, Roy SD, Sopory SK, Bhalla-Sarin N (2005) Cloning and characterization of a mitochondrial glyoxalase II from Brassica juncea that is upregulated by NaCl, Zn, and ABA. Biochem Biophys Res Commun 336(3):813–819

Schaeffer HJ, Forsthoefel NR, Cushman JC (1995) Identification of enhancer and silencer regions involved in salt-responsive expression of Crassulacean acid metabolism (CAM) genes in the facultative halophyte Mesembryanthemum crystallinum. Plant Mol Biol 28(2):205–218

Schaffer BA, Wolstenholme BN, Whiley AW (eds) 2013 The avocado: botany, production and uses

Scheben A, Wolter F, Batley J, Puchta H, Edwards D (2017) Towards CRISPR/Cas crops–bringing together genomics and genome editing. New Phytol 216(3):682–698

Schnable PS (2012) The B73 maize genome: complexity, diversity, and dynamics (November, pg 1112, 2009). Science 337(6098):1040–1040

Shabala S (2013) Learning from halophytes: physiological basis and strategies to improve abiotic stress tolerance in crops. Ann Bot 112(7):1209–1221

Shannon MC, Grieve CM, Francois LE (1994) Whole-plant response to salinity. In: Wilkinson RE (ed) Plant-environment interactions

Shao HB, Chu LY, Jaleel CA, Zhao CX (2008a) Water-deficit stress-induced anatomical changes in higher plants. C R Biol 331(3):215–225

Shao HB, Chu LY, Lu ZH, Kang CM (2008b) Primary antioxidant free radical scavenging and redox signaling pathways in higher plant cells. Int J Biol Sci 4(1):8

Sharma R, Sahoo A, Devendran R, Jain M (2014) Over-expression of a rice tau class glutathione s-transferase gene improves tolerance to salinity and oxidative stresses in Arabidopsis. PLoS One 9(3):e92900

Shi H, Ishitani M, Kim C, Zhu JK (2000) The Arabidopsis thaliana salt tolerance gene SOS1 encodes a putative Na+/H+ antiporter. Proc Natl Acad Sci U S A 97(12):6896–6901

Shinozaki K, Yamaguchi-Shinozaki K (2007) Gene networks involved in drought stress response and tolerance. J Exp Bot 58(2):221–227

Silva EG, Silva AF, Lima JD, Silva MD, Maia JM (2017) Vegetative development and content of calcium, potassium, and sodium in watermelon under salinity stress on organic substrates. Pesq Agrop Brasileira 52:1149–1157

Singh N, Mishra A, Jha B (2014a) Ectopic over-expression of peroxisomal ascorbate peroxidase (SbpAPX) gene confers salt stress tolerance in transgenic peanut (Arachis hypogaea). Gene 547(1):119–125

Singh N, Mishra A, Jha B (2014b) Over-expression of the peroxisomal ascorbate peroxidase (SbpAPX) gene cloned from halophyte Salicornia brachiata confers salt and drought stress tolerance in transgenic tobacco. Mar Biotechnol 16(3):321–332

Singh VK, Mishra A, Haque I, Jha B (2016a) A novel transcription factor-like gene SbSDR1 acts as a molecular switch and confers salt and osmotic endurance to transgenic tobacco. Sci Rep 6(1): 1–16

Singh VK, Mishra A, Haque I, Jha B (2016b) Corrigendum: a novel transcription factor-like gene SbSDR1 acts as a molecular switch and confers salt and osmotic endurance to transgenic tobacco. Sci Rep 6:35128

Singla-Pareek SL, Reddy M, Sopory SK (2001) Transgenic approach towards developing abiotic stress tolerance in plants. Proc Indian Natl Sci Acad Part B 67(5):265–284

Singla-Pareek SL, Reddy MK, Sopory SK (2003) Genetic engineering of the glyoxalase pathway in tobacco leads to enhanced salinity tolerance. Proc Natl Acad Sci U S A 100(25):14672–14677

Singla-Pareek SL, Yadav SK, Pareek A, Reddy MK, Sopory SK (2008) Enhancing salt tolerance in a crop plant by overexpression of glyoxalase II. Transgen Res 17(2):171–180

Skipsey M, Christopher JA, Jane KT, Jepson I, Edwards R (2000) Cloning and characterization of glyoxalase from soybean. Arch Biochem Biophys 374:261–268

Sottosanto JB, Gelli A, Blumwald E (2004) DNA array analyses of Arabidopsis thaliana lacking a vacuolar Na+/H+ antiporter: impact of AtNHX1 on gene expression. Plant J 40(5):752–771

Spindel J, Begum H, Akdemir D, Virk P, Collard B, Redona E, Atlin G, Jannink JL, McCouch SR (2015) Genomic selection and association mapping in rice (Oryza sativa): effect of trait genetic architecture, training population composition, marker number and statistical model on accuracy of rice genomic selection in elite, tropical rice breeding lines. PLoS Genet 11(2):e1004982

Sreeshan A, Meera SP, Augustine A (2014) A review on transporters in salt tolerant mangroves. Trees 28(4):957–960

Sreeshan A, Meera SP, Augustine A (2018) Betaine Aldehyde Dehydrogenase (BADH) gene and free amino acid analysis in Rhizophora mucronata Lam. from Thalassery region of Kerala, India

Srivastava J, Prasad V (2019) Evolution and paleobiogeography of mangroves. Mar Ecol 40(6): e12571

Steudle E (2000) Water uptake by roots: effects of water deficit. J Exp Bot 51(350):1531–1542

Su H, Golldack D, Katsuhara M, Zhao C, Bohnert HJ (2001) Expression and stress-dependent induction of potassium channel transcripts in the common ice plant. Plant Physiol 125(2): 604–614

Sultana S, Khew CY, Morshed MM, Namasivayam P, Napis S, Ho CL (2012) Overexpression of monodehydroascorbate reductase from a mangrove plant (AeMDHAR) confers salt tolerance on rice. J Plant Physiol 169(3):311–318

Sun Q, Gao F, Zhao L, Li K, Zhang J (2010) Identification of a new 130 bp cis-acting element in the TsVP1 promoter involved in the salt stress response from Thellungiella halophila. BMC Plant Biol 10(1):1–12

Taji T, Seki M, Satou M, Sakurai T, Kobayashi M, Ishiyama K, Narusaka Y, Narusaka M, Zhu JK, Shinozaki K (2004) Comparative genomics in salt tolerance between Arabidopsis and Arabidopsis-related halophyte salt cress using Arabidopsis microarray. Plant Physiol 135(3): 1697–1709

Taji T, Komatsu K, Katori T, Kawasaki Y, Sakata Y, Tanaka S, Kobayashi M, Toyoda A, Seki M, Shinozaki K (2010) Comparative genomic analysis of 1047 completely sequenced cDNAs from an Arabidopsis-related model halophyte, Thellungiella halophila. BMC Plant Biol 10(1):1–10

Takeda S, Matsuoka M (2008) Genetic approaches to crop improvement: responding to environmental and population changes. Nat Rev Genet 9(6):444–457

Takhtajan AL (1980) Outline of the classification of flowering plants (Magnoliophyta). Bot Rev 46(3):225–359

Tal M, Shannon MC (1983) Salt tolerance in the wild relatives of the cultivated tomato: responses of Lycopersicon esculentum, L. cheesmanii, L. peruvianum, Solanum pennellii and F1 hybrids to high salinity. Funct Plant Biol 10(1):109–117

Tanaka S, Ikeda K, Ono M, Miyasaka H (2002) Isolation of several anti-stress genes from a mangrove plant Avicennia marina. World J Microbiol Biotechnol 18(8):801–804

Tattini M, Gucci R, Coradeschi MA, Ponzio C, Everard JD (1995) Growth, gas exchange and ion content in Olea europaea plants during salinity stress and subsequent relief. Physiol Plant 95(2): 203–210

Tiwari R, Rajam MV (2022) RNA-and miRNA-interference to enhance abiotic stress tolerance in plants. J Plant Biochem Biotechnol:1–16

Tiwari V, Chaturvedi AK, Mishra A, Jha B (2014) The transcriptional regulatory mechanism of the peroxisomal ascorbate peroxidase (pAPX) gene cloned from an extreme halophyte, Salicornia brachiata. Plant Cell Physiol 55(1):201–217

Tiwari V, Chaturvedi AK, Mishra A, Jha B (2015) Introgression of the SbASR-1 gene cloned from a halophyte Salicornia brachiata enhances salinity and drought endurance in transgenic groundnut (Arachis hypogaea) and acts as a transcription factor. PLoS One 10(7):e0131567

Tiwari V, Patel MK, Chaturvedi AK, Mishra A, Jha B (2016) Functional characterization of the tau class glutathione-S-transferases gene (SbGSTU) promoter of Salicornia brachiata under salinity and osmotic stress. PLoS One 11(2):e0148494

Tomlinson PB (1986) The botany of Mangroves. Cambridge University Press

Tu Anh TT, Khanh TD, Dat TD, Xuan TD (2018) Identification of phenotypic variation and genetic diversity in rice (Oryza sativa L.) mutants. Agriculture 8(2):30

Tudela D, Primo-Millo E (1992) 1-Aminocyclopropane-1-carboxylic acid transported from roots to shoots promotes leaf abscission in Cleopatra mandarin (Citrus reshni Hort. ex Tan.) seedlings rehydrated after water stress. Plant Physiol 100(1):131–137

Tuomainen M, Ahonen V, Kärenlampi SO, Schat H, Paasela T, Švanys A, Tuohimetsä S, Peräniemi S, Tervahauta A (2011) Characterization of the glyoxalase 1 gene TcGLX1 in the metal hyperaccumulator plant Thlaspi caerulescens. Planta 233(6):1173–1184

Udawat P, Mishra A, Jha B (2014) Heterologous expression of an uncharacterized universal stress protein gene (SbUSP) from the extreme halophyte, Salicornia brachiata, which confers salt and osmotic tolerance to E. coli. Gene 536(1):163–170

Udawat P, Jha RK, Sinha D, Mishra A, Jha B (2016) Overexpression of a cytosolic abiotic stress responsive universal stress protein (SbUSP) mitigates salt and osmotic stress in transgenic tobacco plants. Front Plant Sci 7:518

Udawat P, Jha RK, Mishra A, Jha B (2017) Overexpression of a plasma membrane-localized SbSRP-like protein enhances salinity and osmotic stress tolerance in transgenic tobacco. Front Plant Sci 8:582

Umezawa T, Fujita M, Fujita Y, Yamaguchi-Shinozaki K, Shinozaki K (2006) Engineering drought tolerance in plants: discovering and tailoring genes to unlock the future. Curr Opin Biotechnol 17(2):113–122

UN Water and Energy (2014) The United Nations World Water Development Report 2014. UNESCO, Paris

USDA-ARS. 2008. Research databases. Bibliography on Salt Tolerance. George E. Brown, Jr. Salinity Lab. US Dep. Agric., Agric. Res. Serv. Riverside, CA. http://www.ars.usda.gov/Services/docs.htm?docid=8908. Accessed 28 Apr 2010

Vamerali T, Saccomani M, Bona S, Mosca G, Guarise M, Ganis A (2003) A comparison of root characteristics in relation to nutrient and water stress in two maize hybrids. In: Roots: the dynamic interface between plants and the earth, pp 157–167

Veena, Reddy VS, Sopory SK (1999) Glyoxalase I from Brassica juncea: molecular cloning, regulation and its over-expression confer tolerance in transgenic tobacco under stress. Plant J 17(4):385–395

Vinocur B, Altman A (2005) Recent advances in engineering plant tolerance to abiotic stress: achievements and limitations. Curr Opin Biotechnol 16(2):123–132

Volkov V (2015) Salinity tolerance in plants. Quantitative approach to ion transport starting from halophytes and stepping to genetic and protein engineering for manipulating ion fluxes. Front Plant Sci 6:873

Waditee R, Hibino T, Tanaka Y, Nakamura T, Incharoensakdi A, Hayakawa S, Suzuki S, Futsuhara Y, Kawamitsu Y, Takabe T, Takabe T (2002) Functional characterization of betaine/proline transporters in betaine-accumulating mangrove. J Biol Chem 277(21):18373–18382

Walia H, Wilson C, Wahid A, Condamine P, Cui X, Close TJ (2006) Expression analysis of barley (Hordeum vulgare L.) during salinity stress. Funct Integr Genomics 6(2):143–156

Walker RR, Sedgley M, Blesing MA, Douglas TJ (1984) Anatomy, ultrastructure and assimilate concentrations of roots of citrus genotypes differing in ability for salt exclusion. J Exp Bot 35(10):1481–1494

Wang ZL, Li PH, Fredricksen M, Gong ZZ, Kim CS, Zhang C, Bohnert HJ, Zhu JK, Bressan RA, Hasegawa PM, Zhao YX (2004) Expressed sequence tags from Thellungiella halophila, a new model to study plant salt-tolerance. Plant Sci 166(3):609–616

Wang L, Pan D, Lv X, Cheng CL, Li J, Liang W, Xing J, Chen W (2016a) A multilevel investigation to discover why Kandelia candel thrives in high salinity. Plant Cell Environ 39(11):2486–2497

Wang Y, Liu X, Ren C, Zhong GY, Yang L, Li S, Liang Z (2016b) Identification of genomic sites for CRISPR/Cas9-based genome editing in the Vitis vinifera genome. BMC Plant Biol 16(1): 1–7

Wang C, Hu S, Gardner C, Lübberstedt T (2017) Emerging avenues for utilization of exotic germplasm. Trends Plant Sci *22*(7):624–637

Weston AM, Zorb C, John EA, Flowers TJ (2012) High phenotypic plasticity of Suaeda maritima observed under hypoxic conditions in relation to its physiological basis. Ann Bot 109(5): 1027–1036

Whiley AW, Schaffer B, Wolstenholme BN (2002) The avocado. Botany, production and uses. CAB International, Wallingford, UK

Witzel K, Weidner A, Surabhi GK, Varshney RK, Kunze G, Buck-Sorlin GH, Boerner A, Mock HP (2010) Comparative analysis of the grain proteome fraction in barley genotypes with contrasting salinity tolerance during germination. Plant Cell Environ 33(2):211–222

Wong CE, Li Y, Labbe A, Guevara D, Nuin P, Whitty B, Diaz C, Golding GB, Gray GR, Weretilnyk EA, Griffith M (2006) Transcriptional profiling implicates novel interactions between abiotic stress and hormonal responses in Thellungiella, a close relative of Arabidopsis. Plant Physiol 140(4):1437–1450

Wong YY, Ho CL, Nguyen PD, Teo SS, Harikrishna JA, Rahim RA, Wong MC (2007) Isolation of salinity tolerant genes from the mangrove plant, Bruguiera cylindrica by using suppression subtractive hybridization (SSH) and bacterial functional screening. Aquat Bot 86(2):117–122

Wu CA, Yang GD, Meng QW, Zheng CC (2004) The cotton GhNHX1 gene encoding a novel putative tonoplast Na+/H+ antiporter plays an important role in salt stress. Plant Cell Physiol 45(5):600–607

Wu D, Qiu L, Xu L, Ye L, Chen M, Sun D, Chen Z, Zhang H, Jin X, Dai F, Zhang G (2011) Genetic variation of HvCBF genes and their association with salinity tolerance in Tibetan annual wild barley. PLoS one 6(7):e22938

Wu HJ, Zhang Z, Wang JY, Oh DH, Dassanayake M, Liu B, Huang Q, Sun HX, Xia R, Wu Y, Wang YN (2012) Insights into salt tolerance from the genome of Thellungiella salsuginea. Proc Natl Acad Sci U S A 109(30):12219–12224

Wu C, Ma C, Pan Y, Gong S, Zhao C, Chen S, Li H (2013) Sugar beet M14 glyoxalase I gene can enhance plant tolerance to abiotic stresses. J Plant Res 126(3):415–425

Wu P, Yang Q, Wang K, Zhou J, Ma J, Tang Q, Jin L, Xiao W, Jiang A, Jiang Y, Zhu L (2018) Single step genome-wide association studies based on genotyping by sequence data reveals novel loci for the litter traits of domestic pigs. Genomics 110(3):171–179

Wu H, Shabala L, Zhou M, Su N, Wu Q, Ul-Haq T, Zhu J, Mancuso S, Azzarello E, Shabala S (2019) Root vacuolar Na+ sequestration but not exclusion from uptake correlates with barley salt tolerance. Plant J 100(1):55–67

Xie ZM, Zou HF, Lei G, Wei W, Zhou QY, Niu CF, Liao Y, Tian AG, Ma B, Zhang WK, Zhang JS (2009) Soybean Trihelix transcription factors GmGT-2A and GmGT-2B improve plant tolerance to abiotic stresses in transgenic Arabidopsis. PLoS One 4(9):e6898

Xu ZS, Chen M, Li LC, Ma YZ (2008) Functions of the ERF transcription factor family in plants. Botany 86(9):969–977

Xu ZS, Ni ZY, Li ZY, Li LC, Chen M, Gao DY, Yu XD, Liu P, Ma YZ (2009) Isolation and functional characterization of HvDREB1—a gene encoding a dehydration-responsive element binding protein in Hordeum vulgare. J Plant Res 122(1):121–130

Xu R, Wang J, Li C, Johnson P, Lu C, Zhou M (2012) A single locus is responsible for salinity tolerance in a Chinese landrace barley (Hordeum vulgare L.)

Xu R, Wang Y, Zheng H, Lu W, Wu C, Huang J, Yan K, Yang G, Zheng C (2015) Salt-induced transcription factor MYB74 is regulated by the RNA-directed DNA methylation pathway in Arabidopsis. J Exp Bot 66(19):5997–6008

Xue GP, Loveridge CW (2004) HvDRF1 is involved in abscisic acid-mediated gene regulation in barley and produces two forms of AP2 transcriptional activators, interacting preferably with a CT-rich element. Plant J 37(3):326–339

Yadav SK, Singla-Pareek SL, Reddy MK, Sopory SK (2005) Transgenic tobacco plants overexpressing glyoxalase enzymes resist an increase in methylglyoxal and maintain higher reduced glutathione levels under salinity stress. FEBS Lett 579(27):6265–6271

Yamada A, Saitoh T, Mimura T, Ozeki Y (2002) Expression of mangrove allene oxide cyclase enhances salt tolerance in Escherichia coli, yeast, and tobacco cells. Plant Cell Physiol 43(8):903–910

Yamaguchi T, Blumwald E (2005) Developing salt-tolerant crop plants: challenges and opportunities. Trends Plant Sci 10(12):615–620

Yamanaka T, Miyama M, Tada Y (2009) Transcriptome profiling of the mangrove plant Bruguiera gymnorhiza and identification of salt tolerance genes by Agrobacterium functional screening. Biosci Biotechnol Biochem 73(2):304–310

Yang Y, Yang S, Li J, Deng Y, Zhang Z, Xu S, Guo W, Zhong C, Zhou R, Shi S (2015a) Transcriptome analysis of the Holly mangrove Acanthus ilicifolius and its terrestrial relative, Acanthus leucostachyus, provides insights into adaptation to intertidal zones. BMC Genomics 16(1):1–12

Yang Y, Yang S, Li J, Li X, Zhong C, Huang Y, Zhou R, Shi S (2015b) De novo assembly of the transcriptomes of two yellow mangroves, Ceriops tagal and C. zippeliana, and one of their terrestrial relatives, Pellacalyx yunnanensis. Mar Genomics 23:33–36

Yasumoto E, Adachi K, Kato M, Sano H, Sasamoto H, Baba S, Ashihara H (1999) Uptake of inorganic ions and compatible solutes in cultured mangrove cells during salt stress. In Vitro Cell Dev Biol Plant 35(1):82–85

Yin X, Zhao Y, Luo D, Zhang H (2002) Isolating the promoter of a stress-induced gene encoding betaine aldehyde dehydrogenase from the halophyte Atriplex centralasiatica Iljin. Biochim Biophys Acta 1577(3):452–456

Yin XY, Yang AF, Zhang KW, Zhang JR (2004) Production and analysis of transgenic maize with improved salt tolerance by the introduction of AtNHX1 gene. Acta Bot Sin 46(7):854–861

Yin K, Gao C, Qiu JL (2017) Progress and prospects in plant genome editing. Nat Plants 3:17107

Yin S, Han Y, Huang L, Hong Y, Zhang G (2018) Overexpression of HvCBF7 and HvCBF9 changes salt and drought tolerance in Arabidopsis. Plant Growth Regul 85(2):281–292

Yousefirad S, Soltanloo H, Ramezanpour SS, Zaynalinezhad K, Shariati V (2018) Salt oversensitivity derived from mutation breeding improves salinity tolerance in barley via ion homeostasis. Biol Plant 62(4):775–785

Yuan S, Li Z, Li D, Yuan N, Hu Q, Luo H (2015) Constitutive expression of rice microRNA528 alters plant development and enhances tolerance to salinity stress and nitrogen starvation in creeping bentgrass. Plant Physiol 169(1):576–593

Zekri M, Parsons LR (1989) Growth and root hydraulic conductivity of several citrus rootstocks under salt and polyethylene glycol stresses. Physiol Plant 77(1):99–106

Zhang HX, Hodson JN, Williams JP, Blumwald E (2001) Engineering salt-tolerant Brassica plants: characterization of yield and seed oil quality in transgenic plants with increased vacuolar sodium accumulation. Proc Natl Acad Sci U S A 98(22):12832–12836

Zhang GH, Su Q, An LJ, Wu S (2008) Characterization and expression of a vacuolar Na+/H+ antiporter gene from the monocot halophyte Aeluropus littoralis. Plant Physiol Biochem 46(2): 117–126

Zhang J, Liu H, Sun J, Li B, Zhu Q, Chen S, Zhang H (2012) Arabidopsis fatty acid desaturase FAD2 is required for salt tolerance during seed germination and early seedling growth. PLoS One 7(1):e30355

Zhao FY, Zhang XJ, Li PH, Zhao YX, Zhang H (2006) Co-expression of the Suaeda salsa SsNHX1 and Arabidopsis AVP1 confer greater salt tolerance to transgenic rice than the single SsNHX1. Mol Breed 17(4):341–353

Zhen A, Bie Z, Huang Y, Liu Z, Li Q (2010) Effects of scion and rootstock genotypes on the antioxidant defense systems of grafted cucumber seedlings under NaCl stress. Soil Sci Plant Nutr 56(2):263–271

Zhu JK (2002) Salt and drought stress signal transduction in plants. Annu Rev Plant Biol *53*(1): 247–273

Zhu JK (2003) Regulation of ion homeostasis under salt stress. Curr Opin Plant Biol 6(5):441–445

Zhu J, Fan Y, Shabala S, Li C, Lv C, Guo B, Xu R, Zhou M (2020) Understanding mechanisms of salinity tolerance in barley by proteomic and biochemical analysis of near-isogenic lines. Int J Mol Sci 21(4):1516

Zobel RW, Kinraide TB, Baligar VC (2007) Fine root diameters can change in response to changes in nutrient concentrations. Plant Soil 297(1):243–254

Zhu J-K (2000) Genetic analysis of plant salt tolerance using arabidopsis. Plant Physiol 124(3): 941–948

Chapter 11
Plant Conservation Associated with Traditional Knowledge: Past and Future

Athira V. Anand, M. J. Sreedevi, and T. S. Swapna

Abstract World Bank has reported that indigenous people distributed in 90 countries constitute 5% of the global population. They have survived centuries in communion with nature and developed traditional intellect through experience and experimentation. Their awareness of life saving and survival skills is passed on across generations as traditional knowledge. It is the knowledge that is part of the cultural traditions of indigenous communities, having a contemporary application. Plants are inevitable components of traditional knowledge in the areas of herbal medicine, nutrition, cosmetics, perfumery, etc. Currently, interdisciplinary research is being conducted worldwide to develop leads from indigenous facts to marketable ideas and products. Along with bioprospecting, conservation and sustainable use of the bioresources is inevitable for the future generation and the existence of the earth. Measures have been taken at the National and International levels to create awareness of safeguarding nature through legislation and treaties. But the steps are inadequate in conserving the rich biodiversity that holds the solution to all human problems. The focus should be given to the conservation of traditionally significant and RET-listed plant species, while their commercial exploitation in the global market is raising alarms for their management. There is a need to emphasize the importance to safeguard nature, manage its resources and conserve its flora and fauna for the future.

Keywords Traditional knowledge · Indigenous community · Conservation · Ethnomedicine

---

A. V. Anand · M. J. Sreedevi · T. S. Swapna (✉)
Department of Botany, University of Kerala, Thiruvananthapuram, Kerala, India

S. T. Sukumaran, K. T R (eds.), *Conservation and Sustainable Utilization of Bioresources*, Sustainable Development and Biodiversity 30,
https://doi.org/10.1007/978-981-19-5841-0_11

11.1 Introduction

11.1.1 Traditional Knowledge and Flora

Indigenous and local communities' knowledge, cultural practices, myths, and skills are referred to as Traditional knowledge (TK). It is dynamically generated within a community and passed through generations in the form of stories, songs, folklore, proverbs, beliefs, rituals, symbols, community laws, the local language, and agricultural practices. It is considered a communities' culture and practiced, sung, danced, painted, and chanted (https://www.cbd.int/traditional/). This knowledge helps understand and respect natural values. It is not documented or technically recorded. The knowledge covers all the areas, including health, agriculture, horticulture, animal husbandry, and environment. Understandings, in general, are passed to all the members of the community, while specific information remains reserved in persons, like tribal leader, and medicinal and ritual practitioners, who are experienced and well-trained people. In Native American communities, children are selected and trained for this. In Australian communities, traditional owners transfer their role within the family line. In some communities, specific groups are assigned for this purpose (Bruchac 2014).

TK has a huge role in the development of new medicines. For example, *Curcuma longa*, known as turmeric, is widely used by people for wound healing, cosmetic purposes, and as food flavor. It has a number of medicinal properties, like antibacterial and antifungal activity (Moghadamtousi et al. 2014). Now, researchers have developed numerous medicines from the plant. Recently, *Curcuma longa* has been reported to have therapeutic potential against COVID-19 (Emirik 2022). Similarly, *Zingiber officinale, Azadirachta indica, Ocimum sanctum*, etc., are used by local communities of India for treating diverse health issues.

"Arogyapacha" (*Trichopus zeylanicus*) is a herbal medicinal plant endemic to Agasthya hills of Agasthyamala Biosphere Reserve in the border between Kerala and Tamil Nadu, India, widely used by *Kani* tribal community against fatigue. Researchers found that Arogyapacha has several pharmacological properties, like antioxidant (Sharma et al. 1989; Sindhu and Jose 2015), aphrodisiac (Subramoniam et al. 1997), antimicrobial (Vignesh et al. 2016), anti-inflammatory (Kumar et al. 2012), immunomodulatory (Shankar 2017), antitumor (Pushpangadan et al. 1995), anti-ulcer (Shankar 2017), hepatoprotective, anti-diabetic (Sundar Rajan and Velmurugan 2015), etc. Based on the information from the tribal community members, scientists of Jawaharlal Nehru Tropical Botanic Garden and Research Institute (JNTBGRI, Kerala, India) developed and claimed "Jeevani," a herbal medicine used against fatigue and for enhancing immunity. The technology is transferred to Arya Vaidya Pharmacy (Coimbatore) Ltd. for commercialization. This is a model of sharing benefits with the indigenous community, who are the real owners of the knowledge since 50% of the benefits from the commercialization of "Jeevani" are shared with the *Kani* tribes.

TK is pure and effective and not only limited to medicinal remedies but also includes other practices, like agriculture techniques, handicrafts, pottery, fishing, hunting, folklore, etc. Indigenous people conserve natural resources and maintain biodiversity through sustainable use. But, due to the irreplaceable properties of tribal medicines and other forest products, modern people commercially use the knowledge and overexploit them. So, governments have assigned intellectual property rights to TK for protecting them. Several laws and policies have been developed for the rights of indigenous communities as well. Traditional Knowledge Digital Library (TKDL) is a digital library developed in India in 2001 through the combined action of the Council of Scientific and Industrial Research (CSIR) and the Ministry of Health and Family Welfare (India) for documenting TK, chiefly medicinal plants, formulations, and practices. The main objective of this digital library is to protect TK from unlawful patenting and biopiracy. India was also successful in canceling the turmeric and Basmati patents granted by the US patent office and the Neem patent granted by the European Patent Office 1990s (Fredriksson 2021). Scientific documentation of TK and the related plants will help in their conservation and prevent their exploitation.

11.1.2 Human Civilization and the Roles of Plants

Plants are the lungs of nature, and human dependence on plants has an ancient history. Domestication of plants and initiation of agricultural practices paved the way for human civilizations. Evolutionary findings have revealed that human ancestors had unique jaw and gastrointestinal systems that facilitated the consumption of plant materials (Schaal 2019). Also, ancestral humans had certain biochemicals to digest plant material, which proves the evolutionary link between man and plants. Many ancient human civilizations included grains and legumes in their diets (Schaal 2019).

Early men depended on plants for food, therapeutics, clothing, cosmetics, timber, and religious worship. Domestication of plants was a milestone process in human civilization. Thousands of years ago, there were no agricultural crops; only wild plants were present. Then people cultivated those plants that were genetically altered with better yield properties and developed them into crops. Domesticated plants were different from wild ancestors. After domestication and cultivation, people settled into permanent towns (Schaal 2019). While depending on agriculture, they developed several farming tools, food storage systems, and transportation facilities that led to the initiation of other industries.

Ancient people cultivated cereals; they strictly depended on cereals, like rice, wheat, and barley for food. In addition, they used tuber crops, like Cassava, Potato, legumes, vegetables, fruits, etc. Egyptian civilization was built on agricultural wealth. Barley, wheat, beans, chickpeas, Fava beans, onions, garlic, etc., were the cultivated primary food crops in Egyptian civilization (Janick 2000). Rice was domesticated in China. Barley and wheat cultivation is associated with Asian civilizations, too (Jarrige 1986). It is clear that almost all the civilizations are developed on banks of rivers or near rivers to serve the purpose of irrigation. People depended on plants for fodder also. These crops varieties, agricultural strategies, farming tools, and experiences were transferred to many generations as TK.

In ancient times, the curative potential of plants was recognized, and the herbs were widely used for healing in all cultures. They cultivated medicinal plants, viz., ancient Egypt cultivated *Cannabis* (Ibrahim 2017). The opium poppy was cultivated in Egyptian, Mesopotamian, and Persian cultures (Luqman 2011). Sumerians wrote therapeutic records on clay tablets in 5000–3000 BC, the oldest available medicinal record, which revealed that humans used medicinal plants for restoring health (Inoue et al. 2019). Ayurveda is the traditional medicinal system followed by the Indian culture. Each civilization had its own therapeutic knowledge. Based on the leads from TK, several compounds are isolated, and drugs are discovered. Hundreds of such compounds are presently in use in modern medicines, such as Silymarin, Vincristine, Vinblastine, Morphine, Atropine, Digitoxin, Digoxin, and Reserpine, to name a few (Akerele 1993).

Investigations have reported that people in the past used several plants, plant products, and formulations to enhance their beauty. Castor oil (*Ricinus communis* L.), Anise (*Illicium verum* Hook. f), Belladonna (*Atropa belladonna* L.), Cinnamon (*Cinnamomum* Schaeff.), Cardamom (*Elettaria cardamomum*) Myrrh (*Boswellia sacra* Fluck.), and Mustard (*Sinapis alba* L.) were used by Sumerians, Assyrians, and Babylonians to clear the skin (López-Agüero and Stella 2007). Egyptians widely used cosmetics against dry sunlight and winds, consisting of Lilly (Liliaceae), Thyme (Thymus L.), Cedar (*Cedrus libani* A. Rich), and Lavender (Lavandula L.), etc. Ancient people also used body oils and perfumes, in which plants are the essential ingredients (Chaudhri and Jain 2009). Also, plants, such as Myrrh (*Boswellia sacra* Fluck.), Cassia (Cassia L.), Onions (*Allium cepa* L.), and Henna (*Lawsonia inermis* L.), were used for mummification (Abdel-Maksoud and El-Amin 2011). Egyptian people gave a special place to plants in their life, and these aromatic plants were related to their culture and religion. Greek and Roman people also used ointments derived from plants. They applied several facial creams consisting of Cypress (*Cupressus L.*), Cedar (*Cedrus libani* A. Rich), Olive oil (*Olea europaea* L.), Moringa (*Moringa oleifera* Lam.), etc. In Israeli culture, cosmetics have a huge role. In Japan, sunflowers were used to color the eyebrows, and rice powder was used for skin whitening. Chinese people also used several perfumes and cosmetics. Indians chiefly used *Curcuma longa, Santalum album,* and *Azadirachta indica* for beauty enhancing. Skincare products are essential for modern people in their everyday life. Modern cosmetics, like soaps, face creams, sun creams, shampoo etc., contain plant extracts that provide therapeutic effects like antioxidant and anti-aging actions. Green tea (*Camellia sinensis*), *Aloe vera*, Turmeric (*Curcuma longa*), Neem (*Azadirachta indica*). etc., are common in contemporary skincare products (González-Minero and Bravo-Díaz 2018). The perfume industry is mostly based on aromatic plants, such as Rose, Lavender, and Lemon. Flowers, fruits, leaves, woods, seeds, roots, and resins have always been primordial sources for synthesizing aromatic products (Barrales-Cureño et al. 2021). Some of the plants used

Fig. 11.1 Some traditionally used medicinal plants

traditionally by ethnic communities for medicinal and cosmetic purposes are shown in Figs. 11.1a, b and 11.2.

Forests have huge role in human civilization. Ancient men depended on the forest for hunting, collecting food, like honey and fruits, wood, medicinal plants, fodder, etc. Even today, we use several forest products, and forest consists of hidden treasures like unidentified medicinal plants. Ancient civilizations used plant species, like key thatch palm, Florida thatch palm, and some grass species for thatching purposes. Plants were used for dyeing clothes as well. Indus valley civilization, Egyptian civilization, primitive Chinese, and Europeans used various dyes for coloring cloths, cave paintings, etc. Some of the common ancient dye-yielding plants include *Rubia tictorum* L. *Lawsonia inermis*, *Indigofera tinctoria* L., *Crocus sativus* L, and *Curcuma longa* L. (Siva 2007). Tree trunks were extensively used for timber. Oak, Teak, cedar, Mahogany, and woods were employed in construction and carpentry, whereas the trunks of *Taxus, Viburnum,* and *Fraxinus* were used for other purposes, like bow making, arrow making, etc. (Rösch 2012).

Hemidesmus indicus *Ocimum sp.* *Adatoda vasica*

Fig. 11.2 Some traditionally used medicinal plants

11.1.3 Applications of Traditional Knowledge About Plants

TK on biological resources has broad spectrum knowledge-oriented and profit-oriented applications. The role of TK associated with plants in the global market is clearly visible in the way how the multinational companies are marketing their products as derived from ancient knowledge.

Ethnomedicine of indigenous communities is playing a pivotal role in the modern healthcare sector by aiding the discovery of drugs and nutraceuticals. Conventional awareness of herbal medicine is recorded from the ancient Mesopotamian, Egyptian, Chinese, Indian, Greek, and Roman civilizations (Cragg and Newman 2002; Cragg et al. 1994; Duke 1985). World Health Organization (WHO) describes traditional medicine as the health practices, approaches, knowledge, and beliefs that incorporate floral, faunal, and mineral-based medicines, spiritual therapies, manual techniques, as well as exercises, applied singularly or in combination to treat, diagnose, and prevent illnesses or maintain well-being (World Health Organization 2008). The scientific community is well aware of the immense therapeutic potential of plants, which is attributed to the diverse and peculiar bioactive metabolites they contain. In most cases, the medicinal knowledge about plants and healing practices are present only as oral narratives, so research is being conducted to scientifically validate the therapeutic properties of plants and elucidate their bioactive molecules to be used in drug development programs. A study has proved that 60% of the anticancer as well as 75% of the anti-infectious drugs approved during 1981–2002 have a natural origin (Newman et al. 2003). The wisdom of the ethnic communities has contributed many healing principles to modern medicine. Paclitaxel (*Taxus brevifolia*) and Camptothecin (*Camptotheca acuminata*) against cancer, the tranquilizer Reserpine

(*Rauwolfia serpentina*), the cardiotonic Digitoxin (*Digitalis purpurea*), Quinine (*Cinchona officinalis* or *C. ledgeriana*) against malaria are few well-known examples.

One-third of the active components used in modern cosmetic products are extracts of plants (Schmidt 2012). *Curcuma longa* is a reported Ayurvedic medicine, and curcumin isolated from the plant has multiple applications as an anticancer agent, herbal supplement, food flavoring, and coloring agent, as well as in cosmetics. The cosmetic industry of many countries, like India, is feeding on traditional haircare, skincare, and tooth whitening practices. In recent years there has been an increased inclination of people toward anything marketed based on TK. So the multinational companies are relying on the knowledge from our ancestral generation.

Rosmarinus officinalis L., *Cinnamomum verum* L., *Olea europaea* L., *Lawsonia inermis* L., *Eruca sativa* L., *Linum usitatissimum* L., *Avena sativa* L., and *Jasminum grandiflorum* L. are few plants used in the betterment of hair, eyes, skin, and face in the TK of Alexandria, Egypt that could potentially lead to the development of novel cosmetic products (Elansary et al. 2015). *Eclipta alba, Aloe vera, Azadirachta indica, Vernonia cinerea, Carica papaya, Phyllanthus emblica, Santalum album,* etc., are a few commonly used plants in modern cosmetic products that were part of Indian systems of TK (Fig. 11.3).

The traditional ecological knowledge of the ethnic people is the wisdom acquired through experimentation with nature over centuries. The application of conventional farming practices helps in biodiversity conservation, better yield, and food safety. Also, traditional crop improvement techniques find a prominent role in modern agriculture methods. Conventional crop varieties hold more genetic diversity than modern varieties and are more tolerant to adverse environmental conditions. Techniques for the conservation of biodiversity are inspired by the methods used by our ancestors to protect the ecosystem and its resources for generations to come. The modern lifestyle, culture, art forms, yoga, architecture, and interior designing technologies are all drawing inspiration from TK in one way or the other. This is one way that the folks find a connection to nature.

The convention on Biological Diversity on TK had a case study on the plant Hoodia (https://www.cbd.int/abs/infokit/factsheet-tk-en.pdf). According to TK, the indigenous succulent species of Southern Africa, Hoodia (*Hoodia gordonii*), is used by the indigenous San community to eliminate hunger and thirst during long hunting periods. South African Council for Scientific and Industrial Research (CSIR), in 1996, patented the appetite suppressing bioactive compounds present in Hoodia. They went on in agreement with pharmaceuticals companies to exploit the commercial potential of Hoodia as an appetite suppressant in the global anti-obesity drug market. The involvement of media and an NGO has led the CSIR to the initiation of a benefit-sharing agreement involving the San Hoodia Benefit-Sharing Trust. The funds from the commercialization of their TK will be used for the development and education of the community, to support research and protection of the San's TK and heritage along with reinforcing the institutional base of the San community across Southern Africa. This is one of the numerous incidents of application and

Fig. 11.3 Traditionally used medicinal plants

exploitation of TK of ethnic communities to improve human life and benefit the community responsible for the knowledge. Modern-day research in many areas of study, like plant science, animal science, anthropology, archeology, Biotechnology, Bioinformatics, Genomics, environmental sciences, Clinical medicine, etc., uses the TK information to design products or to derive information to improve the life of humans.

11.2 Biodiversity Loss and the Need for Plant Conservation

Biodiversity is the variability of life on earth. There are genetic, species, and ecological diversity that maintains the food chains, ecosystem balance, genetic diversity, natural resources, and providing aesthetic values. A high density of species requires optimum temperature, humidity, soil, etc. The tropical forest region is rich in biodiversity when compared to polar areas. Himalaya, Western Ghats of India,

Madrean pine-oak woodlands, Mexico, Cerrado, Brazil, amazon rain forest, and Pacific Ocean are typical biodiversity hotspots in the world.

Natural and anthropogenic actions are common reasons for biodiversity loss through ecosystem loss, species loss, or area loss. Seasonal change is one natural thing that reduces biodiversity, mainly in winter. Introduction of invasive species, forest fire, landslide, flood, volcano, earthquakes, etc., decreases biodiversity naturally. Unlawful anthropogenic activities are a major threat to biodiversity. Deforestation, fragmentation, habitat loss, herbivory, pollution, hunting, and overexploitation of particular species for medicines, food, and other products also result in biodiversity loss. These will eventually lead to the mass extinction of species. *Trithrinax schyzophylla* is a Palm species seen in Brazilian Atlantic Rainforest, which became extinct due to habitat loss (Glassman 1987). Another Caribbean Palm species, *Coccothrinax barbadensis,* became extinct due to animal grazing (Morici 1997). Overexploitation causes the extinction of particular plant species. For example, plants, such as Orchids, are widely collected for ornamental purposes and are now declining in population. American *Ginseng* is another plant overexploited for its medicinal properties; also, Mahogany is exploited for its wood. Changes in environmental factors, like temperature, light, water, soil nutrients, etc., limit plant population. The natural competition also leads to species extinction. Plants depend on animals for their seed and pollen dispersal, so animal hunting also affects plants' survival. Dodo bird and Calvaria tree mutualism is a well-known example, where overexploitation and extinction of Dodo lead to the extinction of Calvaria tree.

Biodiversity loss leads to the loss of traditionally significant plants and the associated TK. Red data book maintained by the International Union for Conservation of Nature and Natural Resources (IUCN) records threatened living species on earth. This chapter represents categories of life forms that help understand each species' conservation requirement so that measures can be taken accordingly for their management. Governments and non-government organizations in the world have initiated several projects to protect biodiversity and nature.

11.3 Conventional Methods for Conservation of Plants Associated with TK

The TK of the indigenous people is rooted in experiences. They had their own strategies for cultivation, nutrition, and cultural practices that promote the sustainable use of natural resources. The protection of the knowledge, as well as source material, is an equally important job. The Sanskrit manuscript *Matsya-purāṇa* states that "a pond equals ten wells, a reservoir equals ten ponds, while a son equals ten reservoirs, and a tree equals ten sons," which gives a clue about the significance of plant conservation to our ancestors. They had a close connection with nature and worshipped it. Plants were given a prominent place on special occasions, like marriage, birth, death, festivals, etc. Religious myths, folk stories, and taboos helped limit human interference with conservation requiring plants.

11.3.1 Knowledge Bearers

When the curative potential of a plant is known to the public, there are chances of the plant getting exploited commercially, leading to the population decline of the species. Might be foreseeing this, the ancestral generation has sometimes maintained secrecy regarding the miraculous abilities of plants that were limited in population. There used to be selected people in a majority of the ethnic communities who were assigned to carry the legacy and knowledge forward. These selected knowledge bearers are respected in the community as tribal chieftains, traditional healers, etc. Such specialized people are sometimes chosen at a younger age and trained to be experts in Native American groups (Bruchac 2014). Among the *Kani* tribes of Kerala, India, "Plathis" are tribal physicians who are the holder of tribal medicinal knowledge, having the right to transfer the knowledge (Gupta 2002). Similar TK bearers of Australian ethnic communities inherit the role through the family line (Smith and Wobst 2005). In the Maori community, who are the indigenous Polynesians of mainland New Zealand, selected protected understandings are passed on to specific members of the group alone (Smith 2012). They are more skilled than other people in the community due to the knowledge orally passed on to them by their predecessors. These people will have information about the therapeutic and nutritional potential of the surrounding flora and its practical application. Prevention of the exploitation of important plants was successfully achieved by limiting the number of people having know-how on the plants' extraordinary potential.

11.3.2 Sacred Flora

Another unique way used by most of the ethnic communities for plant conservation was to connect them with religious worship. Sacred grooves are observed in many cultural systems around the world, having different theories but serving the same purpose, i.e., the conservation of culturally and ecologically important plants. This includes a small ecosystem consisting mainly of a grove of trees along with other plants that support considerably rich biological diversity.

Present in many countries, like India, Africa, Japan, Italy, Greece, and parts of North America, sacred grooves aim at the conservation of many vulnerable plant species as well because, considering the religious and cultural aspects of the grooves, common people would leave them undisturbed (Fig. 11.3a and b). Sacred forests seen associated with certain ethnic communities are specialized forests mostly associated with religious theories or mythology. These are also protected forests functioning to conserve ethnic plants and maintain ambient ecological conditions of the region. Numerous trees, climbers, shrubs, and herbs are conserved along with animal species, naturally in sacred grooves.

Several sacred trees are of religious significance to indigenous communities. They worship the tree and perform rituals, whereas cutting down the tree or disturbing its habitat would be considered a sin that inspires people to refrain from destroying it. *Aegle marmelos, Ficus benghalensis, Ficus religiosa, Ficus racemosa, Phyllanthus emblica, Prosopis spicigera, Saraca asoca,* and *Azadirachta indica* are a few sacred trees in Indian mythology, similarly are the "*shinboku" in Japan and* "dangsan namu" seen in Korea. There are other sacred plants and plant parts, like *Ocimum tenuiflorum, Nelumbo nucifera, Desmostachya bipinnata* in Indian mythology. All sacred plants possess multiple biological properties, such as antibacterial, antifungal, anti-inflammatory, anticancerous, antipyretic, etc. This concept adopted by indigenous communities is aimed at the conservation of ecologically significant and TK-associated plants (Fig. 11.4a and b).

11.3.3 Domestication and Conservation

Domestication of ethno medicinally and culturally significant plants is another established method to conserve them with people's participation. Since this is done by people having the know-how about the plant's properties, they will conserve it, leading to an increasing the population of the plant. There is a belief in Chinese culture that certain plants, viz., bonsai bamboo, Money plant, Orchids etc., kept in particular areas in the homes, bring happiness, health, and wealth. Oak trees had greater importance in ancient Greek and Roman culture. Indians grow *Ocimum tenuiflorum* and *Azadirachta indica* in their home, both for rituals and herbal home remedies. Studies have proved the role of plants in providing a better quality living environment in the human habitat and advice the incorporation of plants in the interior of living and office paces (Qin et al. 2014). Many countries have initiatives to grow a minimum number of medicinal plants in residential areas. Countless economic and medicinal plant species are conserved through domestication by people.

11.3.4 Selective and Seasonal Harvesting of Plant Parts

Harvesting of the needful plant part alone from the ethno medicinal plants was one way to prevent plant loss and population decline. Also, the seasonal harvesting of plant parts would not disturb the existence of the plant.

(a)

(b)

Fig. 11.4 (**a**) and (**b**) Sacred grooves (*Kavu*) in India serve religious and conservation purposes

11.3.5 Deadwood Collection

Fuelwoods from the forests were extensively used by the ancient population. Then cutting down healthy trees for firewood and deadwood collection were preferred and followed by the communities considering the conservation aspects. Even today, people in rural areas depend on firewood for domestic purposes.

11.3.6 Protection of Plants at the Burial Site

Burial sites, like cemeteries and churchyards, had the potential for the conservation of biodiversity. Studies on the flora of the burial sites across continents have confirmed their role in the conservation of rare and endangered species (Löki et al. 2019). Due to the lack of anthropogenic disturbances, plants would be protected in such areas.

11.4 Modern Methods for Conservation of Plants Associated with Traditional Knowledge

In the modern world, there are lots of methodologies for plant conservation. The methods are formulated through scientific research and authentic data collection of the flora. A lot of effort, initiatives, and the economy are being spent on plant conservation in modern times. Mostly plants are conserved through ex situ and in situ methods. These methods help protect rare, threatened, and endangered TK-associated plants.

11.4.1 In Situ Conservation

This is a method of plant conservation where plants are conserved in their natural habitats. National parks, Wildlife sanctuaries, and Biosphere reserves are common examples of in situ conservation. Evolutionary changes happen in this method of conservation. Not only plants and animals but also tribal communities are safe. Sacred grooves are one of the common examples of in situ conservation.

According to IUCN, a National park is an area to be conserved for ecosystem protection and recreation (Curry 2009). Galapagos Marine Reserve in Ecuador, National Park, Jim Corbett National Park India, North and East Greenland, Grand Canyon National Park, USA, and Manu National Park, Peru are world-famous National Park conserving biodiversity. All anthropogenic activities, such as firewood collection, grazing, hunting, etc., are prohibited in these areas. The Eravikulam National park in Kerala, India, is famous for conserving "Neelakurinji" (*Strobilanthes kunthiana*), the plant which blooms once in 12 years. Numerous RET-listed and TK-associated plant species are conserved in National parks.

In the case of wildlife sanctuaries, human interactions are allowed up to a limit. Some of the wildlife sanctuaries promote ecotourism, and the main objective is the protection of endangered species of flora and fauna. Wayanad wildlife sanctuary, Periyar wildlife sanctuary, and Badra wildlife sanctuary in India are examples. *Chookalikkuni kattunayakka* and *Kumizhi paniya* are the common tribal communities inhabiting inside the wildlife sanctuaries in India, and there they conserve plants associated with their TK.

Biosphere reserves are naturally protected zones documented under UNESCO's Man and the Biosphere (MAB) Program. Nilgiri biosphere reserve, Agasthyamalai biosphere reserve, and Gulf of Mannar are some examples. UNESCO declared Mura-Drava-Danube as the largest biosphere reserve in the world, which lies in Croatia, and Hungary protects 400 plant species along with other animals. Pachmarhi biosphere reserve in Madhya Pradesh, India, conserves endemic flora, such as silver fern, Jamun, Arjun, Sal, and Wild Mango. Nilgiri biosphere region in the Western Ghats, conserves more than 100 endemic plant species. Twenty tribal groups were listed to be inhabiting this reserve forest, viz., *Adujans, Alars, Arandans, Cholanaiks, Irula, Kalanaiks, Kunduradigans, Kurchiyans, Kurumans, Kurumba, Malmuthans, Mullu Kurumbas, Pahinaikas, Paniyans, Pariyans, Tackanadmuppans, Todas, Uralikurulians,* and *Wayanad Kaders* (Hebbar 2018). *Kani* tribes are present in the Agasthyamalai biosphere reserve, India. The presence of indigenous communities in these regions helps in the conservation of their TK and the plants associated with them. Biodiversity hotspots are ecological areas with a crucial amount of diversities and not threatened by human activities, having at least 1500 species of endemic vascular plants. They play a pivotal role in setting conservation priorities and cost-effective methods to preserve biodiversity in terrestrial as well as aquatic ecosystems (Marchese 2015).

Like the Maple tree is the symbol of national pride of Canada, the National tree, National flower, and National fruit are the peculiar icons of a country. Each country has its characteristic flora representing its identity. Declaration of such recognitions on plants and the identification of umbrella species, keystone species, flagship species, etc., support regional plant conservation by drawing public attention to their protection.

11.4.2 Ex Situ Conservation

This is the conservation outside the natural habitat through artificial methods. Several techniques and methodologies are used in ex situ conservation, viz., botanical gardens, nurseries, gene banks, cryopreservation, micropropagation, seed banks, pollen banks, etc. (Jaisankar et al. 2018). Here plants are reserved in man-made habitats similar to their natural ones. Endangered and important plants are protected separately. A number of biotechnological methods are used in ex situ conservation strategies. Both whole plants and parts can be conserved through several methods. Significant benefits of ex situ conservation include the protection of endangered plants, long-term conservation, mass production of rare plants using tissue culture, seeds, and pollens of rare plants that can be stored for the future, etc. Disadvantages associated with ex situ conservation include cost and restricted evolutionary changes. Certain plants nearing extinction, such as Pitcher plants, underground Orchids, etc., could be stored and conserved through ex situ methods. These modern methods are safe and reproducible, and plant propagules can be made available at any time.

The ethno medicinally important plants of the world are protected in botanical gardens. They consist of live plants in artificial habitats, and a variety of medicinal, aromatic, succulent, hydrophytes, cacti, ferns, etc., are protected in the garden. Other than conservation, research activities are done in botanical gardens, like herbarium preparations, taxonomic studies, seminars, awareness programs, etc. People can visit the gardens and become aware of the need for plant conservation. Micropropagation, cryopreservation, seed banks, and gene banks are available in botanical gardens. It is reported that there are more than 1500 botanic gardens around the globe, conserving more than 80,000 plant species (Jaisankar et al. 2018). The Royal botanical garden in Kew is the largest botanical garden in the world.

Jawaharlal Nehru Tropical Botanic Garden and Research Institute are one of the largest botanical gardens in Asia, established by the government of Kerala in 1979 in Thiruvananthapuram district of Kerala, India. It functions under Kerala State Council for Science, Technology and Environment (KSCSTE). JNTBGRI is a conservatory garden for Tropical plant resources. Numerous Rare and endangered plant species are conserved here and propagated through tissue culture. JNTBGRI has several disciplines, including Biotechnology and Bioinformatics, Conservation Biology, Microbiology, Phytochemistry and Phytopharmacology, Ethnomedicine, and Ethnopharmacology. Innovative research works are being conducted in the departments under the supervision of expert scientists. The botanic garden is associated with the local tribal communities through knowledge sharing and benefit sharing, like in the development of anti-fatigue herbal medicine "Jeevani" from the plant *Trichopus Zeylanicus*. Numerous other TK of the Kerala tribal communities associated with plants have been documented and validated by the institute. Conservation of numerous rare, endangered, and threatened (RET)-listed and ethnomedicinally important plants in the country is achieved through the efforts of JNTBGRI.

Micropropagation is the production of plant clones from small aseptic explants aimed at producing a massive number of disease-free plant colonies. Micropropagation plays a remarkable role in plant conservation since rare species could be propagated massively by this method. *Prunus africana*, is an endangered medicinal plant micropropagated and successfully acclimatized (Komakech et al. 2020). Numerous plants of TK have been subjected to micropropagation studies, successfully propagated, and reintroduced back to their natural habitat across the world.

Gene banks preserve genetic material, including seeds, plant parts, pollen, etc. Seed banks, cryobanks, pollen banks, field gene banks, etc., are examples of gene banks. Long-term storage of genetic material without any damage is possible in the controlled conditions of gene banks. These plant parts are safe from any adverse environmental conditions like floods, pests, climatic changes, and other natural calamities. In seed banks, both orthodox and recalcitrant seeds are stored without losing their viability over long periods leading to the conservation of the plants through propagation. Pollen banks are meant for the preservation of pollen grains using cryopreservation techniques. Cryopreservation of plant cells, tissues, and

Fig.11.5 A Miyawaki forest established in Kerala state of India

organs using liquid nitrogen are better to conserve plant samples without losing their viability and genetic modifications. Some rare plants not having viable seeds could be conserved through cryopreservation. This is an efficient technique to conserve endangered ethnomedicinal plants, viz., *Rauwolfia serpentina*, *Digitalis lanata, Atropa belladonna*, *Hyoscyamus* spp., which could be effectively conserved by low temperature storage (Kasagana and Karumuri 2011).

Miyawaki forestation is a method of afforestation developed by Japanese botanist Akira Miyawaki. It is the method of ecological restoration, where native plants are planted over an area to be developed into a dense forest (Fig. 11.4). More than a thousand Miyawaki forests are present in the world. This is an efficient method for plant conservation (Fig. 11.5).

11.4.3 Joint Forest Management (JFM)

It is the conservation of forest through combined efforts of the local community and state forest departments. This program was introduced in India as part of the framework of the National Forest Policy, 1988 by the Ministry of Environment, Forests and Climate Change, Government of India. The village people form a committee, receive adequate training, and take care of nearly located forests. This is a source of income for the villagers, and they can collect forest products. They protect the forest from overgrazing, forest fire, illegal human activities, etc. In India, states like Uttarakhand, Odisha, West Bengal, and Kerala have successfully participated in this program. The government of India has developed many other schemes, such as the national afforestation program (NAP), the national mission for a green India (GIM), and the forest fire prevention and management scheme (FFPM) to protect forests and their flora.

11.4.4 Existing Legislation for Biodiversity Conservation

Overexploitation of medicinal plants used in traditional medicine leads to their extinction. So, several rules and regulations are enacted for their conservation. Therapeutic goods act, Australia (1989), Herbal medicine regulation law No.541, Austria (1989), Natural health products regulations, Canada (2004), Unidad de medicina tradicional, Chile (1992), medicinal products act Estonia (1996), and Traditional Chinese medicine practitioners act, Singapore (2000), etc., are some acts related to medicinal plants in different countries (Ajazuddin and Saraf 2012). The Convention on International Trade in Endangered Species of Wild Fauna and Flora (CITES) is an agreement signed internationally under IUCN to ensure that the trading of animals and plants species does not threaten their survival. Indian Forests Act 1927, Wildlife protection act 1972, Forest Conservation Act 1980, Environment Protection Act 1986, Biological Diversity Act 2002 etc., in India and the Endangered Species Act 1973 in the USA are a few legislation related to Biodiversity conservation at a national level.

11.4.5 Organizations for Plant Conservation

Committed environmentalists in different parts of the world have formed organizations to work toward plant conservation. The world's first international conservation organization was Fauna and Flora International (FFI). FFI protects a lot of ecosystems and threatened areas. Conservation International (CI) is another international organization mainly focusing on major threats to nature and aids governments in the formation of environmental policies.

The International Union for the Conservation of Nature and Natural Resources (IUCN) founded in 1948 is the global environmental organization for the conservation and sustainable use of bioresources, having the largest professional global conservation network. Their major activities include conservation, data collection, analysis, research, education, etc. IUCN also has a significant role in law and policymaking, conservation of the environment, and sustainable use of natural products. The Species survival Commission (SSC) of IUCN is a network consisting of volunteers from every country that provide information to IUCN regarding biodiversity. SSC gives advice to other organizations, agencies, and IUCN members to prevent the loss of biodiversity. World wide fund (WWF) is another international organization associated with biodiversity conservation.

UN has always taken the initiative to bring its member countries together against global environmental issues, like global warming, ozone depletion, unsustainable development, and biodiversity loss. United Nations Environment Program (UNEP) is another international initiative for environmental protection. It was established after the UN conference on the human environment in 1972 Stockholm. United Nations Development Program (UNDP) has conducted several biodiversity and environment protection initiatives as part of its global biodiversity program; their major categories are protecting biodiversity, restoring ecosystems, access, and benefit sharing, clean water and sanitization, protecting and restoring wetlands, sustainable landscape management, and wildlife and protected areas (https://undp-biodiversity.exposure.co/categories). They provide protection to biodiversity, ecosystems, wildlife, traditional people and their knowledge, etc. The international financial institution, World Bank, also focuses on ensuring environmental sustainability and funding for biodiversity conservation. It has developed several safeguard policies to reduce the negative impacts on biodiversity.

United Nations Convention on biological diversity (CBD), also known as biodiversity convention, has three goals, conservation of biological diversity, sustainable use of its components, and the fair and equitable sharing of benefits arising from using genetic resources. It covers ecosystems, species, genetic resources, and biotechnology. International Plant Protection Convention (IPPC), The International Treaty on Plant Genetic Resources for Food and Agriculture, The Convention on Wetlands, The Convention on International Trade in Endangered Species (CITES), and The Convention for the Protection of the World Cultural and Natural Heritage are other biodiversity-related conventions (Pritchard 2005).

The Centre for plant conservation (CPC) is a network of several institutions that collect and manage endangered species in the United States and Canada. Their main aim is to prevent endangered species from extinction and conserve them for future generations. Native Plant Trust is another organization that protects plants from exploitation in the United States. They save native plants, restore them, and educate people regarding conservation. Botanical Garden Conservation International, Kew, London, is in collaboration with 800 botanical gardens in more than 100 countries for plant conservation (https://www.bgci.org/about/about-bgci/). International Plant Genetic Resources Institute (IPGRI) is an international organization for the conservation and use of plant genetic resources supported by the Consultative Group on International Agricultural Research (CGIAR), Rome, Italy. It promotes the sustainable use of genetic resources. In India also, there are many government and non-government agencies associated with biodiversity protection. World Intellectual Property Organization (WIPO) provides intellectual property rights for conserving TK and traditional cultural expressions (TCEs), including medicine, environmental knowledge, art, symbols, and music (https://www.wipo.int/tk/en/). United Nations Educational, Scientific and Cultural Organization (UNESCO) also functions to conserve TK through intangible cultural heritage, various traditions, practices, and skills (https://ich.unesco.org/en/what-is-intangible-heritage-00003). In accordance with the agreement on Trade-Related Aspects of Intellectual Property Rights (TRIPS Agreement), World Trade Organization (WTO) also aims to protect TK. WHO also focuses on the knowledge of indigenous people on traditional medicine and the traditional medicine systems, viz., traditional Chinese medicine, Ayurveda, Unani, Arabic medicine, etc. (World Health Organization, 2002). The food and agriculture organization (FAO) focuses on traditional agricultural knowledge as well. Conference of the CBD Parties and the United Nations Conference on Trade and Development (UNCTAD) are the other organizations connected with TK conservation (Gazizova 2020).

11.5 Future of Plant Conservation Associated with Traditional Knowledge

Conservation, sustainable use, and equitable sharing of the profits derived from the biological resources are topics of concern in modern times (Gadgil 2006). The flora needs to be conserved so that the future generation can breathe. Indigenous people, TK, and conservation are three interrelated topics requiring attention in the future.

11.5.1 International Initiatives

An initiative under the United Nation's Convention on Biological Diversity, named the Global Strategy for Plant Conservation (GSPC), 1999, aims at slowing down the process of extinction of plants around the globe and prevent biodiversity loss by achieving its 16 global targets, where the target 13 is related to the indigenous knowledge about plant resources (CBD 2010). CBD is working toward setting goals to structure global biodiversity policy for the future, even though they have failed in accomplishing most of the targets of the 2011–2020 Strategic Plan for Biodiversity, at a time when the global biological diversity continues to deteriorate (Reyes-García et al. 2022; Green et al. 2019). The decade (2011–2020) was also observed as the United Nations Decade on Biodiversity. The nations should jointly work toward attaining its targets, and this initiative must be taken forward till we do not have plants in the RET category in the future. There are many other national and international organizations aiming at biodiversity conservation. Global Diversity Foundation (GDF) is a non-profit initiative that focuses on the promotion of biocultural diversity through training, documentation, networking, and collaborative research with ethnic groups of Mesoamerica, North Africa, southern Africa, Southeast Asia, etc. (GDF 2018). The organization has the potential to expand its ideologies and work toward plant conservation around the globe in the future.

11.5.2 Ethnobotanical Surveys and Taxonomical Identification

Recognition of the unexplored traditionally significant plants is the first step toward their conservation. There are innumerable plants, part of the knowledge of the ethnic communities across the globe, that remain unknown to the scientific community. Such plants need to be identified and characterized for their bioprospecting and conservation. The population study of the plants would help understand their distribution and current conservation status. Support and guidance must be taken from local indigenous knowledge holders for such ethnobotanical surveys, which will aid in revealing miraculous medicinal plants of ethnic communities along with the discovery of new plant species.

11.5.3 Scientific Documentation of Traditional Knowledge

TK about biological resources is an invaluable asset to every geographical region or community. Scientific documentation of this wisdom is vital for the population. This will help in the transfer of useful information from each community to the global population. Commonly used modes of knowledge transfer like the oral, writing, art forms, folklore, etc., will see a diminishing trend in the near future. Also, according to the World Intellectual Property Organization (WIPO), the oral form of knowledge is not protected by conventional IPR regimes (https://www.wipo.int/pressroom/en/briefs/tk_ip.html). So there is an urgent need to document them in electronic databases as well. The information about medicinal, poisonous, edible, aesthetic, ecologically significant, and other economically important plants could be saved for future use through documentation. So the documentation of TK on bioresources will be effective to prevent biopiracy and avoid conflicts in the future.

India has enacted the Biological Diversity Act, 2002, which has the provisions for the constitution of a Biodiversity Management Committee at local governing bodies and maintenance of People's Biodiversity Registers. This aims at creating an e-database about the traditional healers using the bioresources and the bioresources of the locality (Gadgil 2006). India already has a TKDL, which is a repository of the TK on medicinal plants, and therapeutic formulations in various systems of medicine practiced by the diverse population of the country, viz., Ayurveda, Siddha, Unani, as well as Yoga (Fredriksson 2021). The United Nations Development Program (UNDP) is working with the Ministry of Forestry and Food, India, to promote the traditional healers of the ethnic communities of the country to document their curative knowledge, plants, and practices, which make them responsible for protecting their knowledge. Likewise, every country should document its ethnic plant wealth in the near future. The national biodiversity Authority (NBA) is a statutory body functioning under the Ministry of Environment, Forest and Climate change, under the government of India. It works on issues related to conservation and sustainable use of biological resources along with the reasonable sharing of benefits derived from the application of biological resources (http://nbaindia.org/content/22/2/1/aboutnba.html).

Centre for Plant Biodiversity Research (CPBR) is an Australian government initiative developed jointly by CSIRO Plant Industry and the Director of National Parks (Environment Australia), through the Australian National Botanic Gardens, for new research on conservation, sustainable use, and management of biodiversity for human existence.

Also, the American Museum of Natural History developed a center for biodiversity and conservation in 1993 for conserving our biodiversity and TK. The Intergovernmental Science-Policy Platform on Biodiversity and Ecosystem Services (IPBES) is another independent intergovernmental body established in Panama city, constructed for sustainable use and conservation of biodiversity. Assessments, policy support, building capacity, and knowledge, communications, and outreach are the working areas of IPBES (https://www.ipbes.net/about).

11.5.4 Biodiversity Register for each Province

Like India is maintaining People's Biodiversity Registers, maintenance of a biodiversity register for each province or different geographical area would be a must-have for the conservation of biological diversity of each area in the future. This will help the future generation to understand the flora of each region. The documentation of the TK associated with each plant species would fit the bill, and special focus must be given to flora used by indigenous communities for various purposes. The evidence on the edible, medicinal, toxic plants, as well as the plants having ecological significance will be useful for future research.

11.5.5 Barcoding

The development of QR codes for protected plants in each region will help in educating the layman about the significance of each plant and prevent them from destroying them. Clear information about the TK associated with the plant and its scientific validation will motivate people not to disturb its habitat. This will also inspire the public to grow useful plants themselves. There are plants used in many rituals and art forms of different ethnic groups. They are used as accessories, coloring agents, and as the art of the costume itself. The conservation of the plants will help protect the art forms as well.

11.5.6 Diversifying the Dietary Selections of Humans

Dependence on the limited number of species is one reason for the overuse and decline in the population of traditionally important plants. The use of underutilized plants could play a role here. There are underutilized plant species having the potential to solve the food security issues in the world. The global staple foods are limited to wheat, maize, rice, barley, and several tuber crops. The proportion of plants exploited for edible purposes is very negligible. TK recommends diverse plants for edible purposes, which is one of the food security principles of majority tribal communities (Fig. 11.6). Season-specific plants and parts are advised for consumption based on the deciduous nature of the plants as well as the seasonal heavy metal ion accumulation in the plant parts. Research has reported numerous edible leaves that have nutritional parameters compared with the commonly used vegetables. But there is an urgent need to spread awareness among the population to depend on them as well for their dietary needs. Diversifying the dietary habits of humans to include a greater diversity of plant products would reduce dependence on limited species and will augment their conservation.

Underutilized leafy vegetables, tuber crops, and fruits can be used to overcome the extinction of limited species. *Alternanthera sessilis, Alternanthera philoxeroides, Amaranthus cruentus, Amaranthus spinosus, Ampelygomum malaicum, Diplazium esculentum, Hydrocotyle sibthorpioides, Ipomoea indica, Crataeva magna, Malva verticillata, Oxalis corniculata, Polygonum plebeium, Portulaca oleracea, Trigonella balansae, Typhonium trilobatum*, etc., are common underutilized leafy vegetables in India. We can use these vegetables instead of the most used cultivated ones (Deb et al. 2018).

Zanthoxylum armatum, Salvadora persica, Syzygium cumini, Annona squamosa, Madhuca longifolia, Morus alba, Pyrus pashia, Ziziphus mauritiana, Capparis decidua, Carissa carandas, Manilkara hexandra, Tamarindus indica, Aegle marmelos, Phyllanthus emblica, Date sugar, etc., are some of the underutilized fruits. *Cyamopsis tetragonoloba, Leptadenia pyrotechnica, Prosopis cineraria, Cucumis pubescens* (*Kachri*), and *Cucumis melo* (*Kakri*) are examples of underutilized vegetables (Nandal and Bhardwaj 2014). *Colocasia esculenta, Ipomea batatas, Dioscorea* spp., and *Alocasia* spp. are underutilized tubers with nutritional potential.

11.5.7 Alternative Therapeutic Plants

In the case of exploring the traditional plants for their therapeutic potential also, the number of plants directly employed is less. The genetically closer species of the well-established medicinal plants may also have similar properties and secondary metabolites; only the quantities of individual metabolites would vary. This disparity could be solved by modifying the method for preparation of traditional medicinal

Sauropus androgynus *Cnidoscolus aconitifolius*

Artocarpus hirsutus *Ipomoea sp.*

Fig. 11.6 Some of the traditionally used underutilized plants with nutritional potential

formulations using such alternative plants after research. Exploring other unexplored plants for medicinal use rather than depending on a few plants would help the conservation of the overexploited traditionally known plant species. *Faujasiopsis flexuosa* (Lam.) C. Jeffrey, *Elephantorrhiza elephantina*, *Tithonia diversifolia* (Hemsl) A. Grey, *Glycyrrhiza uralensis*, *Graptopetalum paraguayense*, *Acer saccharum*, *Acer rubrum*, and *Crocus sativus* are some of the unexplored medicinal plants (Adhami et al. 2018).

11.5.8 Sustainable Harvest

Sustainable use of biological resources is mandatory for their conservation. Along with using the plants and products for our purposes, we should save them for future generations as well. TK associated with indigenous communities is interlinked to their nearest forest ecosystem. In many developing as well as underdeveloped countries, people do not have an awareness of the consequences of over-harvesting various forest products. In Brazil, several studies have led to the establishment of legal protocols for the harvest and maintenance of forest products (Hanazaki et al. 2018). India also has several acts and legal rules for the sustainable harvest of forest products. The international leaders and environmentalists should take initiatives to spread awareness regarding the sustainable harvesting of plants of TK to support their conservation.

11.5.9 Implementation of Stringent Legislation

There needs to be stringent legislation to protect the biological diversity on the Earth and prevent any disturbance due to anthropogenic activities. The developments, constructions, as well as expanding population of humans should not harm the environment. Enforcing stronger international wildlife laws for the protection of biological diversity is the need for the future. While doing that, it is necessary to ensure that the freedom and rights of the ethnic communities are not violated. There are national laws in almost every country and international treaties for the conservation of plants in general and the plants associated with TK. International treaties under the UN Convention on Biological Diversity and the International Treaty on Plant Genetic Resources for Food and Agriculture have promoted biodiversity conservation and sustainable use. In order to prevent the loss of biodiversity, transformative change is required in the way how biodiversity policies are designed and executed from international to national scales as well as across sectors (Díaz et al. 2020).

11.5.10 Sustainable Farming Practices Conserving Traditional Crop Varieties

Sustainable farming practices, like "food forests" or "forest gardens," must be promoted among agriculture enthusiasts. Edible forest gardens are multi-strata agroforests having high biodiversity, including perennial plants, ranging from high trees, bushes, small herbs, climbers, soil covers, and tubers (Jacke and Toensmeier 2005). Such systems will help in the conservation and utilization of the local edible plants and products. These are highly productive ecosystems. Biodynamic farming practices, including high biodiversity, are also effective in conserving a variety of edible plants and plant products. The introduction of landraces and other conventionally grown varieties in modern farming will help bring more genetic diversity to crops and improve tolerance to environmental stresses. Such methods will help draw attention to conventional edible plants and prevent their germplasm loss in the future.

11.5.11 Awareness Programs

Conducting awareness programs to educate people regarding the conservation of TK and flora is the need of the hour. The UNICEF must take the initiative toward this cause, and volunteering youth should be selected from every country to spread the idea of conservation. The collaborative working of indigenous communities as well as the bureaucracy is required in the process. People must be educated about the significance of plants in our ecosystem and the harmful effects it could have when plants are eliminated from nature. Community participation is the key to conservation. On-site indigenous communities, well aware of the local biodiversity, could contribute much to the conservation and sustainable utilization of the bioresources. JFM will provide exclusive rights to communities in the forest along with ensuring sustainable use of its products. The practice should begin in schools when children must be educated about practical biodiversity conservation. The traditionally significant plants could be grown and protected in the academic campuses as well. Also, farmers and traditional healing practitioners could be promoted to cultivate and propagate the traditional economic plants (Obute and Osuji 2002).

11.5.12 Establishment of Miyawaki Forests, Including Ethnomedicinal Plants

Initiation and promotion of Miyawaki micro forestation would help conserve our mother Earth by protecting its green cover. It will maintain the floral diversity along with benefitting humans in diverse ways. Miyawaki forestation aims at the conservation of the local plant species in limited areas of land and thereby increasing the forest area of the earth. The outcome of the initiation of the Miyawaki forest in 550 locations in Japan, Malaysia, Southeast Asia, Brazil, Chile and China proved to be effective in quick environmental restoration of degraded zones (Miyawaki 1999). Giving more representation to plants associated with TK in such urban microforests in the future will aid their conservation.

11.5.13 Earth BioGenome Project (EBP)

Another initiative similar to the Human Genome Project, named EBP, aims to sequence and catalog the genomes of eukaryotic species currently on Earth within 10 years. The initiative aims to apply genome sequencing technology, informatics, automation, and artificial intelligence for understanding, utilizing, and conserving biodiversity (https://www.earthbiogenome.org/). The project will also lead to discovering unexplored and unidentified living organisms. It is inevitable to document the plant wealth around the globe and retrieve their genomic as well as metabolome data and store it for future use in electronic databases. Storing the genetic information of the plants associated with TK will also prevent the loss of ethnic knowledge. The results of EBP are expected to address many issues, including the impact of climate change on biodiversity, conservation of endangered species and ecosystems, and the management of ecosystem services (Lewin et al. 2018).

11.5.14 Prevention of Biopiracy

Biopiracy is an illegal practice that the conservation and bioprospecting activities in the future should never follow. The indigenous TK of the natural resources of a geographical region should never be exploited without giving fair compensation to the community which owned the knowledge till then. The international treaties, viz., Nagoya Protocol, 2010 under the UN Convention on Biological Diversity as well as the International Treaty on Plant Genetic Resources for Food and Agriculture have created the legal framework for the unbiased and evenhanded sharing of the profit derived from the utilization of the genetic resources.

Like climate change has become a global existence issue, and the United Nations is coming together for the "Climate summit"; the biodiversity loss will become a matter of concern in the near future unless addressed. There is a need for another space to take the issue of plant conservation associated with TK, like the United Nations Summit on Biodiversity. Global socioeconomic changes along with cultural as well physical genocide will influence the lives of the ethnic and indigenous communities someday, and eventually, they will merge with the general population and disappear from the wild. This will lead to the loss of the invaluable information they have, so documenting them and protecting the knowledge along with the plants are unavoidable for the future. Also, the manifestation of CBD's 2050 vision of "Living in harmony with nature" will help in the conservation of flora (https://www.cbd.int/2011-2020/). The overexploitation of the TK-associated plants could be avoided in the future if bioprospecting of the plants and their conservation go hand in hand.

11.6 Conclusion

Plants are inevitable components of the global economy, and the unfortunate consequence of economic growth is the loss of biodiversity. Conservation of biological resources is vital for our existence, and it is an obligation toward the future generation. The traditional intellect of indigenous communities on the bioresources is passed on across generations and applied by humans in their everyday life. Indigenous communities and TK are a connecting link between the past and the future, like the traditionally used medicinal plants leading to the discovery of efficient therapeutic drugs for the future. Commercial exploitation of such plants leads to a decline in their population. Multinational biotech companies have relied on the plants used by ethnic communities for product development leading to their exploitation. Indigenous communities around the globe are still protecting the biodiversity surrounding their habitat. Several efforts have been made by our ancestors and the modern policymakers for the conservation of plants, but the measures seem to be inadequate for protecting biodiversity. Conservation and documentation of the TK associated with plants deserve prime importance in the future. There is a need to frame efficient strategies to protect plant species from extinction. Every human on the Earth has a responsibility toward Mother Nature and should volunteer in conserving its biodiversity.

References

Abdel-Maksoud G, El-Amin AR (2011) A review on the materials used during the mummification processes in ancient Egypt. Mediter Archaeol Archaeom 11(2):129–150

Adhami S, Siraj S, Farooqi H (2018) Unexplored medicinal plants of potential therapeutic importance: a review. Trop J Nat Prod Res 2(1):3–11

Ajazuddin, Saraf S (2012) Legal regulations of complementary and alternative medicines in different countries. Pharmacogn Rev 6(12):154–160. https://doi.org/10.4103/0973-7847.99950

Akerele O (1993) Nature's medicinal bounty: don't throw it away. World Health Forum 14:390–395

Barrales-Cureño HJ, Salgado-Garciglia R, López-Valdez LG, Reynoso-López R, Herrera-Cabrera BE, Lucho-Constantino GG, Zaragoza-Martínez F, Reyes-Reyes C, Aftab T (2021) Use of secondary metabolites from medicinal and aromatic plants in the fragrance industry. In: Medicinal and aromatic plants. Springer, Cham, pp 669–690

Bruchac M (2014) Indigenous knowledge and traditional knowledge. In: Smith C (ed) Encyclopedia of global archaeology. Springer, New York, pp 3814–3824

CBD—Convention on Biological Diversity (2010) Updated global strategy for plant conservation 2011–2020. https://www.cbd.int/gspc/. Accessed 11 Apr 2018
Chaudhri SK, Jain NK (2009) History of cosmetics. Asian J Pharm 3(3)
Cragg GM, Newman DJ (2002) Drugs from nature: past achievements, future prospects. In: Iwu MM, Wootton JC (eds) Advances in phytomedicine: ethnomedicine and drug discovery, vol 1. Elsevier, Amsterdam, pp 23–37
Cragg GM, Boyd MR, Grever MR, Mays TM, Newman DJ, Schepartz SA (1994) Natural product, drug discovery and development at the National Cancer Institute Policies for International Collaboration and Compensation. In: Adams RP, Miller JS, Goldberg EM, Adams JE (eds) Conservation of plant genes II: utilization of ancient and modern DNA, Monogr Syst Bot Missouri Botanical Garden, pp 221–232
Curry N (2009) National parks. In: Kitchin R, Thrift N (eds) International encyclopedia of human geography. Elsevier, pp 229–235. isbn:9780080449104
Deb P, Prasad BVG, Munsi P, Chakravorty S (2018) Underutilized leafy vegetables: diversity, nutritional security and medicinal benefit in context of eastern India. In: Advances in biodiversity conservation for sustainable development, pp 95–104
Díaz S, Zafra-Calvo N, Purvis A, Verburg PH, Obura D, Leadley P, Chaplin-Kramer R, De Meester L, Dulloo E, Martín-López B, Shaw MR (2020) Set ambitious goals for biodiversity and sustainability. Science 370(6515):411–413
Duke JA (1985) CRC handbook of medicinal herbs. CRC Press, Boca Raton
Elansary HO, Mahmoud EA, Shokralla S, Yessoufou K (2015) Diversity of plants, traditional knowledge, and practices in local cosmetics: a case study from Alexandria, Egypt. Econ Bot 69(2):114–126
Emirik M (2022) Potential therapeutic effect of turmeric contents against SARS-CoV-2 compared with experimental COVID-19 therapies: in silico study. J Biomol Struct Dyn 40(5):2024–2037. https://doi.org/10.1080/07391102.2020.1835719
Fredriksson M (2021) India's traditional knowledge digital library and the politics of patent classifications. Law Critique: 1–19
Gadgil M (2006) Ecology is for the people: a methodology manual for people's biodiversity register. Centre for Ecological Sciences/Indian Institute of Science, Bangalore, 233p
Gazizova AS (2020) Protection of traditional knowledge: the work and the role of international organisations and conferences. Int J Higher Educ 9(8):95–99
GDF—Global Diversity Foundation (2018) Global Diversity Foundation: a world of difference. https://www.global-diversity.org/. Accessed 27 março 2018
Glassman SF (1987) Revisions of the palm genus Syagrus Mart. and other selected genera in the Cocos alliance. University of Illinois Press, Urbana
González-Minero FJ, Bravo-Díaz L (2018) The use of plants in skin-care products, cosmetics and fragrances: past and present. Cosmetics 5(3):50
Green EJ, Buchanan GM, Butchart SH, Chandler GM, Burgess ND, Hill SL, Gregory RD (2019) Relating characteristics of global biodiversity targets to reported progress. Conserv Biol 33(6): 1360–1369
Gupta AK (2002) Value addition to local Kani tribal knowledge: patenting, licensing and benefit-sharing (No. WP2002-08-02). Indian Institute of Management Ahmedabad, Research and Publication Department
Hanazaki N, Zank S, Fonseca-Kruel VS, Schmidt IB (2018) Indigenous and traditional knowledge, sustainable harvest, and the long road ahead to reach the 2020 global strategy for plant conservation objectives. Rodriguesia 69:1587–1601
Hebbar R (2018) Nilgiri biosphere reserve: reflections from the field. Soc Bull 67(3):302–316
http://nbaindia.org/content/22/2/1/aboutnba.html
https://ich.unesco.org/en/what-is-intangible-heritage-00003
https://undp-biodiversity.exposure.co/categories
https://www.bgci.org/about/about-bgci/
https://www.cbd.int/2011-2020

https://www.cbd.int/abs/infokit/factsheet-tk-en.pdf
https://www.cbd.int/traditional/
https://www.earthbiogenome.org/
https://www.fs.fed.us/wildflowers/Rare_Plants/conservation/lawsandregulations.shtml
https://www.ipbes.net/about
https://www.nwf.org/Educational-Resources/Wildlife-Guide/Threats-to-Wildlife/Overexploitation
https://www.wipo.int/tk/en/

Ibrahim V (2017) Cannabis (Marijuana-Hemp) in Ancient Egypt [Internet]. https://www.researchgate.net/profile/Venice_Attia/publication/321420351_Cannabis_marijuana-_hemp_in_Ancient_Egypt/links/5a212c49aca272ab5a623bef/Cannabis-marijuana-hemp-in-Ancient-Egypt. Accessed 11 Sept 2018

Inoue M, Hayashi S, Craker LE (2019) Role of medicinal and aromatic plants: past, present, and future. Pharmacognosy—medicinal plants

Jacke D, Toensmeier E (2005) Edible forest gardens, Ecological design and practice for temperate-climate permaculture, vol II. Chelsea Green Publishing, Hartford

Jaisankar I, Velmurugan A, Sivaperuman C (2018) Biodiversity conservation: issues and strategies for the tropical islands. In: Biodiversity and climate change adaptation in tropical islands. Academic, pp 525–552

Janick J (2000) Ancient Egyptian agriculture and the origins of horticulture. In: International symposium on mediterranean horticulture: issues and prospects, vol. 582, pp 23–39

Jarrige JF (1986) Excavations at Mehrgarh-Nausharo. Pak Archaeol 10(22):63–131

Kasagana VN, Karumuri SS (2011) Conservation of medicinal plants (past, present & future trends). J Pharm Sci Res 3(8):1378

Komakech R, Kim YG, Kim WJ, Omujal F, Yang S, Moon BC, Okello D, Rahmat E, Nambatya Kyeyune G, Matsabisa MG, Kang Y (2020) A micropropagation protocol for the endangered medicinal tree *Prunus africana* (Hook f.) Kalkman: genetic Fidelity and physiological parameter assessment. Front Plant Sci 11:1871

Kumar RS, Perumal P, Shankar BR (2012) Antinociceptive and anti-inflammatory activity of alkaloid fraction of *Trichopus zeylanicus* gaertn. Int J Pharm Pharm Sci 2:1

Lewin HA, Robinson GE, Kress WJ, Baker WJ, Coddington J, Crandall KA, Durbin R, Edwards SV, Forest F, Gilbert MTP, Goldstein MM (2018) Earth BioGenome project: sequencing life for the future of life. Proc Natl Acad Sci U S A 115(17):4325–4333

Löki V, Deák B, Lukács AB, Molnár A (2019) Biodiversity potential of burial places—a review on the flora and fauna of cemeteries and churchyards. Global Ecol Conserv 18:e00614

López-Agüero LC, Stella AM (2007) Aesthetic dermatology through the time. Rev Argent Dermatol 88:227–233

Luqman S (2011) The saga of opium poppy: journey from traditional medicine to modern drugs and nutraceuticals. In: International symposium on Papaver 1036, pp 91–100

Marchese C (2015) Biodiversity hotspots: a shortcut for a more complicated concept. Glob Ecol Conserv 3:297–309

Miyawaki A (1999) Creative ecology. Plant Biotechnol 16(1):15–25

Moghadamtousi SZ, Abdul Kadir H, Hassandarvish P, Tajik H, Abubakar S, Zandi K (2014) A review on antibacterial, antiviral, and antifungal activity of curcumin. Biomed Res Int 2014: 186864

Morici C (1997) *Coccothrinax barbadensis* in Antigua. Principes 41:84–86

Nandal U, Bhardwaj RL (2014) The role of underutilized fruits in nutritional and economic security of tribals: a review. Crit Rev Food Sci Nutr 54(7):880–890

Newman DJ, Cragg GM, Snader KM (2003) Natural products as sources of new drugs over the period 1981–2002. J Nat Prod 66:1022

Obute GC, Osuji LC (2002) Environmental awareness and dividends: a discourse. Afr J Interdiscip Stud 3(1):90–94

Pritchard D (2005) International biodiversity-related treaties and impact assessment—how can they help each other? Impact Assess Proj Apprais 23(1):7–16. https://doi.org/10.3152/147154605781765706

Pushpangadan P, Rajasekharan S, Subramaniam A, Latha PG, Evans DA, Raj RV (1995) Further on the pharmacology of Trichopus zeylanicus. Anc Sci Life 14(3):127

Qin J, Sun C, Zhou X, Leng H, Lian Z (2014) The effect of indoor plants on human comfort. Indoor Built Environ 23(5):709–723

Reyes-García V, Fernández-Llamazares A, Aumeeruddy-Thomas Y, Benyei P, Bussmann RW, Diamond SK, García-del-Amo D, Guadilla-Sáez S, Hanazaki N, Kosoy N, Lavides M (2022) Recognizing indigenous peoples' and local communities' rights and agency in the post-2020 biodiversity agenda. Ambio 51(1):84–92

Rösch M (2012) Forest, wood, and ancient man. IANSA III(2):247–255

Schaal B (2019) Plants and people: our shared history and future. Plants People Planet 1:14–19. https://doi.org/10.1002/ppp3.1

Schmidt BM (2012) Responsible use of medicinal plants for cosmetics. Hortscience 47:985–991

Shankar BR (2017) Anti-ulcer potential of saponin fraction of Trichopus zeylanicus on a various experimental animal model. Int J Green Pharm (IJGP) 11(01)

Sharma AK, Pushpangadan P, Chopra CL, Rajasekharan S, Amma LS (1989) Adaptogenic activity of seeds of *Trichopus zeylanicus* gaertn, the ginseng of Kerala. Anc Sci Life 8(3–4):212

Sindhu C, Jose B (2015) Evaluation of DPPH radical scavenging activity of the leaf, root and fruit extracts of *Trichopus Zeylanicus* from South India. World J Pharm Res 4:1283–1292

Siva R (2007) Status of natural dyes and dye-yielding plants in India. Curr Sci 92(7):916–925. http://www.jstor.org/stable/24097672

Smith LT (2012) Decolonizing methodologies: research and indigenous peoples, 2nd edn. Zed Books, London

Smith C, Wobst HM (eds) (2005) Indigenous archaeologies: decolonizing theory and practice. Routledge, London

Subramoniam A, Madhavachandran V, Rajasekharan S, Pushpangadan P (1997) Aphrodisiac property of *Trichopus zeylanicus* extract in male mice. J Ethnopharmacol 57:21–27. https://doi.org/10.1016/S0378-8741(97)00040-8

Sundar Rajan T, Velmurugan VAS (2015) Anti-diabetic activity of ethanolic extracts of *Trichopus zeylanicus* in streptozotocin induced diabetic rats. World J Pharm Pharm Sci 4:734–740

Vignesh KB, Ramasubbu R, Sasikala N (2016) Analysis of phytochemical constituents and antimicrobial properties of essential oil extracted from the leaves of *Trichopus zeylanicus* ssp. World J Pharm Res 5:499–517

World Health Organization (2002) Traditional medicine in asia (No. Regional publication No. 39). WHO regional office for South-East Asia

World Health Organization (2008) Traditional medicine. http://www.who.int/mediacentre/factsheets/fs134/en/

Chapter 12
Conservation of Landraces and Indigenous Breeds: An Investment for the Future

Emi Mathew and Linu Mathew

Abstract Landraces and indigenous breeds are either animal or plant species developed at a specific location through natural and artificial selection processes over a long period. Due to the changing environment and mixing up with other genotypes, they exhibit an evolving trend. Both landraces and indigenous breeds represent highly resilient gene pools preserving unique features suitable for the specific conditions of their locality, including climate, pests, infections, and traditional practices. Extensive use of pure lines and hybrids with high yield and better quality derived from a small germplasm pool resulted in the loss of potential traits, reducing genetic variability nowadays. To cope with climate change and satisfy the demands of the current scenario, researchers and breeders are continually exploring new sources of genetic diversity.

Conservation of the excellent gene repositories preserved in the landraces and indigenous breeds is the need of the hour in the unpredictable climatic conditions of the world. In situ conservation is a dynamic strategy that allows gradual evolution under an ever-changing scenario. Furthermore, management skills and knowledge linked with these traditional genetic resources are also protected in a co-evolutionary form. Identification, characterization, assessment, documentation, and conservation are essential aspects of preservation. Integrating in situ conservation with ex situ genotyping and phenotyping would help identify the genetic resources to be preserved. Live species ex situ programmes and cryopreservation of genetic materials could be adopted to restore unique genomes. Preservation of genetic resources is crucial for the breeds and races on the verge of extinction to ensure their existence for future generations.

E. Mathew
Bishop Chulaparambil Memorial College, Kottayam, Kerala, India

L. Mathew (✉)
School of Biosciences, Mahatma Gandhi University, Kottayam, Kerala, India
e-mail: linumathew@mgu.ac.in

S. T. Sukumaran, K. T R (eds.), *Conservation and Sustainable Utilization of Bioresources*, Sustainable Development and Biodiversity 30,
https://doi.org/10.1007/978-981-19-5841-0_12

Keywords Landraces of crops · Indigenous breeds · Indigenous cattle breeds · In situ conservation of indigenous breeds · Ex situ conservation · Indigenous traditional knowledge · Gene bank · Cryopreservation of germplasm · Cytogenetic screening · Animal genetic resources

12.1 Introduction

Humanity is at a crossroads for its holdings for future generations (CBD 2020). Biodiversity, the unique feature of life on earth, is currently falling astoundingly (Cardinale et al. 2012). Land-use conversion for agriculture and development, the substitution of local breeds and varieties, rural–urban migration, overexploitation of natural resources, intensive farming, pollution, invasive species, and climate changes are all contributing to this current global trend in genetic diversity loss and genetic resource erosion (EFI 2021).

Diverse genetic resources are the raw elements that people rely on for food, nutrition, livelihood security, ecosystem sustainability (Liu et al. 2018), and bioeconomy. The vast array of genetic resources which execute essential ecosystem functions (Liu et al. 2018) includes wild plants, animals, forest trees, fungi, invertebrates, and microbes, as well as the genetic diversity within them—cultivars, breeds, populations, individuals, and genes (European Union 2018).

During the second half of the twentieth century, a significant shift to high-input intensive monoculture systems occurred in agriculture due to the intense growth of the human population and consumerism (Thanopoulos et al. 2021). Extensive use of pure lines and hybrids with high yield and better quality derived from a small germplasm pool resulted in the loss of potentially valuable traits, reducing genetic variability (Sarbapriya and Aditya 2011).

To cope with climate change and satisfy the demands of the current scenario, researchers and breeders are continually exploring new sources of genetic diversity. Most recent conservation efforts have concentrated on wild species. However, traditional varieties preserved by traditional agricultural techniques contain the majority of gene variability in domesticated species (Villa et al. 2005). Indigenous breeds and landraces are the essential components of agrobiodiversity (Sarbapriya and Aditya 2011). In addition, domesticated species and the variety among and within them have substantial economic and historical importance.

12.2 An Overview of Landraces and Indigenous Breeds

In the wake of agriculture, humans selected wild plants having desirable traits for culturing. These valuable plants were nurtured and multiplied for many years. Nature gradually selected advantageous genotypes appropriate for cultivating on land and gardens. Populations evolved through these combined selection processes are known as landraces. Even though the landraces are endemic to a particular place, they have transported over short or even great distances to new avenues putting them in competition with endemic landraces existing there (Zeven 1998). They could become extinct, replace these endemic varieties, or, more likely, combine with the endemic plants to produce a new landrace. Their phenotype and genotype change in response to each site's environmental conditions and adapt to the specific needs of each time. These adaptations have occurred by changes in the frequencies of phenotypes and thus genotypes for vegetatively propagated crops and the frequencies of alleles for outbreeding crops. New genotypes introduced elsewhere or created by mutation or hybridization can also add to this (Zeven 1998).

Von Rümker (1908) defines landraces as varieties that have been grown since time immemorial in the region for which they bear the name. According to Mansholt (1909), landraces have high stability of their traits and a high resistance capacity to overcome adverse impacts. However, their production capacity is lower than cultivars, and their genetic composition will alter if grown outside their native region (Zeven 1998).

The Merriam Webster dictionary defines a landrace as an animal breed or a plant cultivar that has adapted to its particular habitat, notably via deliberate breeding and agriculture, after being separated from other populations of its species. The word landrace means "country breed". Harlan (1975) gives a rather detailed account of landraces that are made up of a combination of genotypes, all of which are pretty well adapted to the location they developed. But they differ in their specific adaptations to certain environmental conditions. They are usually less productive but stable.

Landraces may be either autochthonous or allochthonous. An autochthonous landrace is one that is native to or has been produced for an extended period (e.g., 100 years or more) inside the agricultural system in which it is present (Zeven 1998). In contrast, an allochthonous landrace is a foreign one imported recently to a particular region. Autochthonous landraces may evolve gradually due to the changing environment and contamination with other genotypes. Allochthonous landrace will most probably develop into a rare type years after its introduction to a particular place through crossbreeding with different genotypes of the locality (Christiaan Struik et al. 2017). Furthermore, natural and artificial selection and migration resulting from seed exchange can all play a role in forming diverse landrace populations and metapopulations (groupings of related populations). As a result, landraces always exhibit an evolving trend contrasting to existing cultivars, planned to be kept stable to type according to the Union for the Protection of New Varieties of Plants' (UPOV) guidelines (Christiaan Struik et al. 2017).

Plant landraces have adapted to conventional basic land preparation, sowing, weeding, and harvesting operations. They are also adaptable to low soil fertility; they are not very demanding, partly since they do not yield much. Landraces have a high level of genetic stability. They are morphologically identifiable; farmers have given them names related to their locality. Distinct landraces differ in soil preferences, planting time, maturity period, height, nutritional value, use, and other characteristics. Past generations of cultivators have left us with dynamic populations in harmony with the environment and diseases. They are the core resources on which future plant breeding must rely. They result from millennia of natural and artificial selections (Harlan 1975; Stalker et al. 2021; Vu et al. 2020).

Crop landraces are excellent sources of unique genotypes that can improve a wide range of crops. They have a moderate edible yield and variable phenology. Landraces include traits for more effective nutrient uptake and use and valuable genes for tolerance to adverse environments, like water stress, salinity, and high temperatures, which help develop improved cultivars. In contrast, modern agricultural practices have resulted in a loss of crop landrace variability during the last few decades. Environmental factors, including genetic erosion and local farming techniques, are the main threats to landrace variety. Specific conservation measures have been developed to address these concerns, and these strategies have an essential role in preserving crop landrace variation. In addition, complete documentation of the problem is necessary to deal with the stress posed by gene flow to crop landraces. The information acquired in its wholeness is worth launching various crop improvement initiatives (Mir et al. 2020).

Because of their great potential to endure biotic and abiotic challenges, landraces or traditional or local varieties are essential crop genetic resources and raw materials for plant breeding (Bellon 2004). Two main approaches of conservation include ex situ (off-site) and in situ (on-site) protection, which are complementary to each other (Brush 1995). In situ conservation of crop species refers to the long-term farming of crop populations on fields, where they have evolved through natural and farmer-selected processes. (Bellon 2004). One of the most convincing arguments for this conservation strategy is that it is dynamic, with crops adjusting to changing environmental conditions, unlike the ex situ approach. In addition to the genetic raw materials of alleles and genotypes, the decentralized management skills and traditional knowledge linked with agricultural genetic resources are protected in a co-evolutionary form with in situ conservation.

Landraces comprise a gene pool repository that enriches biodiversity and maintains and sustainably stabilizes ecosystems. Aside from conserving biodiversity in agriculture, traditional crop cultivation provides humanity with regulatory functions like nutrient cycling, carbon sequestration, soil erosion monitoring, greenhouse gas mitigation, and hydrological process control. However, man has caused an imbalance in biodiversity by overexploiting some plant species while neglecting others, either intentionally or unintentionally, through modern agricultural systems that promote the cultivation of a few high-input and high-yielding species (Azeez et al. 2018). Therefore, landraces may be defined as distinct, recognizable, and dynamic populations adapted to the unique characteristics of a specific location, such as climate, pests, infections, and traditional practices (Villa et al. 2007).

Landraces have origins in a specific geographic area and usually own local name (s). A landrace will show only a moderate yield level but high yield stability, even under adverse conditions. Keeping genetic integrity while preserving genetic diversity (Stalker et al. 2021) is a landmark of landraces. Landrace development is not the result of any institutional breeding programmes or systematic breeding methods. Genetic isolation is primarily a result of geography operating on whatever animals or plants were introduced to a particular location by humans (Stalker et al. 2021; Zeven 1998).

Plant landraces have received greater attention than animal landraces, and the vast bulk of scholarly work on landraces focuses on agricultural botany rather than animal husbandry. Traditional agricultural techniques are linked chiefly with plant landraces. While many landrace animals are involved with farming, others have been used for transportation, companion animals, sports animals, and other non-farming functions. Hence, their geographic distribution may vary. Horse landraces, for example, are less prevalent because they have traveled with people more frequently and continually than most other domestic animals, limiting the prevalence of populations that have been genetically isolated for long (Sponenberg 2000).

The term indigenous breed refers to an animal species native to a specific region that humans have domesticated for breeding and selection purposes. Individuals belonging to a particular breed have similar genotypic and phenotypic features. Gradually these qualities evolve to become acclimated to their specific habitat. Over thousands of years, an indigenous breed becomes adapted to the soil, climate, and sanitary conditions of its locality. Throughout this time, genes for several different traits were evaluated. Gradually, each breed develops its unique pool of genes (Vanishing breeds—India Environment Portal|News, reports, documents, blogs, data, analysis on environment, and development|India, South Asia 1997). As per the definition given by Commission on Genetic Resources for Food and Agriculture, Indigenous breeds, also known as autochthonous or native breeds, are a subgroup of the locally adapted breeds that "originated from, are suited to, and are utilized in a specific geographical location" (CGRFA 2012). Both landraces and indigenous breeds represent highly resilient gene pools preserving unique features well adapted to the specific environment of each locality.

12.3 A Kaleidoscopic World View of Landraces and Indigenous Breeds

Since the shift from a hunting–gathering culture of man to a farming culture, wheat has been vital to civilizations. Continuous selection in wheat farming enabled it to fulfill human requirements and adapt to various environmental conditions. The species of bread wheat (*Triticum aestivum* L.) developed along ancient human migratory routes that traversed Europe and Asia. As a result, domesticated wheat populations have evolved into the so-called landraces adapted to local conditions. Bread wheat was then imported to the New World in the sixteenth century, first in Latin America, then in Northern America and Australia. As Asian landraces are rarely employed in current wheat breeding initiatives, its germplasm comprises a considerable portion of the global diversity that is essentially untapped. Therefore, it is critical to describe these genetic resources to use them adequately (Balfourier et al. 2019).

Wheat landraces can adapt to a variety of environmental conditions. Traditional wheat farmers helped evolve and enhance the on-farm management of unique landraces. However, the creation of low-cost, commercially viable wheat varieties throughout the last century caused a reduction in the genetic diversity of wheat landraces. Despite this, the widespread cultivation of specific landraces has demonstrated their nutritional and agronomic worth to wheat growers compared to current wheat types. Introducing innovative wheat varieties containing wheat landraces is an excellent way to improve yield and stability under stress and climate change. Natural variants provide a great deal of genetic variation and contribute to establishing genetic variation through breeding or genetic engineering (Lodhi et al. 2020).

Until the Green Revolution replaced most traditional folk varieties with a handful of modern cultivars, Indian rice (*Oryza sativa var. indica*) had over 100,000 landraces. Rice variants, such as coloured rice, aromatic rice, medicinal rice, millets, wheat, barley, and maize, are some of the indigenous crops of India. Drought, salinity, and flood resistance are all features of indigenous rice and millet cultivars. For example, Eastern India's Dharical, Dular, and Tilak Kacheri may adapt to various topography, climate, and soil conditions. The residual folk varieties, which are still flourishing on marginal farms, are a testament to local landraces' incredible ability to adapt to various biological and ecological conditions. They also represent a long history of farmers selecting genotypes to suit multiple cultural and gustatory preferences. Furthermore, some of these types contain nutritional qualities documented in folklore and preserved in situ for generations by local communities. The rice gene pool is complemented and supported by various species and environmental variables on-farm (Alvares, and Multiversity (Organization), and Citizens International (Organization) 2017).

Sorghum, pearl millet, maize, barley, and finger millet are the indigenous coarse grains grown in India. Small millets like Kodo millet, little millet, foxtail millet, proso millet, and barnyard millet also come under India's traditional landraces (Renjith 2018b). According to Vedic literature, millets have been grown since the Stone Age and were a staple diet of the Indians. Millets are used in various applications, including food grain, bird/cattle/livestock feed, fuel, and building materials. Millets are C4 plants that are more productive in hot, dry, temperate, tropical, and subtropical conditions with less rainfall and drought-prone circumstances than C3 species, like wheat and rice (Rao 2021).

Indian grain production has more than tripled since the start of the Green Revolution, and the share of Indian grain production contributed by rice and wheat has steadily increased (Rao et al. 2012). The loss of coarse cereals in the Indian diet has significantly reduced iron intake without compensating for other food groups, especially in states where rice has replaced coarse cereals. Along with other measures, increased consumption of coarse grains could lower anaemia prevalence in Indian women (Renjith 2018a).

Diversifying crop production, including indigenous coarse grains, like millets and sorghum, can improve the nutritional quality of food, i.e. similar protein content and higher iron content. It can survive well with less water and energy resources (Davis et al. 2018). Millet cultivation reduces greenhouse gas emissions (Rao et al. 2019).

Compared to rice, alternative grains exhibit significantly high climate resilience without lowering calorie production or more land. Even in the post-green revolution period, certain districts of India have a greater or comparable yield of coarse grains than that of rice. Other nations also can follow such win–win scenarios where food systems achieve numerous nutritional, environmental, and climatic resilience goals using similar multidimensional approaches to food production (Davis et al. 2019).

Indigenous agriculture was developed in Africa by domesticating African plants. The most important crops of Africans as food are sorghum, pearl millet, African rice, yams, oil palm, cowpea, bottle gourd, finger millet, tef, ensete, and noog. The linguistic evidence consistently indicates considerable antiquity of agriculture in sub-Saharan Africa. The African savanna complex did spread out of its hearth (Indigenous African Agriculture 2021).

Every region globally has its traditional food crops extensively consumed by specific people or communities (Eliazer Nelson et al. 2019). African nightshades (*Solanum nigrum*-related species) are essential indigenous green leafy vegetables that rural African populations have long utilized. These vegetables have a lot of genetic variation. The demand for these veggies has steadily increased as people have realized their high nutritional value, therapeutic properties, and health advantages (Ojeivo 2012). Aizen (Mukheit), *Boscia senegalensis,* is another adaptable, resilient crop that grows in plenty in the stretch of Sahel–Sahara savannas of Africa. It yields valuable products and provides year-round shade in extreme drought conditions. So, people of many countries in Africa rely exclusively on Aizen fruits, seeds, roots, and leaves during famine conditions (Lost crops of Africa 2008; Solh et al. 2003) (Akinola et al. 2020). The black walnut (*Juglans nigra L.*), a member of the walnut family (*Juglandaceae*), endemic to North America, is renowned for its uniquely flavoured nuts, which have a long history of usage as a food ingredient. Even though extracting the tree nut is difficult, it offers vital nourishment to indigenous Americans (Vu et al. 2020). Katsura-uri (*Cucumis melo var. conomon*) of Kyoto, Japan is a heritage vegetable now threatened with extinction. Pickles made from immature and mid-ripened fruit have long been famous in Japanese cuisine, but their popularity has waned as consumer demand has declined. Katsura-uri has been given a new strategy to increase its inclusion in the current diet as a functional drink to prevent obesity and diabetes (Sasaki et al. 2017). Aniseed myrtle, *Syzygium anisatum*, is an Australian rainforest tree with leaves that produce an essential oil with an aniseed characteristic aroma. The leaves of this plant, both fresh and dried, are used as a culinary herb. In recent years, aniseed myrtle has been employed as a flavouring agent by the food and beverage industries, and Indigenous Australians have used it for its therapeutic properties (Sultanbawa 2016). Indigenous civilizations have been able to manage food scarcity thanks to agrobiodiversity, embedded in the matrix of specific cultural systems (Alvares, and Multiversity (Organization), and Citizens International (Organization) 2017).

The various indigenous farm animal breeds are solely the result of evolutionary changes. They have adapted to harsh climatic conditions with low management inputs regarding feeds, fodder, and health care. They can more efficiently convert low-quality feeds and hay into animal products and are better adapted to withstand tropical diseases. They are essential components in agriculture (Groeneveld et al. 2010).

Cattle are among the most important livestock species in output and cultural significance. Several indigenous cattle breeds that differ in appearance, performance, and environmental adaption are preserved on all inhabited continents. Thousands of cattle breeds are recognized worldwide, some of which have adapted to the local conditions and others that humans have developed for specific purposes. There are two major categories.

1. *Bos indicus* (or *Bos taurus indicus)* cattle, often known as zebu, are tropical cattle that originated in India, Sub-Saharan Africa, China, and Southeast Asia.
2. *Bos taurus* (or *Bos taurus taurus*) cattle, sometimes known as "taurine" cattle, are suited to colder climates and include nearly all cattle varieties from Europe and northern Asia.

Bos indicus cattle have a more oversized frame and long legs that allow them to quickly cover huge and sparse regions of land in search of food and water. As good foragers, they can adapt to harsh drought conditions, making them a natural choice for humid, tropical environments. Cattle ticks and other parasites are repelled by their smooth, short-haired coats and naturally produced sweat chemicals. Some *Bos indicus* breeds create a substance in their tails that acts as insect repellent when they swat insects. A hump across the shoulders on their back stores fat like a camel. Substantial floppy ears and a dewlap keep them cool. *Bos taurus* cattle, sometimes referred to as "British" breeds, favour milder weather. They have thicker coats to cope with the colder winters and lack the hump that their *Bos indicus* cousins possess. In addition, they mature faster and gain muscle mass more quickly than the *Bos indicus* due to their more petite frame (Australian Cattle|Australian Butchers Guild 2022). More cow species (both wild and domesticated) can be found in various parts of the world, and some of them are so closely related to taurine and indicus cattle (Felius et al. 2014). The noticeable hump at the withers (i.e. the shoulders of a four-legged animal) and the drooping rather than straight ears of indicine cattle are the two most evident morphological distinctions between these groups (McTavish et al. 2013).

In Africa, domestication started with animals and moved much later to crops in the Sahara and Sahel. The agricultural system evolved with a farming village pattern and spread over much of the continent. They developed Livestock herding and pottery production long before any cultivated plant trace (Indigenous African Agriculture 2021). Indigenous humped cattle or zebu cattle (*Bos indicus)* constitute the significant cattle population of Africa (Hanotte et al. 2000). They have adapted to local climatic circumstances that are inappropriate for foreign European breeds (e.g., high temperatures, long periods of drought, vector-borne disease). African indigenous taurine cattle, *Bos taurus* (humpless cattle) are now almost solely found in West Africa. Apart from these two primary cattle breeds, the continent is home to sanga (African humpless *Bos taurus*—humped *Bos indicus* hybrid) and zenga (sanga zebu backcross) (Mwai et al. 2015). Unfortunately, many indigenous African breeds are

Nguni

Sanga is the collective name for the indigenous cattle of Southern Africa. The Nguni (South Africa), Bapedi (Lesotho), Nkone & Mashona (Zimbabwe), Landim (Mozambique), Tswana (Botswana), and Sanga (Namibia) are Sanga ecotypes. Sanga or Sanga phenotypes can also be found in the Afrikaner, Bonsmara, Drakensberger, and Tuli. The cervicothoracic hump is characteristic, as are the lyre-shaped horns with a cylindrical centre. Their body profile suggests a link between laterally horned creatures with a hump on the neck and the shorthorn Brachyceros, which lacks a hump and has European ancestry (Flickr, 2014).

Zenga- Abergele

Crossbreeding between Sanga cattle and zebu created a new cattle breed known as "Zenga", only found in eastern Africa. Abergele, a Zenga breed, is the shortest cattle breed in North Ethiopia (Merha Zerabruk; Vangen, 2011). They are raised in the Abergele lowlands. The Abergele breed is recognised for its ability to adapt to the hotter, drier conditions of the lowlands. The Arado, Fogera, Horro (all of Ethiopia); Jiddu (southern Somalia); Alur, (Democratic Republic of Congo); Nganda (Uganda); Sukuma (Tanzania); all are Zenga breeds indigenous to Africa (Rege, 1999).

White Fulani

White Fulani cattle are a popular beef breed in the Sahel zone of Africa, especially in the areas occupied by the Fulani people. They are predominantly Zebu cattle, although some are Sanga cattle. The White Fulani's coat is often white on black skin, with black ears, eyes, nose, hooves, horn points, and tail tip. In addition, they feature high lyre-shaped horns and either thoracic humps like the Zebu or humps that are transitional between the cervico-thoracic humps and the cervico-thoracic humps like the Sanga (Rege, 1999).

N'Dama

N'Dama cattles are West African breeds. They may be reared in tsetse fly-infested environments since they are trypanotolerant (Hickman, 1991). They are also resistant to ticks and the illnesses they bring and the stomach worm Haemonchus contortus (FAO, 2022).

Fig. 12.1 Indigenous cattle breeds of Africa (All pictures used are under Creative Commons license)

endangered now, and their unique adaptive characteristics may be lost forever. Some of the indigenous African breeds are represented in Fig. 12.1.

Most cattle breeds in Europe are indigenous breeds with a narrow geographic range. However, the breeds have many genetic variations and have significant cultural, historical, socioeconomic, and environmental aspects. Thus, they have a lot of potential. Local cattle is a live legacy, ancient germplasm that shouldn't be lost. These breeds were previously the most popular in their respective regions, but many are labeled endangered. As many breeds became endangered, the hitherto unrecognized "hidden" merits began to be recognized. For decades, European society has recognized the importance of these cow breeds' environmental, social, cultural, commercial, and public qualities. (Hiemstra et al. 2010). Selected indigenous cattle breeds of Europe are shown in Fig. 12.2.

Different domestications of related bovine species also occurred in Asia (Ho et al. 2008). Adaptions of the yak (*Bos grunniens*) to high altitudes (Qiu et al. 2012) have been used in Tibet and neighboring regions. The ranges of the gayal or Mithun (*Bos frontalis*) as well as the domestic Bali cattle (*Bos javanicus*) in Assam and Myanmar overlap with those of the zebu (Payne and Hodges 1997). Because all three species hybridize with taurine and zebu cattle, numerous Asian cattle populations are mixed species and provide distinctive (Payne and Hodges 1997) contributions to the cattle resources (Felius et al. 2014). Figure 12.3 explains some of the indigenous Asian cattle breeds.

Christopher Columbus introduced the first cattle to America on his second journey over the Atlantic in 1493. Cattles of the Canary Islands are from Portuguese and Spanish ancestors brought by early Spanish explorers (Barragy et al. 2002). As a result, these cattle were likely related to Northern African cow breeds. They hence may have an indicine genetic component. The ancestors of modern New World varieties, such as Corriente cattle from Mexico, Texas Longhorns from northern Mexico and the southwestern United States, and Romosinuano cattle from Colombia, were primarily wild Spanish cattle (Rouse 1977). These breeds are more adapted to this terrain than breeds of more recent European origin because of an extended period of natural selection. For example, Texas Longhorns have more immunity to "Texas fever" or "Cattle tick fever", a tick-borne illness caused by the protozoan *Babesia bigemina* (Figueroa et al. 1992). Texas Longhorns also have significantly superior drought resilience (McTavish et al. 2013). American indigenous cattle breeds are shown in Fig. 12.4.

Throughout Australia's history, the livestock has developed from "British" varieties that have prospered in the southern temperate climates but have struggled in the heat of Northern Australia. Australia's breeds have evolved to adapt to diverse conditions around the country. Kerry cattle, a native breed of Ireland, is a landrace population that remains productive in harsh upland regions with poor quality feed, typical of southwest Ireland where it evolved. These cattle were often referred to as the "poor man's cow" due to their ability to produce relatively large quantities of milk on very sparse fodder (Browett et al. 2018). Tropical breeds, such as the *Bos indicus*, were introduced to combat the pest and temperature issues that British varieties, such as the *Bos taurus,* faced. Crossbreeding between British and tropical breeds has resulted in new breeds that combine the best of both worlds: cattle that adapt to tropical environments while still maturing faster and gaining muscular mass quickly (Australian Cattle|Australian Butchers Guild 2022). Australian Braford, Australian Brangus, Australian Charbray, Australian Lowline, Belmont Red, Droughtmaster, Greyman, and Adaptaur are some of the most popular breeds in Australia and some famous Australian breeds are shown in Fig. 12.5.

The United Nations Food and Agricultural Organization (FAO) reports that about 17% of the world's cattle breeds are on the verge of extinction. Many animal breeds are native to the area and have strong agricultural and pastoral histories. More than

Western Finn Cattle

More than any other cattle breed, Finncattle has adapted to the Finnish environment and circumstances. It can be successfully reared as a forest pasture animal. Finncattle are known for their longevity and fecundity. The milk has a greater fat and protein content. However, compared to other breeds, milk output is much lower. Western Finncattle is beige-brown in colour, with white markings or patches. The biggest of the three Finncattle breeds is the Western Finncattle. Bulls stand 140 cm at the withers and weigh 625-800 kg, while cows average 123 cm in height and weigh around 470 kg. Currently, most Finnish native breeds are polled. The Western type was originally predominantly horned (Maki, 2009)

Simmental Cattle

The Simmental, often known as the Swiss Fleckvieh, is a dual-purpose cow breed (milk and meat) from Switzerland. It is named after the Simmental, a valley in Switzerland where the Simme river flows. The Simmental breed is one of the oldest and most extensively dispersed of all cattle breeds in the world, with records dating back to the Middle Ages. It has influenced the development of numerous other well-known European breeds. The Simmental's original coloration has been characterised as "red and white speckled" or "gold and white." Normally, the face is white. The majority of them have pigment around their eyes, which helps to prevent eye disorders caused by harsh sunshine (Department of Animal Science, 2021) .

English Longhorn

The White Longhorn, sometimes known as the British Longhorn, is a beef cow breed with long, curving horns. It originated in northern England. Colours vary, but the cattle are usually finched, with a thick line of white running down the spine, tail, and underside of the belly. The Longhorn was initially a sluggish, heavy draught animal used for ploughing; the milk output was not high, but the milk was rich in fat and was utilized for cheese-making in the eighteenth century, especially in Cheshire (Porter, Lawrence Alderson and Sponenberg, 2016).

Angeln cattle

The Angeln is a cattle breed that originated in Angeln, Schleswig-Holstein, where it was first recorded about 1600. Angeln cattle are red-coloured cattle that were one of the forerunners of the broader Danish Red Cattle breed. Their milk is known for its high milk fat content. Angeln cows produce an average of 7570 kg (16,700 lb) of milk with a fat content of 4.81 per cent(Department of animal science, 2021).

Fig. 12.2 Indigenous cattle breeds of Europe (All pictures used are under Creative Commons license)

Gyr Cattle

The Gir is a Zebu breed that originated in India. It's been utilised to develop other breeds in the area, including the Red Sindhi and the Sahiwal. The Gir has a striking look, with a rounded and domed forehead (the world's only ultraconvex breed), long pendulous ears, and horns that spiral out and back. Gir are speckled in colour, ranging from red to yellow to white. They originated in Gujarat, west India, and have subsequently expanded to neighbouring states (Porter, Lawrence Alderson and Sponenberg, 2016).

Vechur Cattle

The Vechur Cow is a rare Bos indicus bovine breed named after the village Vechoor in Kerala, India. It is the smallest cattle breed in the world, with an average length of 124 cm and a height of 87 cm, and is prized for the increased amount of milk it produces, compared to the amount of food it consumes. The Vechur cow needs very low-maintenance and less feed. [6] Because of the reduced fat globule size, the milk is said to have therapeutic properties and be simple to digest(NDDB, 2020).

ahiwal

Sahiwal is a Zebu cattle breed that is regarded as one of India's greatest milch cow breeds. The name of the breed comes from the Sahiwal region in Pakistan. They were utilised as draught animals also. Today, the Sahiwal is one of India's and Pakistan's greatest dairy breeds. It produces high quality milk. It has been shipped to other Asian nations and Africa, and the Caribbean because of their heat tolerance and excellent milk production. The Sahiwal breed is sold to a broad range of nations and areas due to its distinct traits. The Sahiwal breed came to Australia via New Zealand (Soi, 2021)

Red Chittagong cattle

Red Chittagong cattle (RCC), a native genotype of Bangladesh, are small and reddish in colour, with mature bulls and cows weighing 342 and 180 kg, respectively. from birth weights of 16 and 14 kg. While low mean milk production of 618 L was observed throughout a 228-day lactation, significant amounts of milk protein and fat were also documented, with supplied feed types being usually poor in nutritional value, particularly crude protein. Under farm conditions, however, one out of every five cows produces more than 1,000 L every lactation. Other important characteristics of this breed include resistance to common illnesses and parasites, as well as a high level of adaptability to agro-ecological settings (Gregorini *et al.*, 2021)

Fig. 12.3 Indigenous cattle breeds of Asia (All pictures used are under Creative Commons license)

150 cow breeds have been cataloged in Africa alone (Nierenberg and Sparling 2016). According to a global inventory of animal diversity, many of the world's indigenous livestock breeds are in risk of extinction as commercial varieties take control. According to livestock specialists, their elimination would result in the loss

American Angus

The American Angus is a beef cow breed from the United States. It is descended from the Aberdeen Angus population in Scotland; however, it may simply be black. The black colour, polled head, compact and low-set body and exquisite quality meat are all characteristics of the breed. The Angus is a top-tier beef breed, and for years, purebred or crossbred Angus steers have taken top honours in the world's most prestigious fat-stock events in the United Kingdom and the United States. This breed was first imported to the United States in 1873 (Britannica, 2022).

American Brahman

The Brahman is a zebuine-taurine hybrid beef cow strain from the United States. It was raised in the United States starting in 1885 from cattle that originated in India and were brought from the United Kingdom, India, and Brazil. The Brahman is resistant to parasites and has a strong tolerance for heat, sunshine, and humidity. In addition, Brahman cattle live longer than most other breeds, giving birth to calves as young as 15 years old (Porter, Lawrence Alderson and Sponenberg, 2016).

American Milking Devon

The name American Milking Devon comes from British North Devon cattle that were transported to North America in the 17th century. The American Milking Devon is one of the few true triple-purpose cow breeds in the West, with meat, milk, and draught all valued. The coat is a rich, glossy red hue, and the horns are white, ideally with black ends, on medium-sized cattle. They are greatly valued as oxen because they are energetic, clever, and quite powerful for their size. Physically, milking Devons are tough and capable of surviving on grass (FAO, 2017).

Texas Longhorn

The Texas Longhorn is a cow breed notable for its long horns, which may grow to be over 8 feet long. They are descended from cattle transported to the Americas from the Iberian Peninsula during Christopher Columbus' Second Voyage.The cattle have lived a semi-feral existence on the rangelands for hundreds of years, and they are more tolerant to heat and drought than other European breeds. Texas Longhorns are recognised for their varied colouring, which can be any hue or combination of colours, although dark red and white colouring is the most common (Porter, Lawrence Alderson and Sponenberg, 2016).

Fig. 12.4 Indigenous cattle breeds of America (All pictures used are under Creative Commons license)

of genetic resources that enable animals to resist sickness and drought, especially in developing countries (Hopkin 2007).

Various indigenous breeds of livestock that deserve protection (Nierenberg and Sparling 2016) are mentioned below:

Ankole-Watusi Cattle: A Sanga cattle, found throughout East Africa, initially introduced to the region about 2000 years ago. This breed is distinguished by its conspicuous long horns, which serve to dissipate body heat. These cattle can weigh between 1000 and 1600 pounds when fully matured.

Australian Braford	**Australian Brangus**

The Australian Braford is a beef cattle breed that was developed in Queensland between 1946 and 1952 as part of an effort to develop cattle who were resistant to cow ticks and could endure heat better than other breeds. Braford cattle have a slight hump, loose skin, and a red and white short coat .The breed's genetic base is around 50 percent Hereford and 50 percent Brahman(Corporation, 1989).	Australian Brangus is beef cattle established in the 1950s in Queensland's tropical coastal districts by crossbreeding Brahman and Angus cattle. The breed was developed to have a stronger tolerance to ticks and heat than other cow breeds. The cattle are mostly a sleek black colour. They have a low incidence of eye cancer, which is a concern in many white-faced breeds. Their nose and forehead are large, and their head is medium length (Friend and Denis Bishop, 1978). Brangus are also competent foragers and walkers, and they "do well" in a range of environments.
Australian Charbray	**Australian lowline**

The Australian Charbray was initially developed in the United States in the 1930s, then bred and brought to Australia in the late 1960s. The Charbray is a cross between the American Brahman and the French Charolais, having the American Brahman's hardiness and tick resistance and the French Charolais' lean beef qualities and gentle disposition (*Charbray Society of Australia Ltd - Charbray Breed Attributes.*, 2018). The Charbray lacks the characteristic hump of the Brahman, but it does have the characteristic loose skin and heavy dewlap underneath the throat (Briggs, H.M.; D.M., 1980). It has a huge body and a pale red to cream coat colour.	The Australian Lowline is a heritage breed descended from Aberdeen Angus. It has a gentle disposition, high-quality meat, easy calving, high fertility, breeding endurance, and feed efficiency. These are one of the tiniest cow breeds, but not quite a dwarf, and are predominantly black. This breed is unique in that it may be used in both production and hobby farming (Nicole, 2022).

Fig. 12.5 Famous cattle breeds of Australia (All pictures used are under Creative Commons license)

Brahma Chicken: According to the Livestock Conservancy, Brahma chickens were first produced in the United States from birds imported from China. The Queen of England received a collection of birds from the breeder George Burnham in 1852, sparking a rapid surge in popularity and value. The Brahma, which comes in three hues, lays eggs very effectively in the winter.

Light Sussex Chicken: The Light Sussex Chicken is well known in the United States and the United Kingdom for being a highly productive laying hen. According to research, Roman conquerors may have introduced the Sussex variety to England; it is named after the Sussex area in southern England. The Light Sussex Chicken was created in the nineteenth century when breeders crossed the Sussex chicken with a Mediterranean strain.

Venda Chicken: This chicken is native to South Africa's Limpopo Province, and it produces huge eggs that are usually white, black, or red in hue. The birds, who are skilled scavengers, eat various foods, including grass, kitchen garbage, and the odd small rodent. As a result, the breed produces high-quality eggs and meat.

Ossabaw Island Hog: Spanish Conquistadors left this kind of hog behind on Ossabaw Island, near Savannah, Georgia, over 500 years ago. According to research, the hogs may have originated in the West Indies, a key stop on trade routes between Europe and the Americas. The breed has developed a unique fat metabolic mechanism that permits it to store more fat rich in Omega-3 fatty acids than any other hog.

Gloucestershire Old Spot Pig: In the 1800s, breeders in Gloucestershire, England, produced this pig type. Black dots cover the white coats of this unusual breed, which is thought to be a crossbreed of the now-extinct Gloucestershire and Berkshire pigs. These pigs thrive in the open air and are outstanding foragers and grazers. Pigs were popular in the early twentieth century due to their lean flesh. However, since World War II, they have been classified as endangered since production has shifted to commercial cattle varieties.

Choctaw Hog: These hogs were bred in America and descended from Spanish stock, but they also have a long history with the Choctaw tribe, who retained the hogs after being forced to migrate. The pigs are tiny (about 120 pounds), quick, and athletic. The Choctaw is classified as severely endangered by the Livestock Conservancy. Only a few hundred animals remain, according to population estimates.

St. Croix Sheep: St. Croix sheep are native to the Virgin Islands, and their ancestors are most likely from the Caribbean Hair breed family, a hybrid between two kinds introduced to the Caribbean. The breed has a strong parasite resistance, high heat, and humidity tolerance and is ideally adapted to low-input meat production.

American Buff Goose: The American Buff Goose is a goose that originated in North America, although its exact history is unknown. It is descended from the wild Greylag goose. It yields dark, rich meat, as well as huge white eggs.

American Milking Devon Cattle: The American Milking Devon is a very unusual breed. It was brought to New England by immigrants in 1623 from Devonshire, England. The cattle were quickly adopted across England in the eighteenth and nineteenth centuries. However, when breeders and farmers sought superior beef cattle in the late 1800s, the breed began to lose popularity. Farmers started separating the Beef Devon from the Milking Devon.

American Plains Bison: Weighing in at over 2000 pounds, the American Plains Bison is distinguished by its enormous size and distinctive upper back hump. Despite nearly going extinct at the end of the 1800s, the USDA believes that over 200,000 American Plains Bison survive in private herds in the United States today.

Buckeye Chicken: This chicken has a rich red plumage and prefers free-range life. Buckeye Chicken hens were developed in Ohio and lay medium-sized brown eggs.

Bourbon Red Turkey: The Bourbon Red Turkey was founded in the late 1800s and is named after Bourbon County, Kentucky. The bird is now classified as endangered by the Livestock Conservancy. The flesh from the bird is rich and flavorful. It was popular as a commercial breed in the 1930s and 1940s before being phased out, favouring broad-breasted alternatives.

Cotton Patch Goose: During the 1900s, the Cotton Patch Goose was a multi-purpose bird in the United States. Farmers utilized the birds to weed cornfields and cotton patches, as the name indicates. They provide delectable meat and eggs. During the Great Depression, they were a vital source of income for corn and cotton growers.

Guinea Hog: Guinea Hog: This dark-coloured breed of pig was introduced to the United States as part of the slave trade from West Africa and the Canary Islands. The Livestock Conservancy considers them vulnerable, even though they were historically the most numerous pig species in the Southeast. The guinea hog is a tiny animal that weighs between 100 and 300 pounds.

Gulf Coast Sheep: Immigrants brought this breed of sheep to the New World in the 1500s. In the Deep South, the animals were a major source of wool and meat for domestic usage. The Gulf Coast Sheep has a high parasite resistance. But the rise of antibiotics and more productive sheep breeds is a significant danger to the breed's survival.

Texas Longhorn Cattle: When Spanish settlers established missions in the Southwest, they brought these cattle with them. This unique breed of cow was used for meat, labour, tallow, and skins. According to recent study from the University of Texas, the cows transported from Spain have connections with Middle Eastern and Indian varieties.

Karakul Sheep: The Karakul Sheep are a Central Asian sheep breed, named a high-altitude town where food and water are limited. The pelts of the breed's lambs, commonly made into lambskins, are prized, as is the high-quality wool produced from the coats of older animals.

Kolbroek Pigs: While the actual origins of this pig breed are unknown, they have lived in South Africa for decades and are considered indigenous. Kolbroek pigs are superb foragers, swimmers, and producers of fat in large numbers.

Molo Mushunu Chicken: This chicken is historically grown by the Kikuyu people in the Molo area, in Kenya's Rift Valley. They are famous for their succulent meat and eggs. Their lengthy bodies expand into a featherless head and neck, giving them an unusual look.

Navajo-Churro Sheep: In the sixteenth century, Spanish explorers brought the Navajo-Churro sheep to the southwestern United States. The breed is commonly considered as the New World's first domesticated sheep. The animals are versatile and are utilized for their high-quality wool. The Navajo people adopted the Churro sheep, whose wool they utilized for weaving, as a result of commerce and fighting. The Navajo Sheep Project is now working to revive the breed.

N'Dama Cattle: This breed comes from Guinea's Fouta Djallon mountains. Their meat has a good taste and is minimal in fat. In addition, heat and humidity tolerance is remarkable in this breed. To date, there are roughly seven million N'Dama cattle in West Africa.

Nguni Cattle: Nguni cattle flourish in southern Africa, where community farmers have traditionally relied on them. According to the FAO, few purebred animals are left; many were mated with commercial varieties in the twentieth century. The survival of the Nguni cattle is primarily due to the efforts of rural agricultural communities.

Plymouth Rock Chicken: The Plymouth Rock chicken is the traditional American breed: a crossbreed of Dominiques and Black Javas established in the mid-nineteenth century. The birds have originated in the Boston region, while their exact origin is unknown. Plymouth Rock chickens are dual-purpose chickens, which means they are raised for both meat and eggs.

Red Maasai Sheep: Maasai pastoralists in Kenya and Tanzania are the primary shepherds of the Red Maasai sheep, which are native to West Africa. Sheep are resistant to worms and other illnesses in particular.

Randall Lineback Cattle: Although the actual origins of Randall Lineback cattle are unknown, research shows that they are a cross between Dutch, English, and French breeds. The breed was common in the north-eastern United States, but farmers and breeders crossed the animals with Holsteins in the twentieth century, resulting in the extinction of the breed. For 80 years, the Randall family in Vermont had kept a purebred line of Lineback cattle; today, just a small percentage of that herd exist.

Rhode Island Red Chicken, Old Type: The Rhode Island Red is a reddish-brown bird that is another famous American poultry. Each year, the chickens lay 200–300 eggs. The bird was created in New England as a dual-purpose bird. Hens lay big, spherical eggs.

Royal Palm Turkey: It is a white and black turkey with a red or blue head. Males weigh around 16 pounds on average, while females weigh about 10 pounds. The Royal Palm is classified as endangered by the Livestock Conservancy. In the United States, less than 1000 breeding birds survive.

Tennessee Fainting Goat: These goats have myotonia congenita, a syndrome that causes the goat's muscles to tighten when it is frightened, causing them to fall over. So, they have low body fat due to their condition, resulting in mild meat.

American Rabbit: These rabbits are native to North America and were produced in California. These rabbits were plentiful until the 1950s when they were killed for their meat and fur. The American Rabbit is now the rarest rabbit breed in the United States. The rabbit is classified as critically endangered by the Livestock Conservancy.

The Silver Fox Rabbit: This was created by Ohio breeder Walter B. Garland for its high-quality meat and silver-coloured fur. The rabbits are huge, weighing between 10 and 12 pounds. In the United States, the Silver Fox Rabbit is currently endangered.

Wyandotte Chicken: Breeders called the Wyandotte Chicken after a regional Native American tribe when it was developed in New York in the 1800s. The birds with a large body are poor flyers. The bird was delisted from the Livestock Conservancy's priority list in 2016. It is currently considered a resurgent breed.

Gold and Silver Sebright Chicken: These birds are generally used as ornaments. Both the gold and silver variants have black interwoven feathers. Sebright chickens lay tiny white eggs and are good layers. They are one of the most popular bantam breeds in the United States. John Sebright, the breeder, is the name given to the birds.

Dutch Gold Chicken: The Dutch Gold Chicken is a bantam breed that originated in the Netherlands. The birds most likely came to the Netherlands via Dutch East Indies trading lines. The chickens lay many small eggs, which farmers are supposed to have evolved as a ploy to escape Dutch landlords' insistence that all large eggs be turned over as rent.

12.4 Conservation of Landraces and Indigenous Breeds

The agrobiodiversity heritage and the Traditional Agricultural Knowledge extracted from millennia of experiences are vanishing fast in all countries, owing to forces of innovative development. This Traditional Knowledge documents the consequences of various agronomic and cultural practices that enhance biological and food web variety. It ensures long-term crop yield stability, agroecosystem resilience, and nutritional security for farmers (Alvares, and Multiversity (Organization), and Citizens International (Organization) 2017).

Carlos Seré, Director General of the International Livestock Research Institute in Nairobi, Kenya, claims that local breeds, which make up about 70% of the world's population, are frequently more adapted to their environments than commercially marketed animals with high yields and short-term profitability. For example, red Maasai sheep are naturally immune to intestinal parasites, while Uganda's indigenous Ankole cattle are drought tolerant (Hopkin 2007).

According to Irene Hoffmann, the previous head of the FAO's Animal Genetic Resources programme, diverse indigenous livestock breeds might be necessary for food security in climate change, illness, hunger, drought, and land degradation (Nierenberg and Sparling 2016). As contemporary crop varieties fail to outperform local landraces in marginal environmental conditions, conserving, nurturing, and utilizing ancestral agrobiodiversity and the wealth of local landraces is our most excellent shot for achieving food security in the wake of changing climate disasters (Alvares, and Multiversity (Organization), and Citizens International (Organization) 2017). In addition, farmers and breeders may be better able to adapt and respond to the ever-changing environment if they have access to a broader range of genetic material (Nierenberg and Sparling 2016).

Many farmers' groups and farming and pastoral communities worldwide are working to support the conservation and cultivation of indigenous livestock breeds. Earlier, farmers in Thailand were persuaded to crossbreed their indigenous Kao Lamphun cow with a western breed. Farmers realized after multiple generations that the crossbreed was unable to breed consistently and did not gain weight fast enough following the dry season. Farmers formed a cooperative and petitioned the Government for the return of their Kao Lamphun livestock. The Government consented, and the size of the farmers' herds has now tripled (Keith 1999).

In Vietnam, the Government introduced the I pig, an endangered indigenous breed, to women who maintained pigs to augment their family's income. Farmers preferred larger breeds that produced more meat for the market because the I pig was little. The I pig can live on a diet of sweet potato leaves and rice bran in substandard housing and is disease resistant. It is also easier for women to handle because it is tiny. The women are now alternating litters. One remains a pure breed, while the other is mixed with a giant breed to produce marketable progeny. As a result, the pure breed is preserved, and women's wage levels are raised (Keith 1999).

People are attempting to save cattle in communities all across the world. For example, northern Kenya's Samburu Pastoralists breed cattle, camels, goats, and sheep, shifting from place to place according to rainfall patterns. This approach can aid communal drought resistance while also honouring indigenous peoples' cultural and practical significance. This technique can help community drought resistance while keeping indigenous animal breeds' cultural and practical importance.

Veterinarians are now working on projects to train university students and smallholder farmers on animal husbandry skills and human health in Liberia, Uganda, and Ethiopia. The Turing Foundation provides grants and funds to encourage sustainable livestock and agricultural practices in sub-Saharan Africa, Southeast Asia, and the Pacific Islands. The Africa Centre for Holistic Management advocates cattle use in land restoration and conservation initiatives through on-site training and teaching, thanks to its focus on holistic systems. In addition, Slow Food International's Ark of Taste database archives foods of cultural importance on the verge of extinction.

Crop genetic resources are essential components of biodiversity. However, with the large-scale promotion of monocropping, genetic diversity has largely been lost. Ex situ conservation approaches were widely used to protect traditional crop varieties worldwide. However, this method fails to maintain the dynamic evolutionary processes of crop genetic resources in their original habitats, leading to genetic diversity reduction and even loss of the capacity of resistance to new diseases and pests. Therefore, on-farm conservation has been considered a crucial complement to ex situ conservation (Wang et al. 2016).

According to the FAO, national investment in gene banks and information systems to preserve animal genetic material has increased overall. The Livestock Conservancy, situated in North Carolina, is a national network and resource centre for farmers and breeders. The SVF Foundation, established in Rhode Island, is dedicated to preserving genetic material from uncommon livestock and plant varieties (Nierenberg and Sparling 2016).

The Food and Agriculture Organization of the United Nations published *Voluntary Guidelines for the Conservation and Sustainable Use of Farmers Varieties/ Landraces* on 14 November 2019 to assure crop diversity and farmers' resilience to the loss of plant genetic resources worldwide. These guidelines would help countries develop national plans to conserve their critical crop resources. Proposed recommendations include identifying and documenting existing genetic resources for farming, mapping their potential uses, promoting their preservation, and providing farmers and local communities with support and information, according to national scenarios and preferences. International organizations and instruments, such as the Convention on Biological Diversity (CBD), the International Treaty on Plant Genetic Resources for Food and Agriculture, and the Second Global Plan of Action for PGRFA, supported establishing this systematic approach to conserve farming practices. However, these voluntary guidelines are the first to set out such an approach.

In Europe, landrace variety is a valuable resource for maintaining food security and safeguarding significant economic gains from crop production. Crop wild relatives (CWRs) have sparked a surge in scientific interest, highlighting their potential for crop development, notably mitigating the effects of climate change and contributing to global food security (Biodiversity International, University of Birmingham, National Institute of Agricultural Botany (Regne Unit), Eucarpia 2014). Recent advances in understanding this diversity in the area and planning for their complementary conservation provided a solid foundation for developing a strategic approach to their preservation, based on a range of widely agreed and tested scientific concepts and techniques (Kell et al. 2016).

A European Genetic Resources Strategy was launched by European Forest Institute, enriched by particular strategies for plants, animals, and forests, on 30 November 2021. The initiatives aimed to improve European genetic resource conservation and sustainable usage and the foundation of sustainable agriculture. Genetic resources are also necessary for irreplaceable ecosystem services, like clean water and rich soils. These plans are in response to the European Commission's demand to "create a framework in which the existing mosaic of European, national/ regional entities may join in developing and executing ambitious methods and strategies for managing the crop, forest, and animal genetic resources" (Huyghe 2021; EFI 2021).

The European Genetic Resources Strategy stands out for its ability to translate policy goals into action with a well-set plan. The major three action plans are represented in Figs. 12.6, 12.7, and 12.8. It adds to existing individual agricultural and forest management practices. It provides an integrated framework for increasing agro and forest biodiversity and human and animal genetic resources.

12.5 Conservation Of Indigenous Breeds: Indian Scenario

India is one of the world's 12 mega-biodiverse nations, accounting for around 12% of the worldwide livestock population. The National Bureau of Animal Genetic Resources (NBAGR) of the Government of India aims to conserve and protect indigenous farm animal genetic resources for long-term use and livelihood security. It ensure sustainable management and utilization of livestock and poultry diversity in

Fig. 12.6 Actions proposed for reinforcing conservation and sustainable use of genetic resources (Adapted from Genetic Resource Strategy for Europe EFI 2021)

India through a distinct work strategy, including characterization, documentation, conservation, and evaluation of indigenous breeds (ICAR-NBAGR News Letter 2020).

With a goal of zero non-descript AnGR (Animal Genetic Resources), the country's livestock and poultry population are being surveyed and documented as a primary essential step. They carry out the registration and notification of all sorts of livestock and poultry populations in the country, including breeds, lines, and strains. Conservation is done at three different levels: (1) Conservation of native livestock and poultry species, (2) In situ conservation of threatened breeds, and

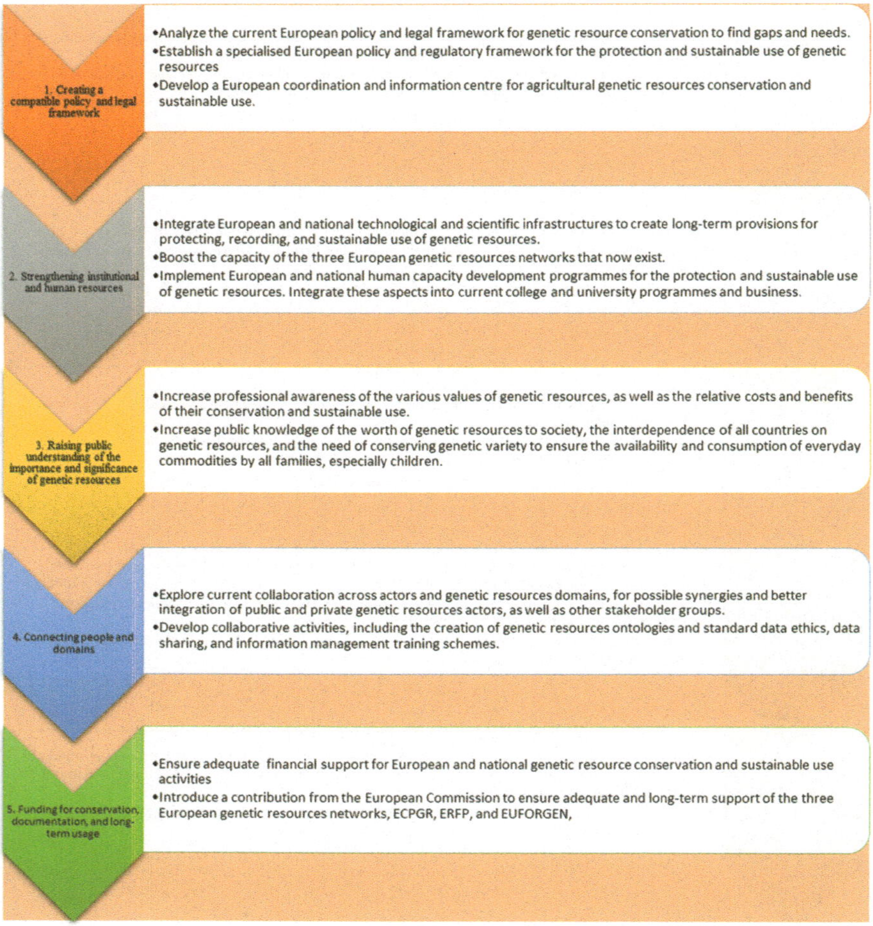

Fig. 12.7 Actions proposed for enabling transformative change for the conservation of genetic resources (Adapted from Genetic Resource Strategy for Europe EFI 2021)

(3) Cryopreservation of germplasm (semen), somatic cells, and DNA of all registered breeds (ICAR-NBAGR News Letter 2020).

Cryopreservation of germplasm, characterization of indigenous, lesser-known populations, identification of bioactive peptides in animal products, and development of assays to detect genetic disorders in bovines are a few projects underway. They use cutting-edge genomic techniques to examine genomic variability and uniqueness of all registered breeds and build a molecular signature of native species for breed standard development based on whole genome-based markers. In addition, transcriptome and metabolome analyses are carried out to assess heat adaptability and sickness tolerance. The establishment of a National Bovine Genome Centre for

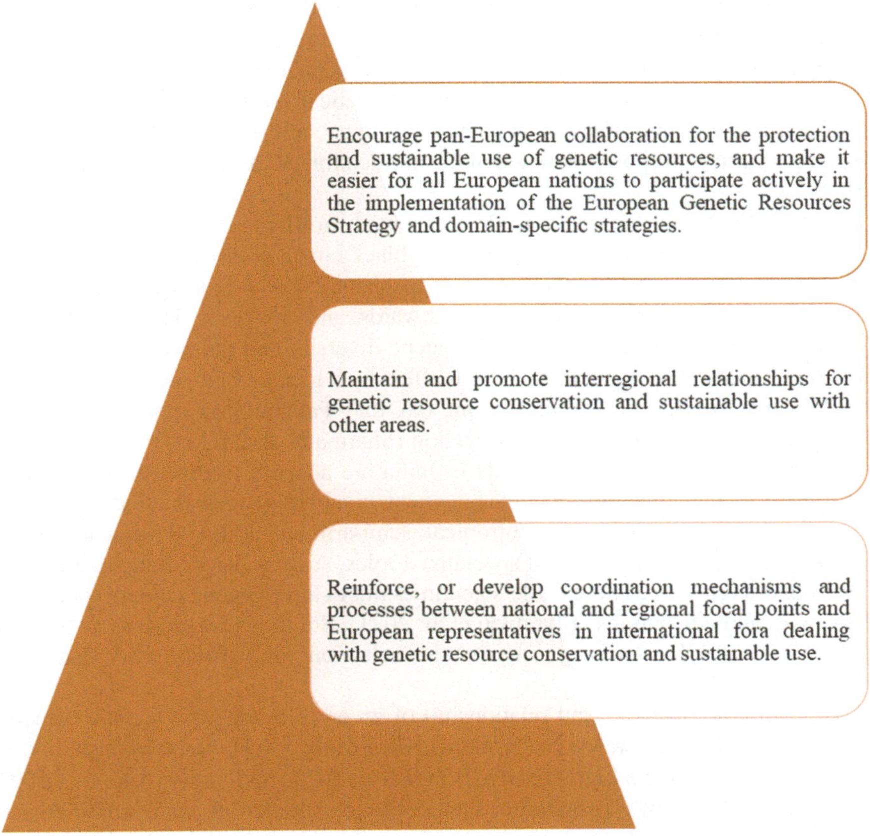

Fig. 12.8 Measures proposed for strengthening international collaboration change for the conservation of genetic resources (Adapted from Genetic Resource Strategy for Europe EFI 2021)

indigenous breeds and the creation and certification of DNA Chips for native cattle, buffalo, and other species for genetic improvement are other objectives of NBAGR.

Another critical action is assessing native AnGR based on social, cultural, economic, and non-marketing values. It is carried out by identifying biomolecules in milk, meat, and other products obtained from native germplasm and their importance for human nutrition and health and developing tests and consulting services to screen genetic disorders in livestock species. Organizing training and sensitization programmes, developing, supporting and assessing policies, and creating databases and other ICT for geo-reference, phenotype, genomics, and survey on AnGR are the capacity-building strategies for the management and conservation of AnGR (ICAR-NBAGR News Letter 2020).

NBAGR is the key entity for registering newly identified animal genetic resources. India is home to nine indigenous dog populations, primarily used for farm guarding and shepherding. However, the characterization of most populations has remained incomplete due to the lack of appropriate methodology specific to dogs (Raja et al. 2016). Rajapalayam and Chippiparai dog breeds of Tamil Nadu and Mudhol Hound of Karnataka are India's first-ever registered dog breeds (Registered Dog—ICAR—National Bureau of Animal Genetic Resources 2020). They were designated as declared breeds for the entire country, allowing them to be kept and reared for various purposes as stated in the notification (ICAR-NBAGR News Letter 2020).

Palamu/Medini is a small-sized goat breed from the Jharkhand state tract (Palamu, Latehar, and Garhwa). They have black coat colour throughout the body. The head shape is slightly convex, the ears are slanted downwards at around 45°, and the horns are straight and oriented backwards and upwards. The Palamu goat possessed a lot of genetic diversity, like gene diversity and the average number of alleles per locus. The population did not affect by a genetic bottleneck over the last 40–80 generations. The findings point out that Palamu goat population has a significant genome in terms of conservation (Sharma et al. 2021).

All the 50 registered cattle breeds in India are adapted to diverse agroclimatic environments and breeding and management practices (Breedslnddb.coop 2022). These breeds flourish in humid subtropical, semi-arid and arid, and tropical wet/dry climates and are suited to several specialized roles, such as dairy, draught, and dual (Dairy & draught) use. These cattle's genomes might have genetic content specific to their tasks and adaptations (Bhardwaj et al. 2021). So, they can serve as a valuable genetic resource for locating genes under selection for various traits that have evolved.

The genomic diversity and relationship of seven different cow breeds, namely, Sahiwal, Tharparkar, Gir, Vechur, Ongole, Kangayam, and Hariana, were studied in random samples using a high-density (Bohra et al. 2022) SNP array. A total of 1993 SNPs with greater power to distinguish these bovine populations were identified and then used to divide each animal's genome into a predetermined number of clusters. Because of its crossbreeding with other breeds, the Vechur population has the highest admixture rate. On the other hand, the Ongole, Kangayam, and Hariana breeds shared only 15% of their genome with other breeds. Even though the studies reveal that all seven breeds emerged from separate ancestors, there was a recent intermixing of these breeds (Dixit et al. 2021).

Several genes associated with immune response, metabolic pathway, glucose and sugar transportation, signalling pathways, cellular processes, cell division, and glycolysis regulation were identified as selection signatures between draught (Kangayam) and dairy breeds. Compared to large-sized cattle, there was considerable variation in the number of QTLs affecting productivity (body weight, growth rate, etc.) and physical features (height) in the short-statured Vechur breed. A QTL/Quantitative Trait Locus is a DNA region associated with a specific phenotype that changes in magnitude through polygenic influences. As a result, they can serve as a valuable genetic resource for finding genes under selection for various qualities that have evolved and determining genetic diversity (Dixit et al. 2021).

In the overlapping regions of north-eastern India, a set of polymorphic SNPs were identified exclusively to distinguish between Indian swamp and riverine buffaloes.

Indian swamp buffaloes had genetic variations that could be identified. The identified probable areas under selective sweeps in swamp buffaloes appear to be related to production and immunological modulation (Kumar et al. 2020). In the Chambal district of Madhya Pradesh, a distinct buffalo population has been discovered in ravine areas. These buffaloes are well adapted to the ravine's scorching climate, scant vegetation, and stony landscape, and they produce a lot of milk. Blood samples were obtained from 11 native "Dang" buffaloes and 11 DNA samples were custom sequenced using Double digest restriction-site associated DNA sequencing (ddRAD) with determining the genetic diversity (ICAR-NBAGR News letter 2020).

Camel is a fascinating part of Indian desert ecosystem. After three decades of reduction, the camel population has reached a critical stage, prompting great anxiety among policymakers. More than a quarter of Indian camels have yet to be identified. It is critical to describe the country's camel germplasm and current diversity to develop a conservation strategy. Sindhis, who live along the Pakistani border, are one such population. Twenty-five microsatellite markers, helpful in measuring genetic diversity, were used to investigate genetic variability and relationships between Sindhi and two Indian camel breeds, Marwari and Kharai. Studies on genetic distances, phylogenetic relationship, correspondence analysis, and clustering method based on the Bayesian approach and individual assignment supported the distinctness of three camel populations. Sindhi camels were readily distinguished from two registered Indian camel breeds. Sindhi appears to be a distinct gene pool, according to the findings. Considerable genetic variation provides a promising outlook for the survival of severely diminishing indigenous camel populations with suitable planning measures to conserve existing genetic variation and avoid inbreeding (Sharma et al. 2020).

Kadaknath is a unique Indian poultry breed that originated in the Jhabua area of Madhya Pradesh (MP). Among India's 19 different chicken breeds, Kadaknath is the only all-black bird. Its meat has energizing and therapeutic effects and is used as traditional medicine. Consumers are becoming more interested in the potential for pharmaceutical advantages. However, scientific data are still needed to back up these statements. As a result, Kadaknath meat was investigated and compared to conventional Cobb-400 broiler meat to identify functional features that may contribute to the lauded benefits. Several in vitro approaches were used to test meat extracts (Sehrawat et al. 2021). More than one in vitro study confirmed Kadaknath meat's enhanced antioxidant capacity. The superior functional characteristic of Kadaknath meat was firmly determined by free radical scavenging assays and metal chelation capabilities. It also had a substantially stronger inhibitory effect on the production of advanced glycation end products, which are linked to age-related and diabetes problems. These experiments added value to the distinctive Indian Kadaknath chicken and its nutritional and functional qualities (Sehrawat et al. 2021).

Barbari is a popular dual-purpose goat breed that has adapted well to the Gangetic plains and is utilized for both meat and milk. Chevon goats from Barbari are of high-quality, and this breed is recommended for commercial goat farming. The Changthangi goat is characterized by high fibre. It thrives in Ladakh, a frigid desert region in the north. The obvious anatomical differences between these two breeds could explain the genetic variations that underpin muscle properties. Therefore, the variations of genetic expression in longissimus thoracis muscles between meat and wool-type Indian goat breeds were analysed. In both Barbari and Changthangi goats, more than 93% of data were of high quality. However, there was a noticeable distinction between the two breeds when it came to the expression of genes related to lipid metabolism. The Changthangi goat's highly interacting genes were mostly linked to muscle fibre type. The research sheds light on the differential expression of genes in Barbari and Changthangi goat breeds, which could be a result of genetic selection or adaptability (Kumar et al. 2021).

If it is no longer possible to simply rebuild a breed population, it is classified as extinct. When there are no breeding males and stored sperm, no breeding females (oocytes), and no embryos left, this scenario becomes extreme (Scherf 2000). Even if no live animals are available, the availability of adequate cryopreserved material could allow for the rebuilding of a breed. Because a tiny number of live creatures reflects a very little amount of genetic information, extinction may occur well before the loss of the last animal, gamete, or embryo (Pimm et al. 2015).

The ICAR-NBAGR germplasm repository is being strengthened by keeping a variety of germplasm types (semen, embryos, DNA, epididymal sperm, and somatic cells). Between July and December 2020, 350 semen doses of Nili Ravi buffaloes were procured for long-term cryopreservation in Gene Bank. The attachment cell culture approach was used to create cell lines from marginal ear tissues from four populations of three species, followed by cell cryopreservation. Mewari and Jalori camels, Halari donkey, and Zanskari horse are the breeds selected. The explant culture technique was used to establish the primary culture. The newly generated indigenous animal cell lines conserve rich genetic resources at the cellular level and provide invaluable materials for genomic, post-genomic, and somatic cell cloning research.

Routine karyotype screening of animals for reproduction is required due to the possibility of spontaneous chromosomal aberrations and the rapid spread of genetic diseases in the population, especially under artificial insemination circumstances. The cytogenetic screening programme, which uses constantly improved diagnostic tools, enables precise karyotype evaluation, chromosomal abnormality detection, and the construction of selection recommendations (Danielak-Czech et al. 2020) .

Recently, cytogenetic screening to detect chromosomal abnormalities was started for bovine males reared for semen production at various centres of NBAGR. Close coordination between breeders and veterinarians may be beneficial in avoiding the financial ramifications of faulty animal reproduction (Holečková et al. 2021).

Establishing breed conservation centres in native breed tracts for in situ conservation and a national gene bank, DNA bank, and somatic cell bank at the institute level for ex situ conservation is a significant step towards conserving economically important and environmentally well-adapted breeds. The NBAGR frequently conducts numerous national and international trainings, exhibitions, and awareness programmes for researchers, field workers, State AH officials, and livestock keepers.

12.6 Conclusion

Restoring and reconstructing are the needs of the hour to protect our invaluable landraces from their vanishing trends. Restoring means safeguarding the landrace gene pool from erosion. It envisages preserving a breed's genetic potential and its administration for future usage. Identification, characterization, assessment, documentation, and conservation are essential aspects of restoring. Furthermore, reconstructing is significant for the species which have a critical existence. In that scenario, the preservation of genetic resources is crucial for tackling the problem and ensuring the existence of indigenous breeds and landraces for future generations.

To achieve a transformative change, we need to establish a worldwide policy and legal framework. Many countries could revise existing model strategies on the go. Such a drive needs empowerment and management of human and institutional resources and collaboration between policymakers and farmers. More communications and awareness of the roles and values of germplasms of landraces and fund mobilization are essential components of this move.

Implementation of restoring and reconstructing strategies will minimize the loss of landraces, contributing to climate-resilient agriculture. In addition, it will preserve precious nutritional resources for future generations. Ensuring the conservation and sustainable use of landraces is a worldwide issue that can be solved by setting strategies with global standards.

References

Akinola R, Pereira LM, Mabhaudhi T, de Bruin FM, Rusch L (2020) A review of indigenous food crops in Africa and the implications for more sustainable and healthy food systems. Sustainability 12(8):3493. https://doi.org/10.3390/SU12083493

Alvares CA, Multiversity (Organization), & Citizens International (Organization) (2017) Chapter 6. Folk rice varieties, traditional knowledge and nutritional security in South Asia. In: Agroecology, ecosystems, and sustainability in the tropics. Studera Press, New Delhi, p 283. https://www.academia.edu/33845663/Folk_Rice_Varieties_Traditional_Knowledge_and_Nutritional_Security_in_South_Asia

Australian Cattle|Australian Butchers Guild (2022). https://www.australianbutchersguild.com.au/supply-chain/on-the-farm/Cattle/

Azeez MA, Adubi AO, Durodola FA (2018) Landraces and crop genetic improvement. In: Rediscovery of Landraces as a resource for the future. https://doi.org/10.5772/INTECHOPEN.75944

Balfourier F, Bouchet S, Robert S, DeOliveira R, Rimbert H, Kitt J, Choulet F, Paux E (2019) Worldwide phylogeography and history of wheat genetic diversity. Sci Adv 5(5):eaav0536. https://doi.org/10.1126/SCIADV.AAV0536

Barragy TJ, Dobiec JF, James F (2002) Gathering Texas gold: J. Frank Dobie and the men who saved the Longhorns, p 259

Bellon MR (2004) Conceptualizing interventions to support on-farm genetic resource conservation. World Dev 32(1):159–172. https://doi.org/10.1016/J.WORLDDEV.2003.04.007

Bhardwaj S, Singh S, Ganguly I, Bhatia AK, Bharti VK, Dixit SP (2021) Genome-wide diversity analysis for signatures of selection of Bos indicus adaptability under extreme agro-climatic conditions of temperate and tropical ecosystems. Anim Gene 20:200115. https://doi.org/10.1016/J.ANGEN.2021.200115

Biodiversity International, University of Birmingham, National Institute of Agricultural Botany (Regne Unit), Eucarpia (2014) International conference on enhanced genepool utilization capturing wild relative and landrace diversity for crop improvement. Cambridge, UK, 16–20 June 2014: book of abstracts. Biodiversity International

Bohra A, Kilian B, Sivasankar S, Caccamo M, Mba C, McCouch SR, Varshney RK (2022) Reap the crop wild relatives for breeding future crops. Trends Biotechnol 40:412–431. https://doi.org/10.1016/J.TIBTECH.2021.08.009

Breedslnddb.coop (2022). https://www.nddb.coop/services/animalbreeding/geneticimprovement/breeds

Browett S, McHugo G, Richardson IW, Magee DA, Park SDE, Fahey AG, Kearney JF, Correia CN, Randhawa IAS, MacHugh DE (2018) Genomic characterisation of the indigenous Irish Kerry cattle breed. Front Genet 9(FEB):51. https://doi.org/10.3389/FGENE.2018.00051/BIBTEX

Brush SB (1995) In situ conservation of landraces in centers of crop diversity. Crop Sci 35(2): 346–354. https://doi.org/10.2135/cropsci1995.0011183x003500020009x

Cardinale BJ, Duffy JE, Gonzalez A, Hooper DU, Perrings C, Venail P, Narwani A, MacE GM, Tilman D, Wardle DA, Kinzig AP, Daily GC, Loreau M, Grace JB, Larigauderie A, Srivastava DS, Naeem S (2012) Biodiversity loss and its impact on humanity. Nature 486(7401):59–67. https://doi.org/10.1038/nature11148

CBD (2020) CBD—Global Biodiversity Outlook 5. Full report. www.emdashdesign.ca

CGRFA (2012) Report of a consultation on the definition of breed categories. http://www.rfp-europe.org/fileadmin/SITE_ERFP/TF_Risk/TFRisk_Tekirdag2011.pdf

Christiaan Struik P, Edward Bradshaw J, Vassilis Papasotiropoulos U, Prohens J, Casañas F, Simó J, Casals J (2017) Toward an evolved concept of landrace. Front Plant Sci 8:145. https://doi.org/10.3389/fpls.2017.00145

Danielak-Czech B, Kozubska-Sobocińska A, Smołucha G, Babicz M (2020) Breeding and economic aspects of cytogenetic screening studies of pigs qualified for reproduction. Animals 10(7):1–14. https://doi.org/10.3390/ANI10071200

Davis KF, Chiarelli DD, Rulli MC, Chhatre A, Richter B, Singh D, DeFries R (2018) Alternative cereals can improve water use and nutrient supply in India. Sci Adv 4(7):eaao1108. https://doi.org/10.1126/SCIADV.AAO1108/SUPPL_FILE/AAO1108_TABLESS1_TO_S10.XLSX

Davis KF, Chhatre A, Rao ND, Singh D, Ghosh-Jerath S, Mridul A, Poblete-Cazenave M, Pradhan N, DeFries R (2019) Assessing the sustainability of post-green revolution cereals in India. Proc Natl Acad Sci U S A 116(50):25034–25041. https://doi.org/10.1073/PNAS.1910935116/-/DCSUPPLEMENTAL

Dixit SP, Bhatia AK, Ganguly I, Singh S, Dash S, Sharma A, Anandkumar N, Dang AK, Jayakumar S (2021) Genome analyses revealed genetic admixture and selection signatures in Bos indicus. Sci Rep 11:21924. https://doi.org/10.1038/s41598-021-01144-2

EFI (2021) Genetic resources strategy for Europe

Eliazer Nelson ARL, Ravichandran K, Antony U (2019) The impact of the green revolution on indigenous crops of India. J Ethnic Foods 6(1):1–10. https://doi.org/10.1186/S42779-019-0011-9/TABLES/2

European Union (2018) A sustainable bioeconomy for Europe: strengthening the connection between economy, society and the environment updated bioeconomy strategy. https://doi.org/10.2777/478385

Felius M, Beerling ML, Buchanan DS, Theunissen B, Koolmees PA, Lenstra JA (2014) On the history of cattle genetic resources. Diversity 6(4):705–750. https://doi.org/10.3390/D6040705

Figueroa JV, Chieves LP, Johnson GS, Buening GM (1992) Detection of Babesia bigemina-infected carriers by polymerase chain reaction amplification. J Clin Microbiol 30(10): 2576–2582. https://doi.org/10.1128/JCM.30.10.2576-2582.1992

Groeneveld LF, Lenstra JA, Eding H, Toro MA, Scherf B, Pilling D, Negrini R, Finlay EK, Jianlin H, Groeneveld E, Weigend S (2010) Genetic diversity in farm animals—a review. Anim Genet 41(1):6–31. https://doi.org/10.1111/j.1365-2052.2010.02038.x

Hanotte O, Tawah CL, Bradley DG, Okomo M, Verjee Y, Ochieng J, Rege JEO (2000) Geographic distribution and frequency of a taurine Bos taurus and an indicine Bos indicus Y specific allele amongst sub-Saharan African cattle breeds. Mol Ecol 9(4):387–396. https://doi.org/10.1046/j.1365-294X.2000.00858.x

Harlan JR (1975) Our vanishing genetic resources. Science (New York, NY) 188(4188):618–621. https://doi.org/10.1126/SCIENCE.188.4188.617

Hiemstra SJ, De Haas Y, Mäki-Tanila A, Gandini G (2010) Local cattle breeds in Europe. In: Local cattle breeds in Europe. Academic Publishers, Wageningen. https://doi.org/10.3920/978-90-8686-697-7

Ho SYW, Larson G, Edwards CJ, Heupink TH, Lakin KE, Holland PWH, Shapiro B (2008) Correlating Bayesian date estimates with climatic events and domestication using a bovine case study. Biol Lett 4(4):370–374. https://doi.org/10.1098/RSBL.2008.0073

Holečková B, Schwarzbacherová V, Galdíková M, Koleničová S, Halušková J, Staničová J, Verebová V, Jutková A, Bagnicka E, Cinelli P (2021) Chromosomal aberrations in cattle. Genes 12:1330. https://doi.org/10.3390/genes12091330

Hopkin M (2007) Local livestock breeds at risk. Nature News. https://doi.org/10.1038/news070903-2

Huyghe C (2021) Sustainable support for plant, animal and forest genetic resources. http://www.genresbridge.eu/genetic-resources-strategy-for-europe/press-release/

ICAR-NBAGR News letter (2020)

Indigenous African Agriculture (2021) 216–235. https://doi.org/10.1002/9780891186342.CH9

Keith H (1999) The value of indigenous animal breeds. https://www.fao.org/NEWS/1999/agr-e.htm

Kell SP, Ford-Lloyd BV, Maxted N (2016) Europe's crop wild relative diversity: from conservation planning to conservation action. In: Enhancing crop genepool use: capturing wild relative and landrace diversity for crop improvement. CABI, Wallingford, pp 125–136. https://doi.org/10.1079/9781780646138.0125

Kumar R, Joel Devadasan M, Surya T, Vineeth MR, Choudhary A, Sivalingam J, Kataria RS, Niranjan SK, Tantia M, Verma A (2020) Genomic diversity and selection sweeps identified in Indian swamp buffaloes reveals it's uniqueness with riverine buffaloes. Genomics 112(3):2385–2392. https://doi.org/10.1016/J.YGENO.2020.01.010

Kumar A, Kaur M, Ahlawat S, Sharma U, Singh MK, Singh KV, Chhabra P, Vijh RK, Yadav A, Arora R (2021) Transcriptomic diversity in longissimus thoracis muscles of Barbari and Changthangi goat breeds of India. Genomics 113(4):1639–1646. https://doi.org/10.1016/J.YGENO.2021.04.019

Liu CLC, Kuchma O, Krutovsky KV (2018) Mixed-species versus monocultures in plantation forestry: development, benefits, ecosystem services and perspectives for the future. Glob Ecol Conserv 15:e00419. https://doi.org/10.1016/J.GECCO.2018.E00419

Lodhi SS, Maryam S, Rafique K, Shafique A, Yousaf ZA, Talha AM, Gul A, Amir R (2020) Overview of the prospective strategies for conservation of genomic diversity in wheat landraces. In: Climate change and food security with emphasis on wheat, pp 293–309. https://doi.org/10.1016/B978-0-12-819527-7.00021-2

Lost crops of Africa (2008) 3:1–354. https://doi.org/10.17226/11879

McTavish EJ, Decker JE, Schnabel RD, Taylor JF, Hillis DM (2013) New world cattle show ancestry from multiple independent domestication events. Proc Natl Acad Sci U S A 110(15):E1398–E1406. https://doi.org/10.1073/PNAS.1303367110/SUPPL_FILE/PNAS.201303367SI.PDF

Mir RA, Sharma A, Mahajan R (2020) Crop landraces: present threats and opportunities for conservation. Undefined, pp 335–349. https://doi.org/10.1007/978-981-15-0156-2_13

Mwai O, Hanotte O, Kwon YJ, Cho S (2015) African indigenous cattle: unique genetic resources in a rapidly changing world. Asian Australas J Anim Sci 28(7):911. https://doi.org/10.5713/AJAS.15.0002R

Nierenberg D, Sparling N (2016) Why we should protect disappearing livestock breeds—CSMonitor.com. https://www.csmonitor.com/Business/The-Bite/2016/0908/Why-we-should-protect-disappearing-livestock-breeds

Ojeivo C (2012) Good agricultural practices for African indigenous vegetables: proceedings of a technical consultation workshop held in Arusha, Tanzania, 7–8 December 2009. Proceedings of a technical consultation workshop

Payne WJA, Hodges J (1997) Tropical cattle: origins, breeds and breeding policies. https://agris.fao.org/agris-search/search.do?recordID=XF2016004516

Pimm SL, Jones HL, Diamond J (2015) On the risk of extinction. Am Nat 132(6):757–785. https://doi.org/10.1086/284889

Qiu Q, Zhang G, Ma T, Qian W, Wang J, Ye Z, Cao C, Hu Q, Kim J, Larkin DM, Auvil L, Capitanu B, Ma J, Lewin HA, Qian X, Lang Y, Zhou R, Wang L, Wang K, Liu J (2012) The yak genome and adaptation to life at high altitude. Nat Genet 44:946–949. https://doi.org/10.1038/ng.2343

Raja KN, Singh PK, Mishra AK, Ganguly I, Devendran P (2016) A new methodology for characterization of dog genetic resources of India. 6(2):87–96

Rao MS (2021) Millet: an alternative food grain. https://doi.org/10.5281/ZENODO.5579567

Rao PP, Pandey L, Jagadeesh E, Deb UK, Jain R, Basu K (2012) Meso-level database coverage and insights village dynamics in South Asia

Rao ND, Poblete-Cazenave M, Bhalerao R, Davis KF, Parkinson S (2019) Spatial analysis of energy use and GHG emissions from cereal production in India. Sci Total Environ 654:841–849. https://doi.org/10.1016/J.SCITOTENV.2018.11.073

Registered Dog—ICAR—National Bureau of Animal Genetic Resources (2020). https://nbagr.icar.gov.in/en/registered-dog/

Renjith D (2018a) Coarse grains better than rice for health, environment—Asia & Pacific. https://www.scidev.net/asia-pacific/news/coarse-grains-better-than-rice-for-health-environment/

Renjith D (2018b) Coarse grains better than rice for health, environment—ICRISAT. https://www.icrisat.org/coarse-grains-better-than-rice-for-health-environment/

Rouse JE (1977) The Criollo: Spanish cattle in the Americas, 1st edn. University of Oklahoma Press, Norman

Sarbapriya R, Aditya I (2011) Impact of population growth on environmental degradation: case of India. J Econ Sustain Dev 2(8). www.iiste.org

Sasaki A, Nakamura Y, Kobayashi Y, Aoi W, Nakamura T, Shirota K, Suetome N, Fukui M, Shigeta T, Matsuo T, Okamoto S, Park EY, Sato K (2017) A new strategy to protect Katsura-uri (Japan's heirloom pickling melon, Cucumis melo var. conomon) from extinction. J Ethnic Foods 4(1):44–50. https://doi.org/10.1016/J.JEF.2017.02.003

Scherf BD (2000) World watch list for domestic animal diversity, 3rd edn

Sehrawat R, Sharma R, Ahlawat S, Sharma V, Thakur MS, Kaur M, Tantia MS (2021) First report on better functional property of black chicken meat from India. Indian J Anim Res 55(6):727–733. https://doi.org/10.18805/IJAR.B-4014

Sharma R, Ahlawat S, Sharma H, Prakash V, Shilpa, Khatak S, Sawal RK, Tantia MS (2020) Identification of a new Indian camel germplasm by microsatellite markers based genetic diversity and population structure of three camel populations. Saudi J Biol Sci 27(7):1699–1709. https://doi.org/10.1016/J.SJBS.2020.04.046

Sharma R, Vij PK, Aggarwal R, Chandran PC, Kamal RK, Dey A, Tantiam S (2021) Characterization of a new potential goat breed (Palamu) from Jharkhand, India. Indian J Anim Sci 90(12). http://epubs.icar.org.in/ejournal/index.php/IJAnS/article/view/113191

Solh M, Amri A, Ngaido T, Valkoun J (2003) Policy and education reform needs for conservation of dryland biodiversity. J Arid Environ 54(1):5–13. https://doi.org/10.1006/jare.2001.0896

Sponenberg DP (2000) Genetic resources and their conservation. In: Bowling AT, Ruvinsky A (eds) The genetics of the horse. CABI Publishing, Wallingford, pp 392–393

Stalker HT, Harold, Warburton M, Harlan JR, Jack R (2021) Harlan's crops and man: people, plants and their domestication. https://www.wiley.com/en-sg/Harlan%27s+Crops+and+Man%3A+People%2C+Plants+and+Their+Domestication%2C+3rd+Edition-p-9780891186335

Sultanbawa Y (2016) Anise myrtle (Syzygium anisatum) oils. In: Essential oils in food preservation, flavor and safety, pp 215–219. https://doi.org/10.1016/B978-0-12-416641-7.00023-7

Thanopoulos R, Chatzigeorgiou T, Argyropoulou K, Kostouros NM, Bebeli PJ (2021) State of crop landraces in arcadia (Greece) and in-situ conservation potential. Diversity 13(11):558. https://doi.org/10.3390/d13110558

Vanishing breeds—India Environment Portal|News, reports, documents, blogs, data, analysis on environment & development|India, South Asia (1997). http://www.indiaenvironmentportal.org.in/content/246/vanishing-breeds/

Villa TCC, Maxted N, Scholten M, Ford-Lloyd B (2005) Defining and identifying crop landraces. Plant Genet Resour 3(3):373–384. https://doi.org/10.1079/PGR200591

Villa TCC, Maxted N, Scholten M, Ford-Lloyd B (2007) Defining and identifying crop landraces. Plant Genet Resour 3(3):373–384. https://doi.org/10.1079/PGR200591

Vu DC, Nguyen THD, Ho TL (2020) An overview of phytochemicals and potential health-promoting properties of black walnut. RSC Adv 10(55):33378–33388. https://doi.org/10.1039/D0RA05714B

Wang Y, Wang Y, Sun X, Caiji Z, Yang J, Cui D, Cao G, Ma X, Han B, Xue D, Han L (2016) Influence of ethnic traditional cultures on genetic diversity of rice landraces under on-farm conservation in southwest China. J Ethnobiol Ethnomed 12(1):51. https://doi.org/10.1186/S13002-016-0120-0

Zeven AC (1998) Landraces: a review of definitions and classifications. Euphytica 104(2):127–139. https://doi.org/10.1023/A:1018683119237

Chapter 13
Conservation and Management of Mangrove Ecosystem in Diverse Perspectives

E. S. Bindiya, P. M. Sreekanth, and Sarita G. Bhat

Abstract The mangrove ecosystem is a dynamic hub of the oceanic environment due to nutrient fluidities, yield, and biodiversity of organisms. It guards the seaside zone against natural catastrophes, confirms pollution reduction, and functions by recycling nutrients. The value of mangroves in providing the forestry and fisheries products to meet the shoreline livelihood and economy is significant. Mangroves are of great ecological significance and socioeconomic implication as a core tropical marine habitat. The mangroves are also one of the world's richest granaries of biological and genetic diversity. They support complex communities where thousands of other species interact, from bacteria to human beings. They provide a valuable nursery habitat for fish and crustaceans, a food source for other faunas. Anthropogenic pressures, development pressures including urbanization and industrialization, and rapid environmental changes have turned tropical and subtropical mangrove forests into one of the Earth's most-threatened ecosystems, causing worldwide loss of coastal livelihoods and ecosystem services. The scientific community finds such an ecosystem as one among the world's most-threatened biomes due to human intervention in the long past and on-going climate change. Many countries have already lost their huge mangrove wealth within the last two decades. Further, the decline of the mangrove cover may cause irreparable damage to the ecosystem to the service of mankind. Now is the time to conserve the precious ecosystem for overall well-being based on in situ or ex situ conservation methods preferable to each species; else, the ecosystem services and other benefits offered by mangroves will be diminished or lost forever.

Keywords Mangrove forests · Geographical distribution · Diversity · Genomic conservation · In situ conservation · Ex situ conservation

---

E. S. Bindiya · P. M. Sreekanth · S. G. Bhat (✉)
Department of Biotechnology, Cochin University of Science and Technology, Cochin, India

S. T. Sukumaran, K. T R (eds.), *Conservation and Sustainable Utilization of Bioresources*, Sustainable Development and Biodiversity 30,
https://doi.org/10.1007/978-981-19-5841-0_13

13.1 Introduction

Mangroves are a group of trees and shrubs which are derived from more than one ancestral stock. Mangroves exhibit different kinds of morphological and physiological adaptations to their habitat, which include tolerance of high salinity, deviating desiccation and saturation of soils across tidal cycles, and exposure of roots to hypoxic, sulfide-rich soils (Friis et al. 2020).

The origin of the word mangrove was well explained by Vanucci (1989). It is said that the word mangrove was derived from West Africa, Senegal, Gambia and Guinea in the fifteenth century. The Portuguese spread these words throughout the world and was later converted into "mangle" and "manglar" by the Spaniards. The English word mangrove is a combination of the Portuguese and Spanish, which means grove made of mangue (Macintosh and Ashton 2002).

Worldwide, natural resources fall under various levels of management and ownership, ranging from private to government ownership (Berkes 2004). Because of the recognized importance of mangroves and the continuing threats to their persistence, actions have been taken internationally for the conservation and sustainable use of wetlands. Protective authorities include the United Nations Forum on Forests (UNFF), Convention on International Trade in Endangered Species, Species of Wild Fauna and Flora (CITES), Convention on Biological Diversity (CBD), United Nations Framework Convention on Climate Change (UNFCCC), Convention on Wetlands of International Importance especially as Waterfowl Habitat (Ramsar Convention), Convention for the Protection of World Cultural and Natural Heritage, and the Convention on the Conservation of Migratory Species of Wild Animals. These agreements have resulted in the protection of large areas of mangrove forests globally. In addition to conservation agreements at the international level, efforts by individual nations to protect or restore these forests have varied from mandated protection by governments to locally initiated efforts. Human use of those resources varies depending on the level of protection. Overall, the global decline of mangroves has slowed, but additional actions need to be taken to ensure their long-term survival (Alongi 2002).

Successful management, conservation, and restoration require the commitment of local, state, and national level government as well as local communities. As patterns of human settlement around the world continue, people are increasingly distanced from nature and therefore may be less concerned about nature conservation (Miller 2005; Zaradic et al. 2009). Additionally, as with many conservation programs, benefits from mangroves might be indirect or poorly understood by local residents, or they may perceive that they are being excluded from access to previously utilized resources (Shackelton et al. 2002).

13.2 Geographical Distribution

Mangroves are usually seen distributed between the Tropic of Cancer and the Tropic of Capricorn on all continents. The biogeographic patterns indicate that a higher species richness is observed in the tropics (especially in Asia) and lower in temperate regions (Koch 1998). Mammals and amphibians follow a similar pattern of highest diversity in the tropics (Wiens et al. 2006; Schreier et al. 2009), and this pattern extends to mangrove forests. They occupy less than 1% of the world's surface and mainly cover an estimated 75% of the tropical coastline worldwide within 123 countries. These forests of the tide collectively make up less than 1% of all tropical forests and less than 0.4% of the total global forest estate.

The first attempt at estimating the total mangrove area in the world was undertaken as part of the Food and Agriculture Organization of United Nations (FAO/UNEP) Tropical Forest Resources Assessment in 1980, where the world total was estimated as 15.6 million hectares. More recent estimates in 2020 are up to 14.8 million hectares. According to the assessment, the most extensive area of mangroves is found in Asia (5.55 million ha—38% of total mangroves), followed by Africa, North and Central America, South America, and Oceania. Mangrove areas were absent in Europe. Indonesia, Brazil, Nigeria, and Australia account for 60% of the total mangrove area found in just ten countries and about 41% of all mangroves. Figure 13.1 represents the region-wise distribution of mangrove forests as per FAO (2020).

13.2.1 Asia

The subregional distribution of mangrove forests is divided into East Asia, South and Southeast Asia, and Western and Central Asia. The major area of mangrove forest is distributed in South and southeast Asia (5.3 million Ha), which includes

Fig. 13.1 The region-wise distribution of mangrove forests as per FAO 2020

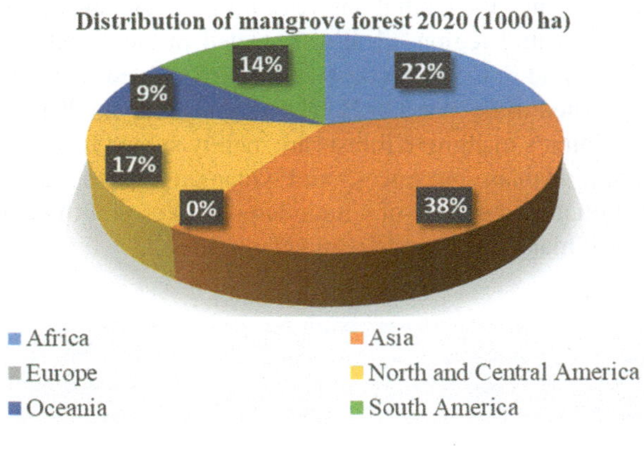

Nepal, Bangladesh, India, Sri Lanka, Pakistan, Myanmar, Vietnam, Thailand, Indonesia, Philippines, Singapore, and Malaysia, which accounts for the majority of mangrove forests.

The total mangrove area in Malaysia is estimated to be approximately 575,000 Ha, of which 60% is found in Sabah, 23% in Sarawak, and the remaining 17% in the Peninsula (Eco-Business 2013) of which 77.8% is considered productive. According to Eco-Business, the vegetated mangrove area in Indonesia is estimated to be 3.5 million Ha. In South Asia alone, the aerial extent of mangrove forests is approximately 1,187,476 Ha, representing ~7% of the global total.

Mangroves in Sri Lanka are restricted to estuaries along the coasts, the extent of which ranges from 15,670 Ha (Edirisinghe et al. 2012). The Puttalam Lagoon–Dutch Bay, Portugal Bay complex, Batticaloa, and Trincomalee are the sites of the largest mangrove patches of Sri Lanka. In Pakistan, approximately 95% (~600,000 hectares) of mangroves are located in the Indus Deltaic swamps of the Sindh Province along the Arabian Sea coastline with Sandspit, Sonmiani (Miani Hor), Kalmat Khor, and Jiwani (Gwadar) as other major areas of distribution and are considered the largest dry-climate mangroves in the world. The total mangrove forest area along the Pakistan coastal line is 86727.86 Ha based on the SPOTXS data (2003).

India has 0.45 million Ha of mangrove. In India, mangroves are found in the States of West Bengal, Orissa, Andhra Pradesh, Tamil Nadu, Andaman and Nicobar Islands, Kerala, Goa, Maharashtra, Karnataka, and Gujarat. Sundarbans in the Gangetic delta of West Bengal is the largest mangrove region in the world and have the largest coverage of mangroves, around 212 million Ha, which consist of 26 plant species. The mangroves cover in Gujarat is about 104,400 Ha (Pandey and Hirose 2013). Pichavram in Tamil Nadu is the second-largest mangrove forest with an area of 1100 Ha, consisting of 12 species (Sundararaju 2019). In Kerala, mangroves are distributed in ten Districts. Among them, Kannur has the highest area of mangrove coverage, up to 755 Ha. Kozhikode and Ernakulam are in second and third positions with coverage of 293 Ha and 260 Ha of mangrove spreads, respectively (Muraleedharan et al. 2009).

The largest continuous mangrove forest in the world, Sundarbans, is located on the border of Bangladesh and India. Mangroves in Bangladesh are located in the Sundarbans, the Chittagong region in the southeast, and the Madhupur tracts in the north-central region, ideally protected in "reserved forests." Mangrove forests are also found in Cox's Bazar and the Noakhali coastal belt. The total mangrove area is estimated to be 599,330 Ha (Islam and Bhuiyan 2018).

China's mangrove forests are mainly spread in areas of Zhejiang, Fujian, Guangdong, Hainan provinces, and Guangxi Zhuang autonomous region. The third national land survey of China demonstrates that the country now has about 27,100 Ha of mangrove forests (Chinadaily 2022).

13.2.2 Africa

East African mangroves are noticed in Mozambique, Tanzania, Kenya, and southern Somalia. The most elaborate mangroves are observed in the Rufiji River Delta in Tanzania and the Zambezi River Delta in Mozambique. There are two general categories of mangroves along the Kenyan and Tanzanian coasts: those found in fringe communities along the open coastline and those found in estuaries and at river mouths. An assessment by Dinerstein et al. (2017) elucidated that 843,000 Ha, or 45%, of the East African mangroves are in protected areas.

West African countries with mangroves include Angola, Benin, Cameroon, Congo, the Democratic Republic of Congo, Equatorial Guinea, Gabon, Gambia, Ghana, Guinea, Guinea Bissau, the Ivory Coast, Liberia, Mauritania, Nigeria, Sao Tome and Principe, Senegal, Sierra Leone, and Togo. Mangroves in West Africa cover more than 2.3 million Ha, which constitutes 13% of mangrove forests worldwide. The most extensive mangrove forests in West Africa are found between Senegal and the south of Sierra Leone; between Ghana and Cameroon; and the Niger Delta (USFS 2014).

The total mangrove cover in South Africa is approximately 1900 Ha (Ward and Steinke 1982; Hoppe-Speer et al. 2013). Mangrove cover starts from Kosi Bay in the north and reaches its southern distributional limit at Nahoon Estuary. The mangroves at Nahoon were once transplanted from Durban Bay in 1969 by Steinke and currently occupy extensive areas on intertidal mudflats in the estuary (Steinke 1999). The largest mangrove forests of South Africa are found in the subtropical areas of St. Lucia and Richards Bay (Naidoo 2016).

13.2.3 American Continents

The Panama Bight Eco-region in the west coast of Central and South America is believed to have one of the best-preserved mangrove ecosystems in the American continent, the widely recognized conservation hotspot. Highly developed mangrove forests are visible along the southern Colombian and northern Ecuadorian Pacific coasts (Castellanos-Galindo et al. 2021). In North America, mangroves are observed from the southern tip of Florida along the Gulf Coast to Texas.

13.2.4 Oceania

Australia owns only 0.9 million Ha of mangrove forest (ABARES, 2019). The mangrove forest is found in all mainland states and the Northern Territory of Australia. They are distributed such that a total of 0.4 million hectares (45%) is found in Queensland, and 0.3 million hectares (39%) is in the Northern Territory.

According to Spalding (2010), New Zealand's mangrove distribution covers about 26,050 Ha. The mangrove trees of New Zealand grow only on the top half of the North Island of New Zealand. Mangroves are found from 34°27′S in the far north of the North Island to 38°05′S at Kawhia Harbour on the west coast and 38°03′ S at Ohiwa Harbour on the east coast (Crisp 1990). Planted mangroves have been observed to establish and survive as far south as 41°13′S on the Hutt River, at the southern end of the North Island (de Lange and de Lange 1994; Horstman et al. 2018). Only one species of mangrove is prominent—*Avicennia resinifera,* although in Australia it is called *Avicennia marina.*

13.3 Biodiversity of Mangrove Ecosystem

Mangroves are living treasures of biodiversity. The term mangrove is often used to refer to both the trees and the community, even though trees are the basic and most visible component of mangrove ecosystems. The mangrove community is an assemblage of a wide range of organisms from different systemic groups, including bacteria, fungi, microalgae, invertebrates, birds, and mammals (Holguin et al. 2001).

On a geological time scale, mangroves are relatively recent and ephemeral coastal features. Even though several sorts of mangroves are located on sheltered and sedimented stretches of reefs and islands, the complex mangrove forest is completely achieved only in estuaries and deltas.

13.3.1 Mangrove Flora and Fauna

There are about 110 species of "mangroves" of which only about 54 species in 20 genera from 16 families constitute the "true mangroves," i.e., species that occur almost exclusively in mangrove habitat (Kibria 2013). Mangroves were thought to have evolved in tropical intertidal environments of the early Tethys Sea between Laurentia and Gondwanaland (King et al. 1990). They are a product of the convergent evolution of angiosperms, where plants from different families from the coastal environments have evolved under similar adaptations that enabled them to colonize and reproduce.

The western center with West Africa, Atlantic, and Pacific South America and the eastern center with East Africa to the Western Pacific are considered as the centers of diversity of the mangroves. The first established angiosperm mangrove vegetation type consisted of *Avicennia* sp., *Rhizophora* sp., and *Hibiscus* sp. in both the centers of diversity (Tomlinson 1986).

Nypa sp. was already existing some 63 million years ago during the Cretaceous era, while *Rhizophora* sp. was recorded to be present only 30 million years ago, as revealed by the fossils. Mangrove trees, with some exceptions, do not always need salt to grow but develop some special adaptation mechanism to tolerate high salt concentration. These adaptations are visibly observed in leaves and roots and in reproduction.

Succulent leaves with a thickened epidermis, waxy cuticle on the upper side with white tomentum on the underside like in xerophytes are sometimes observed in mangrove plants. *Rhizophora* sp. and *Sonneratia* species are good examples of this characteristic leaf pattern. Some leaves are covered with pubescence, as in some *Avicennia* species or scales in *Heritiera* species. They tend to have reduced leaf areas, and sunken stomata like in *Rhizophora, Bruguiera, Ceriops, Avicennia, Aegiceras,* and *Lumnitzera.* Mangroves develop superficial rooting systems. This rooting system serves for shallow anchorage, absorbing water, and oxygen in largely anoxic surroundings.

Instead of tap roots, Avicennia sp. and Sonneratia sp. have radial roots. Small anchoring roots are given off, and vertical pneumatophores or breathing roots protrude above the soil surface. These are slender and flexible in *Avicennia* sp., while strong and woody in *Sonneratia* sp. Only a set of branching prop roots are seen in *Rhizophora* sp. from which a succession of long arched roots, halt-hoops are put out radially, which serve as pneumatophores.

Mangroves can reach far upstream from the sea, wherever an increase in the salinity of the surface mud occurs due to a wedge of heavier seawater that can creep over the bottom. Not all the species of mangrove plants are found in any one mangrove community. According to the decrease in salinity and increase in firmness of the soil from the shoreline to inland, the mangrove community's specific plant species usually occur in zonation. Mangrove swamps are divided into various zones with respect to the dominant tree as a means of zoning. Thus, the zonation can be classified into two categories: (1) From a sheltered tidal area to the beach forest area; (2) From the open sea to the beach forest area.

Rog et al. (2016) identified 464 terrestrial vertebrate species to occur in mangrove forests: 320 mammals, 118 reptiles, and 26 amphibians (including 22 subspecies). This is a fivefold increment in the terrestrial vertebrate species previously reported to use mangroves (excluding birds). Following previous reviews were also added representatives of 14 additional families (Kathiresan and Bingham 2001; Nagelkerken et al. 2008; Luther and Greenberg 2009; Hogarth 2015).

The faunal components of the Mangal belong to a large variety of classes when compared to the floral components. About 75% of the fauna found in mangrove forests is not found in any other zone (Frith et al. 1976). The fauna can be majorly divided into three groups: Tree fauna, ground or surface fauna, and burrow fauna. The trunks, leaves, and roots of mangrove trees support a banding of sessile and mobile animals. The most conspicuous of the insect population in Mangals are the mosquitoes. About 125 species of mosquitoes have been recorded from mangrove areas in Sulawesi, Indonesia, of which *Anopheles sundaicus* carries malaria. The larvae of this species can tolerate water salinity of up to 13%. Colonies of weaver ants, *Oecophylla* sp., fierce predators, live in untidy leaf nests made by weaving five or six leaves together with fine silk thread.

Ciliates and foraminifers are numerous in mangroves, but the meiofauna of the sediment does not exhibit any species diversity. The peculiar peanut worm,

Phascolosoma sp., is able to store oxygen in its coelomic fluid and lives permanently in the soil of the pioneer zone when the tide is out (Göltenboth and Schoppe 2006). The detritus is eaten by a group of small animals such as zooplankton, crustaceans, like penaeid prawns and small crabs, and small fish, particularly fingerlings. The most important litter processors after the crustaceans are isopods, like *Exosperoma* sp., and capitellidae polychaeta, like *Capitellides* sp.

One group of animals that are widely used as indicators of changes in the balance of the mangrove ecosystem is molluscs because of the uniqueness in their distribution. The molluscs, especially class bivalve and gastropod, follow a site-specific distribution pattern that is closely related to the characteristics of their habitat (Vermeij 1973). Molluscs function as predators, detritus eaters, and carcass eaters in the Mangal. They guarantee a balance of energy flow in the ecosystem (Rahardjanto et al. 2020).

The Littorina snails, such as *Littoraria scabra* and *Littoraria carinifera,* reside high in foliage. The tapering, ovoid *Auriculastra* sp., *Vittina* sp., *Cassidula* sp., and *Terebralia* sp. (Fam. Potamididae) are usually found on more muddy parts of the Mangal, whereas the primitive pulmonate snails *Melampus* sp. (Fam. Ellobiidae), *Nerita* sp. (Fam. Neritidae), *Heminerita* sp., and *Truncatella* sp. (Fam. Potamididae) are found on more silty sand (Göltenboth and Schoppe 2006).

The distribution of molluscs associated with the trees depends on the water or tidal level. Thus, the zonation pattern is like branches, mid region, root, and mud. Branches and roots of trees near the high water mark carry small acorn barnacles, *Chthamalus* sp., and oysters, *Crassostrea* sp. At mid-tide level, a small estuarine mussel, *Modiolus* sp. or the bivalve *Enigmonia* sp., which lives on leaves and trunks attached by a stout byssus, may be found. A flat bivalve *Isognomon* sp. may be found among the lower roots. Root holes and clefts hollowed by shipworms shelter amphibious Ellobiidae molluscs, crabs, prawns, nereid, and other polychaete worms. The living roots of *Rhizophora* are invaded by wood-drilling *Sphaeroma* sp. The heteroptera bug *Halobates* sp. skims actively over the surface of stagnant water. The bivalve *Geloina ceylonica* lives buried deep in the mud (Göltenboth and Schoppe 2006).

Besides the snails, the crustaceans are always a prominent and diverse component of the mangrove fauna. The red-clawed fast running Grapsidae *Sesarma* sp. and *Metopograpsus* sp. and the slow moving Xanthidae *Epixanthus* sp. often live under boulders and in holes between roots. The most attractive of tropical mudflat shore crabs belonging to the family Ocypodidae are the fiddler crabs (*Uca* sp.). *Dotilla* sp. or the soldier crab that builds an igloo of mud balls around itself are also seen in the mud. Hermit crabs belonging to *Clibanarius* sp., of the family Paguridae and *Coenobita* sp. of the family Coenobitidae, and mud lobster, *Thalassina anomala* of the family Thalassinidae are usually very common in the Mangal. The king or horseshoe crabs, *Carcinoscorpius rotundicauda,* are living fossils. They are the surviving members of Xiphosura and are ancient marine relatives of the spiders and scorpions which inhabited the oceans some 200 million years ago (Göltenboth and Schoppe 2006).

Drainage channels of the Mangal serve as important habitat for many littoral fish species, reef fishes, and fishes from rivers. Juveniles of the families, particularly Mugilidae, Tetraodontidae, and Gobiidae can be seen. The muddy flats are invaded by the genus *Boleophthalmus* and *Periophthalmus* (mudskippers). *Boleophthalmus boddarti* (Boddart's goggle-eyed goby) is usually found at the seaward edge of the mid-tide area, grazing the algae. Carnivorous *Periophthalmodon schlosseri* (Giant mudskipper) and omnivorous *Periophthalmus chrysospilos* (Gold-spotted Mudskipper) occupy the highest part of the swamps. The banded archerfish, *Toxotes jaculator,* can also be observed "shooting down" insects from above the water at high tide (Göltenboth and Schoppe 2006).

Rana cancrivora is atypical among amphibians in being able to live and breed in modest saline water. The tadpoles are more resistant to salt than the adults, and metamorphosis is delayed until considerable dilution of the water occurs (Göltenboth and Schoppe 2006). Another mangrove frog (*Eleutherodactylus caribe*) was reported in Haiti (Luther and Greenberg 2009).

A variety of reptiles inhabiting the saline water, the ground floor, and the trees can also be visualized. A majority of mangrove-restricted terrestrial reptiles tend to have laterally compressed tails, which are presumed to be helpful in aquatic environments. Water monitor (*Varanus salvator*) and Mangrove monitor (*Varanus indicus*) were the reptiles found inhabiting Philippine Islands and Papua New Guinean and Australian mangroves, respectively. Near-threatened Mangrove terrapin (*Malaclemys terrapin*) was reported in Southern Florida and North America (Luther and Greenberg 2009).

Luther and Greenberg (2009) recognized 11 species of mangrove snakes. The common skink, *Mabuya multifasciata,* is found both on the ground and in the lower root system. While *Cerberus rynchops* (dog-faced water snake), *Homalopsis buccata* (puff-faced water snake), and the crab-eating snake, *Fordonia leucobalia,* are found in shallow water, the cattle snake, *Elaphe radiata,* is a ground snake (Göltenboth and Schoppe 2006). Arboreal snakes are the common cat snake, *Boiga dendrophila,* and the reticulated python, *Python reticulatus.*

Mangrove forests are potentially a very attractive bird habitat. Many of the birds are not restricted to mangroves. But there are 48 species of birds that are restricted to mangroves alone. An additional 18 bird species depend on mangroves for feeding, roosting, or nesting during daily or seasonal migrations but are not considered restricted to mangroves (Luther and Greenberg 2009). The abundance of particularly small prey is very attractive to kingfishers, *Halcyon chloris.* The sweet nectar of flowers attracts both nectarivorous birds, like the brown-throated sunbird, *Anthreptes malacensis,* purple-throated sunbird, *Nectarinia sperata,* and insectivorous birds like the mangrove blue flycatcher, *Cyornis rufigastra.* Residential species confined exclusively to mangroves include Javan coucal (*Centropus sinensis*), Pacific swallow (*Hirundo tahitica*), and mangrove swallow (*Tachycineta albilinea*). The most endangered bird species identified in mangrove communities is the Milky Stork or *Mycteria cinerea.* Kutt (2007) recorded a number of mangrove-obligate species (e.g., Collared Kingfisher, Mangrove Robin, Mangrove Gerygone) more abundantly, and a further suite of species that were mangrove-facultative (e.g.,

Shining Flycatcher, Brown Honeyeater, Dusky Honeyeater, Rufous Fantail, Helmeted Friarbird, Spangled Drongo) in Australian mangroves.

Zonal patterns of the distribution corresponding to the mangrove tree species are exhibited by some birds in different mangroves (Noske 1995, 1996). Prominently exhibited by potentially competing species that occupy largely mutually exclusive zones, such as the seaward fringe, tidal channel, saline flats, or landward fringe. This zonation has been observed in woodpeckers in Malaysia (Noske 1995) and in gerygones and kingfishers in Australia (Noske 1996; Schodde et al. 1982). In Australia the correspondence between plant and bird zonation was more pronounced in the dry season than in the wet season, possibly because of restricted insect richness during the dry season (Noske 1996).

The mangrove forest shares a variety of species of mammals with riverine habitats, including some species of monkeys, like the long-tailed macaque, *Macaca fascicularis,* the Javan lutung, *Semnopithecus auratus;* the Bornean silvered langur, *Presbytis cristata;* and the endemic proboscis monkey, *Nasalis larvatus.* The small-clawed otter, *Aonyx cinerea,* and the fishing cat, *Felis viverrina,* are common residents of the Mangal. Pygmy three-toed sloth (*Bradypus pygmaeus*), Vordermann's pipistrelle (*Pipistrellus vordermanni*), Northern pipistrelle (*Pipistrellus westralis*), Proboscis monkey (*Nasalis larvatus*), Garrido's hutia (*Mysateles garridoi*), and Cabrera's hutia (*Mesocapromys angelcabrerai*) are some of the mangrove-restricted mammalian species (Luther and Greenberg 2009). Fruit bats, particularly the flying fox, *Pteropus vampyrus,* roost in the mangrove forest canopy. Other bats, such as *Eonycteris spelaea,* the cave fruit bat, and *Macroglossus sobrinus,* the long-tongued fruit bat, are very important pollinators of *Sonneratia* sp. (Göltenboth and Schoppe 2006).

Facultative users of mangroves depend for critical stages in their life cycles, using these areas to breed, e.g., the Estuarine Crocodile, *Crocodylus porosus*, and the Sea Krait, *Laticauda colubrina* (Göltenboth and Schoppe 2006). There are also records of species using mangroves to shelter from heat stress, e.g., bat species and kangaroos in Australia (Reef et al. 2014) and to disperse between primary habitats, e.g., the marsh rabbit, *Sylvilagus palustris*, in the United States (Kathiresan and Bingham 2001).

Another unique feature is the existence of swamp tigers in one of the world's famous mangrove forests in India, the Sundarbans. Sundarbans tigers are different from all other Royal Bengal tiger populations in their small, stocky, and muscular frame; coarse and short coats and deep reddish-brown coloration are all adaptations to survive in this hostile habitat. They can swim deceptively fast across a brackish-water channel against a rapidly receding tide (World Atlas 2020). Deer are also regular visitors in some forests, including the Key deer in Florida and the spotted and barking deer in the Sundarbans. Livestock, mainly goats and camels, visit mangroves in many parts of the Middle East, Pakistan, and East Africa. Smaller ones include the hutias of the Greater Antilles (Spalding 2010).

13.3.2 Microbial Diversity

Microscopic investigation of decomposing mangrove leaves reveals a complicated community composed of fungi, bacteria, protozoa, and microalgae (Odum and Heald 1975). In tropical mangroves, bacteria and fungi constitute 91% of the total microbial biomass, whereas algae and protozoa represent only 7% and 2% (Alongi 1988). Microbial communities play crucial roles in mangrove nutrient cycling and plant productivity (Allard et al. 2020). There are only a few studies on the role of microorganisms on the expansion or retraction of mangrove forests linked to sea-level rise (Ward and de Lacerda 2021). Thus, understanding the role of microorganisms and their stability can provide crucial information to protect and restore mangroves in the world of climate change (Alongi and Mukhopadhyay 2015).

The planted mangrove sediment tended to have higher bacterial richness and diversity than the unplanted sediment. Plantation can influence the habitats for microbiota through the unique root exudates of mangroves and oxygen leakage (Tong et al. 2019). The mangrove habitats at geographic sites are unique with factors, like the rate of precipitation, organic carbon, evapotranspiration, and temperature, which could influence bacterial distribution with distinct community structures (Tavares et al. 2021).

Most metagenomic datasets of the core microbiome of mangroves worldwide were dominated by Proteobacteria, mainly by Deltaproteobacteria and Gammaproteobacteria with Planctomycetes, Acidobacteria, Bacteroidetes, and Chloroflexi. Desulfobacterales, which comprise anaerobic members involved in sulfur and carbon cycling as well as in the transformation of methane and nitrogen, is the most abundant order commonly observed in mangrove surface soils (Bashan and Holguin 2002).

Some representatives from non-common phyla, such as Elusimicrobia and Deinococcus-Thermus, as well as Proteobacteria (*Modicisalibacter*) were selected as indicator strains. These genera comprise bacteria capable of reducing nitrate or sulfur compounds, some of them with particular features, such as *Blastocatella* (Elusimicrobia), previously isolated from semi-arid savanna soil; *Truepera* (Deinococcus-Thermus), resistant to ionizing radiation; *Algiphilus* (Proteobacteria), aromatic hydrocarbon degrading; *Thiovulum* (Proteobacteria), rapid swimming bacteria; and *Planktotalea* (Proteobacteria) previously isolated from seawater or marine organisms from cold regions (Parte et al. 2020) and SEEP-SRB1 (sulfate-reducing bacteria cluster).

Even though rare species are non-relevant under a given environmental condition, they may offer a pool of genetic resources under appropriate conditions of climate change or some drastic environmental changes (Jousset et al. 2017). The emergence of novel microbial families Ignavibacteriae (Iino et al. 2010), in Mangalavanam, Kadalundi, and Kallai mangroves (India) and Calditrichaeota (Marshall et al. 2017), and Mucromycota (fungi) in Kadamakkudy mangrove was noticed in a post flood metagenomic analysis of Kerala mangrove sediments (Bindiya and Bhat 2019).

13.3.2.1 Nitrogen-Fixing Bacteria

Nitrogen-fixing bacteria, which constituted members of the genera *Azospirillum*, *Azotobacter*, *Rhizobium*, *Clostridium*, and *Klebsiella,* have been isolated from the sediments rhizosphere and root surfaces of various mangrove species. Several strains of diazotrophic bacteria *Vibrio campbellii*, *Listonella anguillarum*, *V. aestuarianus*, and *Phyllobacterium* sp. were also isolated from the rhizosphere *Rhizophora mangle*, *Avicennia germinans*, and *Laguncularia racemosa* in Mexico. The nitrogen fixing capacity of these bacteria was similar to that of diazotrophic bacteria from the terrestrial environment, such as *Azospirillum* sp. (Bashan and Holguin 2002).

13.3.2.2 Phosphate-Solubilizing Bacteria

In an arid mangrove ecosystem in Mexico, nine strains of phosphate-solubilizing bacteria *Bacillus amyloliquefaciens*, *B. atrophaeus*, *Paenibacillus macerans*, *Xanthobacter agilis, Vibrio proteolyticus, Enterobacter aerogenes*, *E. taylorae*, *E. asburiae*, and *Kluyvera cryocrescens* were isolated from *A. germinans* roots; and three strains *B. licheniformis*, *Chryseomonas luteola*, and *Pseudomonas stutzeri,* from that of *Laguncularia racemosa* (Vazquez et al. 2000). *Penicillium* and *Aspergillus* are very commonly reported as phosphate-solubilizing groups of fungi present in mangroves. *Streptomycetes* is the only known genus of actinomycetes to be identified as a phosphate solubilizer in mangroves (Bashan and Holguin 2002).

13.3.2.3 Sulfate-Reducing Bacteria

In Florida, sulfate-reducing bacteria were the most numerous bacterial group in the rhizosphere of *R. mangle* and *A. germinans* mangroves (Zuberer and Silver 1978). In Goa (India) mangroves, eight species of sulfate-reducing bacteria were isolated which were identified as *Desulfovibrio desulfuricans, Desulfovibrio desulfuricans aestuarii, Desulfovibrio salexigens, Desulfovibrio sapovorans, Desulfotomaculum orientis, Desulfotomaculum acetoxidans, Desulfosarcina variabilis,* and *Desulfococcus multivorans* (Bashan and Holguin 2002).

13.3.2.4 Photosynthetic Anoxygenic Bacteria

Photosynthetic anoxygenic bacteria *Chromatiaceae* (purple Sulfur bacteria) and *Rhodospirillaceae* (purple non-sulfur bacteria) were recovered from Indian mangrove sediments. *Chloronema, Chromatium, Beggiatoa, Thiopedia,* and *Leucothoe* bacteria were the predominant bacterial genera identified in the mangrove ecosystem of Cochin (India). Unidentified species of brown *Chlorobiaceae* were also present (Holguin et al. 2001). Isolates of purple non-sulfur bacteria belonging to ten species, representing four different genera, *Rhodobacter* and *Rhodopseudomonas,* were most common and were identified in mangroves on the coast of the Red Sea in Egypt (Shoreit et al. 1992).

13.3.2.5 Methanogenic Bacteria

A strain of the methanogenic bacterium, *Methanococcoides methylutens* (Mohanraju et al. 1997), and four strains of unidentified thermotolerant methanogenic bacteria have been isolated from the sediment of mangrove forest. Dimethylsulfide-utilizing methylotrophic methanogenic archaeon *Methanosarcina semesiae* was isolated from Tanzanian mangrove sediments (Lyimo et al. 2002). Lyimo et al. (2009) also identified important members from the genus *Methanococcoides* in the mangrove ecosystem both by culture-dependent and culture-independent methods.

13.3.2.6 Fungi

Twenty of the fungal species found in mangrove detritus were identified as ascomycetes (Kohlmeyer et al. 1995) followed by Deuteromycetes and Basidiomycetes. The fungi *Cladosporium herbarum*, *Fusarium moniliforme*, *Cirrenalia basiminuta*, an unidentified hyphomycete, and *Halophytophthora vesicula* were isolated from the dead leaves of *Rhizophora apiculata* (Raghukumar et al. 1994). One of the most notable macroscopic fungi in the mangroves is *Ganoderma* sp. Alias et al. (2010) reported more than 234 species of mangrove and marine fungi prepared as identified and an additional 68 species unidentified.

13.4 Degradation of Mangrove Forest

Anthropogenic pressures due to developmental activities along the coasts, overexploitation and natural calamities have raised concerns about the degradation of mangroves and associated coastal habitats. Major factors which cause the degradation of mangroves around the globe, including land reclamation for agriculture and aquaculture. Increased land-use changes, such as conversion to agriculture, aquaculture, and urban development, have severely degraded the fragile environment of mangroves. Since mangroves are rich in soil nutrients, they are ideally suited for agriculture. Earlier, 90% of mangroves in Kerala were destroyed either for paddy cultivation, coconut orchards, or for land reclamation (Muraleedharan et al. 2009). Mangroves are also being exploited for the production of salt.

13.4.1 Aquaculture

Mangroves are also suited for aquaculture because of their proximity to the sea. Shrimp ponds constructed in the mangroves seriously affect the mangrove ecosystem as the shrimps and other species are fed on specific diets which usually contain certain chemicals. These extra nutrients from food cause eutrophication which decreases the oxygen level and adversely affects the surrounding marine habitats and alters the species distribution. It is reported that shrimp farming alone has caused a loss of 38% of the world's healthy mangroves (Catt et al. 2022). In Kerala massive destruction of mangroves was reported in Kannur district for shrimp farming and Kaippad paddy cultivation (Muraleedharan et al. 2009). In Southeast Asia, 15% of the mangrove areas have been removed for aquaculture.

13.4.2 Coastal Development

Development of the tourism industry, jetties for fisheries landing and mangrove clearance for commercial building and residential areas has severely affected the mangrove population. The introduction of these man-made structures is always associated with the issues of pollution, improper disposal of waste, the emergence of invasive species, altered hydrology, and erosion. Transportation facilities, such as roads and flyovers that have been built within the mangrove areas, forcefully re-routed or blocked the rivers that once flowed through them. This creates the issues of temperature change, salinity, filtration, and sedimentation, which adversely affect the species diversity inside the mangrove ecosystem. In fact, in the past 20 years, more than 700 hectares of mangroves have been cleared for development, yet the current status is still unknown (Catt et al. 2022).

13.4.3 Climate Change

The major impact of climate change along the world's coastlines is sea-level rises and shifting of ocean's chemistry (IPCC 2007; NRC 2011). The rate of these impacts increases at a rate that exceeds the ability of the mangrove ecosystem to adapt to the change. Apart from sea-level rise, increased storminess, altered precipitation regime, and increasing temperature are also threats to the mangrove diversity.

13.4.3.1 Sea-Level Rise

The mangroves are present approximately greater than 1 m above the present low water level. Any increase in sea level may seriously affect the mangroves by altering the hydrological features and related processes. Increase in sea level results in a vertical rise of the water column leading to waterlogging which ultimately kills the mangroves and associated fauna (Ward et al. 2016).

13.4.3.2 Storminess

Frequent storms and hurricanes can significantly influence mangrove productivity. Although some species are resistant to the damage caused by storms, in certain cases, intense storms and cyclones will lead to the uprooting of trees, breakage of branches, and defoliation of the canopy. Extreme storms can also have long-term successional impacts on mangroves by providing a rapid input of non-native sediment, which can increase soil elevation (Ward et al. 2016).

13.4.3.3 Precipitation

The changes in precipitation is often associated with the changes in temperature, which in turn influence the rate of evaporation and transpiration. Changes in rainfall patterns affect the distribution extent and growth rate of mangroves by altering the salinity. When the salinity of the soil gets increased, it will interfere with seedling survival, productivity, and growth of the mangrove system (Ward et al. 2016).

13.4.3.4 Temperature

"Global temperatures are predicted to increase by up to 4.8 °C by 2081–2100 relative to 1986–2005" (IPCC 2013). Alterations in atmospheric temperature substantially influence the composition, phenology, productivity, and finally the pattern of mangrove distribution. Figure 13.2 describes how the temperature is often associated with the ability of CO_2 assimilation in the mangrove ecosystem. Any extreme changes in temperature will collapse the biochemical reactions, which ultimately leads to degradation, vascular embolism and cellular rupture, and the productivity of the mangrove ecosystem will decrease when the temperature exceeds the peak of photosynthesis (Fig. 13.2). The temperature increase is always accompanied with an increased rate of evaporation which ultimately results in increased salinity (Ward et al. 2016).

13.4.4 Deforestation of Mangroves

The emergence of cottage industries in coastline communities' demands for the mangrove timber and charcoal leads to the chopping down of mangroves. Since mangrove wood is suitable as a building material, fencing and fuel, cutting down of

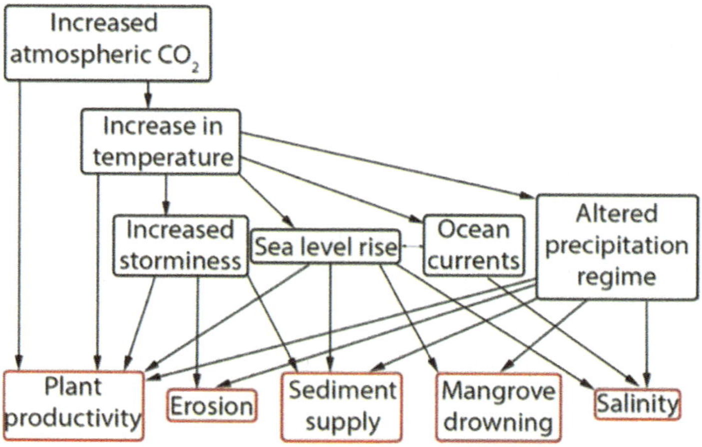

Fig. 13.2 Representation of principal impacting factors of climate change and their influence on mangrove communities

mangroves is proceeding at a higher pace. Many of the coastline communities have turned to charcoal production as their livelihood since mangroves can yield high-quality charcoal (Catt et al. 2022).

13.4.5 Extinction

Approximately 60 species of mangroves have been identified around the world and more than 1 in 6 species of them are under threat. More than 35% of the world's mangrove ecosystems have been destroyed, and 40% of vertebrates endemic to mangroves are threatened with extinction (Catt et al. 2022).

13.5 Conservation of Mangroves: The Need of the Hour

Mangrove ecosystems are considered as the complex ecosystems which form the interface between terrestrial forest and marine ecosystems. From the detailed discussion about the geographical distribution of mangroves (Sect. 13.2) and their biodiversity (Sect. 13.3), we can interpret that along with the structural complexities of mangrove vegetation, it offers a diversified habitat for a wide variety of organisms. Their significant ecological function includes the following:

13.5.1 Stabilization of Shoreline

Mangroves consist of numerous prop roots and other respiratory roots, which form a meshwork under the ground which reduces soil erosion. These complex root systems facilitate the accumulation of debris and prevent the loosening of the soil. Hence, mangroves can be considered as the "coast guards" or "watch dogs" of the shoreline geography and geomorphology (Muraleedharan et al. 2009).

13.5.2 Deep Drainage/Deep Percolation

Mangroves facilitate the movement of surface water into the groundwater flow system which is called the deep drainage or groundwater recharge. Thus, water remains as a part of a shallow water reservoir, which is important for maintaining the water table and influencing the supply of water to surrounding areas. This recharging of groundwater done by mangroves can be considered an environmental service to the community and industries (Muraleedharan et al. 2009).

13.5.3 Regulation of Water Flow

The extensive root system of mangroves can control the excess flow of water entering into them during a flood or heavy rainfall. Mangroves can either store or slow down the flow as it creates several hindrances in the flow path (Muraleedharan et al. 2009).

13.5.4 Sediment Deposition and Nutrient Retention

The structural complexities of mangrove vegetation help slow down the water flow through them, thereby facilitating sediment deposition and nutrient retention. Sediment deposition is always associated with the advantageous removal of toxicants as these molecules are seen attached to the sediment particles. The regulated or controlled flow of water prevents the runoff of nutrients and thus helps in the retention within the system (Muraleedharan et al. 2009).

13.5.5 Carbon Sequestration

The process of capturing carbon dioxide from the atmosphere and its storage is referred to as carbon sequestration. Carbon dioxide is considered a common greenhouse gas that causes global warming, leading to climatic change. Mangroves play a significant role in the biological sequestration of carbon dioxide and are referred to as the "Blue carbon ecosystems". Mangroves are capable of storing three–four times more carbon than terrestrial forests, which are referred to as "Green carbon ecosystems." During the period of growth, mangroves can store and stockpile carbon from 50 metric tons to 220 metric tons per acre. Considering the whole world, mangroves can sequester more than 24 million metric tons of carbon per year (Muraleedharan et al. 2009).

13.5.6 Protection of Habitat and Biodiversity

Mangroves are considered as the most productive marine ecosystems on earth, which provide a unique habitat for many species. They serve as breeding grounds and nursery sites for several species of fish and prawns for spawning and juvenile development (Refer Sect. 13.3) (Muraleedharan et al. 2009).

A diverse range of microbial populations is also found associated with the root system of mangroves. Along with this, fungal endophytes associated with the mesophylls of mangrove leaves are thought to produce novel secondary metabolites of industrial importance.

13.5.7 Gene Bank

Since mangroves are among the most resilient species, they contain more adaptive genes. These genes can be used to modify the genome of crop plants or can be used to impart taste to agricultural products etc. The recombination of salt-tolerant genes of mangroves with crop plants may yield a salt-tolerant plant that can be grown in saline areas and to produce maximum yield (Muraleedharan et al. 2009).

13.5.8 Nutrient Cycling

Different groups of bacteria are usually present in the mangrove ecosystem due to the abundance of carbon and other nutrient contents, where they perform functions, like photosynthesis, nitrogen fixation, and methanogenesis (Das et al. 2007; Holguin et al. 2001). These microbial communities are known to control the chemical environment of the mangrove ecosystem. They stabilize, protect, and nourish coastal water with nutrients and also play a major role in the cycling of nutrients, such as carbon, nitrogen, sulfur, and phosphorus (Alongi 1994). Detritus-supported bacterial biomass routes essential elements through the food web by providing nitrogen and phosphorus to protozoa and metazoa, and eventually to commercially important higher-trophic-level organisms, such as fish and shrimp (Bano et al. 1997).

Mangrove ecosystems are rare and splendid and act as a boundary between land and sea. These ecosystems contribute to the well-being of coastal communities as well as to the environment on a global scale. It acts as a natural coast defense against several natural calamities and disasters.

13.6 Tools for Mangrove Conservation

The mangrove ecosystem is a vital hub of the marine environment because of nutrient fluxes, productivity, and biodiversity of organisms. Despite their importance, mangroves are being degraded at a faster rate because of industrialization, urbanization as well as global climatic change. The conservation of this invaluable treasure of our biodiversity is now considered as the need of the hour. Therefore, the development of effective conservation tools is very much essential to ensure the proper conservation strategies. Although we have well-built conservation policies in legal systems, such as Environmental protection laws and several community efforts and institutional supports, we need to be more practical regarding the conservation. The use of genetic information in mangrove conservation is crucial for its long-term effectiveness. Conservation genetics is an effective tool for highlighting the roles of evolutionary and population genetics for biodiversity conservation (Wee et al. 2019).

13.6.1 Selection of Species to be Conserved

The primary step in the conservation practice is the identification of rare and endangered species of known mangroves which are threatened. Most of the species have a widespread distribution having many congeneric relatives giving them a chance for hybridization and introgression, which leads to perplexes in the proper identification of species to be conserved. Thus, clarification of species identification is important for implementing conservation practices toward priority species and ensuring efficacy. The best example of this is the case study of *Bruguiera hainesii*1 of two critically endangered mangroves, which are a hybrid of *B. gymnorrhiza* and *B. cylindrical*. The International Union for Conservation of Nature (IUCN) prioritizes only the conservation of species, not hybrids (Wee et al. 2019).

13.6.2 Formation of Conservation Units

Since mangroves have a widespread distribution, cross-border conservation is a challenge, especially in implementing protection strategies and enforcing international law. Therefore defining conservation units (CUs) can be helpful for designing reserve networks. The important aspect in defining the CU is genetic connectivity. Genetic connectivity helps identify the geographical regions, metapopulations, and barriers to dispersal.

The identification of CUs based on geographical locations is useful in the genetic evaluation of species present in potential sites. For example, mangroves are present in 274 of over 2000 sites listed under the wetlands of international importance (Ramsar List 2017). The mangrove forest in the Ramsar site has lower rates of deforestation than the global average. The identification of CUs across the borders and prioritizing the region as new Ramsar sites helps conserve the genetic diversity of that particular region. This requires international collaboration and large-scale genotyping (Wee et al. 2019).

13.6.3 Analyzing the Adaptations

Increased anthropogenic activities and global climatic changes have significantly affected the distribution and persistence of mangroves. Because of this, mangroves exhibit differences in the pattern of stress as well as salinity tolerance. Hence, it is essential to understand their molecular mechanisms and physiological responses to environmental stresses and identify the potential evolutionary outcomes under different scenarios. Studies on molecular aspects of adaptive evolution are important in the application of real-time conservation problems. The effective execution of conservation strategies requires the integration of physiological as well as molecular data. With the emergence of NGS, the regulatory mechanisms and phenotypic characteristics for adaptation can be examined through stress tolerance, transcriptome analysis, and epigenetics. By analyzing all this information, managers can potentially identify populations with limited capacity for adaptation and can increase their evolution by providing connectivity to other populations (Wee et al. 2019).

13.6.4 Molecular Tools for Conservation

Molecular techniques used for conservation include different types of markers for the identification of genetic differences between organisms. There are mainly three types of markers, namely, morphological, biochemical, and DNA-based markers (Sahu and Kathiresan 2012).

13.6.4.1 Biochemical Markers and Molecular Markers

Molecular markers can be classified as PCR-based and non-PCR-based markers. Non PCR based markers include; allozymes, Restriction Fragment Length Polymorphism (RFLP) are used to identify genetic diversity within the species. It has been used to study the genomic relationship among 24 mangroves and their associated species.

RAPD is another widely used PCR-based indirect molecular approach for genotypic differentiation and molecular taxonomy among different mangrove species. Mangroves do have inter and intra-generic relations, as proved by RAPD. The RAPD markers alone or along with other molecular markers have widely been used to study genetic variations in mangrove species, including Avicennia, *Excoecaria, Acanthus,* and *Bruguiera*. It is also useful to study the parentage analysis in Rhizophora hybrids (Sahu and Kathiresan 2012). Amplified Polymorphic Length Polymorphism (AFLP) is a PCR-based approach to DNA fingerprinting that provides high levels of resolution to delineate complex genetic structures. AFLP markers have been used to study the genomic relations in three species of Heritiera and the results have shown clear-cut segregation of true mangrove species *H. fomes* from mangrove associate *H. littoralis* and also landrace species *H. macrophylla* (Sahu and Kathiresan 2012). Sreekanth and Anupama (2021) reported the genetic diversity and population structure of *Avicennia marina* from its native distributional range in Kerala, Southern India. Gene diversity within the population is comparatively higher than among the population; the establishment of on-site protection zones for *Avicennia marina* to reduce the impact of human activities would allow its habitats to increase in size through natural regeneration to reach maximum population size. *Avicennia marina* mangrove can be introduced from other populations via appropriate propagation and seedling management to increase the chance of gene exchange and recombination and to improve the level of genetic diversity over time. The comprehensive information on genetic diversity will be baseline data for monitoring the genetic pollution that occurs in response to a particular habitat. The introduction of a species to an ecosystem and the relationship between environmental factors and genetic erosion may not be straightforward. It might be non-linear and site-specific and might involve complex interactions among factors. A more extensive study for conservation and phylogeography of *Avicennia marina* is a high priority. Based on a large number of native distributional areas, population genetic diversity in the entire east coast and west coast of India using genome-wide analysis will be helpful for further predicting the implication of conservation of these species.

Microsatellites, also known as simple sequence repeats (SSRs), are an important class of markers because of their greater abundance and variability. SSRs have been used in population genetic studies in mangroves. Cross-priming studies have shown the polymorphic nature of all the loci in *Kandelia candel* and *Rhizophora stylosa* and proved that the microsatellite markers are ideal for population studies (Sahu and Kathiresan 2012).

13.6.4.2 DNA Barcoding of Mangroves

DNA barcoding is a useful alternative for global biodiversity assessment, providing an accurate identification system for living organisms. DNA barcoding is also considered a powerful method to traditional morphological approaches. The two-locus combination of rbcL + matK has been recommended as the plant barcode and has been approved by the Consortium for the Barcode of Life (CBOL) in 2009 (Sahu and Kathiresan 2012).

13.6.4.2.1 Maturase K (matK) Gene

The matK gene, formerly known as orfK, is considered a valuable gene for studying molecular systematics and the evolution of mangroves. Because of its reasonable size, high substitution rate, evenly distributed codon position variation, low transition and transversion ratio, and the easiness of amplification due to its two flanking coding trnK exons. The 1500 bp matK gene is named according to its possible maturase function and its location within the trnK gene encoding the tRNALys (UUU) (Sahu and Kathiresan 2012).

13.6.4.2.2 Ribulose Bisphosphate Carboxylase Large Subunit (Rbcl) Gene

RbcL gene, located on the chloroplast genome, is an appropriate choice for inference of phylogenetic relationships at a higher taxonomic level. It is a 1350 bp gene which is one of the most widely studied chloroplast DNA genes by means of molecular techniques and is known to exhibit extensive variation above the species level (Sahu and Kathiresan 2012).

13.6.4.2.3 Molecular Phylogenetic Software

Molecular phylogenetics applies a combination of molecular and statistical techniques to infer evolutionary relationships among organisms or genes. Recent advances in sequence analysis methods and computational techniques have revolutionized phylogenetics studies. Mangrove geneticists across the globe routinely employ molecular phylogenetic methods in their research (Sahu and Kathiresan 2012).

13.6.4.3 Next-Generation Sequencing

The development of conservation genomics during the last decades has played a significant role in conservation biology. The emergence of next-generation sequencing technologies made it easier to generate molecular data, which is important in the process of conservation. The next-generation sequencing (NGS)-based data are particularly useful in the areas of species identity, defining CUs, and understanding adaptation (Wee et al. 2019).

13.6.5 In Situ and Ex Situ Conservation Strategies

Increased awareness about the protective, productive, and ecological functions of tropical mangrove ecosystems have highlighted the need to conserve and manage them sustainably. India has taken a great initiative in mangrove forest management. The Sundarbans mangroves of the Bay of Bengal (partly in India and partly in Bangladesh) were the first mangroves in the world brought under scientific management. The area's first management plan was implemented in 1892. The government introduced a scheme for the conservation and protection of mangroves which includes identification of specific mangrove ecosystems for conservation, formulation of a management plan, enhancing research on mangroves, and declaring the participation of state governments, universities, research institutions, and local organizations (Deshmukh and Balaji 1994). The two important strategies adopted for mangrove conservation are as follows:

13.6.5.1 In Situ Conservation of Mangroves

This method helps in the conservation of mangroves within their natural habitat along with the associated animals and plants by creating protected areas such as national parks and wildlife sanctuaries. In situ conservation, which remains as a natural ecosystem, is one of the most recommended methods for conserving forest genetic resources. This involves maintaining the natural areas and executing proper protection measures and maintenance so that the ecosystem is restored in its natural environment itself. In situ conservation not only aims for the conservation of mangroves in their natural habitat but also it provides a site for storing the plant genetic material for future use. In India, 188 permanent preservation plots covering a total area of about 8442 Ha have been established with different forest types. These areas are considered as important "gene sanctuaries," where besides the tree species included, all other associated species can evolve under natural selection pressure (Deshmukh and Balaji 1994).

13.6.5.1.1 Role of Conservation Genomics in Management Actions

Having a clear-cut idea about the genetic variations within a population is crucial for the management of threatened species. The reproductive potential, adaptations, and distribution of a species in a population is directly related to the extent of their genetic diversity within the population. Knowing about this genetic information can be helpful in the decision-making process of conservation, such as prioritization of populations and mixing or excluding certain populations during in situ *and* ex situ management activities (Rossetto et al. 2021).

13.6.5.1.2 Criteria for Sampling

In order to design an effective conservation strategy, certain factors must be considered, which include a well-defined criteria for sampling that can go with the further procedures, like recovery process, analysis, data interpretation, and ultimately the recovery actions. In the case of geographically restricted species, sampling can be done considering a representative species of a population. There are certain special circumstances that we need to consider during sampling. In the case of species that do not have accurate information about evolutionary history and distinctiveness, sampling is needed from closely related taxa or from good quality herbarium specimens collected. The chances for hybridization should also be considered while sampling because it is very crucial for the conservation and management of threatened species. Excluding hybrids from sampling will compromise the purity of the remaining populations and affect the translocation plans (Rossetto et al. 2021).

13.6.5.1.3 Sampling Analysis and Data Interpretation

A perfectly designed sampling procedure can provide data for the analysis of the populations under study and to formulate interpretations which ultimately lead to management approaches. A thoughtful analysis of the sample can lead to productive interpretations. Degradation of the environment can lead to habitat loss, thereby reducing the effective population size. A reduction in population size can lead to stochastic hybridization. When small populations are surrounded by closely related individuals, it leads to interspecific gene flow, which can lead to genetic swamping. Hence, when planning for translocation, understanding the risk of hybridization is very important. Polyploidy also can lead to speciation in plants; it is also important to identify the natural distributions and reproductive boundaries between the species under conservation (Rossetto et al. 2021).

Accelerating genetic diversity through genetic rescue and other mechanisms is an important strategy in the conservation of threatened species recovery. Ideas about the reproductive mechanisms of the species are thus important in taking diversity-related decisions. Small populations which face inbreeding depression and genetic drift are more likely to have less genetic diversity. The less diverse the organism, the more it is prone to extinction. So methods that increase genetic diversity and genetic rescue mechanisms can be used to increase their population size and fitness (Rossetto et al. 2021).

Translocations and other recovery-related methods can be thus adapted according to the kind of species to be conserved. However, conserving a particular species alone can affect the genetic diversity of the entire population.

13.6.5.2 Ex Situ Conservation

This method refers to the conservation of mangroves in the areas outside their natural habitat, such as nurseries and botanical gardens. Ex situ conservation is a protective method applicable for both indigenous species, where in situ conservation is not possible, and for exotic species. In this conservation strategy, genetic diversity cannot be maintained as selected species are being conserved and protected in areas outside their natural habitat. But it can help ensure a fairly good representation of the initial variability. Ex situ conservation of mangroves involved the following:

13.6.5.2.1 Acquisition of Germplasm

Germplasm collection consists of going to the field to search for and collect the genetic variability of the wild species from the natural habitat. The same number should be taken from every plant and in good physical and sanitary conditions. A proper moisture content and temperature should be maintained and controlled to prevent the samples from drying out or rotting (Rahman 2016).

13.6.5.2.2 Storage of Samples

After completing the phytosanitary transactions, the germplasm was taken to the place of conservation, where the samples were checked for sufficient number and viability for conservation. Viable and sufficient samples were collected and stored immediately for raising the nursery. Seedlings selected for plantation should be vigorous and in sufficient numbers to represent the genetic variability, thereby ensuring the continuity of the conserved materials. The seedlings should be planted in such a way that they cannot exchange pollen, thus preventing the populations from losing their original genotype. The exact site where each accession was planted should be recorded on a map. Once germplasm is conditioned and stored in the place of conservation under optimal conditions to ensure its survival, the germplasm has to be managed, starting with characterization and evaluation. Growth and survivability data of planted species should be recorded at regular intervals. Meteorological data were recorded. Data on water salinity, soil salinity, soil pH, sedimentation, soil erosion, and inundation were also recorded. Microsoft Excel programs were used to process all collected data and to prepare tables, charts, and graphs (Rahman 2016).

The Sundarbans mangroves located in the Bay of Bengal were the first mangroves in the world to be put under scientific management. The area's first management plan was implemented in 1892. In 1927, the Indian Forest Act was applied to the mangrove forests of the Sundarbans, which have been declared a reserved area. In Chorao, Goa, 178 ha of the best mangrove area has been declared as a Reserved Forest under the Indian Forest Act, 1927 to protect and conserve the mangrove forests. Subsequently, in 1988, this area was declared a bird sanctuary under the Wildlife (Protection) Act, 1972. Since 1987, the Andaman and Nicobar Administration has banned the extraction of mangrove wood. The plywood industries, power stations, and government steam vessels have since switched over to diesel. Bhitarkanika Wildlife Sanctuary, located at a distance of 120 km from Cuttack in Odisha, is the second-largest mangrove forest in India.

13.7 Conclusion

The insufficiency of proper scientific knowledge in mangrove ecosystems and biodiversity and the ecological functions they are providing basket the efforts to conserve and manage mangroves, which leads to the unscientific and unsustainable exploitation of mangroves. A well-organized team of experts from various disciplines is necessary to sort out the real issues in the degradation and execute proper conservation strategies. This will promote the exchange of valuable ideas regarding the maintenance of mangrove ecosystems.

The lack of knowledge regarding the genetic studies of mangroves is one of the major constraints in the area of conservation. Also a taxonomic study on the species diversity is essential to develop a national database which could be able to monitor the status of species diversity in specific mangrove populations. In order to execute proper management actions, the ecological functions as well as the interdependency of mangroves with other ecosystems should also be considered.

References

Alias S, Zainuddin N, Jones E (2010) Biodiversity of marine fungi in Malaysian mangroves. Bot Mar 53(6):545–554. https://doi.org/10.1515/bot.2010.066

Allard S, Costa M, Bulseco A, Helfer V, Wilkins L, Hassenrück C, Zengler K, Zimmer M, Erazo N, Mazza Rodrigues J, Duke N, Melo V, Vanwonterghem I, Junca H, Makonde H, Jiménez D, Tavares T, Fusi M, Daffonchio D, Duarte C, Peixoto R, Rosado A, Gilbert J, Bowman J (2020) Introducing the mangrove microbiome initiative: identifying microbial research priorities and approaches to better understand, protect, and rehabilitate mangrove ecosystems. mSystems 5(5): e00658–e00620. https://doi.org/10.1128/mSystems.00658-20

Alongi D (1988) Bacterial productivity and microbial biomass in tropical mangrove sediments. Microb Ecol 15(1):59–79. https://doi.org/10.1007/BF02012952

Alongi D (1994) The role of bacteria in nutrient recycling in tropical mangrove and other coastal benthic ecosystems. In: Sasekumar A, Marshall N, Macintosh DJ (eds) Ecology and conservation of Southeast Asian marine and freshwater environments including wetlands, developments in hydrobiology, vol 98. Springer, Dordrecht, pp 19–32. https://doi.org/10.1007/978-94-011-0958-1_3

Alongi D (2002) Present state and future of the world's mangrove forests. Environ Conserv 29(3): 331–349. https://doi.org/10.1017/S0376892902000231

Alongi D, Mukhopadhyay S (2015) Contribution of mangroves to coastal carbon cycling in low latitude seas. Agric For Meteorol 213:266–272. https://doi.org/10.1016/j.agrformet.2014.10.005

Bano N, Nisa M, Khan N, Saleem M, Harrison P, Ahmed S, Azam F (1997) Significance of bacteria in the flux of organic matter in the tidal creeks of the mangrove ecosystem of the Indus River delta, Pakistan. Mar Ecol Prog Ser 157:1–12. https://doi.org/10.3354/meps157001

Bashan Y, Holguin G (2002) Plant growth-promoting bacteria: a potential tool for arid mangrove reforestation. Trees 16(2–3):159–166. https://doi.org/10.1007/s00468-001-0152-4

Berkes F (2004) Rethinking community-based conservation. Conserv Biol 18(3):621–630

Bindiya ES, Bhat SG (2019) KSBB report on Kerala flood. https://www.keralabiodiversity.org/images/2019/November/Flood_Report/Flood_Impact_Metagenomics_Mangroves_Dr_Bindiya_ES.pdf

Castellanos-Galindo G, Kluger L, Camargo M, Cantera J, Mancera Pineda J, Blanco-Libreros J, Wolff M (2021) Mangrove research in Colombia: temporal trends, geographical coverage and research gaps. Estuar Coast Shelf Sci 248:106799

Catt B, Erwig A, Hiebert K (2022) Mangroves | mangroves.elaw.org. https://mangroves.elaw.org

Chinadaily (2022). https://global.chinadaily.com.cn/a/202108/26/WS61277d36a310efa1bd66b690.html

Crisp P (1990) Mangroves in New Zealand: trees in the tide. GP Books

Das S, Lyla P, Khan S (2007) Spatial variation of aerobic culturable heterotrophic bacterial population in sediments of the continental slope of western Bay of Bengal. Ind J Marine Sci 36(1):51–58. http://nopr.niscair.res.in/handle/123456789/14

de Lange W, De Lange P (1994) An appraisal of factors controlling the latitudinal distribution of mangrove (*Avicennia marina* var. *resinifera*) in New Zealand. J Coast Res 10(3):539–548. http://www.jstor.org/stable/4298251

Deshmukh S, Balaji V (1994) Conservation of mangrove forest genetic resources. A training manual. https://idl-bnc-idrc.dspacedirect.org/bitstream/handle/10625/33954/118818.pdf?sequence=1

Dinerstein E, Olson D, Joshi A, Vynne C, Burgess N, Wikramanayake E, Hahn N, Palminteri S, Hedao P, Noss R, Hansen M (2017) An ecoregion-based approach to protecting half the terrestrial realm. Bioscience 67(6):534–545. https://doi.org/10.1093/biosci/bix014

Eco-Business (2013). https://www.eco-business.com/news/malaysias-mangrove-forest-rapidly-depleted-and-degraded/

Edirisinghe EA, Ariyadasa KP, Chandani RP (2012) Forest cover assessment in Sri Lanka. Sri Lanka Forester 34:1–2

FAO (2020) Global forest resources assessment 2020: main report. Rome

Friis G, Vizueta J, Smith E, Nelson D, Khraiwesh B, Qudeimat E, Salehi-Ashtiani K, Ortega A, Marshell A, Duarte C, Burt J (2020) A high-quality genome assembly and annotation of the gray mangrove, *Avicennia marina*. G3 Genes Genomes Genet 11(1):1–7. https://doi.org/10.1093/g3journal/jkaa025

Frith DW, Tantanasiriwong R, Bhatia C (1976) Zonation of macrofauna on a mangrove shore, Phuket Island Thailand. Res Bull Phuket Mar Biol Cent Res Bull 10:1–37

Göltenboth F, Schoppe S (2006) 10 - Mangroves. In: Göltenboth F, Timotius KH, Milan PP, Margraf J (eds) Ecology of insular south-east Asia. Elsevier, Amsterdam, pp 187–214

Hogarth PJ (2015) The biology of mangroves and seagrasses. Oxford University Press

Holguin G, Vazquez P, Bashan Y (2001) The role of sediment microorganisms in the productivity, conservation, and rehabilitation of mangrove ecosystems: an overview. Biol Fertil Soils 33(4): 265–278. https://doi.org/10.1007/s003740000319

Hoppe-Speer S, Adams J, Rajkaran A (2013) Response of mangroves to drought and non-tidal conditions in St Lucia Estuary, South Africa. Afr J Aquat Sci 38(2):153–162. https://doi.org/10.2989/16085914.2012.759095

Horstman E, Lundquist C, Bryan K, Bulmer R, Mullarney J, Stokes D (2018) The dynamics of expanding mangroves in New Zealand. In: Makowski C, Finkl C (eds) Threats to mangrove forests, vol 25. Coastal Research Library, Springer, Cham. https://doi.org/10.1007/978-3-319-73016-5_2

Iino T, Mori K, Uchino Y, Nakagawa T, Harayama S, Suzuki K (2010) *Ignavibacterium album* gen. nov., sp. nov., a moderately thermophilic anaerobic bacterium isolated from microbial mats at a terrestrial hot spring and proposal of *Ignavibacteria* classis nov., for a novel lineage at the periphery of green sulfur bacteria. Int J Syst Evol Microbiol 60(6):1376–1382. https://doi.org/10.1099/ijs.0.012484-0

IPCC Fourth Assessment Report (2007) The physical science basis, 2:580–595

IPCC (2013) Climate Change: the physical science basis. Contribution of Working Group I to the Fifth Assessment Report of the Intergovernmental Panel on Climate Change. Cambridge University Press, Cambridge, 1535 pp

Islam S, Bhuiyan M (2018) Sundarbans mangrove forest of Bangladesh: causes of degradation and sustainable management options. Environ Sustain 1(2):113–131. https://doi.org/10.1007/s42398-018-0018-y

Jousset A, Bienhold C, Chatzinotas A, Gallien L, Gobet A, Kurm V, Küsel K, Rillig MC, Rivett DW, Salles JF, Van Der Heijden MG (2017) Where less may be more: how the rare biosphere pulls ecosystems strings. ISME J 11(4):853–862. https://doi.org/10.1038/ismej.2016.174

Kathiresan K, Bingham B (2001) Biology of mangroves and mangrove Ecosystems. Adv Mar Biol 40:81–251. https://doi.org/10.1016/S0065-2881(01)40003-4

Kibria G (2013) Mangrove forests-its role in livelihoods, carbon sinks and disaster mitigation. RMIT University, Melbourne, VIC, Australia, Technical Report, pp 1–17

King RJ, Adam P, Kuo J (1990) Seagrasses, mangroves and salt marsh plants. In: Clayton MN, King RJ (eds) Biology of marine plants. Longman Cheshire, Melbourne, pp 213–239

Koch F (1998) Biogeography. In: Brown JH, Lomolino MV (eds) Zoosystematics and evolution, 2nd edn. Sinauer Associates, Publishers, p 158

Kohlmeyer J, Bebout B, Vlkmann-Kohlmeyer B (1995) Decomposition of mangrove wood by marine fungi and Teredinids in Belize. Mar Ecol 16(1):27–39. https://doi.org/10.1111/j.1439-0485.1995.tb00392.x

Kutt A (2007) Bird assemblage in a dune-mangrove mosaic, Cairns, Queensland. Aust Zool 34(2):158–164. https://doi.org/10.7882/AZ.2007.013

Luther D, Greenberg R (2009) Mangroves: a global perspective on the evolution and conservation of their terrestrial vertebrates. Bioscience 59(7):602–612. https://doi.org/10.1525/bio.2009.59.7.11

Lyimo T, Pol A, Op den Camp H (2002) Sulfate reduction and methanogenesis in sediments of Mtoni mangrove forest, Tanzania. AMBIO 31(7):614–616

Lyimo T, Pol A, Jetten M, Op den Camp H (2009) Diversity of methanogenic archaea in a mangrove sediment and isolation of a new Methanococcoides strain. FEMS Microbiol Lett 291(2):247–253. https://doi.org/10.1111/j.1574-6968.2008.01464.x

Macintosh DJ, Ashton EC (2002) A review of mangrove biodiversity conservation and management. Centre for Tropical Ecosystems Research, University of Aarhus, Denmark. http://mit.biology.au.dk/cenTER/MCB_Files/2002_Review_WB_MCB_Final.pdf

Marshall I, Starnawski P, Cupit C, Fernández Cáceres E, Ettema T, Schramm A, Kjeldsen K (2017) The novel bacterial phylum Calditrichaeota is diverse, widespread and abundant in marine sediments and has the capacity to degrade detrital proteins. Environ Microbiol Rep 9(4):397–403. https://doi.org/10.1111/1758-2229.12544

Miller JR (2005) Biodiversity conservation and the extinction of experience. Trends Ecol Evol 20(8):430–434. https://doi.org/10.1016/j.tree.2005.05.013

Mohanraju R, Rajagopal BS, Daniels L, Natarajan R (1997) Isolation and characterization of a methanogenic bacterium from mangrove sediments. J Mar Biotechnol 5(2–3):0147–0152

Muraleedharan PK, Swarupanandan K, Anitha V, Ajithkumar C (2009) The conservation of mangroves in Kerala: economic and ecological linkages. Kerala Forest Research Institute, Peechi, pp 8–18

Nagelkerken I, Blaber S, Bouillon S, Green P, Haywood M, Kirton LG, Mayenecke JO, Pawlik J, Penrose HM, Sasekumar A, Somerfield PJ (2008) The habitat function of mangroves for terrestrial and marine fauna: a review. Aquat Bot 89(2):155–185

Naidoo G (2016) The mangroves of South Africa: an ecophysiological review. S Afr J Bot 107: 101–113

National Research Council (2011) Advancing the science of climate change. National Academies Press

Noske RA (1995) The ecology of mangrove forest birds in Peninsular Malaysia. Ibis 137(2): 250–263

Noske RA (1996) Abundance, zonation and foraging ecology of birds in mangroves of Darwin Harbour, Northern Territory. Wildlife Res 23(4):443–474

Odum WE, Heald EJ (1975) The detritus-based food web of an estuarine research: chemistry, biology, and the estuarine system, vol 1. p 265

Pandey C, Hirose, K (2013) International workshop on mangrove conservation in India. In: Proceedings of International Workshop on Mangrove Conservation in India

Parte AC, Sardà Carbasse J, Meier-Kolthoff JP, Reimer LC, Göker M (2020) List of prokaryotic names with standing in nomenclature (LPSN) moves to the DSMZ. Int J Syst Evol Microbiol 70(11):5607–5561

Raghukumar S, Sharma S, Raghukumar C, Sathe-Pathak V, Chandramohan D (1994) Thraustochytrid and fungal component of marine detritus. IV. Laboratory studies on decomposition of leaves of the mangrove *Rhizophora apiculata* Blume. J Exp Mar Biol Ecol 183(1): 113–131

Rahardjanto A, Tosiyana VR, Husamah H, Miharja FJ (2020) Diversity of molluscs in the mangrove forest area of Cengkrong Beach-Trenggalek. In: AIP Conference Proceedings 040075-1-040075-7

Rahman MM (2016) Ex situ conservation of *Rhizophora Mucronata* lam. in the Sundarban mangrove forest of Bangladesh. Int J Bus Soc Sci Res 5(1):103–110

Reef R, Feller I, Lovelock C (2014) Mammalian herbivores in Australia transport nutrients from terrestrial to marine ecosystems via mangroves. J Trop Ecol 30(3):179–188

Rog S, Clarke R, Cook C (2016) More than marine: revealing the critical importance of mangrove ecosystems for terrestrial vertebrates. Divers Distrib 23(2):221–230

Rossetto M, Yap J, Lemmon J (2021) A conservation genomics workflow to guide practical management actions. Glob Ecol Conserv 26:1492

Sahu SK, Kathiresan K (2012) Molecular markers: an intricate tool for new insights in mangrove genetics. Int J Adv Biotechnol Res 3(8):847–863

Schodde R, Mason IJ, Gill HB (1982) The avifauna of the Australian mangroves: a brief review of composition, structure and origin. In: Proceedings of the Australian National Mangrove Workshop, Australian Institute of Marine Science, Cape Ferguson

Schreier BM, Harcourt AH, Coppeto SA, Somi MF (2009) Interspecific competition and niche separation in primates: a global analysis. Biotropica 41:283–291

Shackelton S, Campbell BM, Wollenberg E, Edmunds D (2002) Devolution and community-based natural resource management: creating space for local people to participate and benefit? Nat Resour Perspect 76:1–6

Shoreit AAM, El-Kady IA, Sayed WF (1992) Isolation and identification of purple non-sulfur bacteria of mangal and non-mangal vegetation of Red Sea Coast, Egypt. Bulletin of the Faculty of Science, Assiut University, Egypt

Spalding M (2010) World atlas of mangroves. Routledge

Sreekanth PM, Anupama K (2021) Genetic diversity of mangrove tree species Avicennia marina in eco-geographic regions of Kerala coast, Southern India. Ecol Genet Genomics 20:100094

Steinke TD (1999) Mangroves in South African estuaries. Estuaries of South Africa. pp 119–140

Sundararaju V (2019) Scientific management of mangroves is need of the hour. https://www.downtoearth.org.in/blog/wildlife-biodiversity/scientific-management-of-mangroves-is-need-of-the-hour-64007

Tavares TCL, Bezerra WM, Normando LRO, Rosado AS, Melo VMM (2021) Brazilian semi-arid mangroves-associated microbiome as pools of richness and complexity in a changing world. Front Microbiol 26:2485

Tomlinson PB (1986) The botany of mangroves. Cambridge University Press, London

Tong T, Li R, Wu S, Xie S (2019) The distribution of sediment bacterial community in mangroves across China was governed by geographic location and eutrophication. Mar Pollut Bull 140: 198–203

USFS (2014) West African mangrove ecosystems: developing a basis for a regional perspective. Report compiled by Brooke Czwartacki. https://rmportal.net/library/content/fcmc/publications/mangroves-in-west-africa-a-policy-brief/at_download/file

Vanucci M (1989) The mangroves and us: a synthesis of insights. Indian Association for the Advancement of Science, New Delhi

Vazquez P, Holguin G, Puente M, Lopez-Cortes A, Bashan Y (2000) Phosphate-solubilizing microorganisms associated with the rhizosphere of mangroves in a semiarid coastal lagoon. Biol Fertil Soils 30(5-6):460–468

Vermeij GJ (1973) Molluscs in mangrove swamps: physiognomy, diversity, and regional differences. Syst Zool 22(4):609–624

Ward RD, de Lacerda LD (2021) Responses of mangrove ecosystems to sea level change. In: Dynamic sedimentary environments of mangrove coasts. Elsevier, pp 235–253

Ward CJ, Steinke TD (1982) A note on the distribution and approximate areas of mangroves in South Africa. S Afr J Bot 1(3):51–53

Ward R, Friess D, Day R (2016) Impacts of climate change on mangrove ecosystems: a region by region overview. Ecosyst Health Sustain 2(4):01211

Wee AKS, Mori GM, Lira CF, Núñez-Farfán J, Takayama K, Faulks L, Shi S, Tsuda Y, Suyama Y, Yamamoto T, Iwasaki T, Nagano Y, Wang Z, Watanabe S, Kajita T (2019) The integration and application of genomic information in mangrove conservation. Conserv Biol 33(1):206–209

Wiens JJ, Graham CH, Moen DS, Smith SA, Reeder TW (2006) Evolutionary and ecological causes of the latitudinal diversity gradient in hylid frogs: treefrog trees unearth the roots of high tropical diversity. Am Nat 168:579–596

World Atlas (2020). https://www.worldatlas.com/forests/mangrove-forest.html

Zaradic PA, Pergams OR, Kareiva P (2009) The impact of nature experience on willingness to support conservation. PLoS One 4(10):7367

Zuberer D, Silver W (1978) Biological dinitrogen fixation (acetylene reduction) associated with Florida mangroves. Appl Environ Microbiol 35(3):567–575

Chapter 14
Conservation of RET Plants: Strategies and Plans

V. Mithun, S. Pradeep Kumar, and N. S. Pradeep

Abstract The Earth is rich in biodiversity, rich in valuable flora and fauna. Even so, many destructions are happening through time. Plants, for example, despite their valuable services to mankind, are being ruthlessly destroyed due to development projects and increased dependence. Although many species are threatened by anthropogenic pressure, many are threatened by invasive alien species and climate changes. Thus, many plants are threatened with extinction and are included in the RET group by IUCN. Before taking scientific measures to ensure their conservation and cultivation, it is essential to study their natural distribution and their demographic status. There are a number of strategies adopted by the government and various organizations to protect them. The ex situ and in situ formulas are mostly applicable for RET plant conservation. But there are so many limitations to each of them, and the strategies are specific to plants including in the threatened category. Botanical gardens and seed banks have a major role in this conservation.

Keywords RET plants · Conservation strategies · Ex situ · In situ · Challenges · GSPC · BRAHMS

14.1 Introduction

In the past 50–60 years, the human population has doubled, while the global economy has quadrupled, and around a tenfold increase was recorded in global trade (Diaz et al. 2019). It is recognized that worldwide, tens of thousands of plants are included in the rare category, endangered category, or threatened category. Some of them will become extinct in this century if the present situations continue (Blackmore et al. 2001). Worldwide, many plant species are threatened with extinction because of the spontaneous disappearance of the natural ecosystems for various human activities. More than 50% of the global plant species are endemic to 34 global

V. Mithun · S. Pradeep Kumar · N. S. Pradeep (✉)
KSCSTE-Malabar Botanical Garden and Institute for Plant Sciences, Kozhikode, Kerala, India

S. T. Sukumaran, K. T R (eds.), *Conservation and Sustainable Utilization of Bioresources*, Sustainable Development and Biodiversity 30,
https://doi.org/10.1007/978-981-19-5841-0_14

biodiversity hotspots (GBH), which once covered 15.7% of the Earth's land surface and which are now declined to 2.3%. These areas include most of the endemic species, which face an increasing threat of extinction. The International Union for Conservation of Nature (IUCN) published the Red List of Threatened Plants, first in 1998 (IUCN 1998), now the list includes 5000 species (IUCN 2021). Climate change predictions impacting on plants and their habitats include temperature increases, changes in precipitation patterns, desertification process, rises in sea level, shifts in growing seasons of plants, loss of pollinators and seed dispersers, and increasing frequency and intensity of extreme weather events, such as floods, storms, and droughts (Hawkins et al. 2008). Identifying hotspots or priority areas where biodiversity is most threatened is vital for effective conservation (Myers et al. 2000; Olson and Dinerstein 1998). Unfortunately, as compared with animal conservation, plant conservation is not adequately supported worldwide (Pennisi 2010).

Our collective failure to stem the tide of biodiversity loss has led many to question the effectiveness of our current policies and approaches, and there have been numerous calls for a radical rethink (Bridgewater 2016; Kareiva et al. 2012), although we do not agree on how to proceed. The causes of this situation are complex, including scientific, technical, social, economic, and political factors. As witnessed by the recent creation of the Intergovernmental Science-Policy Platform on Biodiversity and Ecosystem Services (IPBES), the focus is shifting from biodiversity to the social benefits of the goods and services it produces.

It is certain that governments, for whatever reason, will not be able to effectively implement many of the commitments they have made under various international agreements, such as the Convention on Biological Diversity to which they are a party. The inevitable consequence of this is that the various goals set for biodiversity conservation have not been achieved, and the ability of other countries to fill the resulting gaps in achieving global goals has been limited. Despite their importance, they continue to face high biodiversity loss rates (Waldron et al. 2013).

The flora of India is one of the richest in the world according to its wide range of climatic conditions, topology features, and other environmental factors. Conservation of rare, endangered, and threatened (RET) plant species is a serious issue in the field of floristic diversity. Considering the vast extension of the country and the variation in vegetation patterns in different areas and within the same area, a discussion on the floristic diversity and vegetation of the country as a whole would be too general. In India, habitat loss, land-use change, invasion of exotics, pollution, climate change, and unsustainable collection of plants from the wild for commercial purposes have been the primary threats (Rathcke and Jules 1993). Usually, there is an intrinsic relationship between the quantity of native vegetation and the well-being of the plant species that rely on it for their habitat. About 25% of the higher plant species are expected to disappear in the next few decades, and another 25% may be lost by the end of the twenty-first century (Singh and Khurana 2002). Awareness of realizing the importance of floral diversity and the rapid decline has come to notice, which has given an unprecedented impetus for monitoring and conservation (Moza and Bhatnagar 2007).

The progressive degradation of tropical forest ecosystems In many parts of the world has, in addition to undermining biodiversity conservation efforts, negatively impacted human societies and economies at the local, national, and regional scales.

Quantitative assessment of the status and distribution of RET plants is indispensable for proper conservation and management in wild populations. The lack of authentic data on the status, threats, and distribution of RET plants leads to various kinds of disturbances, including the conservation of the natural population.

14.2 RET Plants

The IUCN Red List is a rich summary of information on threats, environmental requirements, and habitats and about conservation measures that can be taken to reduce or prevent extinctions. It is based on an objective system for assessing the risk of extinction of a species based on past, present, and predicted threats. Species evaluation is carried out after a standardized process using strict IUCN Red list categories and criteria, ensuring a high level of scientific documentation, information management, expert review, and justification. The IUCN Red List shows where action should be taken to save nature's building blocks from extinction. It provides a straightforward way to integrate biodiversity requirements into decision-making processes by providing a wealth of useful information about living things. Loss of wildlife reduces our quality of life and basic economic security. Conservation of wildlife protects the biodiversity and ecosystem that provide us with the natural resources we need to live.

The IUCN (The International Union for Conservation of Nature) has been classified the threatened group of plants under nine categories based on mainly three criteria namely: reduction in the population size over the last 10 years or three generations, geographic range in the form of either extent of occurrence or area or occupy both and by the population size numbers. In many instances, the rarity standards assigned have become deviated from the original concept, and threat status is given to a species based on local species/population knowledge by the individuals or assigned institutions. This has resulted in the formation of different threat statuses to a species by different authors. However, the IUCN threat categories are followed for a realistic assessment on red-listed plant species in the country. The plant rarity concept has been mainly classified based on three traits that all the species possess namely: Geographical range/distribution, specificity of the particular habitat, and local population size (Drury 1974).

14.3 Conservation of Biodiversity

India is well known for its biodiversity richness due to the presence of large numbers of flora and fauna. Our cultural diversity has a major role in the conservation of entire biodiversity. Having said that, this flora and faunal diversity are now at a danger level. Global climate change, anthropogenic interventions, entry of invasive alien species into the ecosystem, etc., are the major threats. It is high time that the corresponding government missions and local self-bodies come forward to take the conservation steps for the recovery of RET plants. A systematic species conservation program is the need of the hour.

Identification of the plant species, assessing their natural population status, and studying their geographical distributions are very much necessary before taking the scientific approach or measure toward their conservation. The reduction in the natural population size is considered to be an important threatening factor. It cannot sustain them due to inbreeding and loss of genetic variability. It has become urgent to carry out scientific conservation programs for recovering these species.

Unless immediate measures are taken for the restoration of critically endangered species, they will become extinct. Species recovery is the process through which the decline of a threatened species population is arrested or reversed and threats removed so that the survival of the species population in the wild can be ensured. Many developed countries have initiated and implemented plans to address the restoration of rare, endangered, and threatened species. Species recovery programs' collaboration with international agencies has been carried out successfully in the United States, Australia, United Kingdom, and Canada.

In the case of the United States, there is a special legislation act such as the Endangered Species Act, 1973, (ESA) for carrying out species recovery programs. The Act has provisions for listing the species as "endangered," developing recovery strategies for each species and its corresponding critical natural habitats. So far, 47 species have been stabilized through different recovery processes and have been excluded from the recovery programs. A gradual increase in the population size, habitat restoration, and captive breeding or population stabilization has been achieved through recovery programs.

There are two basic conservation strategies, each of which involves a number of techniques that conservationists can adopt to protect plant genetic diversity. The two conservation strategies are in situ and ex situ. The two protection strategies are fundamentally different in the way they are applied. In-situ conservation involves the designation, management, and monitoring of target taxa where they thrive or are maintained (Fig. 14.1), whereas ex-situ conservation involves the sampling, transfer, and storage of target taxa from the target area (Fig. 14.2). When considering any protection strategies and technologies to be adopted, the Convention on Biological Diversity (CBD) (1992) focuses on ex-situ techniques that are usually considered supplementary supports in-situ techniques and become ideal in the future. The CBD states that each contracting party will take ex-situ conservation actions. Merits and

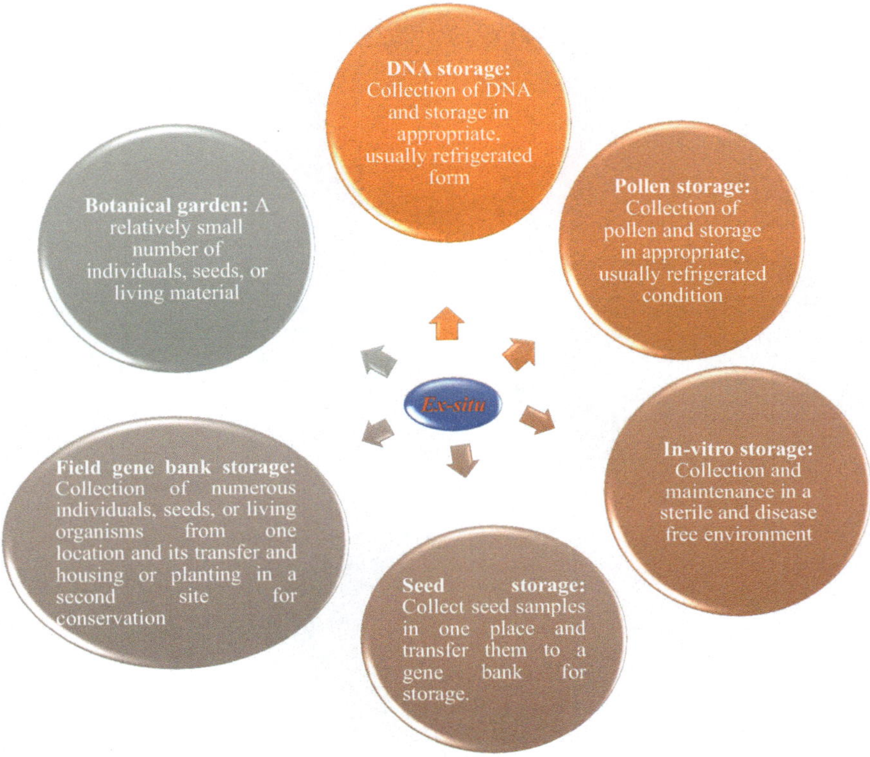

Fig. 14.1 Ex-situ conservation strategies

demerits of both ex-situ and in-situ conservation strategies are expressed in Table 14.1.

14.4 Conservation Challenges

Standardized horticultural practices are meager in the case of RET plants, which deserve special attention to take care of. Ruthless harvesting and unsustainable collection practices lead to the tremendous loss of the natural population of many medicinal plants and are considered one of the major reasons for its rarity. Therefore, it is high time to provide awareness programs to drug collectors for sustainable harvesting of RET medicinal plants.

Certain major challenges exist in plant conservation around the world. It includes mitigating the impacts of major threatening processes associated with the native habitat loss, entry of invasive alien species, raising awareness of the cultural and socio-economic value of plant conservation, dealing with the nationwide loss of

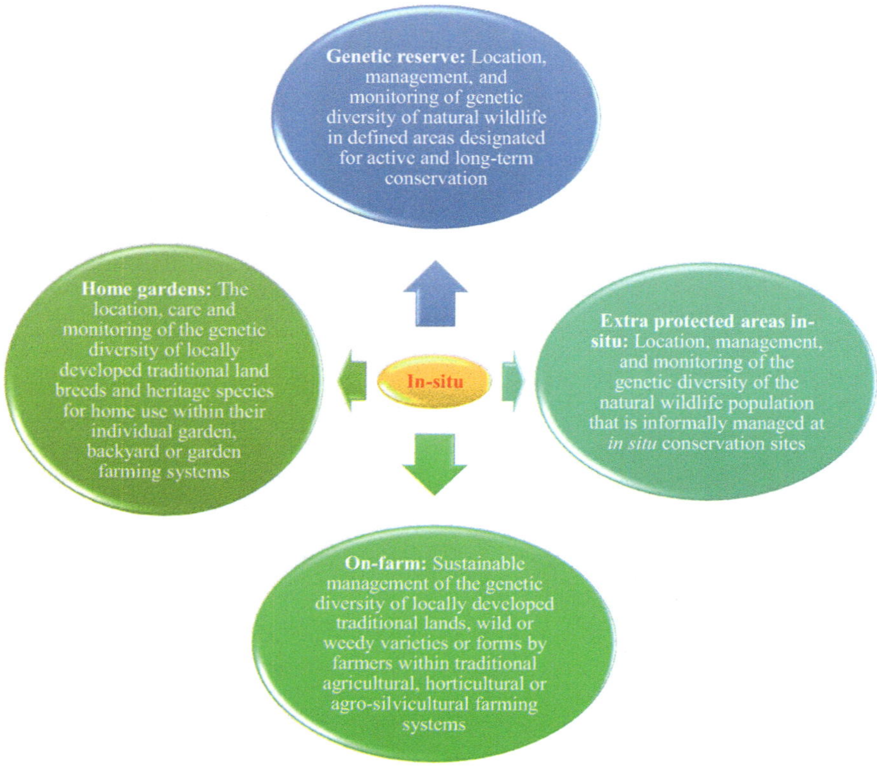

Fig. 14.2 In-situ conservation strategies

taxonomic expertise despite the rich and diverse flora, and improving our understanding of the complete biology of the many plant species targeted for conservation actions.

Over the last few decades, threatening processes have become increasingly clear that targeted active management of reserves, specific threatened plant communities, and threatened plant populations will need to be implemented across the world at levels not previously envisaged. Widespread land clearing and native habitat loss, habitat fragmentation through anthropogenic interventions, invasive alien species, introduced plant diseases, inappropriate fire regimes, and hydrological change are all significant factors in ongoing biodiversity decline.

The significant decline in the natural vegetation throughout the important areas in India is associated to a large degree with widespread land transformation for the construction of buildings, agriculture extension, and other unsustainable construction activities. This, in turn, has, either directly or indirectly, led to major habitat fragmentation, now one of the most important factors affecting the persistence of species and species assemblages in ecosystems. Fragmentation of habitat has been found to result in a wide range of negative impacts on ecological and genetic

Table 14.1 Relative merits and demerits of in-situ and ex-situ conservation strategies

Conservation strategy	Merits	Demerits
In situ	• Target taxa protected along with related taxa within a single reserve • Provides easy access to evolution as well genetic studies • Suitable for intermediate and recalcitrant seeded species • Allows species/pathogen interactions and co-evolution • Appropriate conservation in relation to biotic and abiotic changes	• Materials are not readily available for utilization • Victims of natural and human-mediated disasters, e.g., fire, destruction, urban-industrial development, and climate change • Appropriate management is often not well understood by governments • Requires regular active management and monitoring • Limited genetic diversity can be preserved everywhere
Ex situ	• Medium- and long-term storage is possible • Unlimited genetic diversity can be preserved in a limited location • An easy way to assess resistance to pests and diseases • An easy way to plant breeding, farmers, and other uses • Minor routine maintenance (excluding field gene banks and botanic garden entrances) once the material has been preserved	• It is not possible to store seeds of recalcitrant or intermediate species without the use of in vitro or field gene banks • A practical limitation of storage space for diversity of recalcitrant or intermediate species in field gene banks • Evolution freezes in relation to environmental changes • Genetic diversity is likely to be lost with each regeneration cycle • In vitro storage variations and mutations may be lost

processes in plant populations and has also led to significant degradation of habitat associated with grazing, inappropriate fire regimes, nutrient enrichment, soil disturbances, and a broad range of invasive weeds.

In the case of temperate forests, fire regimes have been identified as an important threat to threatened plant species populations. In temperate regions, particular combinations of fire frequency, fire intensity, and season interrupt their life-cycle processes failing to produce adequate cues that critically affect the phases of germination, development, and reproduction.

Plant diseases are often not highlighted as much as other threats to the flora, but some pathogens are played an important role in the threatening process. Dieback, rust, smut, damping off, etc., are certain plant diseases vitally affecting the plant populations. *Phytophthora* spp. and *Puccinia* spp. are the major pathogens recognized not only because of the impact these are already having on diverse native plant communities and many rare and threatened species but also because of the major difficulties in effective control.

Phytophthora cinnamomi has been listed as one of the world's most destructive invasive species (Lowe et al. 2000), and its historical and predicted future spread into vulnerable Australian plant communities in NSW, Victoria, Tasmania, and Western

Australia is already resulting in significant permanent loss of biodiversity (Shearer et al. 2004; Cahill et al. 2008). Unlike *Phytophthora* dieback, which has been in Australia for many decades, Myrtle rust has only recently arrived. Yet by 2014, it has proved capable of infecting more than 300 native Australian Myrtaceae species. Nearly half of the 2250 native species of Myrtaceae occur in the climatic areas assessed as suitable for the disease making the potential impact on threatened species, plant communities, and in some cases, whole ecosystems significant (Makinson 2014).

The major challenge for effective plant conservation is raising public awareness about the cultural, social, and economic value of conserving rich, unique, and diverse plant species and communities. In many cases, there is far less funding for plants, despite the critical role that plants play in providing food and shelter for animals and perceptions that plants are more easily recovered from decline than animals. This clearly indicates a bias toward animals and a tendency among humans to neither notice nor value plants in the environment (Balding and Williams 2016).

Lack of taxonomic knowledge and the ability to readily identify plant taxa in the field is one of the most significant obstacles to systematic study and effective plant conservation (Coates and Atkins 2001; Wege et al. 2015). Decreasing taxonomic capacity across the country is a growing concern and limitation given the increasing conservation issues requiring some level of taxonomic input. While strategic approaches have been developed recently to deal with taxa that require taxonomic resolution to improve knowledge for conservation management (Wege et al. 2015), there is still an imperative to maintain taxonomic resources to deal with ongoing name changes, species descriptions, and new species discoveries that are critical for the increasing number of species conservation assessments undertaken, as well as and the listing of species of conservation concern.

Limited biological knowledge is one of the important challenges facing RET plant conservation and restoration programs. Interactions between plants and animals are very important in biodiversity conservation. Because protecting the plants is as important as protecting the organisms associated with them. In Australia, rarer Acacias appear to have quite a specific association with symbiotic nodulating bacteria (Thrall et al. 2000), suggesting that the loss of either species will compromise successful plant conservation. In the case of RET species, seed banks may be contingent upon having the corresponding symbiont in the reintroduction or restoration site. Improved knowledge of the specific threatened species and co-dependent species is important for achieving effective conservation outcomes.

14.5 Global Strategy for Plant Conservation

In 2002, the United Nations Convention on Biological Diversity (CBD) adopted The Global Strategy for Plant Conservation (GSPC) program, agreed by the UK and 195 other governments around the world, with targets intended to have been met by 2020. Although it was accepted in September 2020 that the 16-decade-long goals would be missed, significant progress had been made toward achieving them, and this would not have been possible without the GSPC. This set the framework for Plant life's work with Plant Link across the UK and its nations and with our international partners.

14.6 The GSPC 2011–2020 Has 16 Targets Under Five Objectives

- Plant diversity is well understood, documented, and recognized.
- Plant diversity is urgently and effectively conserved.
- Plant diversity is used in a sustainable and equitable manner.
- Education and awareness about plant diversity, its role in sustainable livelihoods, and its importance to all life on Earth are promoted.
- The capacities and public engagement necessary to implement the strategy have been developed.

14.6.1 On Target

- The UK probably has the best documented flora in the world, with nearly 6000 species of flowers, mosses and lichens mapped and monitored. This is an incredible conservation achievement; there are just 557 butterflies, birds, and bees in the whole of Britain.
- Over 75% of our threatened flora is stored in seed banks, and seeds and plants are being used to restore populations in the wild.
- Ninety-six percent of our globally significant Important Plant Areas (IPAs), such as the New Forest IPA, are covered by statutory protection.

14.6.2 Below Target

- Major habitats are not being managed well enough. Just 4% of our woodland, 8% of heathland, and 11% of grasslands are in "favourable" conservation conditions.
- 95% of threatened plants in England and Wales are found on farmland and in woodland, yet current agri-environment schemes are largely failing these species
- Threatened species continue to decline: 72 species, including golden-eye lichen and corn marigold, are not prioritized for action. Some that are, like spreading bellflower and field gentian, continue on the path toward extinction.
- Occurrences of invasive non-natives such as American skunk-cabbage have doubled in the wild, piling pressure on wildlife sites. And they are still on sale.

- Experts are as threatened as the plants and fungi they work hard to conserve. There are more Pandas in British zoos than there are lichenologists employed by agencies, museums, and botanic gardens.

Botanic gardens worldwide have continued to be one of the most active groups in working to refine and implement the GSPC. The botanic garden community immediately began to formulate its ongoing role through the development and implementation of the International Agenda for Botanic Gardens in Conservation (Wyse Jackson and Sutherland 2000).

The international plant conservation scene has undergone an outstanding transformation over the past few years, with the implementation of the GSPC. It has a vital role in stimulating new initiatives and programs at all levels, particularly among certain sectors, especially botanic gardens.

14.7 Role of Botanical Gardens in RET Plant Conservation

A majority of people do not become aware of the role of botanical gardens in RET plant conservation. Most of them just think that botanical gardens only have some recreational values instead of research and development. Botanical gardens are ideal locations for many branches of scientific research. Botanical gardens not only serve as systematic and taxonomic research centers (Dosmann 2006; Stevens 2007), but they also play a vital role as valuable sources of plant ecology data collection such as phenological indication of climate change, plant physiology, plant biochemical changes, and plant growth tactics, and different types of plant–animal interactions in various species (Coates and Dixon 2007; Gratani et al. 2008; Dawson et al. 2009; Primack and Miller-Rushing 2009; Wang et al. 2018). Botanical gardens are good locations for investigations into animal–plant interactions, pollination biology, and ecology, and various seed dispersal patterns. For example, in the 1950s, through the study of seed dispersal in an endangered species, *Taxus Chinensis*, in an ex-situ conservation population introduced into the Nanjing Botanical Garden, researchers were able to propose that any process for the conservation of these Chinese yews should comprise not only conservation of the trees but also conservation of these tree's avian dispersers and habitats for seed germination and seedling growth (Lu et al. 2008; Li et al. 2014).

Plant conservation genetics provides exact tools to propose conservation and successful restoration programs, monitor and measure processes on various parameters, and reduce the extinction risk of threatened plant species in nature (Kramer and Havens 2009). Recent scientific works on the genetic aspects of RET plant conservation have shown that genetic erosion plays an increasing threat to the long-term survival of RET species (Desalle and Amato 2004; Ouborg et al. 2006).

Live plant collections are the major advantage of botanical gardens. In the case of threatened taxa, botanical gardens are given extra care, especially for species that are threatened in the wild. Separate RET conservatories are started there to protect them with additional attention. One of the main objectives of botanical gardens is to establish and support the collection of the endemic taxa and to produce and maintain plant stocks of the desired plants for ex-situ conservation and its sustainable utilization of the plant genetic resource for the future world (Cibrian-Jaramillo et al. 2013).

A basic structure for integrated plant species conservation in a botanical garden includes identification and management of threats affecting the corresponding RET taxon, long-term ex situ or in situ germplasm storage, horticulture, and living collections, research and development information management, conservation priorities, and environmental education (Blackmore et al. 2011). Botanical gardens are also planted and multiplied by the threatened endemic plants for the ex-situ conservation programs (Dosmann 2006). Strategies for conserving the live plants may show some differences among and within botanical garden plant collections (Farnsworth et al. 2006). Living plant conservation collection is very important in the case of conservation of threatened species. In-situ ecosystem management plays a vital role in the conservation of certain endangered plant species in their normal, natural habitats.

In addition to scientific research, botanical gardens of the different countries are conducting various public education, science awareness, and popularization programs (Maunder et al. 2001; Donaldson 2009). Nowadays, citizens engage in science as researchers through citizen science processes organized by various botanical gardens around the globe (Kruger and Shannon 2000). The citizen science process not only uses the citizen as just data collectors but rather considers citizens as scientists (Conrad and Hilchey 2011). Certain NGOs and decision makers are effectively engaging the volunteers to enhance their ability to protect the natural resources, assess the threats and risks to the RET species, and monitor and control the plant's genetic resources and their natural habitats (Silvertown 2009). Using a citizen science program to investigate and document the widespread of invasive plant species in the ecosystem by the local resident may promote both knowledge and behavioral changes in local communities (Jordan and Ehrenfeld 2011).

Local resident people may contribute more valuable information because they have more traditional and local knowledge and innovative ideas from their communities (Cohn 2008). Group actions between volunteers from local communities and scientific researchers from eminent botanical gardens have more potential to collect scientific data and increase the scope of research (Close et al. 2006; Fu et al. 2006; Aguraiuja et al. 2008). Demographics, genetic and ecological responses toward habitat fragmentation, and reproduction studies are included as the part of Citizen Project at botanical gardens (Wagenuis et al. 2007; Donaldson et al. 2002; Wagenuis 2006; Neale et al. 2008). Citizen science programs conducted in botanical gardens should adopt certain basic rules: must follow standardized data collection methods; data collected by the public must be rectified and checked by different experts in various disciplines, and volunteers must receive feedback responses about their contribution to botanical garden's scientific development.

Botanical gardens should play an important role in the development of a plant information database to monitor changes in environmental factors in gardens (Stevens 2007; Paton 2009). Finally, as the globe enters the Anthropocene, a "novel conservation" concept needs to be discussed, and new technologies may offer new opportunities for researchers at botanical gardens for the post-GSPC 2020 (Heywood 2017). As a systematic and scientific botanical garden focusing on science and conservation, it is essential to have a comprehensive collection policy for living collections. To strengthen the capacity and scientific research of botanical gardens:

1. Create specialized gardens by including the native flora and promote interdisciplinary research related to them
2. Improve and develop research facilities based on advanced molecular biology tools and molecular databases
3. Create digitized botanical gardens by incorporating all modern technologies

14.8 Role of Seed Banks in the Conservation of RET Plants

An appropriate approach to preserving a particular plant gene pool requires a comprehensive approach by collaborating the different ex-situ and in-situ methods. Gene banks are a type of bio-repository that protects the genetic material of the species. Svalbard Global Seed Vault is the largest secure seed storage in the world. It is located 1300 km north of Permafrost, Arctic Circle, opened by the Norwegian government in February 2008. In India, the Indian Seed Vault is the largest safe seed bank located in the highlands of the Chang La Valley in Ladakh. It is the second largest seed bank in the world. It was established in 2010 jointly by the National Bureau of Plant Genetic Resources and the Defense Institute of High Altitude Research. There are currently more than 1000 seed banks worldwide, varying in type, size, and focus. Over the past two decades, many botanical gardens have begun to establish seed banks for the conservation of threatened plants. This includes the Millennium Seed Bank project at the Royal Botanic Gardens in Kew, Great Britain. They aim to store 10% of the world's plant diversity, targeting arid tropics, including Great Britain (Smith et al. 1998).

Seed banks play an important role in supporting integrated strategies for plant conservation and sustainable use. The cost-effectiveness of seed banking technology depends on the role that the collection plays in the seed biology of the target species and in long-term conservation and environmental renewal strategies. Seed bank recognizes the opportunity to provide users with environmental restoration materials. Seed banks respond to the growing demand for seeds and provide information on habitat restoration and reintroduction of threatened species. Growing popularity as a tool of the seed bank, several factors can contribute to the conservation of wild tree species. Seed banks provide immediate access to plant samples, allowing researchers and conservation biologists to evaluate their properties (such as new sources of drugs, nutrition, and genes). Perhaps most importantly, the plants preserved in the seed banks are resistant to habitat destruction, diseases, and predators.

Seed gene bank means a place or institution with germplasm is conserved as in the form of seeds. It is very convenient to store the seeds as it has less space than the whole plant. Seeds of all plants cannot be stored in the seed bank at low temperatures. In seed banks, only orthodox seeds can be preserved as germplasms. In the seed banks, there are three types of conservation, viz.:

(i) **Short term**: Working collection is stored for short term (3–5 years) at 5–10 °C.
(ii) **Medium term**: Active collections are stored for medium term (10–15 years) at 0 °C.
(iii) **Long term**: Base collections are conserved for the long term (50 years or more) at −18 °C or −20 °C.

14.8.1 Community Seed Banks

Community seed banks store and manage the seeds, which aims to provide seeds for community members to use for regional or local restoration drives. Seeds are procured from the local community and selected and stored according to the agreed storage system. Community seed banks can take many different forms. For example, seeds can be stored in a shed or in community buildings or pots, on the floor in clay pots, in the family barn, or on the kitchen shelf. Once the seeds are collected from the growers, they are stored in a community seed bank until needed.

14.8.2 BRAHMS for Seed Bank Management

The seed module, developed in collaboration with Millennium Seed Bank in Royal Botanical Garden Kew, has a wide range of queries and research applications for seed collection, storage, testing, and distribution projects. Seed accessions with passport data, processing, weight/count vouchers, new images, calculations, viability/germination tests, storage, and dispatch, are now integrated into updated BRAHMSv.8. The main focus is on promoting good seed management training and the ability to better coordinate seed collection and distribution activities of small and large seed banks.

14.8.3 Important Features

- Seed access records, including all passports and wild origin data, can be managed using seed bank location, including links to vouchers, project and agreement details, seed batch data, cleaning methodology, results, and duplication to other seed banks.

- Seed counting processes, including complete and/or estimates through seed sample weight, X-ray integration, and/or cut-testing analysis to provide customized measurements for each collection.
- Germination test design involves multiple conditions and treatments. It provides test results for germination and viability, including germination rate.
- Provides automatic collections against Millennium Seed Bank Partnership Seed Conservation Standards.
- PDF files, Excel sheets, and other documents can be linked to access, individual plants, species, and other records. Material transfer documents and species-level spreadsheets are examples.
- Easy to download data to other formats such as Excel—for further processing.
- Images of seeds or plants in their original habitat can be attached with available voucher images.
- Projects can expand standard BRAHMS seed data file formats by adding custom fields—which will later become part of your database.
- The Rapid Data Entry (RDE) module is used to capture new seed accessions and test data or to transfer data from other formats such as Excel worksheets.
- You can connect to BRAHMS WebConnect directly from BRAHMS to design your own website and publish a virtual seed bank or online plant catalog. For example, see RBG Kew M SBP Data Warehouse at http://brahmsonline.kew.org/msbp.

One of the most important attitudinal changes needed to ensure successful plant conservation initiatives is to educate the policymakers, funding agencies, and the general public that we are dealing with changes and outcomes across many generations, not the current generation, but recovery with future rewards. However, its success requires a long-term and strategic commitment, including proactive information sought from multiple sources and partners before and during policy development. For example, while major subject matter experts may be useful, they do not need to have the knowledge to make successful ground delivery of programs. It is also important for policymakers and funding agencies developing long-term investment strategies to clearly articulate the risks of not investing partially or withdrawing investment through a program. Successful plant conservation requires investment to attract, train, and retain employees within the long-term framework needed to build the depth of knowledge required to ensure the success of activities and programs.

14.9 National Strategies for Plant Conservation

A limited number of countries have developed national strategies for plant protection, modeled on the GSPC, but with objectives tailored to the national context. China, South Africa, Malaysia, Ireland, and Mexico are examples of this approach.

China is home to a rich variety of flora and fauna, with more than 35,800 species of vascular flora. In 2008, China's Strategy for Plant Conservation (CSPC) was recognized as a joint venture between the Chinese Academy of Sciences, the State Forestry Administration (now National Forestry and Grassland Administration), and the State Environmental Protection Agency (now, Ministry of Ecology and Environment). The CSPC targets were established with the updated GSPC at the global level. China is fully committed to continuing its efforts to protect and preserve and manage its plant diversity sustainably. In 2019, an updated Chinese strategy for Plant Conservation 2021–2030 was launched at an international forum of the GSPC in Dujiangyan, Sichuan Province.

In the case of Mexico, the National Plant Conservation Strategy is with six strategic goals and 33 objectives, a time frame beyond 2020. A coordinator was responsible for each of the six objectives, and a coordination committee was formed to support the implementation of the strategy.

In South Africa, the response to the GSPC was first conducted in 2006 and included an overview of progress against global objectives. As a result, a country-specific strategy was developed to focus on gap areas. A partnership between the Botanical Society of South Africa and the South African National Biodiversity Institute (SANBI) was developed between 2013 and 2015 and laid the foundation for the strategy approved by the Ministry of Environment in 2016. The South African strategy retains the same 16 objectives of the GSPC, but some of the objectives have been modified to suit the national context.

14.10 Multiple Scale Habitat Modeling Approach for Rare Plant Conservation

Habitat loss and modification are the most important causes of the extinction of endangered and threatened plant species (Foin et al. 1998). Habitat assessment should be included in development planning to reduce habitat damage and increase the effectiveness of mitigation efforts (Dale et al. 1998; Cuperus et al. 1999). Given the many contradictions between the conservation of rare species and local economic development, it is essential for development planning to include the distribution of rare plant habitats in evaluating alternative projects to minimize potential land-use conflicts and reduce development project costs.

Multiple scale habitat assessments of rare plants are an important component of conservation and development planning. However, this is challenging due to the lack of synchronization of information about the environment of rare plants, the lack of effective approaches to assess habitat on multiple spatial scales, and the lack of spatial data for relevant environmental attributes and scales. A multiple scale habitat modeling approach was developed to meet this need. Indigenous scale Geographic Information System (GIS) data were used to predict local habitat models based on the spatial distribution of habitat areas of rare plant species and the potential for exposure to potential rare plant habitats. Site-scale models based on the integration of literature and field research have been developed for field surveys and mapping of rare plant habitats to accurately assess the potential and current habitat of specific locations using precise field data on soil and topography, and vegetation structure. The greatest need for assessing the presence and habitat of rare plants is on the landscape scale. Thus, landscape-scale models based on site-scale models but with GIS and remote sensing-based data were developed for spatial and clear assessment of potential and existing habitat. These models can be used as effective tools for conservation planning, monitoring, and maintenance of rare plant habitats, reducing land use conflicts, and reduction of development costs. The processes of model development and application integrate the expanding literature, identify knowledge and data gaps to guide future research, and provide a framework for adopting new information gained in the future to improve habitat assessment.

14.11 Assessment and Reporting of Rare and Endangered Species Through Species Distribution Models

Preventing the depletion of biodiversity around the world is recognized as one of the greatest challenges facing mankind (CBD 2010), and it takes unprecedented effort to achieve this (Wolinsky 2011). Rare species are more likely to become extinct due to smaller geographical ranges, low abundances, and greater sensitivity to environmental changes (Pimm et al. 1995; Lavergne et al. 2005; Broennimann et al. 2005; Lomba et al. 2010). Incomplete information about their distribution, which has long been collected and with limited spatial accuracy, makes the assessment of these species particularly challenging (Engler et al. 2004; Lomba et al. 2010; Gogol-Prokurat 2011; Marcer et al. 2013). Consequently, following the guidelines of the International Union for Conservation of Nature (IUCN), estimating the scope of species distribution is at the heart of most evaluation projects (IUCN 2001).

Species distribution models, SDMs (Guisan and Zimmermann 2000; Guisan and Thuiller 2005), are increasingly used to inform monitoring programs and thus conservation policies (Guisan et al. 2013). These models have been successfully applied to identify new populations of rare and endangered species (Guisan et al. 2006), to prioritize conservation areas (Carvalho et al. 2011), and to assess the implications of global change in species distribution (Thuiller et al. 2005), and to estimate the risk of species extinction (Benito et al. 2009).

SDMs report on rare species and illustrate how the tool applied in such reporting assessments will affect resource allocation decisions for monitoring and conservation. Statistical modeling helps to analyse the potential distribution of targeted species with more accuracy in estimation of their suitable habitat. The effects of climate and land-use changes become more significant in species distribution, and those results should be considered in the next assessment and reporting period. This new framework for future assessments of endangered species and habitats is very important in the context of global change. As a result, scientists and managers dealing with rare and endangered species may be interested in a specific approach.

14.12 Conclusion

On a global scenario, through GSPC, botanical gardens and seed banks follow the proper specific conservation strategy with respect to ecology, habitat, and the targeted taxa's characteristics. By adjoining the scientific advancements with conventional conservation, techniques should produce a better way of RET plant conservations. In the current situation, climate change, other man-made catastrophes, and other anthropogenic effects influence the species distributions and their survival. The developed and developing countries across the globe should implement the country-specific action plans and laws for the protection and conservation of threatened species and their habitats.

References

Aguraiuja R, Zobel M, Zobel K et al (2008) Conservation of the endemic fern lineage Diellia (Aspleniaceae) on the Hawaiian Islands: can population structure indicate regional dynamics and endangering factors? Folia Geobot 43:3–18

Balding M, Williams KJH (2016) Plant blindness and the implications for plant 430 conservation. Conserv Biol 30:1192–1199

Benito BM, Martínez-Ortega MM, Munoz LM, Lorite J, Penas J (2009) Assessing extinction-risk of endangered plants using species distribution models: a case study of habitat depletion caused by the spread of greenhouses. Biodivers Conserv 18(9):2509–2520

Blackmore S, Bramwell D, Crane P, Dias B, Given FT, Leiva A, Morin NR, Pushpangadan P, Raven PH, Samper C, Sarukhan JJ (2001) The Gran Canaria Declaration. BGCI, Richmond UK

Blackmore S, Gibby M, Rae D (2011) Strengthening the scientific contribution of botanic gardens to the second phase of the global strategy for plant conservation. Bot J Linn Soc 166:267–281

Bridgewater P (2016) The Anthropocene biosphere: do threatened species, Red Lists, and protected areas have a future role in nature conservation? Biodivers Conserv 25(3):603–607

Broennimann O, Vittoz P, Moser D, Guisan A (2005) Rarity types among plant species with high conservation priority in Switzerland. Bot Helv 115(2):95–108

Cahill DM, Rookes JE, Wilson BA, Gibson L, McDougall KL (2008) *Phytophthora cinnamomi* and Australias biodiversity: impacts, predictions and progress towards control. Aust J Bot 56:279–310

Carvalho SB, Brito JC, Crespo EG, Watts ME, Possingham HP (2011) Conservation planning under climate change: toward accounting for uncertainty in predicted species distributions to increase confidence in conservation investments in space and time. Biol Conserv 144(7): 2020–2030

CBD (2010) Strategic plan for biodiversity 2011–2020 and the Aichi targets. In: Report of the Tenth Meeting of the Conference of the Parties to the Convention on Biological Diversity

Cibrian-Jaramillo A, Hird A, Oleas N et al (2013) What is the conservation value of a plant in a botanic garden? Using indicators to improve management of *ex-situ* collections. Bot Rev 79: 559–577

Close DC, Messina G, Krauss SL (2006) Conservation biology of the rare species *Conospermum undulatum* and *Macarthuria keigheryi* in an urban bushland remnant. Aust J Bot 54:583–593
Coates DJ, Atkins KA (2001) Priority setting and the conservation of Western Australia's diverse and highly endemic flora. Biol Conserv 97:251–263
Coates DJ, Dixon KW (2007) Current perspectives in plant conservation biology. Aust J Bot 55: 187–193
Cohn PJ (2008) Citizen science: can volunteers do real research? BioScience 58(3):192–197
Conrad CC, Hilchey KG (2011) A review of citizen science and community-based environmental monitoring: issues and opportunities. Environ Monit Assess 176:273–291
Cuperus R, Canters KJ, de Haes HAU, Friedman DS (1999) Guidelines for ecological compensation associated with highways. Biol Conserv 90(1):41–51
Dale VH, King AW, Mann LK, Washington-Allen RA, McCord RA (1998) Assessing land-use impacts on natural resources. Environ Manag 22(2):203–211
Dawson W, Burslem DFRP, Hulme PE (2009) Herbivory is related to taxonomic isolation, but not to invasiveness of tropical alien plants. Divers Distrib 15:141–147
Desalle R, Amato G (2004) The expansion of conservation genetics. Nat Rev Genet 5:702–712
Diaz S, Settele J, Brondízio E, Ngo HT, Gueze M, Agard J, Arneth A, Balvanera P, Brauman K, Butchart SHM (2019) IPBES summary for policymakers of the global assessment report on biodiversity and ecosystem services of the intergovernmental Science-policy platform on biodiversity and ecosystem services. IPBES Secretariat, Bonn, Germany, 2019
Donaldson JS (2009) Botanic gardens science for conservation and global change. Trends Plant Sci 14:608–613
Donaldson JS, Nanni I, Zachariades C et al (2002) Effects of habitat fragmentation on pollinator diversity and plant reproductive success in renosterveld shrublands in South Africa. Conserv Biol 16:1269–1276
Dosmann MS (2006) Research in the garden: averting the collections crisis. Bot Rev 72:207–234
Drury WH (1974) Rare species. Biol Conserv 6:162–169
Engler R, Guisan A, Rechsteiner L (2004) An improved approach for predicting the distribution of rare and endangered species from occurrence and pseudo-absence data. J Appl Ecol 41(2): 263–274
Farnsworth EJ, Klionsky S, Brumback WE (2006) A set of simple decision matrices for prioritizing collection of rare plant species for ex situ conservation. Biol Conserv 128:1–12
Foin TC, Pawley AL, Ayres DR, Carlsen TM, Hodum PJ, Switzer PV (1998) Improving recovery planning for threatened and endangered species. Bioscience 48(3):177–184
Fu Y, Guo H, Chen A (2006) Household differentiation and on farm conservation of biodiversity by indigenous households in Xishuangbanna, China, vol 15. Springer, Netherlands, pp 2687–2703
Gogol-Prokurat M (2011) Predicting habitat suitability for rare plants at local spatial scales using a species distribution model. Ecol Appl 21(1):33–47
Gratani L, Crescente MF, Varone L (2008) Growth pattern and photosynthetic activity of different bamboo species growing in the Botanical Garden of Rome. Flora 203:77–84
Guisan A, Thuiller W (2005) Predicting species distribution: offering more than simple habitat models. Ecol Lett 8(9):993–1009
Guisan A, Zimmermann NE (2000) Predictive habitat distribution models in ecology. Ecol Model 135(2–3):147–186
Guisan A, Broennimann O, Engler R, Vust M, Yoccoz NG, Lehmann A, Zimmermann NE (2006) Using niche-based models to improve the sampling of rare species. Conserv Biol 20(2):501–511
Guisan A, Tingley R, Baumgartner JB, Naujokaitis-Lewis I, Sutcliffe PR, Tulloch AI, Buckley YM (2013) Predicting species distributions for conservation decisions. Ecol Lett 16(12):1424–1435
Hawkins B et al (2008) Plants and climate change: which future? Botanic Gardens Conservation International
Heywood VH (2017) Plant conservation in the anthropocene—challenges and future prospects. Plant Divers 39:314–330

International Union for Conservation of Nature, International Union for Conservation of Nature, Natural Resources. Species Survival Commission, & IUCN Species Survival Commission (2001) IUCN Red List categories and criteria. IUCN

IUCN (1998) The IUCN red list of threatened plants. IUCN, Gland

IUCN (2021) The IUCN red list of threatened species. IUCN, Gland. ISSN: 2307-8235

Jordan RC, Ehrenfeld JG (2011) Knowledge gain and behavioral change in citizen-science programs. Conserv Biol 25:1148–1154

Kareiva P, Marvier M, Lalalz R (2012) Conservation in the anthropocene: beyond fragility and solitude. Breakthrough J. http://thebreakthrough.org/index.php/journal/past-issues/issue-2/conservation-in-the-anthropocene

Kramer AT, Havens K (2009) Plant conservation genetics in a changing world. Trends Plant Sci 14: 599–607

Kruger LE, Shannon MA (2000) Getting to know ourselves and our places through participation in civic social assessment. Soc Nat Resour 13:461–478

Lavergne S, Thuiller W, Molina J, Debussche M (2005) Environmental and human factors influencing rare plant local occurrence, extinction and persistence: a 115-year study in the Mediterranean region. J Biogeogr 32(5):799–811

Li N, An SQ, Liu Z (2014) Fruit consumption and seed dispersal by birds in native *vs. ex-situ* individuals of the endangered Chinese yew, *Taxus chinensis*. Ecol Res 29:917–923

Lomba A, Pellissier L, Randin C, Vicente J, Moreira F, Honrado J, Guisan A (2010) Overcoming the rare species modelling paradox: a novel hierarchical framework applied to an Iberian endemic plant. Biol Conserv 143(11):2647–2657

Lowe S, Browne M, Boudjelas S, De Poorter M (2000) 100 of the world's worst invasive alien species: a selection from the global invasive species database, vol 12. Invasive Species Specialist Group, Auckland

Lu C, Zhu Q, Deng Q (2008) Effect of frugivorous birds on the establishment of a naturally regenerating population of Chinese yew in *ex-situ* conservation. Integr Zool 3:186–193

Makinson R (2014) Myrtle rust—what's happening? Australas Plant Conserv 23:13–15

Marcer A, Saez L, Molowny-Horas R, Pons X, Pino J (2013) Using species distribution modelling to disentangle realised versus potential distributions for rare species conservation. Biol Conserv 166:221–230

Maunder M, Higgens S, Culham A (2001) The effectiveness of botanic garden collections in supporting plant conservation: a European case study. Biodivers Conserv 10:383–401

Moza MK, Bhatnagar AK (2007) Plant reproductive biology studies crucial for conservation. Curr Sci 92(9):1207

Myers N, Mittermeier RA, Mittermeier CG, da Fonseca GAB, Kent J (2000) Biodiversity hotspots for conservation priorities. Nature 403(6772):853–858

Neale JMR, Ranker TA, Collinge SK (2008) Conservation of rare species with island-like distributions: a case study of *Lasthenia conjugens* (Asteraceae) using population genetic structure and the distribution of rare markers. Plant Species Biol 23:97–110

Olson DM, Dinerstein E (1998) The global 200: a representation approach to conserving the Earth's most biologically valuable eco regions. Conserv Biol 12(3):502–515

Ouborg NJ, Vergeer P, Mix C (2006) The rough edges of the conservation genetics paradigm for plants. J Ecol 94:1233–1248

Paton A (2009) Biodiversity informatics and the plant conservation baseline. Trends Plant Sci 14: 629–637

Pennisi E (2010) Tending the global garden. Science 329:1274–1277

Pimm SL, Russell GJ, Gittleman JL, Brooks TM (1995) The future of biodiversity. Science 269: 347–350

Primack RB, Miller-Rushing AJ (2009) The role of botanical gardens in climate change research. New Phytol 182:303–313

Rathcke BJ, Jules ES (1993) Habitat fragmentation and plant-pollinator interactions. Curr Sci 273–277

Shearer BL, Crane CE, Cochrane A (2004) Quantification of the susceptibility of the native flora of the South-West Botanical Province, Western Australia, to *Phytophthora cinnamomi*. Aust J Bot 52:435–443

Silvertown J (2009) A new dawn for citizen science. Trends Ecol Evol 24:467–471

Singh JS, Khurana E (2002) Paradigms of biodiversity: an overview. Proc Natl Acad Sci India-Sect B Biol Sci 68(3):273–296

Smith RD, Linington SH, Wechsberg GE (1998) The Millennium Seed Bank, the convention on biological diversity and the dry tropics. In: Prendergast HDV, Etkin NL, Harris DR, Houghton PJ (eds) Plants for food and medicine. Royal Botanic Gardens, Kew, pp 251–261

Stevens AD (2007) Botanical gardens and their role in ex situ conservation and research. Phyton 46: 211–214

Thrall PH, Burdon JJ, Woods MJ (2000) Variation in the effectiveness of symbiotic associations between native rhizobia and temperate Australian legumes: interactions within and between genera. J Appl Ecol 37:52–65

Thuiller W, Richardson DM, Pysek P, Midgley GF, Hughes GO, Rouget M (2005) Niche-based modelling as a tool for predicting the risk of alien plant invasions at a global scale. Glob Change Biol 11(12):2234–2250

Wagenuis S (2006) Scale dependence of reproductive failure in fragmented Echinacea populations. Ecology 87:931–941

Wagenuis S, Lonsdorf E, Neuhauser C (2007) Patch aging and the S-Allee effect: breeding system effects on the demographic response of plants to habitat fragmentation. Am Nat 169:383–397

Waldron A, Mooers AO, Miller DC, Nibbelink N, Redding D, Kuhn TS, Gittleman JL (2013) Targeting global conservation funding to limit immediate biodiversity declines. PNAS 110(29): 12144–12148

Wang B, Phillips JS, Tomlinson KW (2018) Tradeoff between physical and chemical defense in plant seeds is mediated by seed mass. Oikos 127:440–447

Wege JA, Thiele KR, Shepherd KA, Butcher R, Macfarlane TD, Coates DJ (2015) Strategic taxonomy in a biodiverse landscape: a novel approach to maximizing conservation outcomes for rare and poorly known flora. Biodiversitas 24:17–32

Wolinsky H (2011) Will we wake up to biodiversity? The International Year of Biodiversity has failed to raise awareness or halt decline as economic crises and political interests have sidelined the environment. EMBO Rep 12(12):1226–1229

Wyse Jackson PS, Sutherland L (2000) The international agenda for botanic gardens in conservation. Botanic Gardens Conservation International

Chapter 15
An Outlook on Marine Sponges and Associated Biodiversity Addressing Conservation Strategies

Kizhakke Purayil Anupama, Anet Antony, Shilpa Olakkaran, Reshma Ramarajan, Shanthala Mallikarjunaiah, and Hunasanahally Puttaswamygowda Gurushankara

Abstract Sponges are the pore-bearing animals belonging to the phylum Porifera. They are the oldest living multicellular invertebrates on this planet. Sponges are omnipresent within freshwater and marine ecosystems in the benthic communities of temperate, tropical, and polar habitats, from coastal shores to the sea's abyssal depths. These organisms are sessile, filter feeders, competitors for space, and grow in solitary, colonial, branching, or as thin sheets over a substrate to form a garden-like environment that forms high biomass and diversity in the sea bottom. Sponges potentially create biodiversity hotspots with various associations. They are of commercial importance and provide valuable ecosystem services and functions. Anthropogenic activities and climate changes threaten the community of sponges and their associated biodiversity. There has been no assessment of their current global conservation status. So, it is essential to conserve these ancient organisms with management strategies. This chapter provides sponges' general features, associated biodiversity, significance, threats, and conservation strategies.

Keywords Marine sponges · Conservation strategies · Sponge-associated fauna · Holobiont · Threats

K. P. Anupama · A. Antony · S. Olakkaran · R. Ramarajan · H. P. Gurushankara (✉)
Department of Zoology, School of Biological Sciences, Central University of Kerala, Kasaragod, Kerala, India

S. Mallikarjunaiah
Department of Studies in Zoology, Center for Applied Genetics, Bangalore University, Bengaluru, Karnataka, India

S. T. Sukumaran, K. T R (eds.), *Conservation and Sustainable Utilization of Bioresources*, Sustainable Development and Biodiversity 30,
https://doi.org/10.1007/978-981-19-5841-0_15

15.1 Introduction

Oceans alone cover 65% (335.3 million km^2) of Earth's total area and have an average depth of 12,100 ft. They contain 97% (about 1.35 billion km^3) of the global water and host 99% of life forms (Costello et al. 2010). The largest habitat on Earth, containing almost unexplored biodiversity in the deepest part ever recorded, is located in the western part of the Pacific Ocean, in the Mariana Trench, at around 36,200 ft (Stewart and Jamieson 2019). The vastness of the oceans makes the biodiversity and ecological role of several life forms, but the unexplored deep-sea area accounts for about 95% of Earth's living space (Ramirez-Llodra 2020). Marine ecosystems result from complex interactions between the physiochemical environment and diverse species of organisms (Paulus 2021). The marine biome is broadly divided into two realms—pelagic and benthic. Swimming and floating organisms that live in the pelagic region are Pelagos. The benthic zone is sea bottom sediments, and organisms in this zone are benthos. The ocean seafloor with 4000–6000 m depth is the Abyssal zone, and trenches with 6000–11,000 m depth are the Hadal zone. Many benthos is sessile; some are in burrows, and others are swimming near the bottom. The deep-sea community is associated with benthos and their shared habitats. At the lower continental slope/abyssal plain transition region of the deep-sea, the complex habitats created by large sessile organisms that play an essential role in the marine ecosystem are sponges. Marine biodiversity plays a vital role in providing many ecosystem functions and services that humans derive from the oceans, primary production and nutrient cycling, provisioning services such as food, and cultural services, including tourism (Townsed et al. 2018). The growing human population has impacted 87–90% of the global ocean surface, leading to varying degrees of decline in marine populations at domestic and international levels (Luypaert et al. 2020). It led to global biodiversity loss, suggesting that we are in a biodiversity crisis. Currently, declines in marine populations are ubiquitous, yet their consequences for marine ecosystems. So, it is of utmost importance to study the impact of humans on marine sponges to conserve their populations and associated biodiversity.

15.2 Sponges

The phylum-Porifera (Grant 1836) pore bearer sponges originated during the Cambrian period (Turner 2021). Sponges are freshwater and marine metazoans, the oldest extant animals, and they dominate littoral and tropical reefs; ~75% of benthic regions survived 700–800 million years old (Wulff 2012). They are sessile, filter feeders, competitors for space, grow in solitary, colonial, branching, as thin sheets over substrates from a few centimeters to over 1 m in size, and form extensive garden-like environments characterized by high biomass and diversity in the sea bottom (Przeslawski et al. 2014). Sponges have an aquiferous system that pumps unidirectional water current through the body; a 1 × 10 cm sponge pumps 22.5 L of water daily. The specialized, highly mobile population of cells can differentiate into other cell types (totipotency) and confer plasticity to growth form with siliceous or calcitic spicules, but does not form organs or tissues (Hooper et al. 2002). In general, sponges reproduce both asexually and sexually. Asexual reproduction takes various forms: fragmentation, budding, direct-developing gemmules, and formation of pseudo larvae. Sponges lack differentiated organs and do not possess true gonads. In its place, a significant portion of the sponge body is involved in reproduction. Generally, sponges are hermaphrodites. Sexual dimorphism does not exist in sponges. Most sponges are viviparous, and consequently, the eggs are retained and fertilized internally. However, some sponges are oviparous; either fertilization occurs internally with the zygote eventually being released into the sea, or oocytes are released, and fertilization occurs externally. Within their connective tissue, eggs develop from amoebocytes, and sperm develop from amoebocytes or transformed collar cells. Spermatozoa are shed into the excurrent canals and released into the sea. Two types of short-lived larvae are solid parenchymella larvae and hollow amphiblastula larvae. Sizes range from 50 μm to 5 mm in length. Differential pigmentation of the posterior or anterior pole is not unusual and commonly coincides with areas lacking cilia (Fell 1974).

15.2.1 Habitat

Sponges are globally distributed in the benthic communities of temperate, tropical, and polar habitats, from coastal shores to abyssal depths (Maldonado et al. 2015). Their continued existence in vast numbers is closely linked to the apparent adaptability to environmental conditions and competition for biota. Patterns of sponge distribution depend on the complex physical and chemical features of the sea, the seabed, and currents surrounding the specific location/region. Environmental factors such as wave, light, temperature, sediment load, substrate, high levels of chemicals, and natural or man-made pollution directly regulate sponge assemblage structure and abundance (Wilkinson and Cheshire 1989; Schönberg and Fromont 2012).

15.2.2 Global Status of Sponges

Phylum Porifera comprises four classes: Calcarea, Demospongiae, Hexactinellida, and Homoscleromorpha. In the suborder Spongillida, the order Haplosclerida of the Demospongiae is found in freshwater. About 20% are glass and calcareous species, and the remaining 80% of the known sponges belong to the class Demospongiae. Living freshwater sponges belong to 45 genera in six families for 219 species. Demospongiae have siliceous spicules that are divided into megascleres and microscleres. Class Demospongiae are bath sponges that include carnivorous species from the order Poecilosclerida. Significant numbers of undescribed species often hamper knowledge of sponge species distributions. As of 2022, globally, 9489 sponge (marine + freshwater) species are estimated to be at least 20,000 species likely to be described by the end of this century (De Voogd et al. 2022).

15.2.3 Sponges in India

Indian sponges constitute 30% of world diversity. In 2001, 451 sponge species were recorded in India, with fewer than 3 classes, 17 orders, 65 families, and 169 genera (Pattanayak and Manna 2001). The 15 species which were endemic to Andaman and Nicobar are *Bubaris columnata, Calyx clavata, Cliona kempi, Damiriopsis brondstedi, Discodermia gorgonoides, Haliclona (Gellius) megastoma, Kirkpatrickia spiculophila, Monanchora enigmatica, Poecillastra eccentrica, Thenea andamanensis* of Demospongiae and *Euplectella regalis, Hexactinella minor, Lophocalyx spinosa, Lophophysema inflatum* and *Semperella cucumis* of Hexactinellida. Another study reported that 584 species of sponges are under 288 genera in India. Among them, species from the Gulf of Kachch, 124 species from the Andaman and Nicobar Islands, and 91 species from Lakshadweep were included. Species diversity in the Gulf of Mannar and Palk Bay together constitutes 321 species belonging to 129 genera, of which 257 species are endemic. Almost 67% of the diversity of sponges residing in these areas are from the class Demospongiae (Gunasekaran and Krishnamoorthy 2007). Based on in-situ collection, a study reported 101 species of marine sponges in India's peninsular region. The recorded species of marine sponges belong to 12 orders, 22 families, 5 subfamilies, and 44 genera from 4 subclasses of class Demospongiae and one species from class Calcarea. Most of the recorded species are new to India (George et al. 2020).

15.3 Marine Sponges and Associated Biodiversity

The sponges are known to serve as home to many organisms, including fish and invertebrates. They are significant structural components of seabeds, so understanding sponge-community-associated biodiversity is essential to conserving cryptic marine biodiversity. Sponges' sessile filter-feeding lifestyle constantly exposes them to the microbes in the seawater that form their primary food source; they harbor distinct symbiotic microbial communities. Sponges provide a habitat for many organisms and play an essential role in benthic-pelagic coupling due to their impressive filtering capacity (Folkers and Rombouts 2020). The faunal communities are associated with the specific sponge host; the sponge-associated fauna (SAF) diversity is crucial to understanding communities' specificity of sponge utilization. SAF comprises polychaetes, amphipods, decapods, mollusks, and ophiuroids that use sponges as shelter, nurseries, or food sources (Abdo 2007; Villamizar and Laughlin 1991; Neves and Omena 2003; Fiore and Jutte 2010). Table 15.1 provides the list of organisms associated with sponges, from microbes to chordate levels.

Interactions of sponges with other organisms include competition, predation, and symbiotic associations (Wulff 2006). The symbiotic association of SAF has mutually beneficial, commensal to parasitic or pathogenic. Many species of sponges show various interactions with marine organisms (Table 15.2).

Protection from predators and water currents for suspended food particles are the main advantages of the sponge's host. Some organisms utilize sponges for camouflage. These SAF associations are ubiquitous across sponge species and geography, although these associated communities are less understood and homogenous within sponge species. The factors such as sponge size, internal space/inner hole, large and long-lived sponges, and gross morphological complexity support their higher community abundance.

15.3.1 Shape and Size

Diverse shapes and sizes of sponges include encrusting, massive, erect, and cup forms that offer surfaces as the substrate for the attachment and refuge of other marine taxa (Neves and Omena 2003). Examples: (1) The barnacles have lived in several sponge species, emphasizing the wide geographic range of sponge–barnacle associations (Yu et al. 2017). (2) The host-specific bivalve *Vulsella vulsella*, has been reported to form a mutualistic partnership with the sponge *Spongia* sp., where the sponge provides a habitat for the bivalve to live in, it, in turn, receives exhaled water from the bivalve that increased its own water circulation rates (Tsubaki and Kato 2014). (3) The complex aquiferous canals could offer additional living spaces for some other taxa. The internal space of the sponge is used by multiple species of alpheid shrimps that form colonies with social complexities (Duffy 1996).

15.3.2 Volume

The physical characteristics of sponges' volume influence the patterns of sponge-associated biodiversity (Beepat et al. 2014). Large sponges host higher diversity and abundance of communities (Fiore and Jutte 2010).

15.3.3 Environmental Factors

Habitat type and water depth influence patterns of SAF communities. The abundance and diversity of SAF are determined by biotic interactions such as predation (Abdo

Table 15.1 List of organisms associated with sponge

Sl. no.	Identified taxonomical categories	References
1	**Prokaryotic phylum:** Proteobacteria, Chloroflexi, Cyanobacteria, Crenarchaeota, Acidobacteriota, Bacteroidota, Actinobacteriota, Planctomycetota, Gemmatimonadota, PAUC34f, Dadabacteria, Verrucomicrobiota, Myxococcota, Nitrospirota, Nitrospinota, Desulfobacterota, Deinococcota, Spirochaetota, Latescibacterota, Poribacteria, Bdellovibrionota, Entotheonellaeota, Firmicutes, Methylomirabilota, Thermoplasmatota, and unclassified bacteria ***Prokaryotic class:*** Alphaproteobacteria, Gammaproteobacteria, Cyanobacteria, Nitrososphaeria, Dehalococcoidia, Bacteroidia, Vicinamibacteria, Anaerolineae, Acidimicrobiia, PAUC34f, Planctomycetes, TK17, BD2-11, unclassified Chloroflexi, unclassified Proteobacteria, Thermoanaerobaculia, Dadabacteriia, Verrucomicrobiae, OM190, Rhodothermia, Nitrospiria, AncK6, NB1-j, P9X2b3D02, Polyangia, Phycisphaerae, JG30-KF-CM66, PAUC43f, Bacteriap25, Acidobacteriae, Deinococci, Spirochaetia, Poribacteria, Entotheonellia, Desulfobacteria	Hardoim et al. (2021) Hentschel et al. (2002)
2	**Fungi phylum:** Ascomycota, Basidiomycota, unclassified fungi, Mortierellomycota, Chytridiomycota, Aphelidiomycota, Mucoromycota, Rozellomycota, Kickxellomycota, and unclassified fungi ***Fungi class:*** Agaricomycetes, Sordariomycetes, Tremellomycetes, Dothideomycetes, Eurotiomycetes, unclassified Ascomycota, Microbotryomycetes, Mortierellomycetes, Wallemiomycetes, Saccharomycetes, unclassified Basidiomycota, Lecanoromycetes, Orbiliomycetes, unclassified Aphelidiomycota, Leotiomycetes, Lobulomycetes, Malasseziomycetes, Cystobasidiomycetes, Umbelopsidomycetes, Chytridiomycetes, Dacrymycetes, Agaricostilbomycetes, Rozellomycotina, Kickxellomycetes	Hardoim et al. (2021)
3	**Unicellular eukaryotic phylum:** Diatomea, Florideophycidae, Dinoflagellata, Basidiomycota, Chlorophyta, Phragmoplastophyta, Ascomycota, Protalveolata, Ciliophora, Ochrophyta, Cryptophyceae, Retaria, Euglenozoa, Cercozoa, Prymnesiophyceae, Peronosporomycetes, Aphelidea, Mucoromycota, Holozoa ***Unicellular eukaryotic class:*** unclassified Eukaryota, Mediophyceae, Bacillariophyceae, Dinophyceae, unclassified Diatomea, Nemaliophycidae, Embryophyta, Corallinophycidae, Ulvophyceae, Rhodymeniophycidae, Dothideomycetes, Syndiniales, Agaricomycetes, Malasseziomycetes, Intramacronucleata, Tremellomycetes, Eurotiomycetes, Pedinophyceae, Phaeophyceae, Ustilaginomycetes, Cryptophyceae, unclassified Chlorophyta, Foraminifera, Saccharomycetes, uncultured Trebouxiophyceae, Mamiellophyceae, Microbotryomycetes, Prymnesiophyceae, Clade_VII, unclassified Ciliophora, Chlorodendrophyceae, Chrysophyceae, Dinoflagellata, unclassified Dinoflagellata, Incertae Sedis (Dinoflagellata), Kinetoplastea, unclassified Ascomycota, Peronosporomycetes, Aphelidea, Chlorarachniophyta, Coscinodiscophytina, unclassified Phragmoplastophyta, Thecofilosea, Imbricatea, Prasinophytae, Incertae Sedis (Mucoromycota), unclassified Ochrophyta, Choanoflagellida, Sordariomycetes	Wulff (2006) Hentschel et al. (2002)

(continued)

Table 15.1 (continued)

Sl. no.	Identified taxonomical categories	References
4	**Multicellular eukaryotic phylum:** Cnidaria, Mollusca, Annelida, Arthropoda, Echinodermata, Chordata Phylum: Cnidaria Class: Anthozoa, Scyphozoa, and Hydrozoa Order: Zoanthids, Alcyonacea Genus: P*arazoanthus, Epizoanthus* Phylum: Mollusca Class: Bivalvia and Gastropoda Order: Arcida, Caenogastropoda Genus: Arca Species:*A. noae, siliquariid molluscs* Phylum: Annelida Class: Polychaeta Order: Phyllodocida Genus: Haplosyllis, Branchiosyllis, Species: H. *oculata, spongicola* Phylum: Arthropoda Subphylum: Crustacea Class: Hexanauplia, Malacostraca Subclass: Copepods Order: Isopoda and amphipoda Genus: Panulirus Species: P. *argus* Phylum: Echinoderms Class: Ophiuroidea Order: Ophiurida Genus: Ophiothrix Species: O. *lineata, fragilis*	Rocha et al. (2000) Wulff (2006)
	Phylum: Chordate **Superclass:** Pisces **Fish species associated with Sponges**: *Risor ruber, Elacatinus Figaro, Thalassoma noranhanum, Halichoeres cyanocephalus, Hypleurochilus sp., Starksia brasiliensis, Scorpaenodes tredecimspinosus, Lythrypinus brasiliensis, Lythrypinus sp., Apogon quadrisquamatus, Phaeoptyx pigmentaria, Stegastes pictus, Cephalopholis fulva, Amblycirrhitus pinos, Priolepis dawsoni, Abudefdus Taurus, Acanthurus bahianus, Acanthurus chirurgus, Acanthurus coeruleus, Anisotremus surinamensis, Anisotremus virginicus, Aulostomus maculatus, Balistes vetula, Canthigaster rostrata, Caranx bartholemei, Caranx ruber, Centropyge argi, Cephalopholis cruentatus, Cephalopholis fulvus, Chaetodon capistratus, Chaetodon ocellatus, Chaetodon striatus, Chromis cyanea, Chromis insolata, Chromis multilineata, Clepticus parrae, Coryphopterus personatus, Epinephelus guttatus, Gramma loreto, Liopropoma mawbrayi, Gymnothorax funebris, Haemulon aurolineatum, Haemulon chrysargyreum, Haemulon flavolineatum, Haemulon sciurus, Halichoeres garnoti, Holacanthus ciliaris, Holacanthus tricolor, Holocentrus adscensionis, Holocentrus rufus, Hypoplectrus chlorurus,*	García-Hernández et al. (2018)

(continued)

Table 15.1 (continued)

Sl. no.	Identified taxonomical categories	References
	Hypoplectrus puella, Lachnolaimus maximus, Lutjanus apodus, Lutjanus cyanopterus, Lutjanus jocu, Melichthys niger, Microspathodon chrysurus, Mulloidichthys martinicus, Myripristis jacobus, Neoniphon marianus, Ocyurus chrysurus, Paranthias furcifer, Pomacanthus arcuatus, Pomacanthus paru, Prognathodus aculeatus, Pseudopeneus maculatus, Pterois volitans, Scarus iserti, Scarus taeniopterus, Serranus baldwini, Serranus tabacarius, Serranus tigrinus, Sparisoma aurofrenatum, Sparisoma virides, Sphyraena barracuda, Stegastes leucostictus, Stegastes partitus, Stegastes planifrons, Stegastes variabilis, Thalassoma bifasciatum, Xanthichthys ringens	

2007) and allelochemical attractants (Villamizar and Laughlin 1991). Sponge-derived chemicals serve as attractants to some SAF, such as amphipods (Frith 1976). However, sponge chemistry's influence on SAF communities is poorly understood; more studies are required to elucidate these allelochemical-based interactions.

15.3.4 Holobiont

A holobiont syn. "metaorganisms" is an assemblage of a host and the many other species living in or around it, forming a discrete ecological unit as "nested ecosystems" (Bosch and McFall-Ngai 2011). Marine sponges perfectly demonstrate holobionts as an example of nested ecosystems, given the exceptionally diverse microbial communities housed within them (Thomas et al. 2016). The microbiome provides essential host nutrition, defense, immunity, and development. They mediate the interactions of the holobiont with the surrounding organismal community; through cascading effects, ultimately influencing community structure, ecosystem health, and functioning (Flórez et al. 2015). Different environmental gradients such as geographical distance, season, depth, and habitat harbor species-specific stable microbiomes at different prokaryotic taxonomic levels and prevalence thresholds (Steinert et al. 2017). The sponge microbiome spans the most dominant bacterial symbiont groups such as the phyla Proteobacteria (Gamma and Alphaproteobacteria), Actinobacteria, Chloroflexi Nitrospirae, Cyanobacteria, Candidatus phylum Poribacteria, and Thaumarchaea archaeal group (Thomas et al. 2016). The microbial communities are species-specific but composed of generalist microbes detected in most sponge species from diverse geographic regions and specialists enriched in particular species but are rare or absent in most other species

Table 15.2 Some of the interactions between sponges and their associated organisms

Sl. no.	Interactions	Sponge	Associated organisms	Details
1.	Competition	*Haliclona* sp.	*Acropora nobilis*	*Haliclona* sp. of sponge larvae settling on coral species *A. nobilis* and then killing coral tissue
2.		*Halichondria*	*Zostera marina*	Estuarine species of *Halichondria* on *Z. marina* plant leaf blades appear to be solely due to the relatively rapid growth of the sponge
3.		*Haliclona (Reniera) tubifera*	Tube worm	*H. tubifera* was found to smother neighboring barnacles and tube-dwelling annelids
4.	Predation	*Cliona celata*	Invertebrates species	Boring sponge *C. celata* was preyed on by two gastropod specialists and various shrimp, crabs, a limpet, and sea urchin generalists
5.		*Halichondria panicea*	*Archidoris montereyensis*	*A. montereyensis* species eliminated a population of the intertidal sponge *H. panicea*
6.		*Isodictya*	*Henricia sanguinolenta*	Starfish *H. sanguinolenta* consumes finger sponge species of *Isodictya* according to the size and feeding lesions
7.	Symbiotic associations	*Halichondria panicea*	Scallops	Scallops gain protection from starfish predators when breadcrumb sponge, H. panacea, overgrows their valves; the sponges gain a favorable feeding location by the inhalant feeding currents of their hosts
8.		*Crambe crambe*	*Arca noae*	Sponge *C. crambe* encrusting *A. noae* shells help to inhibit predation on their hosts by a starfish, a non-native invasive snail, and an octopus
9.		*Neofibularia nolitangere*	Polychaetes genus *Haplosyllis*	Caribbean sponge *N. nolitangere* host a dense population of tiny white polychaetes from the genus *Haplosyllis*
10.		*Aplysina cauliformis*	*Haplosyllis spongicola*	Polychaete species *H. spongicola* engulfs the tissue of its host, *A. cauliformis*
11.		Suberitid sponges	Hermit crabs	The sponges encrust gastropod shells inhabited by hermit crabs and then continue to grow, apparently relieving the crabs of the necessity of finding new shells as they grow
12.		*Callyspongia vaginalis*	*Ophiothrix lineata*	*O. lineata* adult brittle stars and tubular Caribbean sponge *C. vaginalis* are mutually beneficial. Brittle stars clean the inhalant surface of sponges as they feed and derive protection from predators on their inedible perches

(continued)

Table 15.2 (continued)

Sl. no.	Interactions	Sponge	Associated organisms	Details
13.		*Halisarca* cf. *dujardini*	*Smittina cervicornis*	Bryozoan *S. cervicornis* overgrown by the thinly encrusting sponge *H. dujardini*. This association helps elevate the sponge above the substratum since sponges in this genus lack skeletal fibers or spicules. The feeding currents appear to be strengthened for both partners by their collaboration

(Erwin et al. 2012). However, evidence suggests that environmental conditions threaten sponge-associated microbial diversity (Erwin et al. 2015).

15.4 Significance

Humans have used sponges for several millennia for commercial purposes. Their significant roles in the ecosystem services and functions are as follows:

1. Primary production and nitrification: Nitrogen metabolism emphasizes ammonia oxidation, carbon metabolism on complex carbohydrates, nitrogen and carbon metabolism utilizing creatinine, vitamin synthesis: thiamine and vitamin B12, secondary metabolism, and carnitine: vitamin BT utilization (Webster and Thomas 2016).
2. Chemical and physical adaptation: Sponges produce a diverse array of secondary metabolites with antiviral, antimicrobial, cytotoxic, allelopathic, and antipredatory effects attributed to the microbiome (Lackner et al. 2017; Wilson et al. 2014). The production of biologically active feeding restraining compounds is common defensive strategy sponges employ to avoid predation (McClintock et al. 2010). Stress proteins, restriction–modification, toxin–antitoxin systems, and clustered regularly interspaced short palindromic repeats (CRISPRs) are in the protection and stress response (Horn et al. 2016). Eukaryotic-like protein (ELPs) domains and lipopolysaccharide modifications are involved in phagocytosis evasion. These defensive functions shield sponge symbionts against incoming foreign DNA, pathogens, and toxins to which they are exposed due to the pumping activity of the host (Slaby et al. 2017). Mobile genetic elements and transposases increased levels of horizontal gene transfer (Gao et al. 2014).
3. Calcification, cementation, and bioerosion: The sponges are important components of coral reefs. Their biomass and ecological tolerance frequently exceed reef-building corals (Rützler 1978). Some sponges could bio-erode and consolidate reef structures (Hooper 1998). Sponges are the major ecosystem engineers on the seafloor, providing habitat for various species (Wulff 2006). They play key ecological roles in substrate modification, nutrient cycling, monitoring benthic

communities, and managing impacted areas (Bell 2008; Schönberg 2008; Xavier and van Soest 2012).

4. Water quality indicators: Sponges are efficient filter feeders and play a critical role in linking the pelagic environment to the benthos through nutrient cycling (Maldonado et al. 2012). They significantly influence water quality and substrate conditions provide nutrition and vital habitat for many other organisms (Wulff 2006). So, sponges are used in biodiversity and impact assessments as potential biological water quality indicators.
5. Secondary metabolites: Sponges have a rich source of bioactive compounds. They maintain symbiotic relationships with microorganisms that bear a structural resemblance to metabolites produced from microbes. More than 5000 chemically diverse compounds have been isolated and structurally characterized, contributing to 30% of marine natural products. The chemical diversity of sponges includes nucleosides, terpenes, peptides, alkaloids, nonribosomal peptides, polyketides, steroids, and macrolides, which have a wide range of biological activities such as antibacterial, anticancer, antifungal, anti-HIV, anti-inflammatory, and antimalarial (Varijakzhan et al. 2021). Commercially sponges' secondary metabolites are used as drugs, anti-biofilm agents, and bath sponges (Stowe et al. 2011; Mehbub et al. 2014).

15.5 Threats to Sponges and Associated Biodiversity

Due to anthropogenic threats, the disturbance in the upper sea surface impacts the deep-sea sponge biodiversity. Major and minor threats to sponges and their associated biodiversity are *Deep-sea mining*: Metal nodules at the seabed are one of the main interests of the industry. Mining these metals disturbs the seabed-containing sponge communities (Oebius et al. 2001). *Cable and pipeline placement*: Telecommunication and electric cable and oil–gas pipelines are installed in the seabed, damaging the sponge grounds (Carter et al. 2009). *Hydrocarbon exploration and exploitation*: Underground stores of oil and gas; its mining threats to sessile sponges, and their associated communities (Bett 2001). *Commercial bottom trawling and other mobile fishing:* Sea bottom trawling and dredges aim to overexploit fishes and attack continental shelves. Deep-sea is highly sensitive and closely linked to the sea surface; hence, changes in Ocean current and temperature impact marine sponges. *Climate change:* Ocean warming, acidification, and the deterioration of water quality-eutrophication, sedimentation, and pollution threatening the marine sponges (Pineda et al. 2017).

15.6 Conservation Strategies

Sponges have received increasing attention in research and management due to the growing scientific and public awareness of their ecological and commercial importance. They generally lack effective management and have been considered a "neglected group." Their significance in the global ecosystem is not much appreciated (Schönberg and Fromont 2012; Van Soest et al. 2012). Hence, a proper conservation strategy is required to provide essential perspectives and directions for sponge populations and their associated biodiversity conservation, maintaining stable marine ecosystems' structure and function.

15.6.1 Species Identification

Difficulty in identifying sponges; many taxonomists broadly categorize them as "sponges" or "filter feeders," which does not allow meaningful ecological or physiological interpretation at the species or genus level. Morphology knowingly underestimates sponge diversity; such discrepancies result from identifications of physical specimens from visual census data. Accurate identification requires experienced taxonomists, and molecular and biochemical profiles, showing concordant variation and being the most efficient sponge biodiversity assessment tool. The OTUs (operational taxonomic units) and MOTUs (molecular operational taxonomic units) are employed to resolve undescribed and cryptic species (Poppe et al. 2010). Proper species identification significantly enhances our understanding of their biodiversity, distributions, and ecology. Also, species databases provide valuable insights into how common or essential sponges are, whether they are likely to be endemic, where sponge hotspots occur, and areas under sampled.

15.6.2 Collection of Regional Baseline Information

To ensure effective management and conservation of sponges, a comprehensive understanding of their diversities, distributions, and biological roles is critical, particularly for regions of high biodiversity and endemism. This knowledge is even more pertinent in the face of rapid global climate change and altered habitat and water quality from increasing coastal development and offshore industries (Przeslawski et al. 2008). We should identify the marine grounds of particular importance for conserving sponge populations based on scientific knowledge. These sponge assemblage data are applied for the future management and monitoring of the region, particularly spatial scale in biodiversity assessments and associated management strategies.

15.6.3 Characterization

The marine environment varies significantly with latitude, ocean current, and bottom topography. It is essential to implement conservation and sustainable utilization measures, considering the characteristics of individual sponge marine areas. The marine environment represents a unique combination of environmental factors such as salinity, pressure, low temperature, and nutrition. However, marine sponge requires remarkable metabolic capabilities to adapt and survive in such condition. The high salt concentration is integral for producing bioactive compounds by sponge-associated fungi. Such an environment has been linked to the osmoregulatory mechanism that signals the production of polyol and amino compounds combined with the increasing concentration of cytoplasmic ions. Many endophytes produce bioactive metabolites involved in the host–endophyte relationship. These metabolites may serve as a source of novel natural products in medicine, agriculture, and industry. It is essential to study endophytic biodiversity, chemistry, bioactivity, and the interaction between sponges and endophytic microorganisms.

15.6.4 Local Knowledge

Participation of various local people and facilitation of coordination is important for conservation and management based on their long history and traditional wisdom.

15.6.5 Sponge Protected Areas

Marine sponges should be designated and managed by law or other effective means to conserve them, supporting the structure and function of marine ecosystems. Areas that would fall under sponges should be designated in various forms, such as National parks. The continuous review should be required for appropriate measures, considering the accumulation of knowledge and changes in the social situation. It is essential to recognize diversity for the constant utilization of the ecosystem services requires maintenance of their ecosystems for our economic activities and social life.

15.6.6 Causes and Factors for Marine Sponge Decline

A knowledge gap will limit understanding of how sponges will respond to environmental degradation. The causes and factors responsible for their decline are identified to promote the conservation of marine sponge populations. Also, actions will be followed with methods and procedures to reduce sponge population decline.

Environmental factors that alter microbiome functioning can lead to changes at the holobiont, community, or even ecosystem level, illustrating the necessity of considering multiple scales when evaluating the functioning of nested ecosystems. Thus, future studies need to target the mechanisms behind host–symbiont interactions and link multiple scales to unravel how the sponge microbiome may alter holobiont functioning under future environmental changes.

15.6.7 Monitoring

Long-term monitoring strategies must collect missing baseline data and identify long-term sponge population trends. It is essential to assess the vulnerability of sponges to global change and other anthropogenic stressors and ensure the maintenance of sponge holobiont functions at the ecosystem level. This will entail multidisciplinary approaches that combine experimental, field, and genomic/transcriptomic data. Sponge culture aspects should be taken up in all the suitable places, including degraded areas. Regularly monitor the activities such as aquaculture, overfishing, mining, and industrialization.

15.7 Conclusion

Sponges play a significant role as water and nutrient's cyclers, biological indicators of water quality, monitoring benthic communities, and managing impacted areas. They provide a habitat for the other taxa. Anthropogenic activities and climate change threaten sponge communities and their associated biodiversity. Since there has been no assessment of their current global conservation status, effective management strategies to conserve sponges are relevant.

References

Abdo DA (2007) Endofauna differences between two temperate marine sponges (Demospongiae; Haplosclerida; Chalinidae) from southwest Australia. Mar Biol 152:845–854

Beepat SS, Appadoo C, Marie DEP et al (2014) Macrofauna associated with the sponge *Neopetrosia exigua* (Kirkpatrick, 1900) from Mauritius. West Indian Ocean J Mar Sci 13: 133–142

Bell JJ (2008) The functional roles of marine sponges. Estuar Coast Shelf Sci 79:341–353

Bett BJ (2001) UK Atlantic margin environmental survey: introduction and overview of bathyal benthic ecology. Cont Shelf Res 21:917–956

Bosch TCG, McFall-Ngai MJ (2011) Metaorganisms as the new frontier. Zoology 114:185–190

Carter L, Burnett D, Drew S et al (2009) Submarine cables and the oceans—connecting the world. UNEP-WCMC Biodiversity Series No 31

Costello MJ, Cheung A, De Hauwere N (2010) Topography statistics for the surface and seabed area, volume, depth, and slope, of the world's seas, oceans, and countries. Environ Sci Technol 44(23):8821–8828

De Voogd NJ, Alvarez B, Boury-Esnault N (2022) World Porifera Database. https://www.marinespecies.org/porifera

Duffy JE (1996) *Synalpheus regalis*, new species, a sponge-dwelling shrimp from the Belize Barrier Reef, with comments on host specificity in Synalpheus. J Crustaccan Biol 16:564–573

Erwin PM, López-Legentil S, González-Pech R et al (2012) A specific mix of generalists: bacterial symbionts in Mediterranean Ircinia spp. FEMS Microbiol Ecol 79:619–637

Erwin PM, Coma R, López-Sendino P et al (2015) Stable symbionts across the HMA-LMA dichotomy: low seasonal and inter-annual variation in sponge-associated bacteria from taxonomically diverse hosts. FEMS Microbiol 91:fiv115

Fell PE (1974) Porifera. In: Giese AC, Pearse JS (eds) Reproduction of marine invertebrates. Academic, New York

Fiore CL, Jutte PC (2010) Characterization of macrofaunal assemblages associated with sponges and tunicates collected of the southeastern United States. Invertebr Biol 129:105–120

Flórez LV, Biedermann PHW, Engl T et al (2015) Defensive symbioses of animals with prokaryotic and eukaryotic microorganisms. Nat Prod Rep 32:904–936

Folkers M, Rombouts T (2020) Sponges revealed: a synthesis of their overlooked ecological functions within aquatic ecosystems. In: Jungblut S, Liebich V, Bode-Dalby M (eds) The Oceans: our research, our future. Springer, Cham

Frith DW (1976) Animals associated with sponges at North Hayling, Hampshire. Zool J Linnean Soc 58:353–362

Gao Z-M, Wang Y, Tian R-M et al (2014) Symbiotic adaptation drives genome streamlining of the cyanobacterial sponge symbiont "*Candidatus Synechococcus spongiarum*". mBio 5:e00079–e00014

García-Hernández JE, Sanchez PJ, Hammerman NM et al (2018) Fish, coral, and sponge assemblages associated with altiphotic and mesophotic reefs along the Guánica biosphere reserve continental shelf edge. Southwest Puerto Rico Front Mar Sci 5:303

George AM, VAN Soest RWM, Sluka RD et al (2020) A checklist of marine sponges (Porifera) of peninsula India. Zootaxa 4885(2)

Gunasekaran S, Krishnamoorthy V (2007) Indian Marine sponges—their biodiversity and conservation. Zoological Survey of India National Symposium on Conservation and Valuation of Marine Biodiversity. pp 1–445

Hardoim CCP, Lôbo-Hajdu G, Custódio MR et al (2021) Prokaryotic, fungal, and unicellular eukaryotic core communities across three sympatric marine sponges from the Southwestern Atlantic Coast are dominated largely by deterministic assemblage processes. Front Microbiol 12:674004

Hentschel U, Hopke J, Horn M et al (2002) Molecular evidence for a uniform microbial community in sponges from different oceans. Appl Environ Microbiol 68:4431–4440

Hooper J (1998) Sponge biodiversity, distribution and biogeography. In: Levi C (ed) Sponges of the new *Caledonian lagoon*. Editions de l'Orstom, Collection Faune et Flore Tropicales, Paris

Hooper JNA, van Soest RWM, Debrenne F (2002) Systema Porifera: a guide to the classification of sponges. Kluwer Academic/Plenum Publishers, New York

Horn H, Slaby B, Jahn M et al (2016) An enrichment of CRISPR and other defense-related features in marine sponge-associated microbial metagenomes. Front Microbiol 7:1751

Lackner G, Peters EE, Helfrich EJN et al (2017) Insights into the lifestyle of uncultured bacterial natural product factories associated with marine sponges. Proc Natl Acad Sci U S A 114:347–356

Luypaert T, Hagan JG, McCarthy ML et al (2020) Status of marine biodiversity in the anthropocene. In: Jungblut S, Liebich V, Bode-Dalby M (eds) The oceans: our research, our future. Springer, Cham

Maldonado M, Ribes M, van Duyl FC (2012) Nutrient fluxes through sponges: biology, budgets, and ecological implications. Adv Mar Biol 62:113–182
Maldonado M, Aguilar R, Bannister R et al (2015) Sponge grounds as key marine habitats: a synthetic review of types, structure, functional roles and conservation concerns. In: Rossi S, Bramanti L, Gori A, Saco O, del Valle C (eds) Marine animal forests. Springer, Cham
McClintock JB, Amsler CD, Baker BJ (2010) Overview of the chemical ecology of benthic marine invertebrates along the Western Antarctic Peninsula. Integr Comp Biol 50:967–980
Mehbub M, Lei J, Franco C et al (2014) Marine sponge derived natural products between 2001 and 2010: trends and opportunities for discovery of bioactives. Mar Drugs 12:4539–4577
Neves G, Omena E (2003) Influence of sponge morphology on the composition of the polychaete associated fauna from Rocas Atoll, northeast Brazil. Coral Reefs 22:123–129
Oebius HU, Becker HJ, Rolinski S et al (2001) Parametrisation and evaluation of marine environmental impacts produced by deep-sea manganese nodule mining. Deep-Sea Res II Top Stud Oceanogr 48:3453–3467
Pattanayak JG, Manna B (2001) Distribution of Marine sponges (Porifera) in India. Proc Zool Soc 54(1):73–101
Paulus E (2021) Shedding light on deep-sea biodiversity—a highly vulnerable habitat in the face of anthropogenic change. Front Mar Sci 8:667048
Pineda MC, Strehlow B, Sternel M et al (2017) Effects of suspended sediments on the sponge holobiont with implications for dredging management. Sci Rep 7:4925
Poppe J, Sutcliffe P, Hooper JNA et al (2010) CO I barcoding reveals new clades and radiation patterns of indo-pacific sponges of the family Irciniidae (Demospongiae: Dictyoceratida). PLoS One 5:e9950
Przeslawski R, Ahyong S, Byrne M (2008) Beyond corals and fish: the effects of climate change on noncoral benthic invertebrates of tropical reefs. Glob Chang Biol 14:2773–2795
Przeslawski R, Alvarez B, Battershill C et al (2014) Sponge biodiversity and ecology of the Van Die-men Rise and eastern Joseph Bonaparte Gulf, Northern Australia. Hydrobiologia 730:1–16
Ramirez-Llodra E (2020) Deep-sea ecosystems: biodiversity and anthropogenic impacts. In: Banet C (ed) The law of the seabed, vol 90. Publications on Ocean Development
Rocha LA, Rosa IL, Feitoza BM (2000) Sponge-dwelling fishes of northeastern Brazil. Environ Biol Fish 59:453–458
Rützler K (1978) Sponges in coral reefs. In: Stoddart DR, Johannes RE (eds) Coral reefs: research methods: monographs on oceanographic methodology. UNESCO Monographs on Oceanographic Methodology
Schönberg CHL (2008) A history of sponge erosion: from past myths and hypotheses to recent approaches. In: Wisshak M, Tapanila L (eds) Current developments in bio-erosion. Springer
Schönberg CHL, Fromont J (2012) Sponge gardens of Ningaloo Reef (Carnarvon Shelf, Western Australia) are biodiversity hotspots. Hydrobiologia 687:143–161
Slaby BM, Hackl T, Horn H et al (2017) Metagenomic binning of a marine sponge microbiome reveals unity in defense but metabolic specialization. ISME J 17:2465–2478
Steinert G, Rohde S, Schupp PJ et al (2017) Host-specific assembly of sponge-associated prokaryotes at high taxonomic ranks. Sci Rep 7:2542
Stewart HA, Jamieson AJ (2019) The five deeps: the location and depth of the deepest place in each of the world's oceans. Earth Sci Rev 197:102896
Stowe SD, Richards JJ, Tucker AT et al (2011) Anti-biofilm compounds derived from marine sponges. Mar Drugs 9:2010–2035
Thomas T, Moitinho-Silva L, Lurgi M et al (2016) Diversity, structure and convergent evolution of the global sponge microbiome. Nat Commun 7:11870
Townsed M, Davies K, Hanley N et al (2018) The challenge of implementing the marine ecosystem service concept. Front Mar Sci 5:359
Tsubaki R, Kato M (2014) A novel filtering mutualism between a sponge host and its endosymbiotic bivalves. PLoS One 9:e108885

Turner EC (2021) Possible poriferan body fossils in early Neoproterozoic microbial reefs. Nature 596:87–91

Van Soest RWM, Boury-Esnault N, Vacelet J et al (2012) Global diversity of sponges (Porifera). PLoS One 7:e35105

Varijakzhan D, Loh JY, Yap WS et al (2021) Bioactive compounds from marine sponges: fundamentals and applications. Mar Drugs 19(5):246

Villamizar E, Laughlin R (1991) Fauna associated with the sponges *Aplysina archeri* and *Aplysina lacunosa* in a coral reef of the Archipielago de Los Roques, National Park, Venezuela. In: Reitner J, Keupp H (eds) Fossil and recent sponges. Springer, Berlin, Heidelberg

Webster NS, Thomas T (2016) The sponge hologenome. mBio 7:e00135–e00116

Wilkinson CR, Cheshire AC (1989) Patterns in the distribution of sponge populations across the central Great Barrier Reef. Coral Reefs 8:127–134

Wilson MC, Mori T, Ruckert C et al (2014) An environmental bacterial taxon with a large and distinct metabolic repertoire. Nature 506:58–62

Wulff JL (2006) Ecological interactions of marine sponges. Can J Zool 84:146–166

Wulff J (2012) Ecological interactions and the distribution, abundance, and diversity of sponges. Adv Mar Biol 6:273–344

Xavier J, van Soest RRM (2012) Diversity patterns and zoogeography of the Northeast Atlantic and Mediterranean shallow-water sponge fauna. Hydrobiologia 687:107–125

Yu M-C, Kolbasov GA, Hosie AM et al (2017) Descriptions of four new sponge-inhabiting barnacles (Thoracica: Archaeobalanidae: Acastinae). Zootaxa 4277:151–198

Chapter 16
In Vitro Conservation of Rare, Endangered, and Threatened Plants

Sangeeth Chandran, A. V. Raghu, and K. V. Mohanan

Abstract The biosphere is very rich in the diversity of fauna and flora. All living organisms depend upon each other and utilize the ecosystem to which they belong for fulfilling their basic needs. Due to many natural as well as anthropogenic reasons, several taxa of plants are being wiped out and many are at the risk of extinction. In this alarming situation, conservation techniques both in situ and ex situ are practiced for the successful recoupment of the losing species diversity. Ex situ conservation techniques are either in vivo or in vitro. In vitro techniques are the most effective tools for the storage of germplasm of various plant species, including rare, endangered, and threatened ones. Different in vitro conservation methods like regular growth and slow growth cultures and cryopreservation techniques can contribute significantly toward the conservation and survival of the existing rare plant species. Long-term viable storage of plant germplasm can be possible by cryopreservation methods. By the combined utilization of both in situ and ex situ methods, the floristic status of our biosphere could be improved and many species saved from complete extinction.

Keywords Diversity · Endangered plants · In situ and ex situ conservation · Tissue culture · Cryopreservation

16.1 Introduction

Plants have been an integral part and vital source of food, medicine, timber, natural products, and other necessities for years and will continue to be so in the future also (Thangavel et al. 2014). According to reports from the World Health Organization, around 80% of the world population is depending on plant systems for medicine

S. Chandran · A. V. Raghu (✉)
KSCSTE-Kerala Forest Research Insititute, Peechi, Thrissur, Kerala, India

K. V. Mohanan
Gregor Mendel Foundation, Calicut University, Malappuram, Kerala, India

S. T. Sukumaran, K. T R (eds.), *Conservation and Sustainable Utilization of Bioresources*, Sustainable Development and Biodiversity 30,
https://doi.org/10.1007/978-981-19-5841-0_16

(Kapai et al. 2010). Due to many biotic and abiotic reasons and combinations of factors such as habitat loss, introduction and colonization of invasive species, climate change, overexploitation, unorganized agricultural practices, industrialization, and urbanization, the diversity of plant species is continuously declining, which will lead to the extinction of many species in the future (Paunescu 2009). Based on the reports of the World Wildlife Fund (WWF) and the International Union for Conservation of Nature and Natural Resources (IUCN), the species extinction rate comes in between 1000 and 10,000 per year. If the current rates of utilization and consumption continue, up to 60,000 higher plant species could become extinct or practically extinct by the year 2050. Once a species goes extinct, the genetic resource associated with it also will lose permanently (IUCN 2012; Maikhuri et al. 2014; Subbaiyan et al. 2015).

16.2 Rare, Endangered, and Threatened Groups

Plants are the foundation of all terrestrial ecosystems and they provide essential support systems for life on earth (Brummitt et al. 2015). Nowadays due to numerous reasons, our rich biodiversity is declining at an alarming rate resulting in the loss of a large number of species. The International Union for Conservation of Nature's (IUCN) Red List of Threatened Species is largely regarded as the most authoritative source of information on the extinction risk of species (Brummitt et al. 2015). They keep a global list of animals, plants, and fungi that tells us if a species is still alive and how likely it is to become extinct in the future. There are seven conservation levels on the Red List: least concern, near threatened, vulnerable, endangered, critically endangered, extinct in the wild, and extinct (IUCN 2012; Kapai et al. 2010; Subbaiyan et al. 2015). Each threat level is represented by a distinct category as given in Fig. 16.1 (Source: IUCN Red list 2000).

According to IUCN, a species is considered extinct when no living members of the species can be found anywhere on the planet. If a species is seen in only a single geographical area and nowhere else, it is said to be endemic to that area. Endangered species are those that are at risk of extinction in the near future. Rare species are that have limited numbers of individuals due to very narrow geographical ranges or low population densities. Vulnerable species are those that are on the verge of becoming endangered in the near future due to population declines across their range. Species classified as endangered, vulnerable, and rare by the IUCN are considered threatened (IUCN 2012; Subbaiyan et al. 2015; National Geographic Society 2012; The National Wildlife Federation 2019).

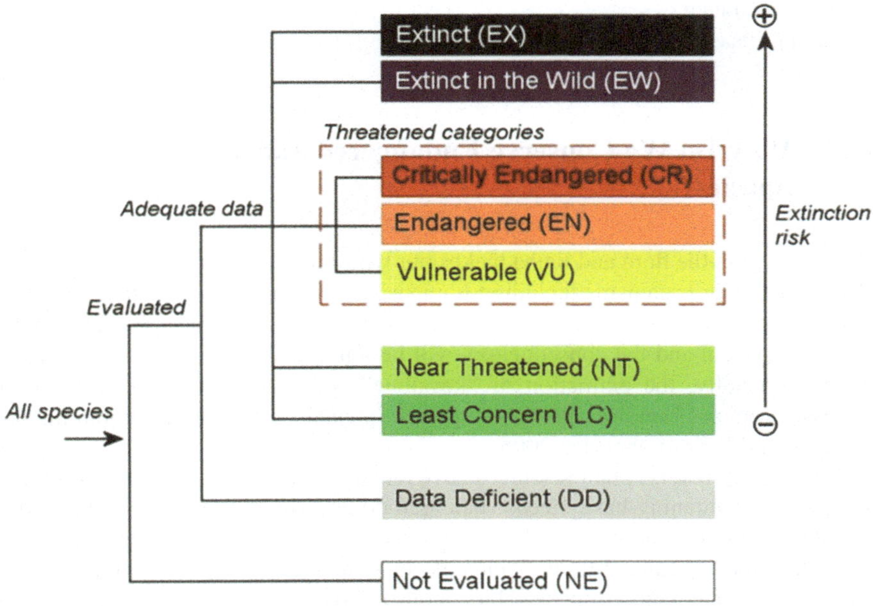

Fig. 16.1 Categories of threatened species

16.3 How Do Species Become Extinct?

Due to some unknown reasons, a few plant species are rare in nature. No intervention is required if their habitat is protected and they continue to reproduce in the wild. Conditions like declining population, limited distribution and fragmentation, small population size, etc., indicate the chances of extinction of a particular species (Kapai et al. 2010).

Instead of a single cause, rarity or extinction of species is resulted from the combination of anthropogenic and evolutionary factors. A few of the major reasons are given below (Paunescu 2009; Kapai et al. 2010; National Geographic Society 2012).

- Habitat destruction
- Fragmentation
- Loss of pollinators
- Reproductive instabilities
- Low seed germination capacity
- Loss in genetic variability
- Overexploitation
- Competition
- Pathogens and diseases
- Environmental conditions

- Climate change
- Natural disasters and calamities.

16.4 Why Do We Conserve Endangered and Threatened Species?

The diversity in the flora and fauna makes our world more attractive and interesting. Mostly all living organisms including humans rely on each other for their survival and existence. Loss of a single species even can cause severe impacts on the balance of the ecosystem and the consequences will be felt at all trophic levels of the food chain. Especially, the extinction of "keystone" species can alter the ecological processes or profoundly alter the species makeup of wildlife communities (US Fish and Wildlife Service 2005).

Throughout history, plant resources have remained an important aspect of human society. After attaining basic needs such as food and shelter, man has been on the lookout for an appropriate remedy for various diseases (Jima and Megersa 2018). Each plant species carry different kinds of phytochemicals which may have medicinal use or not. Unexplored species' biochemistry holds an untapped pool of new and potentially more effective chemicals. Almost 80,000 species of edible plants have been estimated so far. If the less utilized wild species were conserved, they could help to feed the current expanding population. Different wild plant species are also used as environmental quality indicators, for phytoremediation and also as biocontrol agents. If we fail to conserve the existing rare endangered and threatened plants, we will never be going to know how these species might have improved human life (US Fish and Wildlife Service 2005).

The loss of biodiversity at an alarming rate leads to the formation of many species conservation strategies (Engelmann and Engels 2002). Different in situ (conserving species on their natural habitat itself) conservation techniques have been carried out through national parks, biosphere reserves, wildlife sanctuaries, etc. and ex situ (conserving and maintaining living organisms outside their natural habitats) techniques using tissue or cell culture, vegetative propagules, plant seed, pollen, or in the form of a whole plant itself. Both methods are complementary with each other but in certain situations, ex situ is the only feasible method (Ramsay et al. 2000; Paunescu 2009). Cultivation in botanic gardens, seed storage, and in vitro cultivation are the most popular ex situ conservation strategies (Paunescu 2009). Botanical gardens may have the facilities like seed banks, herbaria, greenhouses, nurseries, and research laboratories (Maikhuri et al. 2014; Westwood et al. 2021). The living plant collections of botanical gardens assist ex situ conservation of RET species and provide materials for restoring the RET plants as well as the degraded habitats of each plant species. Currently, botanical gardens contain at least one-third of all known plant species, which includes more than 40% of the threatened species and a considerable number of extinct species in the wild (Westwood et al. 2021). Since ex situ conservation through botanical gardens is time- and space-constrained, additional solutions should be considered. Seed storage appears to be one of the most convenient long-term conservation approaches among the different ex situ conservation methods. However, some species do not produce seeds or produce immature, small, sterile, or recalcitrant seeds that will lose viability within a short period and therefore conventional seed storage methods are not effective (Paunescu 2009).

16.5 In Vitro Conservation

The latest fascinating advances in biotechnology have come up with tools like molecular and in vitro culture technologies for plant genetic resources management and conservation. In vitro tissue culture or systems with controlled environmental conditions offer magnificent opportunities for the production of genetically superior elite populations (Thangavel et al. 2014).

In 1975, Henshaw documented the first in vitro storage of germplasm. In vitro conservation technology enables the management, large-scale production, and international exchange of elite collections of disease-free germplasm in a genetically stable form through tissue culture. Tissue culture has emerged as a well-established method for culturing plant cells, tissues, plant organs, protoplasts, etc. under controlled conditions. Each plant species needs a different kind of growth medium and a specific protocol, which enables the large-scale production of disease-free materials having superior quality. Based on the requirement, cultures can be stored short, medium, or long term by subculturing in various culture media (Rajasekharan and Sahijram 2015; Raghu et al. 2018).

In vitro conservation programs consist of two main steps: (1) in vitro multiplication; (2) in vitro storage. Plant tissue culture is entirely based on the property of totipotency (the ability of a single cell to become a complete plant). Therefore, by the appropriate usage of cryobiological methods, in vitro culture techniques can be used for the preservation and storage of economically important plant species as well as various rare, endangered, and threatened species (Fig. 16.2). Callus, cell suspension, and protoplast cultures are the most suitable methods for the in vitro conservation methods. Periodic subculture of the tissue during long-term storage is not practical because of the chances of microbial contamination and genetic erosion. Hence, callus culture is considered the most suitable culture for long-term germplasm storage. For a few species, tissue culture is the only viable method for the conservation of germplasm (Rajasekharan and Sahijram 2015; Raghu et al. 2018). Tissue culture system has merits such as

- Limited requirement of space and time
- Fast rate of multiplication
- Pathogen-free sample production

Fig. 16.2 In vitro conserved plants

- Minimum chance for genetic erosion by the application of suitable storage conditions
- Limited labor expenses

The development of a successful protocol for in vitro conservation is very important (Rajasekharan and Sahijram 2015). A good in vitro conservation system should meet some requirements as follows:

- Growth and development should be in the minimum range
- Retention of viability and genetic stability of the stored materials as long as possible
- Limited requirement of specialized facilities, materials, and labor input

16.5.1 In Vitro Conservation Strategies

Currently, the use of in vitro approaches for RET plant conservation has global acceptance. Different conservation strategies are applied only after successfully establishing an in vitro culture of the desired plant species and obtaining sufficient multiplication. The maintenance of in vitro cultures can be done by periodic subculture into a fresh medium having a specific composition of chemicals and growth regulators which depend upon the plant species of interest. The conservation techniques can be divided into three subcategories (Rajasekharan and Sahijram 2015; Raghu et al. 2018) such as normal growth cultures, slow growth cultures, and cryopreservation.

16.5.2 Normal Growth Cultures

This is the normal conservation method that can be applied only for a few slow-growing, inherently stable plant species, as well as for species that have no other established methods. This method has only the minimum requirement of low-temperature storage facilities and it avoids the chance of variability induced by stress. Also, this method provides opportunities for further multiplication and global exchange of plant material/culture for extending the plant resources worldwide. However, long-term maintenance of the cultures is time-consuming and needs a high input of labor. There may be chances for the total loss of culture through microbial contamination during each subculture and exchange. Somaclonal variation and morphogenetic capability loss are also possible (Rajasekharan and Sahijram 2015; Raghu et al. 2018).

16.5.3 Slow Growth Cultures

Slow growth method is considered the most accepted and recommended medium-term conservation method which restricts the growth and development of the tissue directly (Raghu et al. 2018). Mostly, this method is applied for the preservation of differentiated plantlets, large callus masses, and also for developing meristem cultures (Johnson 2002). This system offers limited subculture and suppresses the growth, development, and metabolic activities without any tissue damage. The state of slow culture growth can be achieved by altering the culture conditions or by controlling the culture conditions like temperature, oxygen availability, partial oxygen pressure, sucrose, and mineral element concentration or by the addition of any additives that are osmotically active (Engelmann 1991, 2004; Johnson 2002; Rajasekharan and Sahijram 2015; Raghu et al. 2018).

The slow growth approach is used by combining or applying the following criteria in any order (Rajasekharan and Sahijram 2015; Raghu et al. 2018): (1) kind of enclosure; (2) limiting temperature and light; (3) using minimal media; and (4) using growth retardants.

Parameters such as metabolic and physiological state of the explant at the time of storage, types of vessels that we are using and their volume, the closure for the vessels, etc., have a major role in the success rate of slow growth storage (Engelmann 2011). The use of polypropylene caps instead of cotton plugs minimizes the water evaporation rate and extends the storage time (Rajasekharan and Sahijram 2015; Raghu et al. 2018).

Culturing under reduced temperature in combination with controlled light intensity or culturing in the dark itself is the most accepted and recommended storage technique (Engelmann 2011). Temperature below the required level would suppress the metabolic activities of the tissues and subsequently reduce tissue growth (Engelmann 2011; Rajasekharan and Sahijram 2015; Raghu et al. 2018).

In vitro, slow growth methods have been constantly used for the conservation of different species like banana, cassava, potato, coffee, yam, apple, plum, kiwi, etc. (Johnson 2002). Temperate species are normally grown at a temperature between 20 and 25 °C, but at the time of storage, it will be limited up to 6 °C, whereas in the case of tropical plants it comes between a range from 15 to 25 °C (Withers 1991). Tropical plants are cold-sensitive and need to be kept at higher temperatures. The temperature we are using for the storage is not constant for all the species, it may vary from species to species (Engelmann 2011). Therefore, lowering the culture environment temperature below the needed value significantly extends the time intervals between each subculture; however, care must be taken for avoiding temperature below the freezing point and for minimizing the chances of chilling injuries or the denaturation of macromolecules (Johnson 2002).

With or without low-temperature incubation, Osmotica has shown a prominent impact on extending the subculture time interval (Rajasekharan and Sahijram 2015). The concentration of sucrose beyond the normal level in the culture medium causes an inhibitory effect on plant cell growth. Hence the growth of many plant species can be restricted by changing the carbon source of the medium by non-metabolizable, inert sugar alcohol, mostly mannitol and sorbitol. The response of each plant species varies from one another. However, in combination with reduced temperature incubation, it shows a synergistic effect on in vitro conservation of different species (Johnson 2002; Rajasekharan and Sahijram 2015; Engelmann 2011).

Using certain chemicals, simply the overall growth of in vitro culture can be reduced, thereby extending the subculture time interval. Compounds like abscisic acid, *N*-dimethyl-aminosuccinamic acid, maleic hydrazide (MH), trans-cinnamic acid (TCA), chlorocholine chloride (phosphon-D), daminozide (B 995), cycocel

(CCC), etc. showed inhibitory effects on culture growth and thereby prolonged the subculture period for 6–12 months. Usage and selection of appropriate chemicals are species-specific compared with other low-temperature storage techniques; this is the most simple and affordable method of storage. By using specific chemicals, the whole culture can be conserved for a while under normal culture room conditions. However, this method has some drawbacks, such as stunted and abnormal growth, abnormal plant regeneration, development of growth retardant resistant lines, mutation and genetic alterations, etc. (Rajasekharan and Sahijram 2015). There are some alternative techniques that aid in the slow growth conservation of plant materials, such as reduction of oxygen availability to cultures by covering the explants using paraffin, mineral oils, or a liquid medium (Johnson 2002).

16.5.4 Cryopreservation

Cryopreservation is the most accepted method for the long-term storage of living biological materials at extremely low temperatures, typically at or near the temperatures of liquid nitrogen (−196 °C). At this ultra-low temperature, all cell division and other metabolic processes are stopped. Hence the plant sample can be stored for a long period with minimum maintenance (Johnson 2002; Paunescu 2009; Raghu et al. 2018). The availability of successful cryopreservation protocols for endangered and threatened species is very limited (Engelmann 2011). Materials that show natural dehydration, like orthodox seeds, dormant buds, etc., can be cryopreserved directly without any pretreatment. In experimental systems like cell suspensions, embryos, calluses, shoot tips, etc., cellular-free water is present, which causes freezing injury at the time of cryopreservation. Thus, artificial dehydration of the cells is to be done before preservation to avoid ice formation by the crystallization of cellular water (Engelmann 2011; Johnson 2002; Raghu et al. 2018). Classical and new cryopreservation techniques have different methods of operation.

16.5.4.1 Classical Cryopreservation Techniques

Classical cryopreservation techniques involve slow cooling of plant materials to a specific pre-freezing temperature which is followed by sudden immersion into liquid nitrogen. When the cell becomes supercooled, the cell membrane protects the intracellular components from freezing. Further, a decrease in the temperature results in the icing of extracellular fluid and leads to a high concentration of intracellular solutes; thereby, the cell stays supercooled. According to the pre-freezing temperature and cooling rate, intracellular water moves from the cell prior to the solidification of intracellular contents, which is followed by the immersion of tissue in liquid nitrogen. The character of the cell membrane, intracellular solute concentration, etc., are the limiting factors that cause damage to cells at the time of dehydration (Panis and Lambardi 2005; Engelmann 2011; Raghu et al. 2018).

Pregrowth of the sample, cryoprotection, progressive cooling, immersion in liquid nitrogen, storage, thawing and recovery, etc., are the major steps associated with the classical cryopreservation techniques (Johnson 2002; Engelmann 2011).

16.5.4.2 New Cryopreservation Techniques

The new cryopreservation techniques are based on vitrification and are applicable to a wide range of tissues or organs. Vitrification is the only method that can prevent crystallization without the removal of intracellular water from the cells. It is a physical activity of conversion, converting aqueous solutions to a non-crystalline or amorphous state. In this, the cell dehydration process is carried out before freezing by treating the sample with a cryoprotective medium or by air desiccation, which is followed by rapid cooling. Thus, it can avoid all the factors that are responsible for the intracellular ice or crystal formation (Panis and Lambardi 2005; Paunescu 2009; Engelmann 2011).

Comparatively, vitrification-based method is operationally simple and has a wide range of applications. This method was found to be suitable for organs like embryos, shoot tips, etc., which possess wide tissue types and also for several genotypes (Engelmann 2004).

There are seven accepted protocols for vitrification based procedures (Engelmann 2011; Paunescu 2009; Johnson 2002): (1) Encapsulation-dehydration; (2) Vitrification; (3) Encapsulation-vitrification; (4) Dehydration (5) Pregrowth; (6) Pregrowth-dehydration; and (7) Droplet-vitrification.

16.5.4.2.1 Encapsulation-Dehydration

This method is entirely based on the technology developed for artificial seed production (Paunescu 2009). The plant material/tissue is encapsulated in alginate beads which are pregrown in sucrose-enriched liquid medium, and dehydrated under laminar airflow chamber or by using silica gel to around 20% of water content, followed by rapid freezing in liquid nitrogen. The survival and growth recovery of the preserved material are very fast and direct without the formation of callus (Johnson 2002; Panis and Lambardi 2005; Engelmann 2011).

16.5.4.2.2 Vitrification

This is the dehydration method by using different cryoprotectants (chemicals used to protect biological tissue from freezing damage) followed by sudden freezing (Paunescu 2009). Pre-culture of the plant sample on a medium having cryoprotectants, dehydration using vitrification solutions like PVS2 solution, freezing using liquid nitrogen, thawing, cryoprotectant removal, recovery, etc. are the major steps associated with the vitrification process. Successful preservation of embryonic tissues, cell suspensions, somatic embryos, etc. can be done by vitrification (Johnson 2002; Panis and Lambardi 2005; Engelmann 2011).

16.5.4.2.3 Encapsulation-Vitrification

This is formulated by the combination of encapsulation-dehydration and vitrification methods. The specimens are encapsulated in alginate beads and treated with cryopreservative solutions before freezing (Johnson 2002; Paunescu 2009; Engelmann 2011).

16.5.4.2.4 Dehydration

This is the most simple method, which involves the dehydration of specimens and rapid freezing by dipping in liquid nitrogen. Desiccation is usually carried out in a laminar airflow chamber or by using a stream of sterile compressed air or silica gel. Compressed dry air induces rapid drying of samples having relatively high water content without desiccation injury. It is mostly used for freezing embryos of many recalcitrant seeds (Engelmann 2011).

16.5.4.2.5 Pregrowth

In this method, the cells or tissues are cultivated in the medium by exposing them to a cryoprotective substance and then rapid freezing using liquid nitrogen immersion (Engelmann 2011).

16.5.4.2.6 Pregrowth-Dehydration

The samples are pregrown/cultivated using a cryoprotective material and dehydrated either under a laminar airflow chamber or by using silica gel (Engelmann 2011).

16.5.4.2.7 Droplet-Vitrification

Mostly used for freezing apices of apple and potato. The apices pre-treated with vitrification solution are then kept on aluminum foil in the form of small droplets and frozen rapidly by direct immersion in liquid nitrogen (Engelmann 2004).

16.6 Cryopreservation of Rare and Endangered Plants

There are very few reports regarding the successful cryopreservation of rare, endangered, and threatened species. Bunn et al. (2007) reported the preservation techniques of different higher plant groups. Williams and Taji (1987) documented the long-term preservation of rare and threatened Australian native species such as *Boronia edwardsii, B. pilosa, Cheiranthera volubilis, Grevillea biternata, Correa volubilis, Prostanthera eurybioides, P.calycina, P. rotundifolia*, and *Rhagodia spinescens*. The successful preservation of *Eucalyptus gunnii and Grevillea scapigera* was done by Poissonnier et al. (1992) *and* Touchell et al. (1992).

Vitrification of some endangered species of west Australia *(Anigozanthos humilis, A. kalbarensis,* A. *viridis, Conostylis dielsii, C. micrantha* and *C. wonganensis)* was carried out by Turner et al. (2001). Zapartan (1996) and Ibanez and Amo-Marco (1998) proposed cryopreservation techniques for plants such as *Leontopodium alpinum* and *Minuartia valentina,* respectively.

The list of in vitro conserved and cryopreserved plants is given in Tables 16.1 and 16.2.

In this alarming situation, the active and successful application of in vitro conservation technics along with various in situ conservation methods definitely will bring back the depleted populations of rare, endangered, and threatened plants to normal range.

16.7 Conclusion

In the present situation, the conservation of RET plant species is inevitable. The current rate of extinction of species is far higher than the normal rate. If this pattern follows, we will lose numerous important genotypes, and subsequently, it will affect the balance of ecosystems. Different in situ *and* ex situ conservation strategies play a vital role in the preservation of different RET plants. Both the techniques are complementary to each other. In situ conservation is the most effective, cost-efficient technique for long-term storage as well as for exchanging materials globally. Techniques like slow growth, normal growth, and cryopreservation ensure the storage and reuse of plant materials for a long period of time.

Developing conservation protocols can reduce the risk factor for rare and threatened plant species. The preserved plant materials can be utilized for the reproduction of individual plants and thereby preventing the extinction of many plant species.

Table 16.1 In vitro conserved plants

Name	Status	Culture condition	Reference
Arnebia euchroma (Royle) Johnston	Critically endangered	MS medium with 2.5 μM IBA and 2.5 μM BAP promotes somatic embryogenesis and organogenesis	Kapai et al. (2010)
Ceropegia candelabrum L.	Endangered	Propagation by axillary bud multiplication is observed in a medium supplemented with BAP (8.87 μM) and IBA (2.46 μM). Half strength MS medium with IBA promotes in vitro rooting. ½ or ¼ strength of MS medium having 0.23 μM or 0.45 μM 2,4-D induces somatic embryogenesis	Kapai et al. (2010)
Chlorophytum borivilianum Sant. et Fernand.	Endangered	In vitro shoot formation and multiplication are reported on MS basal medium containing 22.2 μM BAP and rooting on B5 medium with 0.57 μM IAA	Kapai et al. (2010)
Gloriosa superba L.	Endangered	Basal medium containing MS salts, B5 vitamins, 9.84 μM 2-iP and 2.32 μM Kn as well as MS medium with 3 mg/L Kn produce the shoot. Half strength MS with 1.0 mg/L IBA or 0.5 mg/L IAA is good for rooting	Kapai et al. (2010)
Ipsea malabarica (Reichb.f.) J.D. Hook	Endangered	Half strength MS medium with 3% sugar and 1.5 mg/L Kn for shoot formation	Kapai et al. (2010)
Picrorhiza kurroa Royle ex Benth.	Endangered	Media containing BAP for micropropagation. MS with 1.0 μM NAA initiates root	Kapai et al. (2010)
Psoralea corylifolia Linn.	Endangered	Regeneration of plant through organogenesis and rooting is observed on normal MS medium with 2% sucrose and either with NAA or IBA	Kapai et al. (2010)
Pterocarpus marsupium Roxb.	Endangered	Cotyledonary node is used for plant regeneration; treatment with IBA (200 μM) followed by the transfer to half-strength semisolid MS medium having IBA (0.2 μM) and phenolic acids is used for rooting	Kapai et al. (2010)
Renanthera imschootiana Rolfe	Endangered	Plant regeneration is observed on MS medium having BAP and NAA. Culturing in MS with 5 μM NAA in conjugation with 1% activated charcoal promotes rooting	Kapai et al. (2010)
Vanda coerulea Griff. ex Lindl.	Endangered	Shoot tip and leaf base culture are used for multiplication and basal medium supplemented with 11.42 μM IAA used for rooting	Kapai et al. (2010)

(continued)

Table 16.1 (continued)

Name	Status	Culture condition	Reference
Coleus forskohlii Briq.	Vulnerable	Multiplication is observed on MS medium and 2 mg/L BAP MS medium with 2.0 mg/L Kn and 1.0 mg/L IAA	Kapai et al. (2010)
Gymnema sylvestre R. Br.	Vulnerable	Combination of 5 mg/L BAP and 0.2 mg/L NAA on MS medium used for shoot generation, half-strength MS having 3 mg/L IBA shows rooting	Kapai et al. (2010)
Holostemma ada-kodien Schult.	Vulnerable	Regeneration of rooted plants is observed on basal MS medium containing 0.2 mg/L BAP	Kapai et al. (2010)
Rauvolfia serpentina Benth. exKurz.	Vulnerable	Propagation is carried out using lateral buds and shoot tips. Culture on half-strength MS with 1.0 mg/L IBA and 1.0 mg/L IAA promotes rooting	Kapai et al. (2010)
Tylophora indica (Burm f.) Merrill	Vulnerable	Somatic embryogenesis and axillary bud multiplication are done using nodal segments; leaf, stem, and petiole explants used for indirect shoot regeneration via callus	Kapai et al. (2010)
Rotula aquatic Lour.	Rare	Woody plant medium supplemented with 6.0 mg/L BAP promotes shoot formation. Half-strength woody plant medium with 0.5 mg/L IAA used for rooting	Kapai et al. (2010)
Aconitum nagarum Stapf	Threatened	MS medium with sucrose (3%) and BA (6 μM) for shoot proliferation; MS medium having sucrose (3%) and NAA (5 μM) promotes rooting	Deb et al. (2018)
Hypericum gaitii Haines	Threatened	Shoot formation is observed on MS medium with sucrose (3%) and BA (1.0 mg/L); for rooting. MS medium supplemented with sucrose (2%) and IBA (1 mg/L) is used	Deb et al. (2018)
Podophyllum hexandrum Royle	Threatened	For shoot proliferation MS medium with sucrose (3%) and combination of GA3 (0.1 μM) and BA (1.0 μM) is used; MS medium having sucrose (3%) and IAA (1.0 μM) shows rooting	Deb et al. (2018)
Vanda bicolor Griff.	Threatened	MS medium with sucrose (3%), NAA and BA (3 μM each) in conjugation with 0.6% activated charcoal is used for both multiplication and rooting	Deb et al. (2018)

(continued)

Table 16.1 (continued)

Name	Status	Culture condition	Reference
Rhododendron macabeanum Watt ex Balf.f.	Threatened	Woody plant medium along with 3% sucrose and 2iP (4 mg/L) shows shoot formation. Woody plant medium conjugated with activated charcoal (0.2%, w/v) promotes rooting	Deb et al. (2018)
Rhododendron wattii Cowan	Threatened	Shooting achieved on woody plant medium containing 3% sucrose along with 2iP (8 mg/L). The same medium conjugated with activated charcoal shows rooting	Deb et al. (2018)
Decalepis arayalpathra (Joseph & Chandra.) Venter	Endangered	Culture of nodal explant on MS medium supplemented with 2.0 of BAP	Seeni and Decruse (2007)
Calophyllum apetalum Willd.	Vulnerable	Culture of nodal explant on MS medium supplemented with 2.0 of BAP	Seeni and Decruse (2007)
Blepharistemma membranifolium (Miq.) Ding Hou.	Threatened	Culture of nodal explant on half strength MS medium supplemented with 2.0 of BAP	Seeni and Decruse (2007)
Calamus travancoricus Bedd. Ex.Becc. and Hook.f.	Threatened	Embryo culture on MS medium supplemented with 0.1 mg/L TDZ	Seeni and Decruse (2007)
Calamus nagbettai Fernandez and Dey	Endangered	Embryo culture on MS medium having 0.1 mgl^{-1} TDZ	Seeni and Decruse (2007)

Table 16.2 Cryopreserved plants

Name	Status	Preservation method	Reference
Buxus hyrcana Pojark.	Critically endangered	Encapsulation-dehydration	Kaviani and Negahdar (2017)
Vanda coerulea Griff. exLindl.,	Endangered	Vitrification	Thammasiri and Soamkul (2007)
Centaurium rigualii Esteve	Endangered	Encapsulation dehydration	Viviani (1997)
Aster altaicus var. *uchiyamae* Kitam	Endangered	Cryopreservation by Droplet-vitrification	Choi et al. (2019)
Mantisia spathulata	Endangered	Dehydration of immature seed	Bhowmik et al. (2011)
Mantisia wengeri	Endangered	Dehydration of immature seed	Bhowmik et al. (2011)
Cosmos atrosanguineus	Endangered	Encapsulation of shoot tip using alginate strips	Wilkinson et al. (2003)
Picrorhiza kurroa Royle ex Benth	Endangered	Cryopreservation of shoot tip by vitrification	Sharma and Sharma (2003)
Valeriana jatamansi Jones	Threatened	Cryopreserved using vitrification	Sharma et al. (2021)
Kaempferia galanga L.	Endangered	Cryopreserved using vitrification	Preetha et al. (2013)
Dioscorea deltoidei Wall.	Endangered	Cryopreservation using the vitrification and encapsulation-dehydration	Mandal and Dixit-Sharma (2007)
Cymbidium eburneum Lindl.	Threatened	Using encapsulation–dehydration	Gogoi et al. (2013)
Cymbidium hookerianum Rchb. f.	Threatened	Using encapsulation–dehydration	Gogoi et al. (2013)
Castilleja levisecta Greenm.	Critically endangered	Using droplet-vitrification method	Salama et al. (2018)
Hypericum perforatum L.	Endangered	Using aluminum foil strips and droplet-vitrification	Yang et al. (2019)
Centaurea ultreiae	Critically endangered	Using vitrification	Mallon et al. (2008)
Aconitum violaceum Jacq	Threatened	By droplet-vitrification method	Rawat et al. (2013)

References

Bhowmik SS, Kumaria S, Tandon P et al (2011) Long-term conservation through cryopreservation of immature seed of *Mantisia spathulata* and *Mantisia wengeri;* two endangered plants of north-east India. Cryo Letters 32(6):498–505

Brummitt NA, Bachman SP, Griffiths-Lee J et al (2015) Green plants in the red: a baseline global assessment for the IUCN sampled red list index for plants. PLoS One 10(8):1–22

Bunn E, Turner S, Panaia M et al (2007) The contribution of in vitro technology and cryogenic storage to conservation of indigenous plants. Aust J Bot 55(3):345–355

Choi CH, Popova E, Lee H et al (2019) Cryopreservation of endangered wild species, *Aster altaicus var. uchiyamae Kitam*, using droplet-vitrification procedure. Cryo Letters 40(2):113–122

Deb CR, Rout GR, Mao AA et al (2018) In vitro propagation of some threatened plant species of India. Curr Sci 114(3):567–575

Engelmann F (1991) In vitro conservation of horticultural species. In: Hortifroid, V International Symposium on Postharvest Physiology of Ornamental Plants; Importance of Cold in Ornamental, vol 298. pp 327–334

Engelmann F (2004) Plant cryopreservation: progress and prospects. In Vitro Cell Dev Biol Plant 40(5):427–433

Engelmann F (2011) Use of biotechnologies for the conservation of plant biodiversity. In Vitro Cell Dev Biol Plant 47(1):5–16

Engelmann F, Engels JMM (2002) Technologies and strategies for ex situ conservation. In: Managing plant genetic diversity. pp 89–103

Gogoi K, Kumaria S, Tandon P et al (2013) Cryopreservation of *Cymbidium eburneum Lindl.* and *C. hookerianum Rchb. f.*, two threatened and vulnerable orchids via encapsulation–dehydration. In Vitro Cell Dev Biol Plant 49(3):248–254

Ibanez MR, Amo-Marco JB (1998) Promotion by phloroglucinol of micropropagation of *Minuartia valentina*, an endangered and endemic Spanish plant. Plant Growth Regul 26(1):49–56

Jima TT, Megersa M (2018) Ethnobotanical study of medicinal plants used to treat human diseases in Berbere District, Bale Zone of Oromia Regional State, South East Ethiopia. Evid Based Complement Alternat Med 2018:1–16

IUCN (2012) IUCN Red List Categories and Criteria, Version 3.1, 2nd edn. IUCN Species Survival Commission, Gland, p 32

Johnson KA (2002) In vitro conservation including rare and endangered plants, heritage plants and important agricultural plants. In: Proceedings of the 7th meeting of the International Association for Plant Tissue Culture and Biotechnology. University of New England, pp 79–91

Kapai VY, Kapoor P, Rao IU et al (2010) In Vitro propagation for conservation of rare and threatened plants of India—a review. Int J Biol Technol 1(2):1–14

Kaviani B, Negahdar N (2017) Propagation, micropropagation and cryopreservation of *Buxus hyrcana Pojark.*, an endangered ornamental shrub. S Afr J Bot 111:326–335

Maikhuri R, Negi VS, Rawat L et al (2014) Notes on conservation of 'RET' plants in India. Curr Sci 06(7):916

Mallon R, Bunn E, Turner SR et al (2008) Cryopreservation of *Centaurea ultreiae* (Compositae) a critically endangered species from Galicia (Spain). Cryo Letters 29(5):363–370

Mandal BB, Dixit-Sharma S (2007) Cryopreservation of in vitro shoot tips of *Dioscorea deltoidea Wall.*, an endangered medicinal plant: effect of cryogenic procedure and storage duration. Cryo Letters 28(6):461–470

National Geographic Society (2012) Endangered species. National Geographic Society. https://www.nationalgeographic.org/encyclopedia/endangered-species/

Panis B, Lambardi M (2005) Status of cryopreservation technologies in plants (crops and forest trees). In: The role of biotechnology in exploring and protecting agricultural genetic resources, vol 5(7). pp 43–54

Paunescu A (2009) Biotechnology for endangered plant conservation: a critical overview. Rom Biotechnol Lett 14(1):4095–4103

Poissonnier M, Monod V, Paques M et al (1992) Cryopreservation in liquid nitrogen of *Eucalyptus gunnii* shoot tips grown in vitro following encapsulation and dehydration. Annales de Recherches Sylvicoles, AFOCEL, pp 5–23

Preetha TS, Hemanthakumar AS, Krishnan PN et al (2013) Shoot tip cryopreservation by vitrification in *Kaempferia galanga L.* an endangered, overexploited medicinal plant in Tropical Asia. IOSR J Pharm Biol Sci 8(3):19–23

Raghu AV, Amruth M, Muhammed Kunhi MK, Raveendran VP, Viswanath S (2018) Prospects in conservation of medicinal plants. KSCSTE-KFRI, pp 13–29

Rajasekharan PE, Sahijram L (2015) In vitro conservation of plant germplasm. In: Plant biology and biotechnology. Springer, New Delhi, pp 417–443

Ramsay MM, Jackson AD, Porley RD et al (2000) A pilot study for the ex situ conservation of UK bryophytes. In: BGCI (ed) EuroGard, pp 52–57

Rawat JM, Rawat B, Agnihotri RK et al (2013) In vitro propagation, genetic and secondary metabolite analysis of *Aconitum violaceum Jacq.*: a threatened medicinal herb. Acta Physiol Plant 35(8):2589–2599

Salama A, Popova E, Jones MP et al (2018) Cryopreservation of the critically endangered golden paintbrush (*Castilleja levisecta Greenm.*): from nature to cryobank to nature. In Vitro Cell Dev Biol Plant 54(1):69–78

Seeni S, Decruse SW (2007) In vitro multiplication and restoration of selected rare, endangered and threatened plants of India. Seed 600:83

Sharma N, Sharma B (2003) Cryopreservation of shoot tips of *Picrorhiza kurroa Royle ex Benth.*, an indigenous endangered medicinal plant, through vitrification. Cryo Letters 24(3):181–190

Sharma S, Parasher K, Mukherjee P et al (2021) Cryopreservation of a threatened medicinal plant, *Valeriana Jatamansi Jones*, using vitrification and assessment of biosynthetic stability of regenerants. Cryo Letters 42(5):300–308

Subbaiyan B, Samydurai P, Prabu MK et al (2015) Inventory of rare, endangered and threatened (RET) plant species in Maruthamalai Hills, Western Ghats of Tamilnadu, South India. Our Nat 12(1):37–43

Thammasiri K, Soamkul L (2007) Cryopreservation of *Vanda coerulea Griff. ex Lindl.* seeds by vitrification. Sci Asia 33:223–227

Thangavel K, Ebbie MG, Ravichandran P et al (2014) Biotechnology and in vitro conservation of medicinal plants. Ann Plant Sci 3(06):734–744

The National Wildlife Federation (2019) Endangered species, National Wildlife Federation. National Wildlife Federation. https://www.nwf.org/Educational-Resources/Wildlife-Guide/Understanding-Conservation/Endangered-Species

Touchell DH, Dixon KW, Tan B et al (1992) Cryopreservation of shoot-tips of *Grevillea scapigera* (Proteaceae): a rare and endangered plant from Western Australia. Aust J Bot 40(3):305–310

Turner SR, Senaratna T, Bunn E et al (2001) Cryopreservation of shoot tips from six endangered Australian species using a modified vitrification protocol. Ann Bot 87(3):371–378

US Fish and Wildlife Service (2005) Why save endangered species

Viviani AB (1997) Cryopreservation of nodal explants of an endangered plant species (*Centaurium rigualii Esteve)* using the encapsulation–dehydration method. Biodivers Conserv 6(4):583–590

Westwood M, Cavender N, Meyer A et al (2021) Botanic garden solutions to the plant extinction crisis. Plants People Planet 3(1):22–32

Wilkinson TIM, Wetten A, Prychid C et al (2003) Suitability of cryopreservation for the long-term storage of rare and endangered plant species: a case history for Cosmos atrosanguineus. Ann Bot 91(1):65–74

Williams RR, Taji AM (1987) Effects of temperature, darkness and gelling agent on long-term storage of in vitro shoot cultures of Australian woody plant species. Plant Cell Tissue Organ Cult 1(2):151–156

Withers LA (1991) In-vitro conservation. Biol J Linn Soc 43(1):31–42

Yang X, Popova E, Shukla MR et al (2019) Root cryopreservation to biobank medicinal plants: a case study for *Hypericumperforatum L.* In Vitro Cell Dev Biol Plant 55(4):392–402

Zapartan M (1996) Conservation of *Leontopodium alpinum* using in vitro culture techiques. Bot. Garden Micropropagation News, Kew, pp 26–19

Chapter 17
Macro- and Micropropagation of Plants for Income Generation

A. S. Hemanthakumar and T. S. Preetha

Abstract The present chapter focused on the development of the macro–micro multiplication system of three economically important plants in silviculture (Rattan palm—*Calamus thwaitesii*), agriculture (Banana cultivars), and medi-culture ('aromatic ginger'—*Kaempferia galanga*) for consistent production, sustainable utilization, and income generation. In *C. thwaitesii,* macro multiplication is through seeds and suckers, while micro multiplication is accomplished by somatic embryogenesis, axenic seedling culture, and shoot tip derived offshoot culture. For somatic embryogenesis, zygotic embryos cultured in Murashige and Skoog (MS) medium supplemented with 7.0 mgL^{-1} 2,4-dichlorophenoxyacetic acid (2,4-D) induced semi-friable calli which transferred in the same medium augmented with 0.5 mgL^{-1} 6-benzylaminopurine (BAP) and 0.2 mgL^{-1} α-naphthaleneacetic acid (NAA) induced ~12 discrete globular embryoids in 6 weeks. The isolated embryoids in hormone-free medium yielded 65% plantlets. The embryoids and axenic shoots thus obtained exhibited maximum shoot induction in a medium supplemented with 0.1 mgL^{-1} Thidiazuron (TDZ). The shoot initials after subculture in medium supplemented with 0.4 mgL^{-1} BAP and 0.1 mgL^{-1} TDZ produced shoot proliferation followed by elongation in basal medium. The elongated shoots produced roots in medium supplemented with 3.0 mgL^{-1} naphthaleneacetic acid (NAA). With this established protocol, ~5940 rooted plantlets could be harvested after 40 weeks from a single embryoid. Similarly, axial shoots were induced from the shoot tip of *C. thwaitesii* suckers on MS medium supplemented with 0.4 mgL^{-1} BAP and 0.1 mgL^{-1} each of TDZ and NAA. The shoot initials obtained were transferred to fresh medium of the same composition for shoot multiplication, and such multiplied shoots were transferred to ½ MS hormone-free medium for shoot elongation. The elongated shoots (5–7 cm) were then transferred to 3.0 mgL^{-1} IBA/4.0 mgL^{-1} NAA

A. S. Hemanthakumar
Biotechnology and Bioinformatics Division, Jawaharlal Nehru Tropical Botanic Garden and Research Institute, Palode, Thiruvananthapuram, Kerala, India

T. S. Preetha (✉)
Department of Botany, University College, Thiruvananthapuram, Kerala, India

S. T. Sukumaran, K. T R (eds.), *Conservation and Sustainable Utilization of Bioresources*, Sustainable Development and Biodiversity 30,
https://doi.org/10.1007/978-981-19-5841-0_17

to raise plantlets. The plantlets thus obtained were hardened in a mist house for 8 weeks, then to 50% shade house for another 16 weeks and the well-established 6-month-old nursery plants reintroduced to selected forest segments exhibited 79–86% field establishment even after 3 years of observation. Thus, the mass multiplication system developed in *C. thwaitesii* ensure a continuous supply of quality planting material to the cane-based cottage industry.

Similarly, the propagation of banana cultivars plays a major role in the agriculture business. It involves macro–micropropagation methods for the supply of quality planting materials. Macropropagation is a very slow method of multiplication that consists of decapitation, decortications, and hardening, which produce only 50–60 shoots per sucker in 4–5 months. Therefore, the expectation of mass multiplication through micropropagation of bananas is very high. In the case of micropropagation of banana cultivars, usually full-strength MS medium supplemented with different concentrations of cytokinins particularly BAP (1–5 mgL^{-1}) is sufficient for culture initiation and multiplication. The shoot bunches thus obtained were transferred to hormone-free MS basal medium for root formation. The hormonal concentration may vary according to the variety of bananas.

Furthermore, macro–micro multiplication methods of the highly priced, medicinal and aromatic plant *Kaempferia galanga* is also demonstrated here for large-scale production, supply, and income generation. Macropropagation is carried out through rhizomes which remain dormant during drought, and they sprout 2–4 plants per rhizome in a year during the onset of spring which is uneconomic. Therefore, micropropagation has given more importance to this species. For micro multiplication, its axillary buds along with basal rhizome part (1 cm) were implanted on MS medium supplemented with 0.5 mgL^{-1} BAP to induce 1–3 shoots within 30 days. Single shoot isolated from such cultures and subcultured on medium supplemented with 4.0 mgL^{-1} BA along with 1.0 mgL^{-1} each of Kinetin and IAA produced an average of 30 shoots after 60 days of culture. Shoot elongation and rhizogenesis were parallel with shoot multiplication irrespective of the hormonal treatments. The roots developed were green and thick, and the number varied from 4 to 8. The rooted plants were hardened in the mist house (28 ± 2 °C, 80 ± 5% RH) for a period of 1–2 weeks and then to a shade net house for 3–4 weeks before transplanting. The plantlets after the short hardening phase transplanted in polybags showed 90–95% survival. Such established plantlets when transferred to the field during southwest monsoon rains facilitated better establishment at 80–90% rate. They showed normal growth like the field-grown plants in their yield and rhizome formation. Thus, the viable highly reproducible propagation systems demonstrated here can be used for the sustained delivery of high-quality planting materials to the beneficiaries thereby creating employment opportunities and income generation to some extent.

Keywords *Calamus thwaitesii* · Banana cultivars · *Kaempferia galanga* · Macropropagation · Micropropagation · Embryoids · Axenic shoot · Reintroduction

17.1 Introduction

Plant propagation is practiced as an art and business than science that caters to the need of a broad range of people, from the small backyard farmers to the large-scale producers. Nowadays, plant propagation practices are considered an important income generative sector that invariably improves the farming community's living standard. The invention of modern tools and techniques such as greenhouses, mist chambers, biotechnology-mediated artificial regeneration, including micropropagation and genetic engineering, has greatly enhanced the qualitative and quantitative production of crop plants which ultimately increased the income, employment opportunity, and export potential in the sector, besides increased household nutritional security of the country populace. The sector has also played a significant role in women empowerment, providing employment opportunities and income generation through macro- and micropropagation of planting material for strengthening the horticulture sector viz. silviculture, agriculture, floriculture, mediculture, etc. Plant propagation is a specialized technique that requires technical expertise and skill. High-quality planting material gives a better rate of field establishment and high-quality yield. The production of quality planting materials is the paramount importance for increasing the yield potential and meeting the increasing demands of quality plant produce, generating additional income, and creating more employment opportunities for both rural and urban youth and women inhabitants.

Propagation through stem cuttings, root cuttings, kiekies, coppice, suckers, and off-shoots in nursery conditions is referred to as macropropagation. It is the most widely used vegetative technique for the mass propagation of tree species. Macropropagation technique is the first step in tree species domestication and offers the opportunity of avoiding the problem of recalcitrant seeds. Macropropagation also facilitates the transfer of genetic potential as well as the non-additive variance of the parent to the new plant and offers the availability of superior individuals in a short period of time for large-scale commercial plantations. One of the promising macropropagation techniques is macro cutting. Compared to other techniques such as micro cutting, grafting, etc., macro cutting is easier, cheaper, faster, and economically more beneficial. Root formation is a critical phase in determining the success rate of macro cuttings. The root formation of cutting is a complex mechanism and is influenced by physiological, environmental, and genetic factors. Though, the mini-cutting propagation system has subsequently been applied with great success in warmer countries for subtropical tree species. These encompass eucalypt clones and interspecific hybrids (Wendling and Xavier 2003; Titon et al. 2006; Saya et al. 2008), teak (Ugalde Arias 2013), *Gmelina arborea,* and others of more local interest (Monteuuis 1993). Propagation by mini-cuttings is nowadays most commonly used for reducing the cost of plants produced by in vitro methods (Thompson 2014; Bonga 2015; Georget et al. 2017). Moreover, macropropagation methods will always face limitations and that can overcome micropropagation techniques (Monteuuis 2016).

The horticulture and plantation sector is widely heralded as a sunrise sector that provides the dynamic tool for improving economic conditions of the farmers and entrepreneurs, creating diversification opportunities with high-value crops, increasing the productivity of land, providing nutritional security, generating employment, ensuring ecological sustainability, and enhancing the export earnings. However, the production and supply of quality planting material is one of the important factors for flourishing the business in the horticulture and plantation sector. It is possible either through conventional or nonconventional means of propagation which is the involvement of science and art in a skillful way. Basic knowledge and skill of it can be a better source of income through commercial nurseries.

Over the years, horticulturists have developed various techniques for the selection of desired types of plants and their multiplication. Recently, interesting developments have taken place in the field of plant multiplication which involves the culture of cells or tissues or plant parts in vitro on a defined medium under aseptic conditions in the laboratory. Traditionally, horticultural plants are multiplied by means of seeds (sexual propagation) or organs other than seeds (asexual or vegetative propagation). Though multiplication by seeds is the cheapest method, it suffers from certain disadvantages. Plants raised from seeds may not repeat the good performance of mother plants. Many horticultural plants take a long time to produce seeds/fruits, and many of them do not produce viable, endospermous, or desired quality seeds. Vegetatively propagated plants do not suffer from these disadvantages. However, vegetative propagation is rather a slow, time- and space-consuming process. Besides, it is usually infected with latent diseases. Some plants are also not amenable to vegetative methods of propagation, for example, date palm, oil palm, coconut palm, papaya, clove, etc. Therefore, scientists started a quest for an alternative method of plant propagation which could overcome these problems. After many trials and errors, plant propagation by tissue culture method, which could overcome the disadvantages of propagation by seeds or vegetative organs, was found commercially successful in the case of orchids. Subsequently, this method has been applied to many other plants. It offers good commercial prospects in ornamental plants, vegetables, and also fruit plants, where the value of the products is high. The technique has reportedly been successful in many species of plants. It has been estimated that more than 350 million plants are being produced annually through this technology. In this backdrop, the present chapter provides macro and micropropagation of few economic plant species/hybrids viz. rattan palm (*Calamus thwaitesii*), banana cultivars, Kacholam (*Kaempferia galanga*) having high demand in silviculture, agriculture, medi-culture industry creating more employment opportunities for both rural and urban youth and women dwellers.

17.2 Propagation of Forest Trees: Silviculture

Propagation of forest trees via seeds is very difficult, particularly for many economically important luxury timbers such as teak (*Tectona grandis*), African teak (*Pericopsis elata*), padauk (*Pterocarpus* spp.), and rose wood (*Dalbergia* spp.); where the seeds are produced only rarely, and at irregular intervals. Furthermore, seed viability and subsequent germination are usually very poor, with the viability of the seed declining rapidly after only a few weeks or months. It is obvious that a far better way to produce forest trees is merely to clone superior elite trees by asexual propagation. For most trees, however, once maturity is attained, vegetative propagation becomes extremely difficult, impractical, or impossible. Propagation of trees from cuttings normally depends on the age of the parent tree, with a rooting capacity of the cutting declining rapidly with the increasing age of the parent plant. Since seed from native desirable trees of commercial value is either in very short supply or is of low viability, there is an urgent need to develop efficient, reliable techniques for asexual clonal propagation of such species.

With plantation crops such as the palms, a similar situation exists. For some palm species, such as date palm (*Phoenix dactylifera*) and sago palm (*Metroxylon* spp.), vegetative propagation is possible using suckers produced from the base of the stem. However, the number of suckers produced is very small, and consequently, the achievable rates of multiplication are low. For other palms, such as oil palm (*Elaeis guineensis*), coconut palm (*Cocos nucifera*), and rattan palm, vegetative propagation using conventional techniques is impossible. Therefore, in such cases, highly successful application of tissue culture propagation techniques is inevitable, and it should be applied to try to solve some of the problems affecting tree propagation and improvement programs.

Micropropagation through the proliferation of axillary buds is a common technique in in vitro multiplication of forest trees. Micro cloning exploits the regeneration capacity of the selected tissue; however, it is difficult to achieve in materials beyond the juvenile stage. As a result, most of the investigations on the development of protocols for clonal propagation are concentrated on juvenile materials. Among various Asian countries, the success in micropropagation of tree species, including bamboos and rattans, differs and depends on their breeding programs. For example, Malaysia has done extensive work in micropropagation of timber species, such as Acacias; the protocols for tissue culture were developed, and the plantlets were tested in the field. In India, the major success has been made in the in vitro propagation of various plant species, including forest trees such as teak, eucalypts, sandal bamboos, and rattans. To some extent, industries have also contributed in the field of tissue culture. The main focus of the industry has been on high-value crops with tremendous export potential, including a few forestry species, such as teak, eucalypts, bamboos, and rattans. However, the involvement of industry is considered essential to commercialize forestry propagules, cost-effective mass multiplication, and supply high-quality products at competitive prices in the market. Indo American hybrid seeds (IAHS) and Southern Petrochemical Industries Corporation (SPIC), which have large micropropagation units in Asia, have successfully adopted the micropropagation technique for mass multiplication of teak, bamboo, eucalypts, etc. Similarly, high-frequency mass multiplication protocols for a few species of bamboos and rattans were developed by Jawaharlal Nehru Tropical Botanic Garden and Research Institute (JNTBGRI), Kerala, India, and also taken up the initiative for mass multiplication and eco-restoration of these species to selected forest segments in the Western Ghats of Kerala as part of institute mandatory activities namely conservation and sustainable utilization of plant genetic resources (Hemanthakumar et al. 2013, 2014).

In spite of these efforts, the continual use of in vitro produced planting stock for large-scale operations is seldom practiced because the significance of integrating tree improvement with micropropagation techniques in producing genetically improved planting stock is still not fully recognized. Besides, for the in vitro clonal propagation program to be successful in plantation forestry, the gain achieved must be much higher. In this backdrop, the first part of the present chapter demonstrates the mass multiplication of quality planting materials of rattan palm for eco-restoration and plantation activities thereby promoting cane/silviculture industry.

17.3 Mass Multiplication and Income Generation of Rattan Palm

17.3.1 Introduction

Rattans constitute unique and versatile group of spiny climbing palms or canes with solid stem but are fast-depleting natural resource in South East Asian countries. They are high-value non-wood forest produce primarily used for making furniture, basket, and handicrafts' items due to its remarkable aesthetic value. In addition, they are important raw materials in the cottage industry contributing significantly to the rural economy in the provinces where they occur. As a high-value fashion-proof commodity, it is much sought after for designer trade goods to grace the homes of urbanites the world over. Since the indigenous rattan resources are overexploited and their availability in nature is limited, the existing resources are insufficient to meet the vertical increase in demand. Predictably, the gap between the rate of production and full capacity utilization in the cane processing units has widened and thereby many of the urban units are already closed down owing to a shortage of raw material. In this background, large-scale production followed by plantation activities is very necessary to meet its demand. However, traditional/macropropagation methods are carried out either through seeds or suckers but seed availability is scarce and conventional propagation through suckers/off-shoots are impractical. Therefore, artificial regeneration is considered the only alternative for the large-scale production and supply of quality planting materials. Besides, rattan cultivation areas are abandoned after shifting cultivation, so this system could also be considered as a possible way of rehabilitating unproductive secondary forests and could help to stabilize shifting cultivation areas to meet the demand. The present chapter demonstrates three efficient multiplication system for one species (three-in-one system) through somatic embryogenesis followed by direct shoot multiplication from embryoids/ axenic shoot cultures and off-shoot derived shoot tip cultures for large-scale production of this economic rattan palm *Calamus thwaitesii*. The plantlets thus multiplied were reintroduced in native forest segments as part of the development of rattan garden which enables sustainable utilization and income generation of rural dwellers from this economic palm.

17.3.2 Embryogenesis

17.3.2.1 Explant Types

Zygotic embryos dissected out of immature green fruits collected from randomly selected rattan palm *C. thwaitesii* growing in the forests of Kollam district of Kerala, India, were used as explants for the induction of somatic embryoids and subsequent production of plantlets through direct shoot multiplication from embryoids and axenic shoots.

17.3.2.2 Surface Sterilization and Embryo Germination

Immature green fruits were separated from the rachis, washed in running tap water for 30 min, and treated with 0.8% (v/v) Teepol under constant stirring for 10 min. They were again washed thoroughly in running tap water followed by rinsing in distilled water. Surface decontamination of the fruits was done by immersion in 70% ethanol and flaming over a spirit lamp inside the laminar air flow hood. After flaming, the embryos were scooped out of the fruits using a sterile surgical blade and the excised embryos were washed in sterile-distilled water to remove remnants of mesocarpic fibres. Finally, the undamaged embryos were transferred to full-strength MS basal media (Murashige and Skoog 1962) for embryo germination. All the cultures were incubated in a culture room maintained at 25 ± 2 °C under 12 h photoperiod and illumination at 50–60 μE m^{-2} s^{-1} provided by cool white fluorescent tubes (Philips India Ltd., Mumbai). Observations on germination of zygotic embryo and shoot formation were periodically recorded at 4- to 6-week interval.

As much as 98% of infection-free embryos were germinated by the surface sterilization procedure described here. The method followed is akin to the embryo culture of a few species of rattans (Barba et al. 1985; Hemanthakumar et al. 2013). Different nutrient media have been used for embryo culture in palms viz. date palm (Tisserat 1983), oil palm (Corley et al. 1976), and coconut palm (Blake and Eeuwens 1982), while in Rattan palms, full-strength MS medium has already been used for the embryo culture of *C. yunnanensis* and *C. ovoideus*. (Chengji and Jiankui 1991).

Since salt concentration of MS formulation is quite high, it may be assumed that embryos of rattans in culture prefer high concentrations of salts for their growth and differentiation. It was observed that the reduction of culture initiation period of embryo up to 2 weeks in MS basal medium and three subcultures of 6 weeks each in fresh medium having the same composition was required for the conversion of leafless seedling into fully grown seedlings with normal leaves.

17.3.2.3 Induction of Somatic Embryoids from Embryogenic Callus Culture

For embryogenic callus formation, the immature zygotic embryos were cultured at regular intervals of 8 weeks in full-strength MS medium supplemented with 1.0–12.0 mgL^{-1} 2,4-D. The white semi-friable calli obtained from the embryos, after 16 weeks of culture, were divided into ~500 mg pieces and subcultured to medium supplemented with varied combinations of BAP (0.2, 0.5, and 1.0 mgL^{-1}) and NAA (0.1, 0.2, 0.5 mgL^{-1}) to induce the embryoids. After 8 weeks, the embryoids were separated and transferred in basal medium for another 8–16 weeks for plantlet development. The calli with the remaining young embryoids were subcultured in basal medium through 2–3 cycles of 8 weeks each to facilitate maturation of the embryoids and from them plantlet formation. Embryoids liberated from the callus mass were subjected to free hand sectioning, stained with a few drops of 2% aqueous toluidine blue and examined under Nikon SMZ 800 stereomicroscope to analyse the route of embryogenesis.

However, zygotic embryos cultured in media supplemented with varying concentrations of 2,4-D (2.0–9.0 mgL^{-1}) induced proliferative mass of calli in 8 weeks, which contained compact and semi-friable portions, as well. Interestingly, the friability increased with increasing concentrations of the auxins and a maximum of 70% of embryos produced white, semi-friable calli after 16 weeks in medium supplemented with higher concentration (7.0 mgL^{-1}) of 2,4-D (Fig. 17.1a). However, 8–12% of the embryos cultured in the presence of this auxin did not respond to callus formation but changed their colour to yellowish brown or brown during the 16-weeks culture period. Nevertheless, a further increase in the concentration of 2,4-D (10.0 mgL^{-1}) resulted in embryo loss (15–32%) due to browning and necrosis. Similar observations were already reported in embryo and tissue cultures of some other rattan palms, where 2,4-D was the most frequently used auxin to produce calli and subsequent normal morphogenesis at lower concentration, within a period of 4–6 weeks (Yusoff 1989; Padmanabhan and Ilangovan 1993; Kundu and Sett 1999; Sett et al. 2002).

The semi-friable portion dissected out of the embryo callus and subcultured for 8 weeks in the medium containing 7.0 mgL^{-1} 2,4-D continued to proliferate into an increasingly semi-friable callus. The calli thus obtained were transformed into embryogenic calli by the subsequent transfer of ~500 mg fresh weight to media fortified with 0.5 BAP mgL^{-1} and 0.2 mgL^{-1} NAA in 4 weeks (Fig. 17.1b) and induced 10–12 embryoids in 6–8 weeks. Stereomicroscopic observation showed that

Fig. 17.1 Embryo-derived embryogenic callus culture: (**a**) Proliferative mass of semi-friable calli; (**b**) Semi-friable calli transformed to embryogenic calli; (**c**) Stereomicroscopic view of embryoids; (**d**) Formation of large embryoids over the browned callogenic tissue; (**e**) Embryoids isolated from callogenic tissue; (**f**) Shoot induction from axenic shoot; (**g**) Mass multiplication through subculture passage; (**h**) Rooted shoots; (**i**) Deflasked plantlets; (**j**) Nursery established plants; (**k**) 3-year-old reintroduced plant in forest segment

certain globular/spherical white-coloured embryoids bodies of embryoids over the embryogenic calli (Fig. 17.1c) enlarged with gradual browning of callogenic tissue (Fig. 17.1d). The longitudinal section of toluidine blue stained chlorophyllous mature embryoid displayed under stereomicroscope showed highly differentiated independent closed vascular system along with leaf primordium and well-organized shoot apex. Similarly, these mature embryoids also showed well-differentiated shoot apex and suspensor in the final stage of embryoid formation. A majority (65%) of 10–12 mm size mature embryoids (Fig. 17.1e) isolated from the callogenic mass were transferred to MS medium, devoid of hormone, developed into well-differentiated shoot system and root system in 4–8 weeks and to independent plantlets in 16 weeks similar to zygotic embryo germination of rattan palms (Hemanthakumar 2010). However, similar observations of somatic embryogenesis were also reported in other economic palms like date palm (Parisa Eshraghi et al. 2005; Sane et al. 2006), oil palm (Scherwinski-Pereira et al. 2010), and coconut palm (Saenz et al. 2006). In fact, the demonstrated ability to produce plantlets from a single zygotic embryo of *C. thwaitesii* confirms the desirability of using embryogenic rather than indirect organogenetic route otherwise reported in general, and in *C. flagellum* (Kundu and Sett 1999) and *C. tenuis* (Sett et al. 2002) in particular.

17.3.3 Induction of Direct Multiple Shoots from Somatic Embryoids/Axenic Shoots

Eight-week-old somatic embryoids and 16-week-old embryoid-derived axenic shoots were separated and implanted into an MS agar nutrient medium fortified with individual concentrations or combinations of plant growth regulators (PGRs), for multiple shoot induction. Among the cytokinins (BAP, 2-iP and Kinetin) tested for the shoot bud induction in embryoid cultures, 0.05–0.1 mgL^{-1} Thidiazuron (TDZ) was the best to induce maximum percentage and number of shoot bud formation. However, 0.1 mgL^{-1} TDZ was the optimal concentration for maximum multiple shoots initiation which produced 6.87 ± 0.23 shoots per embryoid through culture initiation phase (Table 17.1). A significant portion of the shoots (66–95%) becomes fasciated when the concentration of TDZ exceeded 0.2 mgL^{-1}. In addition to individual concentrations of PGRs, combinations were also tried, among them 0.4 mgL^{-1} BAP and 0.1 mgL^{-1} TDZ were the best to obtain 5.17 ± 0.27 shoots per embryoid at 100% efficiency. Overall, the results suggest that 0.1 mgL^{-1} TDZ may be indispensable to obtain justifiable number of shoots from embryoid cultures. However, a combination with other cytokinins was necessary to produce non-fasciated shoots. Unlike, many other palms, rattan embryos preferentially responded well in the presence of TDZ, and this potential can be exploited for commercial purposes, and in *C. thwaitesii* embryoid cultures, the use of TDZ is crucial for successful culture initiation.

Table 17.1 Direct multiple shoot induction and multiplication from zygotic embryos and axenic seedlings in full-strength MS medium supplemented with different concentrations and combinations of PGRs

PGRs (mgL^{-1})		Multiple shoot induction after 6 weeks of culture			
		Embryo		Axenic seedling	
TDZ	BAP	Percentage response	Mean number of shoots*	Percentage response	Mean number of shoots*
0.00	0.00	76.67[c]	1.00 ± 0.00[e]	73.33[c]	1.00 ± 0.00[d]
0.02		86.67[b]	1.00 ± 0.00[e]	86.67[b]	2.00 ± 0.00[c]
0.05		96.67[a]	2.33 ± 0.10[d]	70.00[c]	2.50 ± 0.11[c]
0.1		100.0[a]	6.87 ± 0.23[a]	100.0[a]	5.89 ± 0.27[a]
0.2		86.67[b]	3.83 ± 0.24[c]	100.0[a]	3.87 ± 0.17[b]
0.3		90.00[a]	2.83 ± 0.27[d]	96.67[a]	2.67 ± 0.28[c]
0.4		66.67[d]	2.75 ± 0.19[d]	73.33[c]	2.00 ± 0.17[c]
0.5		60.00[d]	2.00 ± 0.16[d]	66.67[d]	1.71 ± 0.16[d]
1.0		56.67[e]	1.00 ± 0.00[e]	56.67[e]	2.00 ± 0.13[c]
2.0		66.67[d]	1.00 ± 0.00[e]	46.67[f]	2.00 ± 0.18[c]
0.05	0.2	70.00[c]	2.33 ± 0.10[d]	66.67[d]	2.50 ± 1.10[c]
0.05	0.4	73.00[c]	2.50 ± 0.11[d]	70.00[c]	2.71 ± 0.13[c]
0.1	0.2	73.00[c]	3.83 ± 0.27[c]	73.33[c]	2.80 ± 0.16[c]
0.2	0.2	76.67[c]	2.50 ± 0.11[d]	73.33[c]	2.80 ± 0.17[c]
0.1	0.4	100.0[a]	5.17 ± 0.27[b]	73.33[c]	5.60 ± 0.23[a]
0.2	0.4	63.33[d]	3.60 ± 0.18[c]	63.33[d]	3.25 ± 0.19[b]

*Data represent mean ± SE of 15 replicates repeated thrice, recorded after every 6 weeks of culture. Values followed by the same letter in the superscript in a column do not differ significantly based on ANOVA and LSD t-test at $p \leq 0.05$

Whole shoots devoid of roots, dissected out of 16-week-old plantlet raised through embryoids in MS basal medium, have been used as source of explant for multiple shoot induction. In axenic shoot cultures, the basal part (resident meristem of the shoot tip) responded well with differentiation of buds in 6 weeks (Fig. 17.1f). Most of the cytokinins except TDZ tested individually were weak and in the order of priority TDZ alone (0.1 mgL^{-1}) or in combination with BAP (0.4 mgL^{-1}) preferred to obtain maximum shoot initiation in terms of frequency and number of shoots. Although shoot buds were formed *en masse* in the presence of 0.1 mgL^{-1} TDZ, the shoot buds remained more as a proliferating mass than as individual, elongated shoots. Therefore, it is justified that 0.1 mgL^{-1} TDZ is used for direct multiple shoot initiation in axenic seedling-derived whole shoot cultures of *C. thwaitesii*. Hitherto, such individual hormonal concentrations and combinations have been used with certain success for shoot initiation in seedling explant cultures of *Areca catechu* (Hsiang-Chih Wang et al. 2003).

17.3.4 Shoot Multiplication, Elongation, and Rooting

Shoot buds induced after 6 weeks from both embryoids and axenic shoots were transferred, either individually or as a group to MS medium supplemented with the combination of BAP (0.4 mgL^{-1}) and TDZ (0.05–0.2 mgL^{-1}) at 6-weeks interval for further shoot multiplication (Table 17.1). Proliferative mass of shoots achieved through subculture passages in multiplication medium was transferred to hormone-free MS medium at regular intervals of two passages each of 4 weeks to obtain elongated shoots (Fig. 17.1). The underdeveloped shoots were again transferred to fresh medium devoid of hormones for further shoot elongation.

Shoot buds obtained from embryoids and axenic shoot cultures when separated and transferred either individually or as a group to medium containing selected hormonal regimes induced varied rates of shoot multiplication. The number and frequency of shoot multiplication were dependent on the presence of TDZ either alone (0.1 mgL^{-1}) or in combination with 0.4 mgL^{-1} BAP. The rate of shoot multiplication in TDZ as well as in BAP-TDZ was higher than in culture initiation phase. However, unlike TDZ, simultaneous shoot multiplication and elongation were recorded in BAP-TDZ combination with an average of 5940 shoots/explant at 100% efficiency after three subculture passages at 6-weeks period in medium supplemented with 0.4 mgL^{-1} BAP and 0.1 mgL^{-1} TDZ (Fig. 17.1g). The multiple shoots obtained when continuously subcultured in medium supplemented with TDZ alone resulted in reduced length of the shoots. Therefore, the combination of 0.4 mgL^{-1} BAP and 0.1 mgL^{-1} TDZ was preferred as the best PGR combination for shoot multiplication of this species (Table 17.2).

However, shoot elongation was not observed during TDZ-enhanced shoot multiplication phase. Thidiazuron has been shown to induce a high rate of shoot/bud proliferation than other purine-based cytokinins in a number of woody species, including *Fagus sylvatica* (Vieitez and San-Jose 1996), *F. orientalis* (Cuenca et al. 2000), though in many species TDZ-induced buds failed to elongate into shoots

Table 17.2 Shoot multiplication through subculture passages in MS agar medium supplemented with different concentrations and combinations of PGRs

PGRs (mgL^{-1})		Subculture passages*			
TDZ	BAP	I	II	III	Average no. of shoots/explant
0.05		3.00 ± 0.24^{d}	4.50 ± 0.11^{d}	5.25 ± 0.17^{c}	120
0.1		7.87 ± 0.23^{b}	9.50 ± 0.11^{b}	11.89 ± 0.17^{ab}	4158
0.2		4.87 ± 0.17^{c}	5.33 ± 0.16^{c}	5.83 ± 0.10^{c}	300
0.05	0.4	3.71 ± 0.13^{d}	5.00 ± 0.19^{c}	5.75 ± 0.18^{c}	150
0.1	0.4	9.67 ± 0.27^{a}	11.60 ± 0.23^{a}	12.25 ± 0.19^{a}	5940
0.2	0.4	3.25 ± 0.19^{d}	5.71 ± 0.13^{c}	5.80 ± 0.28^{c}	225

*Data represent mean ± SE of 15 replicates repeated thrice, recorded after every 6 weeks of culture. Values followed by the same letter in the superscript in a column do not differ significantly based on ANOVA and LSD t-test at $p \leq 0.05$

Table 17.3 Rhizogenic response of zygotic embryo and axenic seedling derived shoots in MS agar medium supplemented with individual concentrations of auxins

Auxins (mgL^{-1})	Percentage response	Mean no. of roots/shoot*	Mean length of root (cm)*
IBA 0.0	00.00	0.00 ± 0.00	0.00 ± 0.00
1.0	50.00^{d}	1.33 ± 0.10^{b}	8.17 ± 0.39^{c}
2.0	58.33^{d}	1.43 ± 0.11^{b}	8.52 ± 0.52^{bc}
3.0	83.33^{b}	1.60 ± 0.15^{b}	10.0 ± 0.22^{a}
4.0	66.67^{c}	2.00 ± 0.22^{a}	6.63 ± 0.40^{d}
5.0	58.33^{d}	2.14 ± 0.22^{a}	6.44 ± 0.27^{d}
NAA1.0	33.33^{e}	1.25 ± 0.10^{b}	9.14 ± 0.34^{b}
2.0	58.33^{d}	1.43 ± 0.16^{b}	8.98 ± 0.41^{bc}
3.0	96.67^{a}	1.64 ± 0.17^{b}	8.84 ± 0.42^{bc}
4.0	83.33^{b}	2.00 ± 0.17^{a}	6.04 ± 0.31^{d}
5.0	66.67^{c}	2.12 ± 0.21^{a}	5.82 ± 0.30^{e}

*Data represent mean ± SE of 15 replicates repeated thrice, recorded after every 8 weeks of culture. Values followed by the same letter in the superscript in a column do not differ significantly based on ANOVA and LSD t-test at $p \leq 0.05$

(Lyyra et al. 2006). There are many reports on poor elongation of TDZ-induced shoots, which may be consistent with its high cytokinin activity; the concentration and duration of exposure to this compound are critical in this respect (Murthy et al. 1998; Debnath 2005). Shoot elongation was better achieved using liquid or agar medium free of hormones (Singha and Bhatia 1988; Fasolo et al. 1989). Interestingly, irrespective of differential origin of shoot buds (2–3 cm) from source tissue (embryoid, axenic shoot), they all responded alike during the elongation phase. Elongated shoots having a maximum 6–8 cm length could be harvested after two subculture passages of 4-weeks interval, in hormone-free liquid medium. Simultaneous formation of additional shoots during shoot elongation phase may be due to the residual effects of TDZ transferred from the multiplication phase as reported in *Calendula officinalis* (Victorio et al. 2012).

In *C. thwaitesii*, shoot elongation and rooting phase were equally long (8 weeks) and root induction was observed on varied lengths of shoots (4–8 cm) obtained from the same stage of cultures by the transfer of IBA/NAA supplemented media. In fact, among the various concentrations of auxins supplemented individually, 3.0 mgL^{-1}NAA was preferred for better rhizogenic responses at 96% efficiency (Table 17.3; Fig. 17.1h). The observed differences in rooting responses may be related to genetic/physiological constitution and also the ability to accumulate auxins endogenously in this species. Similar results were also reported in other species of rattans viz. *C. simplicifolius* (Zhang Fangqiu 1993) and *C. egregius* (Zeng Bingshan 1997). The process of rooting was slow, often taking 2–3 months before the plants get deflasked. Since most of the plants were single rooted, the ability to form only 1–2 roots may be genetically controlled as in certain species of *Calamus* viz. *C. simplicifolius* (Zhang Fangqiu 1993) and *C. egregious* (Zeng Bingshan 1997).

17.3.5 Hardening and Establishment

Plantlets obtained through embryoid and axenic shoot cultures were weaned away from the flasks, washed repeatedly in running tap water to remove traces of agar, and treated with 0.1% Dithane M-45 for 5 min in order to avoid microbial infection. They were then transferred to polybags filled with potting media of pure river sand and maintained in a mist chamber (28 ± 2 °C and 80 ± 5% RH) specially fabricated for rattan palms (M/s Indo-American Exports Ltd., Bangalore) for hardening. The post-transplantation behaviour of the plantlets was periodically monitored, and data were recorded at regular intervals. After 2 months of hardening, the established plants were transferred to the nursery and reared under diffused light and regular watering for a period of 4 months before the plantlets were reintroduced to selected forest segments of the Western Ghats. Rooted plants (Fig. 17.1i) raised through micro multiplication were established at 96% rate after 8 weeks of hardening in the mist house. The hardy nature of the roots greatly helped the plantlets to establish in the sand medium which is in corroboration with the reports on rapid clonal multiplication of *Morinda umbellata* (Nair and Seeni 2002). Since all the rattan palms employed for cultivation are tropical humid species, the high humidity levels (80 ± 5% RH) maintained in the mist house are very much needed for optimal survival and establishment of the plantlets. However, in order to assure optimal growth of the hardened plants, it was necessary to transfer them to nutrient-rich potting medium consisting of sand: top soil: farmyard manure (3:1:1) to facilitate the continued growth of the hardened plants in the shade net house (Fig. 17.1j).

17.3.6 Eco-restoration

The nursery established plants of 6 months old were transferred to evergreen/semi-evergreen forest segments of the Southern Western Ghats region for field performance study followed by the development of agroforestry system. They were planted near the forest trees to provide shade and support during their growth and establishment. The undergrowth were first thinned out and pits of 30 cm^3 were made at 5-m spacing in the forest floor and were filled with forest soil and leaf mould mixture (1:1). The plants were carefully removed from the poly bags and transplanted into the pits retaining the soil around the roots. Planting was done during southwest monsoon rains. The thick canopy of trees provided shade during their early establishment. As the forest region is infested with wild animals, the field establishment rate of reintroduced plants was monitored periodically by site visits and data on the survival rate of the plants were recorded at regular intervals. The long period (6 months) required for hardening and nursery establishment was due to the slow growth of the in vitro plantlets. This much long period was also required to observe the continuous growth of slow-growing reintroduced palms. The formation of the first new leaf in the reintroduced/translocated plant itself took nearly 2 months, followed by the emergence of 1–2 new leaves within 2–4 months. This kind of encouraging performance of the tissue-cultured plants after reintroduction is also recorded in species such as Blue Vanda (Seeni and Latha 2000) which is attributed to the best health status of the in vitro-derived plants. Long-term observations on the established plants for 3-year period confirmed 81–86% success rate of the embryoid/axenic shoot-derived plantlets (Fig. 17.1k). They continued to grow uniformly at the site of reintroduction, indicating the desirability of using these methods for large-scale production, safe handling and rehabilitating unproductive secondary forest to stabilize shifting cultivation areas which could also be helpful to the conservation of biodiversity and improve socio-economic welfare of the people.

17.3.7 Suckers/Off-Shoot Derived Shoot Tip Cultures of Rattan Palms

17.3.7.1 Collection and Surface Sterilization of Suckers

Three- to six-year-old defoliated suckers of *Calamus thwaitesii*, having maximum number of roots along with a ball of earth were collected from the forest segments of the Western Ghats (Arippa, Thiruvananthapuram Forest Range under Kerala State Forest Department). They were packed in gunny bags without exposure to direct sun, brought to JNTBGRI (Jawaharlal Nehru Tropical Botanic Garden and Research Institute) campus and planted in pure river sand medium in 30 cm diameter pots, reared in the mist house for 2–4 weeks with frequent irrigation. These suckers served as a source of explants for in vitro cloning. Individual shoots were separated from the cluster of suckers and were then washed in running tap water for 30 min. Subsequently, all the leaf sheaths except the innermost three sheaths covering the shoot tip meristem were removed and surface decontaminated by agitating in a mixture of NaClO and Teepol (v/v) (5% + 0.5%) for 20 min followed by washing in running tap water for 30–45 min and finally in distilled water. The subsequent procedures were carried out inside the laminar air flow hood, where the suckers were washed once in sterile-distilled water before removing the remaining protective sheaths. The explant was trimmed followed by exposing the portion of rhizome (1.5 cm) with shoot tip meristem (0.5–0.7 cm) and transferred to 0.1% (w/v) streptomycin sulphate solution for 45 min and rinsed 5 times in sterile-distilled water before sterilizing in 0.1% $HgCl_2$ for 4 min or 5% NaClO for 7 min. Subsequently, the browned basal rhizomatous portion was carefully removed and the remaining portion with intact shoot tip having dormant axillary buds was kept in sterile-distilled water for half an hour to remove exudates before being transferred to the appropriate nutrient medium.

Unlike other palms, the rattans are extremely thorny making it difficult to extract the shoot tips from the offshoot of actively growing underground plant parts which were used as the choice of explant for the standardization of in vitro propagation protocol of *C. thwaitesii*. Underground plant parts often harbour different microbes entailing thorough surface decontamination using rigorous multistep disinfection treatment that offered 80% of contamination-free explant of this species and are comparable with the published report in taro (Tim et al. 1990). However, the treatment of $HgCl_2$ is toxic; therefore, the use of NaClO has the added advantage of low toxicity with a similar effect of $HgCl_2$ (Brondani et al. 2013).

17.3.7.2 Culture Initiation

For culture initiation, the suckers with exposed shoot tips (~0.5–0.7 cm) and rhizome part (~1.5 cm) were inoculated to full-strength Murashige and Skoog (MS) medium (Murashige and Skoog 1962) augmented with required concentrations and combinations of plant growth regulators viz. 0.2–0.4 mgL^{-1} N_6-benzylaminopurine (BAP), 0.05–2.0 mgL^{-1} Thidiazuron (TDZ) and 0.1 mgL^{-1} α-naphthaleneacetic acid (NAA). Subculturing was made every 6 weeks. In certain cases, 2–3 transfers each of 2-week duration had to be followed to minimize exudation and oxidative damage. Cultures were incubated at 24 ± 2 °C under 12 h photoperiod with a photon flux intensity of 50–60 $\mu Em^{-2}s^{-1}$provided by cool white fluorescent tubes. All the cultures were observed at weekly intervals and data were recorded after 4–16 weeks. The infection-free shoot tip explant together with the basal rhizomatous part cultured in MS basal medium invariably responded with the elongation of the existing shoot tip to varied lengths. However, an exogenous supply of cytokinins was sufficient to reverse the longitudinal growth of the apical bud and induce axillary bud proliferation. The inability of conventional cytokinins (BAP, 2-iP) alone or in combination with auxin to induce multiple shoot formation at reasonable levels made it compulsory to try TDZ individually and in combinations. TDZ was very effective to induce multiple shoot formation though the concentration required for optimum bud release and frequencies varied. In this species, relatively high concentration (1.0 mgL^{-1}) of TDZ was required to initiate 2.25 ± 0.23 shoots in 84.45% of the explants in 16 weeks (Table 17.4). However, in the presence of TDZ, apical dominance was completely arrested, and the shoots initiated did not elongate beyond 1.5 cm. Additional shoot bud initiation was axillary as it occurred from the base of the shoot tip particularly in the rhizomatous portion. The inverse relationship between apical bud growth and axillary bud release as observed in this species and stunted growth of even the newly formed axillary shoots could only be related to the known inhibitory influence of TDZ on the longitudinal growth of existing and newly formed shoots (Huetteman and Preece 1993). TDZ is capable of fulfilling the cytokinin and auxin requirements of various regeneration responses (Jones et al. 2007). It can induce new meristem formation, promote shoot development from pre-existing meristems and induce adventitious bud regeneration in a number of species including recalcitrant woody plants (Murthy et al. 1998; Bunn et al. 2005). The results presented in this study are consistent with these observations, as TDZ has played an important role in inducing multiple shoot formation from shoot tips of suckers with greater efficiency than other cytokinins. The indispensability of TDZ for shoot initiation was again confirmed when combinations of

Table 17.4 Direct shoot bud initiation and multiplication from offshoot derived shoot tip culture of *C. thwaitesii* in MS medium augmented with different concentrations of PGRs

PGRs (mgL^{-1})			% response of multiple shoot bud initiation/explant	Mean no. of shoot bud initiation/ explant*	Average no. of shoots/ shoot tip after two subculture passages
TDZ	BAP	NAA			
0.0	–	–	00.00	00.0 ± 0.00	00.00
0.01	–	–	00.00	0.00 ± 0.00	00.00
0.05	–	–	22.22^{e}	1.00 ± 0.00^{b}	04.00^{h}
0.1	–	–	37.78^{d}	1.50 ± 0.13^{b}	04.67^{h}
0.2	–	–	51.11^{b}	1.40 ± 0.20^{b}	04.71^{h}
0.3	–	–	37.78^{d}	1.33 ± 0.13^{b}	05.74^{g}
0.5	–	–	53.33^{b}	1.40 ± 0.12^{b}	06.45^{f}
1.0	–	–	84.45^{a}	2.25 ± 0.23^{a}	16.82^{b}
2.0	–	–	46.67^{c}	1.33 ± 0.13^{b}	05.25^{g}
0.2	0.2	–	22.22^{e}	1.33 ± 0.13^{b}	04.00^{h}
0.1	0.4	–	55.55^{b}	1.80 ± 0.19^{b}	08.41^{e}
0.2	0.4	–	31.11^{d}	1.33 ± 0.13^{b}	05.34^{g}
0.05	0.4	0.1	24.45^{e}	1.50 ± 0.16^{b}	12.25^{c}
0.1	0.4	0.1	86.67^{a}	2.75 ± 0.23^{a}	29.84^{a}
0.2	0.2	0.1	48.89^{c}	1.50 ± 0.13^{b}	16.67^{b}
0.2	0.4	0.1	35.55^{d}	1.40 ± 0.12^{b}	09.28^{d}

*Data represent mean ± SE of 15 replicates repeated thrice, recorded after every 16 weeks of culture. Values followed by the same letter in the superscript in a column do not differ significantly based on ANOVA and LSD t-test at $p \leq 0.05$

cytokinins on shoot initiation were tested. Only combinations involving TDZ proved good for shoot bud initiation from shoot tips. Among the various concentrations and combinations of PGRs tested in this species, maximum of 2.75 ± 0.23 shoots were initiated at 86.67% rate after16 weeks of culture in the presence of 0.4 mgL^{-1} BAP and 0.1 mgL^{-1} each of TDZ and NAA (Table 17.4; Fig. 17.2a). The beneficial effect of NAA for obtaining elongated shoots even in the presence of the cytokinins was evident. In certain species, combinations of cytokinins and auxins act synergistically to obtain normal lengthy shoots during culture initiation/multiplication. Such observations were noticed in *Citrus grandis* (Paudyal and Haq 2000) and *Yucca aloifolia* (Atta-Alla and Van Staden 1997). Shoots induced by higher concentrations of TDZ (2.0 mgL^{-1}) were always distorted or fasciated as observed in the culture. Other cytokinins were weak to initiate buds in shoot tip cultures of this species. TDZ has been used with great success in *Rhododendron* (Preece and Imel 1991) and *Rubus* (Cousineau and Donnelly 1991), wherein the shoots initiated remained stunted. Previous workers have not met with great success in getting in vitro mass multiplication of shoots in *Calamus species* (Yusoff and Manokaran 1985). This is also true with other studies including those by Dekkers and Rao (1989) and Padmanabhan and Ilangovan (1993) working on different species of rattans. Hemanthakumar et al. (2013) published soma plant regeneration from immature embryos of *C. thwaitesii*,

Fig. 17.2 In vitro multiplication of *C. thwaitesii* through suckers/off-shoots: (**a**) Axillary shoot initiation from sucker; (**b**) Shoot multiplication from single shoot initial; (**c**) Shoot elongation; (**d**) Rooting of shoot; (**e**) Deflasked plantlets; (**f**) Nursery established plants; (**g**) Two-year-old reintroduced plant

and the same group demonstrated here the possibility to clone rattan palm from the shoot tips of suckers using TDZ as the principal cytokinin.

17.3.7.3 Shoot Multiplication, Shoot Elongation and Rooting of Shoots

The emergent axillary shoot buds obtained after 12–16 weeks of culture were dissected out from the basal rhizomatous part and transferred to full-strength MS nutrient medium supplemented with different concentrations and combinations of plant growth regulators for shoot multiplication at a regular interval of 6 weeks. The rhizomatous parts free of axillary buds were again transferred to the fresh initiation medium for additional bud emergence. Shoot cultures of 2–3 cm length obtained after subculture passages were isolated either individually or as groups (3–5 shoots) and transferred to full and half strength MS medium free of hormones for further elongation/growth. The individual shoots/shoot clumps (3–5 shoots) of 1.5–7.0 cm length were dissected out carefully and subcultured in full and half strength MS solid or liquid media augmented with auxins IBA and NAA (0–5 mgL^{-1}) for root initiation.

Supplementation of 1.0 mgL^{-1} TDZ in the medium contributed to an increase in the rate of shoot multiplication and helped to obtain 2.67 ± 0.16 shoots at a length of 1.20 ± 0.19 cm in more than 90% shoot buds during the second subculture passage. The repeated subculture passages in TDZ medium greatly helped to achieve an increase in multiplication rate but decreased the shoot elongation and also induced the formation of fasciated shoots (Fig. 17.2). Such shoots did not root even after prolonged period of subculture in hormone-free medium. Similar observation was noticed in peach palm (Graner et al. 2013).

Hence combinations of hormones including TDZ were better desired than the individual concentration of TDZ for shoot multiplication (Table 17.4). A maximum of 5.11 ± 0.30 shoots (Fig. 17.2b) was obtained from a single shoot initial at 93% efficiency with an average of 29.84 shoots harvested after two subculture passages of 8 weeks duration in medium supplemented with 0.4 mgL^{-1} BAP and 0.1 mgL^{-1} each of TDZ and NAA (Table 17.4). Shoot buds multiplied through subculture were transferred to full/half strength MS medium with/without hormones for shoot elongation. Although half strength culture medium devoid of hormones favoured maximum 5.32 ± 0.51 cm long shoots with additional1.33 ± 0.10 shoots after the second subculture passage (Fig. 17.2c). The inhibitory effect of high salts present in full-strength MS medium on shoot elongation and subsequent choice of half-strength nutrient medium for shoot elongation was also reported in *Calophyllum apetalum* (Nair and Seeni 2003).

At least two transfers (each of 6-week duration) of the shoots to fresh medium were required for optimal shoot elongation. Due to the spillover effect from the multiplication stage, marginal number of shoot buds were also induced during the elongation phase as recorded in *Vitis rotundifolia*, *Rhododendron* (Huetteman and Preece 1993) and *Bactris gasipaes* (Almeida et al. 2012). Supplementation of the medium with auxins was essential for rooting of the shoots. However, despite the presence of auxins, shoots that were multiplied in higher concentration of TDZ (2.0 mgL^{-1}) never rooted presumably due to the antirhizogenic influence of this synthetic cytokinin. Therefore, percentage of rooting was significant if the shoots were

Table 17.5 Rooting in shoots of *C. thwaitesii* cultured in MS medium supplemented with various concentrations of auxins

Auxins (mgL^{-1})	Percentage rooting	Mean no. of roots per shoot*	Mean length of root (cm) *
IBA 0.0	00.00	0.00 ± 0.00	0.00 ± 0.00
1.0	00.00	0.00 ± 0.00	0.00 ± 0.00
2.0	48.89^{e}	1.00 ± 0.00	4.24 ± 0.13^{d}
3.0	91.11^{a}	1.78 ± 0.17	8.46 ± 0.45^{a}
4.0	77.78^{b}	1.55 ± 0.15	7.61 ± 0.66^{b}
5.0	48.89^{e}	1.50 ± 0.11	6.85 ± 0.23^{bc}
NAA 1.0	33.33^{f}	1.00 ± 0.00	4.72 ± 0.33^{d}
2.0	57.78^{d}	1.67 ± 0.17	6.28 ± 0.42^{c}
3.0	66.67^{c}	1.43 ± 0.11	7.83 ± 0.39^{b}
4.0	95.55^{a}	1.78 ± 0.17	6.90 ± 0.51^{bc}
5.0	68.89^{c}	1.33 ± 0.10	6.71 ± 0.52^{bc}

*Data represent mean ± SE of 15 replicates repeated thrice, recorded after every 12 weeks of culture. Values followed by the same letter in the superscript in a column do not differ significantly based on ANOVA and LSD t-test at $p \leq 0.05$

treated with higher concentrations of auxins, and maximum of 91.11–95.55% rooting with 1.78 ± 0.17 roots was obtained after 12 weeks in media supplemented with 3.0 mgL^{-1} IBA or 4 mgL^{-1} NAA (Fig. 17.2d), where root length varied from 6.90 ± 0.51 cm to 8.46 ± 0.45 cm (Table 17.5). The remarkable similarity of 1–2 root formation in this species indicates that it may be a predetermined character in the presence of an inducer auxin like NAA, which was more desirable than other auxins for efficient root induction in *Calamus simplicifolius* (Zhang Fangqiu 1993) and *Calamus egregious* (Zeng Bingshan 1997).

17.3.7.4 Hardening and Field Establishment

For hardening and establishment, the plantlets (5–10 cm length) obtained after 12–-16 months were weaned away from the bottles, washed thoroughly in running tap water to remove traces of media adhering to the plantlets and treated with 0.1% Dithane M-45 for 5 min to avoid fungal contamination. They were then transferred to 10-cm diameter earthen pots and also in polybags filled with pure river sand and maintained in the mist house under constant irrigation at 28 ± 2 °C and 80 ± 5% relative humidity (Fig. 17.2). The data on the establishment of the clones were recorded at regular intervals. Three-month-old hardened and mist house established clones transferred to the nursery were maintained under diffused light with regular watering for another 3 months (Table 17.6). After 6–9 months, nursery established plants were planted in selected forest segments and their field performance was assessed.

The rooted plants (Fig. 17.2e) weaned away from the flasks were established in polybags at 85% rate in specially fabricated mist house after 3 months. The established plants in the mist chamber were transplanted into polybags or 4′ pots

Table 17.6 Field performance of 9-months old hardened and established microclones of *C. thwaitesii* in selected forest segments of the Western Ghats

Sites of reintroduction	Total no. of plants reintroduced/ transferred	Number of plants established (%)		
		After 6 months	After 12 months	After 24 months
Palode (Thiruvananthapuram Forest range)	460	390 (84.78)	350 (76.09)	350 (76.09)
Podium (Peppara Wild Life Sanctuary, Thiruvananthapuram Forest range)	350	280 (80.00)	260 (74.29)	250 (71.43)
Aryankavu (Kollam Forest Range)	290	250 (86.21)	230 (79.31)	230 (79.31)

filled with a potting mixture of river sand, top soil and farm yard manure in the 3:1:1 ratio and maintained in 50% shade net house under regular irrigation for another 3 months of nursery establishment (Fig. 17.2f). Such well-acclimatized plants were transferred to natural forest segments during southwest monsoon season for reintroduction and field performance studies. The reintroduced plants got established at 71.43–79.31% rate after 2 years (Fig. 17.2g) and were freed of morphological and growth abnormalities (Table 17.6). The consistently high establishment of the plants recorded in different experimental sites implicates the possibility of replicating the procedure in any tropical region where rattan forms a natural resource.

17.4 Propagation of Crop Plants: Agriculture

Crop diversification in India is generally viewed as a shift from traditionally grown less remunerative crops to more remunerative crops. Food and nutrition security, income growth, poverty alleviation, employment generation, judicious use of land, water and other resources, sustainable agricultural development and environmental and ecological management/improvement have assumed high priority in various countries. Conferring the economy and power of scale to farm families with small holdings is the most serious challenge facing our agriculture (Swaminathan 2010). The Green, White, Yellow and Blue revolutions have been landmarks that have been claimed and recognized the world over. The Green Revolution changed India from a food-deficit to food-sufficient/surplus state.

Diversification of agriculture can affect both the structure and the level of employment. The production of horticultural products offers opportunities for poverty alleviation because it is usually more labour intensive than the production of staple crops. Often, it requires twice as much, sometimes up to four times as much labour than the production of cereal crops. Horticultural production can be made highly profitable to increase employment opportunities, and to bring about increasing commercialization of the rural sector. Fruit and vegetable production is usually lucrative compared to staple crops. Horticultural produce has high value-added and income generation potential, and due to a relative lack of economics of scale (compared to grain production and livestock), their production is attractive, especially for small farmers. The production of fruit and vegetables has a comparative advantage, particularly under conditions where available land is scarce, labour is abundant and markets are accessible.

The rationale for focusing on diversification towards horticultural crops for triggering agricultural development is on account of its contribution to poverty reduction through higher employment generation, higher potential for value addition and for generating foreign exchange, and provision of food and nutrition security through supply of micronutrients and roughages. A large majority of small farmers would be living below the poverty line if they depend only on agriculture as their source of livelihood. In Africa, Asia and Latin America, high-value crop exports are female-intensive industries, with women dominating most aspects of production and processing. The relative profitability of horticultural crops compared to cereals has been shown to be a determining factor for crop diversification into horticultural production in India (Joshi et al. 2003).

Micropropagation industry forms the major backbone of the agri-horti business. The elite quality starter material for the agri/horticulture crops is provided through tissue culture regeneration. Tissue-cultured planting material of fruit crops like banana, strawberries, grapes, papaya, cardamom and ornamental flowers such as gerbera, carnation and orchids is available besides a number of foliage ornamentals. There is a need to give micropropagation industry the status of 'Thrust Sector Industry' for its expansion in days to come (Chandra and Mishra 2010). Considering this background, the second part of this chapter is displaying mass multiplication through macro and micropropagation of banana, a high value food crop promoting agriculture industry in India and elsewhere.

17.4.1 Macro and Micropropagation of Banana Cultivars

17.4.1.1 Introduction

Banana and plantains (*Musa* spp.) are the earliest crop plants from humid tropics and have acclimatized to a broad range of climatic conditions. They are considered as high value, commercial crop, and plantains have remained a staple food of many ethnic groups. Irrespective of their commercial status, banana and plantains are referred as 'Poor man's apple'. Banana is globally ranked fourth, next to rice, wheat and maize in terms of gross value of production. It is a major staple food crop for millions of people as well as provides income through local and international trade. Among the starchy staple food crops, banana ranks third with respect to the total production. Traditional bananas and other species of family *Musaceae* have been the major calorie source of many ethnic tribes of Africa and Pacific Islands. It also finds importance in the diet of Caribbean (Haiti and Dominican Republic) and Latin American countries (Ecuador and Brazil). Presently, banana is grown in around 150 countries across the world on an area of 4.84 million ha producing 95.6 million tonnes. Asia, Africa, and Latin America are the major banana-producing continents. Among the major producers, India alone accounts for 27.43% (26.2 million tonnes) followed by Philippines producing 9.01 million tonnes, and China, Brazil, and Ecuador, with production ranging from 7.19 to 8.21 million tonnes. Production in India, China, and Indonesia has witnessed giant leaps in the last 15 years. In banana trade, Ecuador, Costa Rica, Philippines, and Columbia alone contribute 64% of the world's banana export. While Philippines along with Taiwan derive substantial earnings from banana export, the great bulk of bananas produced in Asian region is traded and consumed in domestic markets. Asian and Pacific consumers are offered a wide choice of banana and plantain cultivars of varying colours (yellow, red, orange, and green), flavours, textures, sizes, and shapes. Many cultivars are consumed fresh as dessert fruit, while a great number of culinary varieties are used in hundreds of recipes based on cooking bananas and plantains.

Traditionally, banana is grown as a perennial crop where the plant is allowed to produce continuous shoots from a subterranean stem. But, the yields fall after 3–5 years and decline rapidly after 10–15 years. The need to shift to cyclic replacement with a new plantation comprising cycles of one crop and one ratoon has been realized only recently in most Asian countries. Natural calamities such as typhoons, floods, droughts, and occasional volcanic eruptions cause devastating losses in banana production. The consequent need for fresh seedlings at regular intervals has led to a very large increase in the demand for clean planting material. Banana is vulnerable to a number of biotic and abiotic stresses which limit its production, particularly among small and marginal farmers with limited resources. Development of tissue culture technology has been the foundation of high-quality, disease-free planting material production at a mass scale, particularly in vegetatively propagated crops. This chapter provides detailed information on macro- and micropropagation procedures of banana seedling for the production and supply of quality planting materials to the beneficiaries.

17.4.1.2 Banana Propagation

Propagation of banana is carried out either sexually through seeds and asexually by means of suckers. Seed propagation is common in wild species, and the seed set, germination, and dormancy depend on the species. In *Ensete*, the only other genus of Musaceae, seed propagation is the only means of perpetuation since sucker production is absent. Seeds are generally brown to black in colour, 2–6 cm in diameter, round or triangular in shape, and mostly compressed in appearance. Fruits of wild species are inedible, being full of seeds that are enveloped in thin mucilaginous pulp. All cultivated commercial bananas are triploid and sterile, excepting a few parthenocarpic AA and AB diploids. Sucker propagation is the only natural means of their perpetuation; artificial methods of propagation viz. macropropagation and micropropagation are included in this chapter.

17.4.1.3 Sucker Propagation

There are two types of suckers, sword suckers—with a well-developed base, pointed tip and narrow leaf blades, as well as water suckers, which are small, less vigorous, broad leaved, and emerge in clumps. A plant produces only 5–20 suckers during its life time of 12–14 months. For accelerating the propagation rate, suckers with growing buds or cut rhizomes called 'bits' and 'peepers' are used. Several good bits, each with a centrally placed germinating eye can be cut from an unbunched rhizome after trimming the roots. Selection of appropriate mother plant for raising new propagules either through in vivo or in vitro methods is important. About 1 kg uniformly sized healthy, true to type pests and disease-free rhizomes or bits, well-trimmed around the growing sprout are the best starting material for propagation.

17.4.2 Macropropagation

Macropropagation is an excellent option for producing low-cost quality planting material. This is a simple method because of the ease of multiplication, saves cost of producing planting material, and has the potential of producing 50–60 shoots per sucker in 4–5 months. Macropropagation is achieved by two methods and could be adopted either in the field conditions (in situ) or in the nursery (ex situ). It involves decapitation, decortication, and hardening.

17.4.2.1 Decapitation

Four- to six-months old plant is headed back, the pseudo-stem is cut down and cross-cuts/incisions are made on the growing meristem so as to stimulate the production of lateral buds. This method results in the production of 10–15 uniform shoots per plant in a short span of time and is highly suitable for small and marginal farmers whose requirements of planting material are relatively small. Suckers of choice varieties can be maintained in a nursery either in sawdust bed or in a big, bottomless concrete pot. The initial planting material should preferably be certified as virus-free and multiplied at farm level under an insect proof net house.

17.4.2.2 Decortication

The pseudo-stem of the mother corm or sword sucker is cut transversely 2 cm above the collar region and then the apical meristem is removed leaving a cavity of 2 cm diameter and 4 cm depth. This is done to overcome the apical dominance. Decapitation and decortication activate the lateral buds giving rise to more side shoots. Generally, the corms that have already flowered give better results than corms that have not yet flowered. Hence, healthy corms left in the field after harvesting are also a potential explant both for in situ and ex situ mass multiplication.

17.4.2.3 Hardening

The lateral sprouts of 8–10 cm length are shifted to pro-trays containing equal parts of cocopeat and vermiculite and after sufficient watering left in a shade net (70% shade) at 85 ± 5% relative humidity. High humidity is achieved by intermittent misting. Sprouts are usually maintained in the pro-trays for a period of 15–20 days and then shifted to polythene bags of size 6′ × 4′ having thickness of 120 gauges for secondary hardening. At this stage, the plants are maintained at 50% shade and 40–50% humidity. Watering is done on alternate days and the plants are ready for field planting in 30–45 days.

17.4.3 Micropropagation for the Supply of Quality Planting Materials of Banana Cultivars

Micropropagation of banana is the practice of rapid multiplication stock plant to produce large number of plantlets under aseptic conditions using modern plant tissue culture techniques. High-frequency mass multiplication, requirement of limited mother stock, product uniformity, season-independent production, saves on labour and energy for transportation, production of disease-free planting material, easy handling of germplasm between laboratories within and across countries, high yield, and returns are the major advantages of banana tissue culture. The earliest reports of in vitro culture of bananas came from Taiwan in the 70s (Ma and Shii 1974; Ma et al. 1978). Till date, protocols have been standardized for in vitro propagation of a wide range of *Musa* species and cultivars belonging to various ploidies and genomes (Sathiamoorthy et al. 1998). Inflorescence, pseudo-stem, suckers, lateral buds, or even small eyes contain a shoot meristem (Jarret et al. 1985; Vuylsteke and De Langhe 1985). It is always better to collect the explants from flowering plants so as to ascertain their trueness to type.

Choice of explant is vital for micropropagation of banana. The well-maintained 60–80 days old virus-free mother plants as well as inflorescence collected from high-yielding healthy plants is used as the source of explant for in vitro multiplication. Success of in vitro culture depends largely on the choice of nutrient medium, including its chemical composition and physical form (Murashige 1974). Several media formulations have been reported for banana shoot tip culture but nearly half of them are modified MS media (Brown et al. 1995). Other popular media include B5 (Gamborg et al. 1968), SH (Schenk and Hildebrant 1972), N6 (Chu et al. 1975), and LS (Linsmaier and Skoog 1975). The culture media vary in both type and concentration of the components, but all have similar basic components of growth regulators, nitrogen, carbohydrates, inorganic macro and micronutrients, vitamins, and organic additives. Generally, the cultures are established on a separate initiation medium, which has a lower concentration of cytokinin than the multiplication medium to which the cultures are subsequently transferred (Jarret et al. 1985; Novak et al. 1989). In banana tissue culture, usually full-strength MS medium supplemented with different concentrations of cytokinin particularly BAP at a range of 1–5 mgL^{-1} is sufficient for culture initiation and multiplication. The hormonal concentration may vary according to the variety of bananas.

17.4.3.1 Culture Initiation

Two types of explants viz. shoot tip of inflorescence as well as suckers were used for culture initiation. Both type of explants were washed thoroughly for half an hour in running tap water to remove mud and dust associated with the explants. Then remove all the spathe of inflorescence/sheath of suckers one by one from the explant except the last 6 spathe/sheath which was retained in the explant. Such types of explants were taken to the laminar air flow hood and washed twice in sterile-distilled water before being treated with 0.1% $HgCl_2$ for 10 min. Then washed four times in sterile-distilled water and remove the remaining spathe/sheath to expose the shoot tip and transferred to the appropriate nutrient medium for culture initiation. Usually, MS medium is supplemented with different concentrations of BAP for culture initiation. Initially, cultures were transferred thrice at 2-day interval in fresh medium for avoiding exudation. Explants devoid of exudation were then transferred to a fresh medium for 15 days to induce shoot initials which were transferred once in the same medium for further shoot initiation, as well as proliferation (Fig. 17.3a). The optimum incubation temperature should be in the range of 24–26 °C. Generally, the light intensity is maintained at 1500–3000 lux. Higher levels of 3000–10,000 lux during later stages improve the survival rate of plantlets upon transfer to soil. Initially, the cultures are maintained at 16 h light/8 h dark cycle, and once after rooting, they are shifted to 14 h light/10 h dark cycle. Decapitation and wounding of shoot tips are carried out to overcome apical dominance and to encourage axillary bud proliferation. But injuring the apical bud through transverse sections, either four or eight cuts, is a much-preferred method. Injuring the explant encourages more production of phenols/exudation, but it can be kept at minimum using antioxidants like citric acid and ascorbic acid.

17.4.3.2 Shoot Multiplication/Proliferation

Shoot initials obtained after 30 days in both the explant types were used for further shoot multiplication/proliferation. Shoot initials were separated either individually or as small groups and subcultured to the fresh medium having the same hormonal composition for further shoot multiplication and each subculture passage lasted for 15–25 days depending on the variety. During 15–25 days after the first subculture,

Fig. 17.3 Micropropagation of Banana: (**a**) Culture initiation; (**b**) Shoot multiplication through subculture passage; (**c**) Shoot elongation and rooting; (**d**) Deflasked plants; (**e**) Primary hardening; (**f**) Hardened plants; (**g**) Poly bag plants ready for sale

the central meristem produces clusters of proliferating buds, and one to three axillary buds get regenerated from the basal parts of explants around the central apical meristem (Fig. 17.3). The number of axillary buds developed during the first and second subculture range from 1 to 12 depending on the genomic constitution of the variety. In general, diploids like Matti, Anaikomban, and Chenkadali produce more buds than commercial cultivars. Among the latter, the number of buds produced during subculture is high in Cavendish (Robusta, Grand Nain—AAA genome) group followed by Plantain (Nendran—AAB genome), and Monthan (ABB genome) types. Subsequent subculture is done by trimming the tip of emerging axillary buds and removal of dead tissue at the base of the explant by gentle scratching. Clusters of proliferating buds develop during the third and fourth subcultures (Fig. 17.3b). For further subculturing, proliferative mass of multiple shoots detached without injury into a smaller group having three to four shoots and cultured to a fresh medium having same hormonal composition to induce enormous number of shoots from the base of each culture initials. After the separation of culture initials from primary explant, it is cut into three to four pieces, and each slice with two to three proliferating clusters is inoculated into individual culture bottles. This subculture cycle is repeated at 3–4-week interval to increase the proliferation rate. During the fourth and fifth subcultures, a single clump contains about 15–25 proliferating shoots. After 6–8 subculture cycles, the proliferated buds were transferred to hormone-free basal solid/liquid medium or medium containing IBA/IAA and activated charcoal for rooting. After a month, the rooted plantlets are ready for hardening (Fig. 17.3c). To minimize somatic variation, the subculturing is restricted to a maximum of eight cycles when each bottle contains 25–30 plantlets with well-developed shoots and roots. Experiments have demonstrated that proliferating shoots can be transferred to polybags (10–20 cm size) having rooting media under greenhouse. This reduces cost and enhances better establishment.

17.4.3.3 Hardening of Micropropagated Plants

Once the plantlets are ready for shifting outside the laboratory, they are carefully acclimatized to adapt to the greenhouse and later to the least protected field conditions. During hardening, the plantlets undergo physiological adaptation to changing external factors like water, temperature, relative humidity, and nutrient supply. The plantlets having 4–8 cm from culture vessels/bottles were deflasked and washed thoroughly in running tap water to remove the agar medium and other dead tissues associated with the plantlets (Fig. 17.3d). It is very important as sucrose in agar encourages microorganisms. Such plantlets were treated with 0.1% Dithane M_{45} for 5 min before being transferred to mini-sand beds at 26 ± 2 °C with $80 \pm 5\%$ RH for 4–6 weeks of primary hardening (Fig. 17.3e). Structures used for primary hardening vary with the climatic conditions. These can be highly sophisticated with UV-stabilized polysheet covering, multiple misting options, thermal shade net and auto-monitoring of light intensity, temperature, and humidity. On the other hand, the structures can be simple with polycarbonate roofing, shade net on all sides with misting facilities. Temperature, relative humidity, and light intensities were monitored manually using a thermometer, hygrometer, and lux meter, respectively. Planting media for primary hardening range from sieved sand augmented with nutrition to mixtures of cocopeat and Soilrite with fine sand in equal proportions. NPK is provided in liquid form on a weekly basis.

17.4.3.4 Secondary Hardening

After primary hardening for 5–6 weeks (Fig. 17.3f), the plantlets were transferred from micropots to polybags. Base substrate is generally soil and sand along with low-cost materials like coir pith, sawdust, or rice husk. Organic manure is either in the form of farm yard manure or poultry manure. Press mud, a by-product of sugar factories, has been found to provide the best substrate for secondary hardening along with soil (Vasane et al. 2006). Plantlets from micropots are dipped in fungicide solution (0.1% bavistin) and planted in polybags containing suitable substrate. Initially, these are maintained in low-light intensity shade nets and 70% RH. The plants are hardened by gradually increasing the light intensity and reducing RH (40%). After 5–6 weeks, the plants become ready for field planting having 3–5 well-developed leaves and a good mass of fibrous roots (Fig. 17.3g).

17.5 Propagation of Medicinal Plants: Medi-culture

Medicinal and aromatic plants possess a pivotal role in Ayurveda, Homeopathy, and Unani medicine. These plants provide raw materials for pharmaceuticals, cosmetics, drugs, and other industries. Farmers collect more income via the cultivation of medicinal and aromatic plants than other commercial and traditional crops. About thousands of native plants have medicinal importance. According to the World Health Organization, 'a medicinal plant is any plant which, in one or more of its organs, contains substances that can be used for therapeutic purposes, or which are precursors for phyto-pharmaceutical semi synthesis'. The use of herbal medicines is safer than modern medicine and reduces the side effects. The global market for herbal medicine and its products is expanding tremendously. In 2004, the market was US$ 62 billion and it will reach about US$ 5000 billion by 2050 (Purohit and Vyas 2004). India is a wealthy country regarding medicinal and aromatic plants, but unfortunately achieved less success in the export of medicinal plant products due to a lack of awareness among Indian farmers about economic values and income generation from medicinal and aromatic products (Singh et al. 2007; Singh 2009; Negi and Sharma 2016).

The production and consumption as well as international trade in medicinal plants, and phytomedicines, are growing quite significantly. In addition, the consumption of herbal medicines is widely spread and increasing. However, harvesting herbal medicine from the wild as a source of raw material is causing loss of genetic diversity and habitat destruction. In spite of herbal treatments for curing different ailments in ancient times, many medicinal plants are being harvested from their wild habitat (Kumari and Priya 2020). Because there is no control over their harvesting, so the agents of traders do harvest them mercilessly, and due to this, several species have become extinct or are on the verge of extinction (Kumari and Priya 2020). Plant tissue culture techniques are said to be a more suitable alternative to help in this alarming problem.

Application of traditional and biotechnological plant-breeding techniques used to improve the genetic level for improving yield and uniformity can be achieved through different methods, including micropropagation (Yushkova et al. 1998). In vitro propagation or tissue culture of plants is very important for the production of high-quality plant-based medicines. The evolving commercial importance of secondary metabolites in recent years resulted in a great interest in secondary metabolism, particularly in the possibility of altering the production of bioactive plant metabolites using tissue culture technology. Different in vitro systems have been used to improve the production of plant chemicals. In vitro plant culture is an important technique for mass multiplication of plants, elimination of plant diseases through meristematic tissue culture technique, plant conservation, and crop improvement through gene transfer (Sarasan et al. 2011). Therefore, prevention of contamination from different sources like bacteria and fungi is necessary for successful culture of medicinal plants by using in vitro propagation. Sustainable production of drugs in the pharmaceutical industry depends on a continuous supply of healthy materials to which plants provide a major contribution (Sahoo et al. 1997). Developing reliable propagation protocols of the economically important medicinal plants through micropropagation is very important for rapid regeneration and quality planting materials production. In this context, third part of this chapter explains macro and micropropagation systems particularly high-frequency in vitro propagation protocol of high-value medicinal plant *Kaempferia galanga* for the consistent production, sustainable utilization and income generation by the supply of quality planting materials to the beneficiaries.

17.5.1 Macro and Micropropagation of Kaempferia galanga *L.*

17.5.1.1 Introduction

Kaempferia galanga L. belonging to family Zingiberaceae is a highly priced, endangered, medicinal, and aromatic plant (Kareem 1997) extensively used in the preparation of ayurvedic drugs, in perfumery, cosmetics, as spice ingredients and it forms an ingredient of more than 59 ayurvedic preparations (Sivarajan and Balachandran 1994). The species known as 'aromatic ginger' is economically important also because of the increased price of its dry rhizomes that are currently having a market value of Rs. 300 per Kg. It has great export potential as the price of essential oil ranges from US$600 to 700 per Kg on the international market and is highly exploited by the local people as well as pharmaceutical industries. The essential oil finds use in perfumery, folk medicine, and curry flavourings. *K. galanga* and its components have numerous pharmacological properties like anti-microbial, amoebicidal, analgesic, anti-tuberculosis, anti-dengue, antioxidant, anti-nociceptive, anti-angiogenic, anticancer, anti-inflammatory, hyper lipidemic, hypo-pigmentary, larvicidal, insecticidal, and mosquito-repellent activities (Kumar 2020). Studies regarding the chemical components in *K. galanga* showed natural compounds such as monoterpenoids, diterpenoids, flavonoids, phenolic glycosides, diary heptanoids, essential oils, and cyclohexane oxide derivatives (Pham et al. 2021).

The plant is probably extinct in the wild, but is now available under cultivation. Interestingly, it is one such species that satisfies the priority criteria for conservation by Cunningham (1997). According to Senarath et al. (2017), *K. galanga* is listed under the threatened category in Sri Lanka and India and in many other Asian countries. The demand for the drug is over 100 MT per year in India and also this species is enlisted in the category of plants prohibited for export under the EXIM policy (Gowthami et al. 2021). Previous studies regarding the feasibility of intercropping *K. galanga* in oil palm plantations showed that it is a profitable intercrop (Josekutty 2003). Moreover, its recognition as a flavouring and perfumery in recent years created a price hike that elevated the crop from restricted cultivation in localized tracts of Kerala to the status of a commercial crop.

17.5.2 Macropropagation of Kaempferia galanga *L.*

K. galanga is a shade-loving plant that needs warm humid climate and survives up to an elevation of 1500 m. Annual rainfall of 1500–2500 mm is needed during the growing period. Rich loamy soil with good drainage is more suitable for cultivation. The plant is propagated by the division of rhizomes. About 840–1700 Kg of rhizome is needed to plant one hectare. Seed rhizomes are stored in a cool dry place under shade or in pits. Smoking of rhizomes preceding planting is beneficial for superior germination (Preetha et al. 2016).

For planting of rhizomes, the land is ploughed and beds of 1 m width, 25 cm height and suitable length are taken and the propagules are planted at 20 × 15 cm spacing. The beds are mulched with green leaves or straw immediately after planting. Irrigation is needed especially during the dry season. Sprinkler irrigation system is better. Organic manuring is recommended. Application of 12 tonnes/acre of FYM or compost and mulching with leaves or straw are beneficial. The field is to be cleared of weeds and earthed up 2–3 times during the early stages as this is a less competitive crop. Incident of attack by insects and pests are not reported in *K. galanga*. Leaf spot and rhizome rot diseases occur particularly during the rainy season, which can be prevented by spraying with 1% Bordeaux mixture (Joy et al. 1998).

The rhizomes are harvested 6–7 months after planting at the drying time that is initiated in the leaves. The rhizomes are dug out, and cleaned by thorough washing to remove the adhering mud and soil particles. They are then sliced into circular pieces of uniform size using a sharp knife and dried for 3–4 days. On the fourth day, heap the rhizomes and keep it overnight. On the next day, it is again spread and dried, subsequently stored in a cool dry place or market it. Prolonged storage sometimes can cause insect and fungus attacks. One of the demerits of the conventional macropropagation is that the rhizomes remain dormant during drought and they sprouts on the onset of spring. Besides, there is a poor seed setting naturally. Even though seeds were obtained after eventual cross-pollination, they are non-viable. The species is annual and about 2–4 plants per rhizome can be obtained in a year. However, it is uneconomic as the rhizomes constitute the commercial product of importance and as a consequence, there is always scarcity of the propagating material too. Considering the demand in economic and medicinal value aspects (>100 MT per year), as well as the propagation problem of the plant, it is necessary to develop a suitable protocol for mass propagation from prevailing elite cultivars. There is an obvious need to develop an efficient and productive regeneration system for rapid multiplication and effective conservation to meet the market demand and replenish these underprivileged taxa.

17.5.3 Micropropagation of K. galanga

17.5.3.1 Shoot culture establishment

Rhizomes with axillary buds collected from the field-grown plants can be used as a source of explants for shoot culture initiation. They need a thorough washing under running tap water to remove the dust and mud followed by removing the outer scales using a scalpel. They were then washed in a liquid detergent (3–4 drops of Teepol, dissolved in 100-mL distilled water) for 20 min and again washed in running tap water. Thereafter several rinses in distilled water, the explants were subjected to surface sterilization with 0.1% (w/v) $HgCl_2$ for 8–10 min succeeded by 4–5 rinses in sterile-distilled water. Axillary buds along with basal rhizome part (1 cm) segmented aseptically were implanted on MS medium (Murashige and Skoog 1962) containing 0.6% (w/v) agar, 3% (w/v) sucrose and supplemented with 0.5 mgL^{-1} BAP. The cultures were incubated for 4–6 weeks in a culture room (26 ± 2 °C) under 12-h photoperiod at a photon flux intensity of 50–60 $\mu Em^{-2} s^{-1}$ provided by cool, white, fluorescent tubes (Philips, India) under 50–60% RH. The surface sterilization procedure affected higher rate of response (76.67%) and the explants produced 1–3 shoots within 30 days. Seasonal variation influenced the rate of contamination. Explants collected during the rainy season are prone to maximum contamination leading to a loss of up to 70% and those collected during summer season showed minimum contamination and loss (30%). The dormant rhizomatous bud transferred to the MS nutrient medium fortified with 0.5 mgL^{-1} BAP responded by breaking the outer thick sheath followed by the emergence of shoot primordium after 15 days of culture (Table 17.7). All the shoots developed were healthy with well-developed leaves and one or two roots (Fig. 17.4a, b).

Table 17.7 Effect of plant growth regulators on shoot multiplication in *K. galanga L*

Growth regulators (mgL^{-1})				% response	Mean number of shoots*	Mean length of shoots*	Mean number of roots*
BA	KIN	NAA	IAA				
1.0	–	–	–	73.33	$2.10 \pm 0.22^{g*}$	3.19 ± 0.15^c	2.9 ± 0.22^e
2.0	–	–	–	76.67	4.89 ± 0.74^e	3.57 ± 0.26^c	3.8 ± 0.25^d
3.0	–	–	–	70.00	5.00 ± 0.38^d	3.07 ± 0.19^c	3.3 ± 0.26^d
4.0	–	–	–	80.00	5.20 ± 0.42^d	3.41 ± 0.24^c	4.1 ± 0.24^c
–	1.0	–	–	66.67	1.40 ± 0.15^h	3.72 ± 0.24^c	2.0 ± 0.21^e
–	2.0	–	–	63.33	1.70 ± 0.14^h	3.10 ± 0.16^c	2.9 ± 0.23^e
–	3.0	–	–	60.00	1.91 ± 0.21^h	3.18 ± 0.12^c	2.8 ± 0.20^e
–	4.0	–	–	63.33	2.71 ± 0.28^g	3.68 ± 0.13^c	3.0 ± 0.26^d
2.0	1.0	–	–	76.67	2.83 ± 0.22^g	3.53 ± 0.17^c	3.3 ± 0.21^d
3.0	1.0	–	–	73.33	3.14 ± 0.26^f	3.54 ± 0.12^c	3.4 ± 0.16^d
4.0	1.0	–	–	66.67	5.75 ± 0.38^d	3.75 ± 0.19^c	3.5 ± 0.17^d
2.0	–	0.5	–	86.67	4.10 ± 0.18^e	3.69 ± 0.12^c	4.6 ± 0.22^c
3.0	–	0.5	–	76.67	5.14 ± 0.31^d	3.58 ± 0.12^c	4.7 ± 0.39^c
4.0	–	0.5	–	76.67	6.14 ± 0.26^c	3.79 ± 0.12^c	4.6 ± 0.26^c
2.0	–	1.0	–	86.67	5.57 ± 0.41^d	3.74 ± 0.13^c	4.8 ± 0.29^c
3.0	–	1.0	–	83.33	5.62 ± 0.47^d	3.43 ± 0.10^c	4.9 ± 0.31^c
4.0	–	1.0	–	86.67	7.40 ± 0.65^b	3.66 ± 0.07^c	5.2 ± 0.29^b
2.0	–	–	0.5	86.67	3.80 ± 0.29^f	3.64 ± 0.10^c	3.8 ± 0.25^d
3.0	–	–	0.5	93.33	4.14 ± 0.39^e	3.82 ± 0.17^c	3.3 ± 0.26^d
4.0	–	–	0.5	90.00	4.80 ± 0.52^e	3.83 ± 0.07^c	3.5 ± 0.22^d
2.0	–	–	1.0	96.67	4.87 ± 0.29^e	4.01 ± 0.13^b	4.6 ± 0.30^c
3.0	–	–	1.0	83.33	5.71 ± 0.44^d	4.17 ± 0.20^b	4.8 ± 0.24^c
4.0	–	–	1.0	80.00	6.00 ± 0.35^c	4.10 ± 0.15^b	4.8 ± 0.29^c
4.0	1.0	1.0	–	86.67	10.60 ± 0.83^a	5.13 ± 0.12^a	5.6 ± 0.37^b
2.0	0.5	0.5	–	83.33	5.40 ± 0.84^d	4.16 ± 0.49^b	4.1 ± 0.85^c
4.0	1.0	–	1.0	76.67	7.90 ± 0.54^b	5.58 ± 0.30^a	5.0 ± 0.30^b
2.0	0.5	–	0.5	73.33	4.40 ± 0.31^e	3.78 ± 0.19^c	6.1 ± 0.41^a

Data represent mean ± SE of 10 replicates repeated thrice; observations were recorded after 30 days of culture. *Means followed by the same letter do not differ significantly based on ANOVA and LSD multiple 't' test at $p \leq 0.05$

17.5.3.2 Shoot Multiplication

Single shoots from established in vitro cultures subcultured on a medium supplemented with a combination of BAP along with auxins (NAA and IAA) increased the shoot multiplication rates and the mean number of shoots varied from 4 to 8 (Fig. 17.4c–e). All the shoots developed were healthy with well-developed green leaves. A positive impact on multiple shoot induction could be noticed when BAP, Kinetin and NAA were used together, where approximately

Fig. 17.4 Micropropagation in *K. galanga*. (**a**, **b**) Shoot bud initiation in MS + 0.5 mgL^{-1} BAP; (**c-e**) Shoot multiplication in MS + 4.0 mgL^{-1} BAP + 0.5 mgL^{-1} NAA after 7, 14 and 20 days; (**f**) Multiple shoots in MS + 4.0 mgL^{-1} BAP + 1.0 mgL^{-1} Kinetin and NAA; (**g–i**) Shoot multiplication during I, II, and III subculture passages; (**j**) Shoot multiplication along with rooting; (**k**) Hardening of plantlets inside the mist chamber (28 ± 2 °C, 85% RH)

Table 17.8 Effect of repeated subculture on shoot multiplication in *K. galanga* L

Growth regulators (mgL^{-1})				Mean number of shoots*			
BA	KIN	NAA	IAA	Subculture I (30 days)	Subculture II (60 days)	Subculture III (90 days)	Subculture IV (120 days)
3.0	–	–	–	5.20 ± 0.42^{d}	13.0 ± 0.62^{e}	11.6 ± 0.52^{g}	9.8 ± 0.39^{h}
4.0	–	–	–	5.75 ± 0.38^{d}	19.8 ± 0.92^{c}	16.1 ± 0.48^{d}	14.3 ± 0.71^{d}
4.0	1.0	–	–	5.00 ± 0.38^{d}	12.0 ± 0.47^{f}	10.3 ± 0.58^{h}	9.3 ± 0.30^{h}
2.0	–	0.5	–	5.57 ± 0.41^{d}	11.7 ± 0.49^{f}	10.1 ± 0.56^{h}	8.9 ± 0.27^{i}
3.0	–	0.5	–	5.62 ± 0.47^{d}	15.1 ± 0.48^{d}	13.1 ± 0.48^{e}	10.6 ± 0.31^{g}
4.0	–	0.5	–	6.14 ± 0.26^{c}	18.3 ± 0.70^{c}	15.9 ± 0.52^{d}	14.5 ± 0.34^{d}
2.0	–	1.0	–	4.87 ± 0.29^{e}	13.0 ± 0.71^{e}	11.5 ± 0.50^{g}	11.0 ± 0.39^{f}
3.0	–	1.0	–	5.71 ± 0.44^{d}	17.5 ± 0.62^{c}	16.2 ± 0.36^{d}	14.9 ± 0.67^{d}
4.0	–	1.0	–	7.40 ± 0.65^{b}	20.7 ± 0.73^{b}	18.4 ± 0.65^{c}	17.1 ± 0.31^{c}
2.0	–	–	0.5	4.10 ± 0.18^{e}	12.8 ± 0.66^{e}	12.0 ± 0.65^{f}	10.1 ± 0.46^{g}
3.0	–	–	0.5	5.14 ± 0.31^{d}	16.3 ± 0.58^{d}	13.9 ± 0.52^{e}	12.4 ± 0.37^{e}
4.0	–	–	0.5	6.00 ± 0.35^{c}	18.0 ± 0.61^{c}	16.4 ± 0.54^{d}	14.5 ± 0.40^{d}
4.0	1.0	1.0	–	10.60 ± 0.83^{a}	30.2 ± 0.76^{a}	27.5 ± 0.52^{a}	24.8 ± 0.39^{a}
4.0	1.0	–	1.0	7.90 ± 0.54^{b}	22.7 ± 0.80^{b}	21.5 ± 0.73^{b}	18.1 ± 0.40^{b}

Data represent mean ± SE of 10 replicates repeated thrice; observations were recorded after 30 days of culture. *Means followed by the same letter in the superscript in a column do not differ significantly based on ANOVA and LSD multiple 't' test at $p \leq 0.05$

8–10 shoots were produced in MS medium supplemented with 4.0 mgL^{-1} BAP along with 1.0 mgL^{-1} each of Kinetin and IAA (Fig. 17.4f; Table 17.7).

17.5.3.3 Scaling up of Shoot Multiplication

Scaling up of shoot multiplication could be achieved through the repeated subculture of the shoot clumps after removing the scales at the sheathing leaf bases. These shoot cultures responded with differentiation of buds from basal part (resident meristem of the shoots) within 1–2 weeks. A relatively higher rate of multiplication was recorded in the second passage compared to the first (Table 17.8). A threefold increase in the number of shoots was achieved during two subculture passages. The shoot clumps with maximum multiplication (~10–11 shoots) obtained during the first subculture in medium amended with 4.0 mgL^{-1} BAP and 1.0 mgL^{-1} each of Kinetin and NAA, produced an average of 30 shoots after 60 days of culture during the second subculture passage in the same media formulation (Table 17.8, Fig. 17.4g–i). Thus, it is possible to produce 30 shoots per explant starting from a single rhizome segment within 60 days of inoculation (Preetha 2012). Subculturing of the shoot clumps in fresh medium of the same composition beyond three cycles resulted in the multiplication of shoots and some of them had wrinkled leaves.

17.5.3.4 Shoot Elongation and Rooting

Plantlets developed in the multiplication medium elongated substantially within 40 days when left in the same medium without subculture during the first subculture period itself hence there was no need for shoot elongation step in the micropropagation in *K. galanga*. Root development was parallel with shoot multiplication irrespective of the hormonal treatments, a noteworthy advantageous feature in almost all zingibers. Small roots were initiated from the basal part of the regenerated shoots after 15–20 days of culture (Fig. 17.4j). The roots were green, thick and the number varied from 4 to 8.

17.5.3.5 Hardening and Field Transfer

The rooted plants weaned away from the culture flasks/bottles needs to be washed thoroughly in running tap water to remove traces of agar, and after that, they were treated with 0.1% Dithane M-45 for 5 min to avoid fungal infection. The deflasked plants should be vigorous and hardy enough as slender and drooping plants had a tendency to get infected. Microplants (7–8 cm) were successfully acclimatized, when they were transferred first to garden soil and pure river sand (1:3) medium, irrigated and hardened in the mist house (28 ± 2 °C, 80 ± 5% RH) for a period of 1–2 weeks and then to a shade net house for 3–4 weeks before transplanting. The plantlets after the short hardening phase transplanted in polybags showed 90–95% survival (Fig. 17.4k). Such established plantlets when transferred to the field during pre-monsoon and monsoon (southwest monsoon) rains facilitated better establishment at 80–90% rate. They grew to mature plants after 3–4 months of transfer and were free of any morphological variations and showed normal growth like the field-grown plants in their yield, rhizome formation, etc. A schematic representation for in vitro micropropagation of *K. galanga* (Preetha 2012) is presented in Fig. 17.5.

17.6 Conclusion

Plant propagation practices are considered as business-oriented art than science which provides to the need of a broad range of people, from small farmers to the large-scale producers. Currently, plant propagation practices are considered an important income generative sector that invariably improved the living standard of the farming community. The invention of biotechnology-mediated artificial regeneration technique has greatly enhanced the easy supply of maximum quality planting materials in a minimum period which ultimately increased the income, employment opportunity and export potential in this sector. It also satisfies the increased household nutritional security of the common people. The sector has also played a significant role in the women empowerment and income generation through

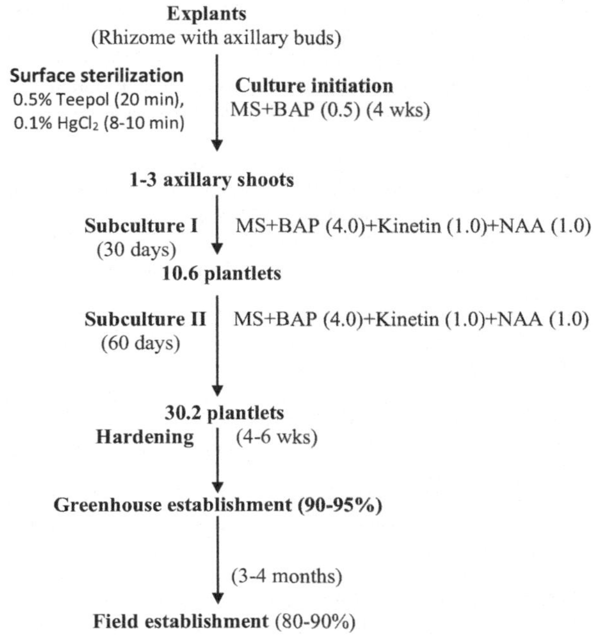

Fig. 17.5 Protocol for in vitro clonal propagation of *K. galanga* L

propagation practices of planting materials and it helps to strengthen the horticulture industry viz. silviculture, agriculture, floriculture, medi-culture, etc. to some extent. This chapter demonstrated conventional and nonconventional propagation practices with special importance to high-frequency in vitro propagation systems of three economically important plants viz. Rattan palm, Banana cultivars and 'aromatic ginger' for mass production, supply and income generation from quality planting materials. High-quality planting material gives a better rate of field establishment and high-quality yield. Hence the production of quality planting materials is the paramount importance for increasing the yield and meeting the increasing demands of elite plant produce which generates additional income and creates more employment opportunities for both rural and urban youth and women dwellers.

References

Almeida M, Almeida CV, Graner EM, Brondani GE, Abreu-Tarazi MF (2012) Pre procambial cells are niches for pluripotent and totipotent stem-like cells for organogenesis and somatic embryogenesis in the peach palm: a histological study. Plant Cell Rep 31:1495–1515

Atta-Alla H, Van Staden J (1997) Micropropagation and establishment of *Yucca aloifolia*. Plant Cell Tissue Organ Cult 48:209–212

Barba RC, Patena LJ, Mercado MM, Lorico L (1985) Tissue culture of rattan (*Calamus manillensis* Wendl.). In: Proceedings of the Second National Symposium on Tissue Culture of Rattan, University Partanian Malaysia

Blake J, Eeuwens CJ (1982) Culture of coconut palm tissues with a view to vegetative propagation. In: Rao AN (ed) Proceedings of COSTE symposium on Tissue Culture of Economically Important Plants, Singapore, pp 145–148

Bonga JM (2015) A comparative evaluation of the application of somatic embryogenesis, rooting of cuttings, and organogenesis of conifers. Can J For Res 45:379–383. https://doi.org/10.1139/cjfr-2014-0360

Brondani GE, Oliveira LS, Bergonci T, Brondani AE, Franca FAM, Silva ALL, Goncalves AN (2013) Chemical sterilization of culture medium: a low cost alternative to in vitro establishment of plants. Sci For Piracicaba 41:257–264

Brown DCW, Finstad KI, Watson EM (1995) Somatic embryogenesis in herbaceous dicots. In: Thorpe TA (ed) In vitro embryogenesis of plants. Kluwer Academic Publishers, Dordrecht, pp 345–416

Bunn E, Senaratna T, Sivasithamparam K, Dixon KW (2005) In vitro propagation of Eucalyptus phylacis L. Johnson and K. Hill., a critically endangered relict from Western Australia. In Vitro Cell Dev Biol Plant 41:812–815

Chandra R, Mishra M (2010) Micropropagation at farmers' door steps. Indian Hortic 55(2):57–60. Published bimonthly, March–April 2010

Chengji Z, Jiankui Z (1991) Plant regeneration in tissue culture of rattan. Acta Bot Yunnaica 13(1): 97–100

Chu CC, Wang CC, Sun CS, Hsu C, Yin KC, Chu CY, Bi FY (1975) Establishment of an efficient medium for anther culture of rice through comparative experiments on the nitrogen sources. Sci Sinica 18:659–668

Corley RHV, Barrett JN, Jones LH (1976) Vegetative propagation of oil palm via tissue culture. In: Malaysian International Agricultural Oil Palm Conference. pp 1–8

Cousineau JC, Donnelly DJ (1991) Adventitious shoot regeneration from leaf explants of tissue cultured and greenhouse grown raspberry. Plant Cell Tissue Organ Cult 27:249–255

Cuenca B, Ballester A, Vieitez AM (2000) In vitro adventitious bud regeneration from internode segments of beech. Plant Cell Tissue Organ Cult 60:213–220

Cunningham AB (1997) An Africa-wide overview of medicinal plant harvesting, conservation and health care. In: Bodeker G, Bhat KKS, Burly J, Vantomme P (eds) Medicinal plants for forest conservation and health care. FAO, Rome, pp 116–129

Debnath SC (2005) A two-step procedure for adventitious shoot regeneration from in vitro-derived lingon berry leaves: shoot induction with TDZ and shoot elongation using zeatin. Hortic Sci 40: 189–192

Dekkers AJ, Rao AN (1989) Some observations on in vitro culture of *Calamus trachycoleus*. In: Rao AN, Yusoff AM (eds) Proceedings of the Seminar on Tissue Culture of Forest Species. Forest Research Institute Malaysia and International Development Research Centre, Singapore, pp 63–68

Eshraghi P, Zarghami R, Mirabdulbaghi M (2005) Somatic embryogenesis in two Iranian date palm cultivars. Afr J Biotechnol 4(11):1309–1312. http://www.academicjournals.org/AJB. ISSN 1684–5315

Fasolo F, Zimmerman RH, Fordham I (1989) Adventitious shoot formation on excised leaves of in vitro grown shoots of apple cultivars. Plant Cell Tissue Organ Cult 16:75–87

Gamborg O, Miller R, Ojima K (1968) Nutrient requirement suspensions cultures of soybean root cells. Exp Cell Res 50:151–158

Georget F, Courtel P, Garcia EM, Hidalgo M, Alpizar E, Breitler J-C, Bertrand B, Etienne H (2017) Somatic embryogenesis-derived coffee plantlets can be efficiently propagated by horticultural rooted mini-cuttings: a boost for somatic embryogenesis. Sci Hortic 216:177–185

Gowthami R, Sharma N, Pandey R, Agarwal A (2021) Status and consolidated list of threatened medicinal plants of India. Genet Resour Crop Evol 2021. https://doi.org/10.1007/s10722-021-01199-0

Graner EM, Oberschelp GPJ, Brondani GE, Batagin-Piotto KD, Almeida CV, Almeida M (2013) TDZ pulsing evaluation on the in vitro morphogenesis of peach palm. Physiol Mol Biol Plants 19:283–288

Hemanthakumar AS (2010) Studies on embryo and tissue culture of three economically important rattan sps. Thesis submitted to Kerala University

Hemanthakumar AS, Preetha TS, Krishnan PN, Seeni S (2013) Utilization of zygotic embryos of an economic rattan palm *Calamus thwaitesii* Becc. (Arecaceae) for somaplant regeneration and cryobanking. 3Biotech 3:195–203

Hemanthakumar AS, Preetha TS, Padmesh P, Krishnan PN, Seeni S (2014) Micro-cloning of an economic rattan palm *Calamus thwaitesii* for eco-restoration programme. Biologia 69(5): 618–624

Huetteman CA, Preece JE (1993) Thidiazuron: a potent cytokinin for woody plant tissue culture. Plant Cell Tissue Organ Cult 33:105–119

Jarret RL, Rodriguez W, Fernandez R (1985) Evaluation, tissue culture propagation and dissemination of 'Saba' and 'Pelipita' plantains in Costa-Rica. Sci Hortic 25:137–147

Jones MPA, Yi Z, Murch SJ, Saxena PK (2007) Thidiazuron induced regeneration of *Echinacea purpurea* L.: micropropagation in solid and liquid culture systems. Plant Cell Rep 26:13–19

Josekutty PC (2003) Techno economic study on intercropping medicinal plants in oil palm plantations. PhD (Hort) thesis, Kerala Agricultural University, Thrissur

Joshi PK, Gulati A, Birthal PS, Tewari L (2003) Agriculture diversification in South Asia: patterns, determinants and policy implications. MSSD Discussion Paper No 57. IFPRI, Washington, DC

Joy PP, Thomas J, Mathew S, Skaria BP (1998) Zingiberaceous medicinal and aromatic plants. Aromatic and Medicinal Plants Research Station, Odakkali, Asamannoor PO, Kerala, India

Kareem AM (1997) Plants in Ayurveda. Foundation for Revitalization of Local Health Tradition (FRLHT), Bangalore, p 82, 240

Kumar A (2020) Phytochemistry, pharmacological activities and uses of traditional medicinal plant *Kaempferia galanga* L.—an overview. J Ethnopharmacol 253:112667. https://doi.org/10.1016/j.jep.2020.112667

Kumari R, Priya (2020) Optimization of concentrations of plant growth regulators for in vitro multiple shoot formation and efficient root induction in *Phyla nodiflora* L. (*Lippia nodiflora* L.) an important medicinal plant. Indian J Sci Res 10(2):39–44

Kundu M, Sett R (1999) Regeneration through organogenesis in Rattan. Plant Cell Tissue Organ Cult 59:219–222

Linsmaier EM, Skoog F (1975) Organic growth factor requirements of tobacco tissue cultures. Physiol Plant 18:100

Lyyra S, Lima A, Merkle S (2006) In vitro regeneration of *Salix nigra* from adventitious shoots. Tree Physiol 26:969–975

Ma SS, Shii CT (1974) Growing banana plantlets from adventitious buds. J Chin Soc Hortic Sci 20: 6–12

Ma SS, Shii CT, Wang SO (1978) Regeneration of banana plants from shoot meristem tips and inflorescence sections in vitro. In: Abstract of the 20th Int Hort Congress, Sydney, Australia, August 15-23, 1978. Abstract No. 1639

Monteuuis O (1993) Current advances in clonal propagation methods of some indigenous timber species in Sabah (Malaysia). In: Davidson J (ed) Proceedings of the regional symposium on Recent Advances in Mass Clonal Multiplication of Forest Trees for Plantation Programmes. UNDP/FAO Regional Project on Improved Productivity of Man-Made Forests Through Application of Technological Advances in Tree Breeding and Propagation (FORTIP), Cisarua, Bogor, Indonesia, 1–8 Dec 1992, pp 168–193

Monteuuis O (2016) Micropropagation and production of forest trees. In: Park YS, Bonga JM, Moon HK (eds) Vegetative propagation of forest trees. Part 1: development and trends in vegetative propagation of forest trees. NIFOS, Korea, pp 32–55

Murashige T (1974) Plant propagation through tissue culture. Annu Rev Plant Physiol 25:135

Murashige T, Skoog F (1962) A revised medium for rapid growth and bioassays with tobacco tissue cultures. Physiol Plant 15:473–497

Murthy BNS, Murch SJ, Saxena PK (1998) Thidiazuron: a potent regulator of in vitro morphogenesis. In Vitro Cell Dev Biol Plant 34:267–275

Nair LG, Seeni S (2002) Rapid clonal multiplication of *Morinda umbellata* Linn. (Rubiaceae), a medicinal liana, through cultures of nodes and shoot tips from mature plant. Phytomorphology 52:77–81

Nair LG, Seeni S (2003) In vitro multiplication of *Calophyllum apetalum* (Clusiaceae), an endemic medicinal tree of the Western Ghats. Plant Cell Tissue Organ Cult 75:169–174

Negi MS, Sharma V (2016) Income generation and economic improvement through cultivation of medicinal plants in Uttarakhand. Asian Reson 5(2):101–105

Novak FJ, Afza R, Van Duran M, Perea-Dallas M, Conger BV, Xiaolang T (1989) Somatic embryogenesis and plant regeneration in suspension cultures of dessert (AA, AAA) and cooking (ABB) bananas. Biotechnology 7:154–159

Padmanabhan D, Ilangovan R (1993) Surgical induction of multiple shoots in embryo cultures of *Calamus gamblei* Becc. RIC Bull 12:8–12

Paudyal KP, Haq N (2000) In vitro propagation of pummel (*Citrus grandis* L. Osbeck). In Vitro Cell Dev Biol Plant 36:511–516

Pham NK, Nguyen HT, Nguyen QB (2021) A review on the ethnomedicinal uses, phytochemistry and pharmacology of plant species belonging to *Kaempferia* L. genus (Zingiberaceae). Pharm Sci Asia 48(1):1–23

Preece JE, Imel MR (1991) Plant regeneration from leaf explants of *Rhododendron* PJM Hybrids. Sci Hortic 48:159–170

Preetha TS (2012) Studies on *in vitro* conservation in *Kaempferia galanga* L. PhD thesis, University of Kerala, India

Preetha TS, Hemanthakumar AS, Krishnan PN (2016) A comprehensive review of *Kaempferia galanga* L. (Zingiberaceae): a high sought medicinal plant in Tropical Asia. J Med Plants Stud 4(3):270–276

Purohit SS, Vyas SP (2004) Marketing of medicinal and aromatic plants in Rajasthan, National Consultative Workshop on Medicinal and Artomatic Plants, held at GBPUAT, Pantnagar

Saenz L, Azpeitia A, Chuc-Armendariz B, Chan JL, Verdeil JL, Hocher V, Oropeza C (2006) Morphological and histological changes during somatic embryo formation from coconut plumule explants. In Vitro Cell Dev Biol Plant 42:19–25. https://doi.org/10.1079/IVP2005728

Sahoo Y, Pattnaik SK, Chand PK (1997) *In vitro* clonal propagation of an aromatic medicinal herb *Ocimum basilicum* L. (sweet basil) by axillary shoot proliferation. In Vitro Cell Dev Biol Plant 33(4):293–296

Sane D, Aberlenc-Bertossi F, Gassama-Dia YK, Sagna M, Trouslot MF, Duval Y, Borgel A (2006) Histocytological analysis of callogenesis and somatic embryogenesis from cell suspensions of date palm (*Phoenix dactylifera*) PMC2803457

Sarasan V, Kite GC, Sileshi Gudeta WS, Philip CS (2011) Applications of phytochemical and *in vitro* techniques for reducing over-harvesting of medicinal and pesticidal plants and generating income for the rural poor. Plant Cell Rep 30:1163–1172. https://doi.org/10.1007/s00299-011-1047-5

Sathiamoorthy S, Uma S, Selvarajan R, Shyam B, Singh HP (1998) Multiplication of virus free banana plants through shoot tip culture. NRCB, Tiruchirapalli, Tamil Nadu, India, p 60

Saya RA, Mankessi F, Toto M, Marien JN, Monteuuis O (2008) Advances in mass clonal propagation of *Eucalyptus urophylla* × *E. grandis* in Congo. Bois For Tropiques 297:15–25

Schenk RU, Hildebrant AC (1972) Medium and techniques for induction and growth of monocotyledonous and dicotyledonous plant cell cultures. Can J Bot 50:199–204

Scherwinski-Pereira JE, da Guedes RS, Fermino PCP, Silva TL, Costa FHS (2010) Somatic embryogenesis and plant regeneration in oil palm using the thin cell layer technique. In Vitro Cell Dev Biol 46:378–385. https://doi.org/10.1007/s11627-010-9279-6

Seeni S, Latha PG (2000) In vitro multiplication and ecorehabilitation of the endangered Blue Vanda. Plant Cell Tissue Organ Cult 61:1–8
Senarath RMUS, Karunarathna BMAC, Senarath WTPSKJG (2017) *In vitro* propagation of *Kaempferia galanga* (Zingiberaceae) and comparison of larvicidal activity and phytochemical identities of rhizomes of tissue cultured and naturally grown plants. J Appl Biotechnol Bioeng 2(4):1–6
Sett R, Kundu M, Sharma P (2002) Regeneration of plantlets through organogenesis in *Calamus tenuis*. Indian J Plant Physiol 7:358–361
Singh KM (2009) Scope of medicinal and aromatic plants farming in Eastern India. SSRN Electron J. https://doi.org/10.2139/ssrn.2019789
Singh KM, Singh U, Singh P (2007) A study on role of ATMA, Patna in development of supply chain for medicinal plants in Patna District, Bihar
Singha S, Bhatia SK (1988) Shoot proliferation of pear cultures on medium containing thidiazuron and benzylamino purine. Hortic Sci 23:803
Sivarajan VV, Balachandran I (1994) Ayurvedic drugs and their plant sources. Oxford and IBH Publishing Co. Pvt. Ltd, New Delhi, p 570
Swaminathan MS (2010) Achieving food security in times of crisis (Review). New Biotechnol 27(5):453–460. https://doi.org/10.1016/j.nbt.2010.08.002
Thompson D (2014) Challenges for the large-scale propagation of forest trees by somatic embryogenesis—a review. In: Park YS, Bonga JM (eds) Proceedings of the 3rd international conference of the IUFRO unit 2.09.02 on Woody plant production integrating genetic and vegetative propagation technologies, 8-12 Sept 2014. Vitoria-Gasteiz, Spain, pp 81–91. http://www.iufro20902.org
Tim WY, John LP, Young I, Kap PL, Fan I, Arditti J (1990) Induction of callus from axillary buds of taro (Colocasia esculenta var. esculenta, Araceae) and subsequent plantlet regeneration. Plant Cell Rep 9:459–462
Tisserat B (1983) Tissue culture of date palms—a new method to propagate an ancient crop and a short discussion of the California Date industry. Principes 27(3):105–117
Titon M, Xavier A, Otoni WC (2006) Clonal propagation of *Eucalyptus grandis* using the minicutting and microcutting techniques. Sci For 71:109–117
Ugalde Arias LA (2013) Teak: new trends in silviculture, commercialization, and wood utilization. In: Cartago CR (ed) International Forestry and Agroforestry, Costa Rica, 552 p ISBN 978-9968-47-716-1
Vasane R, Shailesh RM, Kothari (2006) Optimization of secondary hardening process of banana plantlets (*Musa paradisiaca* L. var. Grand Naine). Indian J Biotechnol 5:394–399
Victorio CP, Lage CLS, Sato A (2012) Tissue culture techniques in the proliferation of shoots and roots of *Calendula officinalis*. Rev Cienc Agron 43(3):539–545
Vieitez AM, San-Jose MC (1996) Adventitious shoot regeneration from *Fagus sylvatica* leaf explants in vitro. In Vitro Cell Dev Biol Plant 32:140–147
Vuylsteke D, De Langhe E (1985) Feasibility of in vitro propagation of bananas and plantains. Trop Agric (Trinidad) 62:323–328
Wang H-C, Chen J-T, Wu S-P, Lin M-C, Chang W-C (2003) Plant regeneration through shoot formation from callus of Areca catechu. Plant Cell Tissue Organ Cult 75:95–98
Wendling I, Xavier A (2003) Miniestaquia seriada no rejuvenescimento de clones de Eucalyptus. Pesq Agropec Bras Brasileira 38(4):475–480
Yushkova EV, Skuratova EV, Velichko NA (1998) Features of the growth of *Catharanthus roseus* L. Callus cultures as related to culture method Biotechnologia

Yusoff AM (1989) Shoot formation in *Calamus manan* under *in vitro*. In: Rao AN, AziahMohd. Yusoff (eds) Proceedings of the seminar on Tissue Culture of Forest Species, Forest Research Institute Malaysia and International Development Research Centre, Singapore, pp 45–49

Yusoff AM, Manokaran N (1985) Seed and vegetative propagation of rattan spp. In: Wong KM, Manokaran N (eds) Proceedings of the rattan seminar. The Rattan Information Centre and Forest Research Institute, Malaysia, pp 13–22

Zeng Bingshan (1997) Tissue culture of *Calamus egregius*. J Central South For Univ 17(4): 563–569

Zhang Fangqiu (1993) A study on Rattan tissue culture. J For Res 6(5):486–492

Chapter 18
In Vitro Secondary Metabolite Production for Sustainable Utilization of Endangered Medicinal Plants

R. Sreelekshmi, T. S. Swapna, and E. A. Siril

Abstract Plant cell cultures represent an inherent source of diverse secondary products or metabolites, which can be applicable in nutraceuticals, pharmaceuticals, cosmeceuticals, or food colors. The valuable products are directly harvested from various plant species from time immemorial. However, thousands of plant species were severely facing extinction and listed as endangered. Besides, the chemical synthesis of valuable and rare metabolites is a very difficult process and not economically feasible. In such cases, plant cell cultures can produce the same compounds as the original or unique ones having economic value. This will result in the development of a reproducible plant cell culture system having enhanced production of desired metabolites. The great interest in in vitro culture techniques is that they may supply a continuous and reliable source of plant metabolites and could support the overproduction of metabolites. This chapter summarizes the applicability of in vitro culture techniques and unique secondary metabolites' production.

Keywords Secondary metabolites · Cell culture · Organ culture · Elicitors · Metabolic engineering

18.1 Why In Vitro Cultures?

Since time immemorial, people have used plants and plant-derived products for nutritional, agricultural, and pharmaceutical purposes. Medicinal plants are a valuable source of drugs because 80% of people in developing countries depend on herbal formulations for their general health care practices, in which over 25% of prescribed medicines are derived from wild medicinal plants (Hamilton 2004). With the higher demand for medicinal plants or their metabolites, the use of medicinal and therapeutic drugs is growing all around the world (Nalawade et al. 2003; Cole et al. 2007). However, ecological, geographical, or seasonal variation influences the

---

R. Sreelekshmi · T. S. Swapna · E. A. Siril (✉)
Department of Botany, University of Kerala, Kariavattom, Thiruvananthapuram, Kerala, India

S. T. Sukumaran, K. T R (eds.), *Conservation and Sustainable Utilization of Bioresources*, Sustainable Development and Biodiversity 30,
https://doi.org/10.1007/978-981-19-5841-0_18

supply of plant raw materials. The extinction of plant species is severely affecting the daily life care practice of our society. Conservation of natural resources is the care and protection of the earth's resources so that they can persist for future generations. Ex-situ and in-situ conservation are the two main modes of biodiversity conservation for protecting endangered plant species. In-situ conservation can be defined as the maintenance of species, genetic, or ecosystem diversity in the natural state in which they occur (Usher 2000). As a result of natural disasters, human involvement, pathogens, pests, or changing environment made in-situ conservation is difficult to maintain. Besides gardens and germplasm banks, in vitro storage offers an alternative source of conserving the endangered plant genotypes in terms of ex vitro methods (Sengar et al. 2010).

Plant tissue culture or in vitro culture system is a well-established method to enhance the industrial importance in the field of conservation, genetic improvement, or production of secondary metabolites. A small piece of tissue (explant) can be used to produce thousands of plants in a short time period, under controlled environmental conditions rather than seasonal variations. Moreover, the techniques reduced the storage space for conserving a large number of plantlets for commercial use, and such plants can easily be established in the field. Compared the conventional whole plant cultivation, the in vitro culture techniques have their own advantages.

When technology buildup, the researchers focused on the production of a specific group of secondary metabolites like alkaloids, flavonoids, anthocyanins, steroids, terpenoids, and quinones (Gupta et al. 2010; Cusido et al. 2014) produced by diverse plants. The production of specific metabolites is often restricted to some genus or species. A majority of these compounds are activated during specific occasions, stressed conditions, or particular developmental stages. The secondary metabolites have no basic role in the sustenance of plant life process, but these valuable chemical compounds play a significant role in the interaction with the environment and also enhance the protective function as defense chemicals under stressed conditions (Radman et al. 2004; Ramakrishna and Ravishankar 2011). In many cases, conventional extraction and isolation of these compounds are very expensive, thus limiting economically sustained production of pure forms, which restricts the wide use of such products. Owing to these efforts on metabolite production, the in vitro production of compounds using organ or cell culture has been focused on in the last few decades. Plant cell tissue and organ culture techniques have been widely used in the biotechnology field to attain overproduction of pharmaceutically valuable products (Verpoorte 2000; Vázquez-Flota et al. 2009; Zhou et al. 2010; Baque et al. 2012a, b; Siddique et al. 2013; Murthy et al. 2014; Dias et al. 2016; Patra and Srivastava 2016; Espinosa-Leal et al. 2018; Maqsood et al. 2018; Krishnan et al. 2020).

For the last three decades, in vitro metabolite production has received due attention in biotechnology and plant science research. The metabolic engineering process on cell culture aimed to flourish various pharmaceutical aspects. The many-fold enhanced production can be targeted by overexpression of key genes in the metabolic pathways. The understanding of such genes is critical for the regulation of targeted phytochemicals and their wide exploitation in medicine, food, and nutraceuticals (Dewick 2002).

18.2 Production of Secondary Metabolites Through In Vitro Techniques

While considering plant cell culture technology, several advantages can be pointed out on in vitro production of metabolites over the field cultivation of endangered medicinal or aromatic plants; such as the number of plant resources, harvesting of target compounds anywhere the world with the control of quality check, production is independent of environmental or geographical variation, production within the limited time periods and the possibility to generate the highly purified form of metabolites (Rao and Ravishankar 2002; Danova 2010; Rather et al. 2012; Sharma et al. 2014; Ramirez-Estrada et al. 2016). However, to enhance the production rate, new technologies were further applied to exclusivity for the value-added secondary metabolites at an industrial level. In vitro culture-based production has become more accepted as a method for the production of valuable metabolites (Murthy et al. 2014; Ramirez-Estrada et al. 2016). The development of different in vitro culture systems got attention to enhance desired metabolic products. The efficient culture system, viz., callus or cell suspension culture, organ cultures, and hairy root cultures deserve specific attention for enhanced production. Plant growth regulators and elicitors could be used to enhance metabolites and play an important role in biosynthesis pathways.

The limited availability of endangered medicinal plants and its slow production of bioactive compounds encourage the necessity of in vitro cell culture techniques to support pharmacological activities. The undifferentiated masses of cells (callus) have the ability to produce programmed metabolites related to specific explant tissue. Callus cultures are mainly concentrated on the sustained and large-scale production of secondary metabolites in pharmaceuticals, food, cosmetics, and related industries (Guern et al. 2012; Georgiev et al. 2013). Cell suspension with elicitation or precursor feeding of many plant species has already been extensively studied regarding the production of secondary metabolites, which could be applicable to achieve enhanced production (Rao and Ravishankar 2002; Ramirez-Estrada et al. 2016). Elicitation and precursor feeding methods enrich better in vitro productivity and lower the production cost. Precursor feeding acts as a strategy that enhances metabolite production at the level of supplying initiator molecules of specific biosynthetic pathways to achieve enhanced production of metabolites.

Elicitation is the most effective biotechnological strategy to achieve stimulated production of secondary metabolites. The specific compound, named elicitor, can stimulate the production of compounds that protect the cell or whole plant system (Klarzynski and Friting 2001; Zhao et al. 2005; Baenas et al. 2014). The term elicitor originally indicated the compound which has the ability to induce phytoalexin in the plant system, but later it is referred to as compounds stimulating any defense responses (Kim et al. 2001; Shams-Ardakani et al. 2005). In another way, the compounds that are pathogenic in origin are termed exogenous elicitors, and the substance produced after a pathogen attack is termed endogenous elicitors (Kim et al. 2001). Exogenous biotic elicitors include microbial enzymes, yeast extract, or

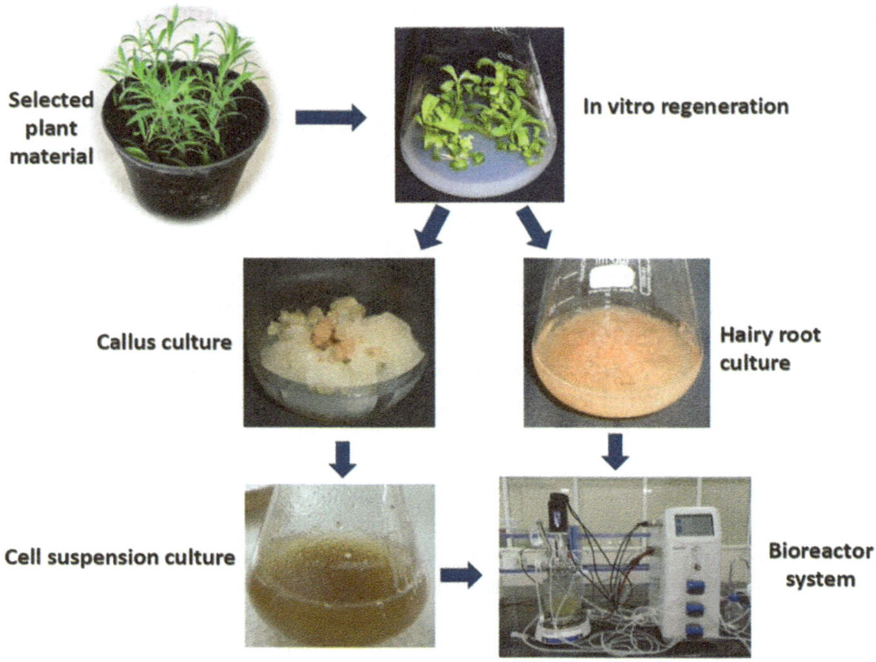

Fig. 18.1 In vitro secondary metabolite production

polysaccharides from microbial cell walls (chitosan and glycanes). According to the nature of the origin, elicitors are classified as biotic or abiotic. The compounds of non-biological origin are considered abiotic elicitors, such as salt or physical factors (Namdeo 2007; Gorelick and Bernstein 2014). Inorganic chemicals like heavy metals are generally used to enhance metabolite production in different culture systems (Trejo-Tapia et al. 2001; Cai et al. 2013; Urdová et al. 2015) with the intention that metal ions modify the secondary metabolism of the plant by inducing a stress response (Verpoorte et al. 2002).

The main research programs focused on producing secondary metabolites go through different culture conditions (Fig. 18.1). At the level of origin, plant material that is capable of producing the desirable compound was regenerated to establish in vitro plantlets for primary conservation. There are various culture techniques such as callus, cell suspension, organ (shoot or root) culture, or hairy root cultures that can be adopted to attain a feasible metabolite production system. Once the best production techniques are identified, the elicitation or precursor feeding experiments can be conducted to precede large-scale production via bioreactor systems.

18.3 Callus and Cell Suspension Culture

Callus, the undifferentiated mass of cells, has the ability on genetic behavior of tissue-specific effective metabolite production. After selecting the most promising elite genotype, the callus initiation works should begin to find out the best-adapted medium. The protocol aimed to produce metabolites on a large scale first requires the production of a high-yielding cell line. Culture conditions play an important role in the growth and development of cell lines. The optimization of culture conditions, including growth regulators, pH, carbon source, light, nitrogen source, salt concentration, temperature, and oxygen, is crucial for regulating the production of desired bioactive compounds. The media constituents in callus or cell cultures are limiting factors for both growth and metabolite production (Zhang and Zhong 1997). Different growth media such as Murashige and Skoog (MS), Gamborg (B5) liquid media, Linsmair & Skoog (LS), N6 liquid medium, and modified media were used to produce the best callus cultures (Coste et al. 2011). The callus is initiated from specific meristematic cells, which possess an undifferentiated, unlimited growth cycle and the ability to synthesize the same compound as the source plant (Ramirez-Estrada et al. 2016). Plant growth regulators (PGRs) are responsible for the proliferation of callus mass from different explant sources. Several PGRs (auxins and cytokinins) or their combination are responsible for the induction of callus. Callus cells are non-differentiated meristematic tissues, small vacuolated and chloroplast lacking nature; however, they can be re-differentiated into entire plants under appropriate growth conditions (Efferth 2019). On the other hand, the single-cell cultures are also generated from the friable callus in a slowly shaken liquid medium.

Cell suspension cultures were initiated from desired friable callus cultures. Before initiating cell suspension, the calli were repeatedly subcultured to establish stable metabolite production. The best friable piece of callus was transferred to the liquid culture medium and kept in a gyratory shaker for specific rotation under standard culture conditions. The unbroken callus or clumps of the developed cultures was removed using sterile standard cell dissociation sieve filters to obtain homogenous cell suspension. For proper culture conditions, different culture media, plant growth regulators, carbon sources, or physical factors like pH, temperature, light, shaking speed, etc. need to be optimized to reach large-scale production (Stafford et al. 1986; Dörnenburg and Knorr 1995; Rao and Ravishankar 2002; Baskaran and Jayabalan 2009; Garcia et al. 2011; Cusido et al. 2014; Masondo et al. 2015). The growth parameters like settled cell volume (SCV), packed cell volume (PCV), total cell count, cell viability, fresh weight (FW), dry weight (DW), and growth index (GI = final fresh weight-initial fresh weight/initial fresh weight) can be recorded at a specific interval to analyze the cell growth. During the initial stage of cell culture growth, the biomass accumulation increases, but the product yield may be relatively low. At the time of the exponential phase, the secondary metabolite production seems to be very low, whereas the primary metabolites were accumulated for biomass production. Later, the primary metabolites become stationary at a stage of culture and from which the secondary metabolites' production register enhancement from the primary metabolites. Furthermore, enzyme activity or biochemical differentiation occurs when growth declines (Payne et al. 1991). This indicates the effectiveness of the stationary phase in the plant culture system (Ramirez-Estrada et al. 2016). The concept of a cell factory was developed later. In a clear strategy named a two-stage culture system, the cultures were maintained in an optimal media composition concentrating biomass production only and then transferred to a specific production medium to stimulate the target molecules. In this stage, the application of elicitors or precursors results enhanced the production of metabolites (Malik et al. 2011).

In vitro metabolite production system gets regulated by changing the type and concentration of PGRs, elicitors, exposure timing, pH, age of culture system, or nutrient level (Namdeo 2007). Different auxins, including α-naphthaleneacetic acid (NAA), 2, 4-dichlorophenoxyacetic acid (2,4-D), picloram, indole-3-acetic acid (IAA), or indole-3-butyric acid (IBA), were supplemented to liquid medium for encouraging the proliferation of callus or cell suspensions. The suspended cell growth kinetics was calculated by plotting the GI at specific time intervals. The commonly used elicitors are yeast extract, pectins, chitosan, xylan, etc. (Yan et al. 2006; Sreeranjini and Siril 2015; Krishnan and Siril 2016a, b), where UV light (Zamboni et al. 2015), heavy metals, nanoparticles, denatured proteins, non-essential media components, and chemicals like methyl jasmonate (MeJ), salicylic acid (SA) can be used as biotic elicitors to enhance the secondary metabolites in plant cell suspension cultures (Kang et al. 2004; Wang et al. 2004). The elicitors and precursors were dissolved in double-distilled water and prepared g L^{-1} stock solution. Various additives at different concentrations were added to the established suspension culture at the end of log-phase through filter sterilization to achieve enhanced production. After optimization of elicitation or precursor feeding, an improved, large-scale production was conducted in bioreactor systems. Further enhancement in the production of the metabolite is reasoned by adopting an alternative method based on the understanding of the metabolic pathway and its manipulation through metabolic engineering approaches.

Generally, the biotic elicitation using compounds having phytoalexinic properties and the ability to regulate the related biosynthetic pathways is considered an excellent method to enhance metabolites production (Ramirez-Estrada et al. 2016). Endogenous biotic elicitors are intracellular proteins and molecules synthesized during stressed or pathogen-attacking conditions, e.g., MeJ or SA (Kim et al. 2001; Sanchez-Sampedro et al. 2005). The elicitor-induced stress in plant cells begins at the plasma membrane. Multiple component signal transduction networks lead to different target responses and finally end in the synthesis of defense molecules or secondary metabolites (Ebel and Cosio 1994; Goel et al. 2011). The mechanism behind the elicitor molecule is recognized to have an important role in the signal transduction process for improving metabolite production by regulating the defense response in plants (Walker et al. 2002; Zhao et al. 2005). Precursors are other enhancing molecules that act as initiators or beginning molecules to precede the biosynthesis route of secondary metabolites and finally reach the increased yield of final products (Rao and Ravishankar 2002). In order to analyze the relationship between hormonal, elicitor, or precursor exposure on changes in metabolic profiling, gene expression studies or transcriptome analysis were conducted on the desired cell line. The ability of plant cell cultures to enhance secondary metabolite production opened new avenues to explore the industrial interest of in vitro metabolites. Bioreactor studies promote the possible commercial production of bioactive compounds by applying plant cell culture techniques. Some major contributions to potent plant metabolites' productions via cell suspension cultures are listed in Table 18.1.

18.3.1 Alkaloids

Cell suspension techniques can be widely utilized for large-scale production of alkaloids, which provide continuous and reliable natural products (Vanisree et al. 2004). Many of the alkaloid source plants are placed under endangered categories and require special attention to ensure the sustained use of such resources in the production of metabolites. The media optimization studies on cell suspension culture indicated 20–30-fold alkaloid increment using high strain cultures (Zhong 2001). Cell suspension cultures of *Vernonia cinerea* reached a maximum biomass and alkaloid content (1.15 mg g^{-1}) on BA, NAA combination medium (Maheshwari et al. 2007). Medicinally and economically important terpenoid indole alkaloids, including vindoline, catharanthine, ajmalicine, bisindoles, vincristine, and vinblastine can be produced via cell suspended cultures of *Catharanthus roseus* (Zhao et al. 2001). The fast-growing cell suspension strains of *Papaver somniferum* produce sanguinarine using a bioreactor system (Park and Yoon 1992). The recent publication forwards the elicitor-induced enhanced production of an efficient alkaloid, berberine in cell suspension cultures of *Tinospora cordifolia* (Pillai and Siril 2022). Cell suspension culture-based production of various alkaloids such as indole alkaloids, quinoline alkaloids, isoquinoline alkaloids, and terpenoid alkaloids from various plant systems was presented and compiled in Table 18.1. A benchtop bioreactor reproduced continuous production of various alkaloid derivatives from high-strain cultures of *Santalum album* and *C. roseus* (Valluri 2009). In terms of metabolic engineering, tropane alkaloid, scopolamine, and its transformed products were successfully biosynthesized (Palazón et al. 2008).

18.3.2 Flavonoids

Flavonoids are highly valued secondary metabolites in the plant kingdom known for their antioxidant, anti-inflammatory, chemopreventive, antitumor, hepatoprotective, antimicrobial, estrogenic, and other types of activities. Flavonoids are especially applied to neurodegenerative, cardiovascular, and other age-related diseases (Kumar et al. 2013; Skrzypczak-Pietraszek et al. 2018). Flavonoids are one of the metabolites

Table 18.1 Bioactive secondary metabolites produced through cell suspension cultures

Plant species	Metabolites	References
Alkaloids		
Aconitum napellus	Aconitine	Hwang et al. (2004)
Coscinium fenestratum	Berberine	Narasimhan and Nair (2004)
Thalictrum rugosum	Berberine	Gugler et al. (1988)
Cinchona ledgeriana	Quinine	Ratnadewi and Sumaryono (2010), Ratnadewi et al. (2013)
Rauvolfia serpentina	Serpentine	Yamamoto and Yamada (1986)
Catharanthus roseus	Ajmalicine Catharanthine,	Zhao et al. (2001); Yahia et al. (1998); El-Sayed and Verpoorte (2002); Zheng and Wu (2004)
Brucea javanica (L.) Merra	Cathinone	Liu et al. (1990)
Flavonoids		
Artemisia absinthium	Flavonoid	Ali et al. (2013)
Cyclopia subternata	Isoflavone	Kokotkiewicz et al. (2012)
Fabiana imbricate	Rutin SC	Schmeda-Hirschmann et al. (2004)
Habenaria edgeworthii	Phenolics	Giri et al. (2012)
Hypericum perforatum	Rutin, kaempferol, luteolin	Pasqua et al. (2003)
Larrea divaricata	Quercetin	Palacio et al. (2012)
Medicago truncatula	Flavonoids, isoflavonoids	Farag et al. (2007)
Morinda citrifolia	Anthraquinones, flavonoids	Deshmukh et al. (2011); Baque et al. (2012a, b)
Petunia hybrida	Anthocyanin	Hagendoorn et al. (1991)
Polygonum hydropiper	Flavonoid	Nakao et al. (1999)
Sophora flavescens	Sophoraflavanone	Yamamoto et al. (1995)
Vitis vinifera	Anthocyanin	Gagne et al. (2007); Qu et al. (2006)
Anthraquinones		
Morinda citrifolia	Anthraquinones	Zenk et al. (1975)
Morinda citrifolia	Rubiadin, lucidin, morindone	Inoue et al. (1981)
Galium mollugo	Anthraquinones	Inoue et al. (1984)

(continued)

Table 18.1 (continued)

Plant species	Metabolites	References
Cinchona ledgeriana	Anthraquinones	Robins et al. (1986)
Gallium verum	Anthraquinones	Banthorpe and White (1995)
Morinda officinalis	Anthraquinones	Xiang and Guo (1997)
Rubia akane	1,2-dihydroxy-AQ, alizarin,	Mizutani et al. (1997)
Ophiorrhiza pumila	Anthraquinones, lucidin	Kitajima et al. (1998)
Morinda elliptica	Anthraquinones	Lajis et al. (2000); Abdullah et al. (1998)
Morinda elliptica	Soranjidiol, morindone	Jasril et al. (2003)
Morinda elliptica	Anthraquinones	Chiang and Abdullah (2007)
Morinda citrifolia	Anthraquinones	Sreeranjini and Siril (2013, 2015)
Oldenlandia umbellata	Anthraquinones	Krishnan and Siril (2016a)
Saponins		
Centella asiatica	Centellosides	Lambert et al. (2011)
Gymnema sylvestre	Gymnemic acids	Bonfill et al. (2011); Veerashree et al. (2012); Chodisetti et al. (2013)
Panax sp., *P. notoginseng*	Ginsenosides	Chodisetti et al. (2015); Biswas et al. (2016)
Calendula officinalis	Oleanolic acid	Biswas et al. (2018)
Solanum lyratum	Solanine, Solanidine, Solasodine	Skrzypczak-Pietraszek et al. (2018)
Asparagus racemosus	Shatavarins	Kuo et al. (2005)
Dioscorea galeottiana D. deltoidei	Diosgenin	Rojas et al. (1999); Pise et al. (2015)

that can be enhanced through elicitation or precursor feeding in various culture systems (Hemashenpagam and Selvaraj 2011; Masoumian et al. 2011; Sonja et al. 2015). In vitro production of flavonoids from callus and cell suspension lines has been accomplished in diverse plant species (Table 18.1). Among various flavonoids, flavones and flavanones have been produced preferably in different in vitro culture systems (Botta and Monache 1995), and the majority of reports showed that isoflavonoid from callus cultures is comparatively higher in herbaceous plants (Luczkiewicz and Głód 2003). One of the flavonoids, anthocyanin, is a well-established product suggested by Zhang et al. (2002) and is produced through the cell suspension culture of *Vitis vinifera*. In precursor feeding, the flavonoids are synthesized from upstream metabolic precursors like phenylalanine, glutamine, or proline via the phenyl propanoid pathway (Rao and Ravishankar 2002). The addition of phenylalanine helps to initiate isoflavones or flavonoids and acts as an upstream metabolic precursor through phenylpropanoid pathway (Zia et al. 2007). Park et al. (1995) studied elicitor-induced isoflavonoid production in the cell suspension system of *Pueraria lobata,* where the addition of yeast extract showed enhancement over control treatment. Application of chitosan (20 mg L^{-1}) resulted in 3.51 mg g^{-1} (2.7-fold) increased flavonoid production in the cell suspension culture of *Andrographis paniculata* (Mendhulkar and Vakil 2013). In *Impatiens balsamina* suspension cultures using chitosan (50 mg L^{-1}) treatment, enhanced production of flavonoids was reported (Kasem 2018). Most recently, a significant increase in total flavonoid content by elicitation treatment in cell suspension cultures of *Dianthus chinensis* was also recorded in our research (unpublished data). There was a negative relationship between increasing the elicitor (yeast extract, pectin, or chitosan) concentration and flavonoid content. Other researchers also made such an observation (Sanchez-Sampedro et al. 2005; Kasem 2018). Chitosan is an exogenous biotic elicitor and is a polymer of B-1,4-glucosamine residue (Chakraborty et al. 2008) and is already reported to induce the production of rosmarinic acid in *Ocimum basilicum* (Kim et al. 2005) and phenylpropanoid enzymes (Chakraborty et al. 2008). George et al. (2008) explained that plant growth efficiency by yeast extract is due to the availability of high amino acid content, which leading jasmonate production and promotes phenolics or flavonoid production (Sanchez-Sampedro et al. 2005; Kasem 2018). The flavonoid pathway-related gene expression was conducted in callus cultures of *Dimocarpus longan* under different blue light conditions, where the genes chalcone synthase (CHS), flavanone 3′-hydroxylase (F3H), dihydroxy flavonol reductase, and leucoanthocyanidin reductase transcript level showed increasing upregulated trends under suspended cell cultures (Li et al. 2018). The increased activity of early phenyl propanoid pathway enzymes is reported in cell suspension cultures of *Cocos nucifera* (Chakraborty et al. 2008), *Scrophularia striata* (Kamalipourazad et al. 2016), *Capsicum chinense* (Kabita et al. 2020), *Pueraria candollei* (Rani et al. 2020), and *Medicago truncatula* (Suzuki et al. 2005). These expression patterns in the phenylalanine feeding experiments correspond with the flavonoid accumulation during the exposure timing. FLS encodes a synthase enzyme, which is believed to synthesize flavanols, including kaempferol, quercetin, and myricetin (Kumar et al. 2018). This relative expression pattern of the flavonoid gene substantiates flavonoid production and enhances it under the influences of phenylalanine feeding. The flavonoid gene expression studies were also used to investigate the specific effect of phenylalanine exposure to the cell suspension culture of *D. chinensis* (unpublished data).

18.3.3 Anthraquinones

Anthraquinones (AQs) are the important group of naturally occurring quinones distributed in higher plants, fungi, and insects, which act as a coloring pigment for these organisms. AQs are industrially used for the manufacture of natural dyes (Lajis et al. 2000; Han et al. 2002) and used as colorants in food, drugs, cosmetics, textiles, or hair dye (Kumar and Sinha 2004). Plants belonging to the families Rubiaceae, Rhamnaceae, Polygonaceae, and Leguminosae are the major source of AQs. Besides being used as dyestuff, various AQ constituents are used as clinically important molecules, especially rubiadin, damnacanthal, alizarin, and purpurin, as they have been used in the formulation of chemotherapeutical drugs, for the treatment of kidney and bladder stones, as a laxative mixture, and as a mild sedative (Huang et al. 2007; Frackowiak et al. 2010). Various constituents of AQs such as alizarin, purpurin, and damnacanthal are currently extracted from two important sources, viz., *Morinda citrifolia* and *Rubia cordifolia* (Han et al. 2002; Sreeranjini and Siril 2013, 2015). However, many other promising candidates for the in vitro production of AQ have been attempted in the past (Table 18.1).

18.3.4 Saponins

Plant cell and tissue and organ culture techniques have been established for the efficient production of saponin and its derivatives. The optimization of culture conditions still continued to enhance the saponin accumulation for a wide range of commercial applications owing to its structural diversity. Some important saponins were reported to have in vitro enhanced production than content in naturally growing plants. Production of ginsenosides from *Panax ginseng*, bacosides from *Bacopa monnieri*, and centellosides from *Centella asiatica* are some examples of efficient in vitro production of saponins (Murthy et al. 2014). Studies on in vitro cell suspended production of some important phytosaponins have been tabulated (Table 18.1). Plant growth regulator auxins are reported to produce steroidal saponin solanine, solanidine, and solasodine in the cell suspension culture of *Solanum lyratum* (Kuo et al. 2012). In vitro culture system stimulated 20 times higher production of saponins (*shatavarins*) in *Asparagus racemosus* (Pise et al. 2015). The influence of abiotic or biotic factors like phosphate, sucrose, light, or fungal elicitation was also reported for the production of diosgenin in the cell suspension cultures of *Dioscorea galeottiana* under dark culture conditions (Rojas et al. 1999). Diosgenin production in *Dioscorea deltoidea* was enhanced by the addition of 2,4-D (Marshall and Staba 1976). Enhanced production of ginsenoside was observed on MeJ (0.2 mM) elicitation in cell suspension cultures of *Panax notoginseng* (Hu and Zhong 2008), whereas 2–4 times ginsenoside production was recorded in cell cultures of *P. ginseng* in response to 0.2 mM salicylic acid elicitation (Hu and Zhong 2007, 2008). Besides, Wang et al. (2005) reported ginsenoside production in both flask and bioreactor conditions, where a maximum of 2.6 and 1.8-fold productions were observed in flask and bioreactor cultures, respectively. *Medicago truncatula* is a better in vitro model plant to study the nature of saponin biosynthesis genes during various elicitation treatments (Suzuki et al. 2002). The expression-related studies on in vitro saponin production were confirmed using key genes including Farnesyl Diphosphate Synthase, Squalene Synthase, Squalene Epoxidase, β-Amyrin Synthase, Cytochrome P450, and Glycosyltransferase (Lambert et al. 2011).

18.4 Hairy Root Cultures

Over the last two decades, the conventional cell suspension culture or organ culture-based metabolite production in several plants that are having property to produce and accumulate metabolite in roots are getting shifted to hairy root cultures. This approach is an effective alternative for the production and enhancement of various bioactive natural metabolites. Hairy roots have been considered a sustainable method for the production of secondary metabolites, because harvesting the root destroys the whole plant in nature. Besides, the highly branched mass of hairy roots has been produced in the absence of external PGRs, without geotropism (Shanks and Morgan 1999). The establishment of hairy root cultures was conducted by infecting the soil bacterium, *Agrobacterium rhizogenes* in which they transfer T-DNA from Ri plasmid into the plant genome. This effective role of *A. rhizogenes* has led to develop a source for root-derived pharmaceuticals (Flores et al. 1999). Well-established hairy root cultures are ginkgolides production from *Gingko biloba* (Ayadi and Tremouillaux-Guiller 2003), serpentine, and ajmalicine from *Catharanthus roseus* and *Rauvolfia micrantha,* respectively (Sudha et al. 2003). Large-scale bioreactor production of ginsenoside from *Panax ginseng* was optimized in various organic nutrient media. These developments indicated the extension of small laboratory-scale techniques to large industrial-level metabolite production and purifications. Hairy root production of camptothecin and podophyllotoxin from *Ophiorrhiza* spp. (Saito et al. 2001; Ya-ut et al. 2011) and *Linum* spp. (Chashmi et al. 2013; Renouard et al. 2018) are the best examples of industrial production of metabolites. Hairy roots *Oldenlandia umbellata* have been found to increase the production of anthraquinones (Krishnan and Siril 2016a, b). In another report, artemisinin present in the aerial part has also been produced from hairy root cultures from *Artemisia annua.*

MeJ derivatives have been reported as the key signaling molecules for in vitro metabolite elicitation. MeJ, chitosan, and vanadyl sulfate are reported as the best elicitors for enhancing the ginsenoside production from hairy root cultures of *P. ginseng* (Palazon et al. 2003). Jasmonic acid and aluminum chloride enhance hyoscyamine and scopolamine production in hairy root cultures of *Brugmansia candida* (Pitta-Alvarez et al. 2000). Furthermore, scopolamine production was enhanced through bacterial elicitors (*Bacillus cereus*, *Staphylococcus aureus*) in hairy roots of *Scopolia parviflora*. Previous reports indicated the efficiency of hairy root culture in improved production of pharmaceutical metabolites (Table 18.2).

Table 18.2 Bioactive secondary metabolites produced through hairy root culture

Plant species	Metabolites	References
Gingko biloba	Ginkgolides	Ayadi and Tremouillaux-Guiller (2003)
Gmelina arborea	Erbascoside	Dhakulkar et al. (2005)
Panax ginseng	Ginsenoside	Palazon et al. (2003)
Camptotheca acuminata	Camptothecin	Lorence et al. (2004)
Rauvolfia micrantha	Ajmalicine, ajmaline	Sudha et al. (2003)
Linum flavum	Coniferin	Lin et al. (2003)
Papaver somniferum	Morphine, sanguinarine, codeine	Le Flem-Bonhomme et al. (2004)
Solidago altissima	Polyacetylene	Inoguchi et al. (2003)
Saussurea medusa	Jaceosidin	Zhao et al. (2004)
Pueraria phaseoloides	Puerarin	Shi and Kintzios (2003)
Linum flavum	Podophyllotoxin	Renouard et al. (2018)
Linum album	Podophyllotoxin	Chashmi et al. (2013)
Ophiorrhiza pumila	Camptothecin	Saito et al. (2001)
Ophiorrhiza alata	Camptothecin	Ya-ut et al. (2011)

18.5 Conclusion

Medicinal plants are the foremost source of life-saving drugs in our society. The demand for herbal medicine in the global market is increasing; meanwhile, the natural biomass material is still harvested. However, thousands of plant species were facing extinction and are listed as endangered. In vitro techniques pave the way for the conservation and production of pharmacologically active compounds. Plant tissue culture techniques like cell cultures or hairy root cultures can be useful for value-added or rare plant-based metabolite production. Thus, in vitro cultures are still more requisite to research to find out the best enhanced production systems at an industrial level. Developing new insight into a biochemical pathway or complex signaling pathway enables researchers to undertake advanced metabolic engineering work or genetic transformation work. The highly specific or controlled in vitro conditions, the cell or organ bio-factories produce high biomass or several enhanced bioactive compounds in a short time period, which enriches drug production industries.

References

Abdullah MA, Ali AM, Marziah M, Lajis NH, Ariff AB (1998) Establishment of cell suspension cultures of *Morinda elliptica* for the production of anthraquinones. Plant Cell Tissue Organ Cult 54:173–182

Ali M, Abbasi BH, Haq I (2013) Production of commercially important secondary metabolites and antioxidant activity in cell suspension cultures of *Artemisia absinthium* L. Ind Crop Prod 49: 400–406

Ayadi R, Tremouillaux-Guiller J (2003) Root formation from transgenic calli of *Ginko biloba*. Tree Physiol 23:713–718

Baenas N, Garcia-Viguera C, Moreno DA (2014) Elicitation: a tool for enriching the bioactive composition of foods. Molecules 19:13541–13563

Banthorpe DV, White JJ (1995) Novel anthraquinones from undifferentiated cell cultures of *Galium verum*. Phytochemistry 38:107–111

Baque MA, Elgirban A, Lee EJ, Paek KY (2012a) Sucrose regulated enhanced induction of anthraquinone, phenolics, flavonoids biosynthesis and activities of antioxidant enzymes in adventitious root suspension cultures of *Morinda citrifolia* (L.). Acta Physiol Plant 34:405–415

Baque MA, Moh SH, Lee EJ, Zhong JJ, Paek KY (2012b) Production of biomass and useful compounds from adventitious roots of high value added medicinal plants using bioreactor. Biotechnol Adv 30:1255–1267

Baskaran P, Jayabalan N (2009) In vitro regeneration of *Psoralea corylifolia* L. through callus cultures. Plant Biotechnol 26:333–336

Biswas T, Kalra A, Mathur AK, Lal RK, Singh M, Mathur A (2016) Elicitors' influenced differential ginsenoside production and exudation into medium with concurrent Rg3/Rh2 panaxadiol induction in *Panax quinquefolius* cell suspensions. Appl Microbiol Biotechnol 100:4909–4922

Biswas T, Pandey SS, Maji D, Gupta V, Kalra A, Singh M, Mathur A, Mathur AK (2018) Enhanced expression of ginsenoside biosynthetic genes and *in vitro* ginsenoside production in elicited *Panax sikkimensis* (Ban) cell suspensions. Protoplasma 255:1147–1160

Bonfill M, Mangas S, Moyano E, Cusido RM, Palazón J (2011) Production of centellosides and phytosterols in cell suspension cultures of *Centella asiatica*. Plant Cell Tissue Organ Cult 104: 61–67

Botta B, Monache GD (1995) *Maclura pomifera* (Osage Orange): *in vitro* culture and the formation of flavonoids and other secondary metabolites. In: Bajaj YPS (ed) Medicinal and Aromatic Plants VIII, vol 33. Springer, Heidelberg

Cai Z, Kastell A, Speiser C, Smetanska I (2013) Enhanced resveratrol production in *Vitis vinifera* cell suspension cultures by heavy metals without loss of cell viability. Appl Biochem Biotechnol 171:330–340

Chakraborty M, Karun A, Mitra A (2008) Accumulation of phenylpropanoids derivatives in chitosan-induced cell suspension culture of *Cocos nucifera*. J Plant Physiol 166:63–71

Chashmi NA, Sharifi M, Yousefzadi M et al (2013) Analysis of 6-methoxy podophyllotoxin and podophyllotoxin in hairy root cultures of *Linum album* Kotschy ex Boiss. Med Chem Res 22: 745–752

Chiang L, Abdullah MA (2007) Enhanced anthraquinones production from adsorbent treated *Morinda elliptica* cell suspension cultures in production medium strategy. Process Biochem 42:757–763

Chodisetti B, Rao K, Gandi S, Giri A (2013) Improved gymnemic acid production in the suspension cultures of *Gymnema sylvestre* through biotic elicitation. Plant Biotechnol Rep 7:519–525

Chodisetti B, Rao K, Gandi S, Giri A (2015) Gymnemic acid enhancement in the suspension cultures of *Gymnema sylvestre* by using the signaling molecules—methyl jasmonate and salicylic acid. In Vitro Cell Dev Biol Plant 51:88–92

Cole IB, Saxena PK, Murch SJ (2007) Medicinal biotechnology in the genus scutellaria. In Vitro Cell Dev Biol Plant 43:318–327

Coste A, Vlase L, Halmagyi A et al (2011) Effects of plant growth regulators and elicitors on production of secondary metabolites in shoot cultures of *Hypericum hirsutum* and *Hypericum maculatum*. Plant Cell Tissue Organ Cult 106:279–288

Cusido RM, Onrubia M, Sabater-Jara AB, Moyano E, Bonfill M, Goossens A, Pedreño MA, Palazon J (2014) A rational approach to improving the biotechnological production oftaxanes in plant cell cultures of *Taxus spp*. Biotechnol Adv 32:1157–1167

Danova K (2010) Production of polyphenolic compounds in shoot cultures of Hypericum species characteristic for the Balkan flora. Bot Serbica 34:29–36

Deshmukh SR, Wadegaonkar VP, Bhagat RP, Wadegaonkar PA (2011) Tissue specific expression of anthraquinones, flavonoids and phenolics in leaf, fruit and root suspension cultures of Indian Mulberry (*Morinda citrifolia* L.). Plant Omics J 4:6–13

Dewick PM (2002) Medicinal natural products: a biosynthetic approach. Wiley

Dhakulkar S, Ganapathi TR, Bhargav S, Bapat VA (2005) Induction of hairy roots in *Gmelina arborea* Roxb. and production of verbascoside in hairy roots. Plant Sci 169:812–818

Dias MI, Sousa MJ, Alves RC, Ferreira IC (2016) Exploring plant tissue culture to improve the production of phenolic compounds: a review. Ind Crop Prod 82:9–22

Dörnenburg H, Knorr D (1995) Strategies for the improvement of secondary metabolite production in plant cell cultures. Enzym Microb Technol 17:674–684

Ebel J, Cosio EG (1994) Elicitors of plant defense responses. Int Rev Cytol 148:1–36

Efferth T (2019) Biotechnology applications of plant callus cultures. Engineering 5:50–59

El-Sayed M, Verpoorte R (2002) Effect of phytohormones on growth and alkaloid accumulation by a *Catharanthus roseus* cell suspension cultures fed with alkaloid precursors tryptamine and loganin. Plant Cell Tissue Organ Cult 68:265–270

Espinosa-Leal CA, Puente-Garza CA, García-Lara S (2018) *In vitro* plant tissue culture: means for production of biological active compounds. Planta 248:1–18

Farag MA, Huhman DV, Lei Z, Sumner LW (2007) Metabolic profiling and systematic identification of flavonoids and isoflavonoids in roots and cell suspension cultures of *Medicago truncatula* using HPLC-UV-ESI-MS and GC-MS. Phytochemistry 68:342–354

Flores HE, Vivanco JM, Loyola-Vargas VM (1999) Radicle biochemistry: the biology of root-specific metabolism. Trends Plant Sci 4:220–226

Frackowiak A, Skibinski P, Gawel W, Zaczynska E, Czarny A, Gancarz R (2010) Synthesis of glycoside derivatives of hydroxyl anthraquinone with ability to dissolve and inhibit formation of crystals of calcium oxalate. Potential compounds in kidney stone therapy 45:1001–1007

Gagne S, Cluzet S, Merillon J, Geny L (2007) ABA initiates anthocyanin production in grape cell cultures. J Plant Growth Regul 30:1–10

Garcia R, Pacheco G, Falcão E, Borges G, Mansur E (2011) Influence of type of explant, plant growth regulators, salt composition of basal medium, and light on callogenesis and regeneration in *Passiflora suberosa* L.(Passifloraceae). Plant Cell Tissue Organ Cult 106:47–54

George EF, Hall MA, De-Klerk GJ (2008) The components of plant tissue culture media II: organic additions, osmotic and pH effects, and support systems. In: George EF, Hall MA, De Klerk GJ (eds) Plant propagation by tissue culture, 3rd edn. Springer, pp 115–173

Georgiev MI, Eibl R, Zhong JJ (2013) Hosting the plant cells *in vitro*: recent trends in bioreactors. Appl Microbiol Biotechnol 97:3787–3800

Giri L, Dhyani P, Rawat S, Bhatt ID, Nandi SK, Rawal RS, Pande V (2012) *In vitro* production of phenolic compounds and antioxidant activity in callus suspension cultures of *Habenaria edgeworthii*: a rare Himalayan medicinal orchid. Ind Crops Prod 39:1–6

Goel MK, Mehrotra S, Kukreja AK (2011) Elicitor-induced cellular and molecular events are responsible for productivity enhancement in hairy root cultures: an insight study. Appl Biochem Biotechnol 165:1342–1355

Gorelick J, Bernstein N (2014) Elicitation: an underutilized tool in the development of medicinal plants as a source of therapeutic secondary metabolites. In: Sparks DL (ed) Advances in agronomy, vol 124. Elsevier, Amsterdam, pp 201–230

Guern J, Renaudin JP, Brown SC (2012) The compartmentation of secondary metabolites in plant cell cultures. Cell Cult Somatic Cell Genet Plants 4:43–76

Gugler K, Funk C, Brodelius P (1988) Elicitor-induced tyrosine decarboxylase in berberine synthetizing suspension cultures of *Thalictrum rugosum*. Eur J Biochem 170:661–666

Gupta S, Chauhan D, Mehla K, Sood P, Nair A (2010) An overview of nutraceuticals: current scenario. J Basic Clin Pharm 1:55–62

Hagendoorn MJ, Poortinga AM, Sang HWWF, van der Plas LH, van Walraven HS (1991) Effect of elicitors on the plasmamembrane of *Petunia hybrida* cell suspensions role of ΔpH in signal transduction. Plant Physiol 96:1261–1267

Hamilton AC (2004) Medicinal plants, conservation and livelihoods. Biodivers Conserv 13:1477–1517

Han YS, van der Heijden R, Verpoorte R (2002) Improved anthraquinone accumulation in cell cultures of *Cinchona robusta* by feeding of biosynthetic precursors and inhibitors. Biotechnol Lett 24:705–710

Hemashenpagam N, Selvaraj T (2011) Effect of arbuscular mycorrhizal (AM) fungus and plant growth promoting rhizo microorganisms (PGPRs) on medicinal plant *Solanum viarum* seedlings. J Environ Biol 32:579–583

Hu FX, Zhong JJ (2007) Role of jasmonic acid in alteration of ginsenoside heterogeneity in elicited cell cultures of *Panax notoginseng*. J Biosci Bioeng 104:513–516

Hu FX, Zhong JJ (2008) Jasmonic acid mediates gene transcription ofginsenoside biosynthesis in cell cultures of *Panax notoginseng* treated with chemically synthesized 2-hydroxyethyl jasmonate. Process Biochem 43:113–118

Huang Q, Lu G, Shen HM, Chung MCM, Ong CN (2007) Anti-cancer properties of anthraquinones from *rhubarb*. Med Res Rev 27:609–630

Hwang SJ, Kim YH, Pyo BS (2004) Optimization of aconitine production in suspension cell cultures of *Aconitum napellus* L. Korean J Med Crop Sci 12:366–371

Inoguchi M, Ogawa S, Furukawa S, Konda H (2003) Production of an allelopathic polyacetylene in hairy root cultures of goldenrod (*Solidago altissima* L.). Biosci Biotechnol Biochem 67:863–868

Inoue K, Nayeshiro H, Inouyet H, Zenk M (1981) Anthraquinones in cell suspension cultures of *Morinda citrifolia*. Phytochemistry 20:1693–1700

Inoue K, Shiobara Y, Nayeshiro H, Inouye H, Wilson G, Zenk MH (1984) Biosynthesis of anthraquinones and related compounds in *Galium mollugo* cell uspension cultures. Phytochemistry 23:307–311

Jasril LN, Mooi LY, Abdullah MA, Sukari MA, Ali AM (2003) Antitumor promoting and antioxidant activities of anthraquinones isolated from the cell suspension culture of *Morinda elliptica*. Asia Pac J Mol Biol Biotechnol 11:3–7

Kabita KC, Sanatombi K, Sharma SK (2020) Efficient enhancement of capsaicinoids biosynthesis in cell suspension cultures of *Capsicum chinense* Jacq. cv. 'Umorok' by elicitors and diferential gene expression analysis of elicited cultures. Plant Cell Tissue Organ Cult 141:145–154

Kamalipourazad M, Sharifi M, Maivan HZ, Behmanesh M, Chashmi NA (2016) Induction of aromatic amino acids and phenylpropanoid compounds in *Scrophularia striata* Boiss. Cell culture in response to chitosan-induced oxidative stress. Plant Physiol Biochem 107:374–384

Kang SM, Jung HY, Kang YM, Yun DJ, Bahk JD, Yang JK, Choi MS (2004) Effects of methyl jasmonate and salicylic acid on the production of tropane alkaloids and the expression of PMT and H6H in adventitious root cultures of *Scopolia parviflora*. Plant Sci 166:745–751

Kasem M (2018) Callus production and suspension elicitation of *Impatiens balsamina* L., plant for enhancing accumulation of phenolics and flavonoids content. J Plant Prod 9:241–248. https://doi.org/10.21608/jpp.2018.35452

Kim CY, Im HW, Kim HK, Huh H (2001) Accumulation of 2,5-dimethoxy-1,4-benzoquinone in suspension cultures of *Panax ginseng* by a fungal elicitor preparation and a yeast elicitor preparation. Appl Microbiol Biotechnol 56:239–242

Kim HJ, Chen F, Wang X, Rajapakse NC (2005) Effect of chitosan on biological properties of sweet basil (*Ocimum basilicum*). J Agric Food Chem 53:3696–3701

Kitajima M, Fischer U, Nakamura M, Ohsawa M, Ueno M, Takayama H, Unger M, Stöckigt J, Aimi N (1998) Anthraquinones from *Ophiorrhiza pumila* tissue and cell cultures. Phytochemistry 48:107–111

Klarzynski O, Friting B (2001) Stimulation of plant natural defenses. C R Acad Sci III 324:953–963

Kokotkiewicz A, Luczkiewicz M, Kowalski W, Badura A, Piekus N, Bucinski A (2012) Isoflavone production in *Cyclopia subternata* Vogel (honeybush) suspension cultures grown in shake flasks and stirres-tank bioreactor. Appl Microbiol Biotechnol 97:8467–8477

Krishnan SRS, Siril EA (2016a) Elicitor and precursor mediated anthraquinone production from cell suspension cultures of *Oldenlandia umbellata* L. Int J Pharm Sci Res 7:3649–3657

Krishnan SRS, Siril EA (2016b) Induction of hairy roots and over production of anthraquinones in *Oldenlandia umbellata* L.: a dye yielding medicinal plant by using wild type Agrobacterium rhizogenes strain. Indian J Plant Physiol 21:271–278

Krishnan SRS, Sreelekshmi R, Siril EA, Swapna TS (2020) Cell and protoplast culture for production of plant metabolites. In: Swapna TS, Shiburaj S, Sabu A (eds) Plant metabolites: methods, applications and prospects. Springer Nature Singapore Pte Ltd, pp 71–88

Kumar JK, Sinha AK (2004) Resurgence of natural colourants: a holistic view. Nat Prod Res 18:59–84

Kumar S, Chashoo G, Saxena AK, Pandey AK (2013) *Parthenium hysterophorus*: a probable source of anticancer, antioxidant and anti-HIV agents. Biomed Res Int 3:810734

Kumar V, Suman U, Rubal, Yadav SK (2018) Flavonoid secondary metabolite: biosynthesis and role in growth and development in plants. In: Yadav S, Kumar V, Singh S (eds) Recent trends and techniques in plant metabolic engineering. Springer, Singapore

Kuo CI, Chao CH, Lu MK (2012) Effects of auxins on the production of steroidal alkaloids in rapidly proliferating tissue and cell cultures of solanum lyratum. Phytochem Anal 23:400–404

Kuo YH, Huang HC, Yang Kuo LM, Hsu YW, Lee KH, Chang FR, Wu YC (2005) New dammarane-type saponins from the galls of *Sapindus mukorossi*. J Agric Food Chem 53: 4722–4727

Lajis NH, Abdullah MA, Ismail NH, Ali AM, Marziah M, Ariff AB, Kitajima M, Takayama H, Aimi N (2000) Anthraquinones from cell suspension culture of *Morinda elliptica*. Nat Prod Sci 6:40–43

Lambert E, Faizal A, Geelen D (2011) Modulation of triterpene saponin production: *in vitro* cultures, elicitation, and metabolic engineering. Appl Biochem Biotechnol 164:220–237

Le Flem-Bonhomme V, Laurain-Matter D, Fliniaux MA (2004) Hairy root induction of *Papaver somniferum* var. album, a difficult-to-transform plant, by *A. rhizogenes* LBA 9402. Planta 218: 890–893

Li H, Lin Y, Chen X, Bai Y, Wang C, Xu X, Wang Y, Lai Z (2018) Effects of blue light on flavonoid accumulation linked to the expression of miR393, miR394 and miR395 in longan embryogenic calli. PLoS One 13(1):e0191444

Lin HW, Kwok KH, Doran PM (2003) Development of *Linum flavum* hairy root cultures for production of coniferin. Biotechnol Lett 25:521–525

Liu KC, Yang SH, Roberts MF, Philipson JD (1990) Production of canthin-6-one alkaloids by cell suspension cultures of *Brucea javanica* (L.) Merr. Plant Cell Rep 9:261–263

Lorence A, Medina-Bolivar F, Nessler CL (2004) Camptothecin and 10-hydroxycamptothecin from *Camptotheca acuminate* hairy rots. Plant Cell Rep 22:437–441

Luczkiewicz M, Głód D (2003) Callus cultures of Genista plants—*in vitro* material producing high amounts of isoflavones of phytoestrogenic activity. Plant Sci 165:1101–1108

Maheshwari P, Songara B, Kumar S, Jain P, Srivastava K, Kumar A (2007) Alkaloid production in *Vernonia cinerea*: callus, cell suspension and root cultures. Biotechnol J 2:1026–1032

Malik S, Cusido RM, Mirjalili MH, Moyano E, Palazon J, Bonfill M (2011) Production of the anticancer drug taxol in *Taxus baccata* suspension cultures: a review. Process Biochem 46:23–34

Maqsood M, Khusrau M, Mujib A (2018) *Catharanthus roseus* (L.) G. Don and *in vitro* techniques: a review. JK Knowl Initiative 2:9–15

Marshall JG, Staba EJ (1976) Hormonal effects on diosgenin biosynthesis and growth in *Dioscorea deltoidea* tissue cultures. Phytochemistry 15:53–55

Masondo NA, Aremu AO, Finnie JF, Van Staden J (2015) Growth and phytochemical levels in micropropagated Eucomis autumnalis subspecies autumnalis using different gelling agents, explant source, and plant growth regulators. In Vitro Cell Dev Biol Plant 51:102–110

Masoumian M, Arbakariya A, Syahida A, Maziah M (2011) Flavonoids production in *Hydrocotyle bonariensis* callus tissues. J Med Plants Res 5:1564–1574

Mendhulkar VD, Vakil MMA (2013) Chitosan and Aspergillus Niger mediated elicitation of total flavonoids in suspension culture of *Andrographis paniculata* (Burm. f.) Nees. Int J Pharm Biol Sci 4:731–740

Mizutani H, Hashimoto O, Nakashima R, Nagai J (1997) Anthraquinone production by cell suspension cultures of *Rubia akane* NAKAI. Biosci Biotechnol Biochem 61:1743–1744

Murthy HN, Lee EJ, Paek KY (2014) Production of secondary metabolites from cell and organ cultures: strategies and approaches for biomass improvement and metabolite accumulation. Plant Cell Tissue Organ Cult 118:1–16

Nakao M, Ono K, Takio S (1999) The effect of calcium on flavanol production in cell suspension cultures of *Polygonum hydropiper*. Plant Cell Rep 18:759–763

Nalawade SM, Sagare AP, Lee CY, Kao CL, Tsay HS (2003) Studies on tissue culture of Chinese medicinal plant resources in Taiwan and their sustainable utilization. Bot Bull Acad Sin 44:79–98

Namdeo AG (2007) Plant cell elicitation for production of secondary metabolites: a review. Pharmacogn Rev 1:69–79

Narasimhan S, Nair GM (2004) Effect of auxins on berberine biosynthesis in cell suspension culture of *Coscinium fenestratum* (Gaertn.) Colebr-a critically endangered medicinal liana of Western Ghats. Indian J Exp Biol 42:616–619

Palacio L, Cantero JJ, Cusido RM, Goleniowski ME (2012) Phenolic compound production in relation to differentiation in cell and tissue cultures of *Larrea divaricata* (Cav.). Plant Sci 193:1–7

Palazon J, Cusido RM, Bonfill M, Mallol A, Moyano E, Morales C et al (2003) Elicitation of different *Panax ginseng* transformed root phenotypes for an improved ginsenoside production. Plant Physiol Biochem 41:1019–1025

Palazón J, Navarro-Ocaña A, Hernandez-Vazquez L, Mirjalili MH (2008) Application of metabolic engineering to the production of scopolamine. Molecules 13:1722–1742

Park JM, Yoon SY (1992) Production of sanguinarine by suspension cultures of *Papaver somniferum* in bioreactors. J Ferment Bioeng 74:292–296

Park HH, Hakamatsuka T, Sankawa U, Ebizuka Y (1995) Rapid metabolism of isofavonoids in elicitor-treated cell suspension cultures of *Pueraria lobata*. Phytochemistry 38:373–380

Pasqua G, Avato P, Monacelli B, Santamaria AR, Argentieri MP (2003) Metabolites in cell suspension cultures, calli, and in vitro regenerated organs of *Hypericum perforatum* cv. Topas. Plant Sci 165:977–982

Patra N, Srivastava AK (2016) Artemisinin production by plant hairy root cultures in gas-and liquid-phase bioreactors. Plant Cell Rep 35:143–153

Payne GF, Bring V, Prince C, Shuler ML (1991) Immobilized plant cells. In: Payne GF, Bringi V, Prince C, Shuler ML (eds) Plant cell and tissue culture in liquid systems. Hanser, pp 179–223

Pillai SK, Siril EA (2022) Exogenous elicitors enhanced berberine production in the cell suspension cultures of *Tinospora cordifolia* (Willd.) Miers ex Hook F & Thoms. Proc Natl Acad Sci India Sect B Biol Sci 92:209–218

Pise MV, Rudra JA, Upadhyay A (2015) Immunomodulatory potential of shatavarins produced from *Asparagus racemosus* tissue cultures. J Nat Sci Biol Med 6:415–420

Pitta-Alvarez SI, Spollansky TC, Giulietti AM (2000) The influence of different biotic and abiotic elicitors on the production and profile of tropane alkaloids in hairy root cultures of *Brugmansia candida*. Enzym Microb Technol 26:252–258

Qu JG, Yu XJ, Zhang W, Jin MF (2006) Significant improved anthocyanins biosynthesis in suspension cultures of *Vitis vinifera* by process intensification. Sheng Wu Gong Cheng Xae Barv 22:299–305

Radman R, Bucke C, Keshavarz T (2004) Elicitor effects on reactive oxygen species in liquid cultures of *Penicillium chrysogenum*. Biotechnol Lett 26:147–152

Ramakrishna A, Ravishankar GA (2011) Influence of abiotic stress signals on secondary metabolites in plants. Plant Signal Behav 6:1720–1731

Ramirez-Estrada K, Vidal-Limon H, Hidalgo D, Moyano E, Golenioswki M, Cusidó RM, Palazon J (2016) Elicitation, an effective strategy for the biotechnological production of bioactive high-added value compounds in plant cell factories. Molecules 21:182

Rani D, Meelaph T, De-Eknamkul W, Vimolmangkang S (2020) Yeast extract elicited isofavonoid accumulation and biosynthetic gene expression in *Pueraria candollei* var. mirifca cell cultures. Plant Cell Tissue Organ Cult 141:661–667

Rao SR, Ravishankar GA (2002) Plant cell cultures: chemical factories of secondary metabolites. Biotechnol Adv 20:101–153

Rather MA, Ganai BA, Kamili AN, Qayoom M, Akbar S, Masood A, Qurishi MA (2012) Comparative GC–FID and GC–MS analysis of the mono and sesquiterpene secondary metabolites produced by the field grown and micropropagated plants of *Artemisia amygdalina* Decne. Acta Physiol Plant 34:885–890

Ratnadewi D, Sumaryono O (2010) Quinoline alkaloids in suspension cultures of *Cinchona ledgeriana* treated with various substances. Hayati J Biosci 17:179–182

Ratnadewi D, Satriawan D, Sumaryono (2013) Enhanced production level of quinine in cell suspension culture of *Cinchona ledgeriana* Moens by paclobutrazol. Biotropia 20:10–18

Renouard S, Corbin C, Drouet S, Medvedec B, Doussot J, Colas C, Maunit B, Bhambra AS, Gontier E, Jullian N, Mesnard F, Boitel M, Abbasi BH, Arroo RRJ, Lainé E, Hano C (2018) Investigation of *Linum flavum* (L.) hairy root cultures for the production of anticancer Aryltetralin Lignans. Int J Mol Sci 19:990

Robins RI, Payne J, Rhodes MJ (1986) The production of anthraquinones by cell suspension cultures of *Cinchona ledgeriana*. Phytochemistry 25:2327–2334

Rojas R, Alba J, Magaña-Plaza I, Cruz F, Ramos-Valdivia AC (1999) Stimulated production of diosgenin in *Dioscorea galeottiana* cell suspension cultures by abiotic and biotic factors. Biotechnol Lett 21:907–911

Saito K, Sudo H, Yamazaki M et al (2001) Feasible production of camptothecin by hairy root culture of *Ophiorrhiza pumila*. Plant Cell Rep 20:267–271

Sanchez-Sampedro MA, Fernandez-Tarrago J, Corchete P (2005) Yeast extract and methyl jasmonate induced silymarin production in cell cultures of *Silybum marianum* (L.) Gaernt. J Biotechnol 119:60–69

Schmeda-Hirschmann G, Jordan M, Gertn A, Wilken D, Hormazabal E, Tapia AA (2004) Secondary metabolite content in *Fabiana imbricate* plants and *in vitro* cultures. Z Naturforsch C J Biosci 5:48–54

Sengar RS, Chaudhary R, Tyagi SK (2010) Present status and scope of floriculture developed through different biological tools. Res J Agric Sci 1:306–314

Shams-Ardakani M, Hemmati S, Mohagheghzadeh A (2005) Effect of elicitors on the enhancement of podophyllotoxin biosynthesis in suspension cultures of *Linum album*. Daru 13:56–60

Shanks JV, Morgan J (1999) Plant 'hairy root' culture. Curr Opin Biotechnol 10:151–155

Sharma P, Kharkwal AC, Abdin MZ, Varma A (2014) *Piriformospora indica* improves micropropagation, growth and phytochemical content of *Aloe vera* L. plants. Symbiosis 64: 11–23

Shi HP, Kintzios S (2003) Genetic transformation of *Pueraria phaseoloides* with *Agrobacterium rhizogenes* and puerarin production in hairy roots. Plant Cell Rep 21:1103–1107

Siddique I, Javed SB, Al-Othman MR, Anis M (2013) Stimulation of *in vitro* organogenesis from epicotyl explants and successive micropropagation round in *Cassia angustifolia* Vahl.: an important source of sennosides. Agrofor Syst 87:583–590

Skrzypczak-Pietraszek E, Piska K, Pietraszek J (2018) Enhanced production of the pharmaceutically important polyphenolic compounds in *Vitex agnus castus* L. shoot cultures by precursor feeding strategy. Eng Life Sci 18:287–297

Sonja GS, Oliver T, Stéphane M, Alain D, Eric L, Claude J, Daniel H (2015) Polysaccharide elicitors enhance phenylpropanoid and naphtodianthrone production in cell suspension cultures of *Hypericum perforatum*. Plant Cell Tissue Organ Cult 122:649–663

Sreeranjini S, Siril EA (2013) Production of anthraquinones from adventitious root derived callus and suspension cultures of *Morinda citrifolia* L. in response to auxins, cytokinins and sucrose levels. Asian J Plant Sci Res 3:131–138

Sreeranjini S, Siril EA (2015) Optimizing elicitors and precursors to enhance alizarin and purpurin production in adventitious roots of *Morinda citrifolia* L. Proc Natl Acad Sci India Sect B Biol Sci 85:725–731

Stafford A, Morris P, Fowler MW (1986) Plant cell biotechnology: a perspective. Enzym Microb Technol 8:578–587

Sudha CG, Obul Reddy B, Ravishanker GA, Seeni S (2003) Production of ajmalicine and ajmaline in hairy root cultures of *Rauvolfia micrantha* Hook f., a rare and endemic medicinal plant. Biotechnol Lett 25:631–636

Suzuki H, Achnine L, Xu R, Matsuda SP, Dixon RA (2002) A genomics approach to the early stages of triterpene saponin biosynthesis in *Medicago truncatula*. Plant J 32:1033–1048

Suzuki H, Reddy MS, Naoumkina M, Aziz N, May GD, Huhman DV, Sumner LW, Blount JW, Mendes P, Dixon RA (2005) Methyl jasmonate and yeast elicitor induce differential transcriptional and metabolic re-programming in cell suspension cultures of the model legume *Medicago truncatula*. Planta 220:696–707

Trejo-Tapia G, Jimenez-Aparicio A, Rodriguez-Monroy M, De Jesus-Sanchez A, Gutierrez-Lopez G (2001) Influence of cobalt and other microelements on the production of betalains and the growth of suspension cultures of *Beta vulgaris*. Plant Cell Tissue Organ Cult 67:19–23

Urdová J, Rexová M, Mučaji P, Balažová A (2015) Elicitation—a tool to improve secondary metabolites production in *Melissa officinalis* L. suspension cultures. Acta Fac Pharm Univ Comen 62:6–50

Usher PJ (2000) Traditional ecological knowledge in environmental assessment and management. Arctic 53:183–193

Valluri JV (2009) Bioreactor production of secondary metabolites from cell cultures of periwinkle and sandal wood. Methods Mol Biol 547:325–335

Vanisree M, Lee CY, Lo SF, Nalawade SM, Lin CY, Tsay HS (2004) Studies on the production of some important secondary metabolites from medicinal plants by tissue culture. Bot Bull Acad Sin 45:1–22

Vázquez-Flota F, Hernández-Domínguez E, de Lourdes Miranda-Ham M, Monforte-González M (2009) A differential response to chemical elicitors in *Catharanthus roseus* in vitro cultures. Biotechnol Lett 31:591–595

Veerashree V, Anuradha CM, Kumar V (2012) Elicitor-enhanced production of gymnemic acid in cell suspension cultures of *Gymnema sylvestre* R. Br. Plant Cell Tissue Organ Cult 108:27–35

Verpoorte R (2000) Pharmacognosy in the new millennium: lead finding and biotechnology. J Pharm Pharmacol 52:253–262

Verpoorte R, Contin A, Memelink J (2002) Biotechnology for the production of plant secondary metabolites. Phytochem Rev 1:13–25

Walker TS, Pal Bais H, Vivanco JM (2002) Jasmonic acid-induced hypericin production in cell suspension cultures of *Hypericum perforatum* L. (St. John's wort). Phytochemistry 60:289–293

Wang YD, Yuan YJ, Wu JC (2004) Induction studies of methyl jasmonate and salicylic acid on taxane production in suspension cultures of *Taxus chinensis* var. *mairei*. Biochem Eng J 19: 259–265

Wang W, Zhang ZY, Zhong JJ (2005) Enhancement of ginsenoside biosynthesis in high-density cultivation of *Panax notoginseng* cells by various strategies of methyl jasmonate elicitation. Appl Microbiol Biotechnol 67:752–758

Xiang H, Guo Y (1997) Studies on the production of anthraquinone by plant cell suspension culture. Huanan Ligong Daxue Xuebao, Ziran Kexueban 25:62–67

Yahia A, Kevers C, Gaspar T, Chenieux JC, Rideau M, Creche J (1998) Cytokinins and ethylene stimulate indole alkaloid accumulation in cell suspensions cultures of *Catharanthus roseus* by two distinct mechanisms. Plant Sci 133:9–15

Yamamoto O, Yamada Y (1986) Production of reserpine and its optimization in cultured *Rauvolfia serpentina* Benth. cells. Plant Cell Rep 5:50–53

Yamamoto H, Ichimura M, Inoue K (1995) Stimulation of prenylated flavanone production by mannans and acidic polysaccharides in callus-cultures of *Sophora-Flavescens*. Phytochemistry 40:77–81

Yan Q, Shi M, Ng J, Wu JY (2006) Elicitor-induced rosmarinic acid accumulation and secondary metabolism enzyme activities in *Salvia miltiorrhiza* hairy roots. Plant Sci 170:853–858

Ya-ut P, Chareonsap P, Sukrong S (2011) Micropropagation and hairy root culture of *Ophiorrhiza alata* Craib for camptothecin production. Biotechnol Lett 33:2519–2526

Zamboni A, Vrhovsek U, Kassemeyer HH, Mattivi F, Velasco R (2015) Elicitor-induced resveratrol production in cell cultures of different grape genotypes (*Vitis* spp.). VITIS-J Grapevine Res 45: 63–68

Zenk MH, el-Shagi H, Schulte U (1975) Anthraquinone production by cell suspension cultures of *Morinda citrifolia*. Planta Med 28:79–101

Zhang Y, Zhong J (1997) Hyperproduction of ginseng saponin and polysaccharide by high density cultivation of *Panax notoginseng* cells. Enzyme Microb Technol 21:59–63

Zhang W, Curtin C, Kikuchi M, Franco C (2002) Integration of jasmonic acid and light irradiation for enhancement of anthocyanins in *Vitis vinifera* suspension cultures. Plant Sci 162:459–468

Zhao J, Zhu W, Hu Q (2001) Selection of fungal elicitors to increase indole alkaloid accumulation in *Catharanthus roseus* suspension cell culture. Enzyme Microb Technol 28:666–672

Zhao D, Fu C, Chen Y, Ma F (2004) Transformation of *Saussurea medusa* for hairy roots and jaceosidin production. Plant Cell Rep 23:468–474

Zhao J, Davis LC, Verpoorte R (2005) Elicitor signal transduction leading to production of plant secondary metabolites. Biotechnol Adv 23:283–333

Zheng Z, Wu M (2004) Cadmium treatment enhances the production of alkaloid secondary metabolites in *Catharanthus roseus*. Plant Sci 166:507–514

Zhong JJ (2001) Biochemical engineering of the production of plant specific secondary metabolites by cell suspension cultures. Adv Biochem Eng Biotechnol 72:1–26

Zhou H, Li M, Zhao X, Fan X, Guo A (2010) Plant regeneration from in vitro leaves of the peach root stock 'Nemaguard' (*Prunus persica*× *P. davidiana*). Plant Cell Tissue Organ Cult 101:79–87

Zia M, Mannan A, Chaudhary MF (2007) Effect of growth regulators and amino acids on artemisinin production in the callus of *Artemisia absinthium*. Pak J Bot 39:799–508

Chapter 19
Enhanced Secondary Metabolite Production for Drug Leads

A. S. Vivek, C. T. Riyas, and T. S. Swapna

Abstract Plant cells produce a versatile array of chemical compounds that develop biochemical networks for survival and proliferation. Primary metabolism entails biochemical processes that result in assimilation, respiration, transport, and differentiation, whereas secondary metabolism constitutes the production of by-products that are plant primary defense mechanisms against pathogens, herbivore attacks, and physical stress such as UV radiation. Secondary metabolites (SMs) are responsible for the flavors, colors, and odors of plants, as well as a source of pharmaceuticals, pesticides, cosmetics, dyes, flavors, and scents. SMs generated from plants have been used as medicine for thousands of years. Currently, bioactive SM are obtained and used either directly or after chemical modification. Because of the constant discovery of their potential functions in healthcare and as lead compounds for future medication development, their pharmacological worth is expanding. This chapter focuses on approaches to improve plant SMs production by using elicitors, as well as plant cell culture, which is seen as a promising alternative for creating bioactive molecules that are difficult to obtain by chemical synthesis or plant extraction. The chapter also concludes that commercial production benefits from in vitro growth of cells with changed and increased metabolic pathways. In comparison to other traditional tactics, metabolic engineering has the potential to lower costs, enhance yield, improve efficacy, maintain metabolic production, and improve quality. This environmentally benign and sustainable technique holds promise for increased metabolite production and desirable de-novo molecules.

Keywords Secondary metabolites · Elicitors · Metabolic engineering · Enhanced metabolite production

---

A. S. Vivek · C. T. Riyas · T. S. Swapna (✉)
Department of Botany, University of Kerala, Thiruvananthapuram, Kerala, India

S. T. Sukumaran, K. T R (eds.), *Conservation and Sustainable Utilization of Bioresources*, Sustainable Development and Biodiversity 30,
https://doi.org/10.1007/978-981-19-5841-0_19

19.1 Secondary Metabolites (SMs)

Photosynthetic and non-photosynthetic organisms (plants, bacteria, and fungi) will produce secondary metabolites through their interaction with terrestrial or aquatic ecological niches (Ramirez-Estrada et al. 2016; Petersen et al. 2020). The environment that intensifies or diverts the particular pathway of primary metabolism (Velu et al. 2018), directly or indirectly, pave the biosynthesis of committed secondary metabolic production via the committed enzyme (Pott et al. 2019). Even the current climatic change and global warming caused secondary metabolic production in organisms, hampering the hazardous effects (Wani et al. 2017). The primary metabolites directly plays role in the growth, reproduction, and repair of the organism. (He et al. 2018). The studies report that the secondary metabolite production in response to primary biosynthetic enzymes is variable from 0 to 95 fold. Such a drastic variation can control by the less multiplexed transcriptional regulation and the unswerving pathways from the primary metabolic pathway to the end product, "secondary metabolite" (Fernie and Stitt 2012; Alseekh et al. 2015).

Researchers are still uncertain how such a predetermined ancestral gene could be a part of the metabolism in fluctuating environment (Kroymann 2011). The intensification of the factors from the niches is stress, and it may be abiotic or biotic in nature. The role and effect of essential abiotic growth factors (light, temperature, humidity, water, micro, and macronutrient availability, heavy metal, pH, and salt content for the growth and existence of the plant) sometimes act vice versa by the variation in the concentration and their intensity (Borges et al. 2017). When the essential factors cross the limit of necessity, it becomes a stress inducer, and through that, the secondary metabolites will produce. Pest attacks, diseases, weeds, or any parasitic attacks are even responsible for the stress induction and successive metabolite production in the organisms and in their surroundings; these living stress inducers are the biotic stress inducers (Bouwmeester et al. 2003; Angessa and Li 2016).

The compound having a size of less than 900 Dalton (Da) is considered the secondary metabolite (Hadacek 2015). Such secondary metabolites aid the duty of the defense mechanism in them (Movahedi et al. 2021). In plants, it may be a defense against herbivores or attacks from the microorganism. Some alkaloid compounds affect ion channels by modulating neuronal signal transduction of the predator (Wink 2015; War et al. 2019). In certain plants, secondary metabolites (SMs) like terpenoids compounds help the plant to increase its pollination rate by their attractive nature to the pollinator as a pheromone (Wink 2018).

The secondary metabolites (SMs) have been exploited by humans unknowingly for millennia through food, medicaments, and feeds for their pets. However, with increasing needs and diseases, the field of research regarding SMs also expanded and flourished with different branches as a growing tree (Sreenivasulu and Fernie 2022). With increasing research, in addition to the basic needs of SMs, luxurious applications also developed (Demain and Fang 2000; Zhang et al. 2021).

The 80% world population still relies on the benefit of secondary metabolites through traditional medicinal practices (Wahid et al. 2020). According to the report of Thirumurugan et al. (2018), out of total isolated secondary metabolites, about 80% of compounds are isolated from plants, and in the case of total isolated antibiotics, plants impart 73.47%, animal based will be about 26.53% (Bérdy 2005).

Since the Primeval period, humans have depended on natural sources like plants, animals, aquatic organisms, and microorganisms for medicines. This type of utilization of natural resources for traditional practice was more successful than other traditional practices with immoral and unscientific bases. Periodically such taboos are eliminated (Velu et al. 2018; Sharma et al. 2021). Paleontological evidence corroborates that the use of medicinal plants was started before 60,000 years (Wen et al. 2010; Velu et al. 2018). The Mesopotamian clay tablets dating back to 2600 BC (Before Christ) is proof of their herbal practice. The *Ebers* Papyrus dating back to 1500 BC is another proof of the herbal use of Egyptian civilization; *Ebers Papyrus* is a well-written form with more than 700 herbal medicines (Saad et al. 2005; Pan et al. 2014). According to the WHO report, traditional medicine (TM) is growing more slowly than previous, especially in Asian countries the TM and its practices have already flourished with national support (Ansari 2021). Traditionally, different countries follow different ways of practices for curing diseases, including Ayurveda, Yoga, Unani, Siddha, and Homeopathy (AYUSH) in India, Traditional Chinese Medicine (TCM), Kampo medicine in Japan, Koryo medicine in Korea, Traditional Medicine of Sri Lanka, and traditional Arabic and Islamic medicine in Arab countries (Ravishankar and Shukla 2008).

Even the natural resources like marine and animal sources are plenty enough. All the written documents predominantly focus on plant-based medicines and remedies. The current research on TM is interdisciplinary with other fields like biotechnology and computer-aided drug discovery (CADD) for the identification and enhanced production of bioactive molecules (Srivastava et al. 2014; Prachayasittikul et al. 2015). Apparently, the new approach of "trans-disciplinarity" increased the data to knowledge conversion, and the traditional knowledge since antiquity has been substantiated by this approach (Yang 2013; Yi et al. 2018).

Using the rapid advancement of modern science and the wide range of modern analytical techniques to identify and isolate the bioactive compound, it is easy to target a lead molecule with its known moiety from an ocean of bioactive molecules in a limited time scale with utmost accuracy (Gurung et al. 2021). This chapter focuses on the enhancement of the production of potential bioactive drug molecules.

19.2 Drug Lead

The genesis of a drug after 12–15 years of gestation period worth billion dollar investments. And the identification of drug lead is based on authentic academic and research data. The target and its drug lead interaction identification also take years of research (Hughes et al. 2011).

The biosynthesis of SMs employed in plants involves a series of pathways like shikimate pathway, acetate pathway, mevalonate pathway, and deoxy-xylulose phosphate pathway. There are a series of reports of the drugs developed from the SMs. The identification of a drug lead can be simply classified into two stages. The first one is the primary assay in which the extract of the plant sample was taken using different polar to non-polar solvents. And the extract is subjected to different activity studies. The plants having positive results were later subjected to the second stage. Here the compounds from the extracts were separated using different chromatographic separation techniques, and individual responsible compound or compounds were identified. Later the compounds were subjected to the structural elucidation and successively evaluated for druggability (Katiyar et al. 2012). A drug development pipeline was structured for the compound having greater druggability, and the drug compound eligible for becoming the medicine was then moved for the commercial production and pharmaceutical industry.

Bellidifolin, fagasterol, epicatechin, gymnemic acid, ginsenoside, kaempferol, lupeol acetate, quercetin, β-sitosterol, and ursolic acid (Srivastav et al. 2019) are few of the compounds identified from various plants. Different in silico studies and drug engineering studies are also employed for the improvement of compound efficacy.

19.3 In Vitro: Secondary Metabolite Production

At present, living organisms are the only significant source for isolating and identifying new compounds with therapeutic use in the drug development pipeline. Currently, organically derived compounds or synthetic forms with modified moiety are used as essential medicaments in the whole world. Drugs, flavoring agents and perfumes, coloring dyes, insecticides, and food additives are a few market forms of secondary metabolites. Most of the medications on the market today are synthetic versions or duplicates of the natural compound. For accuracy and large-scale production, all modern medicines are synthesized artificially instead of isolated from the natural source. The growing commercial value of secondary metabolites has forced and constrained researchers to invest in enhanced secondary metabolite production, especially in recent years, when the idea of tissue culture technologies, in vitro culture, and elicitation to change the production rate of bioactive metabolites (Hussain et al. 2012).

In vitro cell culture ("cell factory") is more cost-effective and high yielding than any traditional strategy in any circumstance in the whole world. The pros of the method is it is the only sustainable and environmentally beneficial approach for producing the valuable compound from critically endangered plant species that are on the verge of extinction. The biotechnological approach of in vitro culture within disciplinary subjects like molecular biology, enzymology, and fermentation technology is the current avenue of research for enhanced metabolic production on a large scale (Abdin 2007; Hussain et al. 2012; Ramirez-Estrada et al. 2016).

Scaling up of secondary metabolite production through in vitro culture consists of two steps as follows:

Biomass production and secondary metabolite synthesis.

19.3.1 Types of In Vitro Culture

1. Callus or suspension culture
2. Shoot culture
3. Root culture (hairy root) culture

An uninterrupted and uniform supply of chemicals from the plant without any human intervention and technology is not possible. Realizing that it was not enough to take chemicals directly from nature and that it was not sustainable, humans began to look for other ways to obtain chemicals, and such research brought them to in vitro synthesis of plants and their products. Such an in vitro synthesis also has the advantage of being able to synthesize chemicals beyond geography and climate (Espinosa-Leal et al. 2018). Nowadays, there are different strategies for scaling up the secondary metabolites through in vitro culture.

An organized structure like shoot, roots are used in compound production, but the unorganized mass of cell called "callus" produced from a sterilized explant with induction of different concentrations of auxins, cytokinins or by their combinations is more reliable than that of any other biomass having well organization (Chandran et al. 2020). The induced callus can be used for the establishment of single-cell suspension culture for enhanced metabolite production using bioreactors. The callus is also used for clonal propagation through a standardized micropropagation procedure (Fig. 19.1).

Fig. 19.1 In vitro propagation in *Gloriosa superba*

Similar to the callus culture, in vitro tuber induction and hairy root culture were suggested for the production of secondary metabolites like colchicine, rosmarinic acid, aconitine anthraquinone, artemisinin, and baicalin; which are relevant in different pharmacological activities. The rosmarinic acid is first isolated from the plant *Rosmarinus officinalis*. With the increasing demand for the target compound, the rosmarinic acid production gets increased using cell and suspension culture. A successful enhancement and elicitation of anthraquinones, phenolics, and coumarin glycosides were achieved from *Polygonum multiflorum* hairy root culture. The commercially relevant indigo dye isolation from *Polygonum tinctorium* was increased with in vitro culture. All the current research on hairy root culture for the enhanced production of SMs is still on a small scale. The large-scale production using bioreactors and implementation of pilot plants of hairy root and cryostorage are still incomplete, and implementation of new technologies is essential (Babich et al. 2020).

Shoot cultures grown in vitro have long been studied as a possible source of high-value compounds. In vitro shoot culture is also a suitable producer molecule of interest, such as lead compounds, antioxidants, and flavoring agents, similar to cell suspension cultures. In vitro shoots preserve the original tissue differentiation of plants; hence they can typically biosynthesize secondary compounds that are not seen in disorganized cell suspensions. On the other hand, large-scale production of shoot cultures is difficult and necessitates the use of specialized bioreactor equipment (Krol et al. 2020).

Different successful organ cultures are reported, like root, root tubers and shoot cultures for the compound isolation, and their metabolic profiling was similar to that of the wild plant (Fig. 19.2). Hairy root culture is another reliable and effective method of in vitro culture to produce the compounds in the root of a plant. The bacteria called *Rhizobium rhizogenes* (*Agrobacterium rhizogenes*) is responsible for the syndrome in plants called hairy root (HR). Hairy roots from a series of plants can produce complex active glycoproteins and other secondary metabolites. The

Fig. 19.2 In vitro tuber induction in *Gloriosa superba*

versatile nature makes the hairy root an excellent tool for custom-made molecules using genetic engineering (Gutierrez-Valdes et al. 2020).

19.3.2 Traditional Strategies

Secondary metabolites are produced from the primary metabolic pathway as a side chain. The yield of the secondary metabolite is determined by the production rate of substrates from core metabolic pathways that are rerouted to secondary biosynthetic pathways. Generally, in a controlled system, the synthesis is influenced by both biotic and abiotic factors such as growth and physiology, temperature, humidity, and intensity of light. The rate of metabolism in in vitro culture is also influenced by the composition of culture media, pH, inoculum density, and culture conditions such as temperature, light density, agitation, and aeration (Fig. 19.3). Hence, for the maximum yield of the targeted compound, these parameters must be standardized. The selection of an appropriate culture medium is also critical in the production of secondary metabolites. Plant growth regulators (PGRs) such as cytokinins, auxins, gibberellins; stress inducers like jasmonates, and salicylates, as well as macro- and micronutrients, vitamins, carbohydrates (sugars), and amino acids, regulate secondary metabolite synthesis in a controlled condition (Chandran et al. 2020).

19.3.3 Elicitors

Elicitors are the biotechnological tool or exploiter of the plant defense mechanism. The synthesis of secondary metabolites gets elevated by enhancing the secondary metabolic pathways by giving stress, while the plant cells are protected from stress. Based on the source, the elicitors could classify as biotic or abiotic (Castro et al. 2021).

Fig. 19.3 Cell suspension culture for secondary metabolite production

19.3.3.1 Biotic Elicitors

Biotic elicitors (BEs) are the elicitors originating from biological sources. The BEs could be categorized as exogenous or endogenous based on their site of origin (Table 19.1). The elicitors like pectin, dextran, chitin, chitosan, and cell wall components like oligo or polysaccharides, Glycopeptides, *N*-acetylglucosamine oligomers, β-glucan, and ergosterol from fungus, bacteria, viruses, or herbivores

Table 19.1 Biotic elicitors for the production of secondary metabolite in plant in vitro cultures

Target compounds	Elicitor
Atropine	*Bacillus cereus*
	Staphylococcus aureus
Azadirachtin	*Claviceps purpurea*
Artemisinin	Mycelia homogenates of *Penicillium chrysogenum*
Cryptotanshinone and tanshinone IIA	Yeast extract
Ginseng saponin	
Tanshinones and	
Phenolic acids	
Artemisinin	
Tropane alkaloids	
Total flavonoids	
Glucotropaeolin	
Plumbagin	
Silymarin	
Glycyrrhizic acid	*Mucor hiemalis*
	Rhizobium leguminosarum
Gymnemic acid	*Aspergillus niger*
	Saccharomyces cerevisiae
	Agrobacterium rhizogenes
	Bacillus subtilis
	Escherichia coli
Hypericin and pseudohypericin	Chitin
	Pectin
	Dextran
Phenylpropanoid and naphtodianthrone	*Fusarium oxysporum*
	Phoma exigua
	Botrytis cinerea
Tanshinone	*Trichoderma atroviride*
Thiophenes	Crude extracts of *Fusarium conglutinans* mycelia
	Polysaccharides from *Bacillus cereus* cells
Rosmarinic acid	Crude extracts of *Phytophthora cinnamon* mycelia
Thiarubrine A	*Protomyces gravidus*
trans-Resveratrol and viniferins	

Source: Wang and Wu (2013); Naik and Al-Khayri (2016)

are exogenous (Chandran et al. 2020; Castro et al. 2021). Out of these elicitors, oligo or polysaccharides are considered to be the finest elicitors for metabolite production since they make signal transduction similar to that of a pathogen attack (Chandran et al. 2020; Patel and Krishnamurthy 2013; Zhao et al. 2005).

Signal transduction in the elicitation of a compound consists of a series of defense mechanisms in the plasma membrane, and all the signaling cascade is complete within a few minutes to hours. It includes the alternation of protein function in the plasma membrane, fluctuation in concentration of cytosolic and nuclear calcium, and the formation of active oxygen species and nitric oxide and reversible phosphorylation of protein (Garcia-Brugger et al. 2006).

Endogenous elicitors are the chemicals produced by the plants themselves to produce secondary metabolites during pathogenic attacks or infections. In plants, cell death and disintegration lead to the release of Reactive oxygen species (ROS), oligosaccharide fragments, and protein fragments, and these groups are also the endogenous elicitors (Yamaguchi and Huffaker 2011; Chandran et al. 2020). The recognition of these molecules not only flags the local area of infection but also boosts the defense mechanism of the plant by acting as a signal to produce the same molecule through a controlled enzymatic pathway, like a positive regulator, even though there may be a neogenesis of a compound happen.

In plants, the elicitors initiate signal transduction responsible for secondary metabolite production. Primarily, the elicitor molecule was received by the receptor of the plasma membrane. And successive depolarization of plasma membrane activates the membrane channels such as K^+/H^+ antiport channels. Cytoplasmic acidification results in the activation of K^+/H^+ and cl^- efflux, which initiates the signaling of secondary metabolite production (Zhao et al. 2005; Narayani and Srivastava 2017). In another way, the elicitors activate G-proteins coupled with the receptor or the mitogen-activated protein kinase (MAPK) cascade. The sequential phosphorylation activates the MAPK cascade. The Ca^{2+}/ion fluxes may be activated due to the increase of cytoplasmic free calcium ions. This also activates Ca^{2+} dependent protein kinases (CDPK), membrane-bound enzymes responsible for defense mechanism, and protein phosphatases result in secondary metabolite production (Fig. 19.4).

Salicylic acid (SA) and methyl jasmonate (MeJA), and jasmonate (JA) are well-recognized compounds for triggering the expression of plant genes committed for diverse metabolic processes, resulting in a broad spectrum of chemicals. In general, MeJA and JA are responsible for the increased production of flavonoids, alkaloids, terpenoids, and phenylpropanoid compounds (Akula and Ravishankar 2011; Baenas et al. 2014). They are also known as "hormones" because they may be employed on plants in a number of ways and stimulate cellular responses at low concentrations distant from their source of production. Their application in different concentrations to different species could induce different compound production: the anthocyanins and phenolic production enhanced in black raspberry fruits by the application of MeJA, abscisic acid (ABA), and anthocyanin content of "Fuji" apples get increased by the application of the *N*-propyl dihydrojasmonate (PDJ), a derivative of JA (Gemma 2000; Baenas et al. 2014).

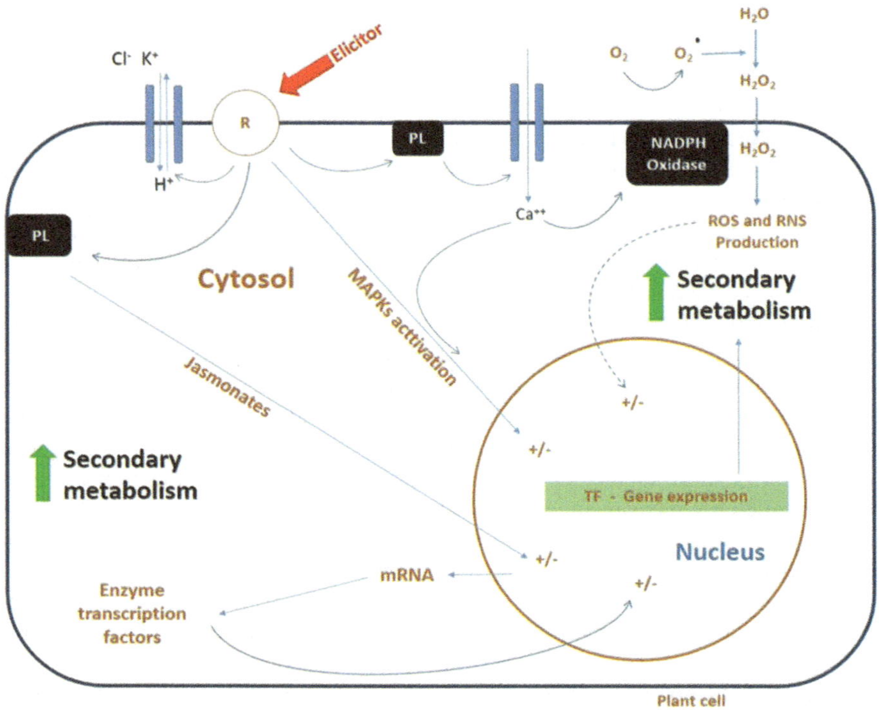

Fig. 19.4 Signal transduction for secondary metabolite production

19.3.3.2 Abiotic Elicitors

Abiotic elicitors are non-biological in origin, and they may be physical or chemical. Cold shock, UV, osmotic, and water stress are a few of the physical stresses, and salts of heavy metals, metal ions, metal oxides, and other inorganic substances are a few of the chemical stresses to induce secondary metabolites and enzymes (Chandran et al. 2020).

Cell suspensions and hairy roots are the usual and widely followed in vitro culture methods for the enhanced metabolite production and jasmonic acid (JA), methyl jasmonate (MeJA), salicylic acid (SA), acetyl salicylic acid (ASA), ethylene (ET) and ethrel (Ethe), heavy metals (HM), and natural or synthetic chemical compounds are the usual and widely following chemical elicitors (Table 19.2). In addition to the quantitative enhancement, the elicitors influence the quality of the secondary metabolite. The elicitors added to an in vitro system simulate the environmental condition with stress factors responsible for activating plant defense mechanisms and successive metabolite production. Di-methyl sulfo-oxide (DMSO), pyrazine 2-carboxylic acid, benzo (1, 2, 3) thiadiazole-7-carboxylic acid-s methyl ester (BION), eugenol, polyamines, morphactin, sodium acetate, sorbitol, benzothiadiazole, saccharin, hydrogen peroxide, silver nitrate, triton-x,

Table 19.2 Abiotic elicitors for the production of secondary metabolite in in vitro cultures

Target compound	Biotic elicitors
Amentoflavone and agathis flavone	Sucrose (30 g L^{-1}) + 2,4 D (5 mg L^{-1}) + Dark
Andrographolide	MeJA (5 μM)
Anthraquinone	MeJA (100 μM)
Baicalein and wogonin	Light (continuous) + 2βmethyl-cyclodextrin (15 mM)
Bergapten	MeJA (100 μM) + 2βMethyl-cyclodextrin (50 mM)
Emodin and physcion	Jasmonic acid (100 μM)
	Salicylic acid (100 μM)
Eugenin	2βMethyl-cyclodextrin (50 mM)
	MeJA (60 mg L^{-1})
Eugenol and methyl eugenol	CuSO4 (25 μM)
Flavonoid	Benzyladenine (8.8 μM) + Glutamine (40 μM)
Gallic acid, myricetin, caffeic acid, catechin and apigenin	Thidiazuron (1 mg L^{-1})
Hydroxycinnamic acid	Coronatine (1 μM)
Isoflavones	2,3,5-triiodobenzoic acid (0.5 mg L^{-1}) + kinetin (5 mg L^{-1})
Phenolic acids and flavonols	Polyethylene glycol (30%)
Phenolic compounds	Methanol (1.5% v/v)
	MeJA (100 μM)
	MeJA (0.5 mg L^{-1})
	Salicylic acid (50 mg L^{-1})
	Salicylic acid (120 μM)
	MeJA (50 μM)
	MeJA (3 μM)
	Salicylic acid (50 μM)
Phenolic compounds Anthocyanins	Naphthalene acetic acid (10 μM)
Phenylpropanoids and naphtodianthrone	Fungal mycelia extracts (50 mg L^{-1})
	Polysaccharides (chitin, dextran and pectin) (100–200 mg L^{-1})
Phthalides (3-Butylidenephthalide)	Light (white)
Plumbagin	Jasmonic acid (50 μM)
Polyphenols	Light quality
Rosmarinic acid	MeJA (11.2 ppm) + Ag + (2.5 ppm)
	Ozone (200 ppb–3 h)
Rosmarinic acid and methoxylated flavonoids (xanthomicrol and cirsimaritin)	SiO2 (100 mg L^{-1})
Salicylic acid, kaempferol, and quercetin	MeJA (0.1 mM)

(continued)

Table 19.2 (continued)

Target compound	Biotic elicitors
Silymarin	MeJA (20–80 mg L^{-1})
	Salicylic acid (20–40 mg L^{-1})
Tanshinone II A	MeJA (200 μM)
Tanshinone II A Rosmarinic acid	Ag + ions (15 μM) Ag + ions (2.5 ppm)
Tanshinone II A Vinblastine and vincristine	AgNO3 (25 μM) AgNO3 (100 μM)
Wogonin and baicalein	2βMethyl-cyclodextrin (15 mM) + light
Xanthone	Acetic acid
β-sitosterol, campesterol, stigmasterol	Benzothiadiazole
Crocin, picrocrocin, and safranal	Polyethylene glycol (PEG)
	Bioregulator prohexadione
Jadomycin	Ethanol
	Ethene
	Inorganic salts: mercuric chloride (HgCl2), copper
	Sulfate (CuSO4), calcium chloride (CaCl2), and
	Vanadyl sulfate (VSO4) [28]
Artemisinin, Protocatechuic, caffeic, *p*-coumaric and chlorogenic acids	As (metal ions)
Total flavonoids (Iristectorigenin A, tectorigenin and tectoridin) Eremophilane type sesquiterpenes and Anhuienoside, Anthocyanins, sinapoyl esters, Hydroxycinnamic acid (5-caffeoylquinic acid and caffeoylquinic acid), cyanidins and flavones	Cu (metal ions)
Ellagic acid, quercetin, kaempferol and epicatechin	Cd(metal ions)
Total alkaloids, phenolic compounds and flavonoids	
Phyllanthin and hypophyllanthin	
Phenolic compounds and anthocyanins	
Chlorogenic acid	
Phenols, flavonoids and lignins	
Total phenols and flavonoids	Ba (metal ions)
p-hydroxybenzoic, *p*-anisic, hesperetic and chlorogenic acids	Ni (metal ions)
Hypericin	Cr (metal ions)
Protocatechuic, *p*-hydroxybenzoic, vanillic, *p*-coumaric, caffeic and gallic acids	
Ferulic acid, salicylic acid, daidzein, vitexin and phenolic acids	Pb (metal ions)
Cynaroside	UV-A/B
Flavonoids	UV-B
Quercetin,	

(continued)

Table 19.2 (continued)

Target compound	Biotic elicitors
Kaempferol,	
and isorhamnetin	
Catharanthine and vindoline	
Anthocyanin	
Catharanthine and vindoline	
Anthocyanin	
trans-Resveratrol	UV-C
Chlorogenic acid,	
Kaempferol, and quercetin	

Source: Patel and Krishnamurthy (2013); Giri and Zaheer (2016); Hashim et al. (2021); Baenas et al. (2014); Wang and Wu (2013)

copper sulphate, ancymidol, cyclodextrins, methylated cyclodextrins, cantharidin, putrescine, glutathione, heavy metal ions (HM), Ca^{+2} cAMP, cantharidin are some other chemical elicitors (Giri and Zaheer 2016; Ho et al. 2020).

The visible light alone and with other elicitors and plant hormones has a significant role in the yielding of medicinally essential metabolites, in particular alkaloids, phenolics, flavonoids, terpenoids, and other volatile organic compounds (VOCs) like ascaridole (de Carvalho et al. 2020; Hashim et al. 2021). By varying light sources, like fluorescent sources, light from light-emitting diodes (LEDs), and ultra violet (UV), the quality and quantity of the metabolites also differ. The UV band in the range is classified into three: UV-A, UV-B, and UV-C. The diversity of UV light is of no great importance in a biological system. Klein et al. (2018) explained the more than 50% increased level of betacyanin and betaxanthin in certain species of *Alternanthera,* during the UV-B irradiation. UV-C band having a wavelength below 280 nm cannot cross the ozone layer. The studies of Schreiner et al. (2012) with the artificial UV-C source resulted in an elevated *trans*-resveratrol content in the callus culture of *Vitis vinifera.*

The visible light spectrum sources like fluorescent and LED are great possibilities as an elicitor. The phenolics, flavonoids, glucuronides, baicalin, wogonoside, verbascoside, rutin, ginsenoside and hypericins (Yu et al. 2005; Ahmad et al. 2016; Fazal et al. 2016; Kawka et al. 2017; Kapoor et al. 2018; Sobhani Najafabadi et al. 2019) have been enhanced in different in vitro cultures grown in a different wavelength of light like blue, white, and green lights. In vitro studies for the enhanced metabolite production in *Stevia rebaudiana, Prunella vulgaris,* and *Scutellaria lateriflora* substantiate that the different spectrum of visible light waves has the property to enhance the secondary metabolites in them. Compared to fluorescent light, LED lights are more promising for elicitation and photomorphogenesis studies (Dutta Gupta and Pradhan 2017). This is because of the wavelength of the wide range of fluorescent light (350–750 nm), and the LED light can be more precisely controlled than the wavelength of ordinary fluorescent light (Dutta Gupta and Pradhan 2017). As well as the hike in the compound production concerning the light intensity, the production of certain compounds like bioflavonoids increases in reduced light (Castro et al. 2021). The study in response to the enhanced metabolite production of *Piper aduncum* and *Epimedium pseudowushanense* also resulted that the increased metabolite production of some flavonoids is observed in low light intensity (Idris et al. 2018).

Salinity is the accumulation of salts. The high salt content in the niche puts stress on the organisms through osmotic, ionic, or nutritional imbalance and affects the normal growth of the organism. For the adaptation of the plant to such a stressed environment, the plant itself diverges or develops specific biochemical mechanisms. Synthesis of new solute molecules brings alteration in the activity of enzymes committed for antioxidant activity and inorganic ion concentration, production of secondary metabolites (Muchate et al. 2019). Out of all modifications, the production of secondary metabolites in response to saline stress can be positively exploited to enhance metabolite production in the in vitro culture.

Salinity has a physiological, biochemical, morphological, and biosynthesis impact on plant natural product profiles by inducing oxidative stress and defense response pathways, which results in the production of reactive oxygen species (ROS), which plays a vital role in enhanced secondary metabolite production in medicinal plants (Gupta and Huang 2014). The molecular and physiological mechanisms of growth and production of SMs on crop species have been studied in the recent decades. Nevertheless, information about medicinal plants, predominantly varying stress and successive defense responses connected with their development, is at present incomplete (Isah 2019).

Metal ions are essential for plants to live a long and healthy life (Singh et al. 2011). Even in plants, the SMs production will get enhanced if the stress is induced by metal ions (Hurmat and Bansal 2020; Anjitha et al. 2021). The metal ions as well as other abiotic elicitors for the production of SMs like flavonoids, phenolics, terpenes, steroids, and alkaloids is listed in Table 19.2.

19.3.3.3 Microbial Elicitors

Microbial cultures and microbial products are nowadays used as an elicitor to produce desirable compounds in in vitro cultures. Proteins, polyunsaturated fatty acids, and oligosaccharides have been found as a few fungal elicitors (Bhaskar et al. 2021). Plant growth-promoting rhizobacteria (PGPR) are a cluster of different bacterial species responsible for eliciting SMs in plants (Prasad et al. 2019; Chamkhi et al. 2021). Compared to other biotic elicitors, shorter growing periods and fewer steps in their manufacture make the bacterial elicitors more optimistic (Table 19.3). Certain bacteria produce elicitors that stress-specific species of plants to elicit SMs, and some have an array of compounds responsible for eliciting particular compounds in different plant species (Chamkhi et al. 2021).

Table 19.3 Bacterial elicitor and bacteria for the production of secondary metabolite in in vitro cultures

Bacterial elicitor and bacteria	Target compound
Gram-negative bacteria	
Peptidoglycan and muropeptides (Agrobacterium rhizogenes)	Saponin and phenolic compounds
	Gymnemic acid
	Glycyrrhizic acid, xanthone, phenolics, and flavonoids
	Phenolics and flavonoids
	Artemisinin
	Phenylpropanoids
	Aryltetralin lignans
	Alizarin and purpurin
Flagellin	Xanthone, quercetin, quercetin, hypericin and hyperforin
Massetolide A and siderophores such as pseudobactin	Phenolic compounds, flavonoids, and anthocyanins
	Hypericin and pseudohypericin
	Hyoscyamine and scopolamine
	cis-Thujone, camphor, and 1,8-cineole
	Terpinene-4-o1, *cis*/*trans* sabinene hydrate, and α-terpineol
	Monoterpenes and phenolics
	Ajmalicine, catharanthine, tabersonine, serpentine, and vindoline
	Gallic acid and ferulic acid
	Isoflavone
	Essential oil yield
Pyocyanin, pyochelin, and Rhamnolipids (Mono-/dirhamnolipids)	Gallic acid and ferulic acid
	Scopolamine
	Hyoscyamine and scopolamine
	cis-Thujone, camphor, and 1,8-cineole
	Stevioside
	Paclitaxel, cephalomannine, 10-deacetyltaxol (Taxus globosa), taxol, and baccatin III (Taxus media)
	Glyceollins (isoflavonoid)
	Benzo[c]phenanthridine (alkaloid)
	Paclitaxel (Taxane)
	Sakuranetin and momilactone A
	HydroXycinnamic acids (caffeic acid, isoferulic acid, *p*-coumaric acid, and sinapic acid)
	Phytosterols (campesterol and β-sitosterol)
	6-Methoxy podophyllotoxin (Aryltetralin-lignans)

(continued)

Table 19.3 (continued)

Bacterial elicitor and bacteria	Target compound
Coronatine derivatives: Coronalon and indanoyl-L isoleucine (In-Ile) Protein (more than 10 kDa)	Total flavonoids, glycyrrhizic acid, and glycyrrhetinic acid
	Diosgenin
	Atropine
	Gymnemic acid
HrpEharpin, HpaXpm, Lipopolysaccharides (LPS) (Lipid A/inner core/glucosamine backbone, peptidoglycan, and muropeptides	Gymnemic acid
Harpin proteins (HarpinXoo)	Phytoalexin
AMEP412 protein, fengycins and surfactins, Acetoin (3-hydroxy-2-butanone), volatile compounds blends	Terpinene-4-o1, *cis/trans* sabinene hydrate, and α-terpineol
	R-terpineol and eugenol
	Pyrethrin
	Diosgenin
	Carbazole alkaloids
	Gymnemic acid
	Secondary metabolites
Dimethyl sulfide, and polysaccharide–protein fractions	Scopolamine
	Phenolic and tanshinones
	Atropine
	Glycyrrhizic acid
PeBA1 protein and lipopeptides	Phenolic compounds
Cyclodextrin	Total taxane
	Silymarin
	trans-Resveratrol
PEBL1 protein	Phenolic compounds
Peptidoglycan (muramyl dipeptide)	Scopolamine
	Bilobalide and ginkgolide A & B Atropine and scopolamine

Source: Chamkhi et al. (2021)

19.3.3.4 Bacterial Coronatine Elicitor

Coronatine is a phytotoxin produced by *Pseudomonas syringae*. They are responsible for eliciting a series of compounds from different plants, like taxol (paclitaxel) from taxus species, isoflavonoids from *Glycine max* species. Coronatine derivatives like indanoyl-L-isoleucine also impart a role in the elicitation of certain compounds like 6-methoxypodophyllotoxin (aryl tetralin-lignans). Tamogami and Kodama (2000) reported that flavonoid phytoalexins sakuranetin and momilactone in the *Oryza sativa* can be enhanced with coronatine (Zhai et al. 2017). In *Lemna paucicostata,* the elevated concentration caffeic acid, isoferulic acid, *p*-coumaric acid, sinapic acid campesterol, and β-sitosterol were reported in the presence of phytoalexins (Kim et al. 2016; Zhai et al. 2017).

19.3.3.5 Bacterial Protein Derivatives Elicitors

Bacterial proteins are also an ideal elicitor for secondary metabolites. PeBL1 protein isolated from *Brevibacillus laterosporus* enhanced the phenolic production in *Nicotiana benthamiana.* Proteins having a size greater than 10 KDa from the *Escherichia coli* are a good enhancer of glycyrrhetinic acid, glycyrrhizic acid, and flavonoids in *Glycyrrhiza uralensis.* A protein complex fraction from *Bacillus cereus* is also a good elicitor of the pharmacologically relevant compound tanshinone in *Salvia miltiorrhiza* (Chamkhi et al. 2021).

Enhanced yield in the elicitation always makes a culture successful. Biotic elicitors like fungal elicitors also have a role in the enhancement of secondary metabolites in commercial in vitro culture experiments. Nowadays, pathogenic fungus is being used as elicitors (Bhaskar et al. 2021; Chamkhi et al. 2021). The fungal cultures in the suspension impart a role of activation of pathogenic response in the plant cell culture, through that induct the enhancement of secondary metabolite production (Wang and Wu 2013). The fungal elicitors are somewhat not much efficient in the case of the symbiotic association because the symbiotic association and successive elicitation are problematic in the synthetic media (Table 19.4).

Table 19.4 Fungal elicitor for the production of secondary metabolite in in vitro cultures

Fungal elicitor	Secondary metabolite
Phytophthora megasperma	Glyceollin
Phytophthora megasperma Alternaria carthami	Polyacetylenes
Verticillium albo-atrum	Phytoalexins
Rhizopus arrhizus	Diosgenin
Altern	Diosgenin
Aria tenuis	*p*-Hydroxybenzoic acid
Pythium aphanidermatum	Anthocyanin
Aspergillus flavus	Sesquiterpenes
Rhizoctonia solani	Artemisinin
Penicillium chrysogenum	Rosmarinic acid
Pythium aphanidermatum	Rosmarinic acid
Aspergillus niger	Thiarubrine A
Protomyces gravidus	Coumarin
Phytophthora parasitica	Hypericin
Yeast	Taxol
Aspergillus niger	Salidroside
Aspergillus niger, Coriolus versicolor, and Ganoderma lucidum	Andrographolide
Ganoderma lucidum and Trichoderma harzianum, Fusarium sp.	Euphol
Gilmaniella sp.	Atractylone gymnemic acid
Aspergillus niger	Glycyrrhizic acid
Mucor hiemalis, Fusarium moniliforme, and Aspergillus niger	Atractylone gymnemic acid

Source: Chamkhi et al. (2021)

Piriformospora indica is an endophytic that currently overcomes the difficulties of axenic culture with fungal elicitors (Smetanska 2009).

19.3.3.6 Algae as Elicitor

Algae or algal products are a promising elicitor nowadays. Marine algae are the significant contributors as elicitors. About 40–69% of the thallus of algae is made of carbohydrates (Arman and Qader 2012; Bhaskar et al. 2021). In addition to the role of elicitation, algae are also a good growth enhancer due to their macro, microelements, and active chemical composition in it. The culture of species from nostocaceae (*Anabaena sps.* and *Nostoc carneum)* is an ideal elicitor of neem cell suspension culture for azadirectin. Microalgae like *Nostoc carneum* are used as an elicitor to enhance pigments in *Carthamus tinctorius* (Bhaskar et al. 2021). Table 19.5 shows a few algal elicitors and their target compounds. 1.2-fold thiophene production increased in the culture of *Tagetes patula* supplemented with *Haematococcus pluvialis*, green algae (Rao et al. 2001).

19.3.4 Co-culture System with Plant, Fungus, and Bacteria

An enhancement strategy in secondary metabolite production by culturing the host plant culture with another living culture system is called a co-culture system (Narayani and Srivastava 2017; Sun et al. 2021). Compared with other elicitation strategies, the co-culture system is a continuous elicitation of SMs. But the difficulty of this system is the simultaneous establishment and management of the host culture and elicitor in a single vessel. The vessel is a combination of two tanks with individual agitation and monitoring system. The two individual tanks are separated by a hydrophilic pyroxylin filter (Narayani and Srivastava 2017).

Table 19.5 Algal elicitor for the production of secondary metabolite in in vitro cultures

Algal elicitor	Secondary metabolite
Anabaena sp.	Azadirachtin
Anabaena cylindrical, Anabaena variabilis	Kinobeon A (Red pigment)
Botryococcus braunii	Vanillin, Vanillylamine Capsaicin
Chlorococcum sp., *Chlorella* sp.	Kinobeon A (Red pigment)
Haematococcus pluvialis	Betalain, Thiophenes
Kappaphycus alvarezii	Picroside-I
Nostoc carneum	Azadirachtin, Kinobeon A (Red pigment)
Nostoc linckia	Kinobeon A (Red pigment)
Spirulina platensis	Capsaicin, Anthocyanin, Betalain, Thiophenes

Source: Bhaskar et al. (2021)

Co-culture of intra- and interspecies was successfully reported in the production of xanthotoxin from root culture of *Ammi majus* and *Ruta graveolens* cell suspension culture. Scopolamine was produced from a co-culture of *Atropa belladonna* and *Duboisia* hybrid shooty teratomas (Mahagamasekera and Doran 1998; Narayani and Srivastava 2017).

There will be a symbiotic relationship in the co-culture of plants and fungus. The enhanced production of podophyllotoxin and 6-methoxypodophyllotoxin was achieved in a co-culture of *Linum album* with fungi; *Piriformospora indica* and *Sebacina vermifera* (Baldi et al. 2008; Narayani and Srivastava 2017). An increased caffeic acid production was achieved by the co-culture of *Ocimum basilicum* (host) and *Rhizophagus irregularis* (fungi) (Srivastava et al. 2016; Narayani and Srivastava 2017). The most essential pharmacologically relevant paclitaxel was observed to be increased 38-fold in a co-culture of *Taxus chinensis* with *Fusarium mairei* (Li et al. 2009; Narayani and Srivastava 2017). The bacterial culture can also be used in co-culture. Tanshinone content in *Salvia miltiorrhiza* hairy root culture gets increased with the co-culture of *Bacillus cereus* (Wu et al. 2007).

Exogenous elicitors originating from fungus, bacteria, viruses, or herbivores (exogenous elicitors) or endogenous elicitors derived from the plant itself (endogenous elicitors) are defined as compounds of biological origin that promote production of metabolites in a live cell system. Their method of action is linked to the activation or inactivation of enzymes or ion channels via receptors. (Castro et al. 2021).

19.3.5 Metabolic Engineering

The production of desired secondary metabolites on a large scale is always a complex and expensive method, and it has been inevitable since the existence of human life. Elicitation in in vitro culture will undoubtedly enhance the undesired compounds with the desired ones. This can be avoided through genetic engineering and synthetic biology (Birchfield and McIntosh 2020). Genetic engineering is a fast-growing field of biotechnology. For the desirable purpose, the genetic code gets altered here. The increased metabolic content can be achieved through this technology. Such a purposeful modification can be achieved through the direct introduction of the gene via gene guns and vectors like *Agrobacterium tumefaciens.* The vector-mediated transformation is restricted to certain dicotyledons only (Verpoorte et al. 2000). Even though *A. tumefaciens* has been used as a successful strategy for the transformation of plants. In the in vitro system, the hairy root resulting from the infestation of *A. tumefaciens* will enhance the secondary metabolites. Such a transformation has tremendous application, especially for enhanced metabolic production. In Genetically modified plants, the desired character gets expressed more, and undesirable get suppressed or less expressed, similarly the unwanted compound production can be avoided in the enhanced metabolite production strategies using genetic engineering. Metabolic engineering not only increase the metabolic level but also can introduce a new compound.

The prime step of metabolic engineering is the mapping of the target biosynthetic pathway. Even though the existing knowledge and confirmation with ^{13}C isotope labeling could reveal the metabolic pathway, targeting is still difficult to attain (Kumar and Prasad 2011; Clendinen et al. 2015). A second step is a computational approach; here, the *in-silico* model of the identified pathway will be designed. "Knock" is one such approach to identify the gene of interest responsible for the upregulation and downregulation of metabolites. The third step is the genetic manipulation of the desired gene using mutagenesis or recombinant DNA (rDNA) technology. Subsequent metabolic flux analysis and their control make sure the end product and its purity. There are a series of approaches that can be executed for the metabolite engineering for desirable from microbes and plants (Kumar and Prasad 2011).

19.3.5.1 Heterologous Expression of Entire Gene Clusters

The transfer of the whole metabolic pathway to another host makes it able for heterologous production of the target compound. In 1997, Hong et al. incorporated a tetrangulol- and tetrangomycin-producing gene to the *Streptomyces rimosus* NRRL 3016 and *Streptomyces strain* WP 4669. Polyhydroxybutyrates (PHBs) are an energy and carbon storage form in many microorganisms in the scarcity (Du and Webb 2011). In 2007, Abd-El-haleem et al. developed a genetically modified *Saccharomyces cerevisiae* that is able to synthesize PHB, by adopting the pathway of *Ralstonia eutropha.*

19.3.5.2 Engineering Regulatory Networks

The metabolite production could up/downregulate using genetic engineering. In general, microbial metabolite production is regulated using regulatory genes responsible for up/downregulation. The regulatory mechanism is a hindrance to the production of the desired compound on a large scale. In metabolic engineering, the regulatory network is modified to produce the metabolite continuously even if it is not necessary, like unrestricted production in *Streptomyces coelicolor* is an example of upregulation mechanism. Correspondingly, increased chromomycin in *Streptomyces griseus* is an example of a downregulation approach. Here the pathway of repressor production is modified to avoid the production of the repressor. Such engineering of networks is not restricted to the field of molecular biology; instead, it flourished by using the benefits of computational biology and in silico approaches. Metabolic engineering is a rational approach to enhanced metabolic production. However, the lack of a well-structured metabolic pathway makes the approach of metabolic engineering more difficult (Kumar and Prasad 2011).

19.3.5.3 Gene Insertion and Deletion

Non-native pathways may induce metabolic imbalance. Such an imbalance can be brought back to the regular mode by deleting or adding one or more genes. Similarly, by adding or deleting one or more genes of a genome, the metabolic pathway can be regulated to produce the molecule of interest.

19.3.5.4 Redirecting Metabolic Pathway

A metabolic pathway can be directed to produce a desirable product by making changes in the pathway. This altered metabolic pathway is the simultaneous result of transcriptomics, metabolomics, proteomics, computational biology, and fluxomics.

19.3.5.5 Stimulation by Precursors

The level of precursors is a limiting factor of secondary metabolites. It may go through either by induction of enzymes responsible for biosynthesis or by regulating the limiting precursor. Usually, the inducers are amino acids and small molecules. In such a simulation, it should be clear that the effect is either due to the elevated precursor supply or due to one or more biosynthetic pathways.

19.3.5.6 Genetic Knockout of Loci

Metabolic engineering can be used to introduce a novel pathway and alter an existing pathway. Knock out (KO) is a promising strategy for enhanced metabolite production even though the KO is supported by a series of "omics" and is not well investigated (McCloskey et al. 2018). CRISPR/Cas9-mediated genome editing in plant in vitro systems is also an efficient technology. Such a mutation can be heritable to the next generations also (Marchev et al. 2020).

19.3.5.7 Quorum Sensing

Gene expression can be regulated as per the cell density in culture due to quorum sensing. The bacteria having quorum sensing will produce chemicals that elevate in concentration as cell density is enhanced is called autoinducers. Such autoinducers are responsible for the altered gene expression in the bacteria (Barnard et al. 2007). It includes a series of activities like symbiosis, competence, virulence, conjugation, motility, antibiotic production, sporulation, and biofilm formation. The quorum sensing may be intra or interspecific (Miller and Bassler 2001). The cell factories are the metabolically engineered cells of plants or microbes that are employed for the overproduction of desired metabolites (Hermann et al. 2018; Navarrete et al. 2020). The high-demand antibiotics and metabolites can be industrially produced using this. One of the significant applications is insulin production from the recombinant *S. cerevisiae* and *Escherichia coli* (Navarrete et al. 2020).

19.3.6 Nanoparticles for Secondary Metabolite

Nanoparticles are particles having a size range of 1–100 nm (Marslin et al. 2017). Currently, they are attaining relevance in different fields of medicine, pharmacology, and enhanced metabolite production. Metallic (Ag, Au, and Fe) and non-metallic (Zn, Ti, Cu) nanoparticles can be used for various purposes. The silver (Ag) nanoparticles attained great interest in the field of in vitro culture and elicitation of secondary metabolites. In in vitro plant culture, the Ag nanoparticles (AgNP) not only impart a single role of elicitor but also induce increased biomass production, shoot proliferation, and callus induction. The vascular system and apoplasts are the transporting channels of the NP. But the membrane transport is size specific to them. The NPs having 40 nm could transfer to plasmodesmata (Rahmawati et al. 2022). There are quite a few reports that the AgNP increases the phenolic content in the in vitro culture of *Caralluma tuberculata* and Sugarcane. Enhancement of capsaicin in *Capsicum* species is another achievement in the application of AgNP (Akash Bhat 2016). In most nanoparticle-mediated target enhancement, the reactive oxygen species (ROS)-mediated signaling is evident and similar to jasmonic acid-induced signaling (Marslin et al. 2017).

19.3.7 Scaling Up of Secondary Metabolites Using Bioreactor

With the increasing demand for pharmaceutical products, the production of secondary metabolites is also increasing. In vitro secondary metabolite production at the flask level is not feasible to keep pace with the uplifting demand. Bioreactors or fermenters with high production rates are the solutions, and currently, existing technologies and research focus on them. The large-scale production also focuses on the cell suspension culture other than somatic embryos and hairy roots due to the ease in different aspects of in vitro culture.

Bioreactors combine different sophisticated systems for a biological conversion. In a global sense, the conversion is for the production of enzymes, microbial products, or for the production of SMs from animal or plant cells. A bioreactor is similar to a chemical reactor, but the bioreactors will be specially designed for the enhancement of a metabolic pathway (Sharma and Shahzad 2013).

A successful bioreactor operation can produce secondary metabolites with low ensured stress, oxygen supply, and nutrient and hormone distribution. However, it is essential to exploit the data and knowledge of metabolic production and morphological differentiation of the candidate suspension to be undertaken on a large scale unless the production will be perilous.

Shikonin, barberine, and ginsenosides are pharmacologically relevant metabolites produced on an industrial scale using bioreactors. In the bio-pharma industry, the scale-up of ginseng, a widely used herbal medicine (Lu et al. 2009), was revolutionary. In addition to optimizing the culture condition of a scale-up, the real-time monitoring of the uneven biomass is also challenging. Regulation of the parameters, such as process timeline, validation, and characterization, is relayed by the desired quality of the product, and its validation from the relevant authorities is essential for an industrial scale-up. In 2004, US FDA had described the designing, controlling, and analysis of Process Analytical Technology (PAT). Automation in the bioprocess makes reliable results. Sensor-mediated monitoring of pH, viscosity, temperature, pressure, redox potential, O_2 and CO_2 supply, biomass and fluid levels, flow measurement, and control of media and gas are some of them.

The production stages of a bioprocess are the biomass production, synthesis, and successive elicitation followed by recovery of the target metabolite. The estimation of target metabolite and biomass production in bioprocessing is employed by different models based on the culture condition, type of culture, and scale-up strategy. In commercial production, the model is also adopted or designed with the feasibility of cost or raw materials and operational costs like growth monitoring, the labor of operation, and cost ease of product recovery. Such a developed or adopted model will be evaluated to identify the best productivity based on the mathematical models and analytical strategies (Isah et al. 2018).

The bioreactors are of different types based on operation: batch culture, fed-batch, chemostat, and continuous. A finite volume of media is the peculiarity of batch culture. However, the fed-batch reactors start with a limited volume of nutrient media, and later the medium is supplied without removing the medium from the system for attaining maximum production. In the chemostat type, the growing organism is supplied with a rich medium constantly, and simultaneously used medium is removed (Sharma and Shahzad 2013). In perfusion culture, the fresh media is supplied to the viable cells constantly, and the product is removed with the spent media while retaining the cells (Stanbury et al. 2017).

Bioreactors can be classified as microbial bioreactors, mammalian cell culture bioreactors, and plant cell or organ culture bioreactors. Nowadays, bioreactors are widely used for the large-scale production of metabolites (Fig. 19.5). Large-scale production using bioreactors have great advantages like control in supply of media, increased working volume, well homogenization, battery control through mechanical and pneumatic steering mechanism, better control in physical environment, enhanced yield, enhanced nutrient uptake compared to other suspension methods, fast yield and easiness in the target compound separation (Sharma and Shahzad 2013).

The conventional bioreactors used in a plant cell or hairy root cultures can be classified into the conventional, gas phase, and hybrid bioreactors. Conventional bioreactors can be classified as Liquid-Phase Reactor, STR (Stirred Tank Reactor), ALR (Air-Lift Reactor), BCR (Bubble Column Reactor) (Fig. 19.5), BTBR

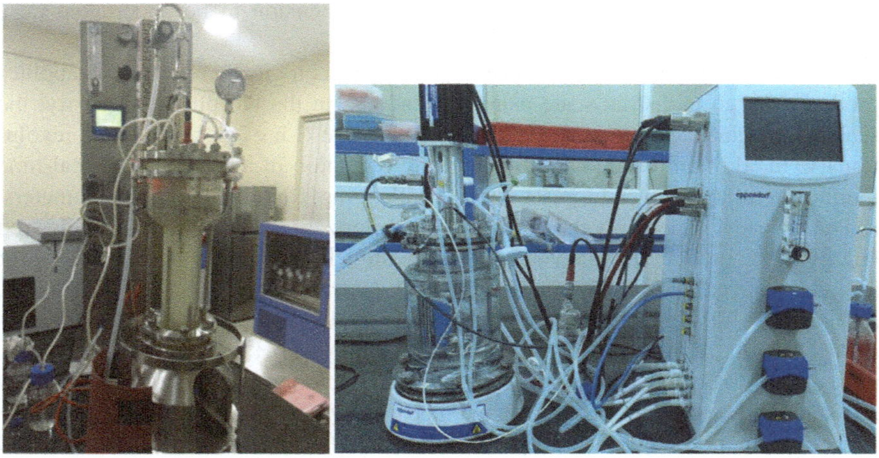

Fig. 19.5 Bubble column and marine impeller bioreactor

(Balloon-Type Bubble Reactor), EFR (Ebb and Flood Reactor), CFR (Convective Flow Reactor), TBR (Turbine Blade Reactor), and RDR (Rotating Drum Reactor). NMR (Nutrient Mist Reactor) is a Gas-Phase Reactor.

For the cell suspension studies and some microbial-related elicitation studies, plastic bags are used instead of permanent vessels. Such containers are made of food-grade plastics and can be reusable, like polyethylene, polystyrene, polyether urethane, and polypropylene (Sharma and Shahzad 2013). Disposable bioreactors can be classified as Wave-Mixed Bag Bioreactors, Stirred Bag Bioreactors, Pneumatically Driven Bag Bioreactors, Box-In-Bag Bioreactors.

19.3.8 Pilot Scale

In addition to the understanding of the cellular and regulatory systems that govern cell physiology, as well as the biophysical and biochemical features of products, the scale-up needs engineering of design and concept. Compared to other scale-up industries, biological products need more attention due to their complexity. In some cases, the system volume may reach up to thousands-to more than a lakh liter. An intermediate pilot-scale production is essential before a commercial establishment considering the financial investment, research and development, and labor investment (Cacciuttolo and Arunakumari 2005).

A pilot-scale is a pre-commercial plant for the scale-up strategy. Laboratory experiment for the enhanced production of SMs is quite easy compared with large-scale production (Dhobale et al. 2018). It is essential for commercial production to establish the technology in a pilot working prototype to evaluate the process control and predetermine the faults on a commercial scale. The pilot plants should be established and investigated to evaluate the process and technology before the establishment of a large-scale to full-scale production. Especially for large-scale production, laboratory data are insufficient to predict the processing of full scale.

Through a successful establishment of a pilot plant, it is possible to develop a standardized procedure of large-scale production, standardize the processing equipment and associated hardware and their quality, identification of factors regulating production rate, and determine the standards for maintain or improve the quality of bioprocessing or the end product. In addition to this, the establishment of the pilot plant will give an understanding of the space required for the infrastructure and associated waste management. As a literal definition, the scale-up is the increase in batch size, the production rate, and the speed of bioprocessing. While doing the research and development of a scale-up technology, there is an array of standardization in each intermediate stage, like pilot plants.

The in vitro culture for secondary metabolite production from plants, hairy root, and suspension culture is more stable. Scale-up of echinacoside, chlorogenic acid, cichoric acid, caftaric acid, and cynarin using *Echinacea angustifolia* is promising proof (Cui et al. 2013). Curacin A is a pharmaceutically relevant compound, and a successful pilot scale was established with a marine cyanobacterium *Lyngbya majuscula.* Pharmacologically relevant ginsenoside was successfully isolated from the adventitious root cultures of *Panax ginseng* C.V. Meyer through large-scale reactor with a capacity of 1–10 tons (Hahn et al. 2003).

19.3.9 Commercial Scale-Up

We are facing a deficit in the availability of certain pharmaceutically important plant metabolites, including nutraceuticals. It is because of the inability to attain a production rate that meets the demand. However, the ongoing biotechnological research and technological revolution could increase the production rate. The establishment of commercial or industrial scale-up with the existing technology can meet the market demand (Cacciuttolo and Arunakumari 2005). And such an increased scale-up will reduce the production cost and thereby reduce product cost (Fig. 19.6).

19.4 Conclusions and Perspectives

The need and production rate of each existing metabolite and metabolites that are de-novo will increase day by day. With the increasing demand, the production strategy should also change. Over the last few decades, the metabolite production for pharmaceuticals has adopted strategies for increasing the metabolic products from the host culture using different elicitors.

In vitro cultivation of cells with modified and enhanced metabolic pathways is beneficial for commercial production. Nowadays, the metabolic engineering and

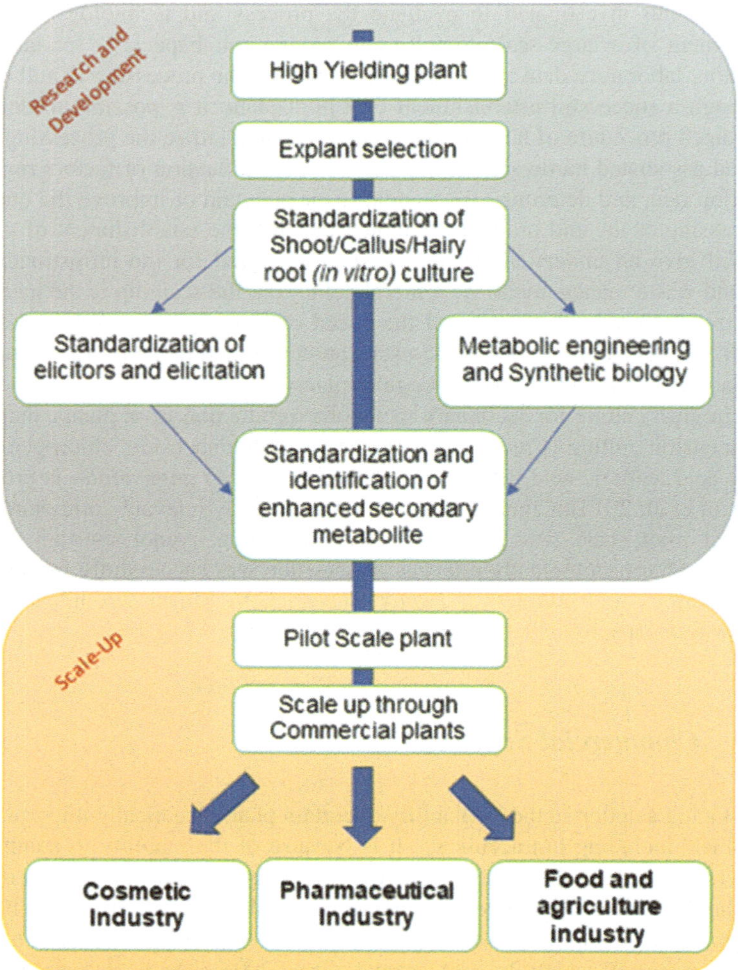

Fig. 19.6 Pipeline for industrial scale production of pharmaceutically important secondary metabolites

synthetic biology approach boom the metabolic production industry. Compared with the traditional strategy, they can provide consistent results in varying external environments and conditions. Compared with other traditional strategies, metabolic engineering and bioreactors can promise reduced cost, increased yield, increased efficacy, uninterrupted metabolic production, and improved quality. This sustainable and eco-friendly approach will be promising for enhanced metabolite production and desirable de-novo compounds.

References

Abd-El-haleem D et al (2007) Biosynthesis of biodegradable polyhydroxyalkanotes biopolymers in genetically modified yeasts. Int J Environ Sci Technol 4(4):513–520. https://doi.org/10.1007/BF03325988

Abdin MZ (2007) Enhancing bioactive molecules in medicinal plants. In: Natural products-essential resources for human. World Scientific Publishing Co. Pvt. Ltd, Singapore, pp 45–57

Ahmad N, Rab A, Ahmad N (2016) Light-induced biochemical variations in secondary metabolite production and antioxidant activity in callus cultures of Stevia rebaudiana (Bert). J Photochem Photobiol B Biol 154:51–56. https://doi.org/10.1016/j.jphotobiol.2015.11.015

Akash Bhat PB (2016) Silver nanoparticles for enhancement of accumulation of capsaicin in suspension culture of CAPSICUM SP. J Exp Scis 7:1–6. https://doi.org/10.19071/jes.2016.v7.3001

Akula R, Ravishankar GA (2011) Influence of abiotic stress signals on secondary metabolites in plants. Plant Signal Behav 6(11):1720–1731. https://doi.org/10.4161/psb.6.11.17613

Alseekh S et al (2015) Identification and mode of inheritance of quantitative trait loci for secondary metabolite abundance in tomato. Plant Cell 27(3):485–512. https://doi.org/10.1105/tpc.114.132266

Angessa TT, Li C (2016) Exploration and utilization of genetic diversity exotic germplasm for barley improvement. In: Exploration, identification and utilization of barley germplasm. pp 223–240. https://doi.org/10.1016/B978-0-12-802922-0.00009-1

Anjitha KS, Sameena PP, Puthur JT (2021) Functional aspects of plant secondary metabolites in metal stress tolerance and their importance in pharmacology. Plant Stress 2:100038. https://doi.org/10.1016/j.stress.2021.100038

Ansari S (2021) Overview of traditional systems of medicine in different continents. In: Egbuna C, Goyal MR, Mishra AP (eds) Preparation of phytopharmaceuticals for the management of disorders. Elsevier, pp 431–473. https://doi.org/10.1016/B978-0-12-820284-5.00017-4

Arman M, Qader SAU (2012) Structural analysis of kappa-carrageenan isolated from Hypnea musciformis (red algae) and evaluation as an elicitor of plant defense mechanism. Carbohydr Polym 88(4):1264–1271. https://doi.org/10.1016/j.carbpol.2012.02.003

Babich O et al (2020) Modern trends in the in vitro production and use of callus, suspension cells and root cultures of medicinal plants. Molecules 25(24):5805. https://doi.org/10.3390/molecules25245805

Baenas N, García-Viguera C, Moreno D (2014) Elicitation: a tool for enriching the bioactive composition of foods. Molecules 19(9):13541–13563. https://doi.org/10.3390/molecules190913541

Baldi A et al (2008) Co-culture of arbuscular mycorrhiza-like fungi (Piriformospora indica and Sebacina vermifera) with plant cells of Linum album for enhanced production of podophyllotoxins: a first report. Biotechnol Lett 30(9):1671–1677. https://doi.org/10.1007/s10529-008-9736-z

Barnard AM et al (2007) Quorum sensing, virulence and secondary metabolite production in plant soft-rotting bacteria. Philos Trans R Soc B Biol Sci 362(1483):1165–1183. https://doi.org/10.1098/rstb.2007.2042

Bérdy J (2005) Bioactive microbial metabolites. J Antibiot 58(1):1–26. https://doi.org/10.1038/ja.2005.1

Bhaskar R et al (2021) Biotic elicitors: a boon for the in-vitro production of plant secondary metabolites. Plant Cell Tissue Organ Cult [Preprint]. https://doi.org/10.1007/s11240-021-02131-1

Birchfield AS, McIntosh CA (2020) Metabolic engineering and synthetic biology of plant natural products—a mini review. Curr Plant Biol 24:100163. https://doi.org/10.1016/j.cpb.2020.100163

Borges CV et al (2017) 'Medicinal plants: influence of environmental factors on the content of secondary metabolites. In: Ghorbanpour M, Varma A (eds) Medicinal plants and environmental challenges. Springer International Publishing AG, pp 259–277. https://doi.org/10.1007/978-3-319-68717-9_15

Bouwmeester HJ et al (2003) Secondary metabolite signalling in host—parasitic plant interactions. Curr Opin Plant Biol 6:358–364. https://doi.org/10.1016/S1369-5266(03)00065-7

Cacciuttolo MA, Arunakumari A (2005) Scale-up considerations for biotechnology-derived products. In: Pharmaceutical process scale-up, pp 129–160

Castro AHF et al (2021) Elicitors as a biotechnological tool for in vitro production of bioactive phenolic compounds. In: Malik S (ed) Exploring plant cells for the production of compounds of interest. Springer International Publishing, Cham, pp 195–226. https://doi.org/10.1007/978-3-030-58271-5_8

Chamkhi I et al (2021) Plant-microbial interaction: the mechanism and the application of microbial elicitor induced secondary metabolites biosynthesis in medicinal plants. Plant Physiol Biochem 167:269–295. https://doi.org/10.1016/j.plaphy.2021.08.001

Chandran H et al (2020) Plant tissue culture as a perpetual source for production of industrially important bioactive compounds. Biotechnol Rep 26:e00450. https://doi.org/10.1016/j.btre.2020.e00450

Clendinen CS et al (2015) An overview of methods using 13C for improved compound identification in metabolomics and natural products. Front Plant Sci 6. https://doi.org/10.3389/fpls.2015.00611

Cui H-Y et al (2013) Scale-up of adventitious root cultures of Echinacea angustifolia in a pilot-scale bioreactor for the production of biomass and caffeic acid derivatives. Plant Biotechnol Rep 7(3): 297–308. https://doi.org/10.1007/s11816-012-0263-y

de Carvalho AA et al (2020) Influence of light spectra and elicitors on growth and ascaridole content using in vitro cultures of Dysphania ambrosioides L. Plant Cell Tissue Organ Cult 143(2):277–290. https://doi.org/10.1007/s11240-020-01892-5

Demain AL, Fang A (2000) The natural functions of secondary metabolites. In: Advances in biochemical engineering/biotechnology. Springer, Berlin, Heidelberg, pp 1–39. https://doi.org/10.1007/3-540-44964-7_1

Dhobale AV et al (2018) Recent advances in pilot plant scale up techniques. Indo Am J Pharm Res 8(4):1060–1069

Du C, Webb C (2011) Cellular systems. In: Comprehensive biotechnology. Elsevier, pp 11–23. https://doi.org/10.1016/B978-0-08-088504-9.00080-5

Dutta Gupta S, Pradhan S (2017) Regulation of gene expression by LED lighting. In: Light emitting diodes for agriculture. Springer, Singapore, pp 237–258. https://doi.org/10.1007/978-981-10-5807-3_10

Espinosa-Leal CA, Puente-Garza CA, García-Lara S (2018) In vitro plant tissue culture: means for production of biological active compounds. Planta 248(1):1–18. https://doi.org/10.1007/s00425-018-2910-1

Fazal H et al (2016) Correlation of different spectral lights with biomass accumulation and production of antioxidant secondary metabolites in callus cultures of medicinally important Prunella vulgaris L. J Photochem Photobiol B Biol 159:1–7. https://doi.org/10.1016/j.jphotobiol.2016.03.008

Fernie AR, Stitt M (2012) On the discordance of metabolomics with proteomics and transcriptomics: coping with increasing complexity in logic, chemistry, and network interactions scientific correspondence. Plant Physiol 158(3):1139–1145. https://doi.org/10.1104/pp.112.193235

Garcia-Brugger A et al (2006) Early signaling events induced by elicitors of plant defenses. Mol Plant-Microbe Interact 19(7):711–724. https://doi.org/10.1094/MPMI-19-0711

Gemma H (2000) Possibility of n-propyl dihydrojasmonate application for thinning fruit, defoliating and promoting the fruit maturation as a cultural technique. Acta Hortic 516:57–66. https://doi.org/10.17660/ActaHortic.2000.516.6

Giri CC, Zaheer M (2016) Chemical elicitors versus secondary metabolite production in vitro using plant cell, tissue and organ cultures: recent trends and a sky eye view appraisal. Plant Cell Tissue Organ Cult 126(1):1–18. https://doi.org/10.1007/s11240-016-0985-6

Gupta B, Huang B (2014) Mechanism of salinity tolerance in plants: physiological, biochemical, and molecular characterization. Int J Genomics 2014:1–18. https://doi.org/10.1155/2014/701596

Gurung AB et al (2021) An updated review of computer-aided drug design and its application to COVID-19. Biomed Res Int 2021:1–18. https://doi.org/10.1155/2021/8853056. Edited by B. Alatas

Gutierrez-Valdes N et al (2020) Hairy root cultures—a versatile tool with multiple applications. Front Plant Sci 11. https://doi.org/10.3389/fpls.2020.00033

Hadacek F (2015) Low-molecular-weight metabolite systems chemistry. Front Environ Sci 3. https://doi.org/10.3389/fenvs.2015.00012

Hahn EJ et al (2003) Adventitious root cultures of Panax ginseng C.V. Meyer and Ginsenoside production through large-scale bioreactor system. J Plant Biotechnol 5(1):1–6

Hashim M et al (2021) Comparative effects of different light sources on the production of key secondary metabolites in plants in vitro cultures. Plants 10(8):1521. https://doi.org/10.3390/plants10081521

He M, He C-Q, Ding N-Z (2018) Abiotic stresses general defenses of land plants and chances for engineering multistress tolerance. Front Plant Sci 9:1–18. https://doi.org/10.3389/fpls.2018.01771

Hermann ATV et al (2018) The use of synthetic biology tools in biorefineries to increase the building blocks diversification. In: Advances in sugarcane biorefinery. Elsevier, pp 41–72. https://doi.org/10.1016/B978-0-12-804534-3.00003-3

Ho T-T, Murthy HN, Park S-Y (2020) Methyl jasmonate induced oxidative stress and accumulation of secondary metabolites in plant cell and organ cultures. Int J Mol Sci 21(3):716. https://doi.org/10.3390/ijms21030716

Hong ST, Carney JR, Gould SJ (1997) Cloning and heterologous expression of the entire gene clusters for PD 116740 from Streptomyces strain WP 4669 and tetrangulol and tetrangomycin from Streptomyces rimosus NRRL 3016. J Bacteriol 179(2):470–476. https://doi.org/10.1128/jb.179.2.470-476.1997

Hughes J et al (2011) Principles of early drug discovery. Br J Pharmacol 162(6):1239–1249

Hurmat RS, Bansal G (2020) Does abiotic stresses enhance the production of secondary metabolites? A review. Pharm Innov 9(1):412–422. https://doi.org/10.22271/tpi.2020.v9.i1g.4315

Hussain MS et al (2012) Current approaches toward production of secondary plant metabolites. J Pharm Bioallied Sci 4(1):10. https://doi.org/10.4103/0975-7406.92725

Idris A et al (2018) Effect of light quality and quantity on the accumulation of flavonoid in plant species. J Sci Technol 10(3). https://doi.org/10.30880/jst.2018.10.03.006

Isah T (2019) Stress and defense responses in plant secondary metabolites production. Biol Res 52(1):39. https://doi.org/10.1186/s40659-019-0246-3

Isah T et al (2018) Secondary metabolism of pharmaceuticals in the plant in vitro cultures: strategies, approaches, and limitations to achieving higher yield. Plant Cell Tissue Organ Cult 132(2):239–265. https://doi.org/10.1007/s11240-017-1332-2

Kapoor S et al (2018) Influence of light quality on growth, secondary metabolites production and antioxidant activity in callus culture of Rhodiola imbricata Edgew. J Photochem Photobiol B Biol 183:258–265. https://doi.org/10.1016/j.jphotobiol.2018.04.018

Katiyar C et al (2012) Drug discovery from plant sources: an integrated approach. Ayu 33(1):10. https://doi.org/10.4103/0974-8520.100295

Kawka B, Kwiecień I, Ekiert H (2017) Influence of culture medium composition and light conditions on the accumulation of bioactive compounds in shoot cultures of Scutellaria lateriflora L. (American skullcap) grown in vitro. Appl Biochem Biotechnol 183(4):1414–1425. https://doi.org/10.1007/s12010-017-2508-2

Kim H-Y et al (2016) Atractylone, an active constituent of KMP6, attenuates allergic inflammation on allergic rhinitis in vitro and in vivo models. Mol Immunol 78:121–132. https://doi.org/10.1016/j.molimm.2016.09.007
Klein FRS et al (2018) UV-B radiation as an elicitor of secondary metabolite production in plants of the genus Alternanthera. Acta Bot Brasil 32(4):615–623. https://doi.org/10.1590/0102-33062018abb0120
Krol A et al (2020) Bioreactor-grown shoot cultures for the secondary metabolite production. In: Plant cell tissue different second. pp 1–62. https://doi.org/10.1007/978-3-030-11253-0_34-1
Kroymann J (2011) Natural diversity and adaptation in plant secondary metabolism. Curr Opin Plant Biol 14(3):246–251. https://doi.org/10.1016/j.pbi.2011.03.021
Kumar RR, Prasad S (2011) Metabolic engineering of bacteria. Indian J Microbiol 51(3):403–409. https://doi.org/10.1007/s12088-011-0172-8
Li Y-C, Tao W-Y, Cheng L (2009) Paclitaxel production using co-culture of Taxus suspension cells and paclitaxel-producing endophytic fungi in a co-bioreactor. Appl Microbiol Biotechnol 83(2): 233–239. https://doi.org/10.1007/s00253-009-1856-4
Lu J-M, Yao Q, Chen C (2009) Ginseng compounds: an update on their molecular mechanisms and medical applications. Curr Vasc Pharmacol 7(3):293–302. https://doi.org/10.2174/157016109788340767
Mahagamasekera MGP, Doran PM (1998) Intergeneric co-culture of genetically transformed organs for the production of scopolamine. Phytochemistry 47(1):17–25. https://doi.org/10.1016/S0031-9422(97)00551-7
Marchev AS, Yordanova ZP, Georgiev MI (2020) Green (cell) factories for advanced production of plant secondary metabolites. Crit Rev Biotechnol 40(4):443–458. https://doi.org/10.1080/07388551.2020.1731414
Marslin G, Sheeba CJ, Franklin G (2017) Nanoparticles alter secondary metabolism in plants via ROS burst. Front Plant Sci 8. https://doi.org/10.3389/fpls.2017.00832
McCloskey D et al (2018) Evolution of gene knockout strains of E. coli reveal regulatory architectures governed by metabolism. Nat Commun 9(1):3796. https://doi.org/10.1038/s41467-018-06219-9
Miller MB, Bassler BL (2001) Quorum sensing in bacteria. Annu Rev Microbiol 55(1):165–199. https://doi.org/10.1146/annurev.micro.55.1.165
Movahedi A et al (2021) Plant secondary metabolites with an overview of populus. Int J Mol Sci 22(13):6890. https://doi.org/10.3390/ijms22136890
Muchate NS et al (2019) NaCl induced salt adaptive changes and enhanced accumulation of 20-hydroxyecdysone in the in vitro shoot cultures of Spinacia oleracea (L.). Sci Rep 9(1): 12522. https://doi.org/10.1038/s41598-019-48737-6
Naik PM, Al-Khayri JM (2016) Abiotic and biotic elicitors–role in secondary metabolites production through in vitro culture of medicinal plants. In: Abiotic and biotic stress in plants—recent advances and future perspectives. InTech. https://doi.org/10.5772/61442
Narayani M, Srivastava S (2017) Elicitation: a stimulation of stress in in vitro plant cell/tissue cultures for enhancement of secondary metabolite production. Phytochem Rev 16(6): 1227–1252. https://doi.org/10.1007/s11101-017-9534-0
Navarrete C et al (2020) Cell factories for industrial production processes: current issues and emerging solutions. Processes 8(7):768. https://doi.org/10.3390/pr8070768
Pan S-Y et al (2014) Historical perspective of traditional indigenous medical practices: the current renaissance and conservation of herbal resources. Evid Based Complement Alternat Med 2014: 1–20. https://doi.org/10.1155/2014/525340
Patel H, Krishnamurthy R (2013) Elicitors in plant tissue culture. J Pharm Phytochem 2(2):60–65
Petersen L-E, Kellermann MY, Schupp PJ (2020) Secondary metabolites of marine microbes: from natural products chemistry to chemical ecology. In: YOUMARES 9—the oceans: our research, our future. Springer International Publishing, Cham, pp 159–180. https://doi.org/10.1007/978-3-030-20389-4_8

Pott DM, Osorio S, Vallarino JG (2019) From central to specialized metabolism: an overview of some secondary compounds derived from the primary metabolism for their role in conferring nutritional and organoleptic characteristics to fruit. Front Plant Sci 10. https://doi.org/10.3389/fpls.2019.00835

Prachayasittikul V et al (2015) Computer-aided drug design of bioactive natural products. Curr Top Med Chem 15(18):1780–1800. https://doi.org/10.2174/1568026615666150506151101

Prasad M et al (2019) 'Plant growth promoting rhizobacteria (PGPR) for sustainable agriculture. In: PGPR amelioration in sustainable agriculture. Elsevier, pp 129–157. https://doi.org/10.1016/B978-0-12-815879-1.00007-0

Rahmawati M et al (2022) Nanotechnology in plant metabolite improvement and in animal welfare. Appl Sci 12(2):838. https://doi.org/10.3390/app12020838

Ramirez-Estrada K et al (2016) Elicitation, an effective strategy for the biotechnological production of bioactive high-added value compounds in plant cell factories. Molecules 21(2). https://doi.org/10.3390/molecules21020182

Rao SR et al (2001) Enhancement of secondary metabolite production in hairy root cultures of beta vulgaris and tagetes patula under the influence of microalgal elicitors. Food Biotechnol 15(1):35–46. https://doi.org/10.1081/FBT-100103893

Ravishankar B, Shukla V (2008) Indian systems of medicine: a brief profile. Afr J Tradit Complement Altern Med 4(3):319. https://doi.org/10.4314/ajtcam.v4i3.31226

Saad B, Azaizeh H, Said O (2005) Tradition and perspectives of Arab herbal medicine: a review. Evid Based Complement Alternat Med 2(4):475–479. https://doi.org/10.1093/ecam/neh133

Schreiner M et al (2012) UV-B-induced secondary plant metabolites—potential benefits for plant and human health. Crit Rev Plant Sci 31(3):229–240. https://doi.org/10.1080/07352689.2012.664979

Sharma S, Shahzad A (2013) Bioreactors: a rapid approach for secondary metabolite production. In: Recent trends in biotechnology and therapeutic applications of medicinal plants. Springer Netherlands, Dordrecht, pp 25–49. https://doi.org/10.1007/978-94-007-6603-7_2

Sharma A, Thakur D, Uniyal SK (2021) Taboos: traditional beliefs and customs for resource management in the Western Himalaya. Indian J Tradit Knowl 20(2):575–581

Singh R et al (2011) Heavy metals and living systems: an overview. Indian J Pharmacol 43(3):246. https://doi.org/10.4103/0253-7613.81505

Smetanska I (2009) Production of secondary metabolites using plant cell cultures. In: Food biotechnology. Springer, Berlin, Heidelberg, pp 187–228. https://doi.org/10.1007/10_2008_103

Sobhani Najafabadi A et al (2019) Effect of different quality of light on growth and production of secondary metabolites in adventitious root cultivation of Hypericum perforatum. Plant Signal Behav 14(9):1640561. https://doi.org/10.1080/15592324.2019.1640561

Sreenivasulu N, Fernie AR (2022) Diversity: current and prospective secondary metabolites for nutrition and medicine. Curr Opin Biotechnol 74:164–170. https://doi.org/10.1016/j.copbio.2021.11.010

Srivastav VK, Egbuna C, Tiwari M (2019) Plant secondary metabolites as lead compounds for the production of potent drugs. In: Phytochemicals as lead compounds for new drug discovery. Elsevier Inc. https://doi.org/10.1016/B978-0-12-817890-4.00001-9

Srivastava P et al (2014) Herbal medicine and biotechnology for the benefit of human health. In: Animal biotechnology. Elsevier, pp 563–575. https://doi.org/10.1016/B978-0-12-416002-6.00030-4

Srivastava S et al (2016) Rhizophagus irregularis as an elicitor of rosmarinic acid and antioxidant production by transformed roots of Ocimum basilicum in an in vitro co-culture system. Mycorrhiza 26(8):919–930. https://doi.org/10.1007/s00572-016-0721-4

Stanbury PF, Whitaker A, Hall SJ (2017) Design of a fermenter. In: Principles of fermentation technology. Elsevier, pp 401–485. https://doi.org/10.1016/B978-0-08-099953-1.00007-7

Sun Y et al (2021) Inducing secondary metabolite production of Aspergillus sydowii through microbial co-culture with Bacillus subtilis. Microb Cell Factories 20(1):42. https://doi.org/10.1186/s12934-021-01527-0

Tamogami S, Kodama O (2000) Coronatine elicits phytoalexin production in rice leaves (Oryza sativa L.) in the same manner as jasmonic acid. Phytochemistry 54(7):689–694. https://doi.org/10.1016/S0031-9422(00)00190-4

Thirumurugan D et al (2018) An introductory chapter: secondary metabolites. In: secondary Metabolites—sources and applications. InTech. https://doi.org/10.5772/intechopen.79766

Tripathi BN, Singh V (2018) Role of metal ions as potential abiotic elicitors of secondary role of metal ions as potential abiotic elicitors of secondary metabolites in plant cell, tissue and organ (April)

Velu G, Palanichamy V, Rajan AP (2018) Phytochemical and pharmacological importance of plant secondary metabolites in modern medicine. In: Bioorganic phase in natural food: an overview. Springer International Publishing, Cham, pp 135–156. https://doi.org/10.1007/978-3-319-74210-6_8

Verpoorte R, Van Der Heijden R, Memelink J (2000) General strategies. In: Metabolic engineering of plant secondary metabolism. Springer Netherlands, Dordrecht, pp 31–50. https://doi.org/10.1007/978-94-015-9423-3_2

Wahid M et al (2020) Pharmacological exploration of traditional plants for the treatment of neurodegenerative disorders. Phytother Res 34(12):3089–3112. https://doi.org/10.1002/ptr.6742

Wang JW, Wu JY (2013) Effective elicitors and process strategies for enhancement of secondary metabolite production in hairy root cultures. In: Biotechnology of hairy root systems, pp 55–89. https://doi.org/10.1007/10_2013_183

Wani SH, Kapoor N, Mahajan R (2017) Metabolic responses of medicinal plants to global warming, temperature and heat stress. In: Medicinal plants and environmental challenges. Springer International Publishing, Cham, pp 69–80. https://doi.org/10.1007/978-3-319-68717-9_4

War AR et al (2019) Plant defense and insect adaptation with reference to secondary metabolites. In: Merillon J-M, Ramawat KG (eds) Co-evolution of secondary metabolites, pp 1–28. https://doi.org/10.1007/978-3-319-76887-8_60-1

Wen SQ et al (2010) Study on natural medicinal chemistry and new drug development. Chin Tradit Herb Drug 41(10):1583–1589

Wink M (2015) Modes of action of herbal medicines and plant secondary metabolites. Medicines 2(3):251–286. https://doi.org/10.3390/medicines2030251

Wink M (2018) Plant secondary metabolites modulate insect behavior-steps toward addiction? Front Physiol 9. https://doi.org/10.3389/fphys.2018.00364

Wu J-Y et al (2007) Enhanced secondary metabolite (tanshinone) production of Salvia miltiorrhiza hairy roots in a novel root–bacteria coculture process. Appl Microbiol Biotechnol 77(3):543–550. https://doi.org/10.1007/s00253-007-1192-5

Yamaguchi Y, Huffaker A (2011) Endogenous peptide elicitors in higher plants. Curr Opin Plant Biol 14(4):351–357. https://doi.org/10.1016/j.pbi.2011.05.001

Yang J (2013) Application of computer-aided drug design to traditional Chinese medicine. Int J Organ Chem 03(01):1–16. https://doi.org/10.4236/ijoc.2013.31A001

Yi F et al (2018) In silico approach in reveal traditional medicine plants pharmacological material basis. Chin Med 13(1):33. https://doi.org/10.1186/s13020-018-0190-0

Yu K-W et al (2005) Ginsenoside production by hairy root cultures of Panax ginseng: influence of temperature and light quality. Biochem Eng J 23(1):53–56. https://doi.org/10.1016/j.bej.2004.07.001

Zhai Q et al (2017) Jasmonates. In: Hormone metabolism and signaling in plants. Elsevier, pp 243–272. https://doi.org/10.1016/B978-0-12-811562-6.00007-4

Zhang S et al (2021) Effects of light on secondary metabolite biosynthesis in medicinal plants. Front Plant Sci 12. https://doi.org/10.3389/fpls.2021.781236

Zhao J, Davis LC, Verpoorte R (2005) Elicitor signal transduction leading to production of plant secondary metabolites. Biotechnol Adv 23(4):283–333. https://doi.org/10.1016/j.biotechadv.2005.01.003

Chapter 20
Modern Ethnobotany and the Development of Drug Leads

Bindu R. Nair, Kumudu Perera, and L. S. Sreeshma

Abstract Ethnobotany has evolved with man and has now developed into a multidisciplinary science termed modern ethnobotany. It involves diverse approaches to identify plants used by several human cultures worldwide, thereby preserving plant diversity, understanding, and interpreting the knowledge. It is hoped that modern ethnobotany could help man understand nature and its processes in greater detail, thereby strengthening his association with nature and enabling him to deal with plants effectively and sustainably. This chapter intends to uncover the essence of modern ethnobotany, more specifically quantitative ethnobotany. Here, both plants and the human races that utilize them judiciously are subjects of study. Documentation and analysis of the collected data enhance the ingenuity of the data. Further, the chapter examines the bioprospecting potential of medicinal plants. An in-depth assessment of plants possessing antidiabetic and antioxidant activities is undertaken. Diabetes mellitus is a chronic disease that occurs due to defective secretion or action of insulin. Phytochemicals targeting various steps such as a decrease in intestinal glucose absorption, increase in insulin secretion, and increase in insulin sensitivity are elaborated in the text. A number of phytochemicals have also been shown to activate major responses of insulin signaling, such as an increase in glucose uptake/decrease in glucose release by the liver and increased glucose uptake by adipose, skeletal, and cardiac muscle tissues. Likewise, inflammation occurs within the living tissue mainly to overcome the harmful effects of toxic principles or tissue damage. Plant-derived compounds have been identified to alleviate inflammation. The active principles having antidiabetic and anti-inflammatory properties and their mechanism of action are discussed. The chapter

B. R. Nair (✉) · L. S. Sreeshma
Department of Botany, University of Kerala, Thiruvananthapuram, Kerala, India
e-mail: bindunair@keralauniversity.ac.in

K. Perera
Department of Biochemistry, Faculty of Medicine, University of Peradeniya, Peradeniya, Sri Lanka

S. T. Sukumaran, K. T R (eds.), *Conservation and Sustainable Utilization of Bioresources*, Sustainable Development and Biodiversity 30,
https://doi.org/10.1007/978-981-19-5841-0_20

concludes on a positive note, suggesting the utility of ethnobotany and why this science should be practiced for the welfare of human society.

Keywords Modern ethnobotany · Methods of data analysis · Antidiabetic plants · Anti-inflammatory extracts · Bioactive molecules

20.1 Introduction to Ethnobotany

Ethnobotany is an amalgamation of the "study of culture" and "study of plants" and is considered "the science of people's interactions with plants" (Nolan 2001). In earlier times, the term "Ethnobotany" was tagged with "plant use by ancient civilizations and primitive people," and therefore was deemed to be of no specific utility to modern human society.

Man evolved into his present form, exploiting plants for food, fiber, dyes, agricultural tools, building materials, ornaments, rituals, and drugs (Rodrigues et al. 2020). However, the importance of co-existence with nature dawned upon man only in the recent past, when he had to face serious repercussions in the form of natural hazards brought about by climatic changes and anthropogenic activities. Knowledge gained through ethnobotany could reinforce our association with nature. Ethnobotany involves diverse approaches to identify plants used by human cultures—thereby preserving plant diversity, understanding and interpreting the knowledge, and enabling him to deal with them effectively and sustainably (Singh et al. 2007; Pandey and Tripathi 2017; Kumar et al. 2021). Just as historical research has helped us to learn and rectify mistakes from the past, ethnobotanical data would help us to widen our perspectives about plants and their beneficial utilization by the different human cultures that exist today. Therefore, ethnobotany is now considered the science of survival, not the science of the ancient or deceased (Pei et al. 2020). For human good, the world may be considered a global village with several sub-communities, living with and protecting mother Nature by sustainably using the plant resources but never exploiting them.

Initially, ethnobotany was considered to be qualitative, simply recording the names of plants and enlisting their utility. However, due to the necessity of times, ethnobotany has evolved into a more focused branch called modern ethnobotany (Gaoue et al. 2017). The father of modern ethnobotany, Richard Evans Schultes, considers this field as an admixture of science and humanities and defines it as "an interdisciplinary field, combining botany, anthropology, sociology, ecology, medicine, economics, ethics, history, chemistry, agriculture, horticulture, forestry, agroforestry, archeology, linguistics, systematics and many other areas of study." It becomes implicit that, over the years, there has been a shift from the raw compilation of data to a procedural and conceptual reorientation, better conceived as academic ethnobotany. Thus, modern ethnobotany has evolved into an analytical, quantitative, experimental, and technology-oriented science and has paved the way for more appropriate and sustainable use of plant resources (Fernandez et al. 2003; Faruque et al. 2018).

20.2 Modern Ethnobotany

The modern outlook considers ethnobotany primarily in the light of applied knowledge. All the data and knowledge gathered as part of ethnobotanical studies are intended to be applied in various aspects. Some of the information gathered has been utilized to uncover or add information to different fields of study, such as traditional phytochemistry, ecology, economic ethnobotany, and even cognitive science (Nolan 2001; Fuller 2013). All these branches find wide application in the present context. Knowledge gained can be used for community development projects, natural resource management, conservation, and also biodiversity prospecting, and ecotourism. The site-specific information collected about plants may be utilized for promoting agro ethnobotany, eco-agriculture, agroforestry and agroecosystems, social forestry, ethnobotanical forest reserves, protection and sustainable use of Non-Timber Forest Produce (NTFP), conservation of wild crop relatives, maintenance of traditional crops and landraces, traditional farming systems and even in the design and development of home gardening efforts attempted by the urban households. Above all, gathering the systematic record of plants in particular localities becomes all the more essential because, with time, the plants, as well as the information givers, may be lost.

Modern ethnobotany projects may be implemented in all the member communities with the help of volunteers who are interested in plant diversity conservation. It goes without saying that the concerned volunteers should develop skills in particular fields of study such as plant identification, economic botany, phytochemistry, pharmacognosy, pharmacology, plant propagation, plant conservation, herbaria preparation, and many more. Realizing the necessity, most universities and colleges have taken steps to conduct recognized and certified courses on ethnobotany and biodiversity conservation in the undergraduate, postgraduate, and research sectors giving special emphasis on practical skills.

20.2.1 Modern Ethnobotany: Quantitative Ethnobotany

One of the applied fields of modern ethnobotany is quantitative ethnobotany. It mainly deals with a systematic approach for plant use documentation.

According to Mallick et al. (2020), India harbors 45,000 plant species and 550 tribal communities. The tribals belong to 227 linguistic groups, and they inhabit varied geographic and climatic zones with diversified plant species, varied cultures, rich traditional knowledge, and wisdom. From the ethnobotanical studies of wild plants, it has been indicated that more than 7000 species have been used for human food at some stage in human history. As research into natural products grows, this framework provides novel insights into how knowledge of ethnomedicinal plants is distributed and might be used (Vedavathy 2002; Yesodharan and Sujana 2007; Jain 2007; Kumar et al. 2012; Souza et al. 2018).

20.2.2 *Quantitative Ethnobotany: Methods of Data Collection and Analysis*

Data collection in any community should first identify the available human resource to be utilized for gaining the required information along with the complete confidence of the stakeholders. In the earlier approach, informants were seen as the objects of research, while in the recent approach, the local community participates in collaborative research, which seeks information on the history of biodiversity and not merely only cataloging it. The International Society of Ethnobiology also emphasizes the importance of collaborative and participatory research. In its Code of Ethics, the issue of participatory research is valued, thus supporting traditional communities in conducting research within their own society, undertaking their own research, recordings, databases, and more for their own use. They propose recommendations, actively participate and reciprocate so that there is a mutual benefit for all parties (Pei et al. 2020).

It is best to describe and demonstrate ways of developing approaches in ethnobotanical research, which may help us to better understand local environmental knowledge, empowering local community members as consultants and collaborators in the research process, and it should be noted that the participation of these community members also increases the chance of success in putting the findings of the research into practice. Members are allowed to actively participate in all phases of the research process, from study design to data interpretation (Pei et al. 2020). The methodology adopted should be briefed to the participants and volunteers identified from among their midst. The group activities for knowledge enhancement should be an enjoyable experience.

Sample questionnaires will likely help in gathering information regarding medicinal plants and their use among ethnic communities. A detailed sample questionnaire developed for identifying the specifications of the study area and its components is given below (Tables 20.1, 20.2, and 20.3).

After initial briefings, activities may be allocated to the volunteers.

Table 20.1 Data collection for the assessment of the community structure

S. no.	Item (Specify the period of data collection)
1	Human population size in the community
2	Gender distribution (male–female ratio)
3	Age strata of the individuals (age groups: children, adolescents, adults, old-aged)
4	Language(s)used and language(s) known
5	Religion(s) practiced
6	Vegetation type and ecosystem (Forest, grassland, desert, tundra)
7	Other flora and fauna (density, number)
8	Climate (at the period when data are collected, season may also be mentioned)
9	Soil composition (alkaline, acidic, neutral)
10	Agricultural details (crops cultivated, livestock maintained, family income source)

Table 20.2 Preparatory activities for data collection

S. no.	Community activity	Methodology (Specify the period of data collection)
1	Selection of participants	Separate groups include (1) traditional medical men, (2) representative samples from the population/community to cross-check data collected (3) volunteers for fieldwork
2	Skilled human resources required	Plant identification, basic knowledge regarding growth and propagation of medicinal plants, awareness about medicinal properties of plants, and herbaria preparation
3	Mode of data collection	Field visits for data and plant collection, semi-structured interviews, group discussions, guided field walks, participant observations, and frequent participatory workshops

Table 20.3 Details of the available plant resources in the selected locality

S. no.	Item (Specify the period of data collection)
1	Name of the plant (if known or to be named if unknown)
2	Habit (herb, shrub, tree)
3	Specificity of habitat (hot and dry areas, cool, water-logged)
4	The geographical location (distance and approach route from a major city in the state)
5	The study area (coordinates—altitude, longitude, latitude)
6	Nature of wild, cultivated, wild relatives of crop plants
7	Information about utility (food, fodder, timber, medicinal)
8	Plant part used (whole, stem, root, leaf, flowers, buds, seeds)
9	Awareness of poisonous nature
10	Availability (aplenty, moderate, or scarce—can be quantified)
11	Longevity
12	Commercial status (availability in the market)
13	The method of preparation, mode of use, dosage, method of administration
14	Record of the present level of threat status
15	Record of conservation efforts being undertaken/or to be undertaken
16	Rate of knowledge transmission to younger generation regarding plants (very poor, poor, good)

First-hand information like plant growth forms, plant part(s) used as edible, availability in nature, method of collection, method of storage, method of processing and preparation, and conservation needs should be carefully recorded.

A detailed sample questionnaire can be developed for tracing the role of plants to man. An example questionnaire is provided.

The raw data collected with the help of participants, inventory interviews, and participatory workshops may be recorded and later analyzed using mathematical calculations, as shown in Table 20.4. Analyses may be performed along with data collection or kept aside for being done later (Silva et al. 2014). All the parameters listed below or a selected few could provide information about medicinal plants and their utility.

Table 20.4 Methods for data recording and analysis

S. no.	Quantitative parameters	Description and calculations
	Distribution of the medicinal plant	
1	Absolute density (AD)	The expected number of plants of that species per square meter (or hectare). The absolute density λk of species k is estimated as the proportion of quarters in which the species is found times the estimated absolute density of all plants **Calculation** $\widehat{\lambda}\ k$ = Quarters with species $k/4n \times \lambda$
2	Relative density of a species (RD)	The percentage of the total number of observations of that species, **Calculation** Relative density (Species k) = $\widehat{\lambda}\ k/\widehat{\lambda} \times 100$
3	Absolute frequency (AF)	The percentage of sites at which a species occurs **Calculation** Absolute frequency = No. of sites with a species/Total number of sites × 100 Higher absolute frequencies indicate a more uniform distribution of a species, while lower values may indicate clustering or clumping
4	Relative frequency (RF) of a species	**Calculation** RF = AF of a species/Total frequency of all species × 100
	A high relative frequency indicates that the species is well distributed along the transect (occurs at many different sampling points). A high relative density indicates that the species appears in a relatively large number of quarters. Consequently, if the relative density is high and the relative frequency is low, then the species must appear in lots of quarters but only at a few points. That is, the species appears in clumps. If both are high, the distribution is relatively even and relatively common along the transect. If the relative density is low (appears in few quarters) and the relative frequency is high (er), then the species must be sparsely distributed (few plants, no clumping)	
5	Absolute cover	The cover or dominance of an individual species is measured by its basal area (BA) or cross-sectional area and is considered mostly in the case of trees where diameter (d), radius (r), and circumference (c) **Calculation** Expressed as the basal area of a tree per hectare The number of trees per species is multiplied by the corresponding mean basal area. The units for cover are m^2/ha
6	Relative cover (relative dominance)	The absolute cover for that species is divided by the total cover times 100 (as percentage value) **Calculation** Relative cover (Species k) = Total BA of species k along with transect/Total BA of all species along transect × 100
7	Importance value (IV)	The IV gives equal weight to the three factors of relative density, cover, and frequency. The importance value can range from 0 to 300 **Calculation** - IV = Relative density (RD) + Relative cover (RC) + Relative frequency (RF)

(continued)

Table 20.4 (continued)

S. no.	Quantitative parameters	Description and calculations
	The usefulness of medicinal plant(s) to the community	
8	Use value (UV)	Quantifies the worth of medicinal plants. If UV score is high for a plant, its use reports are more **Calculation** - UV = *X*/*N* where "*U*" refers to the number of uses mentioned by the informants for a given species, and "*N*" refers to the total number of informants interviewed
9	Informant consensus factor (ICF)	**Calculation**: ICF = Nur − Nt/(Nur − 1), where, "Nur" is the total number of use reports for each disease type and "Nt" is the total number of species used for the treatment of that disease Reveals homogeneity in the documented reports
10	Frequency of citation (FC) and relative frequency of citation (RFC)	**Calculation** FC = (Number of times a particular species was mentioned)/(total number of times all species were mentioned) × 100 **Calculation** RFC index = The number of informants who mentioned the use of the species (FC)/ the total number of informants participating in the survey (*N*). The RFC index ranged from "0" when nobody referred to a plant as useful to "1" when all informants referred to a plant as useful. RFC = FC/N
11	Relative importance index (RI)	**Calculation** RIs = {RFCs(max) + RNUs(max)}/ 2, where RFCs(max) is the relative frequency of citation over the maximum, i.e., it is obtained by dividing FCs by the maximum value in all species of the survey {RFCs(max) = FCs/max(FC)}, and RNUs(max) is the relative number of use-categories over the maximum, obtained dividing the number of uses of the species by the maximum value in all species of the survey {RNUs(max) = NUs/max(NU)} (The RI index theoretically varies from 0, when nobody mentioned any use of the plant, to 1, when the plant was most frequently mentioned as useful in the maximum number of use categories)
12	Direct matrix ranking	The data about the medicinal properties assigned to particular plants are recorded so that the individual plants may be classified and even ranked according to their single or multipurpose use
13	Cochran's Q test	It may be used for documenting the utility of individual plants against disease conditions or cross-checking information collected from various informants in the community

(continued)

Table 20.4 (continued)

S. no.	Quantitative parameters	Description and calculations
14	Jaccard index (JI)	This index is used to compare study data with that of other ethnobotanical studies conducted in more than two study sites. The formula to evaluate the JI index: **Calculation** JI = $c \times 100/a + b - c$, where "*a*" is the recorded number of species of the study area; "*A*," "*b*" is the documented number of species of the area; "*B*" and "*c*" is the common number of species in both area "*A*" and "*B*." In the case of indigenous communities, "*a*" is the number of species reported by an indigenous community "*A*," "*b*" is the number of species cited by the indigenous community "*B*" and *c* is the number of species reported by both "*A*" and "*B*"

Courtesy: Mitchell (2017); Faruque et al. (2018); Silva et al. (2014); Rodrigues et al. (2020); Soejarto et al. (2012); Jima and Megersa (2018)

20.3 Modern Ethnobotany and Bioprospecting

Bioprospecting based on the ethnobotanical approach of medicinal plant research could lead to the production and development of useful healthcare products—phytomedicines, nutraceuticals, and food supplements (Eldeen et al. 2016; Soejarto et al. 2012). Indigenous knowledge and ethnoscience are to be integrated with biotechnological approaches to provide scientific validation of ethnobotanical claims. The potential of the end-products should also be ensured before release.

The following sections highlight how ethnic and indigenous knowledge about medicinal plants has paved the way for scientific validation and bioprospecting for bioactive compounds. Two most common medicinal attributes have been considered here for investigation—Antidiabetic and anti-inflammatory properties.

20.4 . Bioactive Molecules with Antidiabetic Potential from Ethnomedicinal Plants

Diabetes mellitus is a chronic disease that occurs due to defective secretion or action of insulin. In 2021, approximately 537 million adults will suffer from diabetes, and this number is projected to rise to 783 million by 2045 (Dysted et al. 2021). Diabetes is associated with morbidity and mortality mainly due to the associated chronic complications such as cardiovascular diseases, retinopathy, nephropathy, and neuropathy (Simmonds and Howes 2006).

The earliest documentation on the usage of medicinal plants to treat diabetes goes back to 1550 B.C. as written in *Ebers Papyrus*. Medicinal plants remain the main source of medication to treat diabetes in many parts of the world (Simmonds and Howes 2006). A review published using data extracted from 34 reliable journals, from 1990 up to June 2010, has shown antidiabetic plants are widely distributed in six continental regions, while Asia and Africa dominated the global distribution: North America (4.6%), South America (5.4%), Africa (17.9%), Europe (6.6%), Asia (56.9%), and Australia (1.4%). South Asia and East Asia represented the highest distribution in Asia (Chan et al. 2012).

There are more than 1050 antidiabetic plants that are under scientific investigation. Antidiabetic compounds have been isolated from more than 300 of these plants, even though the identification of all the isolated compounds is not completed. Furthermore, active ingredients are not determined in some of the well-known traditional antidiabetic plants (Subramoniam 2016; Simmonds and Howes 2006). Many of the identified compounds have demonstrated multiple therapeutic effects in addition to antidiabetic activities. Some of the antidiabetic plants studied have shown antidiabetic effects through multiple mechanisms (Lankatillake et al. 2019). Active molecules identified from ethnomedicinal plants with a recognized mechanism of action leading to antihyperglycemic effects were discussed here. Some of the well-known antidiabetic plants demonstrated multiple antidiabetic mechanisms of action with no information on the active molecules that were not included here.

20.4.1 *Antihyperglycemic Mechanisms of Active Compounds from Antidiabetic Plants*

A salient feature of uncontrolled diabetes is the persistently elevated blood glucose concentrations which is the basis of the initiation of multiple metabolic changes leading to chronic diabetic complications. Active compounds isolated from a number of antidiabetic plants have demonstrated antihyperglycemic effects through different mechanisms (Subramoniam 2016; Lankatillake et al. 2019; Rosenzweig and Sampson 2021). Mechanisms identified as the targets for antihyperglycemic agents are summarized in Fig. 20.1.

20.4.2 *Phytochemicals That Decline Intestinal Absorption of Glucose*

Some antidiabetic compounds delay postprandial absorption of glucose (Table 20.5) by inhibition of digestive enzymes pancreatic α-amylase, intestinal brush border enzymes α-glucosidase, and sucrase and retardation of glucose absorption by inhibiting sodium glucose co-transporter SGLT-1 (El-Abhar and Schaalan 2014; Saad et al. 2017; Lankatillake et al. 2019). Most of the carbohydrates listed as bioactive molecules in this chapter have inhibited intestinal glucose absorption.

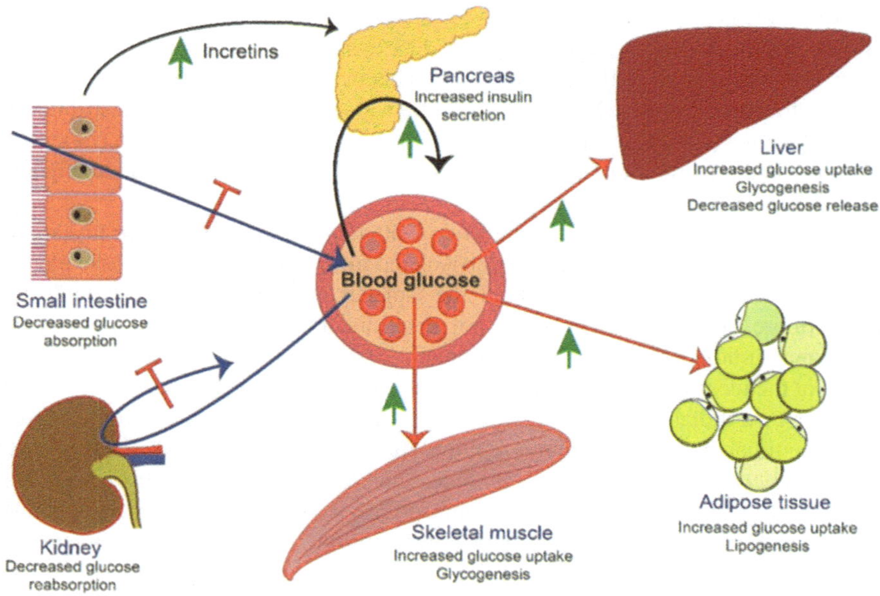

Fig. 20.1 Antihyperglycemic targets identified and mechanism of action

20.4.3 Phytochemicals That Increase the Secretion of Insulin

Normal regulation of blood glucose is primarily achieved by insulin secreted from the β-cells of the pancreas. Tissues such as liver, adipose tissue, and skeletal muscle play major roles in bringing down the blood glucose concentration in response to insulin. Many steps of the insulin signaling pathway are recognized as targets for various antidiabetic plants and phytochemicals (Patel et al. 2012).

One of the modes of increasing pancreatic secretion of insulin includes stimulation of the secretion process via ATP-gated potassium channels and activation of voltage-gated calcium channels. Other modes include inhibition of pancreatic β-cell apoptosis, increase in pancreatic β-cell proliferation, regeneration, and protection of β-cells. Incretin hormones such as glucagon-like peptide-1 (GLP-1) and glucose-dependent insulinotropic polypeptide (GIP) increase glucose-dependent insulin release. Dipeptidyl peptidase-4 (DPP-4), a protease, rapidly degrades circulating incretins. Some compounds stimulate insulin secretion through activation of incretin hormone secretion from the small intestine or inhibition of DPP-4 (Subramoniam 2016). Compounds that have demonstrated one or many of the above effects are shown in Table 20.6 (Oh 2015; Saad et al. 2017; Rosenzweig and Sampson 2021; Shehadeh et al. 2021).

Table 20.5 Phytochemicals that decline intestinal absorption of glucose

Botanical name	Common name	Active compounds
Adhatoda vasica	Malabar nut	Vasicine, vasicinol
Aloe vera	Aloe	Cellulose, mannose, glucomannan
Andrographis paniculate	Creat	Andrographolide
Camellia sinensis	Tea	Catechin
Cinnamomum verum	Cinnamon	Cinnamic acid
Cinnamomum zeylanicum	Cinnamon	L-arabino-D-xylan, cinnzeylanin, cinnzeylanol, D-glucan
Curcuma longa	Turmeric	Curcuminoids
Lagerstroemia speciosa	Banaba	Oleanolic acid, arjunolic acid, asiatic acid, maslinic acid, corosolic acid, 23-hydroxyursolic acid
Mangifera indica	Mango	Mangiferin
Phyllanthus amarus	Carry me seed	Brevifolin carboxylic acid, ethyl brevifolin carboxylate; oleanolic acid, ursolic acid, lupeol
Plantago ovate	Psyllium	Fiber
Rheum emodi	Himalayan rhubarb	Chrysophanol-8-*O*-β-D-glucopyranoside, desoxyrhaponticin, torachrysone-8-*O*-β-D-glucopyranoside
Salacia species	Salacia	Mangiferin, salacinol, kotalanol, kotalagenin 16-acetate
Sesamum indicum	Sesame	(+)-Pinoresinol
Syzygium cumini	Black plum	Betulinic acid, 3,5,7,4′-tetrahydroxy flavanone
Trigonella foenum-graecum	Fenugreek	Guar gum, pectin, mucilaginous fibre

El-Abhar and Schaalan (2014); Saad et al. (2017); Lankatillake et al. (2019)

20.4.4 Phytochemicals That Increase Insulin Sensitivity

A number of bioactive molecules from ethnobotanical plants are reported to increase the insulin sensitivity by activation of the insulin signaling pathway (Simmonds and Howes 2006; Patel et al. 2012). Insulin signaling is initiated with the binding of insulin with the insulin receptor located on the plasma membrane. Subsequent conformational change in the receptor activates its intrinsic tyrosine kinase activity resulting in autophosphorylation of tyrosine residues of the intracellular domains of the receptor. As a response various downstream targets of signaling pathway such as insulin receptor substrates, phosphoinositide 3-kinase and protein kinase B (PKB-Akt) are activated through a cascade of phosphorylation. Some phytochemicals act as agonists of peroxisome proliferator-activated receptor-gamma (PPAR-γ) and increase the expression of insulin-sensitive genes. These events lead to the major responses to insulin; an increase in glucose uptake by the adipose, skeletal, and cardiac muscle tissues, activation of glycogen synthesis in the liver and skeletal

Table 20.6 Phytochemicals that increase secretion of insulin

Botanical name	Common name	Active compounds
Aegle marmelos	Bael	Aegelin, marmesin, marmelosin
Allium sativum	Garlic	Allicin; L-Alliin, Allyl propyl disulfide
Aloe vera	Aloe	Pseudoprototinosaponin AIII, prototino saponins AIII, glucomannan, Aloin, barbaloin, isobarbaloin, aloetic acid, aloe-emodin, lophenol
Anemarrhena asphodeloide	Zhi mu	Mangiferin, mangiferin-7-*O*-β-D-glucoside
Bauhinia variegata	Mountain ebony	Roseoside
Camellia sinensis	Tea	Epigallocatechin-3-gallate
Capsicum annuum	Chilli	Capsaicin
Cassia auriculata	Tanner's Cassia	Bis (2-ethyl hexyl) phthalate
Cinnamomum species	Cinnamon	Cinnamaldehyde, methylhydroxychalcone, trimer procyanidin oligomers
Citrullus colocynthis	Bitter cucumber	Beta-pyrazol-1-ylalanine
Coccinia grandis	Ivy gourd	B-sitosterol, lupeol
Curcuma longa	Turmeric	Curcumin, Ferulic acid
Ephedra distachya	Sea-grape	L-ephedrine
Ficus bengalensis	Banyan	Leucopelargonidin, leucopelargonidin-3-*O*-α-1-rhamnoside
Ginkgo biloba	Kew tree	Ginkgolides
Glycine max	Soybean	Genistein
Glycyrrhizae radix	Licorice	Glycyrrhetinic acid, dihydroxy gymnemic triacetate
Gymnema sylvestre	Gurmar	Gymnemic acids, gourmarin, betaine, choline, trimethylamine, (−) Hydroxycitric acid
Juglans regia	Walnut	Julgon
Momordica charantia	Bitter gourd	Momordicin, vicine
Morus alba	White mulberry	Chrysin, isoquercitrin
Nigella sativa	Black cumin	Thymoquinone
Nymphaea stellate	Blue lotus	Nymphayol
Olea europaea	Olive	Oleuropeoside
Panax ginseng	Ginseng	Ginsenoside
Prunella vulgaris	Selfheal	Jiangtangsu
Psidium guajava	Guava	Strictinin, isostrictinin, pedunculagin
Pterocarpus marsupium	Indian kino	(−)-Epicatechin
Punica granatum	Pomegranate	Punicalagin
Rhizoma coptidis	Huanglian	Berberine
Semen coicis	Coix seed	Coixans

(continued)

Table 20.6 (continued)

Botanical name	Common name	Active compounds
Silybum marianum	Milk thistle	Silymarin
Stevia rebaudiana	Sweetleaf	Steviol, stevioside
Swertia chirayita	Chirata	Swerchirin
Syzygium cumini	Black plum	Quercetin
Teucrium polium	Felty germander	Apigenin
Tribulus terrestris	Puncture vine	Harmane, pinoline
Trigonella foenum-graecum	Fenugreek	4-hydroxyisoleucine
Zea mays	Corn	ZeinH
Zingiber officinale	Ginger	Gingerol

muscle, and a decrease in the hepatic release of glucose (Simmonds and Howes 2006; Saad et al. 2017; Rosenzweig and Sampson 2021).

20.4.5 Phytochemicals That Show Insulin-Like Effects

Modes of insulin-like effects include activation of insulin receptors by receptor tyrosine phosphorylation, stimulation of insulin receptor substrates, and downward signaling pathway leading to cellular responses such as glucose uptake. Compounds that have demonstrated insulin-like effects are shown in Table 20.7 (Patel et al. 2012; Saad et al. 2017; Costa et al. 2020; Tran et al. 2020; Rosenzweig and Sampson 2021). All five peptides listed with antihyperglycemic effects in this chapter have shown insulin-like effects.

20.4.6 Phytochemicals That Inhibit Protein Tyrosine Phosphatases

Function of phosphotyrosine-specific phosphatases (PTP) such as PTP-1B is to neutralize the insulin signaling pathway (Simmonds and Howes 2006; Saad et al. 2017). Inhibition of these phosphatases potentiates insulin signaling (Table 20.8) (Aba and Asuzu 2018; Saad et al. 2017; Subramoniam 2016; Rosenzweig and Sampson 2021).

Table 20.7 Phytochemicals that show insulin-like effects

Botanical name	Common name	Active compounds
Allium sativum	Garlic	*S*-allyl cysteine
Amorpha fruticose	False indigo	Amorfrutins
Angelica giga	Giant angelica	Decursin
Artemisia dracunculus	Tarragon	6-demethoxycapillarisin
Campsis grandiflora	Japanese plum	Ursolic acid
Cinnamomum species	Cinnamon	Cinnamaldehyde, methylhydroxychalcone
Costus igneus	Costus	Insulin-like protein
Echinacea purpurea	Purple coneflower	Hexadeca-2E,9Z,12Z,14E-tetraenoic acid isobutylamide
Eriobotrya japonica	Japanese Plum	Corosolic acid
Gleditsia sinensis	Locust	Aromadendrin
Glycine max	Soybean	Aglycin
Glycyrrhiza foetida	Licorice	Amorfrutins
Lagerstroemia speciosa	Banaba	Penta-galloyl-glucose, corosolic acid
Lavandula angustifolia	Lavender	Ursolic acid
Malus domestica	Apple	Ursolic acid
Mangifera indica	Mango	Penta-galloyl-glucose
Momordica charantia	Bitter gourd	P-insulin (insulin-like polypeptide)
Origanum vulgare	Oregano	Ursolic acid
Panax ginseng	Ginseng	Ginseng oligopeptides, ginsenosides
Pisum sativum	Peas	Vglycin
Pterocarpus marsupium	Indian kino	7-*O*-α-L-rhamnopyranosyl-oxy-4′-methoxy-5-hydroxy isoflavone, pterostilbene
Punica granatum	Pomegranate	Penta-galloyl-glucose, karanjin
Rosmarinus officinalis	Rosemary	Ursolic acid
Salvia officinalis	Sage	Ursolic acid
Symplocos paniculata	Sapphire Berry	Corosolic acid
Theobroma cacao	Cocoa	Gallo-tannins
Thymus vulgaris	Thyme	Ursolic acid
Vitis vinifera	Grape	Gallo-tannins

Patel et al. (2012); Saad et al. (2017); Costa et al. (2020); Tran et al. (2020); Rosenzweig and Sampson (2021)

Table 20.8 Phytochemicals that inhibit protein tyrosine phosphatases

Botanical name	Common name	Active compounds
Artemisia dracunculus	Tarragon	Sakuranetin, 2′,4′-dihydroxy-4-methoxydihydrochalcone, 2′,4-dihydroxy-4′-methoxydihydrochalcone
Berberis vulgaris	Barberry	Berberine
Bombax ceiba	Cotton tree	Lupane
Coptidis rhizoma	Huanglian	Berberine
Crocus sativus	Saffron	Safranal
Eriobotrya japonica	Japanese plum	Corosolic acid
Flemingia philippinensis	Large leaf flemingia	6,8-diprenylorobol
Hydrastis canadensis	Golden seal	Berberine
Lagerstroemia speciosa	Banaba	Corosolic acid
Lophopetalum wallichii	Puen Ta Lei	Lupane
Moraceae family	Mulberry	Morin
Nigella sativa	Black cumin	Nigelladines A–C, nigellaquinomine
Pongamia pinnata	Pongam tree	Karanjin
Psidium guajava	Guava	Psidials B and C
Sophora flavescens	Shrubby sophora	2′-Methoxykurarinone
Sorbus commixta	Japanese rowan	Lupane
Symplocos paniculate	Sapphire berry	Corosolic acid
Theobroma cacao	Cocoa	Gallo-tannins
Vitis vinifera	Grape	Gallo-tannins

20.4.7 Phytochemicals That Increase Uptake/Decrease Release of Glucose by the Liver

Liver contributes to maintaining blood glucose concentration during fed state by increasing the glucose utilization (storage as glycogen and flux through glycolysis) and decreasing the hepatic output of glucose (decreased gluconeogenesis and glycogen breakdown). An increase in glucose utilization by the liver is mediated by induction/increase in activity of glucokinase, phosphofructokinase, and glycogen synthase enzymes and an increase in GLUT2 expression. Downregulation of expression/inhibition of phosphoenol pyruvate carboxykinase, fructose-1,6-bisphosphatase, glucose-6-phosphatase, and glycogen phosphorylase decreases gluconeogenesis and hepatic glycogenolysis. Compounds that have demonstrated one

Table 20.9 Phytochemicals that increase uptake/decrease release of glucose by the liver

Botanical name	Common name	Active compounds
Aegle marmelos	Bael	Aegeline, coumarin
Allium sativum	Garlic	Allyl propyl disulfide, allicin, apigenin, alliin
Artemisia dracunculus	Tarragon	6-demethoxycapillarisin
Camellia sinensis	Tea	Hesperidin, naringin
Cassia auriculata	Tanner's Cassia	Bis (2-ethyl hexyl) phthalate
Cecropia obtusifolia	Trumpet tree	Chlorogenic acid; isoorientin
Mangifera indica	Mango	3β-Taraxerol
Momordica charantia	Bitter gourd	Momordin, momordicine, charantin
Nigella sativa	Black cumin	Thymoquinone
Pterocarpus marsupium	Indian kino	(−)-Epicatechin, pterostilbene
Syzygium cumini	Black plum	Rutin
Trigonella foenum-graecum	Fenugreek	Sotolon, trigonelline, gentianine, carpaine, Momorcharaside A and B

Farzaei et al. (2017); Saad et al. (2017); Bharti et al. (2018); Al-Ishaq et al. (2019)

or many of the above effects are shown in Table 20.9 (Farzaei et al. 2017; Saad et al. 2017; Bharti et al. 2018; Al-Ishaq et al. 2019).

20.4.8 Phytochemicals That Increase Glucose Uptake by Adipose and Muscle Tissues

A major effect of insulin signaling is the predominant increase in glucose uptake by adipose tissue, skeletal muscle, and cardiac muscle. This is accomplished by the translocation of glucose transporter 4 (GLUT4) from intracellular vesicles to the plasma membrane. This event mediates the rapid removal of glucose from the blood, which is utilized by the muscle and adipose tissue to synthesize glycogen and triglycerides, respectively. In addition to translocation, the amount of available GLUT4 is important in maintaining insulin sensitivity.

GLUT4 translocation is also mediated by AMP-activated protein kinase (AMPK), through an insulin-independent signaling pathway (El-Abhar and Schaalan 2014). Compounds that have demonstrated increase in glucose uptake through one or many of the above mechanisms are shown in Table 20.10 (Öberg et al. 2011; El-Abhar and Schaalan 2014; Perera 2016; Saad et al. 2017; Rosenzweig and Sampson 2021; Shehadeh et al. 2021).

Table 20.10 Phytochemicals that increase glucose uptake by adipose and muscle tissues

Botanical name	Common name	Active compounds
Aloe vera	Aloe	Pseudoprototinosaponin AIII, prototinosaponins AIII
Camellia sinensis	Tea	Catechin
Capsicum frutescens	Chilli	Capsaicin
Cassia fistula	Golden shower tree	Catechin
Cecropia obtusifolia	Guarumbo	Chlorogenic acid; isoorientin
Cinnamomum species	Cinnamon	Cinnamaldehyde
Coptidis rhizome	Huanglian	Berberine
Cortex phellodendri	Huang Bai	Berberine
Cryptolepis sanguinolenta	Yellow dye root	Cryptolepine
Curcuma longa	Turmeric	Sesquiterpenoids, curcuminoids
Magnolia officinalis	Magnolia	Honokiol
Lithospermum erythrorhizon	Purple gromwell	Shikonin
Mangifera indica	Mango	3β-taraxerol
Myristica fragrans	Nutmeg	Macelignan
Olea europaea	Olive	Oleuropeoside
Panax ginseng	Ginseng	Ginsenosides
Plumbago zeylanica	White leadwort	Plumbagin
Psidium guajava	Guava	Vescalagin
Pterocarpus marsupium	Indian kino	Vijayoside, pteroside, marsuposide, pterosupol, 7-*O*-α-L-rhamnopyranosyl-oxy-4′-methoxy-5-hydroxy isoflavone
Punica granatum	Pomegranate	Ellagic acid; gallic acid, punicic acid
Syzygium cumini	Black plum	Myricetin
Tinospora cordifolia		Barberin
Trigonella foenum-graecum	Fenugreek	4-Hydroxyisoleucine
Vaccinium macrocarpon	Cranberry	Resveratrol
Vitis vinifera	Grape	Resveratrol

Öberg et al. (2011); El-Abhar and Schaalan (2014); Perera (2016); Saad et al. (2017); Rosenzweig and Sampson (2021); Shehadeh et al. (2021)

20.4.9 Mechanisms Delaying Chronic Complications

Antidiabetic plants have also demonstrated a number of other beneficial effects, including antioxidant activity, antiglycation activity, and antihyperlipidemic effects, which prevent or delay diabetic complications (Simmonds and Howes 2006; Subramoniam 2016).

Most of the phytochemicals related to antidiabetic effect are in the categories of alkaloids, flavonoids, other phenolic compounds, and triterpenoids, and they have shown multiple modes of action leading to antihyperglycemic effects. Other categories included amino acid derivatives, peptides, carbohydrates, coumarins, sterols, and organosulfur compounds.

20.5 Bioactive Molecules with Anti-inflammatory Potential from Ethnomedicinal Plants

Inflammation is a metabolic process that occurs within the living tissue mainly to get relief from the entry or attack of toxic foreign agents or tissue damage. However, inflammatory responses may also be initiated prior to or during many severe disorders, such as rheumatoid arthritis, asthma, chronic inflammatory bowel diseases, type 2 diabetes, neurodegenerative diseases, and cancer. The process of inflammation is very complex.

20.5.1 Anti-inflammation Potential: Methods of Analysis

Anti-inflammatory activity may be evaluated using preliminary screening assays such as inhibition of protein denaturation, proteinase inhibition, and human red blood cell (HRBC), and membrane stabilization activity (Oyedapo and Famurewa 1995). Furthermore, in vitro assays such as Cyclooxygenase (COX) (Walker and Gierse 2010), Lipoxygenase (LOX) (Axelrod et al. 1981), Myeloperoxidase (MPO) (Bradley et al. 1982), and inducible nitric oxide synthase (iNOS) (Salter et al. 1996) assays are also used. The presence of inflammatory cytokines such as TNF-α, IL-1β, and COX 2 can be estimated using ELISA (Huang et al. 2018). In vivo assays include studies on acute (Winter et al. 1962) and chronic anti-inflammatory animal models (male Swiss albino mice) (Tjølsen et al. 1992) and gene expression studies.

The presence and activity of inflammatory cytokine intermediates such as TNF-α, IL-6, IL-8, and ROS have a decisive role in inflammatory responses. Whatever may be the cause of inflammation, the consequent effects have to be curbed and are usually done so with the help of nonsteroidal anti-inflammatory drugs (NSAIDs), glucocorticoids, immunosuppressant drugs, and biologicals (Fürst and Zündorf 2014). Glucocorticoids (GCs) are the most widely used and effective anti-inflammatory drugs. In many chronic immune diseases, GCs decrease the transcription of pro-inflammatory cytokines and chemokines and increase the transcription of anti-inflammatory cytokines. Researchers have proven that inflammation is not only a marker but is also a mediator of disease that was confirmed earlier.

20.5.2 Plant-Derived Compounds with Anti-inflammatory Properties

Innumerable studies describe the utility of plant-derived compounds in alleviating inflammation. However, a majority of the studies cite a clear correlation between the presence of phenolic compounds and antioxidant activity. Phenolic compounds with potent antioxidant activity are also capable of exhibiting anti-inflammatory activity (Gessner et al. 2016; Infante et al. 2016; Rahman et al. 2022). The increased plasma antioxidant capacity of Phenol-rich foods may be explained due to the acceptance of electrons from reactive oxygen species (ROS), thus forming relatively stable phenoxyl radicals. The oxidants are activators of nuclear factor (NF-*k*B), the key regulator of inflammation. Inactive NF-*k*B remains in the cytosol associated with inhibitors. Upon activation, NF-jB target genes, such as proinflammatory cytokines, chemokines, inflammatory enzymes, adhesion molecules, and various other receptors, are induced (Fürst and Zündorf 2014; Gessner et al. 2016).

Many of the plant-derived extracts and compounds are known to affect the inflammation pathway (Tables 20.11 and 20.12). Among the plant-derived compounds, polyphenols were identified to be significant metabolic, and inflammatory modulators due to their capability to influence various cellular and molecular

Table 20.11 Anti-inflammatory activities of some plant extracts/compounds

Sl. no	Scientific name of the plant	Plant part (extract) responsible for anti-inflammatory activity	Phyto compounds responsible for anti-inflammatory activity	Mechanism of inhibition
1	*Balanites aegyptiaca*	Bark	–	COX-1 and 2
2	*Camellia sinensis*	Root	–	COX and LOX
3	*Camellia sinensis*	Leaf	–	TNF-α, IL-6, IL-1β, COX-2, and NO
4	*Litsea quinqueflora*	Leaf	–	COX-2, TNF-α,1β
5	*Rosa canina*	Rose hip	–	COX-1 and 2
6	*Terminalia sericea*	–	Anolignan B isolated from the root	COX-1 and 2

Courtesy-Chattopadhyay et al. (2004); Speroni et al. (2005); Eldeen et al. (2006); Phetcharat et al. (2015); Novilla et al. (2017); Jose and Anilkumar (2020)

Table 20.12 Anti-inflammatory activities of some polyphenolic compounds

Sl. no	Phenolic compounds	Anti-inflammatory activities	Inflammatory markers	Mechanism
1	Apigenin	Inhibiting LPS-induced inflammation	Casp-1, IL-1β	Inhibiting NLRP3 inflammasome activation
2	Catechin	Against monosodium urate-induced inflammation	IL-1β	Modulating NLRP3 inflammasome activation
3	Ellagic acid	Ameliorating monocrotaline-induced pulmonary artery hypertension	IL-1β, IL-2, IL-4, IL6, IL-10, IFN-γ, MIP-1, MDA, NLPR3	Suppressing NLRP3 inflammasome activation
4	Green tea	Decreasing PCB 126 induced oxidative stress	SOD1, GSR, NQO1, GST	Stimulating AhR/Nrf2 pathway
5	Plant polyphenols	Modulating the inflammatory response of human keratinocytes	MCP-1, IL-10, IL-8, IL-6	Impairing phosphorylation EGF induced NFκB via regulating AhR signaling
6	Quercetin	Repair of kidney injury	IL-1β, IL-18	Suppressing NLRP3 inflammasome
7	Resveratrol	Ameliorating hepatic metaflammation	IL-1, TNF-α, IL-6	Inhibiting NLRP3 activation
8	Rutin	Reducing inflammation in pancreas	Casp-1, IL-1β, ASCNLRP3, IL-18, TNF-α	Suppressing NLRP3 inflammasome activation

Courtesy: Hussain et al. (2016); Lautie et al. (2020); Debbarma et al. (2017)

pathway targets (Hussain et al. 2016). Various phenolic compounds act on transcription factors, such as NF-*k*B or nuclear factor-erythroid factor 2-related factor 2 (Nrf-2), to up- or downregulate elements within the antioxidant response pathways. The action of phenols on inflammation has been recorded in several preclinical reports also, such as in vitro, cell-based, and animal studies (Fürst and Zündorf 2014; Gessner et al. 2016; Novilla et al. 2017; Rahman et al. 2022).

Once promising in vitro and in vivo results are obtained for plant-derived compounds, the usual practice is to go for clinical trials. In clinical trials, plant-derived compounds such as curcumin, colchicine, resveratrol, capsaicin, epigallocatechin-3-gallate, quercetin, and berberine have also been tested. Among the tested compounds, curcumin, berberine, and epigallocatechin-3-gallate exhibited high potential comparable even to inflammatory drugs such as glucocorticoids (GCs). Curcumin stood out with a stronger and broader anti-inflammatory profile than even the standard drug, the well-known GC—(Prednisolone disodium phosphate—PLP), inhibiting all tested parameters. Thus, the ability of curcumin to combat numerous inflammatory diseases via multiple pathways was proved (Rahman et al. 2022). The synergistic power of combining traditional knowledge with chemical genomics, metabolomics, and bioassay-guided fractionation would further provide a better understanding of traditional systems of medicine and popularize ethnobotany (Molimau-Samasoni et al. 2021).

20.5.3 Current Concepts and Future Perspectives

Conservation and propagation of medicinal plants for sustainable utilization must go in parallel with the usage of plant species for medicinal purposes, especially when their availability is restricted. Most the plant-derived compounds exhibit molecular diversity, and this diversity can be put to use judiciously.

The beneficial effects of bioactive compounds isolated from several antidiabetic plants are reported, though a lot more remains to be done. It is important to investigate the toxicity over long-term usage of these extracts or compounds as diabetes is a chronic condition that requires long-term medication. Variation in the toxicity due to factors including lack of quality control at the production, variation in the composition of active ingredients, and interaction with other constituents need to be considered. Likewise, simultaneous suppression of various inflammatory response pathways can be considered a more conducive option to alleviate inflammation to a greater extent and to a greater degree than focusing on one pathway at a time. Further, the mechanisms of actions of natural compounds and their analogs in several physiological processes should be clearly understood in order to yield essential insights into their prophylactic and therapeutic uses (Eldeen et al. 2016).

It is a well-known fact that the ultimate choice to ensure the therapeutical or preventive potential of test compounds is clinical trials. However, it is difficult to carry out clinical trials for all the compounds/extracts that show potential results in preliminary screening experiments as they are laborious, time-consuming, and extremely expensive. Thus, most of the research in this area is meager, and there is a big lacuna in this area. The complete scientific validation of proven compounds remains unaddressed. Efforts should be taken to screen out at least the most promising compounds and subjected to clinical trials.

The major challenges faced during the administration of natural products as drugs are poor solubility, proper absorption, difficulty in dosage determination, and bioavailability. Future research should emphasize improving drug-delivery mechanisms as the development of Drug Delivery Systems (DDS) is a promising approach to enhance bioavailability and boost the potency of natural products (Rahman et al. 2022).

20.5.4 Modern Ethnobotany: A Progressive Approach

Multi-disciplinarity of ethnobotany promotes holistic community development as it combines conservation of plant diversity and ensures a healthy way of life. Further, ethnobotanical databases have initiated many plant-based drug discovery programs. Many advantages may be pointed out.

- Recognition of valuable medicinal plants leads to their in-situ conservation (within the locality itself). Organized cultivation of important medicinal plants may lead to their cultivation on a large scale.
- Activities such as propagation and multiplication of potential medicinal plants and the development of new processes and products are likely to result in local income generation. Bioprospecting may lead to drug discovery, the development of nutraceuticals or other value-added products. The value-added end products may promote the development of small-scale cottage industries, ensuring profit sharing for the participants.
- Data collected is documented for posterity. It may be analyzed and utilized for betterment of both urban and rural folk enabling a secure future for all. Information sharing results in knowledge enhancement for both scientists and local inhabitants.
- Activities promote a participatory model wherein the participants acquire specialized skills and learn team management strategies.
- Transparency—The program is very transparent and offers the locals to witness the effective implementation of the processes at all stages.

In short, ethnobotanical research has helped us to understand our cultural heritage and has connected classical botany with the medical sciences. At the same time, caution should be taken during data collection. It often becomes difficult to collect data from a strictly random sample of informants. If informants are not chosen carefully, the data collected may be biased, false, or fabricated. Such activities require a lot of time. This may be due to non-availability of informants as they are engaged with other homely activities or other income-yielding occupations (data collected only from unemployed and elderly persons). Sometimes, people offer information only when they receive incentives for their efforts. Lack of funds may lead to the furnishing of incorrect information. Without proper background knowledge of each unique community, it may not be possible to implement the identified solutions effectively. Social interests should be considered. Antisocial elements in the community may interfere with and hamper the progress of the project.

20.6 Conclusion

Awareness about the importance of ethnobotanical knowledge will prepare man to make effective use of the available plant resources and further conserve them for future use. Knowledge gained from ethnobotanical data will teach the human communities how to address global concerns such as food security for all, hazards caused due to climate change, the importance of biodiversity conservation, and how to use the bioresources sustainably.

References

Aba PE, Asuzu IU (2018) Mechanisms of actions of some bioactive antidiabetic principles from phytochemicals of medicinal plants: a review. Indian J Nat Prod Resour 9:85–96

Al-Ishaq RK, Abotaleb M, Kubatka P et al (2019) Flavonoids and their antidiabetic effects: cellular mechanisms and effects to improve blood sugar levels. Biomolecules 9. https://doi.org/10.3390/biom9090430

Axelrod B, Cheesbrough M, Laakso S (1981) Lipoxygenase from soybeans: EC 1.13. 11.12 Linoleate: oxygen oxidoreductase. In: Methods in enzymology, vol 71. Academic, pp 441–451

Bharti SK, Krishnan S, Kumar A et al (2018) Antidiabetic phytoconstituents and their mode of action on metabolic pathways. Ther Adv Endocrinol Metab 9:81–100. https://doi.org/10.1177/2042018818755019

Bradley PP, Priebat DA, Christensen RD, Rothstein G (1982) Measurement of cutaneous inflammation: estimation of neutrophil content with an enzyme marker. J Investig Dermatol 78(3):206–209

Chan CH, Ngoh GC, Yusoff R (2012) A brief review on anti diabetic plants: global distribution, active ingredients, extraction techniques and acting mechanisms. Pharmacogn Rev 6:22–28. https://doi.org/10.4103/0973-7847.95854

Chattopadhyay P, Besra SE, Gomes A, Das M, Sur P, Mitra S, Vedasiromoni JR (2004) Anti-inflammatory activity of tea (Camellia sinensis) root extract. Life Sci 74(15):1839–1849

Costa IS, Medeiros AF, Piuvezam G et al (2020) Insulin-like proteins in plant sources: a systematic review. Diabetes Metab Syndr Obes 13:3421–3431. https://doi.org/10.2147/DMSO.S256883

Debbarma M, Pala NA, Kumar M, Bussmann RW (2017) Traditional knowledge of medicinal plants in tribes of Tripura in northeast, India. Afr J Tradit Complement Altern Med 14(4):156–168

Dysted MP, Esztergályos B, Gautam S et al (2021) IDF diabetes atlas (Internet), 10th edn

El-Abhar HS, Schaalan MF (2014) Phytotherapy in diabetes: review on potential mechanistic perspectives. World J Diabetes 5:176. https://doi.org/10.4239/wjd.v5.i2.176

Eldeen IM, Elgorashi EE, Mulholland DA, van Staden J (2006) Anolignan B: a bioactive compound from the roots of Terminalia sericea. J Ethnopharmacol 103(1):135–138

Eldeen IM, Effendy MA, Tengku-Muhammad TS (2016) Ethnobotany: challenges and future perspectives. Res J Med Plants 10(6–7):382–387

Faruque MO, Uddin SB, Barlow JW, Hu S, Dong S, Cai Q, Hu X (2018) Quantitative ethnobotany of medicinal plants used by indigenous communities in the Bandarban District of Bangladesh. Front Pharmacol 9:40

Farzaei F, Morovati MR, Farjadmand F et al (2017) A mechanistic review on medicinal plants used for diabetes mellitus in traditional Persian medicine. J Evid Based Complement Altern Med 22:944–955. https://doi.org/10.1177/2156587216686461

Fernandez EC, Sandi YE, Kokoska L (2003) Ethnobotanical inventory of medicinal plants used in the Bustillo Province of the Potosi Department, Bolivia. Fitoterapia 74(4):407–416

Fuller RJM (2013) Ethnobotany: major developments of a discipline abroad, reflected in New Zealand. N Z J Bot 51(2):116–138

Fürst R, Zündorf I (2014) Plant-derived anti-inflammatory compounds: hopes and disappointments regarding the translation of preclinical knowledge into clinical progress. Mediat Inflamm 2014:146832. https://doi.org/10.1155/2014/146832

Gaoue OG, Coe M, Bond M, Hart G, Seyler BC, McMillen H (2017) Theories and major hypotheses in ethnobotany. Econ Bot 71(3):269–287

Gessner DK, Ringseis R, Eder K (2016) Potential of plant polyphenols to combat oxidative stress and inflammatory processes in farm animals. J Anim Physiol Anim Nutr 101(4):605–628

Huang C, Li W, Zhang Q, Chen L, Chen W, Zhang H, Ni Y (2018) Anti-inflammatory activities of Guang-Pheretima extract in lipopolysaccharide-stimulated RAW 264.7 murine macrophages. BMC Complement Altern Med 18(1):1–11

Hussain T, Tan B, Yin Y, Blachier F, Tossou MC, Rahu N (2016) Oxidative stress and inflammation: what polyphenols can do for us? Oxid Med Cell Longev 2016:7432797. https://doi.org/10.1155/2016/7432797

Infante J, Rosalen PL, Lazarini JG, Franchin M, Alencar SMD (2016) Antioxidant and anti-inflammatory activities of unexplored Brazilian native fruits. PLoS One 11(4):e0152974

Jain SK (2007) https://doi.org/10.1002/9780470514634.ch11

Jima TT, Megersa M (2018) Ethnobotanical study of medicinal plants used to treat human diseases in Berbere District, Bale Zone of Oromia Regional State, South East Ethiopia. Evid Based Complement Altern Med 2018:8602945

Jose SM, Anilkumar M (2020) Antibacterial and anti-inflammatory activity of the silver nanoparticles synthesized from the methanolic leaf extract of Litsea quinqueflora (Dennst.) Suresh. Adv Sci Eng Med 12(4):443–449

Kumar S, Jena PK, Tripathy PK (2012) Study of wild edible plants among tribal groups of Simlipal Biosphere Reserve Forest, Odisha, India; with special reference to Dioscorea species. Int J Biol Technol 3(1):11–19

Kumar A, Kumar S, Ramchiary N, Singh P (2021) Role of traditional ethnobotanical knowledge and indigenous communities in achieving Sustainable Development Goals. Sustainability 13(6):3062

Lankatillake C, Huynh T, Dias DA (2019) Understanding glycaemic control and current approaches for screening antidiabetic natural products from evidence-based medicinal plants. Plant Methods 15:1–35. https://doi.org/10.1186/s13007-019-0487-8

Lautie E, Russo O, Ducrot P, Boutin JA (2020) Unraveling plant natural chemical diversity for drug discovery purposes. Front Pharmacol 11:397

Mallick SN, Sahoo T, Naik SK, Panda PC (2020) Ethnobotanical study of wild edible food plants used by the tribals and rural populations of Odisha, India for food and livelihood security. Plant Arch 20(1):661–669

Mitchell K (2017) Quantitative analysis by the point-centered quarter method. http://people.hws.edu/mitchell/PCQM.pdf

Molimau-Samasoni S, Woolner VH, Robichon K, Patel V, Andreassend SK, Sheridan JP, Munkacsi AB (2021) Functional genomics and metabolomics advance the ethnobotany of the Samoan traditional medicine "matalafi". Proc Natl Acad Sci 118:45

Nolan JM (2001) Pursuing the fruits of knowledge: cognitive ethnobotany in Missouri's Little Dixie. J Ethnobiol 21(2):29–54

Novilla A, Djamhuri DS, Nurhayati B, Rihibiha DD, Afifah E, Widowati W (2017) Anti-inflammatory properties of oolong tea (Camellia sinensis) ethanol extract and epigallocatechin gallate in LPS-induced RAW 264.7 cells. Asian Pac J Trop Biomed 7(11):1005–1009

Öberg AI, Yassin K, Csikasz RI et al (2011) Shikonin increases glucose uptake in skeletal muscle cells and improves plasma glucose levels in diabetic Goto-Kakizaki rats. PLoS One 6:1–10. https://doi.org/10.1371/journal.pone.0022510

Oh YS (2015) Plant-derived compounds targeting pancreatic beta cells for the treatment of diabetes. Evid Based Complement Altern Med 2015:629863. https://doi.org/10.1155/2015/629863

Oyedapo OO, Famurewa AJ (1995) Antiprotease and membrane stabilizing activities of extracts of Fagara zanthoxyloides, Olax subscorpioides and Tetrapleura tetraptera. Int J Pharmacogn 33(1):65–69

Pandey A, Tripathi YC (2017) Ethnobotany and its relevance in contemporary research. J Med Plants Stud 5(3):123–129

Patel DK, Prasad SK, Kumar R et al (2012) An overview on antidiabetic medicinal plants having insulin mimetic property. Asian Pac J Trop Biomed 2:320–330. https://doi.org/10.1016/S2221-1691(12)60032-X

Pei S, Alan H, Wang Y (2020) Vital roles for ethnobotany in conservation and sustainable development. Plant Divers 42(6):399

Perera HKI (2016) Antidiabetic effects of Pterocarpus marsupium (Gammalu). Eur J Med Plants 13:1–14. https://doi.org/10.9734/EJMP/2016/23930

Phetcharat L, Wongsuphasawat K, Winther K (2015) The effectiveness of a standardized rose hip powder, containing seeds and shells of Rosa canina, on cell longevity, skin wrinkles, moisture, and elasticity. Clin Interv Aging 10:1849

Rahman M, Rahaman M, Islam M, Rahman F, Mithi FM, Alqahtani T, Uddin M (2022) Role of phenolic compounds in human disease: current knowledge and future prospects. Molecules 27(1):233

Rodrigues E, Cassas F, Conde BE, Da Cruz C, Barretto EHP, Dos Santos G, Ticktin T (2020) Participatory ethnobotany and conservation: a methodological case study conducted with quilombola communities in Brazil's Atlantic Forest. J Ethnobiol Ethnomed 16(1):1–12

Rosenzweig T, Sampson SR (2021) Activation of insulin signaling by botanical products. Int J Mol Sci 22:1–20. https://doi.org/10.3390/ijms22084193

Saad B, Zaid H, Shanak S et al (2017) Anti-diabetes and anti-obesity medicinal plants and phytochemicals: safety, efficacy, and action mechanisms. Springer

Salter M, Duffy C, Garthwaite J, Strijbos PJ (1996) Ex vivo measurement of brain tissue nitrite and nitrate accurately reflects nitric oxide synthase activity in vivo. J Neurochem 66(4):1683–1690

Shehadeh MB, Suaifan GARY, Abu-Odeh AM (2021) Plants secondary metabolites as blood glucose-lowering molecules. Molecules 26:4333. https://doi.org/10.3390/molecules26144333

Silva HCH, Caraciolo RLF, Marangon LC, Ramos MA, Santos LL, Albuquerque UP (2014) Evaluating different methods used in ethnobotanical and ecological studies to record plant biodiversity. J Ethnobiol Ethnomed 10(1):1–11

Simmonds MS, Howes M-JR (2006) Traditional medicines for modern times: antidiabetic plants, first. CRC Press: Taylor & Francis

Singh RK, Singh A, Sureja AK (2007) Sustainable use of ethnobotanical resources. India J Tradit Knowl 6(3):521–530

Soejarto DD, Gyllenhaal C, Kadushin MR, Southavong B, Sydara K, Bouamanivong S, Waller DP (2012) An ethnobotanical survey of medicinal plants of Laos toward the discovery of bioactive compounds as potential candidates for pharmaceutical development. Pharm Biol 50(1):42–60

Souza EN, Williamson EM, Hawkins JA (2018) Which plants used in ethnomedicine are characterized? Phylogenetic patterns in traditional use related to research effort. Front Plant Sci 9:834

Speroni E, Cervellati R, Innocenti G, Costa S, Guerra MC, Dall'Acqua S, Govoni P (2005) Anti-inflammatory, anti-nociceptive and antioxidant activities of Balanites aegyptiaca (L.) Delile. J Ethnopharmacol 98(1–2):117–125

Subramoniam A (2016) Anti-diabetes mellitus plants: active principles, mechanisms of action and sustainable utilization, first. CRC Press: Taylor & Francis

Tjølsen A, Berge OG, Hunskaar S, Rosland JH, Hole K (1992) The formalin test: an evaluation of the method. Pain 51(1):5–17

Tran N, Pham B, Le L (2020) Bioactive compounds in antidiabetic plants: from herbal medicne to modern drug discovery. Biology (Basel) 9:1–31. https://doi.org/10.3390/biology9090252

Vedavathy S (2002) Tribal medicine-the real alternative. Indian J Tradit Knowl 1(1):25–31

Walker MC, Gierse JK (2010) In vitro assays for cyclooxygenase activity and inhibitor characterization. In: Cyclooxygenases. Humana Press, pp 131–144

Winter CA, Risley EA, Nuss GW (1962) Carrageenin-induced edema in hind paw of the rat as an assay for antiinflammatory drugs. Proc Soc Exp Biol Med 111(3):544–547

Yesodharan K, Sujana KA (2007) Wild edible plants traditionally used by the tribes in the Parambikulam Wildlife Sanctuary, Kerala, India. Nat Prod Radiance 6(1):74–78

Chapter 21
Computational Approaches for Identifying Therapeutic Potential of Phytocompounds

Junaida M. Ibrahim, Suveena Sukumaran, and Achuthsankar S. Nair

Abstract Drug discovery is a process that aims at identifying a compound with therapeutic potential in curing and treating disease. Developing methods for diagnosing and treating disease has been a big challenge for pharmaceutical researchers and clinicians. The era of Computer Aided Drug Discovery (CADD) has bought down the time and efforts to a large extent. CADD has contributed largely to different stages of the drug discovery pipeline, mainly in the identification and optimization of lead compounds, leading them to the advanced stages in drug discovery. CADD gives valuable information about the molecular properties, pharmacodynamics, and toxicity, thus helping in the lead optimization phase and improving the efficacy and potency of the drug candidate. This chapter highlights basic steps in the CADD pipeline, two important categories of CADD, i.e., ligand-based and structure-based drug discovery (SBDD and LBDD). Various techniques used in SBDD and LBDD, including molecular docking, molecular dynamic simulation, Quantitative structure-activity relationship (QSAR), pharmacophore modeling, etc., have been briefly discussed. Also, available webservers and tools, significance, applications, and limitations of CADD are also briefly mentioned.

Keywords CADD · QSAR · Molecular docking · SBDD · LBDD · Pharmacophore

21.1 Introduction

The application of computational techniques and methods in the drug discovery and development process is rapidly gaining popularity and appreciation. The use of plant-derived medicines has increased due to their therapeutic value when compared to allopathic medicines, as these phytocompounds exhibit fewer side effects. Also, herbal treatments are independent of any age group, and the sexes are an added

J. M. Ibrahim (✉) · S. Sukumaran · A. S. Nair
Department of Computational Biology and Bioinformatics, University of Kerala, Thiruvananthapuram, Kerala, India

S. T. Sukumaran, K. T R (eds.), *Conservation and Sustainable Utilization of Bioresources*, Sustainable Development and Biodiversity 30,
https://doi.org/10.1007/978-981-19-5841-0_21

advantage (Kapetanovic 2008). According to American hospital records, around 8% of hospital cases are due to the toxicity of the drugs, and more than one lakh people die due to the toxicity. Every year thousands of people lose their lives because of the usage of over-the-counter drugs (Sansgiry et al. 2017). More and more people are turning to medicinal plant-based drugs as the side effects are fewer (Kazemipoor et al. 2012). The plant-derived phytochemicals with putative active substances (e.g., flavonoids, gallates, quercetins, etc.) are considered potent agents for several diseases and health conditions (Sofowora et al. 2013; Tungmunnithum et al. 2018).

Traditional drug discovery and development is time-consuming and cost-intensive and takes an average of 10–15 years to reach the market. A survey conducted by Tufts Center for the Study of Drug Development from 1995 to 2007 showed that only 11.83% of drugs reached the market after the Phase I trials (DiMasi et al. 2016). The rate of success again reduced to 9.6% in the year 2006–2015 (Takebe et al. 2018). The intensified cost and high failure of conventional drug discovery and development encouraged the need for the use of Computer-Aided Drug Design (CADD).

The emerging trends in medicinal plant genomics, evolution, and phylogeny have paved the way for the application of computational software and tools for the initial screening and filtering of active phytocompounds and drug candidates. These dynamic fields are at the intersection of phytochemistry and plant biology and are concerned with the systematics of medicinal plant genomes, the interaction between medicinal plant genomes and the environment, and the correlation between genomic diversity and metabolite diversity, etc. (Hao 2019).

CADD is a key component of multidisciplinary approaches that are being used for the development of the drug. Traditional approaches to drug discovery and development were very complex, time-consuming and costly and were also full of risk of failure during clinical trials. CADD approaches have made significant improvements in compound searching based on similarity, target identification and structure prediction, binding site/cavity prediction and validation, understanding the protein–ligand interaction, screening the vast library of compounds, understanding the dynamics of protein–ligand binding under physiological conditions, predicting the ADMET properties, estimating the biological activity using QSAR, and guiding the necessary changes required in the lead for better efficacy and selectivity. In conventional methods, it is estimated that for every new drug brought to the market, the researchers would typically have employed over 100 screens looking for drug leads (Tiwari and Singh 2022).

The thick and dense forest all over the world holds a good collection of raw materials for the manufacture of drugs. A recent estimation from WHO shows that more than 80% of the people around the world highly depend on medicinal plants, and approximately 21,000 different types of plant species have the potential for being used as medicinal plants. From the pre-historic times, herbal medicines have gained immense acceptance globally (Sofowora et al. 2013). Identification and analysis of such compounds using GC-MS, HPLC, etc., lead to their structural elucidation and further research in the pharmaceutical industry, which in turn have increased the application of CADD in the drug discovery process.

21.2 Drug Design Process

One of the emerging and important applications of bioinformatics in the foreseeable future is in the field of the pharmaceutical industry, especially in medicine and drug discovery (Tang et al. 2006; Kore et al. 2012). Computational methods rely on the fact that any pharmacologically active compound/molecule can interact with targets like proteins/nucleic acids through hydrophobic, hydrogen bond, electrostatic, Van der Waals interactions, etc., which governs the therapeutic potential. The various parameters considered in the drug design are safety and effectiveness of the drug, bioavailability, metabolic stability, fewer side effects, and target-specific distribution and delivery. The powerful computational tools and software have helped discover new/improving potency and effectiveness of the drugs by means of in silico methods. Also, advancements in computational biology and the introduction of several novel technologies in the omics areas have enabled the drug discovery process, mainly for the drug target identification and validation of a more specific one. Other *in slico* screening has also enabled the emergence of new drugs with minimal toxicity through the ADMET (Absorption, Distribution, Metabolism, Excretion, and Toxicity) analysis (Prieto-Martínez et al. 2019). The drug discovery process is very complex and involves the attention, efforts, and collaboration from the interdisciplinary fields. In the drug discovery and developmental process, several factors have to be considered. Some of the important factors are medicinal objective/ disease selection, drug developmental facilities, drug screening facilities, expenses in the entire drug development process, and lack of field experts. A brief description of the basic steps in the drug discovery process is given in the following sections;

21.3 Identification of the Therapeutic Drug Target

The first step in the drug discovery process in both conventional and CADD is the identification of the target, which is involved in the disease/infections that needs to be inhibited or activated to get a particular effect. Usually, the target molecules are proteins which involve in the cell to cell communications, signaling pathways, hormones, nucleic acids, disease biomarkers, etc. Drug targets are usually identified based on the similarity in the bioactive molecules/drugs with known targets and based on available information on molecular interaction and involvement of the proteins in the disease modulation. The two classes of drug targets are potential targets and established targets. The established targets are the molecules with clear scientific knowledge and also available complementary drugs. In the case of potential drug targets, the molecules do not have less or no scientific knowledge and thus does lack complementary drugs targeting them.

Protein Data Bank (PDB) is a high accession structure repository of proteins, which stands as the primary source in retrieving the protein structure for the studies over a few decades. If the structure is available in this database, one can directly download it from there. Due to the limitations of biological techniques, many target proteins do not have structures identified yet (Vyas et al. 2012). In such a scenario, successful computational techniques such as comparative homology modeling, threading, and ab initio modeling can be implemented (Lemer et al. 1995; Vyas et al. 2012; Yousef et al. 2019).

21.3.1 Comparative Homology Modelling

It is widely used for accurate determination/prediction of the three-dimensional structure of a protein based on its amino acid sequence alignment with more similar sequences of proteins having structure (Martí-Renom et al. 2000; Fiser 2004). There are four main steps involved in comparative homology modeling. (1) Fold recognition, which recognizes the similarity between the target and sequence of known template, (2) Aligning the target sequence with template structures, (3) Model building with highly closed sequence templates and (4) validation of the model.

If the target does not have any template structure in the existing databases, ab initio method can be employed. This is applicable when the target protein sequence shares less than 25% identity. Homology modeling only counts the sequence similarity between the target and template, whereas threading takes account of structural information, which covers the secondary structure, solvent accessibility, and pairwise interactions associated with the template. According to this method, the stability of the structure is considered through a global optimization problem to find the dihedral angles (Yousef et al. 2019).

Identifying the binding site of the target is a prerequisite for identifying molecules that bind and interfere with the action of the target. There are several depressions or cavities on the surface of the protein. These are the probable binding sites for the ligand molecule. X-ray crystallographic structures of proteins co-crystallized with substrates/inhibitors or Site-directed mutagenesis are employed for revealing the information about the binding site of the target (Pan et al. 2017). The recognized binding site of the target protein can be obtained through a number of software and servers such as Discovery Studio, CASTp, Metapocket, Q-site Finder, MSPocket, DoGSite Scorer (Huang 2009; Zhu and Pisabarro 2011; Volkamer et al. 2012; Tan et al. 2013; Sun and Chen 2017; Tian et al. 2018).

21.4 Lead Compound Identification and Optimization

The next step is the identification of lead compounds from the vast library of compounds from the available databases. The 3D structure of chemical compounds can be obtained from various chemical databases such as Pubchem, ZINC database, MCULE, ChEMBL, and ChemSpider (Irwin and Shoichet 2005; Laurie and Jackson 2005; Wishart et al. 2008; Gaulton et al. 2012; Kim et al. 2016; Pence and Williams

2010). Researchers can purchase molecules from the databases such as ZINC and PubChem for screening and experimental validation. In this step, the active site or the binding site of the target is identified by predicting the functional domains or motifs from the 3D structure, and the compounds that are complementary and fit stably to the target structures are identified. The molecules showing better interactions are called "hits." The 'hits' are identified by screening the library of natural compounds, commercially available synthetic compounds, combinatorial chemistry libraries, etc. The screening of a huge number of compounds is completed within a short time, so this kind of screening method is called High Throughput Screening (HTS). The 'hits' with the desired pharmacological properties, such as membrane permeability, low toxicity, etc., are identified. These compounds are called lead molecules. In the entire process of drug discovery, the lead optimization step is regarded as the rate-limiting step.

Specific alterations of their chemical structure can also modify the lead compounds by changing the functional groups; this is called in silico drug designing. The general guidelines for assessing the therapeutic index of the selected lead molecules or the drug candidates are identified by applying Lipinski's Rule of 5. According to Lipinski rule of five, a compound is orally active if and only if it satisfies certain rules such as Molecular weight (MW) $\leq$ 500, LogP (partition coefficient between n-octanol and water) $\leq$ 5, number of hydrogen bond acceptor (HBA) $\leq$ 10, number of hydrogen bond donor (HBD) $\leq$ 5 (Daina et al. 2017). The Lipinski's Rule of Five, ADMET (Absorption, Distribution, Metabolism, Excretion, and Toxicity) studies, are used more significantly in filtering the data, especially for docking. Human gastrointestinal absorption (HIA), blood-brain barrier penetration (BBB), Cytochrome P450 inhibition (CYP), and plasma protein binding (PPB) are the commonly used ADMET properties. SWISS ADME, PreADMET prediction, and Molinspiration are some of the online webservers for ADMET prediction.

21.5 Preclinical Studies

The lead compounds having promising activity are subjected to preclinical studies. The preclinical tests are done in animal models for the clinical evaluation of the potential drug candidates. Animal safety studies, carcinogenicity tests, drug elimination, absorption and metabolism studies, drug delivery, drug formulation experiments, dose optimization studies etc., are major studies carried out in the preclinical phase. Certain in vitro studies in non-clinical laboratories can support or enhance preclinical safety and can reduce the number of animal trials. There are several in vitro cell line-based assays that can be explored for addressing ADME, oral bioavailability, permeability, quantifying affinity for the ligand, hepatic metabolic stability, drug-drug interaction etc. (Bajpai and Esmay 2002). This type of in vitro screening might not give definite data on ADME parameters, but it helps rather allows the rank of the compounds with the best characteristics.

21.6 Clinical Trials

In this phase, the pharmacological effects, i.e., the pharmacokinetic and pharmacodynamic effects of the drug candidates in humans, are analyzed and studied. This phase is considered the most costly in the entire drug discovery pipeline. About 90% of the drug candidates subjected to this phase fail and only one-third of the drug candidates develop into drugs. The pharmacological effects of the drug candidates entering into this phase are studied through 4 different phases. Phase 0 in the clinical trials is carried out in a small number (usually 10) of participants with the small dose administration to study the oral bioavailability and shelf life of the drug. More often, this phase is skipped for phase 1. In the first phase, the pharmacological properties, including drug efficacy, safety, and toxicity, are studied and tested in normal and healthy human volunteers. In phase 2 clinical trials, the tests are carried out on the patients suffering from the targeted disease. A larger number of patients, i.e., about 100–1000 patients, are considered in this phase. About 300–3000 people with the specific disease are tested for determining the drug's therapeutic effect; at this stage, the drug is presumed to have some effect. The long-term side effects of a large patient population are studied in phase 4 of the clinical trials. Following successful clinical trials, the drugs are marketed with the FDA approval of the regulatory agencies. The basic steps of the CADD pipeline are summarized in Fig. 21.1.

21.7 Drug Design Methods

The explosion of genomic, proteomic, and structural information has provided hundreds of new targets and bioactive compounds having therapeutic interest. The advancement in analytical techniques, including X-ray crystallography, NMR, etc., and computational power have improved its application in the CADD field also. The

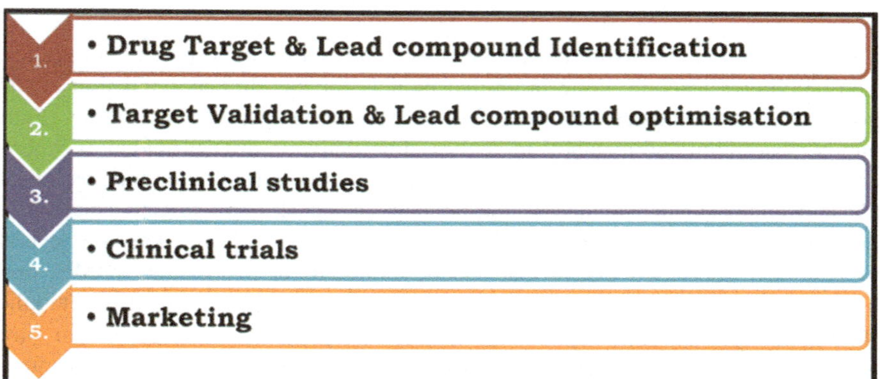

Fig. 21.1 Basic steps in Computer Aided Drug Design (CADD)

two general methods of CADD are Structure-Based Drug Design (SBDD) and Ligand Based Drug Design (LBDD); now a day, a third method, the Hybrid method is also used. These methods are used depending on how much information is available about the drug targets and potential bioactive compounds.

21.7.1 Structure-Based Drug Designing (SBDD)

The SBDD is a rapidly emerging area in which many success stories have been reported in the past years. Several approved drugs have credited their discovery to the tools of CADD. Angiotensin-converting enzyme (ACE) inhibitor captopril and renin inhibitor aliskiren for the treatment of hypertension (Talele et al. 2010; Chang et al. 2015), carbonic anhydrase inhibitor dorolamide for the treatment of cystoid macular edema (Vijayakrishnan 2009), HIV protease inhibitors like saquinavir, ritonavir, and indinavir (Van Drie 2007; Lv et al. 2015; Mahdi et al. 2015), etc. were developed using structure-based molecular docking method. This method is based on the 3D structure knowledge of a biological target obtained from X-ray crystallography/NMR spectroscopy or computed models of a known disease-related drug target (Hoque et al. 2017). Based on the known protein target structure, drug molecules are designed (Fig. 21.2). The designed drug always will be complementary to the target, and thus it stably fits into it. This is done by interactive graphics that identify the complementary structure that has a high affinity to the target structure.

21.7.2 Ligand-Based Drug Designing (LBDD)

This is based on the knowledge of the ligand molecules, mostly natural molecules like plant-based bioactive molecules. This method uses a variety of computational methods to identify novel compounds and design compounds. The new drugs are

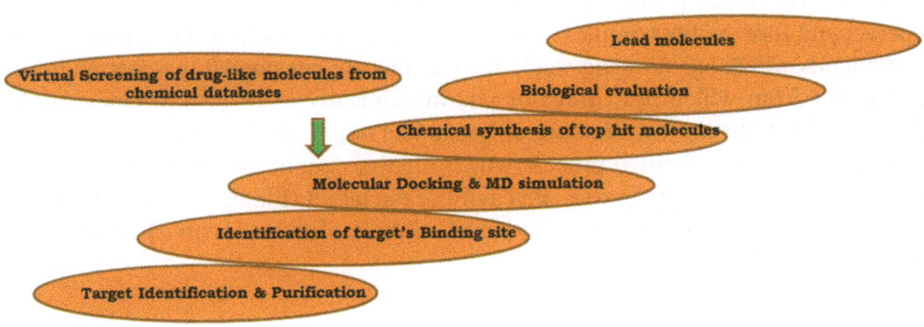

Fig. 21.2 Basic steps in SBDD

designed similar to the structure of a naturally occurring compound or from a drug that is already in use. The designed ligand will have complementarity to its target site, enabling them to bind to its target more effectively (Lin et al. 2020).

21.7.3 *Hybrid Method*

This method is used when the structures of the target, as well as the active molecule, are available. It uses the combination of structure-based and ligand-based drug designing approaches to enhance the success of drug discovery projects. The two main combinations adopted in this method are the interaction-based methods and similarity-docking methods. The combination of LBDD and SBDD methods is used synergistically in the virtual screening of bioactive compounds, ligand profiling, pharmacophore modeling, de novo drug design etc. (Prieto-Martínez et al. 2019).

21.8 Techniques Used in CADD

Since the experimental process and methods are usually laborious, time-consuming, and costly, prediction of drug-target interactions by in silico methods can be employed to derive valuable information.

21.8.1 *Molecular Docking*

Molecular docking is the key tool used in the computational drug discovery process. It is simply defined as binding a small molecule/ligand to a specific site in a target molecule. The binding of the ligand to the target in molecular docking is based on the topological or spatial complementarity of the target site with the ligand and physical-chemical interactions between the ligand and the binding site on the target molecule. The docking analyses are done to identify and screen compounds that can interact and form stable complexes with the target (drug-receptor interactions). The compounds which are showing stable interactions are considered the potential drug candidates from which new drugs or pharmaceutically relevant molecules can be designed. With the known structure of ligand and target, the docking process can be carried out through in silico process which involves confirmation and pose (orientation) of the ligand in the active site cavity of the target (Lin et al. 2020). There are different types of molecular docking based on the molecules involved and docking procedures.

21.8.1.1 Protein-Protein Docking

In this type of docking, the target and ligand will be the proteins. Sometimes the ligand can also be a small peptide rather than a whole protein.

21.8.1.2 Protein-Small Molecule Docking

Here, the ligand will be a small molecule (usually a plant-based bioactive compound, drugs, synthetic compounds, etc.). The complementary ligand molecules are smaller than the protein-protein docking. There are several physical and chemical forces that interact between the two molecules. These forces are used to define various docking scores that measure how good are each docked interactions. These forces include electrostatic, Van der Waals, hydrogen, hydrophobic forces, etc. (Fig. 21.3).

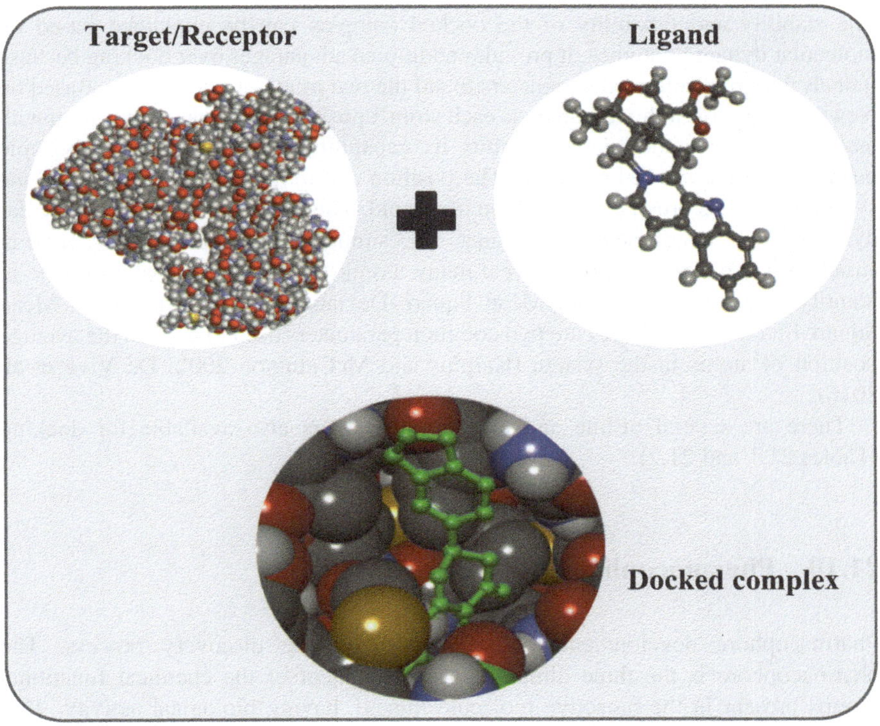

Fig. 21.3 Diagram of molecular docking. The figure depicts the basic process of molecular docking. The target in this figure is Lsa46, an outer membrane protein present in Leptospires, Lsa46 (Ibrahim et al. 2021), which was modeled using ab initio method. The small molecule is a bioactive compound Serpentine present in the plant, The docked image visualized using Discovery studio 2018 is represented here

21.8.1.3 Rigid Docking

In rigid molecular docking, the small ligand molecules are treated as rigid objects that cannot change their spatial shape, including bond angle, bond length, and torsion angles during the docking process.

21.8.1.4 Flexible Docking

In soft/flexible docking, substantial conformational or spatial changes occur within the components during the docking process. However, scoring all possible conformational changes is time-consuming.

21.9 Molecular Dynamics (MD) Simulation

The stability and flexibility of the docked complex can be explained based on molecular dynamics studies. It provides additional advantages over docking because it deals with physiological parameters to aid the real mode of interactions. Based on Newton's laws of motion, it predicts each atom's position in a molecular system with respect to time, and thus, it exhibits its capability in explaining the dynamic behaviour of the docked complex. The position and motion of each particle in the system can be recorded every fraction of second. This is an advantage of molecular dynamics over experimental techniques. MD simulation method is widely used in structure-based drug design to reveal many atomic details, which is impossible to identify in experiments. Root Mean Square Deviation (RMSD) and Root Mean Square Fluctuation (RMSF) are two common parameters used to explain the average position of atoms in the system (Karplus and McCammon 2002; De Vivo et al. 2016).

There are several offline and online web servers also available for docking (Tables 21.1 and 21.2).

21.10 Pharmacophore Development

Pharmacophore development is a part of the drug discovery process. The pharmacophore is the three dimensional arrangement of the chemical functional groups present in the bioactive molecules/ligand, having biological activity. This spatial arrangement of the chemical entities in three dimensional spaces of the ligands is necessary for the appropriate binding at the active site of the target. The features are hydrogen bond donors, hydrogen bond acceptors, aromatic rings, hydrophobic areas, ionizable groups, etc. The pharmacophore screening aims to

Table 21.1 Different tools and webservers used in Molecular docking

SI.	Tool	Description
1	Autodock	Effective in protein-ligand docking. It is based on Lamarckian genetic algorithm, traditional genetic algorithm search and simulated annealing search
2	Autodock vina	Specially designed for protein-ligand docking. It relayed on a sophisticated gradient optimization method
3	CDOCKER	CHARMm-based DOCKER applicable for protein-ligand docking. A module in Discovery Studio
4	Glide	Grid-based ligand docking with energetics for protein-ligand docking
5	DOCK	It evaluates the conformational sampling of small ligand molecules based on the anchor and creates protein-ligand docking search algorithm
6	FlexX	A full automated docking tool for flexible ligands
7	Surflex	To produce putative poses of ligand fragments, it uses a combination of Hammerhead's empirical scoring function and molecular similarity method
8	ZDOCK	Protein-protein docking
9	iGEMDOCK	Protein-ligand docking
10	HEX	Protein-ligand docking
	Docking WebServers	
1	SwissDock	It allows the docking of small molecules to target proteins
2	Hawkdock	It predicts protein-protein docking and analysis
3	HADDOCK	High Ambiguity Driven protein-protein Docking
4	Cluspro	It allows protein-protein docking

Table 21.2 Tools and online web servers used for Molecular Dynamics Simulation

S. I	Tools	Description
1	GROMACS	Groningen MAchine for chemical simulation. It can be used to simulate proteins, nucleic acids, lipids, and carbohydrates
2	NAMD	It is a high-performance biomolecular simulation program. Proteins, lipids, carbohydrates and nucleic acids etc. can be simulated
3	Desmond	It is a powerful dynamics simulation program with speed, accuracy, and scalability. It is applicable to proteins and lipids
4	AMBER	It is an extensively used biomolecular simulation program for the simulation of proteins, nucleic acids, and carbohydrates
5	CHARMM	Chemistry at Harvard molecular mechanics is widely used for the simulation study of Proteins, lipids, carbohydrates, and nucleic acids
6	DL_POLY	A general purpose package for MD simulation that allows large complexity liquids studies. Membranes proteins can be studied with this tool
	Online Web servers	
1	WebGro	It is a fully automated molecular simulation

filter out the features with fewer characteristics (Prathipati et al. 2007; Van Drie 2013; Schaller et al. 2020). All the pharmacophoric features consistently found in the active molecules are taken for the final pharmacophoric model. The pharmacophore model will give better knowledge about the 3D space features of the molecules, and this information can be used for producing better molecules with shape and volumes for proper fitting to the receptor protein structure. The wrong shape prevents the fitting of the ligand molecules with the receptor cavity strongly and stably. Catalyst, Discovery Studio, PHASE, PharmMAPPER etc. are some of the frequently used software for the construction of pharmacophores. There are different types of pharmacophore modeling: such as ligand-based and receptor-based pharmacophore models.

21.10.1 Ligand-Based Pharmacophore (LBP)

In ligand-based pharmacophore, standard pharmacophoric features will be derived from a set of inhibitors against a target of interest with a biological activity represented in terms of inhibitory concentration (IC_{50}) values obtained through the same biological assay (Pommier et al. 2010). This approach comes under three-dimensional (3DQSAR) strategies. There is another type of LBP: i.e., the Common feature pharmacophore approach. It differs from 3DQSAR as there is no limitation in the number of training set compounds and does not require experimental biological activity values obtained in the same biological assay.

21.10.2 Receptor-Based Pharmacophore

In receptor-based pharmacophore, the pharmacophoric features are developed based on the ligand interaction details with active site residues, which in turn are used for querying the databases for filtering out the compounds of similar pharmacophore (Dixon 2010; Vuorinen and Schuster 2015). The three major steps involved in the receptor or structure-based pharmacophore modeling are; (1) construction, (2) subtraction, and (3) optimization. In the first step, the key interacting residues in the active site of the target are identified based on the hydrogen bond and hydrophobic features. This will help to identify possible binding sites at the active site. In the subtraction phase, the possible active sites identified are hierarchically clustered using RMS distance, and only the key residues containing active sites are kept. In the target active site, atoms are added to limit the shape constraint, and prevent clashes with ligand atoms are done to finally optimize the model (Amaravadhi et al. 2014).

21.11 Quantitative Structure-Activity Relationship (QSAR)

In some cases, the lack of target chemical structure information in the databases or the modeled structure diminishes the structure-based drug designing approaches. QSAR gives information about the relationship between chemical structure and biological activity in the form of mathematical expression. Thus the identification of bioactive properties like structure descriptors, biological activity, and physiological properties of the novel synthetic compounds helps avoid the synthesis and testing of the compounds (Melo-Filho et al. 2014). This method is used to correlate the structure of the compounds with their experimental properties. According to QSAR, the structure of the compound must contain features responsible for its physical, chemical, and biological properties. It quantitatively correlates molecular descriptors with functions which include biological activity, toxicity, etc., of a set of compounds. Various methodologies can be adopted for the generation of QSAR. One method required a large number of sufficient data sets with its biological activities obtained from experimental methods, followed by the appropriate division of these compounds into training and test set. The reliability of the predicted final model can be checked with validation methods. Based on the derivatives of the descriptors, QSAR is divided into six different types: (1) 1D-QSAR—It involves the biological activities correlated with logP and pKa. (2) In 2D-QSAR, structural descriptors are connected to biological activity. (3) A correlation of activity with an application of force field calculation of the structure of a set of compounds is considered in 3D QSAR. (4) 4D-QSAR is an extended version of 3D-QSAR, with the addition of ligand configuration collection. (5) Various induced fit models are incorporated in 4D to make it 5D-QSAR. (6) Further addition of certain solvation models in 5D to form 6D-QSAR (Cherkasov et al. 2014).

21.12 De Novo Drug Design

This technique deals with the generation of new molecules with desired and specific drug like properties i.e., entirely new molecules can be designed from scratch. In this method, realistic molecular models of 3D structures are used for predicting new drug candidates having the desired target properties, which are reflected in the target model used (Schneider and Fechner 2005). It can also be used to design or model new chemical classes of compounds that can represent similar substituents to the target using a template that is chemically distinct from previously characterized leads. One of the examples is the mini-proteins of 37–43 residues designed, which target the influenza hemagglutinin and botulinum neurotoxin B (Chevalier et al. 2017).

21.12.1 Fragment-Based Screening

In some scenarios, de novo drug design is less or unsuccessful in generating drug candidates when compared to other methods like large database virtual screening. When a 3D target structure is available, programs for ligand designing such as biochemical and organic model builder (BOMB)/LigMerge/LigBuilder can be used to design ligands by adding substituents into a core structure that can bind to the receptor cavity without utilizing ligand databases. Fragment-based screening is an attractive approach for designing potential drug candidates. In this method, many molecular fragments are screened for finding specific interactions with the receptor-binding cavity. These starting fragments are usually identified from a compound library using sensitive biophysical methods. The fragment-based drug design (FBDD) usually generates a compound starting from a chemical fragment having low binding affinity, low complexity in chemical structures, and low molecular weight, usually less than 300 Da. Then, this compound is modified into drug-like molecules through different strategies. If two fragments similarly bind to the receptor cavity, then these fragments can be connected with linkers to obtain a new drug that could be synthetically feasible (Li 2020). Some of the advantages of FBDD are saving experimental costs, offering diverse hits, and exhibiting multiple ways to develop novel compounds. Using hydroxyethyl amine as the base template structure, novel inhibitors were developed for *Enterococcus faecium* ligase VanA (Sova et al. 2009).

21.13 Artificial Intelligence (AI) in Drug Discovery

Artificial intelligence is a machine learning technique where a computer's ability to learn from the existing data is incorporated into drug discovery processes. Several computational modeling techniques adopted artificial intelligence to predict the biological activities of drug molecules. In addition to that, protein folding, protein-protein interaction, QSAR, virtual screening, de novo design, etc., are some of the specialized areas in drug discovery where artificial intelligence plays its own significant role. The huge collection of big data developed by the high throughput screening cannot be handled by machine learning but can be overcome with deep learning techniques. AI is highly inspiring in identifying preclinical candidates in a cost and time-effective way (Patel et al. 2014).

21.14 Network Pharmacology

The conventional drug discovery approach of 'one drug-one target-one disease' is currently facing many challenges in terms of drug safety, efficacy, and sustainability. The revolution in information technology and multi-target drug development has gained value to network biology and polypharmacology strategies. Network pharmacology (NP) is a newly developing area that uses the combination of network biology and polypharmacology approach. It is a whole system-based approach where it helps to understand the effect of drugs on the molecular interactions in the defected/ailing human cells and human disease networks. As most human diseases are due to the dysfunction of multiple proteins, addressing the multiple targets in the metabolic cascade is more effective for the holistic management of diseases. Also, many bioactive from the traditional knowledge can act and trigger multiple targets rather than a highly specific mono-target (Chandran et al. 2017).

The research strategies (Zhang et al. 2013; Chandran et al. 2017; Zhao et al. 2019) adopted in the NP approach are (1) Gathering information regarding the traditional formulations/botanicals. This information can be obtained from consulting with the Ayurvedic practitioners and researchers and also can be depended on several databases like TKDL, Ayusoft, NLAM, etc. (2) Collecting ingredient information. Literature mining can be done to find out whether someone has retrieved the bioactive information of the formulations. Also, digital databases like PubChem, Universal natural product database, Dr. Duke's phytochemical and ethnobotanical databases etc. can be used to retrieve the chemical constituent information. The collected data can be validated by experimental analysis like LCMS, HPLS-MS/MS, HPLC-NMR, etc. (3) Bioactive data mining. After getting the bioactive information i.e., correlating the digitally acquired ingredients data with the experimental data, a final list of bioactive is made. Then the chemistry of bioactive compounds is studied for finding suitable potential targets. (4) Target data mining and target characterization. The information like diseases, gene annotations, and other related information are available in the Human protein atlas, PDB, Therapeutic target database, OMIM, etc. This knowledge can be explored for finding the potential target or multiple targets. (5) Network construction and analysis. Identification of the connections with the target is the crucial step, and for this, the various data generated so far can be submitted to software/tools for visualizing molecular interaction networks like 'Cytoscape' for making networks and analysis. The pathway analysis can be done by databases like KEGG, DAVID, STRING, etc.

All molecules and their interactions within a living cell can be systematically cataloged with this modern approach. The NP technique can help to improve efficacy and predict unwanted off-target effects in the body. Some of the applications of NP include Network pharmacology evaluations of popular Ayurvedic formulations, Understanding the mechanism of action, evaluation of efficacy, safety, synergistic drug action, drug resistance, and identification of possible substitutes for endangered botanicals. It can help to understand the signaling pathways of disease and discover disease-causing genes. The network pharmacology approach can lead to natural product-based drug repurposing and identification of new adjuvants for cancer chemotherapy, neurodegenerative diseases, etc., from Ayurveda.

21.15 Web Servers and Tools Used in CADD

There are several resources available for CADD, including databases, tools, and servers. Information and details about the target, bioactive, and drug compounds are very crucial in CADD, and many repositories have such stored information. PubChem is a repository of small molecules consisting of millions of bioactive compounds. PDB is a global resource where 3D structures and other critical information about the experimentally determined biological macromolecules are available. DrugBank has about 3000 experimental drugs which can be used for drug repurposing, and also it contains more than 800 FDA-approved drugs. Some other useful databases and servers are given in Table 21.3.

21.16 Significance/Applications and Limitations of CADD in Drug Discovery and Development

21.16.1 Significance of CADD in Drug Discovery

The conventional mode of drug discovery involves the experimental identification and screening of suitable target and ligand molecules. Once a logical hypothesis based on the target and ligand is drawn, several experiments are carried out to find a lead compound has therapeutic efficiency. These drug candidates are tested and evaluated in the laboratories, and approved or successful ones will be passed through preclinical and clinical studies. All these processes are tedious and require enormous resources, infrastructural facilities, expensive, risk-prone trials, multi million dollar investments, etc. CADD searches target-based new and improved drugs through in silico screening and combinatorial chemistry approach helping in the screening of the compounds in a short span. The primary significance of CADD in drug discovery are;

1. The filtration of a large set or huge library of compounds is possible based on the pharmaceutical potential and activity in a short time rather than taking many years. Thus those with predicted potential can be further tested or validated experimentally. The hypothesis is that the reduced number of compounds has more probability of being an active compound.
2. CADD helps in the identification of drugs that can be used against multiple drug targets, i.e., multi-target drug designing and polypharmacology.
3. CADD provides information and insight about lead optimization, pharmacokinetics, and pharmacodynamics properties, including bioavailability, dosage, lethality, ADMET, etc.
4. Alteration of existing drugs so that better therapeutic value can be achieved; is mainly done by adding or removing the functional group in a chemical compound. Also, new chemotypes can be made by joining different fragments.

21.16.2 Application of CADD in Drug Discovery

There are many cases where CADD finds its expertise significant. Amprenavir—a Human Immunodeficiency Virus (HIV) protease inhibitor, was developed through protein modeling and dynamic simulations (Wlodawer and Vondrasek 1998; Clark

Table 21.3 Webservers and offline tools for comparative protein modelling (Webb and Sali 2016)

Databases		
S. I.	Protein sequence databases	Url
1	Ensembl	http://www.ensembl.org
2	Genbank	http://www.ncbi.nlm.nih.gov/Genbank
3	Protein Information Resource	http://pir.georgetown.edu/
4	UniProtKB	http://www.uniprot.org
	Domains and super families	
1	CATH/Gene 3D	http://www.cathdb.info
2	Interpro	http://www.ebi.ac.uk/interpro
3	MEME	http://meme.nbcr.net/meme
5	PRINTS	http://www.bioinf.manchester.ac.uk/dbbrowser/PRINTS/index.php
6	ProDom	http://prodom.prabi.fr
7	ProSite	http://prosite.expasy.org/
8	SCOP	http://scop.mrc-lmb.cam.ac.uk/scop
9	SFLD	http://sfld.rbvi.ucsf.edu/
10	SMART	http://smart.embl-heidelberg.de/
	Protein structures and models	
1	ModBASE	http://www.salilab.org/modbase
2	PDB	http://www.pdb.org
3	Protein Model Portal	http://www.proteinmodelportal.org
4	Swiss Model Repository	http://www.swissmodel.expasy.org/repository/
Alignment		
	Sequence and structure based sequence alignment	
1	AlignMe	http://www.bioinfo.mpg.de/AlignMe
2	CLUSTALW	http://www2.ebi.ac.uk/clustalw/
5	FastA	http://www.ebi.ac.uk/Tools/sss/fasta/
8	GENTHREADER	http://bioinf.cs.ucl.ac.uk/psipred/
10	MAFFT	http://mafft.cbrc.jp/alignment/software/
11	MUSCLE	http://www.drive5.com/muscle
12	MUSTER	http://zhanglab.ccmb.med.umich.edu/MUSTER
13	PROMALS3D	http://prodata.swmed.edu/promals3d/promals3d.php
14	PSI-BLAST	http://blast.ncbi.nlm.nih.gov/Blast.cgi
15	PSIPRED	http://bioinf.cs.ucl.ac.uk/psipred/
16	SALIGN	http://www.salilab.org/salign/
17	SAM-T08	http://compbio.soe.ucsc.edu/HMM-apps/
18	Staccato	http://bioinfo3d.cs.tau.ac.il/staccato/
19	T-Coffee	http://www.tcoffee.org/
	Structure	
4	Mammoth	http://ub.cbm.uam.es/software/mammoth.php
5	Mammoth-mult	http://ub.cbm.uam.es/software/mammothm.php
6	MASS	http://bioinfo3d.cs.tau.ac.il/MASS/

(continued)

Table 21.3 (continued)

Databases		
S. I.	Protein sequence databases	Url
7	MultiProt	http://bioinfo3d.cs.tau.ac.il/MultiProt
8	MUSTANG	http://www.csse.monash.edu.au/~karun/Site/mustang.html
9	PDBeFold	http://www.ebi.ac.uk/msd-srv/ssm/
10	SALIGN	http://www.salilab.org/salign/
11	TM-align	http://zhanglab.ccmb.med.umich.edu/TM-align/
	Alignment modules in molecular graphics program	
1	Discovery Studio	http://www.accelrys.com
2	PyMol	http://www.pymol.org/
3	Swiss-PDB Viewer	http://spdbv.vital-it.ch/
4	UCSF Chimera	http://www.cgl.ucsf.edu/chimera
Comparative modeling, threading and refinement		
3	IntFold	http://www.reading.ac.uk/bioinf/IntFOLD/
4	i-TASSER	http://zhanglab.ccmb.med.umich.edu/I-TASSER/
6	ModWeb	http://salilab.org/modweb/
7	Phyre2	http://www.sbg.bio.ic.ac.uk/phyre2
8	RaptorX	http://raptorx.uchicago.edu/
9	Robetta	http://robetta.bakerlab.org/
10	SWISS-MODEL	https://www.expasy.org/resources/swiss-model
	Programs	
2	Modeller	http://www.salilab.org/modeller/
3	MolIDE	http://dunbrack.fccc.edu/molide/
4	Rosetta@home	http://boinc.bakerlab.org/rosetta/
5	RosettaCM	https://www.rosettacommons.org/home
6	SCWRL	http://dunbrack.fccc.edu/scwrl4/SCWRL4.php
	Quality estimation	
1	ANOLEA	http://melolab.org/anolea/index.html
3	ModEval	http://salilab.org/modeval/
5	PROCHECK	http://www.ebi.ac.uk/thornton-srv/software/PROCHECK/
6	Prosa2003	http://www.came.sbg.ac.at
7	QMEAN	http://www.openstructure.org/download/
9	VERIFY3D	http://www.doe-mbi.ucla.edu/Services/Verify_3D/
10	WHATCHECK	https://swift.cmbi.umcn.nl/gv/whatcheck/
	Methods evaluation	
1	CAMEO	http://cameo3d.org/

2006). Raltitrexed, a thymidylate synthase potential inhibitor against HIV designed through SBDD (Anderson 2003). Norfloxacin—an antibiotic against Urinary Tract Infection (UTI), Topoisomerase II, and IV inhibitor, is an output of SBVS (Batool et al. 2019). A Non-steroidal anti-inflammatory drug (NSAID), flurbiprofen, was discovered through molecular docking studies as Cyclooxygenase-2 (COX-2) inhibitor (Miller et al. 2015; Dadashpour et al. 2015). Isoniazid, an enoyl-acyl-ACP

Table 21.4 Recently approved commercial drugs developed using CADD techniques

Drug name	Target protein	Computational method
2017		
Brigatinib	ALK	Homology modeling and molecular docking
Betrixaban	Serine protease Factor Xa (fXa)	Molecular docking
Copanlisib Hydrochloride	PI3K	SBDD and LBDD
Vaborbactam	B-Lactamase	Molecular Docking and MD Simulation
2018		
Abemaciclib	Cyclin-dependent kinase	SAR and SBDD
Apalutamide	Androgen receptor inhibitor	SBDD and SAR
Dacomitinib	Oral kinase	Combined FBDD and SBDD
Duvelisib	PI3K Kinase	SBDD, LBDD, molecular docking and SAR
Glasdegib Maleate	Hedgehog pathway	SAR
Ivosidenib	Isocitrate dehydrogenase-1 (IDH1)	LBDD and SAR
Larotrectinib Sulphate	Tropomyosin-related kinase	LBDD and SAR
Lorlatinib	Tyrosine kinase	SBDD
Talazoparib Tosylate	Poly (ADP-ribose) polymerase-PARP	SBDD, SAR and Lead Optimization
2019		
Darolutamide	Androgen receptor	SBDD, molecular docking and SAR
Entrectinib	Tyrosine kinase inhibitor	SBDD and SAR
Erdafitinib	FGFR tyrosine	SBDD
Fedratinib Hydrochloride	Tyrosine kinase	SBDD and molecular docking
Selinexor	Nuclear export	SBDD, docking
Zanubrutinib	Bruton's tyrosine kinase inhibitor	SBDD

reductase (InhA) inhibitor as anti-tuberculosis drug obtained through SBVS and pharmacophore modeling (Marrakchi et al. 2000). A carbonic anhydrase inhibitor Dorzolamide is identified as efficient against glaucoma by using fragment-based screening method (Grover et al. 2006). Some of the other recently approved inhibitor drugs that had been developed using computational drug discovery methods (Sabe et al. 2021) are tabulated in Table 21.4.

Both SBDD and LBDD approaches are widely used for the development of a new drug-like compound against COVID-19. So far, a few repurposed approved drugs were undergone clinical trials. The researchers can depend on the structure and sequence of databases for retrieving the information of the targets of SARS-CoV-2 virus. The ligand-based and receptor-based virtual screening and identification of COVID-19 inhibitors are pursued with the aid of CADD (Thodi et al. 2021). Several studies are ongoing involving the three-dimensional modeling of the viral targets. The drug resistance that emerged as a result of viral mutation is the biggest challenge in identifying the antiviral, which can be overcome with the aid of a multiple target approach too. There are several drug-like compounds developed through various methods in CADD. Atazanavir, remdesivir, efavirenz, ritonavir, and dolutegravir are a few FDA approved antiviral drugs showing inhibitory potentiality against SARS-CoV-2 3C-like proteinase (Beck et al. 2020). Sofosbuvir, ribavirin, galidesivir, remdesivir, favipiravir, cefuroxime, tenofovir, and hydroxychloroquine are reached for clinical trial as candidate drugs against SARS-CoV-2 RdRp enzyme (Elfiky 2021). Four compounds, talampicillin, lurasidone, ZINC000000702323, and ZINC000012481889 identified against SARS –Cov-2 Main Protease (M^{pro}). Rubitecan, loprazolam, ZINC000015988935, and ZINC000103558522 are identified as TMPRSS2 inhibitors (Elmezayen and Yelekçi 2021).

21.16.3 Limitations of CADD in Drug Discovery

Professionals with interdisciplinary knowledge are required in the case of CADD, which is a major challenge in the field. Other significant challenges in CADD include; the generation of a vast number of chemical structures for finding better combinations and unstable chemical nature, toxicity, or feasibility of synthesis. There is no single mathematical model available that can be used to standardize the synergetic effect of the drug, as the combined effect of the drug will be much greater than the sum of individual drugs. Synergetic computational models are necessary to study the pharmacological properties, metabolic engineering, signaling pathways, etc. Apart from the availability of good reliable models, quality databases are also required. The quality of data determines the quality of the model. Though there is a number of databases available freely, they are far from the requirement of vigorous drug analysis and discovery. In the computer-based drug discovery process, accurate scientific computing is always important. Numerous algorithms, assumptions, and calculations in a short time have to be taken into account. These greatly diminish the ligand-receptor interactions. Superior software and designing of a synthetically feasible and stale compound are improving, which will be an add-on to the CADD.

21.17 Conclusion

Computer Aided Drug Discovery is an effective and powerful computational technique to develop new therapeutic compounds. The discovery process is an expensive extended process. The appreciable progress of CADD in the last two decades, when coupled with the computational power, made the research community reach its peak in developing quick and effective disease-based solutions at a low cost in a short period of time. It begins with target identification and ends up with its validation before it reaches the market. Preclinical and clinical tests are inevitable to get the approval of the regulatory agencies. The fast-track tools of CADD got attention in identifying the hit compounds, hit to lead, and lead optimization. The increasing databases and tools in the bioinformatics field provide an improved basis for the development of inhibitors with preferred specificity. It is expected that the power of CADD will grow as the technology continues to evolve.

Acknowledgment We would like to extend our sincere gratitude to Prof. (Dr.) P R Sudhakaran, Sir Asuthosh Mukherjee Fellow and Prof. (Dr.) Oommen V Oommen, Emeritus Professor of the Department of Computational Biology & Bioinformatics, the University of Kerala for the valuable feedback, suggestions, contribution, and motivation.

References

Amaravadhi H, Baek K, Yoon HS (2014) Revisiting de novo drug design: receptor based pharmacophore screening. Curr Top Med Chem 14:1890–1898. https://doi.org/10.2174/1568026614666140929115506

Anderson AC (2003) The process of structure-based drug design. Chem Biol 10:787–797. https://doi.org/10.1016/j.chembiol.2003.09.002

Bajpai M, Esmay JD (2002) In vitro studies in drug discovery and development: an analysis of study objectives and application of good laboratory practices (GLP). Drug Metab Rev 34:679–689. https://doi.org/10.1081/dmr-120015690

Batool M, Ahmad B, Choi S (2019) A structure-based drug discovery paradigm. Int J Mol Sci 20: 2783. https://doi.org/10.3390/ijms20112783

Beck BR, Shin B, Choi Y, Park S, Kang K (2020) Predicting commercially available antiviral drugs that may act on the novel coronavirus (SARS-CoV-2) through a drug-target interaction deep learning model. Comput Struct Biotechnol J 18:784–790. https://doi.org/10.1016/j.csbj.2020.03.025

Chandran U, Mehendale N, Patil S, Chaguturu R, Patwardhan B (2017) Network pharmacology. Innov Approaches Drug Discov 2017:127–164. https://doi.org/10.1016/B978-0-12-801814-9.00005-2

Chang C-H, Lin J-W, Caffrey JL, Wu L-C, Lai M-S (2015) Different Angiotensin-converting enzyme inhibitors and the associations with overall and cause-specific mortalities in patients with hypertension. Am J Hypertens 28:823–830. https://doi.org/10.1093/ajh/hpu237

Cherkasov A, Muratov EN, Fourches D, Varnek A, Baskin II, Cronin M et al (2014) QSAR modeling: where have you been? Where are you going to? J Med Chem 57:4977–5010. https://doi.org/10.1021/jm4004285

Chevalier A, Silva D-A, Rocklin GJ, Hicks DR, Vergara R, Murapa P et al (2017) Massively parallel de novo protein design for targeted therapeutics. Nature 550:74–79. https://doi.org/10.1038/nature23912

Clark DE (2006) What has computer-aided molecular design ever done for drug discovery? Expert Opin Drug Discov 1:103–110. https://doi.org/10.1517/17460441.1.2.103

Dadashpour S, Tuylu Kucukkilinc T, Unsal Tan O, Ozadali K, Irannejad H, Emami S (2015) Design, synthesis and in vitro study of 5,6-diaryl-1,2,4-triazine-3-ylthioacetate derivatives as COX-2 and β-amyloid aggregation inhibitors. Arch Pharm (Weinheim) 348:179–187. https://doi.org/10.1002/ardp.201400400

Daina A, Michielin O, Zoete V (2017) SwissADME: a free web tool to evaluate pharmacokinetics, drug-likeness and medicinal chemistry friendliness of small molecules. Sci Rep 7:42717. https://doi.org/10.1038/srep42717

De Vivo M, Masetti M, Bottegoni G, Cavalli A (2016) Role of molecular dynamics and related methods in drug discovery. J Med Chem 59:4035–4061. https://doi.org/10.1021/acs.jmedchem.5b01684

DiMasi JA, Grabowski HG, Hansen RW (2016) Innovation in the pharmaceutical industry: new estimates of R & D costs. J Health Econ 47:20–33. https://doi.org/10.1016/j.jhealeco.2016.01.012

Dixon SL (2010) Pharmacophore methods. In: Reynolds CH, Ringe D, Merz Kenneth M Jr (eds) Drug design: structure- and ligand-based approaches. Cambridge University Press, Cambridge, pp 137–150

Elfiky AA (2021) SARS-CoV-2 RNA dependent RNA polymerase (RdRp) targeting: an in silico perspective. J Biomol Struct Dyn 39:3204–3212. https://doi.org/10.1080/07391102.2020.1761882

Elmezayen AD, Yelekçi K (2021) Homology modeling and in silico design of novel and potential dual-acting inhibitors of human histone deacetylases HDAC5 and HDAC9 isozymes. J Biomol Struct Dyn 39:6396–6414. https://doi.org/10.1080/07391102.2020.1798812

Fiser A (2004) Protein structure modeling in the proteomics era. Expert Rev Proteomics 1:97–110. https://doi.org/10.1586/14789450.1.1.97

Gaulton A, Bellis LJ, Bento AP, Chambers J, Davies M, Hersey A et al (2012) ChEMBL: a large-scale bioactivity database for drug discovery. Nucleic Acids Res 40:D1100–D1107. https://doi.org/10.1093/nar/gkr777

Grover S, Apushkin MA, Fishman GA (2006) Topical dorzolamide for the treatment of cystoid macular edema in patients with retinitis pigmentosa. Am J Ophthalmol 141:850–858. https://doi.org/10.1016/j.ajo.2005.12.030

Hao D-C (2019) Chapter 1—Genomics and evolution of medicinal plants. In: Hao D-CBT-RMP (ed). Academic, pp 1–33

Hoque I, Chatterjee A, Bhattacharya S, Biswas RK (2017) An approach of computer-aided drug design (CADD) tools for in silico pharmaceutical drug design and development. Int J Adv Res Biol Sci 4:60–71

Huang B (2009) MetaPocket: a meta approach to improve protein ligand binding site prediction. OMICS 13:325–330. https://doi.org/10.1089/omi.2009.0045

Ibrahim JM, Shanitha A, Nair AS, Oommen OV, Sudhakaran PR (2021) In silico screening and epitope mapping of leptospiral outer membrane protein—Lsa46. J Biomol Struct Dyn 1–19. https://doi.org/10.1080/07391102.2021.2003247

Irwin JJ, Shoichet BK (2005) ZINC—a free database of commercially available compounds for virtual screening. J Chem Inf Model 45:177–182. https://doi.org/10.1021/ci049714+

Kapetanovic IM (2008) Computer-aided drug discovery and development (CADDD): in silico-chemico-biological approach. Chem Biol Interact 171:165–176. https://doi.org/10.1016/j.cbi.2006.12.006

Karplus M, McCammon JA (2002) Molecular dynamics simulations of biomolecules. Nat Struct Biol 9:646–652. https://doi.org/10.1038/nsb0902-646

Kazemipoor M, Jasimah C, Cordell G, Yaze I (2012) Safety, efficacy and metabolism of traditional medicinal plants in the management of obesity: a. Review 3:288–292. https://doi.org/10.7763/IJCEA.2012.V3.201

Kim S, Thiessen PA, Bolton EE, Chen J, Fu G, Gindulyte A et al (2016) PubChem substance and compound databases. Nucleic Acids Res 44:D1202–D1213. https://doi.org/10.1093/nar/gkv951

Kore P, Mutha M, Antre R, Oswal R, Kshirsagar S (2012) Computer-aided drug design: an innovative tool for modeling. Open J Med Chem 02:139–148. https://doi.org/10.4236/ojmc.2012.24017

Laurie ATR, Jackson RM (2005) Q-SiteFinder: an energy-based method for the prediction of protein-ligand binding sites. Bioinformatics 21:1908–1916. https://doi.org/10.1093/bioinformatics/bti315

Lemer CM, Rooman MJ, Wodak SJ (1995) Protein structure prediction by threading methods: evaluation of current techniques. Proteins 23:337–355. https://doi.org/10.1002/prot.340230308

Li Q (2020) Application of fragment-based drug discovery to versatile targets. Front Mol Biosci 7: 180

Lin X, Li X, Lin X (2020) A review on applications of computational methods in drug screening and design. Molecules 25:1375. https://doi.org/10.3390/molecules25061375

Lv Z, Chu Y, Wang Y (2015) HIV protease inhibitors: a review of molecular selectivity and toxicity. HIV AIDS (Auckl) 7:95–104. https://doi.org/10.2147/HIV.S79956

Mahdi M, Szojka Z, Mótyán JA, Tőzsér J (2015) Inhibition profiling of retroviral protease inhibitors using an HIV-2 modular system. Viruses 7:6152–6162. https://doi.org/10.3390/v7122931

Marrakchi H, Lanéelle G, Quémard AK (2000) InhA, a target of the antituberculous drug isoniazid, is involved in a mycobacterial fatty acid elongation system, FAS-II. Microbiology 146(Pt 2): 289–296. https://doi.org/10.1099/00221287-146-2-289

Martí-Renom MA, Stuart AC, Fiser A, Sánchez R, Melo F, Sali A (2000) Comparative protein structure modeling of genes and genomes. Annu Rev Biophys Biomol Struct 29:291–325. https://doi.org/10.1146/annurev.biophys.29.1.291

Melo-Filho CC, Braga RC, Andrade CH (2014) 3D-QSAR approaches in drug design: perspectives to generate reliable CoMFA models. Curr Comput Aided Drug Des 10:148–159. https://doi.org/10.2174/1573409910666140410111043

Miller Z, Kim K-S, Lee D-M, Kasam V, Baek SE, Lee KH et al (2015) Proteasome inhibitors with pyrazole scaffolds from structure-based virtual screening. J Med Chem 58:2036–2041. https://doi.org/10.1021/jm501344n

Pan L, Gardner CL, Pagliai FA, Gonzalez CF, Lorca GL (2017) Identification of the tolfenamic acid binding pocket in PrbP from Liberibacter asiaticus. Front Microbiol 8:1591. https://doi.org/10.3389/fmicb.2017.01591

Patel HM, Noolvi MN, Sharma P, Jaiswal V, Bansal S, Lohan S et al (2014) Quantitative structure–activity relationship (QSAR) studies as strategic approach in drug discovery. Med Chem Res 23: 4991–5007. https://doi.org/10.1007/s00044-014-1072-3

Pence H, Williams A (2010) ChemSpider: an online chemical information resource. J Chem Educ 87:1123–1124. https://doi.org/10.1021/ed100697w

Pommier Y, Leo E, Zhang H, Marchand C (2010) DNA topoisomerases and their poisoning by anticancer and antibacterial drugs. Chem Biol 17(5):421–433. https://doi.org/10.1016/j.chembiol.2010.04.012

Prathipati P, Dixit A, Saxena KA (2007) Computer-aided drug design: integration of structure-based and ligand-based approaches in drug design. Curr Comput Aided Drug Des 3:133–148

Prieto-Martínez FD, López-López E, Eurídice Juárez-Mercado K, Medina-Franco JL (2019) Chapter 2—Computational drug design methods—current and future perspectives. In: Roy KBT-ISDD (ed). Academic, pp 19–44

Sabe VT, Ntombela T, Jhamba LA, Maguire GEM, Govender T, Naicker T, Kruger HG (2021) Current trends in computer aided drug design and a highlight of drugs discovered via computational techniques: a review. Eur J Med Chem 224:113705. https://doi.org/10.1016/j.ejmech.2021.113705

Sansgiry SS, Bhansali AH, Bapat SS, Xu Q (2017) Abuse of over-the-counter medicines: a pharmacist's perspective. Integr Pharm Res Pract 6:1–6. https://doi.org/10.2147/IPRP.S103494

Schaller D, Šribar D, Noonan T, Deng L, Nguyen TN, Pach S et al (2020) Next generation 3D pharmacophore modeling. WIREs Comput Mol Sci 10:e1468. https://doi.org/10.1002/wcms.1468

Schneider G, Fechner U (2005) Computer-based de novo design of drug-like molecules. Nat Rev Drug Discov 4:649–663. https://doi.org/10.1038/nrd1799

Sofowora A, Ogunbodede E, Onayade A (2013) The role and place of medicinal plants in the strategies for disease prevention. Afr J Tradit Complement Altern Med 10:210–229. https://doi.org/10.4314/ajtcam.v10i5.2

Sova M, Cadez G, Turk S, Majce V, Polanc S, Batson S et al (2009) Design and synthesis of new hydroxyethylamines as inhibitors of D-alanyl-D-lactate ligase (VanA) and D-alanyl-D-alanine ligase (DdlB). Bioorg Med Chem Lett 19:1376–1379. https://doi.org/10.1016/j.bmcl.2009.01.034

Sun J, Chen K (2017) NSiteMatch: prediction of binding sites of nucleotides by identifying the structure similarity of local surface patches. Comput Math Methods Med 2017:5471607. https://doi.org/10.1155/2017/5471607

Takebe T, Imai R, Ono S (2018) The current status of drug discovery and development as originated in United States academia: the influence of industrial and academic collaboration on drug discovery and development. Clin Transl Sci 11:597–606. https://doi.org/10.1111/cts.12577

Talele TT, Khedkar SA, Rigby AC (2010) Successful applications of computer aided drug discovery: moving drugs from concept to the clinic. Curr Top Med Chem 10:127–141. https://doi.org/10.2174/156802610790232251

Tan KP, Nguyen TB, Patel S, Varadarajan R, Madhusudhan MS (2013) Depth: a web server to compute depth, cavity sizes, detect potential small-molecule ligand-binding cavities and predict the pKa of ionizable residues in proteins. Nucleic Acids Res 41:W314–W321. https://doi.org/10.1093/nar/gkt503

Tang Y, Zhu W, Chen K, Jiang H (2006) New technologies in computer-aided drug design: toward target identification and new chemical entity discovery. Drug Discov Today Technol 3:307–313. https://doi.org/10.1016/j.ddtec.2006.09.004

Thodi RC, Ibrahim JM, Surendran VA, Nair AS, Sukumaran ST (2021) Rutaretin1'-(6″--sinapoylglucoside): promising inhibitor of COVID 19 m(pro) catalytic dyad from the leaves of Pittosporum dasycaulon miq (Pittosporaceae). J Biomol Struct Dyn 1–17. https://doi.org/10.1080/07391102.2021.1972841

Tian W, Chen C, Lei X, Zhao J, Liang J (2018) CASTp 3.0: computed atlas of surface topography of proteins. Nucleic Acids Res 46:W363–W367. https://doi.org/10.1093/nar/gky473

Tiwari A, Singh S (2022) Chapter 13—Computational approaches in drug designing. In: Singh DB, Pathak RKBT-B (eds). Academic, pp 207–217

Tungmunnithum D, Thongboonyou A, Pholboon A, Yangsabai A (2018) Flavonoids and other phenolic compounds from medicinal plants for pharmaceutical and medical aspects: an overview. Medicines (Basel, Switzerland) 5:93. https://doi.org/10.3390/medicines5030093

Van Drie JH (2007) Computer-aided drug design: the next 20 years. J Comput Aided Mol Des 21: 591–601. https://doi.org/10.1007/s10822-007-9142-y

Van Drie JH (2013) Generation of three-dimensional pharmacophore models. WIREs Comput Mol Sci 3:449–464. https://doi.org/10.1002/wcms.1129

Vijayakrishnan R (2009) Structure-based drug design and modern medicine. J Postgrad Med 55: 301–304. https://doi.org/10.4103/0022-3859.58943

Volkamer A, Kuhn D, Rippmann F, Rarey M (2012) DoGSiteScorer: a web server for automatic binding site prediction, analysis and druggability assessment. Bioinformatics 28:2074–2075. https://doi.org/10.1093/bioinformatics/bts310

Vuorinen A, Schuster D (2015) Methods for generating and applying pharmacophore models as virtual screening filters and for bioactivity profiling. Methods 71:113–134. https://doi.org/10.1016/j.ymeth.2014.10.013

Vyas VK, Ukawala RD, Ghate M, Chintha C (2012) Homology modeling a fast tool for drug discovery: current perspectives. Indian J Pharm Sci 74:1–17. https://doi.org/10.4103/0250-474X.102537

Webb B, Sali A (2016) Comparative protein structure modeling using MODELLER. Curr Protoc Bioinform 54:5.6.1–5.6.37. https://doi.org/10.1002/cpbi.3

Wishart DS, Knox C, Guo AC, Cheng D, Shrivastava S, Tzur D, Gautam B, Hassanali M (2008) DrugBank: a knowledgebase for drugs, drug actions and drug targets. Nucleic Acids Res 36: D901–D906. https://doi.org/10.1093/nar/gkm958

Wlodawer A, Vondrasek J (1998) Inhibitors of HIV-1 protease: a major success of structure-assisted drug design. Annu Rev Biophys Biomol Struct 27:249–284. https://doi.org/10.1146/annurev.biophys.27.1.249

Yousef M, Abdelkader T, El-Bahnasy K (2019) Performance comparison of ab initio protein structure prediction methods. Ain Shams Eng J 10:713–719. https://doi.org/10.1016/j.asej.2019.03.004

Zhang G, Li Q, Chen Q, Su S (2013) Network pharmacology: a new approach for Chinese herbal medicine research. Evid Based Complement Altern Med 2013:621423. https://doi.org/10.1155/2013/621423

Zhao H, Shan Y, Ma Z, Yu M, Gong B (2019) A network pharmacology approach to explore active compounds and pharmacological mechanisms of epimedium for treatment of premature ovarian insufficiency. Drug Des Devel Ther 13:2997–3007. https://doi.org/10.2147/DDDT.S207823

Zhu H, Pisabarro MT (2011) MSPocket: an orientation-independent algorithm for the detection of ligand binding pockets. Bioinformatics 27:351–358. https://doi.org/10.1093/bioinformatics/btq672

Chapter 22
Antimicrobial Drugs: Possibilities from Medicinal Plants Part A—Antibacterials and Antivirals

Remya Babu, Prakashkumar Raveendran, and Shiburaj Sugathan

Abstract In recent years, the number of microbial infectious diseases has increased considerably. The proliferation of novel viruses and the information of new resistant strains pose a risk to eliminating contagious illnesses. Synthetic antimicrobial drugs are being widely employed to prevent and cure microbial infections; despite this, the growth of antibiotic-resistant microorganisms and decreased antimicrobial efficiency necessitate the development of novel antimicrobials. Since life began, human beings have depended on plant-derived antimicrobials to fight infections. Plant metabolites from medicinal plants are good sources of novel antimicrobials. Extensive in vitro and in vivo research has been undertaken on the antibacterial effect of plant phytochemicals. The current chapter focuses on the antimicrobial activities of medicinal plants and their novel bioactive compounds against bacteria and viruses.

Keywords Antimicrobial drugs · Medicinal plants · Antibacterial · Antifungal

R. Babu
Department of Botany, University of Kerala, Thiruvananthapuram, Kerala, India

Jawaharlal Nehru Tropical Botanic Garden and Research Institute, Thiruvananthapuram, Kerala, India

P. Raveendran
Jawaharlal Nehru Tropical Botanic Garden and Research Institute, Thiruvananthapuram, Kerala, India

S. Sugathan (✉)
Department of Botany, University of Kerala, Thiruvananthapuram, Kerala, India
e-mail: drshiburaj@karalauniversity.ac.in

S. T. Sukumaran, K. T R (eds.), *Conservation and Sustainable Utilization of Bioresources*, Sustainable Development and Biodiversity 30,
https://doi.org/10.1007/978-981-19-5841-0_22

22.1 Introduction

India features an excellent past of the traditional medical system 'Ayurveda,' a singular and distinct health care system thanks to its universe and holistic approach. The indigenous knowledge of medicinal plants has provided a comprehensive source of diverse bioactive molecules of clinical importance. The traditional knowledge about the healing power of many medicinal plants is of centuries old. The scientific evaluations based on Ethnobotanical and Ethnopharmacological studies are possible to tap this traditional knowledge.

The earliest information on the ancient therapeutic system of plants is from Mesopotamia and dated from about 2600 BC; plant-derived substances used were mainly oils, most of which are still used for bacterial infections and inflammations. Since 3000 BCE, civilizations in Egypt, the Middle East, India, and China have complied with specific medicinal herbs and written records. The particular plants to be used and methods for applying them to particular alignments were passed down from generation to generation (Newman et al. 2000). Similarly, the Indian Ayurvedic system has been documented from around 1000 BC (Charaka; Sushruta; and Samhitas with 341 and 516 medicines, respectively), and this system served as the foundation for other systems, and this method was the foundation for the fundamental Tibetan medical treatise, Gyu-Zhi (four tantras), which was translated from Sanskrit in the eighth-century AD (Fallarino 1994). Herbal medications were commonly utilized until the early half of the twentieth century when a trend toward synthetic medicines became popular because they were more effective, patented, and profitable (Tyler 1999). Natural goods, and medical agents produced from plant metabolites, formed an integral part of the healthcare systems of the remaining 20% of the world's population, mostly in developed nations, with natural products accounting for more than 50% of all pharmaceuticals drugs in clinical use (Baker et al. 1995). The emergence of multi-drug resistant bacterial strains with increased antibiotic resistance has recently sparked renewed interest in identifying novel antimicrobial compounds from natural sources for therapeutic and preventative reasons against microbial infections. In constant search for medicinal efficacy of plants and their phytochemicals, as recorded data available now are relatively minor compared to the large variety of plant populations. Microbial sources are the sources of almost all research and natural compounds as antimicrobials. Penicillin's discovery paved the way for antibiotics like streptomycin, aureomycin 459, and chloromycetin to follow. However, soil microbes and fungus create the majority of therapeutically used antibiotics (Trease and Evans 1972). Antimicrobials derived from plants are a substantial unexplored source of medication and need to be investigated further. Antimicrobials from plant extractions have significant therapeutic promises that effectively treat infectious diseases while also avoiding many of the adverse effects of synthetic antimicrobials.

22.2 Antimicrobial Compounds

Plants synthesize a wide array of secondary metabolites as a natural defense mechanism against microbial and insect attacks. Some of these chemicals are hazardous to animals, whereas others are not. Plants have an almost endless potential to produce aromatic compounds, most of which are phenols or their oxygen-substituted counterparts. The database NAPRALERT already included over 88,000 secondary metabolites in 1988, and around 4000 new ones are reported per year (Loub et al. 1985). Many of these bioactive compounds have been employed for food or medical purposes in the form of entire plants or plant extracts. It is well known and experimentally proved that plants produce many secondary metabolites with specific antimicrobial activity. These plant-formed antibiotics are derived from precursors via de novo synthesis in response to defense mechanisms against microbes, insects, and herbivores. The secondary antibacterial metabolites are classified into three large molecular families: phenolics, terpenes, and alkaloids. Some terpenoids give plants their odors; quinones and tannins provide the pigments, and some are responsible for the flavor. Since time immemorial, humans have used medicinal plants or their extracts for various ailments, drugs such as morphine, codeine, reserpine, digoxin, taxol, quinine, and artemisinin.

Phenolic substances are phytochemicals with a phenol structure, defined as an aromatic benzene ring with at least one hydroxyl group (Vermerris and Nicholson 2006). Phenolic substances are often found in the environment. Phytochemicals protect plants from microbial diseases, UV radiation, and external and internal chemical stresses. This vast and varied category of phytochemicals is divided into various subclasses based on chemical structures and plant occurrence. Simple phenols and phenolic acids are some of the primary bioactive phytochemicals with a single phenolic ring. Cinnamic and caffeic acids are phenylpropane-derived molecules with a higher oxidation state. Caffeic acid has significant effective inhibition against bacteria, fungus, and viruses, commonly seen in the herbs tarragon and thyme (Wild 1994). Catechol and pyrogallol are hydroxylated phenols with two and three OH groups, respectively, reported as antibacterial. The number of hydroxyl groups, their location, and highly oxidized phenols is reported as toxic to microorganisms (Geissman 1963). Quinones are ubiquitous, highly reactive colored compounds having an aromatic ring with two ketone substitutes. Reports show that anthraquinone has bacteriostatic and bactericidal activities (Kazmi et al. 1994). Flavones, flavonoids, and flavonols are phenolic structures known to be synthesized by plants in response to microbial infections. One flavonoid catechin can inhibit the cholera toxin produced by *Vibrio cholerae* and has antimicrobial properties against *Streptococcus mutans, Shigella,* and other bacteria (Sakanaka et al. 1992). The polyphenolic compounds found abundantly in plants are tannins in the hydrolyzable and condensed state. Tannins are also reported as having antibacterial and antifungal properties to many microorganisms. Ellagitannin extracts of various berries have significant inhibitory activity against gastro-intestinal pathogenic bacteria such as *Vibrio cholerae, Shigella dysenteriae, Campylobacter* spp., *Salmonella, Listeria,* and *Lactobacillus* strains (Puupponen-Pimiä et al. 2005). Coumarins are discovered in many plants and microorganisms which exhibit a wide range of antimicrobial, antifungal, and antiviral activity. They consist of fused benzene and α-pyrone ring. The coumarin derivatives such as 5-methoxyseslin and its brominated substituents, alloxanthoxyletine, dipetaloloctone, and amino-coumarin-7-amino-4-methylcoumarine have broad-spectrum antibacterial activity and antifungal activity (Liu et al. 2008).

Alkaloids are heterocyclic nitrogen metabolites characterized by their biogenic precursors or carbon skeleton. Indole-type alkaloids have a significant inhibitory effect against some pathogenic Gram-positive, Gram-negative, and acid-fast bacteria and fungi (Mallikharjuna and Seetharam 2009). Saponins are non-volatile compounds naturally seen in plants, which can form soap-like foam when shaken with water because of their polar and non-polar structural elements. The antibacterial activity of a sapogenin moiety depends on its chain length and sugar composition. Saponin can inhibit both Gram-positive and Gram-negative bacteria at different concentrations. Terpenoids/essential oils are the secondary metabolites that give fragrance to the plants, highly enriched in compounds from isoprene units (C_5). They are classified into monoterpenes (C_{10}), sesquiterpenes (C_{15}), diterpenes (C_{20}), squalene (C_{30}), tetraterpenes (C_{40}) based on the number of isoprene units. When these compounds contain an additional oxygen molecule, they are called terpenoids. Phenylpropanoids are synthesized through the mevalonate pathway and terpenoids through shikimic acid pathways. Essential oils are a group of secondary metabolites of volatile compounds from flowers, petals, leaves, stems, fruits, roots, and barks obtained by the steam distillation process. They manifest vigorous antimicrobial activity against many pathogenic and non-pathogenic bacteria and fungi. Curcumin, gingerol, menthol, linalool, geraniol, citral, and cineole are some examples of essential oils which showed a high antibacterial and antifungal activity against a wide range of microorganisms (Pattnaik et al. 1997).

22.3 Antibacterial Drugs from Medicinal Plants

Terminalia sericea is a shrub found abundantly in Africa's tropical and warm temperate regions. Its leaves, root, and bark were reported to have antimicrobial properties (Bruneton 1995; Fyhrquist et al. 2002, 2004). From the roots of *Terminalia* species, pentacyclic triterpenoids have been isolated. After several kinds of research (Eldeen et al. 2006) isolated the bioactive compound Anolignan B which shows a more potent antibacterial activity against both Gram-positive (*Bacillus substilis* and *Staphylococcus aureus*) and Gram-negative bacteria (*Escherichia coli* and *Klebsiella pneumoniae*). Anolignan B, which was first isolated from *Anogeissus acuminata*, was previously active against the HIV-1 reverse transcriptase enzyme, which possesses antiviral medicinal characteristics (Rimando et al. 1994). In traditional Chinese medicine, *Lonicera japonica* Thunb. (*Jinyinhua* in Chinese) commonly used internally for acute rheumatoid arthritis, hepatitis, and upper respiratory tract infections (Sulaiman et al. 2008). Pharmacological studies of these plant extracts also have hepatoprotective, cytoprotective, antimicrobial, antibiotic, antioxidative, antiviral, and anti-inflammatory activities (Chidananda et al. 2010).

The bacteriostatic assay-guided extraction of *Lonicera japonica* leaves yielded five bioactive compounds (Xiong et al. 2013), with significant antibacterial activity against *Staphylococcus aureus* and *Escherichia coli.* These inlude 3,5-di-*O*-caffeoylquinic acid, 4,5-di-*O*-caffeoylquinic acid, luteoloside, 3-*O*-caffeoylquinic acid andd secoxyloganin. Among these, secoxyloganin was isolated for the first time from leaves of L. japonica. The indigenous peoples of South Africa used the roots of *Euclea natalensis* against various bacterial infections (Watt and Breyer-Brandwijk 1962). Six bioactive naphthoquinones were isolated from the root extracts: diospyrin, isodiospyrin, mamegakinone, 7-methyljuglone, neodiospyrin, and shinanolone (Van der Kooy et al. 2006). Among six isolates, neodiospyrin was isolated for the first time, and all six bioactive compounds showed antimycobacterial activity against *Mycobacterium tuberculosis.* The MIC values were compared well to known antimycobacterial drugs, ethambutol, isoniazid, and rifampicin. Next, the traditional medicinal plant *Combretum erythrophyllum* is widely used in South Africa for treating abdominal pains and venereal disease (Hutchings 1996). Seven antibacterial flavonoids were isolated using bioassay-guided fractionation, apigenin, genkwanin, 5-hydroxy-7,4′-dimethoxyflavone, rhamnocitrin, kaempferol, quercetin-5,3′-dimethyl ether, rhamnazin. All the bioactive flavonoids showed antibacterial activity against *Vibrio cholerae* and *Micrococcus faecalis* (Martini et al. 2004). Rhamnocitrin and quercetin-5,3′-dimethyl ether also inhibited *Micrococcus luteus* and *Shigella sonei* (Table 22.1).

The traditional medicine *Warburgia salutaris* acts as an expectorant and is smoked for coughs and colds. Fractionation of the ethyl acetate extract of the stem bark of *W. salutaris* yielded a sesquiterpenoid, muzigadial exhibited antimicrobial activity against Gram-positive bacteria (Rabe and Van Staden 2000). The *Terminalia sericea, Senna petersiana,* and *Anredera cordifolia* are indigenous medicinal plants used by traditional healers to treat illnesses, including sexually transmitted diseases (Ndubani 1997) that showed antibacterial properties against *Bacillus pumilus, Bacillus cereus, Serratia marcescence,* and *Pseudomonas aeruginosa.* The bioactive flavonoid luteolin was isolated as a pure compound from the seed of *Senna petersiana* (Tshikalange et al. 2005). The antimicrobial activity of six tannins isolated from *Vaccinium vitis-idaea* was tested to treat periodontal disease. The clinical pathogens *Actinobacillus actinomycetemcomitans, Porphyromonas gingivalis,* and *prevotella intermedia* were inhibited by procyanidin B-1, procyanidin B-3, proanthocyanidin A-1, cinnamtannin B_1, epicatechin-(4β → 8-epicatechin-(4β → 8, 2β → O → 7)-catechin, and epicatechin-(4β → 6)-epicatechin-(4β → 8, 2β → O → 7)-catechin (Ho et al. 2001) and are helpful for the treatment of tissue damage caused by the generation of ROS, including the treatment of periodontal disease (Fig. 22.1).

Table 22.1 Some bioactive compounds with antibacterial activity

Plant name	Bioactive compound	Chemical constituents	Activity	Organism	References
Terminalia sericea	Anolignan B	2,3-Bis-(4-hydroxy benzyl)1,3-butadiene	Antibacterial and anti-HIV activity	*Bacillus substilis* *Staphylococcus aureus* *Escherichia coli* *Klebsiellae pneumoniae*	Ibrahim et al. (2006)
Lonicera japonica	Phenolic compounds	3,5-di-*O*-caffeoylquinic acid, 4,5-di-*O*-caffeoylquinic acid, luteoloside, 3-*O*-caffeoylquinic acid, secoxyloganin	Antibacterial	*S. aureus* *E. coli*	Xiong et al. (2013)
Euclea natalensis	Naphthoquinones	Diospyrin, isodiospyrin, mamegakinone, 7-methyljuglone, neodiospyrin and shinanolone	Antimycobacterial	*Mycobacterium tuberculosis*	Van der Kooy et al. (2006)
Combretum erythrophyllum	Flavonoid	Apigenin, genkwanin, 5-hydroxy-7,4′--dimethoxyflavone, rhamnazin, kaempferol, quercetin-5,3′-dimethylether, rhamnocitrin	Antibacterial	*Vibrio cholerae*, *Micrococcus faecalis* *M. luteus* *Shigella sonei*	Martini et al. (2004)
Warburgia salutaris	Sesquiterpenoid	Muzigadial	Antibacterial	*S. aureus* *B. subtilis* *S. epidermis* *M. luteus* *E. coli* *K. pneumoniae*	Rabe and Van Staden (2000)
Senna petersiana	Flavonoid	Luteolin	Antibacterial	*Bacillus pumilus*, *B. cereus*, *Serratia marcescence*, *Pseudomonas aeruginosa*	Tshikalange et al. (2005)
Vaccinium vitis-idaea L	Tannin	Procyanidin B-1, procyanidin B-3, proanthocyanidin A-1, cinnamtannin B_1, epicatechin-(4β → 8-epicatechin-(4β → 8, 2β → O → 7)-catechin, and epicatechin-(4β → 6)-epicatechin-(4β → 8, 2β → O → 7)-catechin	Antibacterial	*Actinobacillus actinomycetemcomitans*, *Porphyromonas gingivalis*, *Prevotella intermedia*	Ho et al. (2001)

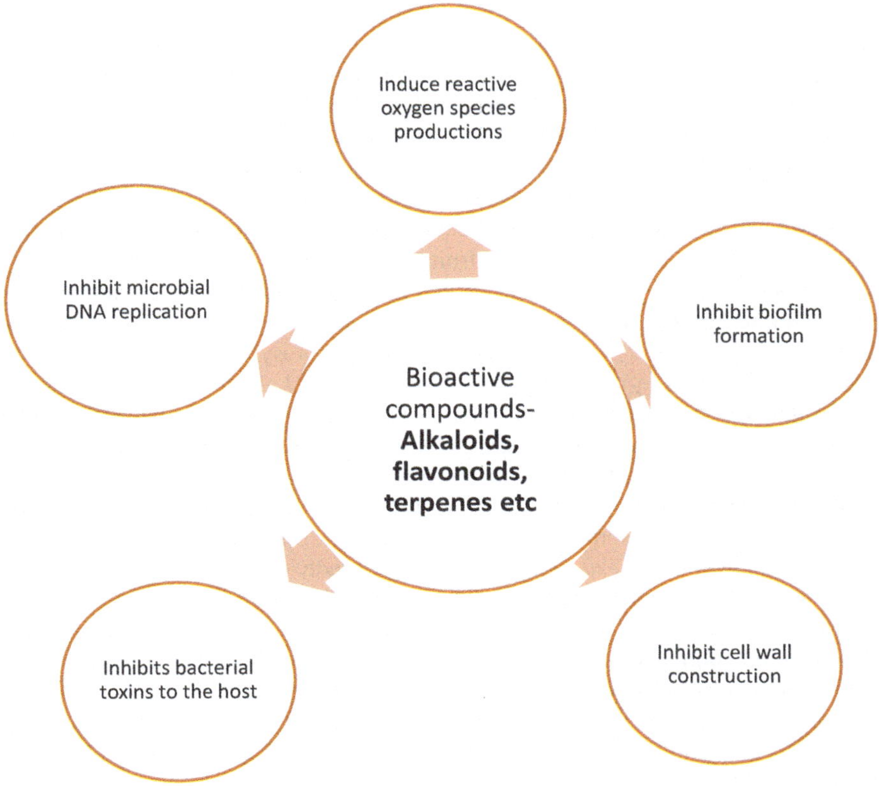

Fig. 22.1 Effects of bioactive compounds on bacteria

22.4 Antiviral Drugs from Medicinal Plants

After the 1918 influenza pandemic, the world battered through the current pandemic, which began in December 2019 and continues to this day, has become the worst epidemic in history. This is the third outbreak produced by the coronavirus through its new variant O-micron in the twenty-second century in India; while writing this chapter about the antiviral drugs and their possibilities by medicinal plants. As of November 13, 2020, 1.30 million people have died due to COVID-19, and 53.76 million have tested positive for the virus (World Health Organisation 2020). It has been discovered that various medicinal plants are already being used to treat respiratory infections.

Traditional Chinese medicinal plant *Forsythiae fructus*, and dried fruits extracts are currently underway as part of the global race to produce viable COVID-19 treatments (Maxmen 2020). It's no surprise that medicinal plant drugs might be used as a powerful weapon against COVID-19. Traditional medicinal drugs derived from these medicinal plants might be used for various purposes, including treating

COVID-19 and providing raw materials for powerful antiviral medications. In 1963 the approval of the first antiviral drug idoxuridine, which inhibits viral DNA synthesis, started the era of antiviral drug development. Since then, more antiviral drugs have been approved to treat viral infections, including HIV, HBV, HCV, and influenza (De Clercq and Li 2016). Medicinal plant extracts have shown to be an excellent option for developing antiviral drugs that can terminate the multiple phases of the virus replication cycle. Plant-based antiviral bioactive compounds, such as anthocyanins, have shown promising benefits in various clinical trials (Mohammadi Pour et al. 2019). *Ribes nigrum* L. fruits extracts were fractionated by column chromatography, and yields were estimated as anthocyanins have antiviral activity against influenza A and B viruses and herpes simplex virus 2. The significant fractions consisted of 3-0-alpha-L-rhamnopyranosyl-beta-D-glucopyranosyl-cyanidin and 3-*O*-beta-D-glucopyranosyl-cyanidin, 3-*O*-alpha-L-rhamnopyranosyl-beta-D-glucopyranosyl-delphinidin and 3-*O*-beta-D-glucopyranosyl-delphinidin showed potent antiviral activity against influenza viruses A and B (Knox et al. 2001). Anthocyanins inhibited viruses through several modes of action (1) prevented viral attachment and entry to cells, (2) suppressed the growth of influenza viruses in vitro (3) inhibited adsorption of influenza viruses A and B onto the cell surface. Asian medicinal plant *Kaempferia parviflora* crude extract showed significant antiviral activity against highly pathogenic avian influenza virus (H5N1). 5-hydroxy-7-methoxy flavone, 5,7-dimethoxyflavone, trimethylapigenin, and tetramethylluteolin were the bioactive compounds isolated from *K. parviflora* bear anti-inflammatory activity and inhibit pathogenic H5N1 influenza virus replication (Sornpet et al. 2017). In the following report, an ancient herbal plant, *Silybum marianum,* native to Southern Europe and Asia, is used to treat liver and gallbladder disorders, including hepatitis, cirrhosis, and hepatoprotection agent.

Hepatitis C is a primary chronic disease that slowly leads to cirrhosis and hepatocellular carcinoma development. Available drugs had considerable side effects and were partially effective. The compound silymarin, derived from the milk thistle from *Silybum marianum,* contains seven flavonolignans (silybin A, silybin B, isosilybin A, isosilybin B, silychristin, isosilychristin, silydianin), and one flavonoid (taxifolin). Biochemically fractionated by high-performance liquid chromatography of extract, silybin A, silybin B, and isosilybin A, isosilybin B reported significant anti-NF-kB and anti-HCV activity (Polyak et al. 2007). Human immunodeficiency virus 1 (HIV-1) causes the majority of HIV infections. Currently, using anti-HIV drugs produces drug-resistant viruses. Finding new novel plant-based bioactive compounds that are useful for preventing transmission of HIV and treatment of patients were necessary.

Reported showed that flavonoid compounds have a significant anti-HIV activity which inhibits reverse transcription, entry of virus, the integrase, and protease activity with low toxicity. The herbal plant *Securigera securidaca* is an annual herb in eastern folk medicine used against various diseases, and its seed extraction has been detected with cardenolides, steroidal and pentacyclic compounds (Table 22.2).

The novel compounds Kaempferol and kaempferol-7-*O*-glucoside isolated have potent anti-HSV activity, and a further study reported that the compounds showed

Table 22.2 Bioactive compounds with antiviral activity

Plant name	Bioactive compound	Chemical constituents	Organism	References
Ribes nigrum L	Anthocyanin	3-0-alpha-L-rhamnopyranosyl-beta-D-glucopyranosyl-cyanidin and 3-*O*-beta-D-glucopyranosyl-cyanidin, 3-*O*-alpha-L-rhamnopyranosyl-beta-D-glucopyranosyl-delphinidin and 3-*O*-beta-D-glucopyranosyl-delphinidin	Influenza virus A and B, Herpes simplex virus 2	Knox et al. (2001)
Kaempferia parviflora		5-hydroxy-7-methoxyflavone, 5,7-dimethoxyflavone, trimethylapigenin and tetramethylluteolin	Avian influenza virus (H5N1)	Sornpet et al. (2017)
Silybum marianum	Flavonolignans Flavonoid	Silybin A, silybin B, isosilybin A, isosilybin B, silychristin, isosilychristin, silydianin Taxifolin	HCV	Polyak et al. (2007)
Securigera securidaca	Flavonoid	Kaempferol and kaempferol-7-*O*-glucoside	HCV-1 HIV	Behbahani et al. (2014)
Green tea	Polyphenol	Epigallocatechin gallate (EGCG)	Zika virus HIV, HCV, HSV	Carneiro et al. (2016)
	Flavonoid	Herbacetin, rhoifolin and pectolinarin	SARS CoV-2	Jo et al. (2020)
	Phenol Flavonol Flavanone Alkaloid Phenylpropyl alkaloid	Gallic acid, Quercetin, Naringin, Capsaicin and Psychotrine	SARS CoV-2	Alrasheid et al. (2021)

significant anti-HIV-1 reverse transcriptase activity. This could be considered a new anti-HIV potential drug for the early treatment of HIV infection (Behbahani et al. 2014). The first Zika virus was reported, which causes Zika fever in 1947 in Uganda. The first outbreak was observed in Brazil in May 2015, with increased microcephaly cases reported in affected newborns. No approved drugs and vaccines are available for treating viral infections. Green tea contains a polyphenol (−) epigallocatechin gallate (EGCG) antiviral compound against many viruses, including human immunodeficiency virus (HIV), influenza virus, herpes simplex virus (HSV), and hepatitis C virus (HCV) (Calland et al. 2012). The report showed that EGCG could inhibit Zika virus entry on host cells and has a virucidal effect on a minimum concentration of >5 μM (Carneiro et al. 2016).

Coronaviruses are single-stranded RNA viruses with large envelopes, which are highly diverse, cause respiratory tract infections, and can infect both animals and humans. Based on the information available, the bioactive drug compounds from the plants' extracts targets are viral RNA-dependent RNA polymerase, receptors seen in the cell membrane of SARS-CoV-2 (ACE-2), and the spike proteins depend on the viral structure and its replication process. The flavonoids herbacetin, rhoifolin, and pectolinarin have antiviral activity against SARS-CoV by effectively blocking its enzymatic activity by inhibiting its 3C-like protease (Jo et al. 2020). Another recent study aimed to assess the bioactive drug from medicinal plants, antiviral inhibitors for COVID-19, using the molecular docking method. The essential tool for drug discovery is the computational technique used to estimate the affinity between two molecules, such as ligand-protein and the protein-protein interaction, and help predict the toxic side effects of compounds. The bioactive compounds isolated from medicinal plants are Gallic acid (−17.45), Quercetin (−15.81), Naringin (−14.50), Capsaicin (−13.90), and Psychotrine (−13.5), which have a potential inhibition than that of chloroquine (Alrasheid et al. 2021). These compounds might be used to prevent COVID-19 infection, leading to the development of a natural medicinal drug against COVID-19. The life-threatening coronavirus infections, several studies are ongoing to produce effective vaccines against the unpredictable changes of COVID variants. Is evidence that new variants can evade immunity produced by vaccines or previous infections. The researchers and scientists are analyzing the idea of modifying the vaccines that are already being used worldwide. Extensive studies and research investigations are needed for new drug discovery and development of safe, effective, and low-cost antiviral compounds from medicinal plants and herbs.

22.5 Conclusions

Plants have always been a source of healing for humanity. A multi-prolonged approach will be needed to curb the rise of emerging and resistant infectious diseases, including developing novel therapeutic compounds or formulations. Researchers are looking for novel compounds with antimicrobial action and synergism with already available antimicrobial agents. As a result, medicinal plants are becoming a viable alternative for infection therapy.

References

Alrasheid AA, Babiker MY, Awad TA (2021) Evaluation of certain medicinal plants compounds as new potential inhibitors of novel coronavirus (COVID-19) using molecular docking analysis. In Silico Pharmacol 9(1):1–7

Baker JT, Borris RP, Carté B, Cordell GA, Soejarto DD, Cragg GM, Tyler VE (1995) Natural product drug discovery and development: new perspectives on international collaboration. J Nat Prod 58(9):1325–1357

Behbahani M, Sayedipour S, Pourazar A, Shanehsazzadeh M (2014) In vitro anti-HIV-1 activities of Kaempferol and kaempferol-7-O-glucoside isolated from Securigera securidaca. Res Pharm Sci 9(6):463

Bruneton J (1995) Pharmacognosy, phytochemistry, medicinal plants. Lavoisier Publishing

Calland N, Albecka A, Belouzard S, Wychowski C, Duverlie G, Descamps V, Séron K (2012) (−)-Epigallocatechin-3-gallate is a new inhibitor of hepatitis C virus entry. Hepatology 55(3): 720–729

Carneiro BM, Batista MN, Braga ACS, Nogueira ML, Rahal P (2016) The green tea molecule EGCG inhibits Zika virus entry. Virology 496:215–218

De Clercq E, Li G (2016) Approved antiviral drugs over the past 50 years. Clin Microbiol Rev 29(3):695–747

Eldeen IM, Elgorashi EE, Mulholland DA, van Staden J (2006) Anolignan B: a bioactive compound from the roots of *Terminalia sericea*. J Ethnopharmacol 103(1):135–138

Chidananda SR, Bharathi A, Jianping Z, Troy JS, Ikhlas AK (2010) Qantitative determination of phenolic acids in *Lonicera japonica* thunb. Using high performance thin layer chromatography. J Liq Chromatogr Relat Technol 34(1):38–47

Fallarino M (1994) Tibetan medical paintings. Herbalgram 31:38–34

Fyhrquist P, Mwasumbi L, Hæggström CA, Vuorela H, Hiltunen R, Vuorela P (2002) Ethnobotanical and antimicrobial investigation on some species of Terminalia and Combretum (Combretaceae) growing in Tanzania. J Ethnopharmacol 79(2):169–177

Fyhrquist P, Mwasumbi L, Hæggström CA, Vuorela H, Hiltunen R, Vuorela P (2004) Antifungal activity of selected species of Terminalia, Pteleopsis, and Combretum (Combretaceae) collected in Tanzania. Pharm Biol 42(4–5):308–317

Geissman TA (1963) Flavonoid compounds, tannins, lignins, and, related compounds. In: Comprehensive biochemistry, vol 9. Elsevier, pp 213–250

Ho KY, Tsai CC, Huang JS, Chen CP, Lin TC, Lin CC (2001) Antimicrobial activity of tannin components from Vaccinium vitis-idaea L. J Pharm Pharmacol 53(2):187–191

Hutchings A (1996) Zulu medicinal plants: an inventory. University of Natal Press

Jo S, Kim S, Shin DH, Kim MS (2020) Inhibition of SARS-CoV 3CL protease by flavonoids. J Enzyme Inhib Med Chem 35(1):145–151

Kazmi MH, Malik A, Hameed S, Akhtar N, Ali SN (1994) An anthraquinone derivative from Cassia italica. Phytochemistry 36(3):761–763

Knox YM, Hayashi K, Suzutani T, Ogasawara M, Yoshida I, Shiina R, Azuma M (2001) Activity of anthocyanins from fruit extract of Ribes nigrum L. against influenza A and B viruses. Acta Virol 45(4):209–215

Liu X, Dong M, Chen X, Jiang M, Lv X, Zhou J (2008) Antimicrobial activity of an endophytic Xylaria sp. YX-28 and identification of its antimicrobial compound 7-amino-4-methylcoumarin. Appl Microbiol Biotechnol 78(2):241–247

Loub WD, Farnsworth NR, Soejarto DD, Quinn ML (1985) NAPRALERT: computer handling of natural product research data. J Chem Inf Comput Sci 25(2):99–103

Mallikharjuna PB, Seetharam YN (2009) In vitro antimicrobial screening of alkaloid fractions from Strychnos potatorum. E-J Chem 6(4):1200–1204

Martini ND, Katerere DRP, Eloff JN (2004) Biological activity of five antibacterial flavonoids from Combretum erythrophyllum (Combretaceae). J Ethnopharmacol 93(2–3):207–212

Maxmen A (2020) More than 80 clinical trials launch to test coronavirus treatments. Nature 578(7795):347–349

Mohammadi Pour P, Fakhri S, Asgary S, Farzaei MH, Echeverria J (2019) The signaling pathways, and therapeutic targets of antiviral agents: focusing on the antiviral approaches and clinical perspectives of anthocyanins in the management of viral diseases. Front Pharmacol 10:1207

Ndubani P (1997) Knowledge about and herbal treatment of sexually transmitted diseases among the Goba of Chiawa, Zambia. Central Afr J Med 43(10):283–287

Newman DJ, Cragg GM, Snader KM (2000) The influence of natural products upon drug discovery. Nat Prod Rep 17(3):215–234

Pattnaik S, Subramanyam VR, Bapaji M, Kole CR (1997) Antibacterial and antifungal activity of aromatic constituents of essential oils. Microbios 89(358):39–46

Polyak SJ, Morishima C, Shuhart MC, Wang CC, Liu Y, Lee DYW (2007) Inhibition of T-cell inflammatory cytokines, hepatocyte NF-κB signaling, and HCV infection by standardized silymarin. Gastroenterology 132(5):1925–1936

Puupponen-Pimiä R, Nohynek L, Hartmann-Schmidlin S, Kähkönen M, Heinonen M, Määtt-ä-Riihinen K, Oksman-Caldentey KM (2005) Berry phenolics selectively inhibit the growth of intestinal pathogens. J Appl Microbiol 98(4):991–1000

Rabe T, Van Staden J (2000) Isolation of an antibacterial sesquiterpenoid from Warburgia salutaris. J Ethnopharmacol 73(1–2):171–174

Rimando AM, Pezzuto JM, Farnsworth NR, Santisuk T, Reutrakul V, Kawanishi K (1994) New lignans from Anogeissus acuminata with HIV-1 reverse transcriptase inhibitory activity. J Nat Prod 57(7):896–904

Sakanaka S, Shimura N, Aizawa M, Kim M, Yamamoto T (1992) Preventive effect of green tea polyphenols against dental caries in conventional rats. Biosci Biotechnol Biochem 56(4): 592–594

Sornpet B, Potha T, Tragoolpua Y, Pringproa K (2017) Antiviral activity of five Asian medicinal plant crude extracts against highly pathogenic H5N1 avian influenza virus. Asian Pac J Trop Med 10(9):871–876

Sulaiman MR, Zakaria ZA, Moin S, Somchit MN, Chai Y (2008) Antinociceptive and anti-inflammatory effects of Lonicera japonica thunb. Res J Pharmacol 2:38–42

Trease G, Evans W (1972) Pharmacognosy. Univ. Press, Aberdeen, Great Britain, pp 161–163

Tshikalange TE, Meyer JJM, Hussein AA (2005) Antimicrobial activity, toxicity and the isolation of a bioactive compound from plants used to treat sexually transmitted diseases. J Ethnopharmacol 96(3):515–519

Tyler VE (1999) Phytomedicines: back to the future. J Nat Prod 62(11):1589–1592

Van der Kooy F, Meyer JJM, Lall N (2006) Antimycobacterial activity and possible mode of action of newly isolated neodiospyrin and other naphthoquinones from Euclea natalensis. S Afr J Bot 72(3):349–352

World Health Organization (2020) Weekly epidemiological update - 17 November 2020. https://www.who.int/publications

Vermerris W, Nicholson R (2006) Phenolic compound biochemistry. Springer, Netherlands

Watt JM, Breyer-Brandwijk MG (1962) The medicinal and poisonous plants of Southern and Eastern Africa (No. 581.96 W38)

Wild R (1994) The complete book of natural and medicinal cures. Complement Health Pract Rev 3: 139

Xiong J, Li S, Wang W, Hong Y, Tang K, Luo Q (2013) Screening and identification of the antibacterial bioactive compounds from Lonicera japonica Thunb. leaves. Food Chem 138(1): 327–333

Chapter 23
Antimicrobial Drugs: Possibilities from Medicinal Plants Part B—Antifungals

Thahiya Naushad, Irfan Türetgen, and Shiburaj Sugathan

Abstract Natural products, especially plant-derived ones, are the source of diverse chemical structures with potential bioactive molecules. The fast-growing Antimicrobial Resistance (AMR) problem is identified as one of the top ten global public health threats. The increasing prevalence of antimicrobial resistance has urged the discovery of novel molecules with a new spectrum of bioactivity or activity on alternative drug targets. Traditional screening of natural compounds for bioactivity is a cumbersome process. But the advances in Computational or in silico tools possibility of high thorough screening methods ease the identification of bioactive metabolites as possible drug leads.

Keywords Antimicrobial resistance · Plant metabolites · Antifungal · Medicinal plants · Zingiberaceae

23.1 Introduction

23.1.1 The Discovery of Miracle Drug

The Coronavirus pandemic has been affecting humanity since December 2019 with a series of COVID-19 waves. Preventing infectious diseases is one of the biggest

T. Naushad
Department of Botany, University of Kerala, Thiruvananthapuram, Kerala, India

Fundamental and Industrial Microbiology Division, Department of Biology, Faculty of Science, Istanbul University, Istanbul, Turkey

I. Türetgen
Fundamental and Industrial Microbiology Division, Department of Biology, Faculty of Science, Istanbul University, Istanbul, Turkey

S. Sugathan (✉)
Department of Botany, University of Kerala, Thiruvananthapuram, Kerala, India
e-mail: drshiburaj@karalauniversity.ac.in

S. T. Sukumaran, K. T R (eds.), *Conservation and Sustainable Utilization of Bioresources*, Sustainable Development and Biodiversity 30,
https://doi.org/10.1007/978-981-19-5841-0_23

challenges facing the scientific world. Approximately, 50% of deaths in tropical countries are due to infectious diseases and are hiked after the Coronavirus outbreak. The principal causative organisms are bacteria, viruses, fungi, and helminths. Contagious diseases are always a significant threat to public health. In the early periods, the mortality rate due to infectious diseases was relatively high. Between 1347 and 1352, Black Death (Plague) killed around two-thirds population in Europe (Alfani and Murphy 2017). Plague is a bacterial disease caused by *Yersinia pestis*. Another ruinous disease recorded in history is smallpox, caused by the Variola virus. It is estimated that the death rate of smallpox is much higher than the death rate due to all the wars that occurred in the world. The infections like cholera, flu, Severe Acute Respiratory Syndrome Coronavirus (SARS-CoV), and the Middle East Respiratory Syndrome Coronavirus (MERS-CoV) have also been listed among the diseases that afflict humans in the past.

With the discovery of antibiotics and vaccines, there has been a steady decline in the mortality rate caused by the mentioned diseases. Serendipity happened in Alexander Flemming's lab and opened a new hope to the scientific world. And then, the "miracle drug" called antibiotics was used worldwide to treat infections. But in fact, antibiotics have been widely used by people in the prehistoric era. Osteology studies conducted in ancient populations revealed that they had been exposed to certain antibiotics. According to a survey carried out by Bassett et al., human skeletons from old Sudanese Nubia demonstrated the presence of trace amounts of tetracycline in their bones dated back to 350–550 CE. It was documented that the exposure to this kind of antibiotics in their diet might have been a reason for a reduced rate of infections in such populations (Bassett et al. 1980; Aminov 2010). It has also been reported that ancient Egyptians used moldy pieces of bread and larval therapy to treat wound infections. Proteolytic enzymes produced by maggots have bactericidal properties, and these help in granulation tissue formation and fibroblast development and thereby hasten the wound healing process (Pahor 1992).

23.1.2 Antimicrobial Resistance (AMR) and the Need for Novel Drugs

Before Alexander Flemming, in 1870, Sir John Scott Burdon-Sanderson detailed the bactericidal inhibition in culture fluid covered in mold. After this, scientists like Joseph Lister, Dr. John Tyndall, and Ernest Duchesine have also observed antimicrobial activities in their studies. The commercialization of penicillin in 1942 instigated new insights into microbiology. Since then, humankind has been depending on antibiotics to cure infections. But despite that, the misuse of antimicrobial drugs by humans results in the development of antimicrobial-resistant strains. The World Health Organisation (WHO) declared Antimicrobial Resistance (AMR) as one of the top ten global public health threats facing humanity. The increasing prevalence of antimicrobial resistance has reduced patient treatment options and increased mortality and morbidity. Antimicrobial-resistant organisms cause more than two million infections in the United States of America, and there are approximately 23,000 deaths reported annually.

In contrast, antimicrobial-resistant organisms cause approximately 25,000 deaths in Europe (Marston et al. 2016). The hike in antimicrobial resistance poses an extraordinary threat to human life, and therefore developing new antimicrobial drugs has become a compelling situation for the scientific community. Plants are a promising source for the discovery of novel antimicrobials.

23.1.3 Medicinal Plants: The Rich Source of Bioactive Compounds

The use of plants for human health has been documented for many years. Plants are so closely associated with human life. Human beings have depended on plants from the beginning of their life on earth. One of the most notable benefits of plants is their ability to treat diseases. Many people around the globe rely on medicinal plants as primary health care even without knowing their bioactive compounds. To give an example, Willow tree bark and leaves were used by ancient Egyptians to reduce infection. Besides, various types of pain were relieved by willow and its extracts.

Researchers discovered that salicylic acid found in Willow extracts was a significant component of acetylsalicylic acid, commonly known as aspirin. Cancerous cells are also inhibited by these extracts (Mahdi et al. 2006). As is known, plants are rich sources of bioactive compounds produced as their secondary metabolites. Bioactive compounds are phytochemicals extracted from different plant parts and are not required for daily functioning. Plants use these compounds as attraction, protection, and signaling molecules. Many plant bioactive compounds like glycosides, glucosinolates, saponins, anthraquinone glycosides, flavonoids, proanthocyanidins, tannins, terpenoids, resins, lignans, alkaloids, furocoumarins, etc. have been documented (Fig. 23.1). Medicinal and poisonous plants are the rich source of these compounds (Bernhoft 2010).

Worldwide, medicinal plants play a pivotal role in developing human culture. These plants are an integral part of traditional medicines. Approximately 85% of conventional drugs are plant-based (Wilson and Peter 1988). It is estimated that several ethnic communities in India use 7500 different species of plants as medicine. They have been using these plants to treat wounds and multiple diseases like abdominal disorders, jaundice, liver problems, etc. (Datta et al. 2014). Nowadays, scientists working on drug development are exploring ethnobotanical knowledge to incorporate traditional medicines into modern research. Many bioactive compounds in these medicinal plants can be utilized for pharmacopoeial, non-pharmacopoeial, and synthetic drug development. The rich source of bioactive compounds present in medicinal plants should be more studied to innovate novel antimicrobial drugs. In the current review, we discuss the antimicrobial potentiality of medicinal plants with a particular focus on plants belonging to the Zingiberaceae family.

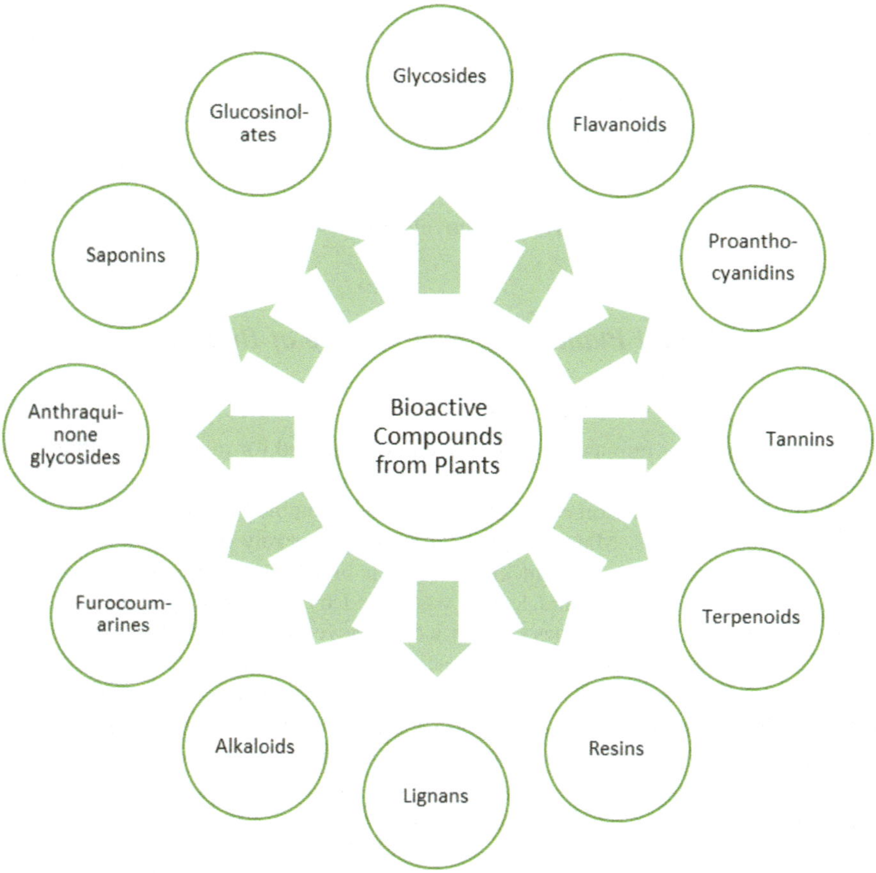

Fig. 23.1 Bioactive compounds from plants

23.2 Antifungal Potential of Medicinal Plants

Fungi permeate the environment, so fungal diseases are common and complex to treat compared to bacterial infections. Fungi can cause many illnesses in humans, including meningitis, asthma, lung infection, bloodstream infection, dermatological infections, vaginal candidiasis, ringworm, etc. Furthermore, opportunistic fungal infections are more common in AIDS patients and are a leading cause of morbidity and mortality (Garbino et al. 2001). As a home remedy, people include yogurt and probiotics in their diet to evict the fungus from the body. These foods contain good bacteria which fight against the infected fungus (Sharma et al. 2022). There have been numerous studies reported on the antifungal activity of plant extracts which can be further explored to develop antimycotic medications. Ethnobotanical information and ethnopharmacological expedition play a significant role in identifying novel antifungal agents.

Among the 14 plants studied by Webster et al., aqueous extracts of *Fragaria virginiana, Epilobium angustifolium,* and *Potentilla simplex* exhibited potential antifungal properties. *Fragaria virginiana* inhibited the growth of *Candida krusei, C. glabrata,* and *C. lusitaniae.* The root extracts of *Epilobium angustifolium* were effective against the yeasts *Candida glabrata, Candida lusitaniae,* and *Saccharomyces cerevisiae.* In contrast, the stem and leaf extracts of *Potentilla simplex* were good at suppressing the growth of *Candida glabrata* (Webster et al. 2008). Many plant essential oils have been used as antisepsis, pain reliever, or even to treat a wide range of infections in folk medicine. Essential oils are odiferous oils extracted from different plant parts. The essential oils extracted from *Cymbopogon martini, Eucalyptus globulus, and Cinnamomum zylanicum* have shown antimycotic activity against the human pathogens *Aspergillus fumigatus* and *Aspergillus niger* at 0.025% (v/v) (Bansod and Rai 2008). It has also been documented that *Foeniculum vulgare* (Fennel) seed essential oil retarded the growth of *Cladosporium cladosporioides, Phomopsis helianthi,* and *Trichophyton mentagrophytes.* The presence of compounds like anethole, pinene, and fenchone in essential oil is related to the antimycotic properties (Mimica-Dukić et al. 2003). Furthermore, the essential oil extracted from the aerial part of *Foeniculum vulgare* has shown inhibitory and fungicidal activities on *Acremonium sclerotigenum*, a human and plant pathogen (Mafakheri and Mirghazanfari 2018).

Buwa and van Staden screened 13 plants used to treat venereal diseases in South Africa. Ethanolic extracts of *Bersama lucens* and *Harpephyllum caffrum* were most effective against *Candida albicans* with a MIC value of 0.78 mg/mL. The biological activity of *Harpephyllum caffrum* is attributed to the phenolic compounds present in the extract. (Buwa and Van Staden 2006). In another study, stem and leaf aqueous extracts of plants *Ziziphus joazeiro* Mart. and *Caesalpinia pyramidalis* Tul. exhibited antifungal activity against a wide range of fungal pathogens like *Trichophyton rubrum, Candida guilliermondii, Candida albicans, Cryptococcus neoformans,* and *Fonsecaea pedrosoi* (Cruz et al. 2007). A further study conducted in Saudi Arabia disclosed the antimycotic ability of medicinal plants, which are widely used across the country. The study documented that the fungus *Trichophyton rubrum* was inhibited by the methanol extracts of *Cymbopogon citratus* DC. Stapf. (Lemongrass), *Lantana camara* L., *Nerium oleander* L. by 85–90%, while the ethyl acetate extracts exhibited 80–85% inhibition. Compared to these extracts, the percentage inhibition of aqueous extracts was less (Bokhari 2009).

Denture stomatitis, an oral infection caused by *Candida albicans,* can be treated with clove (*Syzygium aromaticum*) and cinnamon (*Cinnamomum verum*) oils loaded with emulgel (Emulgel is a topical drug formulated with dual release technology, that contains both emulsion and gel). The enhanced acceptability in taste without side effects makes it a promising alternative to existing allopathic medications. The MIC values of clove oil and cinnamon oil were 512 μg/mL and 64 μg/mL, respectively (Iyer et al. 2022). Another study with *Acalypha wilkesiana*, a medicinal plant used in traditional medicine to treat skin diseases like pityriasis versicolor and seborrheic dermatitis, was scientifically studied by Sherifat et al. The leaf extracts of two varieties of *Acalypha wilkesiana* (Macrophylla and Hoffmanii) were found to be effective against fungi *Candida albicans* ATCC 10231, *Trichophyton rubrum* ATCC 2188, and *Malassezier furfur* ATCC 14521. Compared to ethyl acetate and hexane extracts, ethanol extract exhibits a more potent inhibition, possibly due to the ability of ethanol to extract more antifungal components of plants than the others (Sherifat et al. 2022). The anticandidal property of aqueous leaf extract of *Acalypha wilkesiana* (Macrophylla variety) has also been reported with a MIC value of 31.2 μg/mL (Lumpu et al. 2014).

The plants used in Indian traditional medicines were also documented for their antimycotic properties (Table 23.1). For example, *Solanum xanthocarpum* (Indian nightshade) treats infections and degenerative diseases. The methanolic extract of *Solanum xanthocarpum* was effective against *Aspergillus fumigatus, A. flavus,* and *A. niger*. The antifungal potentiality of *Datura metel* against these three fungal strains has also been documented (Dabur et al. 2004). *Solanum xanthocarpum* is also known for its anticancer activity against lung cancer and leukemia (Kumar and Pandey 2014).

The prevalence of fungal infections has increased considerably in recent years. Immunocompromised and AIDS patients are very much prone to fungal infections. The rising number of cases and antifungal resistance necessitates the development of new therapeutic approaches. All the studies mentioned here disclosed that most medicinal plants used in traditional medicines have high antimycotic potential. However, more research is required before it can be developed into a therapeutic medication.

23.3 Zingiberaceae: The Potential Source of Antimicrobials

Pants belonging to the Zingiberaceae family are most abundantly dispersed in the tropics, especially in Southeast Asia. About 50 genera and 1300 species of plants were identified from this family. Most of these plants are characterized by aromatic leaves and fleshy rhizomes. They have been used as medicines, spices, flavoring agents, and dyes (Habsah et al. 2000). They are well-documented antimicrobial properties. For example, *Alpinia rafflesiana* exhibited a wide range of activities against *Bacillus subtilis,* Methicillin-resistant *Staphylococcus aureus, Pseudomonas aeruginosa* (Habsah et al. 2000).

Curcuma longa (turmeric) is considered one of the spices that have been cultivated for centuries. Rhizome powder of *Curcuma longa* is used as a primary flavoring agent in Asian countries. In a study conducted by Apisariyakul et al., turmeric oil was effective against dermatophytes (fungi that cause dermatophytosis) like *Trichophyton rubrum* and *T. mentagrophytes, Epidermophyton floccosum,*

Table 23.1 Medicinal plants with antifungal properties

Plant	Part used	Fungi	Extract	Reference
Fragaria virginiana	Leaf	*Candida krusei, Candida glabrata, Candida lusitaniae*	Aqueous	Webster et al. (2008)
Epilobium angustifolium	Root	*Candida glabrata, Candida lusitaniae, Saccharomyces cerevisiae*	Aqueous	Webster et al. (2008)
Potentilla simplex	Stem Leaf	*Candida glabrata*	Aqueous	Webster et al. (2008)
Cymbopogon martini	Leaf	*Aspergillus fumigatus Aspergillus niger*	Essential oil	Bansod and Rai (2008)
Eucalyptus globulus	Leaf	*Aspergillus fumigatus Aspergillus niger*	Essential oil	Bansod and Rai (2008)
Cinnamomum zylanicum	Bark Leaf	*Aspergillus fumigatus Aspergillus niger*	Essential oil	Bansod and Rai (2008)
Foeniculum vulgare Mill.	Seed	*Cladosporium cladosporioides, Phomopsis helianthi, Trichophyton mentagrophytes*	Essential oil	Mimica-Dukić et al. (2003)
Foeniculum vulgare	Aerial	*Acremonium sclerotigenum*	Essential oil	Mafakheri and Mirghazanfari (2018)
Bersama lucens	Bark	*Candida albicans*	Ethanol	Buwa and van Staden (2006)
Harpephyllum caffrum	Bark	*Candida albicans*	Ethanol	Buwa and van Staden (2006)
Ziziphus joazeiro Mart.	Leaf Stem	*Trichophyton rubrum, Candida guilliermondii, Candida albicans, Cryptococcus neoformans, Fonsecaea pedrosoi*	Aqueous	Cruz et al. (2007)
Caesalpinia pyramidalis Tul.	Leaf Stem	*Trichophyton rubrum, Candida guilliermondii, Candida albicans, Cryptococcus neoformans, Fonsecaea pedrosoi*	Aqueous	Cruz et al. (2007)
Cymbopogon citratus DC. Stapf.	Stalk Leaf	*Trichophyton rubrum*	Methanol Ethyl Acetate Aqueous	Bokhari (2009)
Lantana camara L.	Leaf Flower	*Trichophyton rubrum*	Methanol Ethyl Acetate Aqueous	Bokhari (2009)
Nerium oleander L	Leaf	*Trichophyton rubrum*	Methanol Ethyl Acetate Aqueous	Bokhari (2009)
Syzygium aromaticum	Flower bud	*Candida albicans*	Oil	Iyer et al. (2022)
Cinnamomum verum	Bark	*Candida albicans*	Oil	Iyer et al. (2022)

(continued)

Table 23.1 (continued)

Plant	Part used	Fungi	Extract	Reference
Acalypha wilkesiana	Leaf	*Malassezia furfur, Candida albicans, Trichophyton rubrum*	Ethyl acetate Hexane Ethanol Aqueous	Lumpu et al. (2014) Sherifat et al. (2022)
Solanum xanthocarpum	–	*Aspergillus fumigatus, Aspergillus flavus, Aspergillus niger*	Methanol	Dabur et al. (2004)
Datura metel	–	*Aspergillus fumigatus, Aspergillus flavus, Aspergillus niger*	Methanol	Dabur et al. (2004)

Microsporum gypseum, and *Sporothrix schenckii.* Turmeric oil was also influential in inhibiting the growth of *Trichophyton*-induced dermatophytosis in Guinea pigs. Besides, the study investigated that curcumin (a primary bioactive compound on *Curcuma longa* known for its antioxidant property) has no antifungal against the isolates mentioned above (Apisariyakul et al. 1995).

The results from another study indicate that the essential oil of *Alpinia conchigera* Griff. can inhibit the growth of Gram-positive bacteria (Qamaruz Zaman et al. 2022). *Stahlianthus thorelii* Gagnep. is a plant widely seen in China, Thailand, India, and Vietnam. This plant plays a significant role in traditional medicine to treat health issues like hemorrhage, rheumatism, abnormal menstruation, digestive problems, and joint pains. Tuan et al. reported the antimicrobial activity of essential oil extracted from the leaf and rhizome of *Stahlianthus thorelii* for the first time. *Staphylococcus aureus, Bacillus subtilis,* and *Aspergillus niger* were inhibited by rhizome oil with a MIC range between 150 and 200 μg/mL. In contrast, leaf oil was effective only against *Bacillus subtilis* at MIC 200 μg/mL (Tuan et al. 2022).

Eighteen species of six genera (*Alpinia, Costus, Curcuma, Hedychium, Vanoverberghia, Zingiber*) of medicinal plants in Taiwan, belonging to the family Zingiberaceae, were exposed to the antimicrobial and antioxidant studies. Almost all the ethanolic extracts of plants showed inhibition against *Escherichia coli, Salmonella enterica, Staphylococcus aureus,* and *Vibrio parahaemolyticus.* However, the extracts of *Alpinia japonica, Alpinia kawakamii,* and *Alpinia kusshakuensis* were resistant to all the mentioned bacteria (Chen et al. 2007).

Emerging and re-emerging viral infections constitute a significant threat to public health in the contemporary world. Antiviral drug discovery strategies for such disorders could use medicinal plants with broad antiviral effects. Zingiberaceae plants are a promising source for immunomodulating and antiviral medicines (Mbadiko et al. 2020). For example, *Curcuma longa* is known for its broad spectrum of antiviral activity. These plant extracts can inhibit Human Immunodeficiency Virus (HIV) 1 and 2, Herpes Simplex Virus Type1, Feline Infectious Peritonitis Virus, Coxsackieviruses, Hepatitis virus, Human papillomavirus, Chikungunya virus, Epstein-Barr virus, Dengue virus, Human Norovirus, SARS-COVID 1 and so on (Ichsyani et al. 2017; Baikerikar 2017; Khaerunnisa et al. 2020; Mbadiko et al.

2020). The plant extract of *Aframomum melegueta* has antiviral activity against the Avian pox virus, Fowlpox virus, HIV, etc. (Onwuatuegwu et al. 2017). Studies conducted on antiviral activities of *Zingiber officinale* revealed that the rhizome extracts could be used against rotavirus, H1N1, H5N1, Herpes (Type 1 and 2), Rhinovirus, Hepatitis B, and SARS Covid 1 (Singh et al. 2018; Imo and Za'aku 2019; Mbadiko et al. 2020).

The antimicrobial activity of medicinal plants is attributed to rich bioactive compounds in extracts. Most medicinal plants contain phenols and flavonoids as the main bioactive compounds. Depending on the variety and growth conditions, the contents of the bioactive compounds may differ. Zingiberaceae's major bioactive compounds are gingerol, curcumin, curcumenol, and eugenol (Chumroenphat et al. 2019). Periodontal disease-causing microorganisms like *Porphyromonas gingivalis* ATCC 53978, *Porphyromonas endodontalis* ATCC 35406, and *Prevotella intermedia* ATCC 25611 were inhibited by the ethanolic and n-hexane extracts of *Zingiber officinale* Roscoe. The analytical studies on extracts revealed that [10]-gingerol and [12]-gingerol is the bioactive compounds inhibiting the growth of these anaerobic Gram-negative bacteria associated with oral disease (Park et al. 2008). Curumenol from the plant *Curcuma aeruginosa* has the ability to inhibit *Escherichia coli* and *Salmonella typhi* (Rahayu and Sugita 2018). Curcumin, a polyphenolic compound present in the rhizome of *Curcuma longa* L., exhibits broad antimicrobial activities. Researchers have documented it as antibacterial, antiviral, antifungal, and antimalarial agents. It also shows synergistic antibacterial activity with other antibiotics. For example, the synergistic action of curcumin with ampicillin reduced the MIC value of ampicillin. Curcumin has also been documented for its antiviral properties and has been found effective against HIV, HSV (Type 1 and 2), influenza, etc. (Zorofchian et al. 2014; Mazumder et al. 1995). A recent study indicates that a combination of curcumin derivatives, azithromycin, and antiviral inhibitor N3 is a promising therapeutic approach for treating COVID-19 (Alves et al. 2022). Monoterpenes like 1,8-cineole, α-pinene, α-terpinene, β-pinene, and flavonoid, isopanduratin A from Zingiberaceae plant extracts, are known to retard the bacterial growth by inhibiting cell wall or cell membrane (Prasad et al. 2019).

23.4 Conclusion

In conclusion, medicinal plants are rich in secondary metabolites, making them favorable resources for novel antimicrobials. Researchers study many medicinal plants, and still, more is commenced. The optimum potential of medicinal plants as therapeutic drugs needs to be explored as the progression of multidrug resistance becomes a public threat.

References

Alfani G, Murphy T (2017) Plague and lethal epidemics in the pre-industrial world. J Econ Hist 77(1):314–343

Alves DR, da Rocha MN, Passos CCO, Marinho MM, Marinho ES, de Morais SM (2022) Curcumins and its derivatives as potential inhibitors of New Coronavirus (COVID-19) main protease: an in silico strategy. Res Soc Dev 11(1):e6511124334

Aminov RI (2010) A brief history of the antibiotic era: lessons learned and challenges for the future. Front Microbiol 1:134

Apisariyakul A, Vanittanakom N, Buddhasukh D (1995) Antifungal activity of turmeric oil extracted from *Curcuma longa* (Zingiberaceae). J Ethnopharmacol 49(3):163–169

Baikerikar S (2017) Curcumin and natural derivatives inhibit Ebola viral proteins: an in silico approach. Pharm Res 9(1):S15

Bansod S, Rai M (2008) Antifungal activity of essential oils from Indian medicinal plants against human pathogenic *Aspergillus fumigatus* and *A. niger*. World J Med Sci 3(2):81–88

Bassett EJ, Keith MS, Armelagos GJ, Martin DL, Villanueva AR (1980) Tetracycline-labelled human bone from ancient Sudanese Nubia (AD 350). Science 209(4464):1532–1534

Bernhoft A (2010) A brief review on bioactive compounds in plants. In: Bioactive compounds in plants-benefits and risks for man and animals, vol 50. pp 11–17

Bokhari FM (2009) Antifungal activity of some medicinal plants used in Jeddah, Saudi Arabia. Mycopath 7(1):51–57

Buwa LV, Van Staden J (2006) Antibacterial and antifungal activity of traditional medicinal plants used against venereal diseases in South Africa. J Ethnopharmacol 103(1):139–142

Chen IN, Chang CC, Ng CC, Wang CY, Shyu YT, Chang TL (2007) Antioxidant and antimicrobial activity of Zingiberaceae plants in Taiwan. Plant Foods Hum Nutr 63(1):15–20

Chumroenphat T, Somboonwatthanakul I, Saensouk S, Siriamornpun S (2019) The diversity of biologically active compounds in the rhizomes of recently discovered Zingiberaceae plants native to North Eastern Thailand. Pharm J 11(5)

Cruz MCS, Santos PO, Barbosa AM Jr, De Mélo DLFM, Alviano CS, Antoniolli AR, Trindade RC (2007) Antifungal activity of Brazilian medicinal plants involved in popular treatment of mycoses. J Ethnopharmacol 111(2):409–412

Dabur R, Singh H, Chhillar AK, Ali M, Sharma GL (2004) Antifungal potential of Indian medicinal plants. Fitoterapia 75(3–4):389–391

Datta T, Patra AK, Dastidar SG (2014) Medicinal plants used by tribal population of Coochbehar district, West Bengal, India—an ethnobotanical survey. Asian Pac J Trop Biomed 4:S478–S482

Garbino J, Kolarova L, Lew D, Hirschel B, Rohner P (2001) Fungemia in HIV-infected patients: a 12-year study in a tertiary care hospital. AIDS Patient Care STDs 15(8):407–410

Habsah M, Amran M, Mackeen MM, Lajis NH, Kikuzaki H, Nakatani N, Ali AM (2000) Screening of Zingiberaceae extracts for antimicrobial and antioxidant activities. J Ethnopharmacol 72(3): 403–410

Ichsyani M, Ridhanya A, Risanti M, Desti H, Ceria R, Putri DH, Dewi BE (2017) Antiviral effects of *Curcuma longa* L. against dengue virus in vitro and in vivo. In: IOP Conference Series: Earth and Environmental Science, vol 101(1). p 012005

Imo C, Za'aku JS (2019) Medicinal properties of ginger and garlic: a review. Curr Trends Biomed Eng Biosci 18:47–52

Iyer MS, Gujjari AK, Paranthaman S, Abu Lila AS, Almansour K, Alshammari F, Gowda DV (2022) Development and evaluation of clove and cinnamon supercritical fluid extracts-loaded emulgel for antifungal activity in denture stomatitis. Gels 8(1):33

Khaerunnisa S, Kurniawan H, Awaluddin R, Suhartati S, Soetjipto S (2020) Potential inhibitor of COVID-19 main protease (Mpro) from several medicinal plant compounds by molecular docking study. Preprints 2020:2020030226

Kumar S, Pandey AK (2014) Medicinal attributes of *Solanum xanthocarpum* fruit consumed by several tribal communities as food: an in vitro antioxidant, anticancer and anti-HIV perspective. BMC Complement Altern Med 14(1):1–8

Lumpu N, Manienga K, Bumoyi M (2014) Antibacterial and antifungal screening of extracts from six medicinal plants collected in Kinshasa-Democratic Republic of Congo against clinical isolate pathogens. J Pharmacogn Phytother 6(3):24–32

Mafakheri H, Mirghazanfari SM (2018) Antifungal activity of the essential oils of some medicinal plants against human and plant fungal pathogens. Cell Mol Biol 64(15):13–19

Mahdi JG, Mahdi AJ, Bowen ID (2006) The historical analysis of aspirin discovery, its relation to the willow tree and antiproliferative and anticancer potential. Cell Prolif 39(2):147–155

Marston HD, Dixon DM, Knisely JM, Palmore TN, Fauci AS (2016) Antimicrobial Resistance. JAMA 316(11):1193–1204

Mazumder A, Raghavan K, Weinstein J, Kohn KW, Pommier Y (1995) Inhibition of human immunodeficiency virus type-1 integrase by curcumin. Biochem Pharmacol 49(8):1165–1170

Mbadiko CM, Inkoto CL, Gbolo BZ, Lengbiye EM, Kilembe JT, Matondo A, Mpiana PT (2020) A mini review on the phytochemistry, toxicology and antiviral activity of some medically interesting Zingiberaceae species. J Complement Alternat Med Res 9(4):44–56

Mimica-Dukić N, Kujundžić S, Soković M, Couladis M (2003) Essential oil composition and antifungal activity of *Foeniculum vulgare* Mill. obtained by different distillation conditions. Phytother Res 17(4):368–371

Onwuatuegwu JTC, Abraham OAJ, Umeoduagu ND (2017) In vitro antiviral activities of *Aframomum melegueta* leaf extracts on Newcastle disease virus (NDV), fowl pox virus (FPV) and infectious bursal disease virus (IBDV). IDOSR J Sci Technol 2(2):33–45

Pahor AL (1992) Ear, nose, and throat in ancient Egypt. JLO 106(8):677–687

Park M, Bae J, Lee DS (2008) Antibacterial activity of [10]-gingerol and [12]-gingerol isolated from ginger rhizome against periodontal bacteria. Phytother Res 22(11):1446–1449

Prasad MA, Zolnik CP, Molina J (2019) Leveraging phytochemicals: the plant phylogeny predicts sources of novel antibacterial compounds. Future Sci OA 5(7):FSO407

Qamaruz Zaman F, Ridzuan R, Abdelmageed AHA (2022) Chemical composition, antioxidant and antimicrobial activities of the essential oils from rhizomes and leaves of *Alpinia conchigera* Griff. (Zingiberaceae). J Essent Oil Bear Plants 24(6):1311–1322

Rahayu DUC, Sugita P (2018) Antibacterial activity of curcumenol from rhizomes of Indonesian *Curcuma aeruginosa* (Zingiberaceae). Rasayan J Chem 11(2):762–765

Sharma S, Kaur H, Singh J, Yadav R (2022) A short review on liquisolid technology in antifungal drugs. World J Pharm Sci 144–150

Sherifat KO, Itohan AM, Adeola SO, Adeola KM, Aderemi OL (2022) Antifungal activity of *Acalypha wilkesiana*: a preliminary study of fungal isolates of clinical significance. Afr J Infect Dis 16(1):21

Singh P, Srivastava S, Singh VB, Sharma P, Singh D (2018) Ginger (*Zingiber officinale*): a Nobel herbal remedy. Int J Curr Microbiol Appl Sci Special Issue-7:4065–4077

Tuan NH, Quang LV, Tung NT, Ngoc NB, Khanh PN, Averyanov LV (2022) Chemical composition and antibacterial properties of essential oil extracted from the leaves and the rhizomes of *Stahlianthus thorelii* Gagnep. (Zingiberaceae). J Essent Oil Bear Plants 24(6):1365–1372

Webster D, Taschereau P, Belland RJ, Sand C, Rennie RP (2008) Antifungal activity of medicinal plant extracts; preliminary screening studies. J Ethnopharmacol 115(1):140–146

Wilson EO, Peter FM (1988) Screening plants for new medicines. In: Biodiversity. National Academies Press (US)

Zorofchian MS, Abdul Kadir H, Hassandarvish P, Tajik H, Abubakar S, Zandi K (2014) A review on antibacterial, antiviral, and antifungal activity of curcumin. Biomed Res Int 2014:186864

Chapter 24
Marine Resources and Sustainable Utilization

Rakhie Narayanan, K. Sreelekshmi, and Keerthi T R

Abstract Sustainable development and sustainable resource management are two important concepts put forward by the modern world to maintain ecological balance and development. The ocean is the sink of all the wastes produced by humans, and dumping this into oceans causes the deterioration of the ecosystem and biodiversity. Ocean pollution also contributes to climate change, the heat generated in oceans causes marine heat waves, and it is at record levels. It is high time to take initiatives for the protection as well as sustainable utilization of resources from the ocean. Most coastal countries have policies for protecting their territorial sea by establishing protected areas and putting restrictions on developmental activities. Several organizations, including the UN have put forward several goals for achieving sustainability. Integrated ocean management is a strategy for a balanced exploration and utilization of ocean resources with the cooperation of government authorities, the public, and international collaborations. Being responsible consumers of marine resources is the major path to attaining sustainability.

Keywords UNCLOS · Policy instruments · SDG14 · IOM · ICM

24.1 Introduction

Oceans and seas are an inevitable part of economic growth, providing food and vast scope of job opportunities. When demand doesn't meet the supply, exploitation of resources will happen, which leads to a huge loss and imbalance in the biosphere. The biotic and abiotic resources found in oceans and the bottom together are called marine resources that were being utilized by humans from historical times. The major marine resources are included in Fig. 24.1.

R. Narayanan · K. Sreelekshmi · K. T R (✉)
School of Biosciences, Mahatma Gandhi University, Kottayam, India
e-mail: keerthi@mgu.ac.in

S. T. Sukumaran, K. T R (eds.), *Conservation and Sustainable Utilization of Bioresources*, Sustainable Development and Biodiversity 30,
https://doi.org/10.1007/978-981-19-5841-0_24

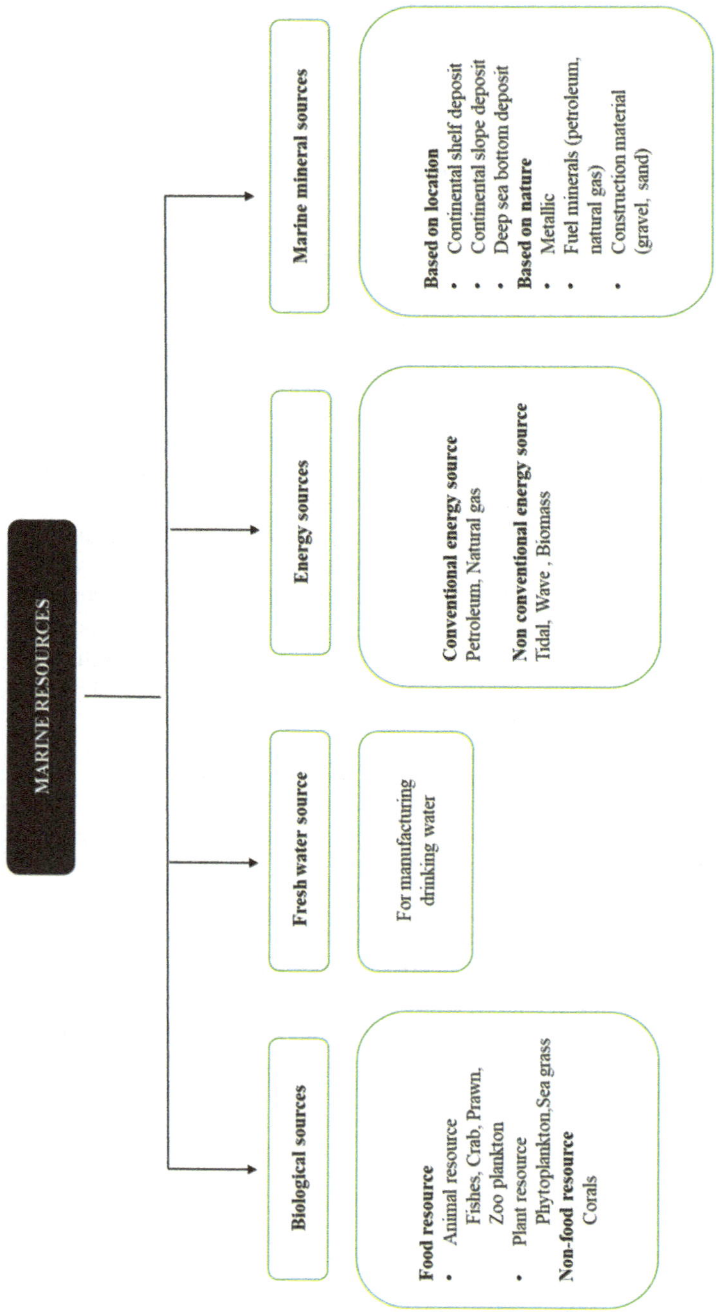

Fig. 24.1 The major marine resources

It is estimated that more than three billion people depend on marine sources for their livelihood, among which the Fisheries industry is the most prominent. Marine resources are renewable, and hence it was not a matter of overexploitation until the modernization of traditional methods with the advancement in science and technology. Human activities have a huge impact on the marine ecosystem and thereby on the whole globe. Plastic pollution is a major threat. Single-use plastic bottles and plastic bags are non-degradable. When it reaches the sea, it causes habitat degradation and ultimately, ecosystem degradation. The other land-originated pollutants include agricultural run-offs, untreated sewages, accidental oil spills, etc. The excess carbon emission causes ocean warming, acidification, and oxygen loss. According to the United Nations, eutrophication will increase to 20% by 2050. Ocean act as a carbon dioxide and heat buffering system. It absorbs 30% of CO_2 emitted by human activities (https://www.conserve-energy-future.com/causes-and-effects-of-ocean-pollution.php). The UN has declared 2021–2030 as the 'Decade of Ocean Science for Sustainable Development', aiming to reduce the pressure on the marine ecosystem as well as utilize the ocean for development activities without causing any impact on the marine environment. The challenges for sustainable utilization and development include:

1. Understand and beat marine pollution
2. Protect and restore ecosystem and biodiversity
3. Sustainably feed the world population
4. Develop a sustainable and equitable ocean economy
5. Unlock ocean-based solutions to climate change
6. Increase community resilience to ocean hazards
7. Expand the Global Ocean Observing System
8. Create a digital representation of the ocean
9. Skills, knowledge and technology for all
10. Change humanity's relationship with the ocean (https://www.oceandecade.org/challenges/)

24.2 Sustainability of Marine Food Resources

Illegal human activities are the main reason for the depletion of marine resources. The unsustainable fish catching techniques like bottom trawling, cyanide fishing, dynamite fishing, ghost fishing, and bycatch not only cause the depletion of the specific fish group but also other species are getting reduced in the marine ecosystem. Other illegal and uncontrolled human activities that cause unsustainability in marine food resources include recreational fisheries, tourism development, heavy marine traffic, global climate change and acidification of oceans, etc. (Gonzalez 2016).

24.2.1 How to Ensure Sustainability in Wild Fisheries?

It is complicated to maintain sustainability in fisheries because many commercial species are overfished before the efforts taken by any organization or governmental bodies, and the demand for seafood is also increasing. The rational use of all the strategies and equipment used in fishing and trawling is the best and most reliable method to retain the abundance of all the edible and non-edible organisms in the marine system. The efficiency, effectiveness, and specificity of fishing gear is the most important concern. The catchability of commercial fishing vessels is increasing with technological creep. On average, the catchability increases at a rate of 3.2% per year (Eigaard et al. 2014). With the improvement in gear design, the efficiency of engines, deck equipment, and catch handling procedure, along with the increase in the harvest of fish, the bycatch (harvest/capture of other species of fish or shellfish rather than the specific species for which the fishing gear is set) and habitat degradation were also significantly increased. In professional as well as recreational fisheries, regulations and control must be put in place to avoid the unnecessary usage of trawling nets and gears. It is very important to have knowledge of the spawning season and reproductive patterns of commercial fishes. The trawling must not disturb the growth and development of fries. Trawling of commercial fishes must be done based on their landing size (the minimum amount of fish that is required for commercial selling). The land size of each fish depends on the species and reproductive potential. A periodical assessment of the abundance of each species is required to know whether it has a minimum landing size. An effective statistical and technical system is required to control landings in the fish market. It is actually the management of people than the management of fish resources. Responsible fish consumption and giving awareness to children and young people about the depletion and deterioration of marine food resources and the importance of maintaining the ecosystem for the future are some of the qualitative approaches for sustainability (Gonzalez 2016).

24.3 Marine Mineral Resources

Oil, natural gas, and other mineral resources are plentiful in the marine environment. Tin, diamonds, sulfur, sand, gravel, calcium carbonate, heavy mineral sand etc., are some of the marine mineral sources. Even though deep-sea mineral deposits are vast, commercial exploitation is not feasible due to the high cost of processing, especially from the high sea. Most of the production of ocean minerals takes place in shallow, nearshore waters. The only production from the deep sea is oil and natural gas.

The UNCLOS had put forward a royalty provision concerning the continental shelf mineral excavation development. It states that the production occurring on the continental shelves beyond 200 nautical miles is subjected to pay to International Seabed Authority, which shares with UNCLOS as equal. The payment begins only in the sixth year of production at 1% of the production value, which reaches 12% by the 12th year and remains fixed (Hoagland et al. 2001).

24.4 Policy Instruments for Conservation and Sustainable Use of Marine Resources

The policy instruments for maintaining a sustainable marine ecosystem include regulatory, economic, information, and voluntary approaches (https://www.oecd-ilibrary.org/sites/b9c62341-en/index.html?itemId=/content/component/b9c62341-en#section-d1e5875). Regulatory approaches set out the legal territory/environment in which ocean industries operate within the national exclusive economic zone. Economic instruments continuously create incentives/revenue for actors by putting a price on externalities. Information and voluntary approaches are intended for producers and consumers.

24.4.1 Regulatory Policy Instruments

24.4.1.1 Marine Spatial Planning

Marine spatial planning (MSP) is the process of allocating temporal and spatial space for industries and other economic activities by the government. Eigaard et al. (2014) explained the steps in MSP and it includes:

1. Identify needs and establish authority
2. Obtaining financial support
3. Organizing the process through pre-planning
4. Organizing stakeholder participation
5. Defining and analyzing the existing condition
6. Define and analyze the future condition
7. Preparing and approving the spatial management plan
8. Implementing and enforcing the spatial management plan
9. Monitoring and evaluating performance
10. Adapting the marine spatial management process

One of the major hurdles in establishing MSP is the lack of data regarding the benthic habitat, biodiversity of marine system etc. Huge investments are needed to collect and study such data. One of the examples of a well-established MSP is Seychelles in the Indian Ocean. It is an integrated ecosystem-based sustainable practice.

24.4.1.2 Marine Protected Areas

Marine Protected Areas (MPAs) are one of the policy instruments adopted to minimize the deterioration of marine habitats and to protect vulnerable marine species with defined management objectives. Human activities are strictly regulated in MPAs, and it helps in long-term conservation and maintains consistency in the marine population. The deteriorated ecosystem will re-establish naturally with time. MPAs can be established in a variety of ocean habitats like the open ocean, coastlines, estuaries, and sometimes freshwater habitats also (protected areas in the great lake of the USA) and can be called marine reserves, marine parks, marine conservation zone, marine sanctuaries, etc. MPA is a conservative strategy, gaining popularity, the total covering in MPAs increased from 2.1% in 2000 to 17.3% in 2019.

Other regulatory instruments include a ban on single-use plastics in the marine environment, dynamite fishing, Regulations on vessel size, catch limit, regulatory standards for fishing gear etc.

24.4.2 Economic Instruments

The economic instruments that foster the sustainable marine system include:

- Taxes are imposed on polluters for dumping waste and other pollutants into the sea
- Fees and charges like fishing license fees or entry fees to an MPA
- Tradable permit: right to access or harvest certain resources is allocated only to permit holders. They are limited in number and can be traded between permit holders.
- Subsidies: for the production of certain things, the government provides subsidies, but it can have a negative impact if they increase production (like extraction for construction). Subsidies are intended to lower the cost of sustainable economic activities.
- Payment for ecosystem services: The payment given to ecosystem service providers by the users. For example, the 'blue carbon payments' for mangrove restoration.
- Biodiversity offsets: offsetting is a process of recreating the destroyed ecosystem due to unavoidable developmental activities. The main aim is no net loss of biodiversity; the recreation creates more habitat and thereby more rich biodiversity.

24.4.3 Information and Voluntary Approach

The information instrument gathers marine information between government, society and business. It can be done through voluntary agencies like the Marine Stewardship Council and Friend of Sea, which implements standards and certifications for sustainable marine utilization.

24.5 International Treaties and Commissions for Sustainable Utilization of Marine Resources

All nations have their own policies and strategies for conserving and utilizing marine resources within their territory. It depends on biodiversity and marine ecosystem mode without disturbing the economy and commercial development.

24.5.1 United Nations Convention on the Law of the Sea (UNCLOS)

The UNCLOS was signed in 1982 by 117 states and entered into force in 1994 to define territories for states, regulate the overexploitation of seabed, and improve the economic benefit from marine resources without overexploitation. The main principles underlying UNCLOS include:

1. Sovereignty over resources: the state or citizens of the state should utilize the resources within its territorial sea and exclusive economic zone.
2. Precautionary action states that the release of potentially harmful substances to the marine environment should be prohibited.
3. Common heritage: as the name suggests, this principle implies that everyone owns the common space, universal interest has priority over national interest, the economic benefit from the common space of the ocean must be shared among all the states, and the exploration of common space must be for peaceful purpose, scientific exploration must be done without any treatment to the environment.
4. Environmental conservation: all states are responsible for taking necessary measures for the conservation of the marine ecosystem.
5. Sustainable development: economic development must be in a way that both present and future requirements must meet.
6. International cooperation: cooperation among states is required for the conservation and management of living and non-living resources, scientific studies, settlement of disputes, pollution control, and technology transfer.

24.5.2 Sustainable Development Goal 14

Sustainable Development Goal 14 or SDG14 is one of the 17 sustainable development goals established by the United Nations in 2015 for the conservation and utilization of marine resources (https://www.un.org/sustainabledevelopment/oceans/). There are ten goals which include:

14.1. By 2025, prevent and significantly reduce marine pollution of all kinds, in particular from land-based activities, including marine debris and nutrient pollution

14.2. By 2020, sustainably manage and protect marine and coastal ecosystems to avoid significant adverse impacts, including by strengthening their resilience and taking action for their restoration in order to achieve healthy and productive oceans

14.3. Minimize and address the impacts of ocean acidification, including through enhanced scientific cooperation at all levels

14.4. By 2020, effectively regulate harvesting and end overfishing, illegal, unreported, and unregulated fishing, and destructive fishing practices and implement science-based management plans in order to restore fish stocks in the shortest time feasible, at least to levels that can produce maximum sustainable yield as determined by their biological characteristics

14.5. By 2020, conserve at least 10% of coastal and marine areas, consistent with national and international law and based on the best available scientific information

14.6. By 2020, prohibit certain forms of fisheries subsidies that contribute to overcapacity and overfishing, eliminate subsidies that contribute to illegal, unreported, and unregulated fishing and refrain from introducing new such subsidies, recognizing that appropriate and effective special and differential treatment for developing and least developed countries should be an integral part of the World Trade Organization fisheries subsidies negotiation

14.7. By 2030, increase the economic benefits to Small Island developing States and least developed countries from the sustainable use of marine resources, including through sustainable management of fisheries, aquaculture, and tourism

14.8. Increase scientific knowledge, develop research capacity, and transfer marine technology, taking into account the Intergovernmental Oceanographic Commission Criteria and Guidelines on the Transfer of Marine Technology, in order to improve ocean health and to enhance the contribution of marine biodiversity to the development of developing countries, in particular, small island developing States and least developed countries

14.9. Provide access for small-scale artisanal fishers to marine resources and markets

14.10. Enhance the conservation and sustainable use of oceans and their resources by implementing international law as reflected in UNCLOS, which provides the legal framework for the conservation and sustainable use of oceans and their resources.

Table 24.1 Current status of SDG 14 (https://sdg-tracker.org/oceans#targets)

SDG no	Target	Current status
14.1	Reduce marine pollution	Reduction in beach litter by 6% Increased eutrophication (2020)
14.2	Protect and restore marine ecosystem	A few countries like China, Malaysia, Singapore and Germany adopted ecosystem-based approaches to manage marine area
14.3	Reduce ocean acidification	Average pH of the marine water was reduced from 8.11 to 8.05 at the end of 2019
14.4	Maintain the fish stocks within sustainable levels	34.15% of total commercial fishes were overexploited and remaining were equal to or above the maximum sustainable yield (2017)
14.5	Protection of marine areas with specific interest	Percentage of protected area increased to 2.74% of total marine area (2018)
14.6	Combat illegal, unregulated and unreported fishing	Data from only some countries are known. Other countries implemented instrumentation and technology for regulating fishing (2022)
14.7	Increasing income from sustainable fisheries	The economic benefit from sustainable fisheries was lower than in previous periods (2011–2019)
14.A	Increase scientific knowledge on the marine system through research and develop technologies for sustainability	Not much progress in ocean research funding and thereby slow growth in research and technology
14.B	Support small-scale fisheries	Most of the coastal countries promote small-scale fisheries
14.C	Implementing UNCLOS	A few countries succeeded in implementing UNCLOS

The current status of SDG 14 is included in Table 24.1.

24.6 Integrated Coastal and Ocean Management

The coastal area management or integrated management of coastal or ocean resources begins with the 'interface between land and sea' or the area or place where the land surfaces and the surface of the sea meet and interact (Yusuf and Sultan 2001). It is a unique place in our global geography. Economically coasts are very important sites for port and harbor facilities that provide large financial benefits. The coasts are highly valued and greatly attractive as sites for resorts and as vacation destinations. The combination of fresh water and saltwater in coastal estuaries creates some of the most productive and richest habitats on earth, resulting in the bounty of fishes and other marine species, including cone shells, sharks, and horseshoe crabs are very useful in the formulation of medicines for different diseases such as cancer, muscle diseases, chronic pain, etc. (Thiruvenkatasamy and Rahman 2017).

Approximately eight million populations are located on the coast. It is evaluated that 90% of sewage is directly dumped into the sea in developing countries, and half the world's coastal wetlands have disappeared. About 75% of global fisheries are fully utilized or overfished, and 60% of coral reefs are threatened. Brander-Smith et al. (1990), say that the climate change report of the Intergovernmental panel, 2007 shows significant warming, sea-level rise, increase in the frequency of storms, changes in precipitation patterns, ocean acidification, and wind patterns are affecting each region differently. So, these climate changes cause greater threats to biodiversity (Thiruvenkatasamy and Rahman 2017).

The management of multiple uses and expectations from ever more crowded oceans and coasts is a major challenge for developed and developing countries alike. With increasing resource and user conflicts over sectoral and political boundaries, there is a need to take a more comprehensive, holistic and Integrated Coastal Management (ICM), which is a process for accommodating multiple uses in coastal and ocean areas to achieve sustainable development of coastal and ocean areas. Similarly, ecosystem Based management (EBM) includes the full range of interactions within an ecosystem (including social system), rather than focusing on individual uses, ecosystem, or species services, with a focus on maintaining ecosystem service functions. The two practices will both be needed in concert to address the huge challenges facing the world's coastal and ocean areas.

The 1992 UN Conference on Environment and Development (UNCED) and the 2002 World Summit on Sustainable Development (WSSD) recognized the importance of integrated ocean and coastal management (IOCM). Chapter 17 (the Ocean chapter) of Agenda 21 presented a complete blueprint for the management of the world's oceans, emphasizing that the oceans can no longer be managed as they had been traditionally through a sector, use-by-use approach. Instead, as chapter 17 put it, approaches that are "integrated with content and anticipatory in ambit" must be adopted. The 2002 Johannesburg Plan of Implementation (JPOI) of the World Summit on Sustainable Development (WSSD) followed with a more specific target and timetables calling for "the application by 2010 of the ecosystem approach" and promotion of integrated coastal and ocean management at the national level. WSSD also provided encouragement and assistance to countries in developing ocean policies and mechanisms for integrated coastal management (Global Forum 2002).

24.6.1 Integrated Coastal and Ocean Management Concept and Guidelines

The term ICM has been replaced by various other terms such as Integrated Area Management (IAM), Integrated Coastal and Ocean Management (ICOM), and Integrated Coastal Zone Management (ICZM). Although there are some differences in concepts, generally, all terms refer to the same approach to management.

ICM can be defined as "a continuous and dynamic process by which decisions are taken for the sustainable use, development, and protection of coastal and marine areas and resources" (Cicin-Sain et al. 1998). The goals of ICM are to attain sustainable development of coastal and marine areas; to reduce exposure of coastal areas and their populations to natural hazards; and to maintain essential ecological processes, life support systems, and biological diversity in coastal and marine areas. International guidelines for ICM are represented in Table 24.2.

Table 24.2 International guidelines for ICM

Year	organization	Guidelines
1992	UN (United Nations)	Agenda 21, chapter 17
1993	OECD (Organization for Economic Cooperation and Development) World Bank IUCN (International Union for Conservation of Nature)	Coastal Zone Management: Integrated Policies Guidelines for integrated Coastal Zone Management Cross-Sectoral, Integrated Coastal Area Planning (CICAP): Guidelines and principles for Coastal Area Development
1995	UNEP (United Nations Environment Programme)	Guidelines for Integrated Management of Coastal and Marine Areas: With Special Reference to the Mediterranean Basin
1996	UNEP (United Nations Environment Programme)	Guidelines for Integrated Planning and Management of Coastal and Marine Areas in the Wider Caribbean Region
1998	FAO (Food and Agricultural Organization)	Integrated Coastal Management and Agriculture, Forestry, and Fisheries
1999	UNEP (United Nations Environment Programme) EC Council of Europe	Conceptual Framework and Planning Guidelines for Integrated Coastal Area and River Basin Management Towards a European Integrated Coastal Zone Management (ICZM), Strategy: General Principles and Policy Options European Code of Conduct for Coastal Zone
2000	CBD (Convention on Biological Diversity)	Review of existing Instruments Relevant to Integrated Marine and Coastal Area Management and Their Implementation for the Implementation of the Convention on Biological Diversity
2004	CBD (Convention on Biological Diversity)	Integrated Marine and Coastal Area Management (IMCAM) Approaches for Implementing the Convention on Biological Diversity

24.6.2 Principles of Integrated Coastal Management

Integrated coastal management includes (1) a set of both practical and technical principles; (2) a management strategy that underlines adaptation and feedback; and (3) the use of particular approaches, methods, and techniques.

Two types of principles for guiding ICM can be identified: principles based on agreed international norms for environment and development that have originated from the Earth Summit and key international agreements and principles specifically related to the special character of coasts and oceans.

24.6.2.1 Principles Related to Environment and Development

The Rio Declaration on Environment and Development (Rio Declaration) is a group of 27 principles to guide national and international actions on environment, development, and social issues approved by all nations participating in the Earth Summit conference. Some of these principles are new; others represent the repetition of principles already established in international law; still, others represent changes in established principles of international law (Van Dyke 1996). Overall, they provide a broad set of norms to guide nations in the pursuit of sustainable development. Some of the major principles are described below;

24.6.2.1.1 Principle of Interrelationship and Integration

The Rio Declaration indirectly states that the principle of interrelationship and integration forms the backbone of sustainable development and is the underlying theme of the Rio Declaration and Agenda 21. It means that we must address the interrelationships, or interdependence, among issues and sectors and between environment and development. In contrast to past thinking and past practices, environmental protection and development cannot be considered as separate activities; each one must incorporate the other (Cicin-Sain et al. 1998).

24.6.2.1.2 Inter- and Intragenerational Equity Principles

The principle of intergenerational equity reflects the view that as members of the present generation, we hold the earth in trust for future generations and therefore, we should not prevent the options of future generations (WCED 1987). The principle of intragenerational equity refers to the obligation to take into account the needs of other members of society, especially regarding the distribution of the benefits of development.

24.6.2.1.3 Principle of the Right to Develop

This principle relates to the basic right to life of every human being as well as the right to develop his or her potential so as to live in dignity. It is the first principle stated in the Rio Declaration (Cicin-Sain et al. 1998).

24.6.2.1.4 Environmental Safeguards Principle

This principle relates to the prevention of environmental harm through anticipatory measures to prevent harm rather than through post hoc efforts to repair it or provide compensation for it. Environmental safeguards together with the precautionary principle and with two other Rio principles—the need for states to enact and implement effective environmental legislation and the principle of prevention of transboundary (across frontiers) environmental harm (Cicin-Sain et al. 1998).

24.6.2.1.5 Precautionary Principle

Lack of scientific certainty is no reason to postpone action to avoid potentially serious or irreversible harm to the environment. Principle 15 of the Rio Declaration reads, in part, "Where there are threats of serious or irreversible damage, lack of full scientific certainty shall not be used as a reason for postponing cost-effective measures to prevent environmental degradation" (Cicin-Sain et al. 1998).

24.6.2.1.6 "Polluter Pays" Principle

The Polluter-Pays Principle, as defined in paragraph 4 of the "Guiding Principles", states that the polluter should bear the expenses of preventing and controlling pollution "to ensure that the environment is in an acceptable state". The principle was originally developed by the Organization for Economic Cooperation and Development (OECD) to ensure that firms paid the full costs of controlling pollution and were not supported by the state. The principle is intended to apply within states rather than between states. Principle 16 of the Rio Declaration brings the "polluter pays" approach beyond a strictly developed country context; it calls on national authorities to "endeavor to promote the internalization of environmental costs and the use of economic instruments, taking into account the approach that the polluter should, in principle, bear the cost of pollution" (Cicin-Sain et al. 1998).

24.6.2.1.7 Transparency Principle and Other Process-Oriented Principles

The transparency principle requires that decisions be made in an open, transparent manner, with full public involvement. This principle goes hand in hand with a number of related principles: encouragement of participation by all major groups, including women, children, youth, indigenous peoples and their communities, local authorities, and others; the public's right to access to environmental information; and the importance of conducting environmental impact assessments to help ensure informed decision making and to provide for public participation and access to information.

Integrated Coastal Management is also conducted by principles related to the special character of oceans and coasts and to the public nature of the oceans and the use of coastal ocean resources.

24.6.2.2 Principles Related to the Special Character of Oceans and Coasts (Cicin-Sain et al. 1998)

- Coastal and ocean systems need special planning and management methods due to their high productivity, great mobility, and interdependence.
- The substantial interactions within the land-water boundary require recognizing and managing the whole system.
- Activities well inland can significantly affect coastal resources.
- The boundary of land and water (e.g., beaches and dunes) that helps as buffers against erosion and sea-level rise should be conserved.
- Disturbances of the natural longshore drift system should be reduced.
- The biodiversity of rare and fragile ecosystems and endangered/threatened species should be protected.
- Efforts to stabilize the coast should be "designed with nature" using, e.g., special vegetation instead of physical structures for erosion control.

24.6.2.2.1 Principles Related to the Public Nature of the Oceans and to the Use of Coastal Ocean Resources

- Since ocean resources are part of the public area, management must be guided by a stewardship ethic, fairness, and equity.
- Historically based claims of native peoples should be recognized.
- ICM is the future to promote the coexistence of multiple uses in an area; in case of irreconcilable conflicts, protecting renewable living resources and their habitats should have priority over the exploitation of non-living, non-renewable resources.
- New coastal developments that are marine dependent should have priority over those that are not.

24.6.3 Functions of ICM (Karthick et al. 2015)

The main functions of integrated coastal management are presented in Table 24.3.

24.6.4 Application of ICM

The implementation of the river basin and sub-basin management in Houag Champi (Lao People's Democratic Republic) has made full use of the ICM concepts and working modality. Integrated river basin management (IRBM) and ICM are complementary and mutually reinforcing in natural resources management across a landscape from the river to the coast.

The ICM method was also accepted in developed countries like Singapore and Japan. While evaluating the implementation of integrated urban coastal management (IUCM) in Singapore, the conclusion was that "the ICM system remains fully relevant to a highly urbanized city that has all along been developed through sectoral management". Realizing the usefulness and effectiveness of integrated management, several local governments in Japan have initiated an integrative, comprehensive management approach with measurable success (PEMSEA 2015).

The major goal of an ICM system is to attain environmental, social, and economic sustainability in the coastal and marine areas. Many local governments implementing ICM programs are addressing selected sustainable development challenges such as disaster risk reduction and management, biodiversity loss, pollution, access to and security of fresh water supplies, and sustainable use of marine resources. As such, these local governments are directly contributing to Sustainable Development Goals (SDGs) such as clean water and sanitation (SDG 6), sustainable cities (SDG 11), climate action (SDG 13), life on land and below water (SDG

Table 24.3 Functions of integrated coastal management

Functions	Activities
1. Area planning	Plan for present and future uses of coastal and marine areas; provide a long-term vision
2. Promotion of economic development	Promote proper uses of coastal and marine areas (e.g., marine aquaculture, ecotourism)
3. Stewardship of resources	Protect the ecological base of coastal and marine areas; conserve biological diversity; ensure sustainability of uses
4. Conflict resolution	Integrate and balance existing and potential uses; address conflicts among coastal and marine uses
5. Protection of public safety	Protect public safety in coastal and marine areas typically prone to significant natural, as well as human-made hazards
6. Proprietorship of public submerged lands and waters	As governments are often outright owners of specific coastal and marine areas, manage government held areas and resources wisely and with good economic returns to the public

14, 15), and partnerships (SDG 17). The ICM system can also carry out other relevant, sustainable development targets which are critical in a local government context, including reduction of poverty and hunger (SDG 1, 2), gender equality (SDG 5), sustained, inclusive and sustainable economic growth (SDG 8), healthy lives and wellbeing at all ages (SDG 3), and inclusive and equitable quality education (SDG 4) (Thia-Eng and Bonga 2018).

The ICM system plays a key role in achieving global targets of international conventions and protocols. For example, the inclusion of "habitat protection, restoration, and management" in the common framework for Sustainable Development of Coastal Areas enables local action in addressing some of the concerns of biodiversity conventions, particularly in achieving the Aichi Biodiversity Targets (ABTs). The application of the ICM system to achieve most of the 13 ABTs is outlined in the "Practical Guidance on Implementing Integrated Coastal Management in the Context of Achieving Aichi Biodiversity Targets" (CBD 2015). Similarly, the inclusion of "pollution reduction and waste management" enables local ICM initiatives to implement the provisions of UNEP's Convention on Global Actions for Land-based Pollution (GPA). The inclusion of "food security and livelihood management" contributes to addressing FAO's Convention on Responsible Fisheries and the inclusion of "disaster management", to meeting commitments of the climate change convention.

References

Brander-Smith D, Therrien D, Tobin S (1990) Public review panel on tanker safety and marine spills response capacity – final report

CBD (2015) Integrated coastal management for the achievement of the aichi biodiversity targets (Issue 76)

Cicin-Sain B et al (1998) Integrated coastal and ocean management: concepts and practices

Eigaard OR et al (2014) Technological development and fisheries management. Rev Fish Sci Aquacult 22(2):156–174. https://doi.org/10.1080/23308249.2014.899557

Global Forum (2002) 'Promoting Integrated Oceans Governance', 17

Gonzalez JA (2016) Sustainability of marine food resources—an ecological and fishery approach. J Environ Health Sci 2(3):1–5. https://doi.org/10.15436/2378-6841.16.1027

Hoagland P, Jacoby J, Schumacher ME (2001) 'The law of the sea'. In: MER Marine Engineers Review. p 47. https://doi.org/10.1006/rwos.2001.0415

Karthick S et al (2015) Integrated coastal area management. Int J Eng Adv Technol 4:2249–8958

PEMSEA (2015) PEMSEA Accomplishment Report 2014–2015

Thia-Eng C, Bonga D (2018) A functional integrated coastal management system towards achieving sustainable development objectives. In: Local contributions to global sustainable agenda: case studies in integrated coastal management in the East Asian Seas Region. CMC, pp 525–552

Thiruvenkatasamy K, Rahman F (2017) Integrated coastal and ocean management: a global overview. J Eng Appl Sci 12(Special issue 2):6260–6268

Van Dyke JM (1996) The Rio principles and our responsibilities of ocean stewardship. Ocean Coast Manag 31(1):1–23

Yusuf A, Sultan M (2001) 'The Maritime Commons: Digital Repository of the World'. An evaluation of integrated coastal and ocean management as a means for sustainable development: a case study: the environmental and socio-economic impact of shrimp farming in Bangladesh

Chapter 25
Traditional Knowledge and Its Sustainable Utilization

S. Rajasekharan, Vinodkumar T. G. Nair, M. Navas, T. C. James, and K. Murugan

Abstract Traditional knowledge (TK) refers to the tradition or culture of respective countries of the world. TK is sometimes specifically focused on information belonging to different ethnic communities of the world, especially that of tribal people. It is often restricted to location-specific knowledge of common people, including ethnic communities residing in a particular region/country, and the knowledge is confined to the genetic and non-genetic resources available within their surroundings. The importance of TK is highlighted by the fact that more than 80% of the livelihood needs of the world's poor directly or indirectly depend upon the use of biological resources and associated TK. TK is getting eroded rapidly because of the changing lifestyles of the people. In the present review, Sector-wise details of Traditional knowledge, sources of TK, TK, and Health Traditions in India, Methodology adopted for the systematic documentation of TK, Prior Informed consent was developed by the Division of Ethno medicine and ethnopharmacology, JNTBGRI, Kerala state of India, selected few case Studies in the Oral Health Tradition, treatment procedures prevailing among the tribal as well as folk communities to combat various health issues, passport Script Data of knowledge provider, ethnomedicine and its scope in developing novel herbal drugs and nutraceuticals, ethnomedical research, sustainable utilization of TK and medicinal Plants, ethnopharmacology and drug development—JNTBGRI's contribution, phytoconstituents as nutraceuticals, food and medicinal plants used by the tribes of Kerala having immunoenhancing properties, challenges and opportunities of TK,

S. Rajasekharan
Kerala State Biodiversity Board, Thiruvananthapuram, Kerala, India

V. T. G. Nair · M. Navas
Jawaharlal Nehru Tropical Botanic Garden and Research Institute, Thiruvananthapuram, Kerala, India

T. C. James
Research and Information System for Developing Countries, New Delhi, India

K. Murugan (✉)
CISSA Phytotech, Thiruvananthapuram, Kerala, India

S. T. Sukumaran, K. T R (eds.), *Conservation and Sustainable Utilization of Bioresources*, Sustainable Development and Biodiversity 30,
https://doi.org/10.1007/978-981-19-5841-0_25

current scenario of protection of TK were discussed. There is an urgent need to document such valuable information for the welfare and betterment of posterity.

Keywords Traditional knowledge · Phytoconstituents · Protection · Health Traditions · Sustainable utilization

25.1 Introduction

25.1.1 Definition of Traditional Knowledge

Traditional Knowledge (TK), Indigenous Knowledge (IK), Traditional Environmental Knowledge (TEK), and Local Knowledge (LK) generally refer to long-standing traditions and practices of certain regional indigenous or local communities. TK also encompasses the wisdom, knowledge, teaching, and experience of these communities mostly transmitted orally from generation to generation. The definition of TK has been widely debated and discussed on various national and international floors and is yet to arrive at a common consensus hence a universal definition could not be provided because of its nature, characteristics use, etc. TK, IK, TEK, and LK are relative terms coined by policy makers, researchers, or scholars belonging to different traditions from time to time according to the origin and utility of genetic resources which is associated with biodiversity, including plant, animal, and micro-organism diversity. Apart from this, TK associated with non-genetic resources also exists in different cultures, which include martial arts, magico-religious cures, yoga, etc. The definition of TK is sometimes specifically focused on the knowledge belonging to different ethnic communities of the world, especially that of the tribal people. Since TK is restricted to location-specific knowledge of common people, including ethnic communities residing in a particular region/country, the knowledge is confined to the genetic and non-genetic resources available within their surroundings.

Ethnomedical surveys and systematic documentation of traditional knowledge conducted among different tribal communities of Kerala are very relevant even today, as there is immense potential and knowledge base still remaining with the tribal communities, particularly those living in the interior forest patches. The systematic documentation of traditional knowledge could give new insights for conservation and sustainable utilization of biodiversity, which will definitely give new leads for further research that can be validated and developed into new medicines, functional foods, products for agro-chemical industries, etc., ensuring benefit sharing to the knowledge providers. This will also help create new programs and projects for the Local Governing Bodies/Grama panchayaths and also generate awareness among local people and tribal communities on the importance of conservation of forests, biodiversity, and traditional knowledge, particularly related to plants used for food and medicine.

25.1.2 Significance

- Ethnomedicine/Traditional knowledge is directly linked with the cultural heritage of the people and has been handed over from generation to generation.
- Ethnomedical knowledge is getting eroded rapidly because of the changing lifestyles of people. So there is an urgent need to document such valuable information for the welfare and betterment of posterity.
- Bio-prospecting of plants used for food and medicine based on Ethnomedicine/ Traditional/Folkpractices is an important area of research due to the possibility of developing diverse novel products which are essentially required for taking care of human health.
- The role of Ethnomedical leads is important in evaluating the medicinal and food value of plants which may help develop therapeutically active formulations with commercial viability.
- Technology transfer of these products will further ensure their commercialization and benefit-sharing with the knowledge providers.

25.1.3 Thrust Areas of Traditional Knowledge

The sector-wise areas of Traditional Knowledge can be classified into different segments such as Art and Culture, Agriculture, Animal Husbandry, Architecture, Biodiversity Conservation and Utilization, Eco-friendly Practices, Fisheries, Forest and Wild life Management, Health Care, Medicinal plants and Food Plants, Rural Technology, etc. Traditional Knowledge and Folk Practices Related to Art and Culture can be further classified into Songs, Dance forms, Drama, Murals and Paintings, Dyes and Natural colours, Music and musical instruments, Martial arts, Ornaments and costumes, Customs and beliefs, Tribal customs, and beliefs, Taboos and religious practices (Table 25.1). Traditional Knowledge and Folk Practices Related to Health Care include health and hygiene, prevention of diseases, method of diagnosis, disease management, customs, rituals and religious practices, healing techniques, etc. (Rajasekharan et al. 2012c) (Appendix 1).

25.1.4 Different Sources of Traditional Knowledge

Based on source, Traditional Knowledge (TK) could be divided into (1) the knowledge related to plants used for food, (2) Knowledge related to plants used in the branches of science like Ayurveda, Yoga, Naturopathy, Unani, Siddha, Sowa-Rigpa and Homoeopathy (AYUSH), and (3) Knowledge related to plants used in indigenous medicine. There are diverse sources for each category of knowledge, which is represented in Fig. 25.1.

Table 25.1 Sector-wise analysis of Traditional knowledge

Part 1-primary data	Art and culture, Agriculture, Animal husbandry, Architecture, Biodiversity conservation and utilization, Eco-friendly practices, Fisheries, Forest and wildlife management, Health care, Medicinal plants and food plants, Rural Technology, Miscellaneous
Sub sector-1 Traditional knowledge related to art and culture	Songs, dances, Drama, Murals and paintings; Dyes and natural colours; Music and musical instruments; Martial arts; Ornaments and costumes, Customs and beliefs, Tribal customs and beliefs, Taboos and religious practices
Sub sector-2 Traditional knowledge related to agriculture	Common agricultural practices including selection of seeds, selection of sites, pre-treatment of seeds before sowing; Propagation and multiplication, Prevention of soil erosion, Desalination/reclamation of soil, Timing for sowing the seeds, weeding, application of manure, harvesting; Pest control, Irrigation, Intercropping, Water harvesting/watershed management/maintaining ecological balance, Combating unexpected and sudden changes in climate, Enhancing soil fertility, Breeding and hybridization, Post-harvest technology; Storage of harvests, Processing harvested goods; Religious, ritual and spiritual practices concerning agriculture
Sub sector-3 Traditional knowledge related to animal husbandry	Selection of appropriate breeds, Feed and fodder, Breeding and delivering, Prenatal and postnatal care, Enhancement of milk production, Animal health care, and disease management, Plants used in veterinary medicine; Use of milk, other animal products, Taboos, and beliefs in rearing and keeping animals
Sub sector-4 Traditional knowledge related to architecture	Structure and architecture of dwelling places, Places of worship, Places of public assembly, commercial establishment, etc., Traditional construction techniques including masonry, carpentry, varnishing, etc.
Sub sector-5 Traditional knowledge related to biodiversity conservation and utilization	Collection of non-wood forest product, Non-destructive and renewable extraction technique, Timing and periodicity of extraction, Conservation and utilization of rare, endangered of plants, Conservation of frequently extracted species and threatened species, Plants yielding dyes and pigments; Plants yielding useful gums and resins, Aromatic plants and plants having insecticidal properties
Sub sector-6 Traditional knowledge related to eco-friendly practices	Topography and terrain of areas, Self-protection and safe guarding against wild animals and reptiles, Prediction of sudden climatic changes and natural calamities, Ethnic food and natural drinks, allergic and poisonous plants
Sub sector-7 Traditional knowledge related to fisheries	Seasonal availability and migration of fish in the sea and river waters; Weather forecasting and fishing; Manufacture of fishing nets and fishing equipment, Manufacture of fishing boat; Breeding, egg laying,

(continued)

Table 25.1 (continued)

	hatching, etc., Pisciculture; Poisonous, toxic and edible fish, Fish harvesting, Fish processing and preservation, Fish poisoning plants, Vernacular name, credibility, and other properties; Medicinal properties
Sub sector-8 Traditional knowledge related to forest and wildlife management	Flora and fauna of the forest, Wild edible and medicinal plants, Wild edible and toxic plants; Mushrooms Concentration, migration, and movement of wild animals
Sub sector-9 Traditional knowledge related to health care	Health and hygiene, Prevention of diseases, Method of diagnosis, Disease management, Custom, rituals and religious practices, Healing techniques, Prenatal and postnatal care, Animal products; Metals and minerals
Sub sector-10 Traditional knowledge related to medicinal and food plants	Medicinal plants, Wild edible plants, Food plants, Single drug remedies, Compound drugs; Selection and collection of drugs, processing preparation, storage and administration of food and medicine
Sub sector-11 Traditional knowledge related to rural technology	Traditional crafts such as treatment of coconut husk, preparation of fibre, coir and coir products; Bamboos, reeds, and their products; Hand loom cotton cloth, dyeing and printing of textiles using natural colours; Medicated cloth, Tanning of leather; Potteries and clay products
Sub sector-12 Traditional knowledge related to miscellaneous	Any other relevant information based on above-mentioned sectors or other areas. TK is related to plant and animal health care

25.1.5 Traditional Knowledge and Health Traditions in India

India is one of the mega diversity countries having rich cultures and practices related to food and medicine. Classical and Oral Health Traditions are the two main areas in the medicare system. The Classical system includes highly organized, classified and codified systems like Ayurveda, Siddha, Unani, Homoeopathy, Yoga and Naturopathy, and Sowa-rigpa, which all come under the AYUSH system of medicine. The oral system or non-codified system includes Tribal medicine and Folk medicine with the service of bone setters, local mid-wives, *visha vaidyas* (traditional toxicologists in Kerala, India), *marma/varma* (neuromuscular therapy in Ayurveda) practitioners, *kalari* (martial art form in Kerala, India) practitioners, local traditional *vaidyas* (traditional healers) and farmers (Rajasekharan et al. 2012b).

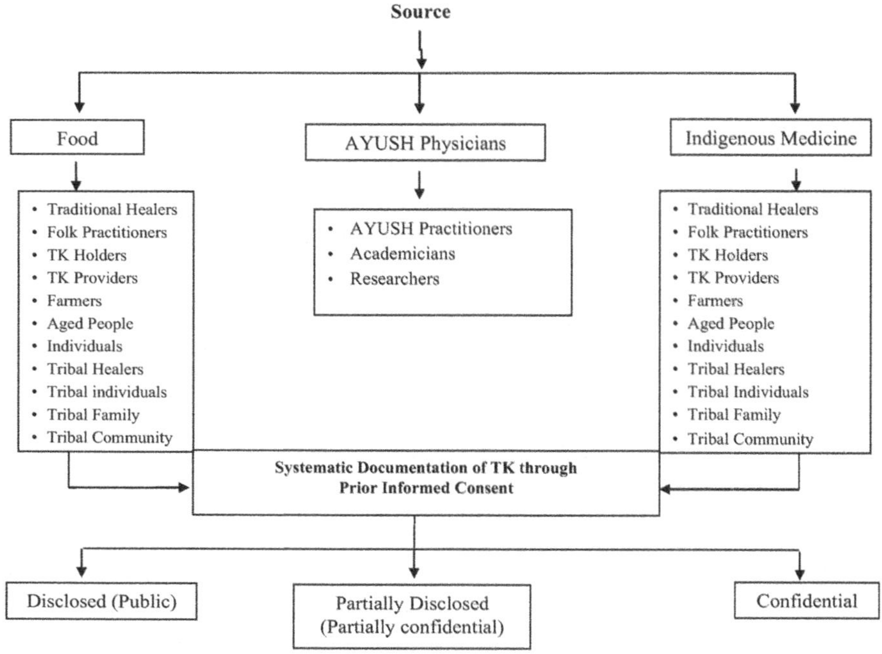

Fig. 25.1 Sources of category of knowledge

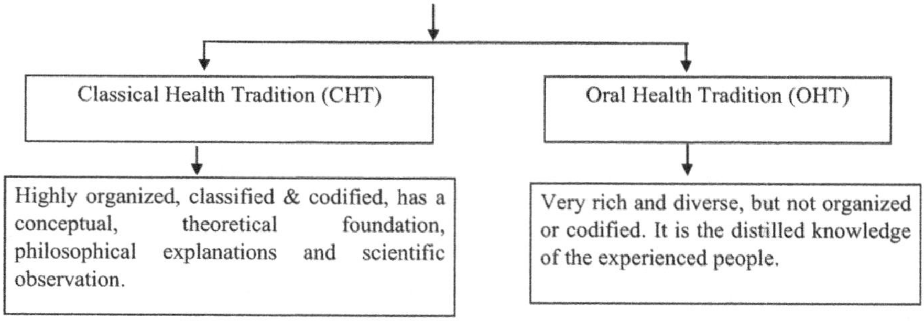

Fig. 25.2 TK related to traditional medicine in India

25.1.6 Health Traditions in India

TK related to Traditional Medicine in India as well as the Source of TK related to Traditional Medicine in India is represented in the Figs. 25.2 and 25.3.

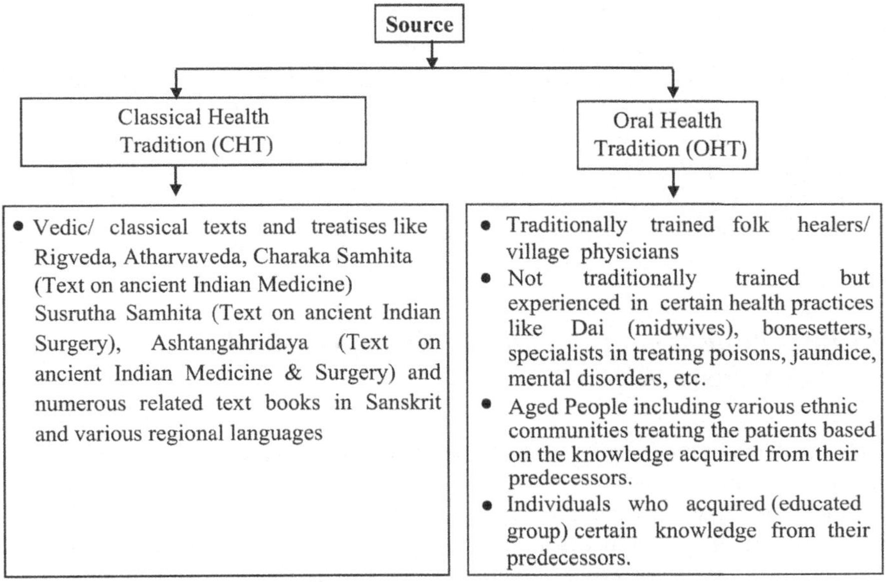

Fig. 25.3 Source of TK related to traditional medicine in India

25.1.7 Systematic Documentation of Traditional Knowledge

TK associated with biodiversity includes both codified (documented) as well as undocumented (orally transmitted information/expressions of folklore) information which are accepted as cultural components of biodiversity. If these are not preserved and documented, they are likely to be lost in the modern materialistic world. Hence, there is amble scope and urgent necessity to carry out systematic documentation of Traditional Knowledge (TK) associated with biodiversity. It is essentially required for protecting the knowledge under a *sui-generis* system by establishing different viable models. Systematically documented TK will provide valuable information that is highly beneficial to the research workers for developing new processes, products, patenting, technology transfer, commercialization, benefit-sharing, etc. It is also possible to make TK cost-effective and helpful for developing appropriate technology. This will further lead to the integration of TK into the planning process for the various developmental activities at the grassroots level. Implementation of an effective legal framework to check bio-piracy is also highly essential while implementing sector-wise programmes related to biodiversity and associated TK.

Sector-wise thrust areas are identified for the documentation of TK includes—TK related to art and culture, agriculture, animal husbandry, architecture, biodiversity conservation and utilization, eco-friendly practices, fisheries, forest and wildlife management, health care, medicinal and food plants, rural technology, etc. Several countries have adapted existing Intellectual Property systems to the need of TK holders through *Sui generis* measures for TK protection. These take different forms in various countries. A database of official insignia of Native American Tribes prevents others from registering these insignia as trade marks in the United States of America. New Zealand's trademark law has been amended to exclude trademarks that cause offence, and this applies especially to indigenous Maori symbols. India's Patent Act has been amended to clarify the status of TK within patent law (Rajasekharan et al. 2012a; Shravan 2012). The Chinese State Intellectual Property Office has a team of patent examiners specializing in traditional Chinese medicine (WIPO publication No.920).

25.1.8 Protection of Traditional Knowledge Under Sui Generis *Rights*

In some communities and counties, the judgment has been made that even adaptations of existing IP rights systems are not sufficient to cater to the holistic and unique character of TK subject matters. This has led to the decision to protect TK through *Sui generis* rights. What makes an IP system a *Sui generis* one is the modification of its features so as to properly accommodate the special characteristics of its subject matter and the specific policy needs which lead to the establishment of a distinct system. Few national experiences in using *Sui generis* IP rights for protecting TK are narrated here. A more detailed analysis of national laws is presented in the latter part of this chapter (Rajasekharan et al. 2012c; Rajasekharan and Latha 2011)

1. The *Sui generis* regime of Peru was established by Law No.27811 of 2002 (WIPO,920E)
2. The Biodiversity Law No.7788 of Costa Rica (WIPO, 920E)
3. Portugal's *Sui generis* Decree-Law No.118 of April 20, 2002 (WIPO, 920E)
4. The Act on Protection and Promotion of Traditional Thai Medicinal Intelligence, B.E.2542. (WIPO, 920E)

25.1.9 Methodology Adopted for the Systematic Documentation of TK

The methodology adopted for the systematic documentation of TK starts with groundwork and planning to access and benefit sharing and further planning, which is depicted in Fig. 25.4.

Methodology

↓

Ground work and planning for the implementation of the project

↓

Establishing contact with the Gramapanchayath

↓

Interaction with elected local body members and TK providers of Gramapanchayath

↓

Team building and vision development

↓

Interactive session with TK holders/providers including

↓

Question and answer session

↓

List out the name of TK providers/holders through interactive session

Organization and planning of field visit

↓

Interview / meeting with individuals/ TK holders after obtaining prior informed consent (pic)

(With the aid of voice recorder and photography with short video clippings)

↓

Field level TK documentation in the prescribed format as per the guidelines

↓

Data pooling and analysis

↓

Preparation of electronic database

↓

Digitisation of TK

↓

Video documentation of selected case studies based on traditional and folk practices

↓

Establishment of data bank

↓

Availability of the data for the end users (through signing of contractual agreement)

↓

Guidelines for the knowledge providers and end users

↓

Access and benefit sharing (ABS)-including publication of case studies

↓

Integration of TK into the planning process

Fig. 25.4 Methodology adopted for the systematic documentation of TK

25.1.10 Prior Informed Consent

Jawaharlal Nehru Tropical Botanic Garden and Research Institute (JNTBGRI) in Thiruvananthapuram district of Kerala, India, is a Conservatory Botanic Garden of tropical plant resources in India and Kerala state in particular. The institute established in 1979, is a prestigious research centre recognized as a 'National Centre of Excellence' in ex situ conservation and sustainable utilization of tropical plant diversity. A Prior Informed consent procedure was developed by the Division of Ethno medicine and Ethnopharmacology, JNTBGRI for the systematic documentation of TK associated with plants used for food and medicine; which was used for the documentation of TK associated with *Arogyapacha (Trichopus zeylanicus)* (Appendix 1).

25.2 Case Studies

In the Oral Health Tradition, there are several treatment procedures prevailing among the tribal as well as folk communities to combat various health issues. Kerala is endowed with a rich and varied cultural heritage rooted in Traditional Knowledge systems. One of the areas where Traditional Knowledge has made a tremendous contribution to the betterment of the quality of life is in the primary health care sector. Apart from Ayurveda, there is a strong and vibrant oral health tradition existing in the countryside, villages, and tribal habitats. The practitioners of the oral health tradition cater to the needs of the local communities, often at no cost or at easily affordable prices, as they make use of the locally available herbs, minerals, and animal products for the preparation of medicines.

In the past, most of the houses in Kerala had maintained a 'minigene pool' of plants used for food and medicine. It included different local varieties of Paddy, Yam, Colocasia, Tubers, Horse gram, Pepper, Turmeric, Sesamum, Arrowroot, different varieties of grains, etc. They also conserved a number of medicinal plants within their surroundings, which are still continued to be utilized in the form of 'home remedies for common ailments'. Among many Folk/Traditional food and Medicinal plant practices documented, 20 selected case studies are described below.

Case studies were performed during the field level survey by Rajasekharan and his team, JNTBGRI in 2005 from different parts of Kerala, India. A pipeline for bio prospecting of ethnomedicine or TK is narrated in Fig. 25.5.

25.2.1 Marunnu Kanji (Medicated Gruel)

'*Marunnu kanji*' is a kind of medicated gruel preparation. There are 25 ingredients (8 fresh medicinal plants and 16 dried medicinal plants and cow's milk) in the preparation of medicated gruel (Table 25.2).

Method of preparation: One kilogram of rice is cooked with 2.5 L of cow's milk and 1 L of decoction prepared from ingredients No.2–18. At the end of the preparation, expressed juice obtained from ingredients No. 19–24 is mixed with the gruel, and a sufficient quantity of rock salt and honey is added.

Quantity prescribed: 250 mL for an adult.

Indications: It is exclusively recommended to enhance appetite, immunity, liver function and cure disorders in the alimentary canal. An average of 10–15 thousand

Bioprospecting of ethnomedicinal and Traditional Knowledge (TK)

↓

Ethnomedicine/ TK

↓

Ethno medico Botanical exploration investigation

↓

Preparation of database

↓

Prediction of database and guidelines for sharing knowledge

↓

Legal implication related to the protection of TK

↓

Section and collection of plant materials for preclinical study

↓

Extraction

↓

Bio-assay

↓

Isolation of compounds and structure elucidation

↓

Pre-clinical study (In vitro- in vivo) (Including pharmacognosy, Pharmacology, Toxicology)

↓

Product development

↓

Filing of patent

↓

Clinical trial

↓

Development of drugs

↓

Technology transfer, Commercialization and benefit sharing

Fig. 25.5 A pipeline for bio prospecting of ethnomedicine or TK

Table 25.2 Ingredients of medicated gruel

Sl. no.	Botanical name	Local name	Parts used	Quantity
1.	*Oryza sativa* L.	Pachari	Grain	1 kg
2.	*Aegle marmelos* Corr.	Koovalam	Root	15 g
3.	*Premna serratifolia* L.	Munha	Root	15 g
4.	*Oroxylum indicum* Vent.	Vayyazhantha	Root	15 g
5.	*Stereospermum suaveolens* (Roxb)DC.	Paathiri	Root	15 g
6.	*Gmelinia arborea* Roxb.	Kumbil	Root	15 g
7.	*Desmodium gangeticum* (L.) DC.	Orila	Root	15 g
8.	*Pseudathria viscida* (L.) Wt. and Arn.	Moovila	Root	15 g
9.	*Solanum indicum* L.	Cheru vazhuthana	Root	15 g
10.	*Solanum virginianum* L.	Velvazhuthana (Kantakarichunda)	Root	15 g
11.	*Tribulus terrestris* L.	Njerinjil	Root	15 g
12.	*Sida alnifolia* L.	Bala/Kurunthotti	Root	15 g
13.	*Ricinus communis* L.	Avankku	Root	15 g
14.	*Jatropha curcas* L.	Kodiyavanakku	Root	15 g
15.	*Vigna vexillata* (L.) A.Rich	Kattuzhunnu	Seed	15 g
16.	*Pueraria montana* var. *lobata* (Willd.) Sanjappa and Pradeep	Kattupayar	Seed	15 g
17.	*Cyperus rotundus* L.	Muthanga	Tuber	15 g
18.	*Boerhavia diffusa* L.	Thazhuthama	Leaves	25 g
19.	*Biophytum sensitivum* (L.) DC	Mukutti	Whole	25 g
20.	*Cardiospermum halicacabum* L.	Valli uzhinja	Whole	25 g
21.	*Emilia sonchifolia* (L.) DC. ex DC.	Muyalcheviyan	Whole	25 g
22.	*Cissus quadrangularis* L.	Ellodiyan	Stem	25 g
23.	*Euphorbia hirta* L.	Nilampala	Whole	25 g
24.	*Justicia gendarussa* Burm. f.	Kuzhalvathakodi	Leaves	25 g
25.	Cow's milk-			2.5 L
26.	Water			4 L[a]
27.	Expressed juice[b]			0.5 L

[a]Ingredients No.2–18 prepared in the form of decoction by adding 4 L of water and reduced into 1 L
[b]0.5 L expressed juice is to be obtained from ingredients No. 19–24

people consumed this medicated gruel in the function organized to commemorate their ancestors during the year 2011.

25.2.2 Mukkudi *(Medicated Butter Milk)*

Mukkudi, is a polyherbal formulation consisting of 28 ingredients. It contains 12 fresh leaves of medicinal plants and 16 raw drugs. Major ingredients are presented in Table 25.3.

25.2.2.1 Method of Preparation and Administration

This is a unique kind of polyherbal preparation popularly known as 'Mukkudi'[1] formulated by the traditional Ayurvedic physicians of Kerala. It is an admixture of decoction of raw drugs, a paste of fresh leaves, and curd (prepared from cow's milk). It is to be administered in the morning on an empty stomach.

Indications: To prevent and cure diseases occurring during the rainy season, especially diarrhoea and dysentery. It also enhances appetite and is an effective remedy for piles.

Dosage: 250 mL for adults. An average of 300–400 patients/individuals were administered this treatment during special occasions.

25.2.3 Attayum Kuzhambum *(Leech for Bloodletting and Application of Medicated Oil)*

The history of Sri Maruthorvattom Sree Dahanwanthari temple is directly linked with the local ruler Sri Puthiyaikkal Thamburan, Sri Velludu Namboothiri, and Ashtavaidyan Chirattamon Mooss.

The term *Atta* means Leech and *Kuzhambu* denotes the medicated oil. The *kuzhambu* is popularly known as *'Narayana thailam'*. It is one of the healing practices prevalent in the temple to cure *Vatarogas* (rheumatic disorders) since time immemorial. This practice was introduced in the temple by Chirattamon Mooss, one of the members of the Ashtavaidya family (Physicians and Surgeons, who specialized in the eight specialities of Ayurveda in Kerala), who migrated from Central Kerala to Maruthorvattom, Alapuzha District of Kerala. Leech must have been used to perform the bloodletting treatment. At present, only medicated oil is dispensed from the temple for treatment (Table 25.4).

Indications: This medicated oil is administered orally and externally to cure Gout, Joint pain, Arthritis and Lumbago. An average of 200–250 patients received this oil during the occasion.

---

[1] '*Mu*' means three, which denotes the decoction (raw drugs) + paste (fresh leaves) + curd: *Kudi* means drink – a kind of drink blended with three items ('Mukkudi').

Table 25.3 Ingredients of medicated butter milk

Sl. no.	Botanical name	Local name	Parts used	Quantity
1.	*Oxalis corniculata* L.	Puliyarila	Whole fresh	10 g
2.	*Phyllanthus amarus* Schumach. and Thonn.	Keezharnelli	Whole (fresh)	10 g
3.	*Mangifera indica* L.	Nattu Mavu	Tender leaves (fresh)	10 g
4.	*Aegle marmelos* Corr.	Koovalam	Leaves (fresh)	10 g
5.	*Murraya koenigii* (L.) Spreng.	Kariveppila	Leaves (fresh)	
6.	*Zingiber officinale* Roscoe	Inchi	Rhizome (fresh)	10 g
7.	*Emilia sonchifolia* (L.) DC. ex DC.	Muyal cheviyan	Whole (fresh)	10 g
8.	*Cardiospermum halicacabum* L.	Valli uzhinja	Whole (fresh)	10 g
9.	*Centella asiatica* (L.) Urb.	Kudangal	Whole (fresh)	10 g
10.	*Piper nigrum* L.	Kurumulaku	Fruit (dried)	6 g
11.	*Punica granatum* L.	Mathalam	Fruit rind (dried)	6 g
12.	*Cyperus rotundus* L.	Muthanga	Tuber (dried)	6 g
13.	*Myristica fragrans* Houtt.	Jathi	Seed aril (dried)	6 g
14.	Prepared from cow's milk	Curd		1 L
15.	Induppu (rock salt)			Sufficient quantity
16.	Honey			Sufficient quantity

25.2.4 Formulations from Mr. Vaidyamadom Cheriya Narayanan Namboothiri: A Legend and an Eminent Traditional Ayurvedic Physician of Kerala

Mr. *Vaidyamadom* Cheriya Narayanan Namboothiri, aged 84, is an authority on classical Ayurvedic text, popularly known as '*Ashtangahridayam*', written by Vagbhatta. The *Vaidyamadom* family has in its possession two of the exclusive three copies of the Palm leaf manuscript of well-known commentaries of '*Ashtangahridayam*' and '*Ashtagasangraham'* by *Indu,* the disciple of Vagbhatta (personal communication of Shri *Vaidyamadom* Cheriya Narayanan Namboothiri). He wrote a commentary in Sanskrit on '*Ashtangahridayam'* and was published by Chaukhambha Sanskrit Book publisher Banaras.

According to *Vaidyamadom* Cheriya Narayanan Namboothiri, Vagbhatta first visited the house of *Alathiyur Nambi* (one of the late *Ashta Vaidyan* of Kerala, Appendix 2). Then he visited several places in Kerala and disseminated the knowledge and treatment techniques to the members of higher sections of the society based on his own text '*Ashtangahridayam*'. This includes *Alathiyur Nambi, Karathol Nambi, Choondal Nambi, Olassa Mooss, Vayaskara Mooss, Chirattamon Mooss,*

Table 25.4 Ingredients of Narayana thailam[a] (Medicated oil)

Sl. no.	Botanical name	Local name	Parts used	Quantity
1.	*Gmelinia arborea* Roxb.	Kumbil	Root	5 kg
2.	*Aegle marmelos* Corr.	Koovalam	Leaves (fresh)	5 kg
3.	*Stereospermum suaveolens* (Roxb.) DC.	Pathiri	Root	5 kg
4.	*Oroxylum indicum* (L.) Kurz	Palakapayyani	Root	5 kg
5.	*Solanum indicum* L.	Cheru vazhuthana	Root	5 kg
6.	*Sida alnifolia* L.	Kurunthotti	Root	5 kg
7.	*Boerhavia diffusa* L.	Thazhuthama	Leaves	5 kg
8.	*Premna serratifolia* L.	Munha	Root	5 kg
9.	*Merremia tridentata* (L.) Hallier f.	Prasarani	Stem and leaves	5 kg
10.	*Withania somnifera* (L.) Dunal	Amukkuram	Root	5 kg
11.	*Solanum virginianum* L.	Kantakari	Root	5 kg
12.	*Abutilon indicum* (L.) Sweet	Vellurppam	Root	5 kg
13.	*Mucuna pruriens* (L.) DC.	Naikurana	Root	5 kg
14.	*Santalum album* L.	Chandanam	Stem	1 kg
15.	*Anethum graveolens* L.	Chathakuppa	Fruit	1 kg
16.	*Nardostachys jatamansi* (D. Don) DC.	Manchi	Root	1 kg
17.	*Acorus calamus* L.	Vayambu	Rhizome	1 kg
18.	*Valeriana wallichii* DC.	Thakaram	Root	1 kg
19.	*Saussurea costus* (Falc.) Lipsch.	Kottam	Root	1 kg
20.	*Elettaria cardamomum* (L.) Maton	Elathari	Root	1 kg
21.	*Desmodium gangeticum* (L.) DC.	Orila	Root	1 kg
22.	*Pseudathria viscida* (L.) Wt. and Arn.	Moovila	Root	1 kg
23.	*Alpinia galanga* (L.) Willd.	Aratha	Rhizome	1 kg
24.	*Cedrus deodara* (Roxb. ex Lamb.) G.Don	Devadaram	Heart wood	4 kg
25.	Bitumen	Kanmadam	Rock exudation (purified)	200 g
26.	Rock salt	Induppu		1 kg
27.	Gingelly oil	Nallenna		3 tin (30 kg)
28.	*Asparagus racemosus* Willd.	Sathavari	Tuberous root	25 kg
29.	Cow's milk			70 L

[a]Medicated oil is prepared according to the Ayurvedic Pharmacopoeia

Pulamanthole Mooss, Kuttanchery Mooss, Elayadath Thaikkat Mooss, Pazhanellipurath Thaikkat Mooss, Parappur Moos and Vaidya Madam (Famous Ayurvedic Physicians of Kerala) (Appendix 2).

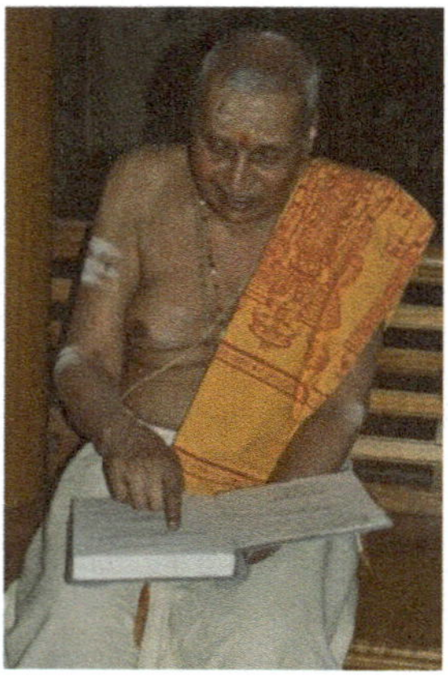

Fig. 25.6 Shri Vaidyamadom Cheriya Narayanan Namboothiri

Shri Vaidyamadom Cheriya Narayanan Namboothiri is a resident of Mezhathoor, Trithala of Palakkad district of Kerala state. He studied Ayurveda through Gurukula tradition, once prevalent in Kerala before the introduction of formal Ayurvedic education in the state. He is considered as one of the living legends and an exponent of Ayurveda.

During the field study, authors met Shri Vaidya Madom in his residence at Mezhathoor. He agreed to share his vast experience in the field of Ayurveda especially for the treatment of Chronic disorders like diabetes, cancer, peptic ulcer, *vataroga* (rheumatic disorders/arthritis), etc. He explained the new therapeutic efficacy of three medicinal plants which are not mentioned in Ayurveda (Fig. 25.6).

One such plant is described here.

25.2.5 'Erachikootti' *(Muscle Toner)*

The local term '*Erachi'* means flesh and '*kootti'* denotes that, therapeutically, the plant acts as a toner of muscle. This is one of the medicinal plant species described by him to enhance the functioning of 'mamsadhathu'—muscle (One among the 'sapta dhatus'—tissues mentioned in Ayurveda). Based on the meaning of the local name *Erachikooti,* he has given a new Sanskrit name to the plant as 'Mamsavardhini' (enhances the function of the circulatory and musculoskeletal system) and also helps to promote the strength of the muscles. He acquired this knowledge from a local healer in Kasaragod district, situated in the northern part of Kerala. According to him, the bark of this plant is having anti-cancer activity. Information about this plant was also recorded by our research team while conducting the systematic documentation of traditional knowledge related to plants used for food and AYUSH and indigenous medicine in Kasaragod district of Kerala State. The process of botanical identification is in progress. He also revealed the

Table 25.5 Therapeutic efficacy of some medicinal plants revealed by Shri Vaidya Madom

Sl. no.	Local name	Botanical name	Sanskrit name	Parts used	Uses/indication
1	Erachikootti	Identification is in progress	Mamsavardhini[a]	Bark	Anti-cancer, muscular impairments
2	Kudajadri	Identification is in progress	Kudajadri[b]	Leaves	In leukaemia, to increase platelets and haemoglobin. And also arrest the uterine bleeding
3	Unknown	Identification is in progress (thorny plant with red flower)	–	Leaves	Anti-diabetic
4	Cherkkuru	*Semicarpus anacardium*	Bhallatakom	Seed	Leukaemia
5	Poocha meesa	*Orthosiphon aristatus* (Blume) Miq.	–	Stem and leaf	Diabetes
6	Kadukka	*Terminalia chabula* Ritz.	Abhaya	Fruit	TB, cough
7	Neeraral	Identification is in progress	Sunishanna		To arrest uterine bleeding (whole plant in the form of paste with butter(10–15 g) for 4 days in empty stomach), anti-diabetic
8	Kayunni	*Eclipta alba*	Bringaraj	Whole plant	Luecoderma

[a]Given name by Shri Vaidya Madom Cheriyanarayanan Namboothiri
[b] Place of collection

therapeutic efficacies/uses of many underutilized medicinal plants in Ayurveda (Table 25.5).

25.2.6 *Changampally Tradition of Thirunavaya-1*

'Changampally tradition' is famous for its traditional Kalari arts (martial arts) and '*Marma Chiktsa*' (traditional orthopaedics) the art and science of prevention, diagnosis and treatment of diseases and abnormalities of the musculo skeletal system (Fig. 25.7).

According to the 'Changampally tradition', diagnosis of the disease is based on pulse reading and examination of anatomically/physiologically important vital points (*marma*) of the body. The tradition further shows their ability to link between the human body and Nature while designing the treatment protocol by considering the environmental health and availability of location-specific medicinal plants. They strongly believed that the abundant distribution of certain medicinal plants in a

Fig. 25.7 Dr. Kunhalan Gurukkal (Right) interacting with Dr. S Rajasekharan in his Medicinal Plant Garden

particular location is a signal of impending diseases that these plant species can cure. Interestingly, Dr. Kunhalan Gurukkal, the present custodian of Changampally tradition, revealed his experience with an abundant distribution of Castor oil plants in a particular area near Kunnamkulam in Thrissur district. He has predicted the impending incidence of jaundice. He claimed that his prediction was accurate. According to him, in the local health tradition, the tender leaves of the Castor Oil plant (Ricinus communis) are considered an effective remedy for the treatment of jaundice administered either alone or in combination with other medicinal plants. The study carried out by our team also documented more than 50 information on *Ricinus communis* exclusively for treating jaundice.

Changampally tradition has 800-year-old history, and it is claimed that they are the descendants of Tulu Brahmins who were brought to Thirunavaya (Malabar region of Kerala, located near the bank of *Bharathapuzha* river) from Southern-Karnataka by the Samoothiries, the rulers of the Malabar for treating the wounded soldiers in the war field and also the wounded fighters participating in the Mamangam festival (festival of '*kalaripayattu' which is a traditional martial arts form).* Later they got converted to Islam and settled in Thirunavaya. The famous Changampally Kalari located at Thirunavaya ('*Nalpatheeradi Kalari*') was famous for its traditional Kalari–Marma treatment and also gave birth to many famous Kalari Gurukkals (Masters of Kalari).

The knowledge of Changampally tradition was orally transmitted from generation to generation through the two traditional streams. One is the knowledge acquired from their forefathers, mainly the knowledge of Ayurveda, and the second stream is from the 'Guru Parampara' (transfer of oral tradition knowledge from master to disciples). The Guru of Kunhalan Gurukkal was a Sufi Saint called Mushoor Mullakkoya Thangal. It may be the reason that they are effectively applying spiritual healing while treating various diseases. Currently, he is engaged in developing new herbal medicine for treating non-communicable diseases like Cancer, Diabetes, Hyper tension, Obesity, Hypercholesterolemia, Alzheimer's, Osteoarthritis, Osteoporosis, Inter vertebral Disc Prolapse (IVDP), etc. Dr. Kunhalan Gurukkal is researching for the last 35 years for developing—anti-cancer drugs, and he has come out with 55 new and novel polyherbal formulations. He strongly believes that treatment must always be service-oriented in nature and not fully utilized for commercial purposes. On his campus, he has also conserved a number of rare and important medicinal plants. He also disclosed some of the therapeutic efficacy of medicinal plants which are not mentioned in Ayurveda. He disclosed the new therapeutic efficacy of *Plumbago indica* to minimize the side effect occurring due to radiation therapy in cancer patients.

25.2.7 Preparation of Kooman Kudukka Appam (Steamed Food Article)

This is a piece of tribal information initially documented by Shri. Vasudevan Vaidyar, aged 47 residing at Edavanna Gramapanchayath in Malappuram district. It is a unique, steamed food article prepared by using the green fruit shell of a tree, locally known as Kooman Kudukka. The term Kooman denotes the name of a bird locally known as '*Uppan*' (Greater Coucal or Pheasant Crow); Kudukka means the shell of a fruit tree. It shows the relationship between the bird and fruit, which is directly linked with the sustenance of the bird. The bird is fond of eating nutritive tissue present inside the mature green shell. Probably, this may be the reason that the local tribal people called this tree '*Kooman Kudukka Maram*'.

A study related to the taxonomical identity of the tree is in progress. In the past, the tribes used this shell for the preparation of an ethnic food article known as '*Kooman Kudukka Appam*' (Table 25.6).

Method of preparation: Ingredients are mixed well and filled in the four grooved fruit shells of Kooman Kudukka. Then it is placed in a steam cooker for a period of 10–15 min. Subsequently, the material placed in the fruit shell is removed and then transferred to the plate. This sweet steam cake is very tasty and palatable (Fig. 25.8).

Indication: It is recommended for anaemia and palpitation, relives stress and tension and is also a remedy for insomnia.

Table 25.6 Ingredients of 'Kooman Kudukka Appam'

Sl. no.	Botanical name	Local name	Parts used	Quantity
1	*Bambusa bambos*(L.) Voss	Mulayari	Dried seeds	250 g
2	*Adhatoda beddomei* C.B. Clark	Chittadalodakam	Fresh leaves juice	30 mL
3	*Cuminum cyminum* L.	Nalla Jeerakam	Dried fruit	5 g
4		Jaggery		200 g
5	Identification in progress	Kooman Kudukka	Green (mature) fruit shell with the presence of nutritive tissues (pale yellow coloured portion—the carpel wall present in each locule)	Quantity available in the fruit shell

Fig. 25.8 *Kooman Kudukka Appam*

25.2.8 Herbal Preparations by Veerammal Vaidyar (Irular Tribe)

Veerammal aged 54, (Fig. 25.11) is a tribal healer from the Irular tribal community, residing at *Vattalakki ooru* (Tribal settlement), Sholayur Gramapanchayath of Palakkad District and practising Tribal medicine for the last 25 years. She is a specialist in treating cancer and hypertension. Veerammal is practising in her residence and also has maintained a good Medicinal plant garden. For clinical practice, she collects raw drugs from nearby forest areas. About 50 very important medicinal plants, including herbs, shrubs, and trees, are growing in her garden. According to her, genuine drugs are highly essential for treating cancer,

Fig. 25.9 Ethnic food preparation of Irular Tribe of Sholayur in Kerala state of India

Table 25.7 Therapeutic efficacy of some medicinal plants revealed by Veerammal Vaidyar

Sl. no.	Tribal name	Botanical name	Parts used	Uses/indication
1	Mathalatha	*Punica granatum* Linn.	Tender leaves (fresh)	Memory enhancer, hypertension
2	Kathivettucheddi	*Chromolaena odorata* (L.) R. King and H. Robinson	Whole plant (fresh)	Cut wounds
3	Karamaram	Identification in progress	Bark (powder)	Chest pain
4	Melakulukki	*Glycosmis pentaphylla (Retz.) DC.*	Leaf, Bark and Root (powder—external use)	Body pain, cramp (external application)
5	Eswamaram	–	Bark, tender leaf (expressed juice)	Fever
6	Sinkamaram	–	Tender leaves (expressed juice)	Fever
7	Munnacheera	*Amaranthus spinosus* Linn.	Bark	Snake bite (external application and oral administration)
8	Garudakodi	*Aristolochia indica* Linn.	Whole plant	
9	Pithae	–	Leaf (fresh)	
10	Kilikkathakka	–	Tuber (fresh)	

hypertension, fever, jaundice pre-hepatic, hepatocellular, and post-hepatic, etc. (Figs. 25.9, 25.10, 25.11, 25.12, 25.13).

She has also revealed the therapeutic efficacy and uses of some medicinal plants (Table 25.7).

Veerammal (Figs. 25.9, 25.10 and 25.11) has developed a medicinal plant preparation with around 100 ingredients for treating Cancer and tumours. She partially disclosed some of the ingredients of the product (Fig. 25.12). They are Peechimba (*Nothapodytes foetida* (Wight) Sleumer), Koduveli (*Plumbago indica* L.), Nenmenivaka (*Albizia lebbeck* (L.) Benth.), Melakulukki (*Glycosmis*

Fig. 25.10 (**a**) Food article prepared from Ragi (*Eleusine coracana* Gaertn)

Fig. 25.11 Veerammal Vaidyar (Irular Tribal healer)

pentaphylla (Retz.) DC.), Erigontha(*Toddalia asiatica* (L.) Lam.) represented in Fig. 25.13.

25.2.9 An Absorbent Technique for Snake-Bite Using 'Visha Kallu': A Kind of Medicated Stone

'*Visha kallu*' is a kind of medicated stone potentiated with anti-poisonous properties. Preparation of the stone is a very lengthy and complex procedure. Ingredients required for the preparation of the stone are:

1. Pebbles from the river (Vellaram kallu)—50 g approximately
2. Expressed juice of the leaves of *Ocimum tenuiiflorum* (Krishna thulasi), *Anisomeles malabarica* (Perum thumba), *Leucas aspera* (Thumba), *Piper betle* (Vettila)—20 mL. each approximately and paste of Santalum *album* (Chandanam)—50 g. approximately.

Thina ari (*Setaria italica* (L.) P. Beauv.)

Karamaram (*Canthium sp.*)

Munha (*Premna sp.*)

Karuvelam (*Acacia nilotica* (Benth.) Brenan

Nechikarimba – *Murraya paniculata* (L.) Jack

Pillai thenki (*Crinum asiaticum* L.)

Fig. 25.12 Plants used by Irular Tribe for food and medicine

Method of preparation: The pebbles are ground well and mixed with the said ingredients, and prepared in the form of paste. The paste is then covered with seven leaves of *Aristolochia tagala* (*Valiya arayan*) and placed on a rock and roasted using the wood of *Chukrasia tabularis* (*Chuvanna akil*), *Santalum album* (Chandanam), *Ocimum tenuiiflorum* (*Krishna thulasi*), and Camphor (*Karpooram*). The roasted material is again covered with a paste of termite soil and soil content obtained from the trees. The above materials are again covered with the leaves of *Aristolochia indica* or *Aristolochia tagala* and then kept under low fire. When the essence of the soil is fully absorbed by the mass, the soil is removed. The mass is then transferred into the water for 1 h. It is then covered with the pounded mass of the stem bark of

Peechimba (*Nothapodytes foetida* (Wight) Sleumer) **Melakulukki (*Glycosmis pentaphylla* (Retz.) DC.**

Fig. 25.13 Ingredients of the anticancer formulation developed by Veerammal Vaidyar

Fig. 25.14 Medicated stone potentiated with anti-poisonous properties

Alstonia venenata (‘*Analivegam*’) or *Pittosporum nilgherrensis* (‘*Analivegam/ Analivenga*’) and again kept over a low fire for half an hour and later stored for a period of 1 week. After 1 week, the mass is again covered with a paste prepared with the ashes of coconut shell or *Kunstelaria keralensis* (‘*Kariveppila valli*’) and sun-dried for 3–4 days. The dried material is then shaped in the form of a stone and kept in the ashes of cow dung or dried leaves of *Cannabis sativa* (‘*Kanchavu*’) or *Nicotiana tabacum* (‘*Pukayila*’) to maintain the potency of the stone (Fig. 25.14).

Mode of Application: The tribal healers directly apply the stone on the bitten part, which will stick automatically and absorb the poison from the body. During the process, they chant ‘mantras’ to propitiate Lord Siva. When the absorption is complete, the stone falls down on its own.

Preservation Technique: After this process, the stone is to be transferred and stored in cow’s milk for 2 h for necessary purification. Then it is taken out, dried and kept in the ashes of cow dung for re-use. In this manner, one stone can be used at least 20 times. This kind of practice is prevalent among the Malavedan tribe and traditional Ayurvedic physicians of Kerala. The claim is yet to be scientifically validated.

25.2.10 Malavedan Tribe from Pathanamthitta District

25.2.10.1 ‘Amrithapala’: *Decalepis arayalpathra* (Joseph and D Chandras.) Venter

(Syn: *Janakia arayalpathra* Joseph and Chandrasekharan) Amrithapala, a rare and endemic plant species found in the southern forests of the Western Ghats region of Kerala, is used by the local ‘Kani’ tribe as an effective remedy for peptic ulcers, cancer-like afflictions, and as a rejuvenating tonic. Search made in Ayurvedic literature indicates that the plant may be ‘Thampra rasasyani’ mentioned in the ‘Oushadha Nighantu’ of Thayyil Kumaran Krishnan (1906).

Method of Preparation and Mode of Administration: The expressed juice of the pounded mass of the fresh tuberous root of Amrithapala is mixed with an equal quantity of the expressed juice of fresh coconut kernel. The mixture is then boiled for some time to attain a semi-solid form which after cooling is administered at a dose of 10–15 g twice daily for 15–30 days to cure all kinds of peptic ulcers and cancer-like afflictions. It is also recommended as a stamina booster and blood purifier (Knowledge provider: Lakshmikutti, Mottamoodu tribal settlement, Kallar).

Patents Awarded: A process to prepare a herbal preparation for cancer from *Janakia arayalpathra* and *Trichopus zeylanicus* leaf—Patent No.193609 dated 22/09/2006. *Inventors*: A. Subramonium, S. Rajasekharan, P. Pushpangadan, V. George and G. Sreekandan Nair. File No; MAS/650/2001.

1. A process for the preparation of an anti-cancer drug from the roots of *Janakia arayalpathra*—Patent No. IN191799 dated January 2004. *Inventors*:

Table 25.8 Curative effects of some medical plants

Sl no.	Botanical name	Local name	Parts used	Disease treated
1.	*Drynaria quercifolia*	Marappanna	Rhizome	Jaundice
2.	*Moringa oleifera*	Muringa	Tender leaves	Conjunctivitis
3.	*Pterospermum heyneanum*	Elluttipatta	Bark	Bone fracture
4.	*Aristolochia tagala*	Valia arayan	Root	Snake bite
5.	*Alangium lamarkii*	Ankolam	Root	Dog bite
6.	*Chenopodium ambrosioides*	Cheriya Peechambam	Whole plant	Rabies
7.	*Eclipta prostrata*	Kaithonni	Whole plant	Asthma
8.	*Cephalandra indica*	Koval	Root	Epilepsy/ insanity
9.	*Luffa cylindrica*	Kattupeechedu	Fruit fiber	Jaundice
10.	*Lawsonia inerms*	Mailanchi	Root	Jaundice

A. Subramonium, S. Rajasekharan, P. Pushpangadan, V. George, P.G. Latha and V.V. Asha. File No: 947/DEL/2000 filed by DST

25.2.11 Fumigation Therapy for Treating Piles

This is a novel method of fumigation therapy exclusively applied to cure piles. In this treatment, the patient is seated on a specially designed hollow-shaped wooden stool. The stool has an inner space where the ingredients of a specially designed formulation with three raw drugs consisting of dammar, *Achyranthes aspera,* etc., including a small piece of flesh removed from the head of the forest cane turtle (Kareelama). The flesh is removed without killing the animal. The ingredients are fumigated inside this special stool, and the patient suffering from piles or hernia is made to sit on it for 5–10 min. This is a unique treatment developed by the tribes (Muthuvan, Malaarayar, Mannan, and Kurichiyar) of the Vellaramkuthu colony of Kuttampuzha, Ernakulam district. Such a kind of treatment is neither described in Ayurveda nor Modern medicine.

25.2.12 Ethnomedical Investigation of **Drynaria quercifolia** *(L.) J. Smith. (Polypodiaceae)*

Folk healers of Kerala are specialized in treating with ‘Ottamoolies’ (single drugs) for specific disorders. Typical examples of medicinal plants used against some disorders are given in Table 25.8.

The use of *Drynaria quercifolia* for treating chronic jaundice and also as a liver tonic was found very promising. In view of the great importance of hepato-protective herbal drugs, detailed pharmacognostic and ethno-pharmaco-logical investigations on this plant were reported.

Enumeration of Ethnomedical Data

Name:	*Drynaria quercifolia* (L.)
Family:	Polypodiaceae/Drynariaceae
Order:	Filicales
Subclass:	Leptosporagiatae
Class:	Filicinate
Division:	Pteridophyta
Habit:	Stoloniferous creeping epiphytic herb
Habitat:	Growing on rocky surface and trees
Vernacular name:	Marappanna Kizhangu (Kani tribe), Anappacha (Malavedan tribe), Tudiyanpara (Pathinaikan tribe)

The therapeutic potentialities of *Drynaria quercifolia* have not been mentioned in ancient Ayurvedic classics. It is recorded in the Wealth of India raw materials that the rhizome is bitter and astringent. Aqueous extracts possess anti-bacterial properties. "In Malaya, the fronds are used in the form of a poultice for swellings."

Recent studies revealed that this is used by the tribes of Western Ghats to cure liver diseases. A pounded mass of the stolon (5–10 g) along with the previous day's rice gruel or with the juice of coconut kernel is administered thrice daily for 4 days by the Kani tribe for chronic Jaundice. A salt-free diet along with a dip bath in river water is also advised. The Malavedan tribe uses medicated oil prepared from the rhizome or a combination with leaves to be applied externally over the scalp to cure insanity and to get a cooling effect. Pathinaikan tribe uses the stolon (rhizome) for edible purposes (Rajasekharan and Latha 2011).

25.2.13 Tribal Medicine Presented by Rengi Amma (Mudugar Tribe), Palakkad District of Kerala, India

Rengi Amma aged 65, belongs to Mudugar tribal community, residing at North Omala Ooru, Jellipara, Agali Gramapanchayath of Palakkad District (Fig. 25.15). She started her practice at the age of 30 and studied the healing techniques from her grandfather, late Kuppa Moopan, who died at the age of 90. She usually accompanied her grandfather as and when he went inside the forest for grazing cattle. He taught her to identify numerous medicinal plants available in the forest for treating various ailments. At present, she is practising in his residence at Jellipara and also working as a visiting healer in the Muddha Mooppan Centre for Tribal Medicine Development Hospital at Mathur village in Palakkad district. As a tribal healer, she is a specialist in treating Cancer, Peptic ulcers, Diabetes, Joint pain, Osteo-arthritis, Intestinal colic, Urinary calculi, and Cataract. She is also practising special massage, 'Uzhichil', with medicated oil. Medicinal plants required for the treatments are

(a) (b) (c) (d)

Fig. 25.15 (**a**) Rengi Amma vaidyar (Mudugar Tribe). (**b**) Rengi Amma explaining about the uses of medicinal plants. (**c**) Preparation of medicated oil at her residence (**d**) Interaction with Rengi Amma by Dr. Rajasekharan

collected from the forest and prepared in the form of powder, decoction, medicated oil, etc. Some of the important information documented from Rengi Amma is presented in Table 25.9.

25.2.14 Traditional Knowledge Presented by Chelli Vaidyar (Irular Tribe), Sholayur in Palakkad District of Kerala, India

Chelli Vaidyar, aged 60, is a tribal healer from the Irular tribal community, residing at Vadakkekadampara, Sholayur Gramapanchayath of Palakkad District of Kerala. She studied various tribal healing techniques from Sri Bodha Mooppan who was a famous tribal leader and healer among the Irular tribal community at Attappady, Sholayur. He died at the age of 108 years. She acquainted the knowledge of examining the patients and the diagnosis method through pulse reading from her grandfather. She is an expert in identifying medicinal plants available in the forest areas of Attappady.

Table 25.9 Traditional knowledge associated with some plants revealed by Rengi Amma

Sl. no.	Tribal name	Botanical name	Parts used	Uses/indication
1	Pooveru	*Rauvolfia serpentina* (L.) Benth. Kurz (used red and white flowered *Rauvolfia)*	Root (fresh)	Prepared in the form of paste (3–5 g) administered orally along with water twice daily to control hypertension and to cure intestinal colic
2	Nannari Juliveru Chuvannulli	*Hemidesmus indicus* (L.) R. Br. *Toddalia asiatica* (L.) Lam. *Allium cepa* L.	Tuberous root (fresh) Root (fresh) Bulb (fresh)	Paste prepared from the *Allium cepa* (2 bulb) + 10 g root of *Hemidesmus* and *Toddalia* and administered orally in the morning, empty stomach for a period of 2 months to cure peptic ulcer. Restriction—avoid fish, meat, egg and spicy food
3	Muthali maram	Not identified. A small tree, growing inside the forest, leaf round-shaped, not aromatic and root contains mucilage	Tender root (dried)	Prepared in the form of powder (5 g), administered orally, twice daily, before meals, for a period of 1–2 months to control diabetes
4	Chunigaduravalli Chuvannulli	Not identified. A climber, leaves small round shaped *Allium cepa* L.	Root and leaves (fresh) Bulb (fresh)	Prepared in the form of paste and applied over cheek and neck to cure oral cancer
5	Ambukana	Not identified (A small tree)	Root and leaves (fresh)	Prepared in the form of paste (3 g) and administered orally along with 30 mL water and also applied the paste over the affected part to cure tumour
6	Kavalai Elakkai Kurumulaku Veluthulli	*Dioscorea oppositifolia* L. *Elettaria cardamomum* (L.) Maton *Piper nigrum* L. *Allium sativum* L.	Stem (dried) Seed (dried) Fruit (dried) Bulb (fresh)	10 g stem powder of *Dioscorea* + 1 g *Elettaria* + 2 g *Piper* + 1 g *Allium* boiled in 1.5 L of water and whole water is consumed during day time to control the obesity

Chelli Vaidyar is currently practising Tribal medicine at her residence. She is also a consultant tribal healer of Muddha Mooppan Centre for Tribal Medicine Development Hospital at Mathur village, Palakkad District of Kerala, India. She is treating a number of diseased conditions like rheumatic disorders, chronic head ache, diabetes, hypertension, bronchial asthma, peptic ulcer, hypercholesteremia, etc.

(a) (b) (c)

Fig. 25.16 (**a**) Chelli Vaidyar. (**b**) Ponnuchami. (**c**) Interaction with Chelli Vaidyar

She is a specialist in managing infertility problems in both males and females. According to Chelli Vaidyar, the therapeutic efficacy of medicinal plants collected from the forest is better than the plants collected from cultivation (Fig. 25.16 a, b & c).

At present Chelli Vaidyar's son Ponnu Swamy, aged 30, is helping her with the collection of medicinal plants required for the treatment from the nearby forest areas. He is also assisting her with the preparation of tribal medicine in the form of medicated oil, powder, pills and medicated decoction. Ponnu Swamy is a specialist in traditional massage known as '*uzhichi'l*. It is considered as a treatment for nervous disorders to enhance peripheral blood circulation and problems related to traumatic injury. She is not willing to disclose most of the information related to her treatment because she strongly believes that, once it is revealed to outsiders, the medicine may not produce the desired effect. Therefore, she has kept most of the information secret, and some of the information was orally passed on to her son.

Some of the relevant information documented from Chelli Vaidyaris presented in Table 25.10.

25.3 Ethnomedicine and Its Scope in Developing Novel Herbal Drugs and Nutraceuticals

The term 'ethno' is derived from the Greek word *ethnos*, meaning 'nation' and has in combination with the suffix 'ology' gave us the term 'ethnology' which means science of races and people, their relation to one another, distinctive physical and other characteristics. Ethnomedicine denotes plants, animal products, and minerals used by indigenous communities of a particular region or country for healthcare/ medicinal purposes other than those mentioned in the classical health traditions of respective cultures (Rajasekharan et al. 2012d). Ethnomedicine is directly linked with Traditional Knowledge.

'Acharya Charaka', who wrote the famous medical treatise 'Charaka Samhitha' (Textbook of Ayurvedic Medicine) was the first to describe the concept of

Table 25.10 Some of the relevant information documented from Chelli Vaidyaris

Sl. no.	Tribal name	Botanical name	Parts used	Uses/indication
1.	Churuli	*Diplazium esculentum* (Retz.) Sw.	Root (fresh)	Expressed juice obtained from the root (60 mL), administered orally in the morning during the sunrise. This medicine is administered immediately after the menstrual period for 7 days and used as an antifertility agent. After the administration of the medicine, sexual intercourse should be avoided for next 3 months
2.	Kattuparuthi Chanthamalli Goolitholi Kattumullanki	*Cochlospermum religiosum* (L.) Alston Yet to be identified Yet to be identified Yet to be identified	Leaf and root (dried) Leaf (dried) Bark (dried) Leaf and root (dried)	All four ingredients taken in equal quantity are made into a powder (5–8 g) and administered orally along with goat milk (120 mL) on empty stomach, once daily for a period of 2 months to cure Leucorrhoea
3.	Dhanti maram Venga Puli Kudalanga Pavakka	Yet to be identified *Pterocarpus marsupium* Roxb. *Tamarindus indica* L. Yet to be identified *Momordica dioca* Roxb. Will.	Bark (dried) Bark (dried) Bark (dried) Fruit (dried) fruit (dried)	Take all five ingredients in equal quantity and prepare them in the form of powder (5–8 g) and administer orally along with lukewarm water, twice daily for a period of 4–5 months to control diabetes

Ethnomedicine. '*Oushadi namarupabhyam janatehiya japavane, Avipaschavia gopasca ye canye vanacarine*'; In this Sanskrit text, Charaka stressed the importance of correct identification of a plant as most important for designing treatment protocol. He further stated that simply confirming the name of a plant does not help in the treatment, and therefore, one should be well versed with morphological characters. To acquire such knowledge of plants, including their therapeutic uses, etc., one should interact/inter-associate with the hill tribes, cowherds, and sages of the forest. These views of Charaka have been well appreciated and considered as the basic concept of Ethnomedical research even in the modern era (Rajasekharan et al. 2012d).

25.4 Ethnomedical Research

Ethnomedical data collected through systematic documentation is classified based on (1) General information on bioresources (food and medicine) (2) Local identity (3) Botanical details (4) Utility-based categorization (4) Parts used (5) Collection/Harvesting (6) Ingredients (Single and Poly Herbal Formulations (7) Method of Preparation and Administration (8) Method of diagnosis and treatment (9) Social information. All these components are essential to execute Ethnomedical Research, right from identification of plants species (both Botanical and Local) to scientific validation such as Pre-clinical studies including Ethnopharmacological, Phytochemical investigation, Drug design and formulation, Standardization, and Clinical trial followed by Product development, Technology transfer, Commercialization, and Benefit-sharing.

Kerala is the first State in India to initiate Ethnomedical Research, and the responsibility was entrusted to Jawaharlal Nehru Tropical Botanic Garden and Research Institute (JNTBGRI) (Rajasekharan and Latha 2011).

25.5 Sustainable Utilization of TK and Medicinal Plants

25.5.1 Scope of Developing Novel Herbal Drugs and Nutraceuticals

According to WHO Guidelines, Herbal medicines include herbs, herbal materials, herbal preparations, and finished herbal products. In some countries, Herbal Medicines may contain, by tradition, natural organic or inorganic active ingredients which are not of plant origin (e.g. animal and mineral materials).

Nutraceuticals are products having both nutritional and therapeutic value. Generally, it is defined as a food product that acts through its preventive, promotional, corrective, and curative properties and provides health and medical benefits. The term nutraceutical was originally defined by Dr. Stephen L. DeFelice, Founder and Chairman of the Foundation of Innovation Medicine (FIM), Crawford, New Jersey (Rajasekharan et al. 2012d; Huang 2018).

The concept of nutraceuticals is described in Ayurveda. Several treatises on Ayurveda elaborate on the importance of food products, food, and its intake according to season. 'Acharya Charaka' stated that "man is the product of food". Therefore, the quality of man basically depends on the quality of the food/food products consumed. Most of the Rasayanas (Polyherbal formulations having rejuvenating properties) mentioned in Ayurveda are nutraceuticals in nature and are recommended for intake as food (e.g. 'Chyavanaprasa') to maintain the health of a healthy individual (Swastha) and also prescribed as drugs in many diseases. The term 'Chyavana' (trademark name) denotes the name of the physician who formulated the product, and 'Prasa' means consumed like food according to the digestive power of an individual.

The third-world nations are rich in biodiversity and indigenous knowledge particularly Traditional Ethnomedical Practices. Among the Asian countries, India and China are the two major stakeholders in herbal product development and trade, especially Ethnomedical practices and Phytopharmaceuticals. In current allopathic medicine, there are a large number of plant-derived drugs. We know that the modern allopathic system has produced miraculous results. But we also know that there is a great deal that is not known about how systems function together. There are energy flows in the system, and there are linkages as well as positive and negative responses along with feedback systems. In a sense, Ayurveda, Siddha, and Unani have aimed at a holistic approach to treatment (Tasnia et al. 2021). We have developed these systems, but over a period of time, we have allowed them to be subservient to or below in standing status of the modern allopathic system. This is not to say that the modern allopathic system is not good and not powerful. But the other systems have a place of honour, and we have to ensure that they are given support, particularly with the wealth of the natural resources that exist in developing countries like India. We have to understand the body as a whole system and healthcare must be available at a low cost to the millions of people in such countries.

25.6 Ethnopharmacology and Drug Development: JNTBGRI's Contribution

At Jawaharlal Nehru Tropical Botanic Garden and Research Institute (JNTBGRI), the Ethnomedicine Division was started in 1992 with a view to conducting detailed ethnopharmacological studies of medicinal plants based on Traditional Knowledge. Therefore, systematic documentation of ethnomedical information on medicinal/food plants is essential to obtain leads for developing novel herbal/modern drugs for treating diseases like cancer, liver disorders, osteoporosis, diabetes, Alzheimer's disease, malaria, etc. This will also help to enrich the Ayurvedic pharmacopoeia (Table 25.11). In this context, TBGRI has initiated ethnopharmacological studies of selected medicinal plants, and a brief summary of the investigations is presented (Rajasekharan 2013; Vansh 2020).

25.7 Phytoconstituents as Nutraceuticals

The sources of nutraceuticals include ginseng, spirulina, *Gingko biloba*, amino acids, glucosamine, chondroitin, and *Aegle marmelos* that are to be formulated and consumed or administered internally under the supervision of a qualified medical practitioner and intended for the specific dietary management of a disease or

Table 25.11 Research highlights on medicinal plants of Kerala obtained from ethnomedical leads in JNTBGRI, palode

Sl. no	Name of the plant	Leads obtained (Tasnia et al. 2021)
1.	*Artemisia pallens*	Anti diabetic
2.	*Barringtonia racemose*	Analgesic, anti inflammatory
3.	*Caesalpenia pulcherrima*	Anti diabetic
4.	*Capparis brevispina*	Hepatoprotective
5.	*Commiphora caudata*	Anti-inflammatory, analgesic
6.	*Cyclea peltate*	Hepatoprotective
7.	*Drynaria quercifolia*	Analgesic, anti inflammatory
8.	*Decalepis arayalpatra*	Antitumor, antiulcer
9.	*Elephantopus scaber*	Anti cancer
10.	*Epipremnum pinnatum*	Analgesic, anti inflammatory
11.	*Epipremnum pinnatum*	Anti lipidperoxidative
12.	*Evolvulus nummularius*	Aphrodisiac
13.	*Ficus religiosa*	Analgesic, anti inflammatory
14.	*Ficus religiosa*	Anti lipidperoxidative
15.	*Hedyotis corymbosa*	Hepatoprotective
16.	*Helminthostachys zeylanica*	Hepatoprotective
17.	*Heracleum candolleanum*	Anti-inflammatory
18.	*Hibiscus hispidissimus*	Hepatoprotective
19.	*Holostemma adakodien*	Aphrodisiacs
20.	*Ixora coccinea*	Anti cancer, Hepatoprotective
21.	*Janakia arayalpathra*	Anti cancer
22.	*Jasminum azoricum*	Anti-hypertensive
23.	*Justicia gendarussa*	Analgesic, anti inflammatory
24.	*Kaempferia rotunda*	Wound healing, anti-inflammatory, analgesic
25.	*Mucuna pruriens*	Aphrodisiacs

(continued)

Table 25.11 (continued)

Sl. no	Name of the plant	Leads obtained (Tasnia et al. 2021)
26.	*Pisonia alba*	Anti diabetic
27.	*Pittosporum neelgherense*	Hepatoprotective
28.	*Plumbago rosea*	Anti diabetic
29.	*Psoralea corylifolia*	Anti cancer
30.	*Rhaphidophora pertusa*	Anti lipidperoxidative
31.	*Rhinacanthus nasuta*	Hepatoprotective
32.	*Ricinus communis*	Hepatoprotective
33.	*Saraca asoca*	Hepatoprotective
34.	*Sida acuta*	Hepatoprotective
35.	*Spilanthes ciliata*	Hepatoprotective
36.	*Thespesia populnea*	Hepatoprotective
37.	*Tinospora cordifolia*	Anti diabetic
38.	*Trichopus zeylanicus ssp. travancoricus*	Aphrodisiac, Adaptogenic anti fatigue, immunomodulatory, anti hepatotoxic Anti genotoxic, anti stress, aphrodisiac, anti allergic, Adaptogenic, anti gastric ulcer, anti oxidant, DNA protecting activity
39.	*Utleria salicifolia*	Hepatoprotective
40.	*Wattakaka volubilis*	Analgesic, anti inflammatory, anti lipidperoxidative
41.	*Withania coagulans*	Anti diabetic
42.	*Wrightia tinctoria*	Antidermatophytic

condition for which distinctive nutritional requirements, based on recognized scientific principles, are established by medical evaluation. The nutraceutical-containing phytosterols are effective in lowering LDL cholesterol (Shruti et al. 2010; Rajasekharan and Latha 2012).

Recent research, however, has enabled scientists to group phytonutrients into classes on the basis of similar protective functions as well as individual physical and chemical characteristics of the molecules. Each class offers a unique kind of protection for the body. All classes of phytonutrients need to be consumed to keep the body healthy. Some of them are Terpenes, Carotenoids, Limonoids, Phenols, Flavonoids, Isoflavones, Glucosinolates, Allylic Sulfides, Indoles, Tocotrienols, Tocopherols, Lipoic acid, and Ubiquinone (Rajasekharan and Latha 2012).

25.7.1 Merits of Nutraceuticals

Plants constitute a major source of nutraceuticals. Since a new molecule is difficult to discover and more expensive and risky than ever before, many pharmaceutical companies are now trying to discover nutraceuticals because there is undoubtedly a very huge and growing market. The belief among consumers is that these "food-like substances" are either harmless or least toxic as compared to conventional pharmaceuticals. Increased healthcare costs with conventional pharmaceuticals, recent legislation, and scientific discoveries have attracted people toward nutraceuticals. Inappropriate dietary habits are seen as contributing to the leading cause of death due to coronary heart disease, certain types of cancer, etc. in the present human generation. The role of nutraceuticals in treating these conditions is thus speculated. The emergence of diet-disease relations has led to the search for specific constituents of plants, animals, and minerals having a beneficial role in our mental and physical health. Nutraceuticals are gaining popularity as people are relying on them for safeguarding their health and avoiding side effects associated with drugs as well. As public knowledge in this field has evolved, manufacturers have sought to fulfil their appetite for these products resulting in exploding markets. Japan, USA and UK are the world leaders in the global nutraceutical market. There are around 5000 established nutraceutical products globally (Rajasekharan and Latha 2012). Long use history, better patient tolerance as well as public acceptance are the reasons for the success of nutraceuticals. Renewable sources, ease of cultivation and processing, environmental friendliness, and local availability are the other merits of the manufacturing and availability of nutraceuticals.

25.7.2 Dietary Antioxidants

Antioxidants are substances, which retard or prevent deterioration, damage or destruction caused by oxidation. Fortunately, the body has an army of antioxidants for damage limitation. Antioxidants form an integral part of the nutraceutical market. During the last few years, research has confirmed that many of the common diseases and ailments of the twenty-first century (CVS, diabetes, cataracts, high blood pressure, infertility, respiratory infection, and rheumatoid arthritis) are associated with tissue deficiency and/or low dietary levels of compounds called antioxidants (Rajasekharan et al. 2013).

Increased dietary intake of long-chain omega-3-fatty acids such as eicosapentaenoic acid (EPA) and docosahexaenoic acid (DHA) built up in algae and plankton have demonstrated many cardioprotection effects, including antiarrhythmic, blood triglyceride-lowering, and anti-thrombotic effects as well as improving endothelial relaxation and inhibiting both atherosclerosis and inflammation. α–linolenic acid is another omega-3-fatty acid found in plant foods and vegetable oils (Table 25.12). Additionally, vegetable oils yield high levels of

Table 25.12 List of some nutraceuticals with reported bioactivity

Nutraceuticals	Reported activity/use (Espín et al. 2007; Nasri et al. 2014)
Liquorice	Expectorant, treatment of peptic ulcer
Isoflavones in soyabeans	Reduces cholesterol, prevention of osteoporosis
Phosphatidylcholine	Natural treatment of liver disease
Ginger	Carminative, antiemetic, treatment of dizziness
Kambocha tea	Improve diet, relief from arthritis and menstrual cramps
Glucosamine-sulphate, Chondroitin—sulphate, vit C, vit D, vit E, zinc, selenium, copper	Osteoarthritis
Lycopene, resveratrol, beta -1, 3-glucans, soya	Management of cardiovascular disease
Green tea	Treatment of cancer
Carotenoids, lycopene and lutein	Prostate cancer
Antipain, leupeptin	Urokinase inhibitor (prostate cancer)
Fenugreek (*Trigonella foenum-graceum*)	Anti-diabetic, anti-cancer
Noni (*Morinda citrifolia*)	Relief from blood pressure, muscle pain

omega-6-fatty acid linoleic acid. α–linolenic acid is a precursor of DHA and EPA. However, the limited conversion effectiveness of dietary α–linolenic acid to DHA strongly suggests that the dietary intake of EPA and DHA is a potent way to enhance the levels of circulating omega-3-fatty acids in human serum.

Nutraceuticals, functional foods, designer foods, and medicinal foods are now becoming popular in developed countries, although the very concept had its origin in oriental medicine, particularly the traditional medicine systems of India. The Indian folklore medicine and other local health tradition practices and formulations offer ample opportunities for the nutraceutical or health food industries. Scientific validation of the folklore knowledge and resources is, therefore, a promising area for R & D in natural product development (Rajasekharan et al. 2013).

The contemporary lifestyle categorized by growing stress levels, poor eating habits, and a lack of exercise has increased the incidence of health disorders globally. Obesity, blood sugar, digestive disorders, and heart ailments have become more common than the common cold (Espín et al. 2007). The use of nutraceuticals as an attempt to accomplish desirable therapeutic outcomes with reduced side effects, as compared with other therapeutic agents, has met with great monetary success. The preference for the discovery and production of nutraceuticals over pharmaceuticals is well seen in pharmaceutical and biotechnology companies (Rajasekharan et al. 2013).

Based on the repository of data on Traditional Knowledge (TK), there is a plan to establish a National facility of plants used for food and medicine for the development of Ayurveda, Modern, Herbal medicines, nutraceuticals, functional foods for Non-Communicable Diseases (NCDs), infectious diseases (US$ USD 69.2 Billion), Viral diseases (Global antiviral drugs market size was valued at USD 41.1 billion in the year 2015 and is estimated to reach a value of USD 69.2 billion by 2025, growing at a CAGR of 4.0%.), Nutrition deficiency (India's nutraceutical industry is set to double in size to Rs 26,764 crore (approx USD 4 billion) by 2020), etc. There is a huge domestic and international market for medicines/nutraceuticals, especially for metabolic disorders. For example, Global Liver Disease Treatment Market is expected to garner $19,536 million by 2022, registering a CAGR of 11.72% during the forecast period 2016–2022. In India, Diabetes Market is expected to grow to US$ 7441.6 million by 2023 from US$ 4778.7 million in 2016 (Rajasekharan et al. 2017a).

The medicinal plant wealth of Kerala has been the focus of extensive studies by several scholars/healers. The foremost classic work *Hortus Indicus Malabaricus* by Hendrik Adriaan Van Rheede, was written in Kerala. It was published in 1678 and 1703 in 12 volumes with illustrations of 742 plants (534 medicinal species) and formed the main reference work o7 Karl Linnaeus on the plants of Malabar. The book's content is mainly based on 'Keralaramam' written by Mr. Itty Achuthan, a famous traditional physician of Ayurveda who lived in Cherthalla of Allapuzha district of Kerala Malayalam. Appu Bhat, Vinayaka Pandit and Ranga Bhat were the Konkani Brahmin scholars who scrutinized the part of the description included from Sanskrit text (Singh 2013). The classic work was translated from Latin to English by Prof. K S Manilal of Calicut University and was published by the University of Kerala in 2004.

25.8 Wild Medicinal and Edible Plants Used by the Tribes of Kerala Having Immuno-Enhancing Properties

1. ***Pellionia heyneana*** Wedd., (Nilampatti) belongs to the family Urticaceae. They are Subshrubs with arched, angled stems; densely lineolate, distichous leaves found in Evergreen and semi-evergreen forests of Palakkad, Kollam, Idukki, Pathanamthitta, Malappuram, Kannur, Thiruvananthapuram, Thrissur and Wayanad districts of Kerala, India. The plant is used by Cholanaikkans and Malai Pandaram tribes (Fig. 25.17). The medicinal parts are leaf and entire plant used as Immuno booster, hepatoprotective, antioxidant and stamina booster.
2. ***Neurocalyx Calycinus*** (R. Br. ex Bennett) Robins, (Pachachedi). The plant belonging to the family Rubiaceae is used by Mala Kuravans tribal community of Kerala, India. Leaf and roots are the useful parts. They are large endemic pubescent herbs having leaves all simple, whorled at tip, oblanceolate, acute at apex, having rusty puberulus. Distributed along Evergreen forests of Palakkad, Kollam, Idukki, Pathanamthitta, Malappuram, Kannur, Thiruvananthapuram

Fig. 25.17 Immuno boosting wild medicinal plants

and Wayanad districts of Kerala, India. The application includes Immuno booster, wound healing, analgesic, anti-inflammatory and antioxidant action.

3. ***Caesalpinia Bonduc*** (L) Roxb. (Caesalpinioideae) is a climbing plant with stems up to 15 m long, usually armed with robust prickles. The seeds are often sold in local markets with vernacular names as 'Kalamchikuru', 'Kaalanchi', 'Kazhinch–Kai'. Seed, Leaf, and Root contain bitter substances phytosterenin, bonducin, saponin, phytosterol, fixed oil, starch, and sucrose. Seeds also contain α, β, γ, δ and ζ caesalpins (Rajasekharan et al. 2017b). The plant is used against Malaria, Diabetes, Stomach disorders, Rheumatism, Cough, Fever, Headache, Jaundice, Diarrhoea, Skin eruptions, Asthma, and Internal blood clots. Other application includes Immuno and stamina booster,

hepato protective, adaptogenic/stabilization of physiology, antimicrobial, antidiabetic. *C. bonduc* is used by Malai Pandaram.

4. ***Abrus Precatorius*** **L.** (Chuvannakunni, Kakani, Kunni, Kunnikuru, Gunj) belongs to the family Fabaceae and are woody perennial twining shrubs having puberulent young stems. Leaves are even-pinnate; with leaflets 12–16 pairs which are oblong to elliptic, base and apex obtuse with lower surface sparsely pubescent. The flowers are pink. The plant is distributed along deciduous forests and also on the plains and is used by Kannikars (tribal community). The parts used are leaf, root and seeds. The application includes Immuno and stamina booster, antimalarial, anti-inflammatory, antidiabetic, antioxidant and skin diseases (Rajasekharan et al. 2018).
5. ***Bauhinia Variegata*** L. (Chuvannamandaram, Malayakathi, Kovidaram, Mandaram, Kongu), belongs to the family Fabaceae, and it is a medium-sized tree with dark brown, nearly smooth bark with pubescent young shoots. The plant parts used are stem bark, flower, and roots. They are applied by Kurumbar and Kannikars tribals as Immuno and stamina booster, antioxidant, anticancer, hypolipidemic, hepatoprotective, anti-inflammatory, nephrotic protection, antiulcer, wound healing.
6. ***Andrographis Elongata*** (Vahl) Anders. (Acanthaceae). These are erect herbs up to 2 m long with divaricate branches with stem stout, tetragonous and pubescent. Leaves show an opposite arrangement. The plant is observed in moist deciduous forests of Western Ghats. The plant is used by Malai Pandaram tribes as Antipyretic, antidiabetic, antimicrobial, Immuno and stamina booster, hepatoprotective, anticancer, antiviral, antihepatitis, antivenom against snake bite. The parts used are the leaf, seed, and entire plant (Rajasekharan et al. 2018).
7. ***Aristolochia indica*** **L.** (Eshwaramulla, Garudakodi, Iswaramooli, Kadalivegam, Karalakam, Karalayam, Karalvegam, Karanavalli, Karandavalli, Kudukkamooli, Urithoongi, Urikizhangu). This plant belongs to the family Aristolochiaceae and is twining perennial herbs with slender branchlets and simple, alternate Leaves. They are seen along with the degraded moist deciduous forests, also in the plains growing along fences. The plant is used by the Kannikars tribes of Kerala, India. The parts used are roots and leaves as Immuno and stamina boosters and as medicine for fever and headache with anticancer and antioxidant potential.
8. ***Zingiber nimmonii*** Dalz. (Mala-inchi) belonging to the family Zingiberaceae, has a small rhizome with purplish-lilac inside and is strongly aromatic. Roots are fleshy, bearing fleshy root tubers. The plant is seen in moist deciduous forests and also in the plains of Western Ghats. A rhizome is used by Malai Pandaram tribes as Immuno and stamina booster, anticancer, antioxidant and antiviral medicines.
9. ***Dioscoria Oppositifolia*** L. (Kaachil, Kanjirakizhangu) in the family Dioscoreaceae is a fast-growing twining vine that has escaped from cultivation. It can survive in different habitats and environmental conditions but is most commonly found at the edges of rich, mesic bottomland forests, along stream banks and drainage ways and near fencerows. Initial infestations are generally

Fig. 25.18 Immuno boosting wild edible plants

associated with human-caused disturbances, such as near old homesites and along roadways, and from these areas, they can easily spread to nearby riparian swaths and undisturbed habitats. It can tolerate light levels ranging from full sun to full shade, but mostly grows at intermediate light levels along forest edges and is typically found in silty loam soils. It also prefers soils that are relatively rich in nitrogen. *D. oppositifolia* propagules are dispersed primarily by gravity but may be dispersed further by water or by animals. Both the tuber and bulbils are edible. It is used by Kannikars and Kattunayakans tribes of Kerala, India and the tubers are rich in dietary fibres, protein, carbohydrates and minerals. They enhance stamina and immunity andwork against fatigue, gaining body weight etc (Fig. 25.18).

10. ***Opuntia vulgaris*** Mill. (Chappathikally, Pattanathukally) in the family Cactaceae is a lesser-known edible species used by Hill Pulayas community in Kerala, India. They are subshrubs with fleshy branches and areoles woolly having spines to 2.5 cm. Leaves are subulate or absent. Flowers are solitary, sessile and fruit is berry. The plant is seen in dry deciduous forests, also in the plains Idukki, Pathanamthitta, Wayanad and Palakkad in Kerala, India. Used as Immuno booster, energy booster and for vitality.

11. ***Glycosmis pentaphylla*** (Retz.) DC, (Kurumpannal, Panal, Panchi), belongs to the faily Rutaceae. They are large shrubs to small trees up to 4 m tall. The plant is observed as undergrowth in disturbed evergreen forests up to 1400 m in the Western Ghats of Kerala, India. They are lesser known underutilized species. Useful parts are Fruit, leaf and root used by Malai Pandaram and Kannikars tribal communities. The plant is harvested from the wild, mainly for local use as a food and medicine (Immuno booster, antidiabetic, analgesis, anticancer, antiviral, antibacterial, antioxidantvitality, energy booster, antitrust).
12. ***Termitomyces Microcarpus*** (Berk. and Broome) Heim, is an agaric fungus of Lyophyllaceae. It is an edible species found in Africa and Asia, where it grows in groups or clusters in deciduous forests near the roots of bamboo stumps associated with termite nests. The species contain the largest amount of total amino acids as well as essential amino acids. It is used as immuno booster, stamina booster, nutraceutical, and antifatigue by all tribes of Kerala.
13. ***Baccaurea Courtallensis*** (Weight) Mull.Arg (Mootapalam), of Euphorbiaceae is a wild-growing fruit of the Western Ghats of India. It grows as an understory plant in moist evergreen forests from North Kannada to South Kerala and the adjoining western parts of Tamil Nadu up to an altitude of 900 m. *B. courtallensis* is an endemic tree growing up to 10 m tall, having grey bark usually smooth or scaly; blazes light orange. It bears edible fruits and leaves. The plant is used by tribes Kannikars and Malai Pandaram in Kerala, India. Though the fruits are sour, they are eaten by local people. These are also used for pickling. The plant could enhance stamina, immunity and has antiviral, antibacterial, antioxidant, anti-inflammatory, antihyperlipidemic properties (Rajasekharan et al. 2018).

25.9 Challenges and Opportunities of TK

Traditional knowledge is an invaluable wealth in every geographical region. There are a lot of challenges and opportunities associated with the conservation and application of knowledge. Systematic documentation of Genetic Resources and associated TK will help in the protection and conservation, thereby preventing data loss. Preparation of a biodiversity Database and the documentation of TK in digital form, which is a challenging task, could play a prominent role in positive protection (Technical report 2013). If scientifically done, it will be useful for the future generation in the application of TK for product development. Lead-based scientific validation of TK and Product development are important areas of research in the biodiversity-rich areas of the world. There is a need to raise awareness regarding the role of Prior Informed Consent for the protection of TK and implement customary laws on the protection of TK. Prior Informed Consent has an inevitable role in the application of TK since it helps in knowledge sharing and benefit-sharing. Framing of legal protection for the TK holders and legitimate users may help to prevent biopiracy and data loss which helps in Access and Benefit Sharing (ABS). Formulating an International dimension for the protection and benefit-sharing associated with genetic resources will help to prevent international conflicts regarding biopiracy and patenting. Capacity building programmes for the TK holders/Tribal communities should be a responsibility of the scientific fraternity and governments should take necessary action towards achieving this. Learning from existing case studies carried out so far at the National and International levels and integrating TK into a comprehensive development framework will help in the betterment of human living conditions in many ways.

25.10 Current Scenario of Protection of Traditional Knowledge

The first major international treaty that took serious note of the protection of Traditional Knowledge (TK) was the Convention on Biological Diversity (CBD), 1992. TK is often associated with biological resources and traditional lifestyles. The CBD talks about both in situ and ex situ conservation measures, and so far as TK is concerned, its approach is to ensure that the benefits arising out of the use of TK are "relevant to the conservation of biological diversity and the sustainable use of its components" are shared equitably (Preamble). The treaty obligates member countries to respect, preserve and maintain such knowledge and to promote its wider application with the approval and involvement of the holders of the knowledge, along with encouraging the equitable sharing of the benefits arising out of the utilization of the knowledge [Article 8 (i)]. The three objectives of the CBD are (1) the conservation of biological diversity, (2) the sustainable use of its components, and (3) the fair and equitable sharing of benefits arising from the utilization of genetic resources. Towards this, it encourages the exchange of information, including indigenous and traditional knowledge (Article 17). This approach has the advantage of seeing that TK does not remain static but has its place in the knowledge that humanity exploits to create wealth through sustainable use of biological diversity (Shruti et al. 2010).

The CBD is an international treaty with one of the largest number of parties (196 parties). Still, the fulfilment of the obligations that it bestows on member countries is voluntary. It does not have any mandatory enforcement clauses leading to penalties, unlike the case with the World Trade Organisation (WTO) treaties. Among the WTO agreements, the Agreement on Trade-Related Aspects of Intellectual Property Rights (TRIPS Agreement), 1994 is conspicuous by its omission of TK protection from the ambit of IPRs. It is, however, a minimum obligations treaty, and member states are free to extend higher protection for IPRs, as per Article 27. Many countries have enacted national legislation granting rights to traditional knowledge holders of their country.

A major lacuna in the CBD was that it had not made provisions for the protection of genetic resources and associated traditional knowledge in countries, which are not the countries of origin, that is, in those countries that got them imported either legally or illegally. The obligations applied to domestic materials and knowledge. This situation was effectively not enabling the fight against bio-piracy and the steps to ensure fair and equitable benefit sharing arising out of the access to GR and TK. To address this challenge, a protocol, namely, the Nagoya Protocol on Access to Genetic Resources and the Fair and Equitable Sharing of Benefits Arising from their Utilization to the Convention on Biological Diversity (hereafter Nagoya Protocol) was adopted by the Conference of the Parties in its tenth meeting on 29th October 2010 in Nagoya, Japan. This was the result of 6 years of negotiations.

The Protocol unambiguously stated that it applies to TK associated with GR and to the benefits arising from the utilization of such knowledge (Art.3). The Nagoya Protocol recognized the symbiotic relationship between the tangible Genetic Resources (GR) and the intangible TK. This is an essential innovation since other IPRs, except the protection of Geographical Indications, do not refer to the interrelationship between the tangible and the intangible. It also recognizes the diversity of circumstances in which TK is held or owned and also the different ways such as oral, documented, or in any other form in which TK is maintained. The Protocol highlights the need to identify rights holders even within a community.

The most important obligation so far as TK is concerned that the Protocol imposes on member countries is that they should ensure that TK associated with GR is accessed with prior and informed consent (PIC) of the holders and with mutually agreed terms established (Art.7). This article, however, specifies the holders as "indigenous and local communities" (Tasnia et al. 2021). This may raise doubts about the protection of TK that is trans-community and nation-wide such as Ayurveda, which is even trans-national. The Protocol, of course, refers to the TK "shared by one or more indigenous and local communities in several Parties" and provides that the Parties concerned "shall endeavour to cooperate, as appropriate, with the involvement of the indigenous and local communities" (Art. 11). The Protocol makes it obligatory to give due consideration to the customary laws, community protocols and procedures of the local communities (Art. 12). The overall approach of the Protocol is to empower and enable indigenous and local communities to control and benefit from their TK. This stress on indigenous and local communities is natural since the Nagoya Protocol is the first international instrument that has relevance to those communities that were negotiated after the adoption of the UN Declaration on the Rights of Indigenous Peoples in September 2007 and are viewed as "a significant step in mainstreaming indigenous rights as a cross-cutting issue in international negotiations" (Tasnia et al. 2021).

The most important provision of the Protocol is that the contracting parties should ensure that their nationals comply with the domestic legislation and regulatory requirements of provider countries related to access and benefit-sharing of TK associated with GR. This is a major step forward from the CBD and safeguards the interests of TK holders in provider countries. The Protocol provides for several instrumental mechanisms such as the establishment of National Focal Points and

Access and Benefit-Sharing Clearing Houses to serve as national contact points for information, grant of access, and cooperation on issues of compliance. The Protocol, which came into force on 12th October 2014 has received 132 ratifications or accessions so far (Tasnia et al. 2021).

The World Intellectual Property Organisation (WIPO), the specialized agency of the United Nations Organisation dealing with Intellectual Property Rights (IPR) matters, has been discussing for more than two decades the protection of traditional knowledge. It is doing this through a forum known as the WIPO Inter-Governmental Committee on Intellectual Property and Genetic Resources, Traditional Knowledge and Folklore (IGC), established in 2000. Advancements in technology, especially the emergence of biopharma, have led to the patenting of traditional knowledge without proper acknowledgment and having access and benefit-sharing agreement. In the 1990s, with the coming into force of the Agreement on Trade-Related Aspects of Intellectual Property Rights, many countries in the South felt that while technology in the North gets protected under the patent regime and accessing which high rates of royalties had to be paid whereas the knowledge especially relating to biological resources which the South had in plenty was not so protected, and individuals and corporations could access and appropriate the same with minor modifications under the patent regime, like the turmeric patent in the United States of America (Tasnia et al. 2021). (USA). The issue caught the attention of the WIPO Standing Committee on Patents and finally led to the establishment of the IGC. The current discussions in the IGC are based on draft texts for one or more legal instruments for the protection of TK, Traditional Cultural Expressions, and Access to Genetic Resources. The WIPO General Assembly extended the mandate of the IGC for the biennium 2022/2023 in October 2021 with the objective of finalizing one or more legal instruments, as the case may be. So far as TK is concerned, the current text before the IGC is 'The Protection of Traditional Knowledge: Draft Articles Facilitators' Rev. (June 19, 2019)' (Tasnia et al. 2021). This text contains a large number of square bracketed words and phrases on which consensus has to emerge.

The draft text includes articles on definition (Art.1), objectives (Art. 2), beneficiaries (Art.3), the scope of protection (Art 4), development of national TK databases (Art 5), sanctions, remedies, and exercise of rights (Art. 6), disclosure requirement (Art. 7), administration of rights (Art. 8), exceptions and limitations (Art. 9), term of protection (Art. 10), formalities (Art. 11), transitional measures (Art. 12), relationship with other international agreements, non-derogation (Art. 14), national treatment (Art. 15), and transboundary cooperation (Art. 16). Most of these provisions are standard provisions in IPR treaties. What is of material significance are those relating to definition, objectives, beneficiaries, the scope of protection, and remedies.

Article 1 on Definition Reads as Under:

Traditional Knowledge refers to knowledge originating from indigenous [peoples], local communities and/or [other beneficiaries] that may be dynamic and evolving and is the result of intellectual activity, experiences, spiritual means, or insights in or from a traditional context, which may be connected to land and environment, including know-how, skills, innovations, practices, teaching, or learning.

What one observes in this and generally throughout the text is that TK is by default associated with 'indigenous' people. This term has been in use to distinguish the people in the present nation-states who were there before the European colonial settlers. In a way, this approach may limit the scope of TK of the inherited knowledge of certain communities (Tasnia et al. 2021). Countries like India, China, Bangladesh, Sri Lanka, etc., which have large systematised traditional medicine knowledge in Ayurveda, Chinese medicine, etc. may find themselves squeezed out unless these systems are accepted as alternative medicine systems Allopathy. Even then, protection of the knowledge which the systems have, which have never enjoyed patent rights, from misappropriation may also be problematic. The systems which have tested their knowledge and medicine through long public use will not be able to obtain patents since the knowledge is already in the public domain.

Article 2 on Objectives has Three Alternatives:

Of these, the first one clearly states the objective as

"To provide effective, balanced and adequate protection relating to intellectual property against:

(a) Unauthorized and/or uncompensated uses of traditional knowledge; and
(b) The erroneous grant of intellectual property rights over traditional knowledge, [while supporting the appropriate use of traditional knowledge]".

In the second and third alternatives, the stress is on the appropriate use of TK; they differ in that while the second one speaks of protection within the intellectual property system, the third one speaks of protection within the patent system. Since none of the existing IPRs accommodates TK, how IP protection will be extended is an issue. Should it be a *sui generis* law? Alternately, whether any of the existing IP instruments be tweaked to suit the requirements of TK is a question. As regards the third option within the patent system is concerned, how the single size fits all approach of the patent will be adjusted to TK is a question. Novelty and inventiveness are two fundamental features to make an invention patentable. How these features will be adapted to TK for a patent regime is not clear in the third alternative.

These issues loom up when we look into Article 3, which proposes the eligibility criteria. This Article has two by and similar large alternatives and a third one that mentions the subject matter of protection, and all are in square brackets. The first one talks of protection for TK,

(a) created, generated, received, or revealed by indigenous [peoples], local communities and/or [other beneficiaries] and developed, held, used, and maintained collectively by them [in accordance with their customary laws and protocols];
(b) linked with, and is an integral part of, the cultural and social identity and traditional heritage of indigenous peoples, local communities and/or [other beneficiaries]; and
(c) transmitted between or from generation to generation, whether consecutively or not.

This alternative also has a provision that countries may make the prior existence of TK for a reasonable term a condition for eligibility. The second alternative is materially not much different. The alternative Article 3 links the TK to the definition of beneficiaries proposed in Article 4 and also prescribes existence for a 50-year period or two generations for the eligibility of the TK. The major issue, here also, is the definition of 'indigenous people' or 'local community.' The proposed instrument nowhere defines it. In countries like the USA, where the indigenous people are clearly distinguishable from the European or subsequent to colonization settlers, this is not an issue, but in a country like India, China, or Indonesia, where there are no large-scale settlements during the last four or five centuries, this poses a problem. The Constitution does not speak of 'indigenous people'. There are, however, groups who refer to themselves as '*Adivasis*' (original dwellers), implying they are the people eligible for protection of TK. Most Indians trace their origins to the humans who migrated to parts of this country long back, and as of now, neither the '*Adivasis*' nor the others can say with any documentary evidence when their forefathers initially settled in this country. That being so, the caveat that India's representative proposed while passing the UN Declaration on the Rights of Indigenous Peoples in 2007 remains valid that all Indians at the time of India's independence from Great Britain in 1947 were indigenous (James et al. 2021; James 2022). In the absence of that, there is the possibility of most of the TK in India falling outside the proposed treaty since stretching the 'local community' to include the entire population of more than 130 crore looks neither desirable nor rational.

Article 9 is about exceptions and limitations. The current text contains three alternatives. The first alternative is almost like the exceptions proposed for patents, copyrights, etc. in other treaties and reads as under:

In complying with the obligations set forth in this instrument, the Member States [as in special cases] [should] adopt justifiable exceptions and limitations necessary to protect the public interest, provided such exceptions and limitations shall not unreasonably conflict with the interests of beneficiaries nor unduly prejudice the implementation of this instrument.]

The second alternative has detailed provisions for general exceptions and specific exceptions. It presents a list of activities to be included in exceptions and limitations. The third alternative is only an enabling clause permitting members to provide for exceptions and limitations. The exceptions and limitations depend on the scope of protection and rights and can be meaningfully examined only after finalizing those issues. All that can be said at this stage is that "they will have to be drafted keeping in cultural advancement without destroying heritage or denying fair and equitable benefits to the holders in case of commercial exploitation" (Rajasekharan and Latha 2012).

25.11 National Laws

Many countries have enacted legislation for the protection of TK. An overview is presented below:

25.11.1 ARIPO

The Swakopmund Protocol on the Protection of Traditional Knowledge and Expressions of Folklore[2] (2010) applies to the countries that are members of the African Regional Industrial Property Organization (ARIPO).[3] This Protocol defines TK as "any knowledge originating from a local or traditional community that is the result of intellectual activity and insight in a traditional context, including know-how, skills, innovations, practices, and learning, where the knowledge is embodied in the traditional lifestyle of a community, or contained in the codified knowledge systems passed from one generation to another." It goes on to clarify that the term is limited to a specific technical field and may include agricultural, environmental, or medical knowledge and knowledge associated with genetic resources.

As per the Protocol, the owners of TK, namely, "the local and traditional communities, and recognized communities within such communities, who create, preserve and transmit knowledge in a traditional and intergenerational context". The owners of TK are granted the exclusive right to authorize the exploitation[4] just like a patent. The Protocol also provides an exception that the "protection shall not be prejudicial to the continued availability of traditional knowledge for practice, exchange, use and transmission of the knowledge by its holders within the traditional context."[5]

25.11.2 Bhutan

Bhutan has included certain provisions in the Biodiversity Act of Bhutan 2003[6] for the protection of TK. As per this TK means "any knowledge, innovation and practices of local communities relating to the use, properties, values and processes

[2]Available at wipo.int/tk/en/databases/tklaws/articles/article_0044.html.

[3]The present members of the Organization are Botswana, Kingdom of Eswatini, The Gambia, Ghana, Kenya, Kingdom of Lesotho, Liberia, Malawi, Mauritius, Mozambique, Namibia, Rwanda, Sao Tome and Principe, Sierra Leone, Somalia, Sudan, Uganda, United Republic of Tanzania, Zambia and Zimbabwe.

[4]Section 7.

[5]Section 11.

[6]Available at wipo.int/tk/en/databases/tklaws/articles/article_0044.html.

of any biological and genetic resources or any part thereof" (Article 51). Those who desire to access TK for non-customary use need to obtain the PIC of the traditional owners of the TK (Articles 36–38).

25.11.3 Brazil

Law No. 13.123 of May 20, 2015 (Access and Benefits Sharing of Genetic Resources and Associated Traditional Knowledge)[7] deals with the protection of and access to traditional knowledge associated with genetic resources in Brazil. The protection is against illegal use and exploitation. (Article 8). Traditional users are exempted from the law (Art. 8). Other access and use are subject to PIC and ABS. The law is in fulfilment of obligations under the Nagoya Protocol.

25.11.4 Chile

Law No. 19.039 on Industrial Property (Consolidated Law approved by Decree-Law No. 3)[8] is the law on TK in Chile. The law guarantees that the protection afforded by industrial property rights shall be granted while safeguarding and respecting national TK.

25.11.5 Costa Rica

As early as 1998, Costa Rica had, in its law No. 7788, regulated access to TK and also provided for equitable distribution of benefits from the exploitation of TK to its holders. The definition used in this law for knowledge is that it is "a dynamic product generated by society over time and by different mechanisms, and includes that which is produced by traditional means or generated by scientific practice" (Art.7.6). The law also extends automatic community rights over TK associated with biodiversity without any registration (Art. 82). It also recognizes and accepts the continuous evolution of TK.[9]

---

[7] Available at wipolex.wipo.int.

[8] Available at wipo.int/tk/en/databases/tklaws/articles/article_0005.html.

[9] The Biodiversity Law of Costa Rica available at https://www.endangeredearth.com/wp-content/uploads/es_laws/Costa-Rica-Biodiversity-Law-of-Costa-Rica.pdf.

25.11.6 Egypt

Egypt's Law on the Protection of Intellectual Property Rights, Law No. 82, 2002[10] says that where an invention involves, inter alia, traditional medicinal, agricultural, industrial, or handicraft knowledge, it should have been acquired in a legitimate manner.

25.11.7 Ethiopia

The law titled Access to Genetic Resources and Community Knowledge, and Community Rights Proclamation No. 482/2006[11] defines community knowledge as "knowledge, practices, innovations or technologies created or developed over generations by local communities on the conservation and use of genetic resources" (Art. 2). The local community has the right to regulate access to their community knowledge, an inalienable right to use their genetic resources and community knowledge, and the right to share from the benefit arising out of the utilization of their genetic resources and community knowledge. (Art. 6) The right to regulate access includes the right to give PIC. (Art. 7).

25.11.8 Peru

Law No. 27811 of 24 July 2002, introducing a Protection Regime for the Collective Knowledge of Indigenous Peoples derived from Biological Resources[12] regulates protection of TK in Peru. The Act defines collective knowledge as "the accumulated, transgenerational knowledge evolved by indigenous peoples and communities concerning the properties, uses and characteristics of biological diversity" Art.2). The law is restricted to indigenous peoples' rights that existed before the formation of the Peruvian State. The other articles of the law provide for PIC and ABS.

25.11.9 Philippines

The Philippines also protect the traditional rights of the indigenous communities, as different from the Filipinos who form the majority.[13] Implementing Rules and

[10] Available at wipo.int/tk/en/databases/tklaws/.

[11] Available at wipo.int/tk/en/databases/tklaws/articles/article_0009.html.

[12] Available at wipo.int/tk/en/databases/tklaws/.

[13] The Indigenous Peoples Rights Act of 1997 (Republic Act No. 8371).

Regulations of Republic Act No. 10055 (Joint Administrative Order No. 02-2010)[14] deals with IPR applications directly based on any biodiversity, genetic resources or materials, traditional knowledge, and indigenous knowledge, systems, and practices to which the RDI has had access to before the filing of the IPR application.

25.11.10 South Africa

South Africa's National Environmental Management Biodiversity Act, 2004[15] regulates traditional knowledge protection. This law talks about the "customary utilisation or knowledge of indigenous biological resources by an indigenous community, in accordance with written or unwritten rules, usages, customs or practices traditionally observed, accepted and recognized by them, and includes discoveries about the relevant indigenous biological resources by that community" (Art. 1). It is essentially about bioprospecting, for which permits have to be issued (Art.4) and research other than bioprospecting is exempt (Art.5).

25.11.11 Thailand

Protection and Promotion of Traditional Thai Medicinal Intelligence Act, B.E. 2542 (1999)[16] have provisions for the protection of traditional medicinal knowledge. The Act defines traditional Thai medicine as "the medicinal procedures concerned with examination, diagnosis, therapy, treatment or prevention or promotion and rehabilitation of the health of humans and animals, obstetrics, traditional Thai massage, and also includes the production of Traditional Thai drugs and the invention of medical devices on the basis of knowledge or text that has been passed from generation to generation" (Sect. 3). The law grants rights to those who have registered their intellectual property rights on traditional Thai medical intelligence under the Act. It is a right of sole ownership of the production of the drug and over the research, distribution, improvement, or development of formulas on traditional Thai drugs.

The picture that emerges is that there is no uniform approach among countries toward the protection of TK. The issues start with the definition of the term TK itself. That is one reason why WIPO IGC has not yet been able to finalize the definition of TK. Then comes the issue of the nature of protection. Some may prefer to have what is termed defensive protection against biopiracy or misappropriation of TK, and some argue for positive protection along the lines of IPR. Often, the positive protection is linked with the protection of biological resources, as in India. What

[14] Available at wipo.int/tk/en/databases/tklaws/articles/article_0156.html.

[15] Available at wipo.int/tk/en/databases/tklaws/articles/article_0022.html.

[16] Available at wipo.int/tk/en/databases/tklaws/articles/article_0024.html.

about knowledge relating to textiles or handicrafts, or toys? Will they be covered under the biological resource or genetic resource associated with traditional knowledge?

25.11.12 *India: A* Sui generis *Law?*

India does not yet have any specific law protecting TK. However, it has taken many measures, such as the creation of a Traditional Knowledge Digital Library (TKDL) as a defensive mechanism. The database has information on the formulations mentioned in the classical texts of Ayurveda, Siddha, Unani, Sowa Rigpa, and Yoga which will remain prior knowledge for purposes of a patent application. The digital library is in the five most commonly used languages in patent offices, namely, English, French, German, Japanese and Spanish. The formulations have been converted into patent application abstract format and are also based on Traditional Knowledge Resource Classification. At present, more than 4,00,000 formulations/ practices are available in the TKDL database[17]. An amendment made to the Patents Act, 1970 specifically bars any invention which, in effect, is TK or aggregation or duplication of known properties of the traditionally known component from being patented.[18]

Under the Biological Diversity Act, 2002, the Biodiversity Management Committees constituted at local levels across India's states and union territories have compiled 265,458 Peoples' Biodiversity Registers. The PBRs contain documentation of knowledge of individuals and communities concerning biodiversity and its uses. The TKDL and PBRs are important repositories of TK. They serve as documentary proof of the existence of the knowledge and the holder of that knowledge and serve as defensive mechanisms against piracy and misappropriation of knowledge.

The Biological Diversity Act 2002 also contains detailed provisions for the protection of TK associated with biological resources. They remain regulated and are governed by ABS and PIC mechanisms as required by CBD. An interlinkage between the National Biodiversity Authority (NBA) and the Patent Office has also been established whereby, for any invention based on biological resources or associated TK, the applicant has to submit a no-objection certificate from the NBA.

The National Intellectual Property Rights Policy 2016[19] contains specific directions regarding TK. The promotion of TK is part of the vision statement itself. Objective 3 of the Biological Diversity Act, which is about the legal and legislative

[17] http://www.tkdl.res.in/tkdl/langdefault/common/Abouttkdl.asp. Accessed on 28 January 2022.

[18] Section 3(p) of the Patents Act, 1970.

[19] Government of India, Ministry of Commerce and Industry, Department for Promotion of Industry and Internal Trade. (2016). National Intellectual Property Rights Policy. 12th May 2016.

framework that balances the interests of the right owners with the larger public interest, stresses the importance of protecting traditional medicinal knowledge, which exists in diverse forms in the country. It also points out that the Biological Diversity Act 2002 provides a mechanism for regulation access and ensures fair and equitable sharing of benefits arising from the use of biological resources and associated traditional knowledge. The Policy also states that activities for the promotion of TK have to be conducted with the effective participation of holders of TK. A specific commitment made in the Policy is that the ambit of TKDL will be expanded while exploring the possibility of using it for further R & D. Documentation of oral knowledge is to be done with due care for preserving the integrity of the knowledge and without compromising the traditional ways of life of communities. The TK holders have to be given support for furthering the knowledge systems.

Many attempts were made in the past to draft a *sui generis* legislation for the protection of TK in India. These are mostly remained at departmental level discussions in the Ministry of Human Resource Development (1998–1999), Ministry of Commerce and Industry, and the Office of the Controller General of Patents, Designs and Trade Marks. An attempt was mooted by the Ministry of AYUSH with the Jawaharlal Nehru University, New Delhi[20]. A draft seems to have been made by the IPR Centre of Cochin University of Science and Technology. Shri Shashi Tharoor, Member of Parliament of India also moved a private member's bill. But no legislation has been enacted so far in India.

The Department Related Parliamentary Standing Committee on Commerce, in its 161st Report (presented to the Rajya Sabha and Lok Sabha of the Indian Parliament on 23rd July 2021), has suggested various measures for the protection of TK. These include:

(a) Review Section 3(p) of the Patents Act for including traditional knowledge under patents.
(b) Incorporate provisions to investigate claims of patents to prevent the misuse or exploitation of TK of the country.
(c) Incorporate provisions under the Geographical Indications of Goods Act for registering traditional knowledge and traditional cultural expressions as Geographical Indications if the description of a product or process is closely linked to that of traditional knowledge of a specific geographical location.
(d) Make the creators and holders of TK, especially tribal communities, forest dwellers, artisans, and craftsmen, aware of the novelty or inventive steps involved in traditional expressions or work.
(e) Mobilize the creators or communities practising TK to claim IPRs wherein the Government should play the role of a joint owner, thereby restricting their misappropriation and exploitation.

[20] Personal knowledge of the author, having been associated with these efforts.

The Report, however, does not suggest any means to overcome the patentability criteria of novelty and inventiveness for TK, which is already existing knowledge. The Indian Patents Act also includes oral knowledge as well as both documented and undocumented information in the prior art. The Committee's recommendation that India should engage at the international level for the protection of TK is certainly welcome (Singh 2019).

25.12 Conclusion

Traditional Knowledge (TK) related to plants used for medicinal purposes is getting depleted rapidly. A well-designed mechanism is to be brought out to codify the remaining TK at the earliest; otherwise, this valuable knowledge will be lost forever. Therefore, systematic documentation of TK is highly essential to save the TK existing in the oral tradition. This will further lead to the establishment of Traditional Knowledge-based R & D activities to develop novel herbal products and nutraceuticals for the benefit of ailing humanity, respecting the Intellectual Property Rights (IPR), and strictly following the Access and Benefit Sharing (ABS) mechanism. Alternatively, this knowledge can also be effectively utilized for enriching Ayurvedic Pharmacopoeia as it was practised in the past by ancient Ayurvedic scholars.

Appendix 1: Prior Informed Consent Procedures

BIODIVERSITY AND ASSOCIATED TRADITIONALKNOWLEDGE **HOW TO DOCUMENT** 1. PRIOR INFORMED CONSENT (PIC)[a] **MODEL OF PRIOR INFORMED CONSENT FORM DEVELOPED BY JNTBGRI** I. Has it been explained to you, that you have complete discretion on Whether to disclose or not, the IK you hold?			
Yes		No	
(This is the most important part of the PIC. The right of the holders of Tk to refuse to divulge has to be informed to them. This should be explained to them in a language known to them Answer to these shall depend on the answer to Q.NO.4 In case, the knowledge belongs to the community or traditional healers care shall be taken to check who is authorized to give the consent. Enquiries are regards the customary Law of the community abs whether they are followed shall be made (Rajasekharan et al. 2012c)			

[a] Input by Adv. Yeshwanth Shenoy

HOW TO DOCUMENT 1. PRIOR INFORMED CONSENT (PIC) **MODEL OF PRIOR INFORMED CONSENT FORM DEVELOPED BY JNTBGRI** II. Were the advantages and disadvantages of disclosure of TK made known to you?

(continued)

Yes		No	

(The advantages and disadvantages of disclosure of TK note to them are to be made clear to them. They should be made known in short bio-piracy, the IP possibility and Public Domain. There may be various kinds of TK holders with varying exposure to the nitty-gritty of the IP world or the issue of bio piracy)

HOW TO DOCUMENT

1. PRIOR INFORMED CONSENT (PIC)

MODEL OF PRIOR INFORMED CONSENT FORM DEVELOPED BY JNTBGRI

III. Were you explained in brief the Biological Diversity Act and Your rights under the Act?

Yes		No	

(The passing of the Biological Diversity Act brought in some kind of relief to the knowledge holders. They can claim statutory rights if the value of the TK they hold. Therefore, informing them of their rights under the Act would be helpful for them to decide better what is good or not for them)

HOW TO DOCUMENT

1. PRIOR INFORMED CONSENT (PIC)

IV. Is the knowledge disclosed by you exclusively known to you?

Yes		No	

(If NO, who else knows about the Disclosed Knowledge?

- All members of the Community
- Locally known
- Traditional healers of your community

If known only the Traditional Holders of the Community, a brief explanatory note:

____

____

(This question will helping in knowing the stake holders. If the person disclosing has exclusive knowledge, it is easier, but to find a TK exclusively may not happen since TK in its very nature is shared. Therefore, it has to be ascertained if the TK is of a community or is of a particular group from the community)

HOW TO DOCUMENT

1. PRIOR INFORMED CONSENT (PIC)

V. Whether you approve that the knowledge disclosed by you be published in a scientific journal or other journal or any other media with the intention of inviting enquiries from commercialization/research?

Yes		No	

(If there is something interesting in the disclosed TK, it might well be worth writing about to get more information or may be some research organization would be interested in conducting a study. With the present IP law in the country, any kind of information that goes out has to be weighed well lest it might fall in the public domain)

HOW TO DOCUMENT

1. PRIOR INFORMED CONSENT (PIC)

(continued)

VI.* Whether you approved this knowledge to be protected by the existing intellectual Property Rights?

Yes		No	

(This is very interesting aspect. Even if TK contributes to a invention, TK in itself may not quality for it. At the same time section 6 makes it mandatory to get the approval form the NBA. This is again related to the PIC as envisaged under section 20 Of the Act, What happens if ever the holders answer the question in negative will be interesting to find out)

HOW TO DOCUMENT

1. PRIOR INFORMED CONSENT (PIC)

VII.* Whether you approve of research/commercialization of the knowledge with improvements?

Yes		No	

(Scientific research on TK is very important to authenticate the TK commercialization of the TK is the key for the benefit sharing)

VIII.* What kind of benefit sharing is preferred?

(a) Monetary. (b) Non-Monetary

If Non-Monetary, what kinds of Benefit Sharing are you interested in?

(This is a question to ascertain what kind of benefits weak the stakeholders like to get from the disclosed TK if they are ascertained to have any commercial value)

HOW TO DOCUMENT

1. PRIOR INFORMED CONSENT (PIC)

I/We have been explained every question in a language known to me/us. All answers are given by me/us after understanding the questions completely and the implications thereof. All information given are the lo the best of my/our knowledge and belief

Place Signature

Date Name

Contact Address

*(Answer to these shall depend on the answer to Q.NO.4 in case the knowledge belongs to the Community or Traditional Healers, due care shall be taken to check who is authorized to give the consent? Enquiries as regards the Customary Low of the Community and whether they are followed shall be made)

Appendix 2

The term *Ashta Vaidya* is misinterpreted on many occasions as eight families of physicians. According to the legends, after the creation of Kerala, Sage *Parasurama* permitted 64 *Brahmin* families to migrate and establish their colonies in the land of Kerala. For the welfare of these families, he constituted 18 'SABHAMATAS' (centre for higher learning). For administrative convenience, he brought these centres under three main faculties namely 'Sastra Sabhamata' (faculty of holistic science), 'Sannyasa Sabhamata' (faculty of spiritual learning) and 'Karmy Sabhamata' (faculty of Vedic studies). Apart from this, to deal with the health care of the people, Sage *Parasurama* trained a selected group of individuals who later became the scholars in *Ashtagas* (eight specialities) of Ayurveda and this group was named *Ashta Vaidyas. Ashta Vaidyas* of Kerala mastered all the eight specialities of Ayurveda namely *Kayachikitsa* (General medicine), *Shalyatantra* (General surgery), Shalakyatantra(Ophthalmology and *Otorhinolaryngology*), *Kaumarabhritya* (Paediatrics, Obstetrics and Gynaecology), *Agadatantra* (toxicology), *Rasayanatantra* (Gerontology/Rejuvenation therapy), *Vajeekaranatantra* (Rejuvenation of the reproductive system) and *Bhutavidya* (Psychiatry). They were also known as 'Poornavaidyas' (Complete physicians)

Appendix 3: Selected Case Studies with Passport Script Data of Traditional Knowledge

Case Study No: AYUSH/Erna/Puth/07-07	**Disclosed**	✔	**Partially Disclosed**	**Undisclosed**
Name of the Gramapanchayath (Local governing body) Surveyed:	**Puthenvelikkara, Ernakulam District, kerala, India**			
Date of Survey:	**25-02-2020**			

Practice Title	Preparation of Mullayari Thoran
Country	Kerala, India
Domain	Health
Technology	Traditional Practices
Bearers of Knowledge	Local Health Tradition (Community)

Summary of Information/Claim/Practice

Ingredient(s): Uluva (*Trigonella foenum*-graecum L.) – 1/2 teaspoon + Uzhunu (*Vigna mungo* (L.) Hepper) – 1/2 teaspoon + Mula (*Bambusa arundinacea* Willd.) – 250 g + Kaduk (*Brassica juncea* (L.) Czern. and Coss.) – 1/2 teaspoon + Thengu (*Cocos nucifera* L.) – One cup + Kodapuli (*Garcinia gummi-gutta* (L.) Robs. – 20 g

Method of Preparation and Administration:

The bamboo rice is cooked in water and water is drained off. Then it adds to the fried Trigonella, Vigna and Brassica in coconut oil along with Malabar tamarind, grated coconut and dried chilli. It then heats up for 10 min

Restrictions: Nil

Dosage: Nil, Nil

Adjuvant: Nil

Knowledge Provider	Kamalam Bhatt

Documented by: Dr. Vinodkumar T G Nair and his team, (Recorded by Sruthy C R and Vishnu M R)

Passport Script Data of Knowledge Provider

Name and photo	Address	Age	Sex	Traditional occupation	Present occupation	Educational qualifications (s)
Kamalam Bhatt 	Eluparambil House Manjerikkunnu Puthenvelikkara P O	72	F	Not mentioned	Not mentioned	4th Standard

Passport Script Data of Plant Species Used

Scientific name	Tribal/local name
Trigonella foenum-graecum L.	Uluva
Vigna mungo (L.) Hepper	Uzhunu
Bambusa arundinacea Willd.	Mula
Brassica juncea (L.) Czern. and Coss.	Kaduk
Cocos nucifera L.	Thengu
Garcinia gummi-gutta (L.) Robs.	Kodapuli

Selected Case Studies

Case Study No: AYUSH/THIR/ARUV/03-01	**Disclosed**	✔	**Partially Disclosed**	**Undisclosed**
Name of the Gramapanchayath (Local governing body) Surveyed:		**Aruvikkara, Thiruvananthapuram District, Kerala, India**		
Date of Survey:		**07-10-2021**		

Practice Title	**Single Drug for Asthma**
Country	Kerala, India

(continued)

Domain	Health
Technology	Traditional Practices
Bearers of Knowledge	Local Health Tradition (Individual Knowledge)

Summary of Information/Claim/Practice
Ingredient(s): Valli-pala (*Tylophora asthmatica* (L. f.) Wight and Arn.)
Method of Preparation and Administration:
The leaves are steamed in the coal and juice is extracted. This juice is consumed and the leaf is placed on the chest
Restrictions: Nil
Dosage: Juice pf 7 leaves, Once a day
Adjuvant: Nil

Knowledge Provider	Prathapachandran

Documented by: Dr. Vinodkumar T G Nair and his team, (Recorded by Dr. Vinodkumar T G and his team). 2020

Passport Script Data of Knowledge Provider

Name and photo	Address	Age	Sex	Traditional occupation	Present occupation	Educational qualifications (s)
Prathapachandran 	Maruthamcode, Backside of Keltron, Aruvikkara P O	61	M	Marmachikilsa	Marmachikilsa	Post Graduate

Passport Script Data of Plant Species Used

(a) Scientific Name: *Tylophora asthmatica* (L. f.) Wight and Arn.
(b) Tribal/Local Name: Valli-pala

Name of the plant	Part(s) used	Part condition
Tylophora asthmatica (L. f.) Wight and Arn.	Leaf	Fresh

(c) Habit:

Herb		Shrub		Tree		Climber	✔	Twiner		Others	

(d) Indication(s):

What Kind of benefit Sharing is preferred?	Monetary	✔	Non-monetary	

References

Espín JC, García-Conesa MT, Tomás-Barberán FA (2007) Nutraceuticals: facts and fiction. Phytochemistry 68(22–24):2986–3008. https://doi.org/10.1016/j.phytochem.2007.09.014. PMID: 17976666

Huang D (2018) Dietary antioxidants and health promotion. Antioxidants (Basel) 7(1):9. https://doi.org/10.3390/antiox7010009

James TC (2022) International discussions on indigenous people and India. Research and Information System for Developing Countries, New Delhi

James TC, Namrata P, Apurva B (2021) In-depth study on protection of traditional knowledge, traditional cultural expressions and plant genetic resources. Research and Information System for Developing Countries, New Delhi. ISBN: 81-7122-160-2

Nasri H, Baradaran A, Shirzad H, Rafieian-Kopaei M (2014) New concepts in nutraceuticals as alternative for pharmaceuticals. Int J Prev Med 5(12):1487–1499

Rajasekharan S (2013) Traditional and folk practices—contemporary relevance and future practices. In: Proceedings of Kerala Environment Congress, Centre for Environment and development in association with KSBB & KSCSTE, Thiruvananthapuram

Rajasekharan S, Latha PG (eds) (2011) Traditional and folk practice—contemporary relevance and future prospects. Number 1. JNTBGRI, Palode, Thiruvananthapuram

Rajasekharan S, Latha PG (eds) (2012) Traditional and folk practice—contemporary relevance and future prospects. JNTBGRI, Palode, Thiruvananthapuram. ISBN 978-81-920098-5-8

Rajasekharan S, Latha PG, Nair VTG (2012a) Concepts on drug development based on traditional knowledge/Ayurveda substantiated with modern science and technology. JNTBGRI, Palode, Thiruvananthapuram. ISBN 978-81-920098-8-9

Rajasekharan S, Latha PG, Nair VTG (2012b) Learning from traditional knowledge—'systematic documentation of traditional knowledge related to plants used for food and AYUSH & indigenous medicine'. JNTBGRI, Palode, Thiruvananthapuram. ISBN 978-81-920098-6-5

Rajasekharan S, Pushpangadan P, Latha PG, Nair VTG (2012c) Learning from traditional knowledge—intellectual property, traditional knowledge: community prospective on access and benefit sharing—ABS Kani model of access and benefit sharing. JNTBGRI, Palode, Thiruvananthapuram. ISBN 978-81-920098-7-2

Rajasekharan S, Pushpangadan P, Latha PG, Nair VTG (2012d) Concept of drug development based on traditional knowledge/Ayurveda substantiated with modern science and technology Number 2. JNTBGRI, Palode, Thiruvananthapuram

Rajasekharan S, Nair VTG, Navas M, Latha PG (2013) Traditional and folk practices of Kerala—case studies. J Tradit Folk Pract 1(1):50–68

Rajasekharan S, Vinodkumar TG, Navas (2017a) Traditional/folk practices of Kerala—case studies—part 4: traditional food articles. J Tradit Folk Pract 5(01):37–42

Rajasekharan S, Vinodkumar TG, Navas M, Latha PG (2017b) Traditional/folk practices of Kerala—case studies—part V: healing art of tribal communities of Kerala. J Tradit Folk Pract 5(02):94–103

Rajasekharan S, Vinodkumar TG, Navas M (2018) Traditional/folk practices of Kerala case studies—part VI: healing art of tribal/folk communities of Kerala. J Tradit Folk Pract 6(01): 98–102

Shravan K (2012) Traditional knowledge and Patent Strategy. JIPR 17(5):430–436

Shruti S, Archana M, Pradeep M, Suresh PV, Shivaprasad HN (2010) *In vivo* immunomodulatory activities of the aqueous extract of bonduc nut *Caesalpinia bonducella* seeds. Pharm Biol 48(2): 227–230. https://doi.org/10.3109/13880200903085474

Singh NG (2013) Traditional knowledge systems, international law and national challenges: marginalization or emancipation? Eur J Int Law 24(4):1205–1221. https://doi.org/10.1093/ejil/cht077

Tasnia KC, Chowdhury NS, Fatema IB (2021) Traditional and pharmacological reports of the Genus Baccaurea. A review. Am J Biomed Sci Res 11(6):ID.001683. https://doi.org/10.34297/AJBSR.2021.11.001683

Technical Report (2013) Project on Systematic Documentation of Traditional Knowledge Related to Plants used for Food and AYUSH & Indigenous Medicine. Dept. of AYUSH, Govt. Of India

Vansh M (2020) Phytochemicals effective in lowering low-density lipoproteins. J Biol Eng Res Rev 7(1):16–23

WIPO publication No.920 (E) on Intellectual Property and Traditional Knowledge, Booklet No. 2, pp 20–21. https://www.wipo.int/edocs/pubdocs/en/tk/920/wipo_pub_9

Correction to: Conservation and Sustainable Utilization of Bioresources

Swapna Thacheril Sukumaran and Keerthi T R

Correction to:
S. T. Sukumaran, K. T R (eds.), *Conservation and Sustainable Utilization of Bioresources*, Sustainable Development and Biodiversity 30, https://doi.org/10.1007/978-981-19-5841-0

This book published under the series 'Sustainable Development and Biodiversity' was inadvertently published without the volume number. Volume number 30 has been updated with this correction.

The updated version of the book can be found at
https://doi.org/10.1007/978-981-19-5841-0

S. T. Sukumaran, K. T R (eds.), *Conservation and Sustainable Utilization of Bioresources*, Sustainable Development and Biodiversity 30,
https://doi.org/10.1007/978-981-19-5841-0_26

The manufacturer's authorised representative in the EU is Springer Nature Customer Service Centre GmbH, Europaplatz 3, 69115 Heidelberg, Germany. If you have any concerns regarding our products, please contact ProductSafety@springernature.com

Printed and bound by CPI Group (UK) Ltd, Croydon, CR0 4YY

30/09/2026

02225245-0020